Physics

Physics: A Focused Introduction provides a thorough overview of the principles and practical applications of introductory undergraduate physics. Featuring a streamlined narrative, it encourages active engagement between fundamental concepts and fully worked examples.

Major topics such as motion (kinematics, forces, and conservation of momentum and energy), thermodynamics, waves, optics, electricity and magnetism are covered, in addition to the foundations of modern physics.

Key features:

- More than 200 fully worked examples and 1,000 end-of-chapter problems
- Learning Objectives at the start of each chapter
- An appendix which introduces key mathematical content ranging from scientific notation and the metric system to geometry, trigonometry, and techniques for solving systems of equations
- Calculus is introduced and used as a tool to clarify the physics; examples and problems that require proficiency in calculus are clearly marked
- Supported by modular lab activities developed at the University of Mount Union, detailed lecture notes, and a comprehensive solution manual available to instructors

Written in a relaxed and engaging style, this textbook is suitable for students of science and engineering who require a concise introduction to the field, as well as students of both algebra-based and calculus-based courses.

Colin Campbell is an Associate Professor of Physics and Data Science at the University of Mount Union, USA, where he teaches at all levels of the undergraduate curriculum. He earned his B.S. from Westminster College and his Ph.D. from the Pennsylvania State University. He has received multiple honors for his teaching, including the University of Mount Union's Distinguished Teaching Award.

Textbook Series in Physical Sciences

This textbook series offers pedagogical resources for the physical sciences. It publishes high-quality, high-impact texts to improve understanding of fundamental and cutting-edge topics, as well as to facilitate instruction. The authors are encouraged to incorporate numerous problems and worked examples, as well as making available solutions manuals for undergraduate and graduate level course adoptions. The format makes these texts useful as professional self-study and refresher guides as well. Subject areas covered in this series include condensed matter physics, quantum sciences, atomic, molecular, and plasma physics, energy science, nanoscience, spectroscopy, mathematical physics, geophysics, environmental physics, and so on, in terms of both theory and experiment.

Understanding Nanomaterials
Malkiat S. Johal

Concise Optics: Concepts, Examples, and Problems
Ajawad I. Haija, M. Z. Numan, W. Larry Freeman

A Mathematica Primer for Physicists
Jim Napolitano

Understanding Nanomaterials, Second Edition
Malkiat S. Johal, Lewis E. Johnson

Physics for Technology, Second Edition: With Applications in Industrial Control Electronics
Daniel H. Nichols

Time-Resolved Spectroscopy: An Experimental Perspective
Thomas Weinacht; Brett J. Pearson

No-Frills Physics: A Concise Study Guide for Algebra-Based Physics
Matthew D. McCluskey

Quantum Principles and Particles, Second Edition
Walter Wilcox

Electronic Conduction: Classical and Quantum Theory to Nanoelectronic Devices
John P. Xanthakis

Physics: A Focused Introduction
Colin Campbell

For more information about this series, please visit:
www.crcpress.com/Textbook-Series-in-Physical-Sciences/book-series/TPHYSCI

Physics
A Focused Introduction

Colin Campbell

CRC Press
Taylor & Francis Group
Boca Raton London New York

CRC Press is an imprint of the
Taylor & Francis Group, an **informa** business

Front cover image: Colin Campbell

First edition published 2026
by CRC Press
2385 NW Executive Center Drive, Suite 320, Boca Raton FL 33431

and by CRC Press
4 Park Square, Milton Park, Abingdon, Oxon, OX14 4RN

CRC Press is an imprint of Taylor & Francis Group, LLC

ISBN: 978-1-032-94587-3 (hbk)
ISBN: 978-1-032-94588-0 (pbk)
ISBN: 978-1-003-57156-8 (ebk)

DOI: 10.1201/9781003571568

Typeset in Latin Modern font
by KnowledgeWorks Global Ltd.

Publisher's note: This book has been prepared from camera-ready copy provided by the authors.

Access the Support Material: www.routledge.com/9781032945873

Contents

Part I

The Fundamentals of Motion

We begin with the study of motion on the scale of everyday life: how do we describe it, and what causes it? This is sometimes referred to as the study of *mechanics*; we begin here in part because it speaks to our curiosity about things we can readily experience. Why does a thrown ball take the path that it does? Why does an apple fall to the ground but the Moon does not? What determines the distance a car will travel when the driver slams on the brakes?

To answer these questions, we will develop important concepts and hone our ability to think carefully about the world around us. Many of these concepts – including vectors, forces, and energy – will be very important as we examine other branches of physics in the remainder of this book.

1

Describing Motion on a Line

Our goal in this chapter is to *describe* simple situations where objects move in one dimension (that is to say, a straight line). This is referred to as the study of one-dimensional **kinematics**. (After discussing two-dimensional kinematics in Chapter 2, we will turn to the study of what *causes* things to move as they do in Chapter 3.)

Learning Objectives

After reading this chapter, you should be able to:

- Understand the meaning and importance of a *coordinate system* for approaching problems in physics.

- Be able to explain, and mathematically define, *position*, *velocity*, and *acceleration*.

- Analyze and draw motion diagrams and motion graphs.

- Understand where the *one-dimensional kinematic equations* come from, and use them to analyze one-dimensional motion with constant acceleration.

- Account for observers moving relative to one another when analyzing motion.

1.1   Position, Velocity, and Acceleration

As a first example, consider a ball that is dropped off of a tall building and allowed to fall until it strikes the ground. Suppose we start a stopwatch the instant the ball is dropped, and to be specific, suppose also that the building is 100 meters tall. Table 1.1 shows the data and Figure 1.1 visualizes the position of the ball at half-second intervals as it falls.[1]

It is often convenient to show this information on a position vs. time plot, as on the right side of the figure. If you imagine the left side of the figure as a series of snapshots all laid overtop one another, then all we're doing here is spreading them out, side by side, in order of increasing time. Referring to either side of the figure, we might say something like: "When we start the stopwatch, the ball is 100 meters above the ground". Notice that there are three components to this statement: the time (here we are discussing the instant where $t = 0$ seconds), a reference point (the ground), and the object's position relative to that reference point (100 meters above it). We call the reference point the **origin**.

Of course, there is no reason why we must identify the ground as the origin. Another obvious choice would be to measure the ball's position from the top of the building: "When we start the stopwatch, the ball is even with the top of the building". A third choice might

---

[1]For now we're ignoring the effects of air resistance. We'll discuss how air resistance changes things in Chapter 3.

DOI: 10.1201/9781003571568-2

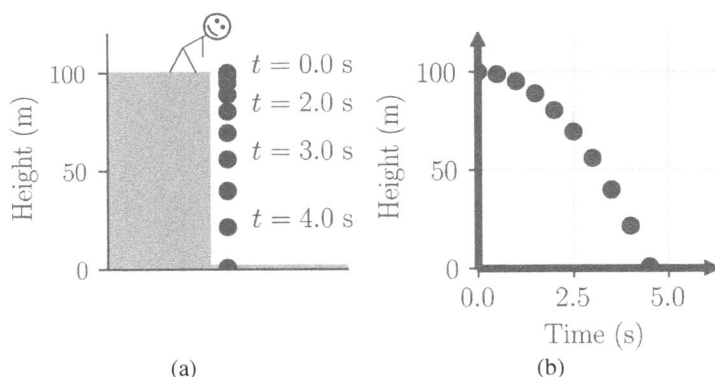

(a)   (b)

FIGURE 1.1
(a) The height of a ball at half-second intervals as it falls from a 100-meter-tall building.
(The stick figure is not to scale!). (b) The same information shown on a **position-time**
graph.

be where some observer is standing on the ground, off to the side of the building: "When we
start the stopwatch, the ball is *this* distance to my right and *this* distance above the ground".
In principle we can choose whatever origin we like when we set out to analyze the motion of
an object. The motion of the object, after all, does not depend on our choice!

Whatever our choice for origin, it is important to
note that the position is not simply a number: we
must also specify a direction (we specified that at
$t = 0$ seconds, the ball is 100 meters *above the origin*).
Quantities that require both an amount (i.e. a *mag-*
nitude) and a direction are **vectors**, while quantities
that require only a magnitude (like temperature) are
scalars.[2] When we work algebraically with vectors,
we draw an arrow over them, like so: $\vec{x}$. Figure 1.2
draws three position vectors, one for each of the three
origins described above, for the ball mid-flight. Note
that our choice of origin influences the length and
direction of the position vector.

To be mathematically precise in describing a vector, we need to define an **axis**. For instance, if we are
on the roof and looking down at the ground as the
ball falls, we might say "down toward the ground is a
positive distance away from the origin, and up toward the sky is a negative distance", or
we might prefer the opposite choice and describe *up* as positive and *down* as negative. We
also need to *label* the axis; in this chapter we'll generally use the label x. Taken together,
the origin and the orientation of our axis determine our **coordinate system**. Choosing a
coordinate system is often the first step in a physics problem: it defines the framework we
use to mathematically describe the world around us.[3]

TABLE 1.1
A ball's height as it falls.

Time (s)	Height (m)
0.0	100.0
0.5	98.8
1.0	95.1
1.5	90.0
2.0	80.4
2.5	69.4
3.0	55.9
3.5	40.0
4.0	21.6
4.5	0.8

---

[2]Scalars only require a magnitude, but it is worth keeping in mind that they can still be *negative*. For
instance, electric charge is a scalar, and the electric charge of an electron is a negative (the electric charge of
a proton is, in contrast, positive).
[3]Some problems may specify a coordinate system for you, but this is usually just for the convenience of
the person or program grading your work.

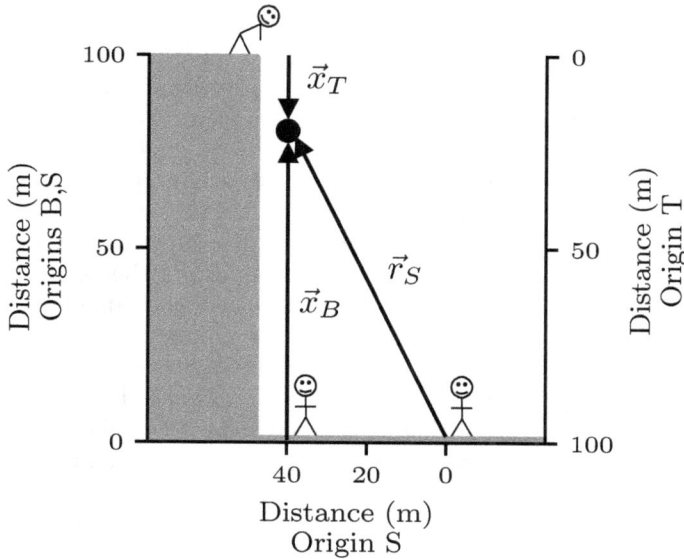

FIGURE 1.2

Three different choices of origin (the top of the building, the bottom of the building, and from an observer standing on the ground and off to the side) and a position vector for each. Notice the scales define different choices for the positive direction: the person at the top of the building defines *down* as positive (the right scale), while the people at the bottom define *up* as positive (the left scale). Meanwhile, the person off to the side also defines a horizontal scale such that *left* is the positive horizontal direction.

Let's consider a specific example. If we use the roof as our origin and call *down* the positive direction, then when the ball is halfway to the ground, we would describe the position like so:

$$\vec{x} = 50 \text{ m}\hat{x}$$

Here $\hat{x}$ is read "x hat" and is simply a label: the distance is along the axis we've labeled x.[4] When we are concerned only with the magnitude of a vector (in this example, 50 m), we use the absolute value symbol, $|\vec{x}|$, or alternatively just drop the arrow: x.

As we noted above, the observer standing off to the side needs to describe the position with *both* a distance above the ground *and* a distance to the side. Thus we need **two** axes, which we'll often call x and y. Generically we describe such a vector like so:

$$\vec{r} = r_x\hat{x} + r_y\hat{y}$$

Here we use a new symbol $\vec{r}$ to avoid implying the vector is purely along the x axis, as in the first case. In this chapter we will restrict ourselves to situations where the motion can be described with a single axis (this is so-called *linear* motion). We'll consider more complicated motion that occurs on a *plane* in Chapter 2.[5]

---

[4]If we made the other choice and called down the *negative* direction, then the ball in the same position would be labeled $\vec{x} = -50 \text{ m}\hat{x}$.

[5]While we could avoid using the word "vector" or the "$\hat{x}$" notation in this chapter and just talk about positive and negative numbers, it will be helpful to get used to these ideas now. I should also mention that these "hat" quantities are formally called **unit vectors**, i.e. vectors with length 1. Note also that some other books use $\hat{i}$ and $\hat{j}$ instead of $\hat{x}$ and $\hat{y}$; the meaning is the same.

Example 1.1 One-Dimensional Coordinate Systems   ⋆
Consider the ball in Figure 1.1. For an origin at the top of the building and an origin
on the ground, there are a total of *four* possible one-dimensional coordinate systems.
Draw each of them, and for each, draw a position vector to the ball mid-flight.

The position vector is always drawn from the origin of our coordinate system
to the position of the ball, as shown in Figure 1.3. The vectors on the top left and
bottom right coordinate systems are in the negative direction, while the other two
are in the positive direction. The top two vectors are of the same length and are
shorter than the bottom two vectors, which are also of the same length.

It is worth noting that we don't *have* to use the ground or the top of the building
as the origin. We've already seen an example where an observer is standing off to the
side of the building and might choose their location as the origin. Another example
is an observer with their head sticking out of a window 60 m above the ground: they
may fix the origin at that height. In any case, nothing about *the ball* has actually
changed – just our choice of origin. This all serves to underscore how important it is
to carefully choose your coordinate system and to be clear in your choice when you
relay your work to others!

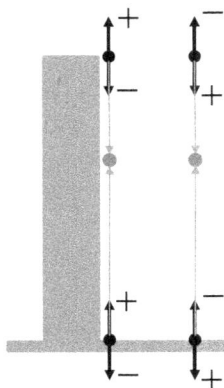

FIGURE 1.3
Vectors (in gray) showing the position of a ball using four different coordinate
systems. The ball is shown twice just so we can see the coordinate systems
without overlapping.

Once we have a coordinate system in place, we're able to numerically describe the position
of the ball at some instant. Now let's turn to describing how that position changes over
time. We label a position at some initial instant t_i as $\vec{x}_i$ and its position at some later (final)
instant t_f as $\vec{x}_f$. We use the Greek capital letter delta, Δ, to mean "change in", and we
mathematically write the change in position, or the **displacement**, as

$$\Delta\vec{x} = \vec{x}_f - \vec{x}_i \tag{1.1}$$

Visually, the displacement vector $\Delta\vec{x}$ runs from the initial position $\vec{x}_i$ to the final position
$\vec{x}_f$ (Figure 1.4).

Now, some time $\Delta t = t_f - t_i$ elapsed as the ball made this progress, and we define the
ball's average **velocity** during this time to be[6]

---

[6]We will typically drop the "avg" subscript to reduce clutter.

$$\vec{v}_{\text{avg}} = \frac{\Delta \vec{x}}{\Delta t} = \frac{\vec{x}_f - \vec{x}_i}{\Delta t} \tag{1.2}$$

It is important to be able to analyze what equations like this are telling us. First, the units: the average velocity is defined in terms of a length (with the SI unit of meters) divided by a time (with the SI unit of seconds), so the SI units for a velocity is a meter per second, or "m/s".

Next, note that because Equation (1.2) involves vectors, it has something to say about *magnitudes* and *directions*. Let's consider magnitudes first. For instance, if we imagine holding $\Delta \vec{x}$ fixed while increasing Δt, then the right hand side of the equation decreases (because the denominator increases). Because the left hand side must equal the right hand side (this is the definition of an equation), it follows that $\vec{v}$ decreases: it takes longer to cover the same distance.

Equation (1.2) also says that the *direction* of $\vec{v}$ is the same as the direction of $\Delta \vec{x}$. Suppose you are standing 5m from the origin of some coordinate system (Fig. 1.5). If you start walking in the positive direction (that is, with a positive velocity), then $\Delta \vec{x}$ is positive (a moment later you are 6m from the origin, then 7m, etc.). If, on the other hand, you walk toward the origin (that is, with a *negative* velocity), then $\Delta \vec{x}$ is negative: for example, a moment later you are 4m from the origin, so $\Delta \vec{x} = (4 - 5)\,\text{m}\hat{x} = -1\,\text{m}\hat{x}$. We summarize all of this by saying that *the position vector grows/shrinks in the direction of the velocity*.

It is an excellent idea to perform this kind of analysis whenever you encounter a new equation: you don't really understand an equation until you think through what it says.

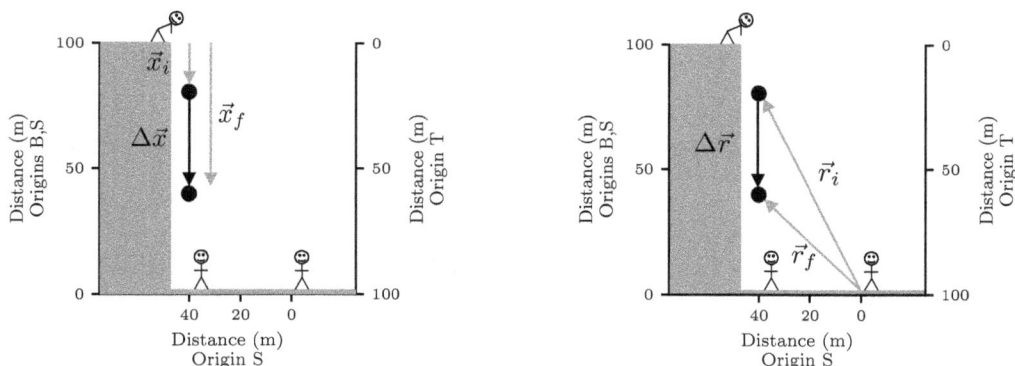

FIGURE 1.4

Two choices of coordinate systems for analyzing the ball as it drops. (Left) The origin (T) is at the top of the building and *down* is the positive direction. We use $\vec{x}$ to indicate a vector aligned with a vertical axis labeled x. (Right) The origin (S) is at the foot of the observer on the ground and off to the side; *up* is the positive vertical direction and *left* is the positive horizontal direction (because these are two-dimensional vectors we use the symbol $\vec{r}$). In each panel, two positions are shown with black dots and the initial and final position vectors are shown in gray. The displacement vectors are shown in black. Note that the length and direction of the displacement vectors are the same in each panel, though someone using the coordinate system on the left panel would say the vector points in the *positive* direction (along their one axis) while someone using the coordinate system on the right panel would say it is pointing in the *negative* direction along their vertical axis.

Before we proceed to some examples, let's consider the limiting case when the two time points t_i and t_f become arbitrarily close to the same instant. In this case our equation for

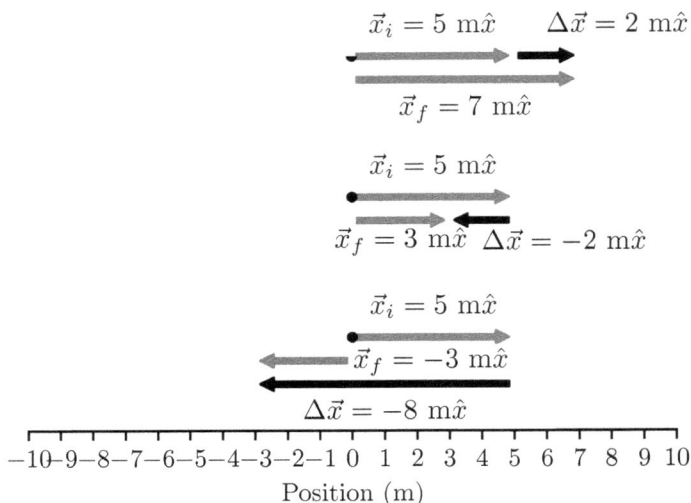

$$\vec{x}_i = 5 \text{ m}\hat{x} \qquad \Delta\vec{x} = 2 \text{ m}\hat{x}$$

$$\vec{x}_f = 7 \text{ m}\hat{x}$$

$$\vec{x}_i = 5 \text{ m}\hat{x}$$

$$\vec{x}_f = 3 \text{ m}\hat{x} \quad \Delta\vec{x} = -2 \text{ m}\hat{x}$$

$$\vec{x}_i = 5 \text{ m}\hat{x}$$

$$\vec{x}_f = -3 \text{ m}\hat{x}$$

$$\Delta\vec{x} = -8 \text{ m}\hat{x}$$

$$-10\;-9\;-8\;-7\;-6\;-5\;-4\;-3\;-2\;-1\;\;0\;\;1\;\;2\;\;3\;\;4\;\;5\;\;6\;\;7\;\;8\;\;9\;\;10$$

Position (m)

FIGURE 1.5
The change in position (that is, the displacement) for something that is initially +5m from the origin. (Top) The displacement (and therefore the average velocity) points in the same direction as the initial position, so the position vector increases in length. (Middle) The velocity points in the opposite direction, so the position vector decreases in length. If this process goes on long enough, as in (Bottom), the position vector passes through the origin and reverses direction to align with the velocity vector.

average velocity becomes an exact equation for the velocity *at some instant*. Appropriately, we refer to this as the **instantaneous velocity**.

When we're talking about infinitesimally small changes, we're invoking the language of calculus (Section A.5); specifically, we say that *instantaneous velocity is given by the derivative of position with respect to time.* In one-dimensional situations we mathematically express this like so (note that the Δ symbols have been replaced by the letter d):

$$v(t) = \frac{d}{dt}x(t) \tag{1.3}$$

Depending on whether or not you've studied calculus, you might opt to skip the calculus-based examples and problems in this book (they are clearly marked). Of course, if you *have* studied calculus, you should take advantage of these examples and work some of these problems; it will only enrich your understanding. And if you move on to study advanced physics, your future self will thank your present self! That said, having a basic sense for what a derivative *is* makes it easier, not harder, to understand many of the concepts we will develop in this book.

Example 1.2 Average Speed and Velocity   ⋆

Suppose you jog 1.0×10^2 yd (1 yd = 3 ft) in a straight line (in the $+x$ direction, say) and it takes you 4.0×10^1 s. Then, you take an additional 6.0×10^1 s to move halfway back to your starting point. (a) What is your average velocity across the entire trip? (b) Is this the same as the total distance you moved divided by the total time you spent on your trip? (c) Convert your answer to meters per second *and* miles per hour. Are the numbers used in this example realistic for a person jogging?

We sketch the motion in Figure 1.6, noting your position as you start with t_1 (i.e. the first time we have), your second position (when you're done with the first part of your jog) with t_2, and your final position with t_3. You walk for a total of 4.0×10^1 s $+ 6.0 \times 10^1$ s $= 1.00 \times 10^2$ s and your displacement (from start to end, i.e. from time 1 to time 3) has magnitude 5.0×10^1 yd. Thus your average velocity has a magnitude

$$|\vec{v}_{13}| = \frac{|\Delta\vec{x}_{13}|}{\Delta t_{13}} = \frac{5.0 \times 10^1 \text{ yd}}{1.00 \times 10^2 \text{ s}} = 5.0 \times 10^{-1} \text{ yd/s}$$

Meanwhile the total **distance** you've covered is 1.50×10^2 m: you walk 1.00×10^2 m, then turn around and walk half that distance again. The distance you travel divided by the time elapsed is a scalar quantity called the average **speed**:

$$\text{average speed} = \frac{\text{distance}}{\text{time}} = \frac{1.50 \times 10^2 \text{ yd}}{6.0 \times 10^1 \text{ s}} = 2.5 \text{ yd/s}$$

This certainly isn't the same thing as your average velocity! (If you hadn't turned around, though, they *would* be the same.)

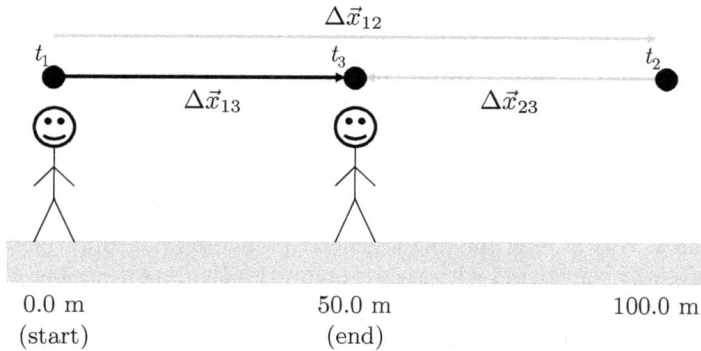

FIGURE 1.6
The motion diagram has three positions. The displacement vectors from time 1 to time 2 and, separately, from time 2 to time 3 are shown in gray. The overall displacement is from the first position (time point 1) to the last (time point 3) and is shown in black.

We can convert from yd/s to m/s:

$$2.5\frac{\text{yd}}{\text{s}} \times \frac{0.9144 \text{ m}}{1 \text{ yd}} = 2.3 \text{ m/s}$$

And to go to miles per hour:

$$2.3\frac{\text{m}}{\text{s}} \times \frac{1 \text{ mi}}{1609 \text{ m}} \times \frac{3600 \text{ s}}{1 \text{ hr}} = 5.1\frac{\text{mi}}{\text{hr}}$$

This is a reasonable value for someone jogging: running is usually considered to be above around 6 mph (elite sprinters can exceed 20 mph!) and walking is usually 4 mph and slower.

This example, where we consider someone backtracking, emphasizes how different the *average* speed can be compared to the *average* velocity. That being said, if we shrink the time interval then this difference becomes less pronounced, and (as we shall see) in the limit where the time difference is arbitrarily small the *instantaneous* speed becomes equivalent to the magnitude of the *instantaneous* velocity. (It is worth emphasizing that speed, like distance, is a *scalar*, while velocity, like displacement, is a *vector*.)

Let me make one more comment before we move on. Figure 1.7 looks awfully similar to Figure 1.6, except $\Delta \vec{x}$ has been replaced with a $\vec{v}$. Why can we label a vector as a *velocity* when we are connecting dots that represent the *position* of an object? A vector indicating a change in position is a *displacement*, not a velocity; the two quantities don't even have the same units!

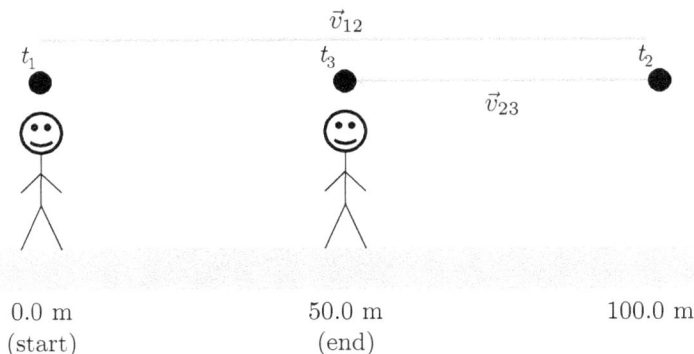

FIGURE 1.7
The same motion diagram with the average velocities $\vec{v}_{12}$ and $\vec{v}_{23}$ included.

Let's address this by first making a general point. In analyzing the world around us, we will often be confronted with vector quantities that are described with different units. For instance, when talking about the motion of charged particles, we'll think about the particle's position, velocity, acceleration, and the existence of magnetic and/or electric fields at the particle's location – each of these five quantities is a vector, and each has different units!

The way to reconcile this with the fact that the length of a vector conveys its magnitude is to realize that *it only makes sense to compare the lengths of vectors with the same units.* Drawing an electric field vector that is twice as long as a velocity vector doesn't inherently mean anything: you could also draw an accurate diagram where the relative sizes are reversed. (In such cases the vectors need to be labeled with their magnitudes, so you know what you're dealing with, or otherwise you need to know the magnitude from information provided outside of the diagram.) What matters is the length of a vector compared to the length of other vectors of the same type: you can correctly say things like "*this* velocity has twice the magnitude as *that* velocity because it is twice as long on the page".

Let's now turn to Figure 1.7. The average velocity is related to the corresponding displacement according to $\vec{v} = \Delta \vec{x}/\Delta t$, and because the time interval Δt is the same for every successive pair of positions in a motion diagram, knowing $\Delta \vec{x}$ is sufficient to determine the relative size of $\vec{v}$. For example, if $\Delta \vec{x}_{12}$ is twice the value of $\Delta \vec{x}_{23}$, then we know that $\vec{v}_{12}$ is twice the value of $\vec{v}_{23}$ (thus we say that the two vectors have *proportional* magnitudes).

The fact that Δt is the same *for successive points* in a motion diagram is precisely why I didn't draw a velocity vector from point 1 to point 3: that would imply that the magnitude of the average velocity over the whole trip, $\vec{v}_{13}$, is the same as the magnitude of the average velocity for the second part of the trip, $\vec{v}_{23}$. But as you can check for yourself, this isn't the case: the reason is because even though the magnitude of the displacements ($|\Delta x_{13}| = |\Delta x_{23}|$) are the same, the time intervals aren't ($\Delta t_{13} \neq \Delta t_{23}$), so the velocities don't have equal magnitudes ($|\Delta x_{13}|/\Delta t_{13} \neq |\Delta x_{23}|/\Delta t_{23}$).

I've gone into the weeds a bit here, but the point is simply that the velocity points in the same direction as the displacement, so it can be convenient to indicate the direction (and relative magnitude) of the average velocity between two successive points on a motion diagram by connecting them with a vector and labeling it as the average velocity. As we shall see as we proceed, including velocity vectors in this way can be useful when we consider the *acceleration* of an object.

Example 1.3 Approaching Friends ★★

You and a friend are facing each other from an initial distance of 1.00×10^2 m. At the same instant, you begin walking toward one another: you move at 2.00 m/s and your friend moves at 1.00 m/s. How far have each of you moved after 5.00 s, and what is your separation at that point? How far will each of you have moved when you walk into one another, and when does that occur?

It is worth noting that we could approach this problem several different ways. For instance, you could generate a data table of each person's position in some small time interval, and proceed until they collide. Or, you could make a graph with the position on the vertical axis and time on the horizontal axis: you have enough information to represent the position as a function of time as a line for each walker, and the collision occurs when the lines intersect (see Problem 1.14).

Here, we will take an algebraic approach. We can use Equation (1.2) and algebraically solve for $\Delta \vec{x}$ to determine how far you will both have moved:

$$\Delta \vec{x} = \vec{v}\Delta t$$

We should be careful about *signs*; if you are walking in the *positive* direction then your friend is walking in the *negative* direction. Thus

$$\Delta \vec{x}_{\text{you}} = (2.00 \text{ m/s})\,(5.00 \text{ s}) = 1.00 \times 10^1 \text{ m}\hat{x}$$

and

$$\Delta \vec{x}_{\text{friend}} = (-1.00 \text{ m/s})\,(5.00 \text{ s}) = -5.00 \text{ m}\hat{x}$$

So between the two of you, you've covered a total of 15.00 m and there are 85.00 m left between you.

Clearly more time will pass before you collide! To work out when this occurs, we'll invoke Equation (1.1) as well as Equation (1.2) for you and your friend:

$$\vec{x}_{f,\text{you}} - \vec{x}_{i,\text{you}} = \vec{v}_{\text{you}}\Delta t$$
$$\vec{x}_{f,\text{friend}} - \vec{x}_{i,\text{friend}} = \vec{v}_{\text{friend}}\Delta t$$

Now note that if *you* start at 0.0 m then your friend starts at 1.00×10^2 m. If we have the same $\vec{x}_f$ at the same time, we find that we have two equations and two unknowns:

$$x_f - 0 \text{ m} = (2.00 \text{ m/s}) \Delta t$$
$$x_f - 1.00 \times 10^2 \text{ m} = (-1.00 \text{ m/s}) \Delta t$$

(In the first line I've algebraically canceled the $\hat{x}$ from both sides.) If we solve the first equation for x_f and insert into the second, we find

$$(2.00 \text{ m/s}) \Delta t - 1.00 \times 10^2 \text{ m} = (-1.00 \text{ m/s}) \Delta t$$
$$(3.00 \text{ m/s}) \Delta t = 1.00 \times 10^2 \text{ m}$$
$$\Delta t = 33.3 \text{ s}$$

Now that we know how long each person walks, it isn't too challenging to find *where* they are:

$$x_f = (2.00 \text{ m/s}) \Delta t = (2.00 \text{ m/s}) (33.3 \text{ s}) = 66.6 \text{ m}$$

So you walked 66.6 m and your friend walked $1.00 \times 10^2 \text{ m} - 66.6 \text{ m} = 33$ m. Figure 1.8 summarizes the motion.

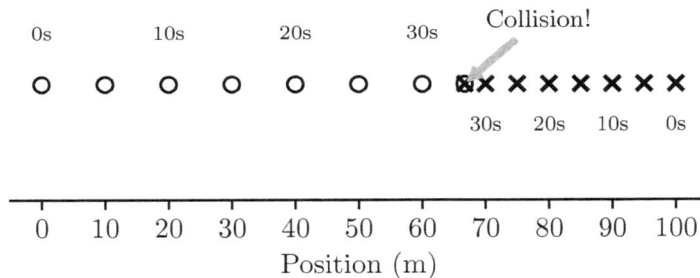

FIGURE 1.8
You (circles) and your friend (crosses) begin 1.00×10^2 m apart, walk toward one another, and collide near the 67 m mark after about 33 s.

At least in retrospect, this shouldn't be too shocking of a result: you are walking twice as fast as your friend, so you cover twice the distance. By the same logic, if you were walking, say, three times as fast as your friend, then you should cover *triple* the distance, which means you'd meet 75 m from your starting point: split the total distance into four equal distances ($100 \text{ m}/4 = 25$ m), so three can be assigned to you ($3 \times 25 = 75$ m) and one (25 m) to your friend.

Just as velocity measures the rate at which an object's *position* is changing (in $\frac{\text{m}}{\text{s}}$), we define the average **acceleration** to measure the rate at which an object's *velocity* is changing (in $\frac{\text{m/s}}{\text{s}}$, or $\frac{\text{m}}{\text{s}^2}$):

$$\vec{a}_{\text{avg}} = \frac{\Delta \vec{v}}{\Delta t} = \frac{\vec{v}_f - \vec{v}_i}{\Delta t} \tag{1.4}$$

Similarly, the *instantaneous* acceleration is the time derivative of the velocity:

$$a\left(t\right) = \frac{d}{dt} v\left(t\right) \tag{1.5}$$

In one-dimensional motion, acceleration measures the extent to which the object is "speeding up" or "slowing down", where **speed** is simply the distance covered (without regard to direction) per unit time.[7] Importantly, everything we said about *position and velocity* surrounding Equation (1.2) applies also to *velocity and acceleration*. Thus, for instance, *the velocity vector grows/shrinks in the direction of the acceleration*. A good example of this is a car accelerating from rest. If we set the origin on top of the car's initial position and set the positive direction in the direction of the car's motion, then the motion diagram will look something like Figure 1.9, where we've sampled the time points at 0.50 s intervals and used an acceleration of 6.0 m/s². The fact that the acceleration vectors are the same (they have the same magnitude and point in the same direction) means that the velocity changes by the same amount in every equal time interval.

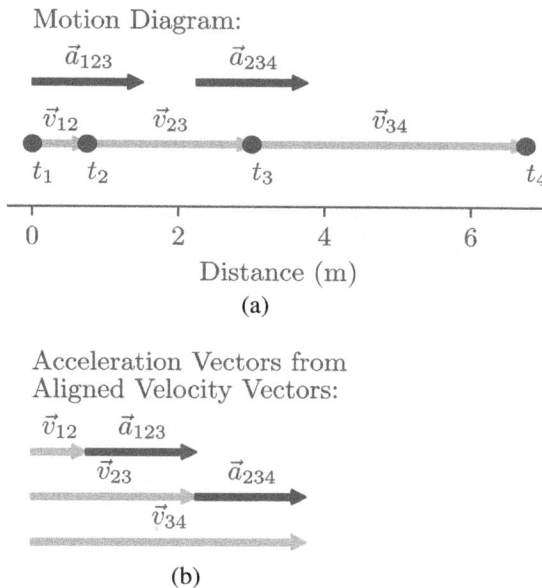

Motion Diagram:

(a)

Acceleration Vectors from
Aligned Velocity Vectors:

(b)

FIGURE 1.9

(a) The motion diagram for a car accelerating from rest. (b) To determine the acceleration vectors, we draw the velocity vectors with a common starting point. (Unlike velocity vectors, this is naturally the case for position vectors because position vectors all start at the origin, as in Figure 1.4). Once the velocity vectors are drawn with a common starting point, each acceleration vector is drawn from the end one velocity vector to the end of the following velocity vector (e.g. $\vec{a}_{123}$ runs from the end of $\vec{v}_{12}$ to the end of $\vec{v}_{23}$), just as *velocity* vectors are drawn from one *position* vector to the next. In the motion diagram at the bottom, each acceleration vector is normally placed next to the middle position used for calculating the acceleration (e.g. $\vec{a}_{123}$ is placed next to the car's position at t_2, which is equivalent to the position where the constituent velocities $\vec{v}_{12}$ and $\vec{v}_{23}$ meet).

---

[7]More generally, acceleration measures how quickly the velocity vector is changing. You might think this means the same thing, but remember vectors have a direction as well as a magnitude: we will see that when an object is moving in a circular path it can have a constant *speed* but still accelerate because the *direction* of the velocity vector is changing as the object turns.

Another example is tossing a ball into the air. The velocity is initially up, and as it slows the velocity shrinks to 0 m/s. Evidently the acceleration points down, opposite the velocity. Then, the ball picks up speed as it falls back toward your hand; the acceleration still points down, now in the same direction as the velocity.

In fact, if we ignore complicating factors like rocket propulsion or air resistance,[8] then every object that is released near the surface of the Earth accelerates down toward the surface of the Earth at a constant rate called **the acceleration due to gravity**:

$$g = 9.8 \frac{\text{m}}{\text{s}^2} \tag{1.6}$$

We'll get in to these considerations and why g has this value in Chapter 3. Example 1.4 considers the "ball tossed up" scenario in more detail.

Example 1.4 Ball Toss Motion Diagram   ⋆

Table 1.2 shows the height of a ball as it is tossed into the air: we set the origin at the initial height of the ball, and *up* is the positive direction of our coordinate system. (a) Show that the velocity and acceleration values shown in the table follow from the time and position values. (b) Draw an annotated motion diagram that includes velocity and acceleration vectors. (c) Make a graph of the position vs. time, velocity vs. time, and acceleration vs. time (with time on the horizontal axis in each case). Connect your data points with straight lines to guide the eye.

TABLE 1.2

Time label	Time (s)	x (m)	v (m/s)	a (m/s^2)
t_1	0.00	0.00		
t_2	0.51	3.83	7.5	
t_3	1.02	5.10	2.5	-9.8
t_4	1.53	3.83	-2.5	-9.8
t_5	2.04	0.00	-7.5	-9.8

(a) Note first that $\Delta t = 0.51$ s; each row of the time column increases by this amount. To calculate an average velocity, we need Δt and two adjacent positions to calculate a displacement. The *first* velocity we can calculate uses the first two positions:

$$\vec{v}_{12} = \frac{\vec{x}_2 - \vec{x}_1}{\Delta t} = \frac{(3.83 - 0.00) \text{ m}\hat{x}}{0.51 \text{ s}} = 7.5 \frac{\text{m}}{\text{s}}\hat{x}$$

Rounded to two significant digits, this is the first velocity reported. Working our way down the column, each velocity uses the position to its left and one above (note that we get 1 velocity for each successive pair of positions, which is why there is always one less average velocity to report than there are positions reported).

Once we have the average velocities between each successive pair of positions, we can calculate the average accelerations between each successive pair of velocities. For instance, the first average acceleration is given by

$$\vec{a}_{123} = \frac{\vec{v}_{23} - \vec{v}_{12}}{\Delta t} = \frac{(2.49 - 7.51) \frac{\text{m}}{\text{s}}\hat{x}}{0.51 \text{ s}} = -9.8 \frac{\text{m}}{\text{s}^2}\hat{x}$$

---

[8]For instance, if you drop a feather and a nail, they'll certainly accelerate at different rates because air resistance has a much stronger effect on the feather.

where I've included one additional significant figure in the calculation for the velocities to avoid rounding error when determining the acceleration. Repeating the process as we work down the column shows a consistent value for the acceleration with a magnitude equal to g, as we expect from Equation (1.6). (The negative sign comes from the fact that we called *up* positive and the acceleration points *down*, toward the surface of the Earth.)

The motion diagram requested in (b) is shown in Figure 1.10. Note, for instance, that the acceleration vectors are all drawn with the same length, as we expect from the values shown in the table. You're invited to draw the acceleration vectors by first drawing the velocity vectors with a common starting point, as in Figure 1.9, in Problem 1.16.

For part (c), we need to make a series of graphs using the values from the table. The result is shown in Figure 1.11, where I've reported velocities and accelerations with the corresponding average time (e.g. $\vec{v}_{12}$ uses the average of t_1 and t_2, while the first acceleration uses t_2 since it is directly in the middle of the time interval used to calculate $\vec{a}_{123}$).

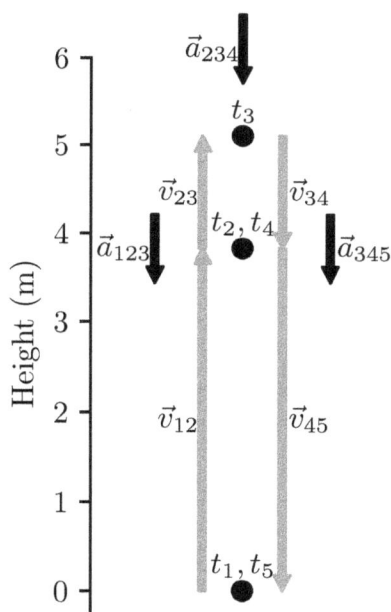

FIGURE 1.10
Motion diagram.

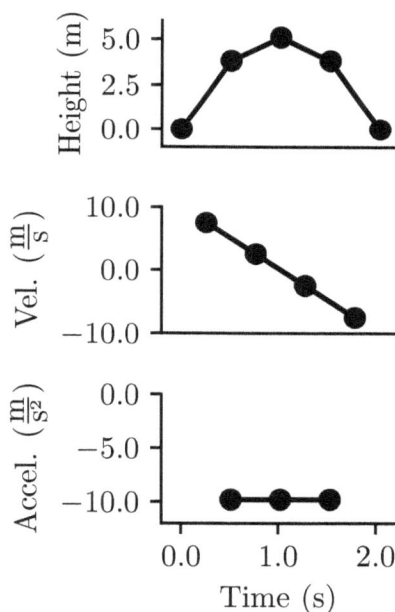

FIGURE 1.11
Velocities and accelerations with the corresponding average time.

1.2 Motion Graphs

Now that we have defined position, velocity, and acceleration, we can talk in more depth about **motion graphs** like the one shown in the right panel of Figure 1.1 and more recently in Figure 1.11. While the horizontal axis of a motion graph always represents *time*, the vertical axis can show *position*, *velocity*, or *acceleration*. The central question is this: given one motion graph for an object (say, velocity vs. time), what can we say about the others? Answering this question involves analyzing graphs and ideas from calculus; you may wish to read Sections A.4 and A.5 if you haven't already.

To proceed, recall that if some section of a graph is a line, then the slope of that line is given simply by

$$\text{slope} = \frac{\text{change in vertical coordinate}}{\text{change in horizontal coordinate}} = \frac{\text{rise}}{\text{run}}$$

If we have a graph of, say, position vs. time $(x(t))$, then slope $= \frac{\Delta \vec{x}}{\Delta t}$, which is equivalent to the average velocity. In the general case, though, $x(t)$ can be varying in complex ways for whatever time interval Δt we consider. It is more precise to shrink Δt down to an infinitesimally small interval, in which case $\frac{\Delta \vec{x}}{\Delta t}$ becomes the derivative of $x(t)$, i.e. $\frac{d}{dt}x(t)$.

The derivative of $x(t)$ evaluated at a certain instant is the slope of the line tangent to the curve at that point (Fig. A.20). Moreover, according to Equation (1.3), the derivative of $x(t)$ is the instantaneous velocity. Thus *the velocity at any instant is given by the slope of the line tangent to the position vs. time curve.* Analogously, the *acceleration* at any instant is given by the slope of the line tangent to the *velocity* vs. time curve.

In Figure 1.12 we demonstrate this for the case where $x(t)$ starts at the origin (i.e. $x_i = 0$) and has a constant nonzero slope. Because the slope (velocity) is constant, $v(t)$ is a horizontal line with a y-intercept given by the slope of the $x(t)$ curve. Similarly, because $v(t)$ is constant, its slope (which is the acceleration) is 0 m/s^2.

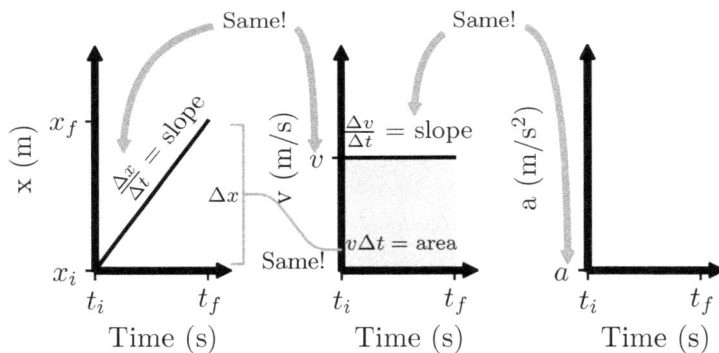

FIGURE 1.12
The position vs. time, velocity vs. time, and acceleration vs. time graphs are all related. We can look at the *slope* of one graph to determine the value on the graph to its right: the slope of the x-t graph at any instant is equal to the velocity, and the slope of the v-t graph at any instant is equal to the acceleration. We can also look at the *area* bounded by a graph to determine how the value on the graph to the left changes. For example, the shaded area on the v-t graph s equal to the displacement: $\Delta x = v\Delta t$. (Applying the same logic to the a-t and v-t graphs, we see that because $a = 0$ m/s^2, the a-t graph bounds no area and therefore the velocity is constant.)

Notice that in this analysis we've analyzed slopes to go from $x(t)$ to $v(t)$ to $a(t)$. We can reverse the direction of the analysis. To see how, consider the $v(t)$ curve in Figure 1.12:

we have a constant velocity with magnitude v. The area between this line and the time axis is a rectangle with area A given by

$$tA = v\Delta t$$

But the right hand term in the above equation is the change in position ($\vec{v} = \frac{\Delta \vec{x}}{\Delta t} \Rightarrow \vec{v}\Delta t = \Delta \vec{x}$). In the *general* case, the velocity could be changing significantly for any finite time interval Δt, so just as we argued above it pays to shrink Δt down to an infinitesimally small interval, in which case the displacement is given by the *integral* of $v(t)$ with respect to time (Fig. A.21).

If you're unfamiliar with integral calculus, don't panic. The key point is that an integral measures a *cumulative effect*: in this case, the cumulative effect of the instant-by-instant displacements that are determined by the velocity. Mathematically, we express the displacement as an integral of velocity over time like so:

$$x(t) = \int v(t)\, dt + x_i \tag{1.7}$$

In other words, *the change in position is given by the area between the time axis and the velocity curve in a velocity vs. time graph*. Analogously, *the change in velocity is given by the area between the time axis and the acceleration curve in an acceleration vs. time graph*:

$$v(t) = \int a(t)\, dt + v_i \tag{1.8}$$

How do these principles apply to the motion graphs shown in Figure 1.12? Well, start with the acceleration curve: the acceleration is 0, so there is no area and therefore no change in the velocity.[9] We don't see directly what the velocity *is*, but we know that in the velocity vs. time graph the velocity will be a horizontal line.

If you repeat this analysis for the velocity vs. time graph, the area is a simple rectangle, as we noted above. The result is a position vs. time graph with a displacement given by $\Delta x = v\Delta t$. The position vs. time graph is *linear* because this equation is linear in time.[10]

We've spend quite a bit of time analyzing Figure 1.12, where there is *zero* acceleration. It is more interesting to consider the case where there is *nonzero* acceleration. Let's next consider the case where the acceleration is *constant* (but nonzero), as in the bottom of Figure 1.13. Here the velocity changes linearly in time (remember that acceleration is related to velocity in the same way that velocity is related to position). To make things simple, suppose $\vec{v}_i = 0$ m/s. What does the position-time graph look like now?

Well, in this case the velocity-time graph forms a triangle that starts at the origin; the displacement is given by its area: $\Delta \vec{x} = \frac{1}{2}(\Delta t)(\Delta \vec{v})$. But because $\vec{v}_i = 0$ and $\vec{v}_f = \vec{a}\Delta t$, we can write instead

$$\Delta \vec{x} = \frac{1}{2}(\Delta t)(\vec{a}\Delta t - 0) = \frac{1}{2}a(\Delta t)^2$$

This means that the position varies *quadratically* with time, and the position-time graph is a *parabola*. We show the motion graphs for this situation on the bottom of Figure 1.13. For comparison, a stripped-down version of Figure 1.12 is shown on the top.

---

[9]Formally, $v = \int 0\, dt = v_i$ where v_i is the constant of integration: the initial velocity. Thus the velocity after the period we're considering is the same as the initial velocity. In other words, the velocity doesn't change.

[10]Formally, $x = \int v\, dt = vt + x_i$ where x_i is the constant of integration: the initial position.

Before we move on, let me emphasize the importance of Figure 1.13. When objects are dropped near the surface of the Earth, they accelerate toward the surface of the Earth with a constant magnitude of $g = 9.8\frac{m}{s^2}$. Thus the bottom panels of Figure 1.13 shows the general behavior of anything dropped near the surface of the Earth.[11] Beyond that, these "flat $\rightarrow$ linear $\rightarrow$ parabolic" mathematical relationships will come up again in other contexts (for instance, in relating *energy* and *force*). It is well worth your time to make sure you follow the arguments presented here.

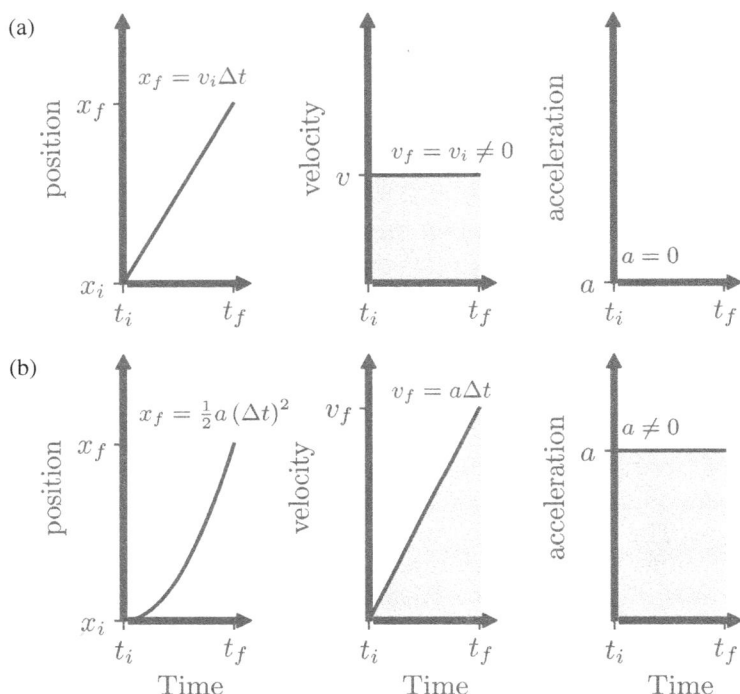

FIGURE 1.13
(a) This replicates Figure 1.12, where the velocity is a constant nonzero value and, therefore, the acceleration is 0 and the position increases linearly with time ($\Delta \vec{x} = \vec{v}\Delta t$; note also that here we suppose $\vec{x}_i = 0$). (b) The parallel motion graphs for the case where the acceleration is a constant nonzero value and, therefore, the velocity increases linearly with time ($\Delta \vec{v} = \vec{a}\Delta t$; we suppose $\vec{v}_i = 0$) and the position (again, supposing $\vec{x}_i = 0$) increases quadratically with time ($\Delta \vec{x} = \frac{1}{2}a\left(\Delta t\right)^2$).

We'll analyze these relationships in some detail using calculus in Example 1.10, after considering a sequence of examples that can be analyzed geometrically. First, we present one more set of motion graphs for constant (nonzero) acceleration in Figure 1.14. All of the same arguments we've been developing (e.g. the velocity is given by the tangent of the position-time graph) apply here, as well–see if you agree.

---

[11]Assuming we call "down toward the Earth" the positive direction – since the acceleration is positive in Figure 1.13. The next example considers the case where we call *up* the positive direction, which means acceleration due to gravity is negative.

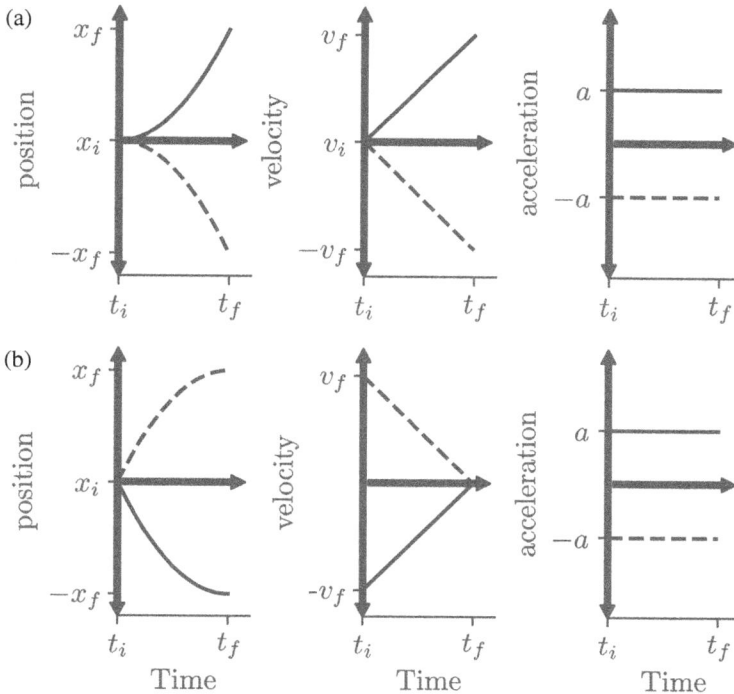

FIGURE 1.14
Two sets of motion graphs for constant acceleration. In both part figures, the solid curves correspond to a positive constant acceleration and the dashed curves correspond to a negative constant acceleration, and in all cases we set $x_i = 0$ m. (a) The velocities are "positive and increasing" (solid) or "negative and decreasing" (dashed). (b) The velocities are "positive and decreasing" (dashed) or "negative and increasing" (solid).

Example 1.5 A Car Accelerating ⋆
A certain car is advertised as being able to go from 0 to 6.0×10^1 mph (27 m/s) in 9.5 seconds.

(a) What is the average acceleration during this time?

(b) Assuming the car accelerates at this average rate the entire time, make a plot of the acceleration (on the vertical axis) vs. the time (on the horizontal axis). Repeat for velocity vs. time.

(c) Determine the total distance traveled by the car from your plot in (b).

First, we must choose our coordinate system. The natural choice is to set the origin at the car's initial position (so $x_i = 0\hat{x}$), and choose the positive direction to be the direction of the car's acceleration (so the velocities and accelerations are all positive).

(a) The average acceleration is just $\vec{a} = \frac{\Delta \vec{v}}{\Delta t} = \frac{(27-0)\frac{m}{s}\hat{x}}{9.5 \text{ s}} = 2.8\frac{m}{s^2}\hat{x}$.

(b) The acceleration vs. time and velocity vs. time plots are as shown in Figure 1.15.

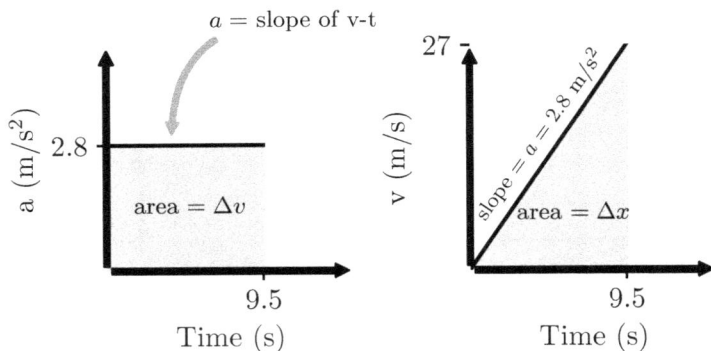

FIGURE 1.15
The motion graphs for Example 1.5.

The plots are labeled with some key bits of analysis; of particular note is the fact that the area under the acceleration vs. time plot corresponds to the change in velocity. This follows directly from the equation defining acceleration, from which we find $\Delta\vec{v} = \vec{a}\Delta t$ (the shaded rectangle in the acceleration vs. time plot has an area equal to its length multiplied by its width, which is just the right hand side of the equation). Meanwhile, if we plot any two quantities against one another, then the slope is just "the rise over the run", or $\frac{\text{change in vertical axis}}{\text{change in horizontal axis}}$. For a velocity-time graph, this corresponds to $\frac{\Delta\vec{v}}{\Delta t}$, which is the acceleration.

(c) From the velocity equation $\vec{v} = \frac{\Delta\vec{x}}{\Delta t}$ we find $\Delta\vec{x} = \vec{v}\Delta t$, which graphically corresponds to the area under the velocity-time curve. From the above diagram we see that the area is given by a triangle with a base of 9.5 seconds and a height of 27 m/s, from which we find $\Delta\vec{x} = \frac{1}{2}(\text{base} \times \text{height}) = \frac{1}{2}(9.5\text{ s})\left(27\frac{\text{m}}{\text{s}}\right)\hat{x} = 130\text{ m}\hat{x}$.

Example 1.6 Slamming on the Brakes   ★
The same car from Example 1.5 is advertised as having a 6.0×10^1 mph to 0 mph stopping distance of 130 ft (4.0×10^1 m). What is its average acceleration? How long does this take?

We choose our coordinate system such that the car is initially at the origin and moving in the positive direction. Here the car is going from a higher speed to a lower speed (to a complete stop, in fact), which is to say it is slowing down and the acceleration is negative:

$$\vec{a} = \frac{\Delta\vec{v}}{\Delta t} = \frac{\vec{v_f} - \vec{v_i}}{\Delta t} = \frac{(0 - 27)\frac{\text{m}}{\text{s}}\hat{x}}{\Delta t}$$

We can't solve this equation directly because there are two unknown variables. If you're faced with such a scenario and are unsure of how to proceed, think through what else you know about the situation. Here, for instance, we can look to the other part of the question and describe the velocity-time graph, as shown in Figure 1.16.

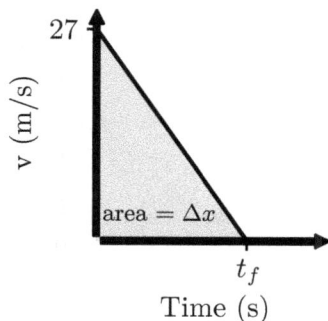

FIGURE 1.16
Velocity-time graph.

As in Example 1.5, the area bounded by the time axis and the velocity curve corresponds to the displacement, which we know in this problem is 40 m $= \frac{1}{2}\left(27\frac{m}{s}\right)\Delta t$ (because the area of the v-t curve is a right triangle). Solving for Δt yields

$$\Delta t = \frac{2\left(4.0\times 10^1 \text{ m}\right)}{27\frac{m}{s}} = 3.0 \text{ s}$$

Then, returning to the equation for acceleration, we find

$$\vec{a} = \frac{(0-27)\frac{m}{s}\hat{x}}{3.0 \text{ s}} = -9.0\frac{m}{s^2}\hat{x}$$

Example 1.7 Racers' Position and Velocity Graphs   ⋆
Consider a 1.0 km race between a turtle and a hare, where the turtle quickly accelerates to its top speed of 0.50 m/s. The hare, being magnanimous (or perhaps overconfident), decides to take a nap. If it can quickly accelerate to its top speed of 2.0×10^1 m/s once it wakes up, then how long can the hare sleep if it is to beat the turtle? Make a plot of the position (on the vertical axis) vs. the time (on the horizontal axis) for the turtle and the hare. Repeat for velocity vs. time.

From the equation for velocity, $\vec{v} = \frac{\Delta \vec{x}}{\Delta t}$, we find $\Delta t = \frac{\Delta \vec{x}}{\vec{v}}$. Thus $\Delta t_{\text{turtle}} = \frac{1.0\times 10^3 \text{ mm}}{0.50 \text{m/s}} = 2.0 \times 10^3$ s and $\Delta t_{\text{hare}} = \frac{1.0\times 10^3 \text{m}}{2.0\times 10^1 \text{m/s}} = 5.0 \times 10^1$ s. It follows that if the hare sleeps for $\left(2.0 \times 10^3 - 5.0 \times 10^1\right)$ s, or 32.5 minutes, it will tie the turtle. Any time less than this and the hare will win. The position-time and velocity-time graphs are shown in Figure 1.17.

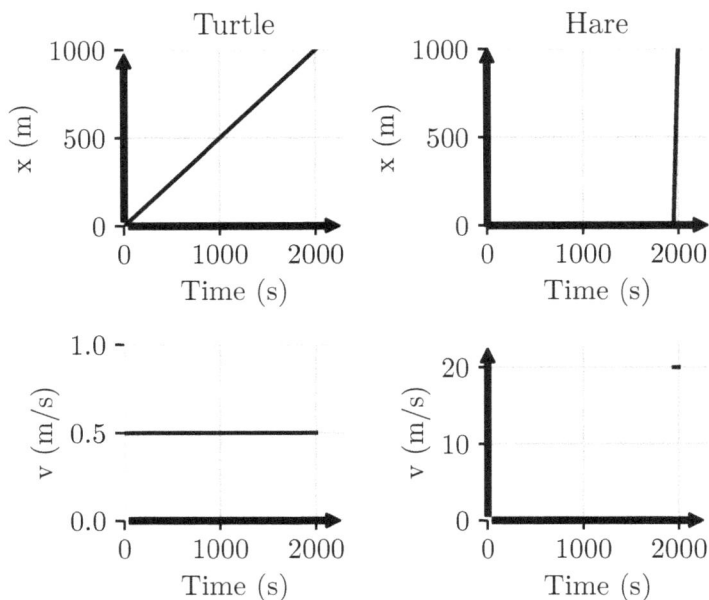

FIGURE 1.17
Motion graphs for the turtle and hare. They have the same displacement, but the turtle achieves it with a slow velocity over a long time, while the hare achieves it with a high velocity over a short time. The time scale highlights the fact that the hare leaps to the finish line almost instantly, at the very end of the race!

Example 1.8 Velocity vs. Time for a Unicyclist   ⋆
Figure 1.18 shows a sample velocity vs. time graph for a unicyclist. Generate the corresponding acceleration vs. time graph. What is the unicyclist's total displacement? What is the unicyclist's total change in velocity?

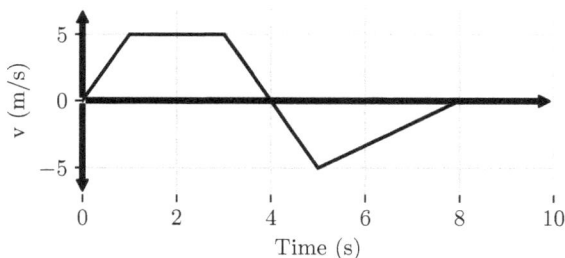

FIGURE 1.18
A sample velocity vs. time graph.

The value on the acceleration-time graph corresponds to the slope of the velocity-time graph. With this in mind, we can construct the acceleration-time graph, as shown in Figure 1.19.

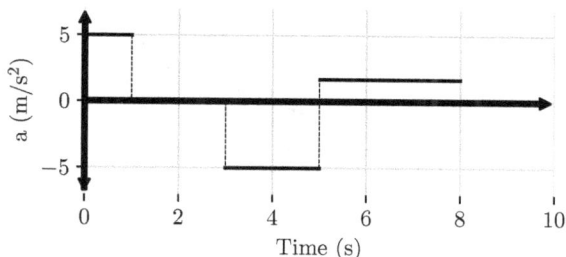

FIGURE 1.19
The acceleration vs. time graph corresponding to Figure 1.18.

The total displacement is the area bounded by the velocity curve and the time axis. Importantly, if some component of the area is below the time axis, the velocity is negative (the unicyclist is moving backward), and so too is the displacement. Referring to the velocity-time graph, the area is split up into rectangles and triangles. Computing the areas from left to right and combining them, we find

$$\Delta x = \frac{1}{2}\left(1 \text{ s} \times 5\frac{\text{m}}{\text{s}}\right) + \left(2 \text{ s} \times 5\frac{\text{m}}{\text{s}}\right) - \frac{1}{2}\left(1 \text{ s} \times 5\frac{\text{m}}{\text{s}}\right) - \frac{1}{2}\left(1 \text{ s} \times 5\frac{\text{m}}{\text{s}}\right) - \frac{1}{2}\left(3 \text{ s} \times 5\frac{\text{m}}{\text{s}}\right)$$
$$= 5 \text{ m}$$

The total change in velocity, meanwhile, can be computed directly from the velocity-time plot: at the beginning ($t = 0$s) the velocity is 0 m/s, and at the end ($t = 8$ s) the velocity is 0 m/s. Thus the unicyclist started at rest and ends at rest, and there is no overall change in velocity. We could also determine this by calculating the total area bounded by the time axis and the acceleration curve on the acceleration-time graph. These are all rectangles, and working from left to right we find

$$\Delta v = \left(1 \text{ s} \times 5\frac{\text{m}}{\text{s}^2}\right) + \left(2 \text{ s} \times 0\frac{\text{m}}{\text{s}^2}\right) - \left(2 \text{ s} \times 5\frac{\text{m}}{\text{s}^2}\right) + \left(3 \text{ s} \times \frac{5 \text{ m}}{3 \text{ s}^2}\right) = 0\frac{\text{m}}{\text{s}}$$

as expected.

Example 1.9 Position, Velocity, and Acceleration Graphs ⋆⋆
Figure 1.20 shows a partially filled set of motion graphs. Each of the time periods (separated by vertical dashed lines) are filled in with just *one* of the position, velocity, and acceleration graphs. The object finishes at $x = 8.0$ m when $t = 6.0$ s. (a) Fill in the missing information on the graphs.

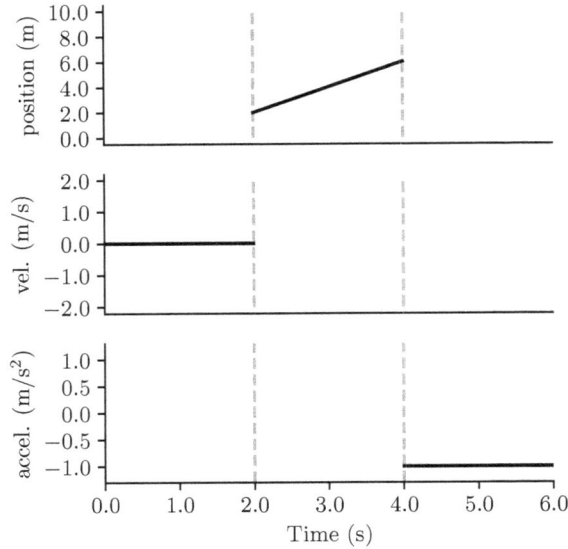

FIGURE 1.20
Partially filled set of motion graphs.

Let's begin with the leftmost region ($t = 0.0$ s to $t = 2.0$ s). The velocity 0.0 m/s, which means the position isn't changing. Thus the position marked at $t = 2.0$ s, $x = 2.5$ m, must be the position for the entirety of the first two seconds. Meanwhile, the acceleration for the first two seconds must be 0.0 m/s^2 if the velocity is *any* constant value.

That does it for the first two seconds. What about the second region, from $t = 2.0$ s to $t = 4.0$ s? The position increases linearly from 2.0 m to 6.0 m, so we can calculate the magnitude of the velocity:

$$v = \frac{\Delta x}{\Delta t} = \frac{(6.0 - 2.0)\,\text{m}}{2.0\text{s}} = 2.0 \text{ m/s}$$

Note that we find a **discontinuity**: the velocity jumps from 0.0 m/s in the first region to a nonzero value at $t = 2.0$ s. This is not physically possible, but we can consider this to be a simplification for the case of a very rapid acceleration surrounding $t = 2.0$ s. In any event, the velocity is a constant value for $t = 2.0$ s to $t = 4.0$ s (the position-time graph has a constant slope), so the acceleration must be 0.0 m/s^2 for this interval.

Finally, for the last two seconds, we see that the acceleration is a constant negative value of -1.0 m/s^2. This is the value of the slope of the velocity-time graph. If we start with the 2.0 m/s we found for $t = 4.0$ s, then over the ensuing 2.0 s the velocity will drop linearly to 0.0 m/s. The position, meanwhile, will increase *at a decreasing rate*: the velocity is positive but decreasing, so the position increases rapidly (relatively speaking) at the beginning of this interval, then decreases to a standstill at the end of the interval (when the velocity reaches 0.0 m/s; see the dashed curves on the bottom of Figure 1.14 for comparison). As indicated in the problem statement, it finishes at $x = 8.0$ m. Figure 1.21 shows the complete motion graphs.

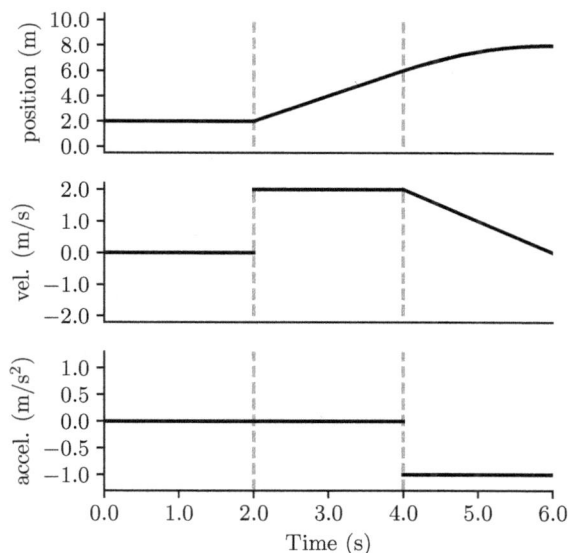

FIGURE 1.21
The complete motion graphs.

Example 1.10 Position, Velocity, and Acceleration with Calculus $\star\star \int dx$
(The "$\int$" and "dx" labels respectively indicate that the example includes integral and differential calculus.)

(a) At $t = 0$ s, a particular object has an arbitrary initial position and initial velocity. If the acceleration is given by $a(t) = a$ (that is, the acceleration is a constant value a), determine $v(t)$ and $x(t)$. Check your answers by finding $v(t)$ from $x(t)$ and $a(t)$ from $v(t)$.

(b) Suppose an object's position as a function of time is given by $x(t) = A\sin(\omega t)$. Determine the velocity and acceleration at an arbitrary time t.

(a) Adapting equation (1.7) from "position and velocity" to "velocity and acceleration", we have

$$v(t) = \int a(t)\, dt$$
$$v(t) = \int a\, dt$$
$$v(t) = at + C$$
$$v(t) = v_i + at$$

In the last step we have noted that the constant of integration corresponds to the initial velocity, i.e. $C = v_i$.

To acquire $x(t)$, we have

$$x\left(t\right) = \int v\left(t\right) dt$$

$$x\left(t\right) = \int \left(v_i + at\right) dt$$

$$x\left(t\right) = v_i t + \frac{1}{2}at^2 + C$$

$$x\left(t\right) = x_i + v_i t + \frac{1}{2}at^2$$

Once again, in the last step we noted that the constant of integration corresponds to the initial position, i.e. $C = x_i$.

To check our work, we'll take the derivative of our expressions for $x\left(t\right)$ and $v\left(t\right)$. Starting with equation (1.3), we have

$$v\left(t\right) = \frac{dx\left(t\right)}{dt}$$

$$v\left(t\right) = \frac{d}{dt}\left(x_i + v_i t + \frac{1}{2}at^2\right)$$

$$v\left(t\right) = v_i + at$$

as expected. Similarly,

$$a\left(t\right) = \frac{dv\left(t\right)}{dt}$$

$$a\left(t\right) = \frac{d}{dt}\left(v_i + at\right)$$

$$a\left(t\right) = a$$

as expected. We will obtain these important "constant acceleration" equations for position and velocity in the next section by taking a more cumbersome algebraic and geometric approach.

(b) Because we are given $x\left(t\right)$, we'll take its derivative to find $v\left(t\right)$ and then take the derivative of $v\left(t\right)$ to find $a\left(t\right)$. To begin, we have

$$v\left(t\right) = \frac{dx\left(t\right)}{dt}$$

$$v\left(t\right) = \frac{d}{dt}\left(A\sin\left(\omega t\right)\right)$$

$$v\left(t\right) = A\omega\cos\left(\omega t\right)$$

We repeat this process to determine the acceleration:

$$a\left(t\right) = \frac{dv\left(t\right)}{dt}$$

$$a\left(t\right) = \frac{d}{dt}\left(A\omega\cos\left(kt\right)\right)$$

$$a\left(t\right) = -A\omega^2\sin\left(\omega t\right)$$

Integrating these equations to check our results is left as an exercise (Problem 1.33).

If we compare this to $x\left(t\right)$, you'll notice $a\left(t\right) = -\omega^2 x\left(t\right)$. This turns out to be an immensely useful result; we will see it when discussing diverse topics including piano strings, planetary motion, and light.

1.3   Kinematic Equations

With the exception of Example 1.10, our arguments so far have relied heavily on drawing and analyzing motion graphs. While this is certainly a useful tool for solidifying our understanding of these concepts, we would like a more precise approach (it is difficult to draw and measure velocity-time graphs with an area of exactly 2.75 m, for instance). For now we will restrict ourselves to the case where acceleration is constant, meaning that the motion graphs at the bottom of Figure 1.13 are relevant. In that figure, however, we assumed the initial velocity was $0\frac{m}{s}$. If we relax that requirement, then from the definition of velocity we find

$$\Delta\vec{v} = \vec{a}\Delta t$$
$$\vec{v}_f - \vec{v}_i = \vec{a}\Delta t$$

or, finally,

$$\vec{v}_f = \vec{v}_i + \vec{a}\Delta t \tag{1.9}$$

In words, this equation tells us the final velocity of some object, given its initial velocity, acceleration, and the time during which it accelerates.[12]

As we've seen, velocity is related to position in the same way as acceleration is related to velocity. Thus, if we perform the same algebra to the equation that defined velocity that we did above to the equation that defines acceleration, we find $\vec{x}_f = \vec{x}_i + \vec{v}\Delta t$. Importantly, however, we're considering the case where velocity is not necessarily constant (only acceleration is), so it isn't immediately clear what $\vec{v}$ actually means here.

To think this through, remember that the quantity $\vec{v}\Delta t$ corresponds to the displacement, $\Delta\vec{x}$. The displacement is just the area bounded by the time axis and the position axis on a position-time plot, so we can determine an appropriate expression by looking at a generic velocity-time graph under constant acceleration (Figure 1.22). The area under the curve can be divided into a rectangle and a triangle, and we see

$$\Delta\vec{x} = \text{area of rectangle} + \text{area of triangle}$$
$$\Delta\vec{x} = \vec{v}_i\Delta t + \frac{1}{2}\left(\vec{a}\Delta t\right)\Delta t$$

Plugging in the definition of displacement ($\Delta\vec{x} = \vec{x}_f - \vec{x}_i$), we find

$$\vec{x}_f = \vec{x}_i + \vec{v}_i\Delta t + \frac{1}{2}\vec{a}\left(\Delta t\right)^2 \tag{1.10}$$

We now have two equations (1.9 and 1.10) and a total of 6 variables.[13] Combining equations to eliminate variables (a standard exercise in algebra) allows us to derive new equations. For instance, both equations rely on time, but it is feasible that we might have a situation where we do not know (and do not care) how much time has elapsed and would prefer to analyze the situation without requiring it as a variable. If we solve Equation (1.9) for time

$$\Delta t = \frac{\Delta\vec{v}}{\vec{a}} = \frac{\vec{v}_f - \vec{v}_i}{\vec{a}}$$

---

[12] If you read Example 1.10, you'll recognize this as automatically falling out of a calculus-based analysis of velocity in the context of constant acceleration.

[13] Again, we derived this more straightforwardly using calculus, in Example 1.10.

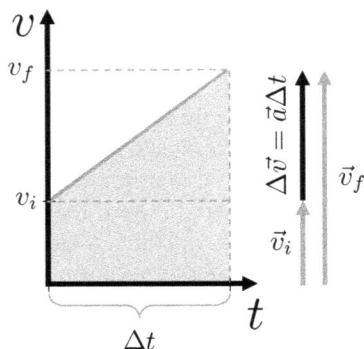

FIGURE 1.22
The shaded area corresponds to the displacement $\Delta \vec{x}$ under constant acceleration, where the initial velocity is not 0 (in contrast to Figure 1.13). The vector addition to the side demonstrates Equation (1.9).

and substitute it into Equation (1.10), we are confronted with some messy algebra:[14]

$$\vec{x}_f = \vec{x}_i + \vec{v}_i \frac{\vec{v}_f - \vec{v}_i}{\vec{a}} + \frac{1}{2}\vec{a}\left(\frac{\vec{v}_f - \vec{v}_i}{\vec{a}}\right)^2$$

Subtract $\vec{x}_i$ to the left and factor out $1/\vec{a}$ on the right:

$$\vec{x}_f - \vec{x}_i = \frac{1}{\vec{a}}\left(\vec{v}_i\left(\vec{v}_f - \vec{v}_i\right) + \frac{1}{2}\left(\vec{v}_f - \vec{v}_i\right)^2\right)$$

Multiply $2\vec{a}$ to the left hand side:

$$2\vec{a}\Delta\vec{x} = 2\vec{v}_i\left(\vec{v}_f - \vec{v}_i\right) + \left(\vec{v}_f - \vec{v}_i\right)^2$$

Distribute terms on the right:

$$2\vec{a}\Delta\vec{x} = 2\vec{v}_i\vec{v}_f - 2\vec{v}_i^2 + \vec{v}_f^2 + \vec{v}_i^2 - 2\vec{v}_i\vec{v}_f$$

Cancel like terms on the right:

$$2\vec{a}\Delta\vec{x} = \vec{v}_f^2 - \vec{v}_i^2$$

Or, finally,

$$\vec{v}_f^2 = \vec{v}_i^2 + 2\vec{a}\Delta\vec{x} \tag{1.11}$$

Phew! We have now derived the **one-dimensional kinematic equations**, which I summarize below.

$$\vec{v}_f = \vec{v}_i + \vec{a}\Delta t \qquad \text{(no } \Delta\vec{x})$$
$$\vec{x}_f = \vec{x}_i + \vec{v}_i\Delta t + \frac{1}{2}\vec{a}\left(\Delta t\right)^2 \quad \text{(no } \vec{v}_f) \tag{1.12}$$
$$\vec{v}_f^2 = \vec{v}_i^2 + 2\vec{a}\Delta\vec{x} \qquad \text{(no } \Delta t)$$

---

[14]If you're not particularly comfortable solving algebraic equations, it is well worth your time to carefully go through these steps.

The parenthetical note after each equation is a reminder of what kinematic term each equation *omits*. In a simple kinematics problem, the operative question is often, "what do I neither know nor care to know?" This guides you to the appropriate equation, which can then be solved directly. Of course, more complicated problems require more than this, as the next sequence of examples shows.

Example 1.11 Kinematics of a Car Stopping ★

A car is approaching a stop sign. When it is 45 m from the sign, it is traveling at 35 mph. What average acceleration is needed if it is to come to a complete stop just in front of the sign?

We set our coordinate system on top of the car's initial position and call the positive direction its direction of motion. (Fig. 1.23; notice that the horizontal separation is not to scale!) As you can check for yourself, converting the speed from mph (miles per hour) to SI units yields 15.6 m/s. We should report this to two significant figures (16 m/s) but I'm noting an extra digit so we can avoid rounding error in the following calculations. In the figure we've noted the kinematic variables along the corresponding positions (note that the acceleration deals with the behavior *between* the two positions, which is why it is drawn in the middle of the figure). We've indicated the quantities we *don't* know with a question mark.

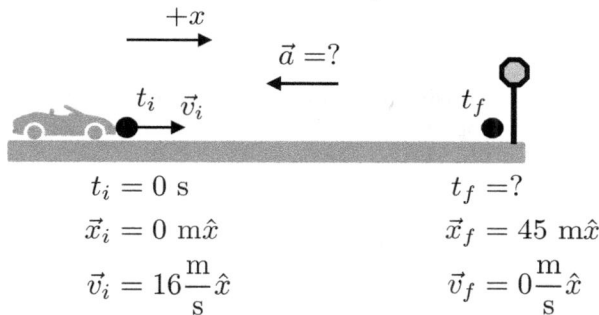

$$t_i = 0 \text{ s} \qquad\qquad t_f = ?$$
$$\vec{x}_i = 0 \text{ m} \hat{x} \qquad\qquad \vec{x}_f = 45 \text{ m} \hat{x}$$
$$\vec{v}_i = 16 \frac{\text{m}}{\text{s}} \hat{x} \qquad\qquad \vec{v}_f = 0 \frac{\text{m}}{\text{s}} \hat{x}$$

FIGURE 1.23

We can also list the kinematic variables and solve for the acceleration using the "no Δt" equation:

$$\vec{a} = ?$$
$$\vec{x}_i = 0 \text{ m}$$
$$\vec{x}_f = 45 \text{ m}$$
$$\vec{v}_i = 15.6 \frac{\text{m}}{\text{s}}$$
$$\vec{v}_f = 0 \frac{\text{m}}{\text{s}}$$
$$\Delta t = ?$$

$$\vec{v}_f^2 = \vec{v}_i^2 + 2\vec{a}\Delta\vec{x}$$

$$\vec{a} = \frac{\vec{v}_f^2 - \vec{v}_i^2}{2\Delta\vec{x}}$$

$$\vec{a} = \frac{\left(0\frac{\text{m}}{\text{s}}\right)^2 - \left(15.6\frac{\text{m}}{\text{s}}\right)^2}{2\left(45 - 0\right)\text{m}}\hat{x}$$

$$\vec{a} = -2.7\frac{\text{m}}{\text{s}^2}\hat{x}$$

The negative value follows from the fact that we're calling the direction of motion (to the right in the figure) positive. In order to slow down to rest, the acceleration must point opposite the velocity, in the negative direction. If you instead called *left* the positive direction, you'd find the acceleration is 2.7 m/s² (Problem 1.41).

Example 1.12 Kinematics of a Car's Acceleration ★★

The car from Example 1.11 slows down and stops at the stop sign, as we've already discussed. Suppose it stays stopped in front of the stop sign for 3.0 s. Then, it begins to accelerate at 3.5 m/s². How far past the stop sign will it be 15.0 s after it originally began braking? Make motion graphs for the entire duration of motion.

Let's first make a drawing of the situation. We have *four* distinct positions to worry about: the two from Example 1.11, a third that accounts for the 3.0 s the car is stopped, and a fourth for the end of our analysis, when the car is farther down the road. So while before we used i and f as our subscripts (initial and final), that meaning is now ambiguous. So, we'll switch to subscripts 1, 2, 3, and 4 (Fig. 1.24).

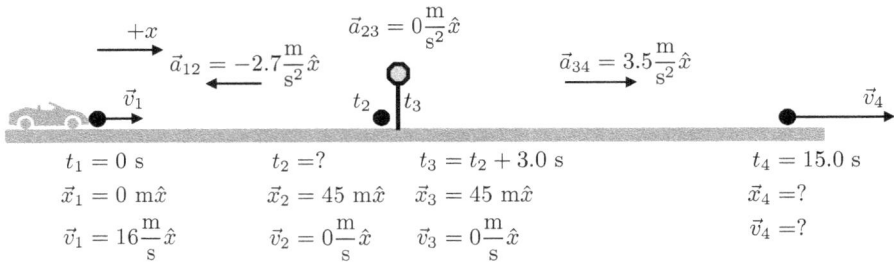

$$t_1 = 0 \text{ s} \qquad t_2 = ? \qquad t_3 = t_2 + 3.0 \text{ s} \qquad t_4 = 15.0 \text{ s}$$
$$\vec{x}_1 = 0 \text{ m}\hat{x} \qquad \vec{x}_2 = 45 \text{ m}\hat{x} \quad \vec{x}_3 = 45 \text{ m}\hat{x} \qquad \vec{x}_4 = ?$$
$$\vec{v}_1 = 16\frac{\text{m}}{\text{s}}\hat{x} \qquad \vec{v}_2 = 0\frac{\text{m}}{\text{s}}\hat{x} \quad \vec{v}_3 = 0\frac{\text{m}}{\text{s}}\hat{x} \qquad \vec{v}_4 = ?$$

FIGURE 1.24

To determine the final position $\vec{x}_4$, we first need to determine how long it took for the car to get to the stop sign:

$$\vec{v}_2 = \vec{v}_1 + \vec{a}_{12}\Delta t_{12}$$

$$\Delta t_{12} = \frac{\vec{v}_2 - \vec{v}_1}{\vec{a}_{12}}$$

$$t_2 - 0 \text{ s} = \frac{(0 - 15.6)\frac{\text{m}}{\text{s}}}{-2.70\frac{\text{m}}{\text{s}^2}}$$

$$t_2 = 5.77 \text{ s}$$

I'm keeping additional significant figures here to avoid rounding error as we proceed. Now we know $t_4 = 15.0$ s, so the interval $\Delta t_{34} = (15.0 - 5.77 - 3.0) = 6.23$ s. We can now find the final position x_4 using the "no v_f" equation for the "3 to 4" interval:

$$\vec{x}_4 = \vec{x}_3 + \vec{v}_3\Delta t_{34} + \frac{1}{2}\vec{a}_{34}\Delta t_{34}^2$$

$$\vec{x}_4 = (45 \text{ m}\hat{x}) + \left(0\frac{\text{m}}{\text{s}}\hat{x}\right)(6.23 \text{ s}) + \frac{1}{2}\left(3.5\frac{\text{m}}{\text{s}^2}\hat{x}\right)(6.23 \text{ s})^2$$

$$\vec{x}_4 = 113 \text{ m}\hat{x}$$

This is the ending position using the same origin as in the previous example. To obtain the distance *past the stop sign*, we need $\Delta\vec{x}_{34} = (113 - 45) \text{ m}\hat{x} = 68$ m. The motion graphs are shown in Figure 1.25.

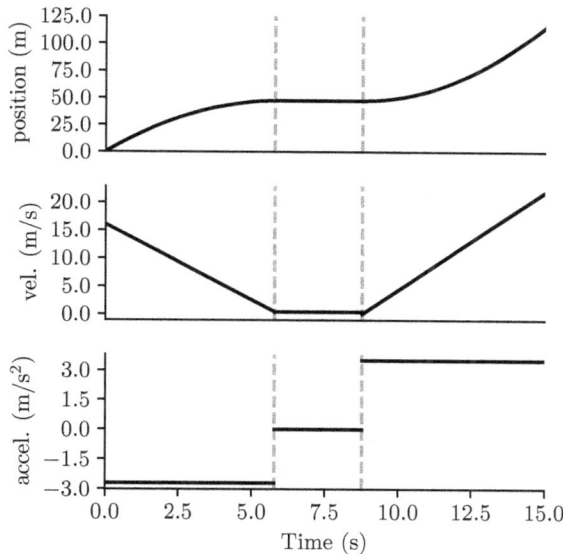

FIGURE 1.25
Motion graphs.

Example 1.13 Sledding on a Hill ★

A child on a sled moves down a steep hill with an acceleration of $4.0\frac{m}{s^2}$. If she starts from rest, how long will it take to reach a speed of $20.0\frac{m}{s}$? How far will she travel in this time?

We choose our coordinate system such that the child is initially at the origin and moves in the positive direction. We're here only considered with two instances, so we'll use i and f as the subscripts:

$\vec{a} = 4.0\frac{m}{s^2}$	The "no $\Delta\vec{x}$" equation allows us to solve for Δt:	Then either other equation allows us to solve for Δx:
$\vec{x}_i = 0$ m		
$\vec{x}_f = ?$	$\vec{v}_f = \vec{v}_i + \vec{a}\Delta t$	$\vec{v}_f^2 = \vec{v}_i^2 + 2\vec{a}\Delta\vec{x}$
$\vec{v}_i = 0\frac{m}{s}$	$\dfrac{\vec{v}_f - \vec{v}_i}{\vec{a}} = \Delta t$	$\Delta\vec{x} = \dfrac{\vec{v}_f^2 - \vec{v}_i^2}{2\vec{a}}$
$\vec{v}_f = 20.0\frac{m}{s}$	$\dfrac{(20.0 - 0)\frac{m}{s}}{4.0\frac{m}{s^2}} = \Delta t = 5.0$ s	$\Delta\vec{x} = \dfrac{\left(20.0^2 - 0^2\right)\left(\frac{m}{s}\right)^2}{2\left(4.0\frac{m}{s^2}\right)}\hat{x}$
$\Delta t = ?$		$= 5.0 \times 10^1$ m$\hat{x}$

"Whoa whoa", you might object. "Where's the figure?" I left the figure off here to encourage you to try to picture this in your mind's eye before seeing a drawing, but never fear. Flip ahead to Figure 1.27 on page 33 and you'll see a diagram for a very similar situation (the acceleration $\vec{a}$ here points in the direction of g_x in that figure).

Example 1.14 Pinching a Falling Ruler ⋆⋆

Consider an experiment where your friend holds a ruler vertically while you come just shy of pinching the bottom with your thumb and forefinger. Your friend releases the ruler and you try to pinch it before it falls entirely below your fingers. Objects in free-fall near the surface of the earth accelerate downward at a rate of $g = 9.8 \frac{m}{s^2}$. If the ruler is 0.30 m long, how much time, t, do you have to react? Where will you pinch the ruler if your reaction time is $\frac{1}{2}t$? Support your work with a motion diagram.

We choose our coordinate system such that our fingers are at the origin, and down is the positive direction. This means that the acceleration due to gravity (which points down toward the Earth) is positive in our coordinate system. The maximum reaction time corresponds to grabbing the ruler at the very top, when the bottom of the ruler is 0.30 m below our fingers. Thus our list of kinematic variables is:

List:

$$\vec{a} = 9.8 \frac{m}{s^2}$$

$$\vec{x}_i = 0 \text{ m}$$

$$\vec{x}_f = 0.30 \text{ m}$$

$$\vec{v}_i = 0 \frac{m}{s}$$

$$\vec{v}_f = ? \frac{m}{s}$$

$$\Delta t = ?$$

We can solve for Δt directly using the "no $\Delta \vec{v}_f$" equation. The algebra becomes simpler because the first two terms go to 0, and we find

$$\vec{x}_f = \vec{x}_i + \vec{v}_i \Delta t + \frac{1}{2}\vec{a}(\Delta t)^2$$

$$\vec{x}_f = \frac{1}{2}\vec{a}(\Delta t)^2$$

$$\left(\frac{2\vec{x}_f}{\vec{a}}\right)^{\frac{1}{2}} = \Delta t$$

$$\left(\frac{2(0.30 \text{ m})}{9.8 \frac{m}{s^2}}\right)^{\frac{1}{2}} = \Delta t = 0.25 \text{ s}$$

For the second question we use half of this value to solve for $\vec{x}_f$, which we can compute using the same equation:

$$\vec{x}_f = \frac{1}{2}\vec{a}(\Delta t)^2 = \frac{1}{2}\left(9.8 \frac{m}{s^2}\right)\left(\frac{0.25 \text{ s}}{2}\right)^2 \hat{x}$$

$$\vec{x}_f = 0.075 \text{ m} \hat{x}$$

Apparently if our reaction time is cut in half, we will grip the ruler much less than halfway down its length (a quarter of its length, to be precise; Fig 1.26). We could have deduced this directly, without any numbers, by considering the following:

$$(\vec{x}_f)_1 = \frac{1}{2}\vec{a}(\Delta t)^2 \quad \text{and} \quad (\vec{x}_f)_2 = \frac{1}{2}\vec{a}\left(\frac{\Delta t}{2}\right)^2$$

Factoring out the factor of $\left(\frac{1}{2}\right)^2$ in the equation for $(\vec{x}_f)_2$, we find

$$(\vec{x}_f)_2 = \frac{1}{4}\left(\frac{1}{2}\vec{a}(\Delta t)^2\right) = \frac{1}{4}(\vec{x}_f)_1$$

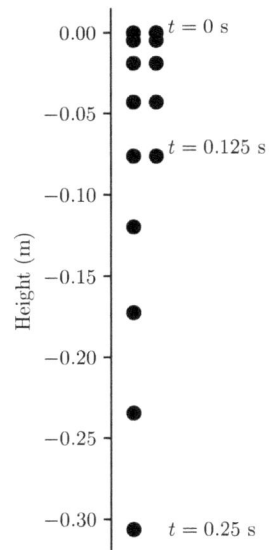

FIGURE 1.26

Example 1.15 Acceleration on a Ramp ★★

For a frictionless surface that is inclined relative to the horizontal at an angle θ (Fig. 1.27), the acceleration due to gravity obeys

$$a = g_x = g \sin \theta \tag{1.13}$$

directed down the ramp, as we will show rigorously in Example 2.3. If a frictionless surface is 75 cm long, at what angle should it be positioned if an object that starts from rest on the top side is to be moving at 2.0 m/s when it reaches the bottom?

We are not here concerned with time, so our attention should be drawn to the equation that omits it:

$$\vec{v}_f^2 = \vec{v}_i^2 + 2\vec{a}\Delta\vec{x}$$

$$v_f^2 = v_i^2 + 2\left(g \sin \theta\right) \Delta x$$

$$\frac{v_f^2 - v_i^2}{2g\Delta x} = \sin \theta$$

$$\sin^{-1}\left(\frac{v_f^2 - v_i^2}{2g\Delta x}\right) = \theta$$

$$\sin^{-1}\left(\frac{(2.0 \text{ m/s})^2 - (0\text{m/s})^2}{2\left(9.8 \text{ m/s}^2\right)(0.75 \text{ m})}\right) = \theta = 16°$$

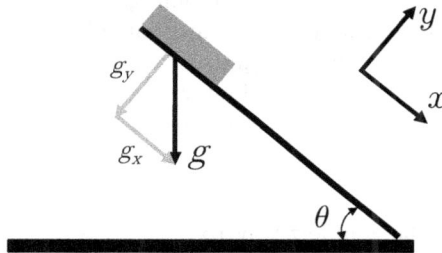

FIGURE 1.27

Example 1.16 Up, then Down ★★

A child stands at the railing of a building of height h and throws a ball vertically upward with a speed v. Develop an expression for the time elapsed until the ball strikes the ground. The acceleration (toward the ground) of a ball in free fall has a magnitude g.

First, note that this is an *algebraic* problem, in that no numbers are given. Our answer will also be algebraic. This can be daunting, at first, if you like to plug in numbers as you work on a problem. However, a competent student of physics needs to be comfortable with algebra, and problems such as these are excellent practice. Just as importantly, arriving at a mathematical expression is more informative than arriving at a number: you can see more clearly how various quantities are related (we saw this at the end of the preceding example, as well). Obtaining such an understanding is at the core of physics.

To begin, let's draw a diagram, shown in Figure 1.28.

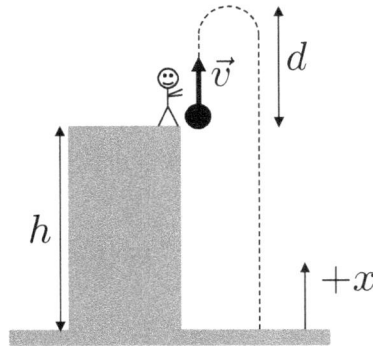

FIGURE 1.28
Schematic of a ball thrown vertically from the top of a building.

Here we've shown the ball moving a bit to the side just so we can see clearly that it moves up to some maximum height and then falls back down to the ground. We know the ball will reach some maximum height (a distance above the edge of the building that we've labeled d). Thus we can think of the ball as having two distinct parts: it travels up, then it travels down.

A key insight is that at the instant that the ball is at its maximum height, its velocity is 0 m/s. Otherwise it would either still be moving up (and not yet be at its maximum height) or moving down (and already be below its maximum height). Thus, if we analyze *just* the upward part of the motion, then we know its *final* velocity is 0 m/s. Likewise, if we analyze *just* the downward part of the motion, then we know its *initial* velocity is 0 m/s. With this in mind, let's list the kinematic variables for the upward part of the motion. Of course, first we need a coordinate system: we'll set the origin level with the ground and choose up as the positive direction. Then:

UP:

$$\vec{a} = -g\hat{x}$$
$$\vec{x}_i = h\hat{x}$$
$$\vec{x}_f = (h + d)\,\hat{x}$$
$$\vec{v}_i = v\hat{x}$$
$$\vec{v}_f = 0\frac{\text{m}}{\text{s}}\hat{x}$$
$$\Delta t_{\text{up}} = ?$$

We can use either of the kinematic equations involving time, because we know everything except time. Perhaps the simplest way to proceed is:

$$\vec{v}_f = \vec{v}_i + a\Delta t_{\text{up}}$$
$$0 = v - g\Delta t_{\text{up}}$$
$$\Delta t_{\text{up}} = \frac{v}{g}$$

This says the time up depends just on the velocity and the acceleration. Because the velocity is in the numerator, it says that it will take more time to reach its maximum height if the velocity is larger. Likewise, if the acceleration due to gravity was greater, the ball would slow down more rapidly and take less time to reach its maximum height.

We can proceed in the same way for the second half of the motion:

DOWN:

$$\vec{a} = -g\hat{x}$$
$$\vec{x}_i = (h + d)\,\hat{x}$$
$$\vec{x}_f = 0 \text{ m}$$
$$\vec{v}_i = 0\frac{\text{m}}{\text{s}}\hat{x}$$
$$\vec{v}_f = ?$$
$$\Delta t_{\text{down}} = ?$$

Now we don't know $\vec{v}_f$, so unless we wish to solve for it we should appeal to kinematic equation that involves time but omits final velocity:

$$\vec{x}_f = \vec{x}_i + \vec{v}_i \Delta t + \frac{1}{2}\vec{a}\,(\Delta t)^2$$

$$0 = (h + d) + (0)\,\Delta t_{\text{down}} - \frac{1}{2}g\,(\Delta t_{\text{down}})^2$$

$$\frac{1}{2}g\,(\Delta t_{\text{down}})^2 = (h + d)$$

$$\Delta t_{\text{down}} = \left(\frac{2\,(h + d)}{g}\right)^{\frac{1}{2}}$$

This says the time down depends just on its initial height and gravity: the scenario is identical to simply dropping a ball from a height $h + d$. The time increases the higher the ball is, and it would decrease if the acceleration due to gravity increased.

Now that we've analyzed the time up and the time down, we can easily determine the total time from when it is thrown to when it strikes the ground:

$$\Delta t = \Delta t_{\text{up}} + \Delta t_{\text{down}}$$

$$\Delta t = \frac{v}{g} + \left(\frac{2\,(h + d)}{g}\right)^{\frac{1}{2}}$$

Example 1.17 A Bottle Rocket ★★

A bottle rocket is sitting on the ground when it is fired, causing it to accelerate upward at a rate of 4.0 m/s² for 8.0 s. How long is it in the air, and what maximum height above the ground does it reach?

We sketch the motion and note the kinematic variables that we know (or can easily determine) in Figure 1.29. Note that once the rocket has stopped firing, the rocket will continue moving upward under the influence of gravity. In the Figure, the location marked 1 corresponds to the initial position of the rocket, the location marked 2 corresponds to the instant the rocket stops firing, location 3 is the highest point, and location 4 corresponds to the point of impact on the ground. (The path down is shown to the side just to avoid overlap.) We'll set the origin on the ground and call up the positive direction, as usual.

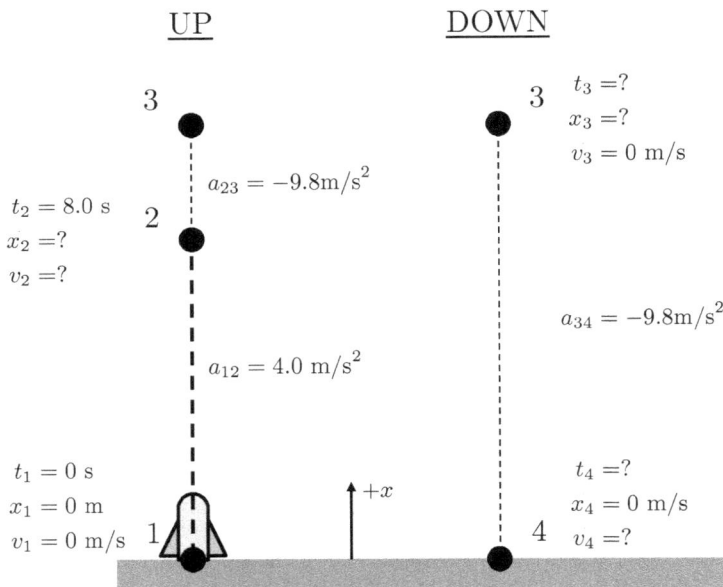

FIGURE 1.29

We'll begin by analyzing the motion of the rocket for the first 8.0 seconds, when the rocket is accelerating upward.

ROCKET ON:

$$\vec{a}_{12} = 4.0\frac{\text{m}}{\text{s}^2}\hat{x}$$

$$\vec{x}_1 = 0 \text{ m}\hat{x}$$

$$\vec{x}_2 = ?$$

$$\vec{v}_1 = 0\frac{\text{m}}{\text{s}}\hat{x}$$

$$\vec{v}_2 = ?$$

$$\Delta t_{12} = 8.0 \text{ s}$$

We can calculate $\vec{v}_2$:

$$\vec{v}_2 = \vec{v}_1 + \vec{a}_{12}\Delta t_{12}$$

$$v_2 = (0 \text{ m/s})\,\hat{x} + \left(4.0 \text{ m/s}^2\right)(8.0 \text{ s})\,\hat{x}$$

$$v_2 = 32\frac{\text{m}}{\text{s}}\hat{x}$$

We can also find the altitude when the rocket shuts off:

$$\vec{x}_2 = \vec{x}_1 + \vec{v}_1\Delta t_{12} + \frac{1}{2}\vec{a}_{12}\left(\Delta t_{12}\right)^2$$

$$x_2 = (0 \text{ m}) + (0 \text{ m/s})(8.0 \text{ s}) + \frac{1}{2}\left(4.0 \text{ m/s}^2\right)(8.0 \text{ s})^2$$

$$x_2 = 128 \text{ m}$$

(I divided out the $\hat{x}$ and indicate the magnitude of x_2.) If we round x_2 to the appropriate number of significant figures we should report 130 m, but I am leaving in the "calculator value" to avoid rounding error for future calculations.

Once the rocket shuts off the rocket will be in free fall; the acceleration will be -9.8 m/s^2. Thus the rocket will slow down until it reaches a maximum altitude (the positive velocity decreases to 0), then pick up speed as it falls back to the ground.

Importantly, the kinematic equations work for any interval with *constant acceleration*, so we're free to compare "location 2" with "location *3 or 4*", or we can compare "location 3" to "location 4". (We can't compare locations 1 or 2 with a later location, though, because the acceleration *changes* from 4.0 m/s^2 to -9.8 m/s^2 as we pass location 2.)

We could proceed a number of different ways. Let's first work out the maximum height by comparing the instant the rocket shuts off (location 2) to the instant it reaches the maximum height (location 3):

Note that I provided the calculated values with additional significant figures to avoid rounding error, and I've asserted that the velocity at the maximum height is 0. We'll solve for $\vec{x}_3$ by avoiding Δt_{23}:

$$\vec{a}_{23} = -9.8\frac{\text{m}}{\text{s}^2}\hat{x}$$

$$\vec{x}_2 = 128 \text{ m}\hat{x}$$

$$\vec{x}_3 = ?$$

$$\vec{v}_2 = 32.0\frac{\text{m}}{\text{s}}\hat{x}$$

$$\vec{v}_3 = 0\frac{\text{m}}{\text{s}}\hat{x}$$

$$\Delta t_{23} = ?$$

$$\vec{v}_3^2 = \vec{v}_2^2 + 2\vec{a}_{23}\Delta x_{23}$$

$$\frac{\vec{v}_3^2 - \vec{v}_2^2}{2\vec{a}_{23}} = \vec{x}_3 - \vec{x}_2$$

$$\frac{\vec{v}_3^2 - \vec{v}_2^2}{2\vec{a}_{23}} + \vec{x}_2 = \vec{x}_3$$

$$\frac{(0 \text{ m/s})^2 - (32 \text{ m/s})^2}{2\left(-9.8\text{m/s}^2\right)} + 128 \text{ m} = x_3 = 180 \text{ m}$$

Where I've rounded the final answer to the appropriate number of significant figures. To find the time until the rocket hits the ground, we'll compare the instant the rocket shuts off (subscript 2) to the instant the rocket hits the ground (subscript 4):

Here we'll sidestep the unknown velocity $\vec{v}_4$:

$$\vec{a}_{24} = -9.8\frac{\text{m}}{\text{s}^2}\hat{x}$$

$$\vec{x}_2 = 128 \text{ m}\hat{x}$$

$$\vec{x}_4 = 0 \text{ m}\hat{x}$$

$$\vec{v}_2 = 32.0\frac{\text{m}}{\text{s}}\hat{x}$$

$$\vec{v}_4 = ?$$

$$\Delta t_{24} = ?$$

$$\vec{x}_4 = \vec{x}_2 + \vec{v}_2\Delta t_{24} + \frac{1}{2}\vec{a}\left(\Delta t_{24}\right)^2$$

$$0 = (\vec{x}_2 - \vec{x}_4) + \vec{v}_2\Delta t_{24} + \frac{1}{2}\vec{a}\left(\Delta t_{24}\right)$$

$$0 = (128 \text{ m}) + (32.0 \text{ m/s})\left(\Delta t_{24}\right) + \left(-4.9 \text{ m/s}^2\right)\left(\Delta t_{24}\right)^2$$

We have a quadratic equation with (temporarily suppressing units) $a = -4.9$, $b = 32$, and $c = 128$, which we can solve by applying the quadratic formula:

$$\Delta t_{24} = \frac{-b \pm \left(b^2 - 4ac\right)^{1/2}}{2a}$$

$$\Delta t_{24} = \frac{-32 \pm \left(32^2 - 4\left(-4.9\right)\left(128\right)\right)^{1/2}}{2\left(-4.9\right)}$$

$$\Delta t_{24} = 9.3 \text{ s}$$

Where in the last line I have reintroduced the units and I have neglected the second solution provided by the quadratic equation because (as you can check for yourself) the second solution is *negative*, which is not physically viable.

Now it is straightforward to calculate the time at impact:

$$\Delta t_{24} = t_4 - t_2 \implies t_4 = t_2 + \Delta t_{24} = 8.0 \text{ s} + 9.3 \text{ s} = 17.3 \text{ s}$$

Thus the rocket spends 17.3 s in flight and reaches a maximum altitude of 180 m. It is worth noting that we could have analyzed the motion of the rocket once it is in free fall using the method of Example 1.16. In the context of *this* example, that approach entails separately working out the time to the maximum height and then the time from the maximum height back to the ground. There are additional steps but the advantage is that you don't need to apply the quadratic formula. Which approach is *better* is largely one of taste.

1.4 Visually Combining 1D Vectors: The "Tail to Tip" Method

Because vectors in this chapter can be represented simply by positive and negative numbers (being one-dimensional), the mathematics of vector addition and subtraction is quite simple. Nonetheless, it is worth our time to describe how we can add these vectors visually, because it (1) helps to make it clear what we're doing and (2) is a good prelude for combining vectors in two-dimensional kinematics. The rule for adding vectors visually is simple: place the beginning of one vector (the "tail") at the end of the other (the "tip"). The vector sum then runs from the beginning of the first vector to the end of the last (thus, this is often called the "tail to tip method"). For vector subtraction, one simply reverses the direction of the negative vector. An example of both vector addition and vector subtraction is shown in Figure 1.31, and Example 1.18 provides some additional practice. In Section 1.5 we will apply this technique to help us understand situations where objects are moving *on other moving objects*, like someone running across a moving train.

Example 1.18 Tail to Tip Vector Addition   ⋆
Consider the vectors $\vec{A} = 5\hat{x}$ and $\vec{B} = -3\hat{x}$. Calculate (a) $\vec{A} + \vec{B}$, (b) $\vec{A} - \vec{B}$, and (c) $\vec{B} - \vec{A}$. In each case demonstrate the resulting vector using the tail to tip method.

Calculating the numerical results is straightforward. For (a) we have

$$\vec{A} + \vec{B} = (5 + (-3))\,\hat{x} = 2\hat{x}$$

For (b),

$$\vec{A} - \vec{B} = (5 - (-3))\,\hat{x} = 8\hat{x}$$

And for (c),

$$\vec{B} - \vec{A} = (-3 - 5)\,\hat{x} = -8\hat{x}$$

These are shown with the tail to tip method in Figure 1.30.

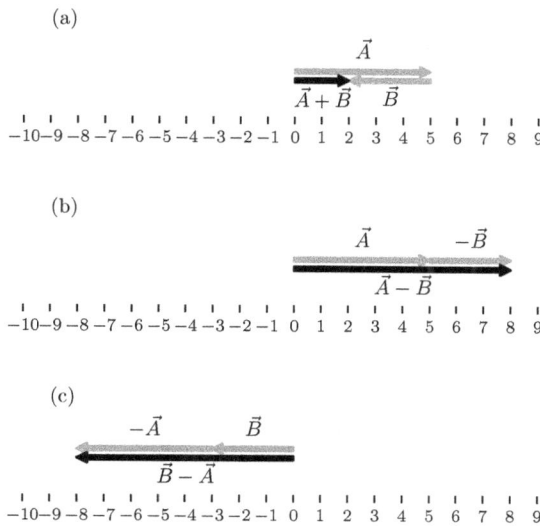

FIGURE 1.30
The vectors being combined are shown in "tail top tip" in gray, and the resulting vector, which runs from the beginning (tail) of the first vector to the end (tip) of the second vector is shown in black. The vertical spacing is used only to avoid overlap in the vectors, and the number lines below the vectors are included to indicate scale.

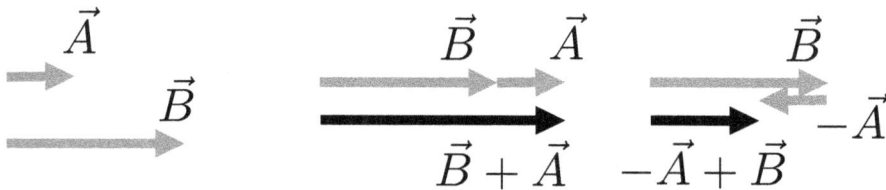

FIGURE 1.31
Visually adding and subtracting one-dimensional vectors with the tail-to-tip method.

1.5   Relative Motion

Suppose person A is sitting on the ground, watching a train pass by from left to right at some constant speed v. If person B is sitting on the train, then from their point of view the train isn't moving at all: as time passes, the front of the train does not move any farther from or closer to person B (Figure 1.32). This is an example of *relative motion*, which comes into play when two observers are moving at different velocities.[15]

---

[15]You may have heard of Einstein's theories of special and general relativity. Relativity is used in the same sense there as it is here, though Einstein's theories are relevant for speeds approaching the speed of light and aren't important for our present purposes, where we're talking about objects moving at everyday speeds.

To formalize this, let's consider a specific example. Suppose the train is moving at a speed of 2 m/s from left to right according to person A. Suppose also that person B stands up and starts walking to the back of the train at a speed of 1 m/s relative to the train (suppose there are marks on the floor of the train, spaced every 1 meter, and person B sees that he passes one of these marks every second). We denote the velocity of person B relative to (i.e. from the perspective of) person A as $\vec{v}_{BA}$. The value of $\vec{v}_{BA}$ is determined by combining the velocity of person B relative to the train ($\vec{v}_{BT}$) and the velocity of the train relative to person A ($\vec{v}_{TA}$). Specifically:

$$\vec{v}_{BA} = \vec{v}_{BT} + \vec{v}_{TA} \tag{1.14}$$

If person A calls "to the right" the positive direction in their coordinate system, then

$$\vec{v}_{BA} = -1\frac{\text{m}}{\text{s}}\hat{x} + 2\frac{\text{m}}{\text{s}}\hat{x} = 1\frac{\text{m}}{\text{s}}\hat{x}$$

which you may have realized intuitively. Now, if person B appears to be moving to the right at 1 m/s from person A's perspective, then person A appears to be moving to the left at 1 m/s from person B's perspective. In other words,

$$\vec{v}_{BA} = -\vec{v}_{AB} \tag{1.15}$$

Equations (1.14) and (1.15) can be generalized to any sort of one-dimensional relative motion.

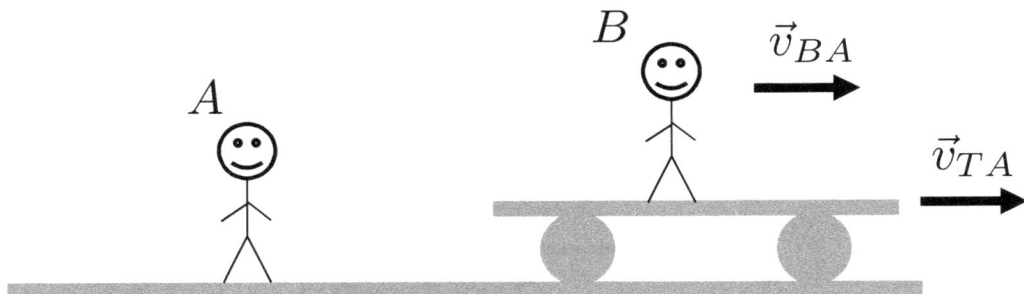

FIGURE 1.32
Person A is standing on the ground, observing person B sitting on a train. Both the train and person B have the same velocity according to person A, while person B observes the train as having no velocity at all.

Example 1.19 Relative Velocities in a Hot Air Balloon   ⋆
Suppose a hot air balloon is moving up at a speed of 10.0 m/s. A passenger in the balloon leans over the edge and throws a ball down at a speed of 5.0 m/s relative to the balloon. What is the speed of the ball relative to an observer on the ground? Explain with a drawing and describe the subsequent motion of the ball.

We call up the positive direction and denote the ball, balloon, and ground with b, B and G, respectively. Then

$$\vec{v}_{bG} = \vec{v}_{bB} + \vec{v}_{BG}$$

$$\vec{v}_{bG} = -5.0\frac{m}{s}\hat{x} + 10.0\frac{m}{s}\hat{x}$$

$$\vec{v}_{bG} = 5.0\frac{m}{s}\hat{x}$$

The balloon is rising faster than the ball is thrown (relative to the balloon), so it is still moving up relative to a ground-based observer, just at a slower speed. We can perform the same calculation with a diagram (Figure 1.19).

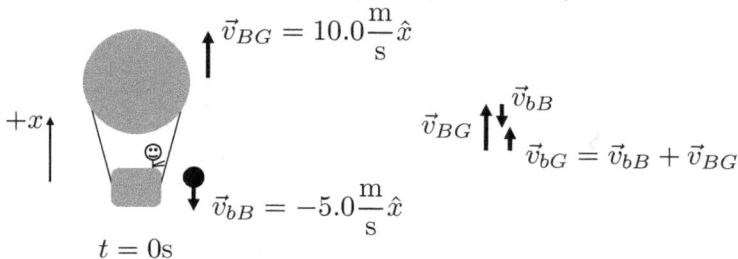

FIGURE 1.33
Vector addition for a ball thrown down from a rising hot air balloon.

On the right, we draw the vectors and their sum with the tail to tip method (the vectors are separated horizontally a bit just so you can easily distinguish them). As we saw mathematically, the result is positive. Thus the motion of the ball is similar to if you stood on stationary ground and threw it up (it will increase in height, then eventually fall back to the ground).

Example 1.20 Collision on a Train   ★★
In a movie, two cars experience a head-on collision. Car A is on a train moving at 15. mph to the left. Car A's speedometer reads 40. mph and car B's speedometer reads 30. mph, as shown (Recall that the decimal after the digits indicates all figures to the left are significant). What is B's speed relative to A just before the collision?

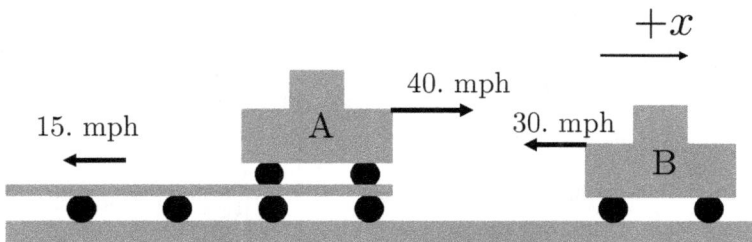

FIGURE 1.34
An ultra-realistic diagram of two cars colliding.

The vectors on the diagram show the velocity of B relative to the ground, the velocity of A relative to the train, and the velocity of the train relative to the ground.

We'll choose the coordinate system such that right is the positive direction. To start, note

$$\vec{v}_{BA} = \vec{v}_{B,\text{ground}} + \vec{v}_{\text{ground},A}$$
$$\vec{v}_{BA} = \vec{v}_{B,\text{ground}} + (\vec{v}_{\text{ground, train}} + \vec{v}_{\text{train},A})$$

Now note that if the train is moving to the left from the ground's perspective, then the ground is moving to the *right* from the train's perspective (that is, $\vec{v}_{\text{ground,train}} = -\vec{v}_{\text{train,ground}}$, as in Equation (1.15)). Similarly, if A is moving to the right relative to the train, then the train is moving to the left relative to A. Plugging in, we find

$$\vec{v}_{BA} = \vec{v}_{B,\text{ground}} + \vec{v}_{\text{ground, train}} + \vec{v}_{\text{train},A}$$
$$\vec{v}_{BA} = (-30. + 15. - 40.)\,\text{mph}\hat{x}$$
$$\vec{v}_{BA} = -55\,\text{mph}\hat{x}$$

Example 1.21 Multiple Drops from a Balloon   ⋆ ⋆ ⋆

A hot air balloon is experiencing an upward acceleration a. 10.0 seconds after leaving the ground, a passenger drops a ball over the edge. 1.50 seconds after that, she throws a second ball straight down with an initial speed (relative to the balloon) of 12.0 m/s. The two balls strike the ground at the same time and with the same velocity. What is a?

This problem forces us to track multiple objects simultaneously, so we need to be particularly cautious as we proceed. As usual, it is best to begin with a careful drawing. In fact, let's make several drawings, corresponding to the key instants of the motion (Figure 1.35).

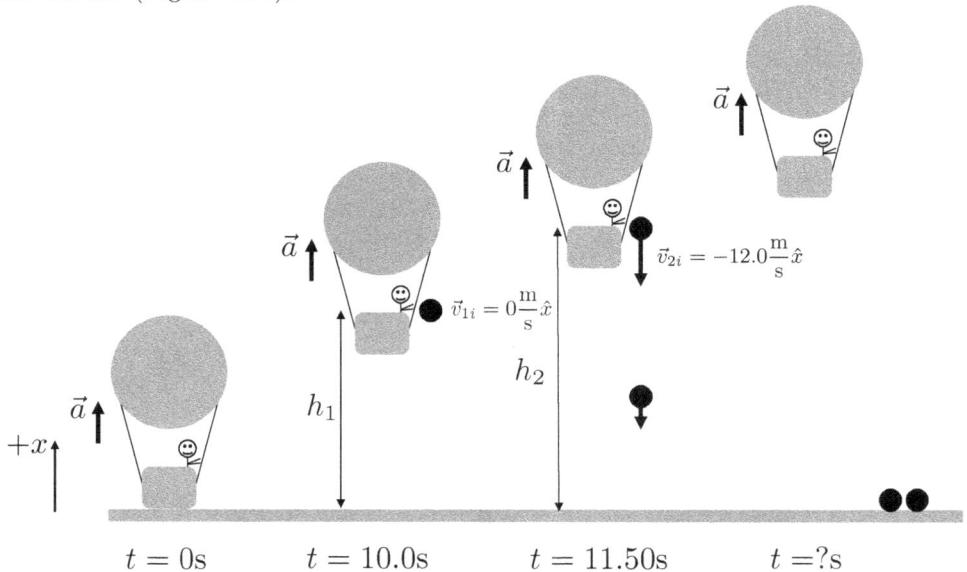

FIGURE 1.35
Balls falling from a rising hot air balloon.

We'll choose our coordinate system such that the origin is level with the ground, and up is the positive direction. Let's begin by mathematically expressing the fact that the impact velocities of the balls will be the same:

$$\vec{v}_{1f} = \vec{v}_{2f}$$
$$v_{1i} - g\Delta t_1 = v_{2i} - g\Delta t_2$$

In the second line I've applied the kinematic equation for v_f that omits $\Delta \vec{x}$ (we're told something about time so it seems reasonable to make this choice). We don't have much direct information about any of these parameters, but let's work with this equation as we proceed. Ultimately we want to obtain an expression for the acceleration of the balloon in terms of known quantities.

As our next step, note that $\Delta t_2 = \Delta t_1 - 1.50$ s: the second ball is thrown 1.50 s after the first, but they strike the ground at the same instant. Let's insert this into our "master" equation:

$$v_{1i} - g\Delta t_1 = v_{2i} - g\left(\Delta t_1 - 1.50 \text{ s}\right)$$

This eliminates one of our unknown variables. Let's next consider the balloon during its first 10.0 seconds. It will rise to some height that we've labeled h_1 in the above diagram, and it will have some upward velocity which is equivalent to the initial velocity of the first ball (the passenger doesn't throw the ball, but until the instant it is dropped, the ball is still locked in motion with the balloon and passenger, so it has the same velocity at the instant that it is released). Considering the kinematic equation $\vec{v}_f = \vec{v}_i + a\left(\Delta t\right)$ and applying it to the balloon's ascent, we can note that the initial velocity (of the balloon) is 0 m/s. Thus $v_{1i} = a\left(10.0 \text{ s}\right)$.

Inserting into our master equation, we find

$$a\left(10.0 \text{ s}\right) - g\Delta t_1 = \vec{v}_{2i} - g\left(\Delta t_1 - 1.50 \text{ s}\right)$$

Now we can make a similar argument for the velocity of the second ball when it is thrown. At this instant (at $t = 11.50$ s) the *balloon* is moving up with a speed $a\left(11.50 \text{ s}\right)$, but the second ball is thrown down with a speed of 12.0 m/s relative to the balloon. Thus the initial velocity of the second ball (relative to the ground) the instant it is thrown is $v_{2i} = a\left(11.50 \text{ s}\right) - 12.0\frac{\text{m}}{\text{s}}$. Once again we can insert this result into our master equation:

$$a\left(10.0 \text{ s}\right) - g\Delta t_1 = a\left(11.50 \text{ s}\right) - 12.0\frac{\text{m}}{\text{s}} - g\left(\Delta t_1 - 1.50 \text{ s}\right)$$

Notice we have a term $-g\Delta t_1$ on both sides of our equation. Thus these algebraically cancel and we're left with an equation with numbers and a, the variable we set out to identify (Recall $g = 9.8$ m/s^2). It remains to finish off the algebra:

$$a\left(10.0 \text{ s}\right) = a\left(11.50 \text{ s}\right) - 12.0\frac{\text{m}}{\text{s}} + g\left(1.50 \text{ s}\right)$$
$$\left(1.50 \text{ s}\right)a = g\left(1.50 \text{ s}\right) - 12.0\frac{\text{m}}{\text{s}}$$
$$a = 1.8\frac{\text{m}}{\text{s}^2}$$

Phew!

1.6 Problems for Chapter 1

1.1 Position, Velocity, and Acceleration

⋆ **Problem 1.1.** If the fastest you can safely drive is 65 mph (miles per hour), what is the longest time you can stop for dinner if you must travel 541 mi in 9.6 hr total?

⋆ **Problem 1.2.** Figure 1.36 shows part of a motion diagram. Make your own copy of this diagram and add in the *preceding* position (#1) and the average velocity $\vec{v}_{12}$ if the acceleration during the motion is directed

   a. to the left

   b. to the right

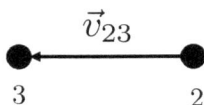

FIGURE 1.36
Problem 1.2

⋆ **Problem 1.3.** Consider the motion diagram shown in Figure 1.37. Draw in the velocity and acceleration vectors. What is the sign (positive, negative, or zero) of the position, velocity, and acceleration of the particle? Provide an example of a physical situation described by this motion diagram.

⋆ **Problem 1.4.** Repeat Problem 1.3 for the motion diagram shown in Figure 1.38.

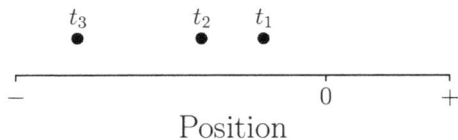 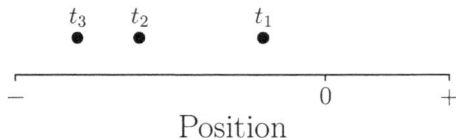

FIGURE 1.37
Problem 1.3

FIGURE 1.38
Problem 1.4

⋆ **Problem 1.5.** Repeat Problem 1.3 for the motion diagram shown in Figure 1.39.

⋆ **Problem 1.6.** Repeat Problem 1.3 for the motion diagram shown in Figure 1.40.

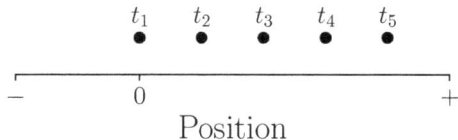 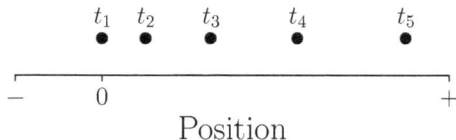

FIGURE 1.39
Problem 1.5

FIGURE 1.40
Problem 1.6

⋆ **Problem 1.7.** Repeat Problem 1.3 for the motion diagram shown in Figure 1.41.

★ **Problem 1.8.** Repeat Problem 1.3 for the motion diagram shown in Figure 1.42.

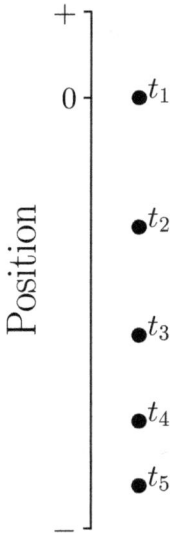

FIGURE 1.41
Problem 1.7

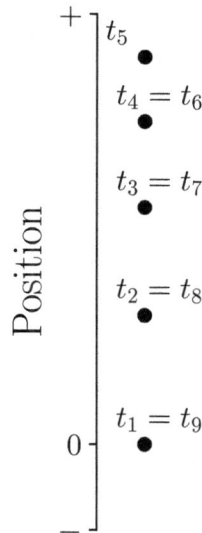

FIGURE 1.42
Problem 1.8

★ **Problem 1.9.** Repeat Problem 1.3 for the motion diagram shown in Figure 1.43.

★ **Problem 1.10.** Repeat Problem 1.3 for the motion diagram shown in Figure 1.44. Here the object is moving along a ramp. *You* choose the origin and which direction along the ramp is positive. Indicate your choice when providing your answer.

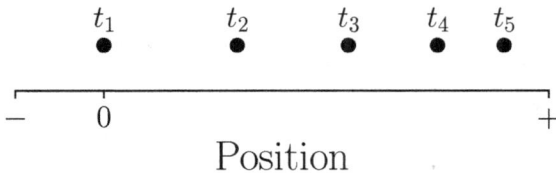

FIGURE 1.43
Problem 1.9

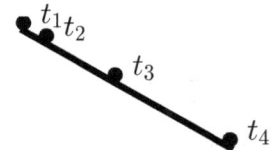

FIGURE 1.44
Problem 1.10

★ **Problem 1.11.** You step out of your house at 9:05 AM and run down the street to a friend's apartment building, 6.00×10^2 m to the east. You arrive at 9:07 AM. You realize you dropped your wallet along the way, turn around, and run back 5.00×10^2 m to get it. You arrive at your wallet at 9:09 AM. What is your (a) average velocity and (b) average speed during the entire run, expressed to 3 significant figures?

★ **Problem 1.12.** You watch a bug wandering along your desk. Bored, you place a ruler nearby and observe that it walks from the 35 cm mark to the 15 cm mark, back to the 30 cm mark, then over to the 10 cm mark. The entire epic journey lasted 180 s.

a What distance did it travel?

    b What was the bug's average speed?

    c What was the bug's average velocity?

★★   **Problem 1.13.** You and your friend are facing each other 10.0 m apart. You both say "I am standing at the origin of my coordinate system, and I am facing in the positive direction". Suppose you're driving a toy car toward your friend. At a certain instant, you observe it as having a position $\vec{x}_1$, velocity $\vec{v}_1$, and acceleration $\vec{a}_1$ according to *your* coordinate system. Your friend observes it as having a position $\vec{x}_2$, velocity $\vec{v}_2$, and acceleration $\vec{a}_2$ according to *their* coordinate system. Determine three equations that allow you to express $\vec{x}_1$ in terms of $\vec{x}_2$, $\vec{v}_1$ in terms of $\vec{v}_2$, and $\vec{a}_1$ in terms of $\vec{a}_2$.

★★   **Problem 1.14.** Consider the situation described in Example 1.3.

    a. Make a data table of your position and your friend's position for 5 second intervals until you find that you've passed each other–which means you've *passed* the point of the collision. How precisely can you describe the instant of the collision from your table?

    b. Determine an equation of motion $x_{\text{you}}(t)$ and $x_{\text{friend}}(t)$. Plot them on one graph and mathematically determine the time where they collide.

★   **Problem 1.15.** Suppose you are analyzing a video of your friend jogging down a hallway. At $t_0 = 0.0$ s, $t_1 = 1.0$ s, and $t_2 = 2.0$ s your friend is respectively at $x_0 = 0.0$ m, $x_1 = 2.0$ m, and $x_2 = 3.0$ m.

    a. Make a complete motion diagram, including velocity and acceleration vectors.

    b. What is your friend's average velocity during the first second ($\vec{v}_{01}$)? What about during the second second ($\vec{v}_{12}$)? What is their average velocity over the entire time interval?

    c. What is your friend's average acceleration during the two seconds you analyzed?

★   **Problem 1.16.** Consider the situation described in Example 1.4. Carefully draw each of the velocity vectors so they start from a common location (use a ruler and make sure their lengths are proportional to their actual values), and demonstrate that the acceleration vectors all have the same magnitude (and point down) as in Figure 1.9 (though here the analysis will be *vertical* rather than horizontal).

★   **Problem 1.17.** An object is moving to the left and is slowing down. Which way does the object's velocity vector point? Which way does the object's acceleration vector point? Support your answer with a motion diagram.

★   **Problem 1.18.** An athlete runs 2.5 km in a straight line, then turns around and runs back to where they started. Suppose the first 2.5 km took 1.0×10^3 s and the second 2.5 km took 1.2×10^3 s.

    a. What was their speed during the first 2.5 km? Express your answer in both m/s and mph.

    b. Repeat (a) for the second 2.5 km.

    c. What was the runner's average velocity during the entire trip?

★   **Problem 1.19.** Consider an object that, at t_0, has a downward velocity $\vec{v}_1$.

    a. Draw a motion diagram (including at least three dots) in the case that the acceleration $\vec{a}$ points in the same direction as the velocity. (You don't have numerical values to work with here, so all that matters is the relative spacing of the dots.)

    b. Repeat (a) in the case where the acceleration points opposite the velocity.

    c. In words, describe what is different, and why, between your diagrams for (a) and (b).

★★   **Problem 1.20.** At $t = 0.0$ s, a car is moving at 10.0 m/s due East. It continues like this for 5.0 s, then over the course of the following 5.0 s it slows down at a constant rate until it comes to rest. Draw a complete motion diagram with dots at 1.0 second intervals.

★★   **Problem 1.21.** A car starts from rest, accelerates at 5.0 m/s² for 4.0 s, coasts at a constant speed for 3.0 s, then slows at a rate of 3.00 m/s² until it comes to rest. Draw a complete motion diagram.

★ **Problem 1.22.** Draw a complete motion diagram of a rocket uniformly accelerating from rest such that it reaches a speed v at height h. Include at least 5 points in your diagram.

★ **Problem 1.23.** You throw a ball up, allow it to reach some maximum height h, and then catch it at the same position it was thrown. Draw a complete motion diagram. (Hint: the acceleration is constant and downward, toward your hand.)

1.2 Motion Graphs

★ **Problem 1.24.** The velocity of a particle is shown in Figure 1.45. If the particle is located at $x = 0.0$ m at $t = 0.0$ s, where is it at (a) $t = 4.0$ s and (b) $t = 6.0$ s?

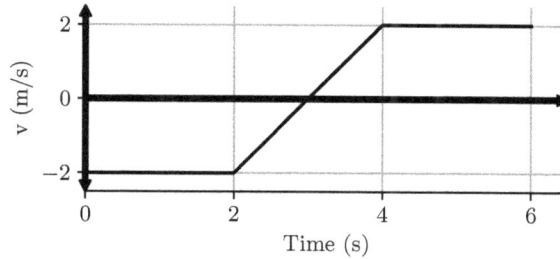

FIGURE 1.45
Problem 1.24

★★ **Problem 1.25.** Figure 1.46 shows a position-time graph. Draw the corresponding velocity-time graph and acceleration-time graph. How fast and in what direction is the object moving at $t = 3.0$ s?

★★ **Problem 1.26.** Figure 1.47 shows a velocity-vs-time graph for a particle that is located at $x = 6.0$ m at $t = 0$ s. What is the particle's (a) position, (b) velocity, and (c) acceleration at $t = 2.0$ s?

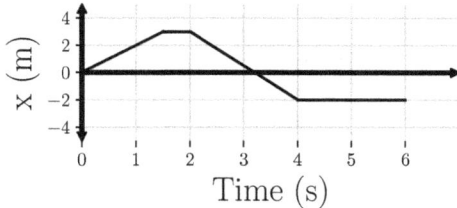

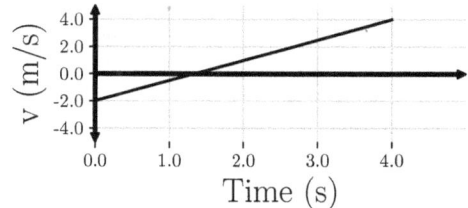

FIGURE 1.46
Problem 1.25

FIGURE 1.47
Problem 1.26

★★ **Problem 1.27.** Consider the 40-yard dash, an athletic event where players begin from rest and then run 40 yards as fast as they can. (For this problem you may express distances in units of yards rather than in meters.)

a. Table 1.3 shows a participant's position at 0.5 second intervals for the first 4 seconds of their 40-yard dash. Use this data to make a position-vs-time graph.

b. What was the sprinter's average velocity during the sprint?

c. When during the sprint does the sprinter have an acceleration that is (i) positive, (ii) negative, and (iii) very close to 0? If you think the answer is "never" for any of these categories, say so.

★ **Problem 1.28.** Figure 1.48 shows position-vs-time graphs for two objects as they move along the same axis. When during the motion (if ever) is the speed of object B (a) *smaller than*, (b) *greater than*, and (c) *equal to* the speed of object A? (d-e) Repeat where we are concerned with the *velocity* rather than the *speed*. (You'll have to make some estimates here–explain your reasoning in words!)

TABLE 1.3
Problem 1.27

Time (s)	Distance (yd)
0	0.0
0.5	2.0
1.0	5.0
1.5	9.0
2.0	14
2.5	19
3.0	25
3.5	30
4.0	36

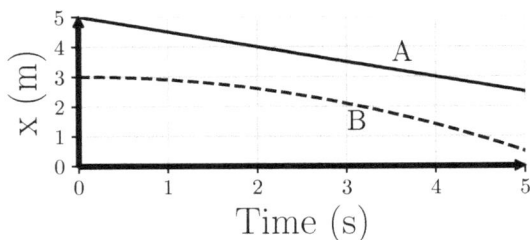

FIGURE 1.48
Problem 1.28

★★ **Problem 1.29.** Figure 1.49 shows a velocity-time graph. Draw the corresponding position-time graph and acceleration-time graph. Assume the object is at $x = 5.0$ m at $t = 0.0$ s. What is the object's acceleration at $t = 3.5$ s?

★★ **Problem 1.30.** Figure 1.50 shows an acceleration-time graph. Draw the corresponding velocity-time graph and position-time graph. Assume the object is at $x = 0.0$ m with $v = 1.0$ m/s at $t = 0.0$ s. What is the object's position and velocity at $t = 6.0$ s?

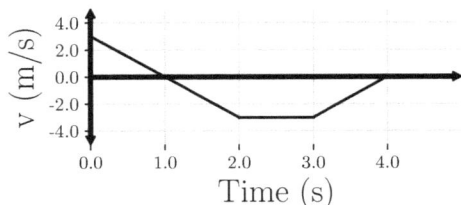

FIGURE 1.49
Problem 1.29

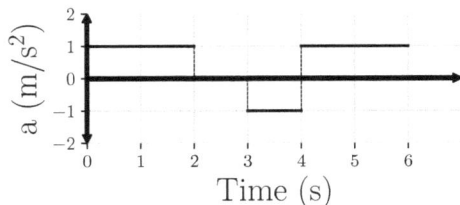

FIGURE 1.50
Problem 1.30

★★ **Problem 1.31.** Figure 1.51 shows motion graphs for a single object. The first phase of the motion is shown in the position-time graph, the second phase of motion is shown in the velocity-time graph, and the third phase of motion is shown in the acceleration-time graph.

a. Sketch the missing sections of each of the three graphs. Assume the velocity is continuous.

b. In a few sentences, describe how the object's velocity changes from the beginning of the motion to the end. (The axes are not labeled with numbers, so this is all qualitative–your response won't have any numerical values.)

★★ **Problem 1.32.** Repeat Problem 1.31 for the motion graphs shown in Figure 1.52.

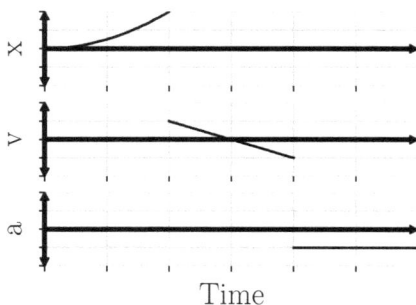

FIGURE 1.51
Problem 1.31

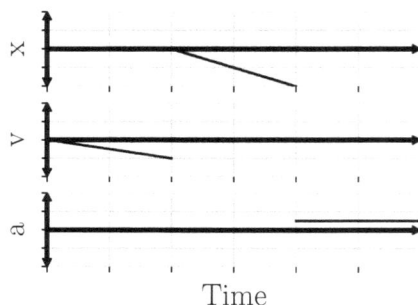

FIGURE 1.52
Problem 1.32

Problem 1.33. Integrate the equations for $a(t)$ and $v(t)$ in part (b) of Example 1.10 to verify that you obtain the expected equations for $v(t)$ and $x(t)$, respectively. What values must you assign the constants of integration? What do they mean, physically?

1.3 Kinematic Equations

Problem 1.34. A certain car is advertised as being able to accelerate from 0 mph (miles per hour) to 60.0 mph in 5.0 seconds. What is the car's average acceleration?

Problem 1.35. A bus is traveling at 10.0 mi/hr and must stop to pick up more passengers. If it takes 10.0 seconds to stop, what is its acceleration?

Problem 1.36. A clown car traveling at 10.0 km/hr on a straight level road undergoes an acceleration of 1.00 m/s^2 for 10.0 seconds. How far do the clowns travel in 10.0 s?

Problem 1.37. You and a friend are racing your bikes on a level bit of road. You "improve your chances" by starting with a velocity of 2.0 m/s and after traveling 23 m you cross the finish line moving with a velocity of 6.0 m/s. What's your average acceleration?

Problem 1.38. If you drop a pebble over a 75.0-m-tall cliff, what velocity will it have when it strikes the ground? (Assuming, as usual, that we neglect air resistance!)

Problem 1.39. Near the end of an action movie, a hero dramatically jumps from a building into a net below just as the building explodes. If the fall took 4.6 seconds, (a) how far did he fall and (b) what speed did he have on impact? Ignore air resistance.

Problem 1.40. Figure 1.53 shows a velocity vs. time graph for a particle moving along the x-axis. At $t_0 = 0.0$ s its position is $x_0 = 2.0$ m. At $t = 3.0$ s, what is the particle's position?

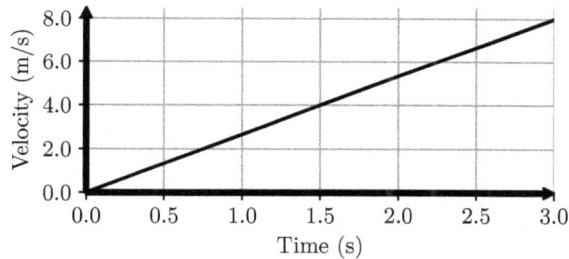

FIGURE 1.53
Problem 1.40

Problem 1.41. Repeat Example 1.11 but set the positive direction of your coordinate system to be *opposite* the car's initial direction of motion. Briefly comment on your answer compared to the answer found in the Example.

Problem 1.42. A block slides past you traveling at 1.5 m/s. If the block is slowing down at a rate of 0.50 m/s^2, how far from you will the block be once it stops moving?

Problem 1.43. You are at a train station, standing next to the train at the front of the first car. The train starts moving with a constant acceleration and 5.0 s later the back of the first car passes you. Each train car is 25 m long.

a. What is the acceleration of the train?

b. How fast is the fourth car traveling when it fully passes you?

c. How long does it take for all four cars to fully pass you?

d. Sketch the position vs time graph for the entire motion of the train. Explain the shape in the context of your responses to (a)–(c).

★★ **Problem 1.44.** In a game show, two contestants have a boat race from a starting line to a finish point 10.0 m away from the start. The position vs. time graph of the contestants are shown in Figure 1.54.

   a. Give an estimated numerical answer for the location where the contestants have the same velocity. Explain your logic.

   b. What is contestant A's velocity?

   c. If contestant B starts from rest and arrives at the finish line in 4.0 s, what is her average acceleration?

   d. How fast is contestant B traveling when she arrives at the finish line?

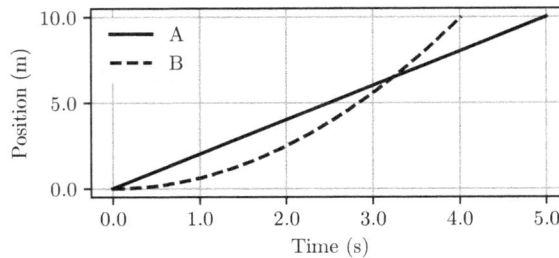

FIGURE 1.54
Problem 1.44

★ **Problem 1.45.** On another planet, the vertical motion of a ball's vertical motion is measured to obey $y(t) = -4.5t^2 + 2.0t + 1.0$ where the units are such that y is in meters. What is the acceleration due to gravity on this planet?

★ **Problem 1.46.** You drop a ball from rest on the moon, where the acceleration due to gravity is about 1/6 of that on the Earth. What is the velocity of the ball when it strikes the ground 1.0 m below?

★ **Problem 1.47.** An airplane that is flying level needs to accelerate from a speed of 2.00×10^2 m/s to a speed of 2.40×10^2 m/s while it flies a distance of 1.20 km. What must be the acceleration of the plane?

★ **Problem 1.48.** A block is traveling up a frictionless ramp with an angle of 20.0° above the horizontal with an initial speed of 10.0 m/s. What is the velocity of the block after 5.00 seconds?

★ **Problem 1.49.** A rocket powered hockey-puck has a constant acceleration of 2.0 m/s² when fired. It is released from rest on a large frictionless table. After traveling 4.0 m the rocket stops firing, but the puck continues to coast for another 2.0 m.

   a. How long was the rocket firing from release to the 4.0 m mark?

   b. What is the total time from release of the puck to the end of the coast?

   c. Draw a full motion diagram for the motion of the puck for the entire motion (acceleration & coast). Be sure to clearly label the start and end of the motion in question.

   d. What is the speed of the puck after traveling the 4.0 m?

★ **Problem 1.50.** In a review session I once made up the following question (and drew Figure 1.55, not to scale) on the spot: my colleague Bob Ekey is riding on a unicycle holding a cup of coffee, screaming, as he zooms past a concerned student. If Bob starts with a velocity of 10.0 m/s and is traveling with a speed of 20.0 m/s after 2.5 s, then what is the angle of the ramp with respect to the horizontal? Assume Bob isn't pedaling so his acceleration is purely due to gravity.

FIGURE 1.55
Problem 1.50

★ **Problem 1.51.** A car traveling at +30.0 m/s runs out of gas while traveling up a 20.0° slope. After 10.0 seconds, what is the car's velocity?

★ **Problem 1.52.** A block sliding up a 10.0° incline takes 3.3 s to stop. What was the initial velocity of the block? Ignore friction.

★ **Problem 1.53.** A baseball pitcher throws a 85 mph fastball into a padded wall. The padding compresses by 4.0 cm as the ball comes to rest. What was the ball's acceleration while it compressed the padding? If a thicker, softer padding was used such that the compression was over 8.0 cm rather than 4.0 cm, what would the new acceleration be? (Can you answer the second part without repeating all of your calculations from the first part?)

★★ **Problem 1.54.** A ball is thrown straight up from a height of 1.8 m above the ground. The initial velocity of the ball is 12.0 m/s. Someone catches the ball when it first reaches a height of 4.0 m above the ground. How long after the ball was thrown was it caught? What was the ball's velocity when it was caught?

★★ **Problem 1.55.** In Problem 1.54 we supposed the ball is caught *on the way up*. Now suppose that the ball passes the person and then they catch it *on the way down*. How long after the ball was thrown will the ball be caught in this situation? What is the ball's velocity?

★★ **Problem 1.56.** A student standing on the ground throws a ball straight up. The ball leaves the student's hand with a speed of 15.0 m/s when the hand is 2.00 m above the ground. The student catches the ball 3.12 seconds later.

  a. What maximum vertical altitude above the launch point does the ball reach?

  b. At what vertical distance above the ground does the student catch the ball?

  c. What is the velocity of the ball right before the student catches it?

  d. Sketch the velocity vs. time graph for the entire motion of the ball. Explain the shape in terms of your responses to (a)–(c).

★ **Problem 1.57.** A rocket ship in deep space is traveling at −10.0 m/s and is caught in a tractor beam which causes the ship to slow. If the ship has a displacement of +10.0 m after 1.0 minute, what is the acceleration of the ship?

★ **Problem 1.58.** A car starts from rest and accelerates with a constant acceleration of 1.00 m/s² for 3.00 s. The car continues for 5.00 s at constant velocity. How far has the car traveled from its starting point?

★★ **Problem 1.59.** A predator spots me from a distance of 100.0 meters away and attacks, accelerating from rest at 3.5 m/s² up to its top speed of 25.0 m/s.

  a. How far has the predator traveled before it reaches its top speed?

  b. How long does it take for the predator to catch me, assuming I stay in the same spot?

  c. How long does it take for the predator to catch me, assuming I start running away from the predator (at the same time that the predator begins running) and quickly reach my top speed of 3.0 m/s?

★★ **Problem 1.60.** A child is hiding behind a couch in the middle of a 3.6-m-wide room when a sibling wielding a foam dart gun enters with a speed of 1.50 m/s. The defenseless child moves toward an open door on the opposite side of the room in 1.4 s. The sibling slows down at 0.30 m/s² and begins firing a dart every 0.25 s starting 0.50 s after they enter the room. How far apart are they when the child makes it to the door? How many darts did the sibling fire up until that instant?

★★ **Problem 1.61.** A student's hand is 1.5 m above the ground when she throws a ball straight up. It leaves her hand with a speed of 2.0 m/s.

  a. How long does it take to return to her hand?

  b. How long would it take to strike the ground if she moved her hand out of the way?

★★ **Problem 1.62.** You launch a ball of mud vertically upward and it hits the ceiling 0.500 seconds later. What is the launch velocity of the ball, if the distance from launch to the ceiling is 1.00×10^2 in?

★ **Problem 1.63.** In a movie, a ring is dramatically dropped into a lava pit from a narrow ledge. It takes 7.0 seconds for it to land in the lava pit. How far above the pit is the ledge?

★ **Problem 1.64.** Suppose you and a friend get into street racing. You are in a car that can accelerate at a rate of 11.0 m/s² and your friend is in a car that can accelerate at a rate of 10.0 m/s². If you start accelerating at the same instant, how far apart will the cars be after you've driven 50.0 m?

★★ **Problem 1.65.** You are driving along State Street at constant velocity and the light turns red ahead of you. It takes you 0.50 seconds to react before your car brakes with constant acceleration.

  a. Draw a full motion diagram for the motion of the car from the instant the light turns red to when you come to rest. Be sure to clearly label the start and end of the motion in question.

  b. If you travel 10.0 m before you start to brake, how fast were you traveling before the light turned red?

  c. Once applying the brakes, it takes you 5.0 seconds to come to rest. What is the acceleration of the car during the braking?

  d. How far did the car travel during the 5.0 second braking experience?

★★★ **Problem 1.66.** A rocket accelerates (from rest) straight up for 25 s at a rate of 35 m/s², at which point it runs out of fuel and the rocket stops firing. Take $g = 9.8$ m/s² to be the acceleration due to gravity.

  a. What is its altitude when the rocket stops firing?

  b. What is the maximum altitude?

  c. What is the total flight time from launch until impact with the ground?

★★ **Problem 1.67.** A professional baseball pitcher can throw a fastball at about 90.0 miles per hour.

  a. Suppose the pitcher's throwing motion brings the ball from rest through a horizontal displacement of 1.00 m. How long does it take for the pitcher to complete the throwing motion?

  b. Home plate is 60.5 ft from the pitching mound. How long does the ball take to reach home plate from the instant the pitcher begins his throwing motion?

★★ **Problem 1.68.** A rock is dropped from a height of 4.0 m above the surface of a pond. Once it hits the water, the rock sinks at a constant velocity (the same as the velocity with which it impacted the surface) and sinks to the bottom 4.5 s after it was dropped. How deep is the pond?

★★ **Problem 1.69.** A car is driving at 20. m/s in the $+x$ direction when the driver, who is originally driving in the right lane, observes a group of deer straight ahead. The driver swerves into the left lane and, at $t = 0.0$ s, begins accelerating at a rate of 1.5 m/s² for 3.0 s. It then travels at a constant speed for 2.0 s before changing back into the right lane.

  a. How fast is the car traveling at $t = 3.0$ s?

b. How far does the car travel in the left lane?

c. Suppose that a truck is traveling in the $-x$ direction at a steady 20. m/s. If the truck is 1.0 km in front of the car at $t = 0.0$, how much distance is there between the car and the truck when the car changes back into the right lane?

Problem 1.70. A bead is able to slide without friction along a thin rod of length L. Take the acceleration due to gravity to have a constant magnitude g.

a. The rod is held vertically and the bead is released from rest from the top. How long does it take to reach the bottom?

b. How would your answer to (a) change if the rod's length was $2L$?

Problem 1.71. Suppose a super hero's friend is pushed off of an 85.0-m-tall tower by a villainous villain (I encourage you to come up with your own names for these characters and label your diagram accordingly). 2.0 seconds after the friend begins to fall, the hero jumps downward with some initial speed v in an effort to catch the friend before they hit the water below.

a. How long does it take for the friend to reach a position 20.0 m above the water?

b. What must v be if the hero is to be 20.0 m above the water at the same time as the friend?

c. Assuming v has the value you found in (b), what speed will the friend and the hero have when they collide 20.0 m above the water?

1.5 Relative Motion

Problem 1.72. Some airports have moving walkways: you step on and the walkway carries you forward. Suppose the walkway moves (relative to the ground) at 1.0 m/s, and your jogging speed is 2.0 m/s. What is your speed, relative to the ground, if you get on the walkway and jog (a) in the same direction as the walkway or (b) the wrong way, opposite the direction of the walkway? Include a vector diagram to support your work.

Problem 1.73. Two cars are driving toward each other. Driver A's speedometer indicates their speed (relative to the ground) is 40 mph. Driver B's speedometer indicates their speed is 60 mph. What is driver B's speed relative to driver A? What is driver A's speed relative to driver B? Include a vector diagram to support your work.

Problem 1.74. Three sprinters are all in a line moving in the same (call it the positive) direction. Sprinter A (in the back) is running at 2.7 m/s, Sprinter B (in the middle) is running at 3.5 m/s, and Sprinter C (in the front) is running at 4.0 m/s. What is the velocity of each sprinter according to each of the other sprinters? (You'll have 6 answers here!)

Problem 1.75. You are standing on the ground as a train passes by; suppose it is moving from left to right at a speed of 15 m/s. Much to your surprise, you see me standing on top of the last car of the train, facing left. As I pass you, I throw a ball.

a. If *you* determine the speed of the ball to be 5.0 m/s to the *left*, then what was the speed of the ball as I threw it from *my* perspective?

b. Repeat (a) in the case where you determine the speed of the ball to be 5.0 m/s to the *right*.

c. Repeat (a) in the case where you determine the speed of the ball to be 0.0 m/s.

Problem 1.76. You're standing on a stationary train and drop a ball from shoulder height. The position on the floor of the train where the ball lands is marked with a delightful sticker. Then, the experiment is repeated while the train is moving at a steady speed v. Is the new sticker closer to the front of the train, the back of the train, or is it on top of the original sticker? Explain.

Additional Problems for Chapter 1

★★ **Problem 1.77.** Consider the motion diagram in Figure 1.56. A position is indicated every 10.0 s and the object moves to the left (in the negative direction) for the duration shown.

   a. Make a table of the position of the object at every time.

   b. Calculate the object's velocity between every adjacent pair of positions identified in (a).

   c. Calculate the object's acceleration between every adjacent pair of velocities identified in (b).

   d. Draw the motion diagram and label the velocities and accelerations using your results from (b) and (c).

   e. Make a position-vs-time graph, velocity-vs-time graph, and acceleration-vs-time graph for the object.

   f. During which frame-to-frame interval does the object have the greatest speed? How do you know?

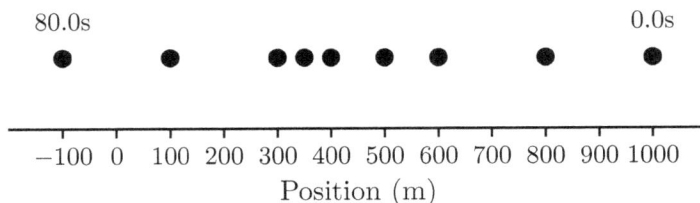

FIGURE 1.56
Problem 1.77

★★ *dx* **Problem 1.78.** Consider the position function $x(t) = 5(t-3)^2 + 6$ where x is measured in meters and t is measured in seconds. Determine the object's velocity and acceleration as functions of time. If these equations accurately describe the motion of some particle in the time interval $t = 0.0$ s to $t = 5.0$ s, what is the particle's maximum position? What is its minimum position?

★★ *dx* **Problem 1.79.** Repeat Problem 1.78 for the position function $x(t) = t^3 - 6t^2 + 4t + 12$.

★★ ∫ **Problem 1.80.** In the text we considered cases where an object's acceleration is 0, a constant value, or (in Example 1.10) sinusoidally. Now consider the case where the acceleration varies linearly in time, i.e. $a(t) = bt$ where b is some positive constant. Determine the corresponding equations for the object's velocity as a function of time and position as a function of time. Interpret the physical meaning of the constants of integration.

★★ **Problem 1.81.** A cart is resting at the bottom of a long track that has been elevated at an angle of 15° above the horizontal. The cart is given a shove such that it has an initial velocity of 0.50 m/s up the incline.

   a. How far up the incline does it reach before beginning to slide back down?

   b. How long does it take for the cart to return to the bottom of the track?

   c. Sketch the position vs. time, velocity vs. time, and acceleration vs. time graphs for the cart.

★ **Problem 1.82.** Consider the situation described in Problem 1.81, but now determine what initial velocity the cart needs to have if it is to travel 0.80 m up the ramp before stopping.

2

Describing Motion on a Plane

In Chapter 1 we restricted ourselves to motion that occurred on a straight line. In this chapter we will see that the techniques we developed in Chapter 1 extend straightforwardly to motion that occurs in two dimensions (for example, a ball thrown as two people play catch).

Learning Objectives

After reading this chapter, you should be able to:

- Set up a two-dimensional coordinate system.

- Analyze 2D vectors and motion diagrams. In particular, calculate and describe (mathematically and graphically) position, displacement, velocity, and acceleration in two-dimensional coordinate systems.

- Recognize that motion in two dimensions can be analyzed as a pair of one-dimensional problems *linked by time*.

- Apply the one-dimensional kinematic equations to two-dimensional motion.

- Relate the linear kinematic variables (x, v, and a) to their angular counterparts (θ, ω, and α).

- Use motion graphs and the angular kinematic equations to analyze the motion of objects traveling in a circular path.

2.1   Two-Dimensional Vectors and Motion Diagrams

In Chapter 1 we saw that a position vector points from the origin of our coordinate system to the location of the object that we are describing. If the object is moving, then the displacement vector points from the initial position to the final position; mathematically we expressed this as $\Delta \vec{x} = \vec{x}_f - \vec{x}_i$. The exact same logic holds if we are describing motion that occurs on a plane rather than on a straight line, like if we consider a ball being tossed across the room (Fig. 2.1).

We will return to this example, but to get a handle on describing two-dimensional vectors we will start with a simpler example: consider a top-down view of a person standing on a flat field, and suppose they walk 30.0 meters east. This is a one-dimensional problem, and if we set up our coordinate system so the positive direction is east, we can say their displacement $\Delta \vec{x}$ is 30.0 m east. Let's introduce a slightly different way of saying the same thing:

$$\Delta \vec{x} = 30.0 \text{ m} \hat{x} \tag{2.1}$$

DOI: 10.1201/9781003571568-3

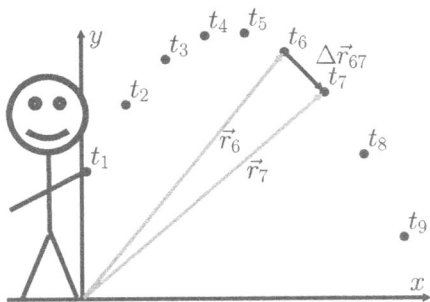

FIGURE 2.1
An example of a two-dimensional motion graph showing a ball being tossed across the room. The origin is on the ground below the person's hand, two example positions vectors are shown in gray, and the corresponding displacement vector is shown in black.

The symbol $\hat{x}$ (read "x hat") is called a **unit vector**; it is simply a shorthand way of assigning a direction to the vector's magnitude (which is 30.0 m in this example). To see why this is useful, consider a second scenario where the person instead walked 40.0 m north. To be clear that this is a distinct situation from moving east, we'll set up a new axis with the variable y instead of x; the displacement, $\Delta \vec{y}$, is 40.0 m north. Alternatively, using our new notation, we can say instead

$$\Delta \vec{y} = 40.0 \text{ m} \hat{y} \tag{2.2}$$

We can combine these two one-dimensional scenarios into a single two-dimensional scenario: suppose our walker moves 30.0 m east, then 40.0 m north. The displacement is still defined to be the change in position, which we can define in terms of a displacement along the x axis and a displacement along the y axis (Figure 2.2):

$$\Delta \vec{r} = 30 \text{ m} \hat{x} + 40 \text{ m} \hat{y} \tag{2.3}$$

Note that we're using $\Delta \vec{r}$ to represent displacement to differentiate it from our coordinates. That is, the displacement is not necessarily just in the x dimension or just the y dimension; rather, there is a *component* in the $\hat{x}$ direction and a *component* in the $\hat{y}$ direction.[1] The approach of having two perpendicular axes like in Figure 2.2 is referred to as a **Cartesian coordinate system** after René Descartes (1596-1650), a giant of science, mathematics, and philosophy.

Describing a two-dimensional vector by saying "some amount along the x axis and some amount along the y axis" is commonly referred to as the **component method**. A second method, called the **magnitude-angle method**, summarizes a vector with its magnitude and the angle formed by the vector and the positive x axis.[2] Using our example from Figure 2.2, we can describe the displacement like so:

$$\Delta \vec{r} = 50.0 \text{ m at } 53.1° \text{ above the } \hat{x} \text{ axis} \tag{2.4}$$

---

[1] In the scenario we are discussing, our walker actually moved along the x and y vector components shown in Figure 2.2. However, it is important to recognize that this isn't required: the displacement only depends on the starting position and the ending position, so as long as we end at the same position, the displacement $\Delta \vec{r}$ is the same regardless of the route we take to get there.

[2] The Greek symbols θ ("theta") and ϕ ("phi") are commonly used as variables to represent angles.

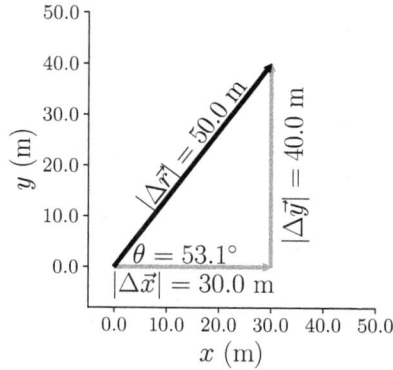

FIGURE 2.2

A two-dimensional displacement vector $\Delta \vec{r}$ with a 30.0 m x component and a 40.0 m y component. The vector can be fully described in terms of these components or in terms of its overall magnitude, 50.0 m, and the angle it forms with the $+x$ axis, 53.1°.

Both Equations (2.3) and (2.4) give complete descriptions of the same vector. It is important to become comfortable converting back and forth between these two methods: they both have their advantages and you'll use them both frequently. To carry out these conversions, you'll need to make use of right-triangle trigonometry and the Pythagorean Theorem (Section A.3). Example 2.1 demonstrates the conversion for this situation, and Examples 2.2–2.3 demonstrate the analysis of two-dimensional vectors in some situations we will see frequently as we proceed.

Example 2.1 Cartesian and Polar Vector Components ⋆

Verify that the vector described by Equation (2.4) is equivalent to the vector described by Equation (2.3).

When we represent a vector and its components, the vector is always the hypotenuse of a right triangle and the components are the other two sides. The Pythagorean Theorem allows us to use the components provided in Equation (2.3) to determine the magnitude of the displacement vector (that is, the hypotenuse):

$$c^2 = a^2 + b^2$$
$$|\Delta \vec{r}|^2 = (30.0 \text{ m})^2 + (40.0 \text{ m})^2$$
$$|\Delta \vec{r}| = 50.0 \text{ m}$$

So far, so good–the magnitude specified in Equation (2.4) agrees with what Equation (2.3) says it must be. It remains to determine the value of θ. This requires a bit of trigonometry:

$$\tan(\theta) = \frac{40.0 \text{ m}}{30.0 \text{ m}}$$
$$\tan^{-1}(\tan(\theta)) = \tan^{-1}\left(\frac{40.0 \text{ m}}{30.0 \text{ m}}\right)$$
$$\theta = \tan^{-1}\left(\frac{40.0}{30.0}\right) = 53.1°$$

If you're rusty with your trigonometry, note that in the second line above we've applied an *inverse trigonometric function* to "undo" a trigonometric function: $\tan^{-1}(\tan(\theta)) = \theta$, and likewise for sin with $\sin^{-1}$ and cos with $\cos^{-1}$. You can think of trig functions and their inverses as paired in a similar way to multiplication and division: if we have $2x$ and want x, we divide by 2 to "undo" the multiplication by 2 and arrive at x.

Now that we've checked both the magnitude and the direction, it is clear that the two Equations (2.3) and (2.4) do indeed define the same vector.

Example 2.2 Velocity Components   ⋆

A ball is kicked such that as it leaves the kicker's foot, it has a speed of 3.0 m/s at an angle of 32° above the ground. What is the ball's horizontal velocity? What is its vertical velocity?

The problem of reducing a vector from the magnitude-angle form to the component form is often referred to as **vector decomposition** (going the other way is often referred to as **vector composition**). To begin, we set up our coordinate system so the ball starts at the origin and its velocity has components in the positive x direction and the positive y direction, as shown in Figure 2.2.

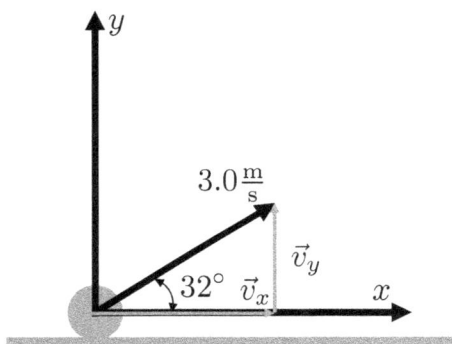

FIGURE 2.3

Note that we know the hypotenuse of the triangle and one of the interior angles. Sine deals with the hypotenuse and the side of the triangle *opposite* the angle, which is $\vec{v}_y$ in this case. Thus

$$\sin(32°) = \frac{v_y}{3.0\frac{m}{s}}$$

$$v_y = \left(3.0\frac{m}{s}\right)\sin(32°) = 1.6\frac{m}{s}$$

Similarly,

$$\cos(32°) = \frac{v_x}{3.0\frac{m}{s}}$$

$$v_x = \left(3.0\frac{m}{s}\right)\cos(32°) = 2.5\frac{m}{s}$$

We can summarize by writing

$$\vec{v} = 2.5\frac{m}{s}\hat{x} + 1.6\frac{m}{s}\hat{y}$$

Example 2.3 Components of $\vec{g}$ on a Ramp ⋆

An object is sliding on a surface inclined above the horizontal by some angle θ. How much of the acceleration due to gravity is parallel to the surface? (Recall that we considered a related problem in Example 1.15).

We sketch a diagram in Figure 2.4. Because the angles inside a triangle add to $180°$, it follows that

$$90° + \theta + \phi_1 = 180° \implies \phi_1 = 90° - \theta$$

However, the two angles ϕ_1 and ϕ_2 add to $90°$, so we have

$$\phi_1 = 90° - \theta$$
$$90° - \phi_2 = 90° - \theta$$
$$\phi_2 = \theta$$

Thus the piece of $\vec{g}$ parallel to the surface, which we've aligned with the x axis, is

$$g_x = \pm g \sin(\theta) \tag{2.5}$$

The $\pm$ indicates this quantity can be positive or negative. It always points *down the ramp*, but that is only a positive quantity if we call *down the ramp* the positive direction, as we've done here. If *up the ramp* is the positive direction, then $g_x = -g\sin(\theta)$.

This is a useful result, because it allows us to analyze motion on an incline plane in *one* dimension! In contrast, if we used the usual "up is $+y$, right is $+x$" convention, the motion would be two dimensional since it moves both right and down. We will see in Chapter 3 that things become more complicated when we include effects such as friction, but for now you can think of Equation (2.5) as a way to determine the acceleration of an object on an incline plane when it is being subjected only to gravity.

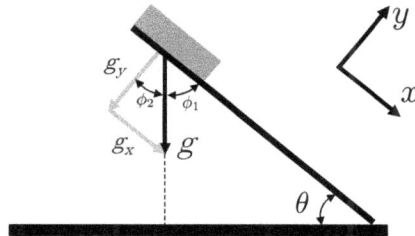

FIGURE 2.4

When we added one-dimensional vectors, we were essentially just adding positive and negative numbers. For instance, if we have two one-dimensional vectors $\vec{p} = 5\hat{x}$ and $\vec{q} = -7\hat{x}$, then their sum is given by

$$\vec{p} + \vec{q} = 5\hat{x} + (-7\hat{x})$$
$$\vec{p} + \vec{q} = (5 - 7\hat{x})$$
$$\vec{p} + \vec{q} = -2\hat{x}$$

and their difference is given by

$$\vec{p} - \vec{q} = 5\hat{x} - (-7\hat{x})$$
$$\vec{p} - \vec{q} = (5 + 7\hat{x})$$
$$\vec{p} - \vec{q} = 12\hat{x}$$

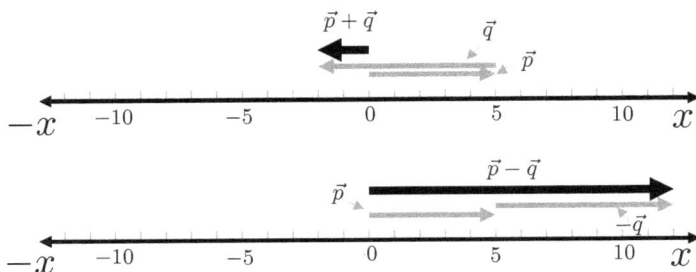

FIGURE 2.5

The tail to tip method of visually combining vectors in one dimension. The vectors are shifted vertically just for visual clarity.

We show the tail-to-top method applied to these cases in Figure 2.5.

In the algebra of this example, the $\hat{x}$ serves as a label and we simply combine the magnitudes of the vectors. This label is important in two dimensional vector analysis, because *we treat the x and y components independently*. For example, if we have two two-dimensional vectors $\vec{p} = 5\hat{x} + 2\hat{y}$ and $\vec{q} = -7\hat{x} - 2\hat{y}$, then their sum is given by

$$\vec{p} + \vec{q} = (5\hat{x} + 2\hat{y}) + (-7\hat{x} - 2\hat{y})$$
$$\vec{p} + \vec{q} = (5 - 7)\,\hat{x} + (2 - 2)\,\hat{y}$$
$$\vec{p} + \vec{q} = -2\hat{x}$$

And their difference is given by

$$\vec{p} - \vec{q} = (5\hat{x} + 2\hat{y}) - (-7\hat{x} - 2\hat{y})$$
$$\vec{p} - \vec{q} = (5 + 7)\,\hat{x} + (2 + 2)\,\hat{y}$$
$$\vec{p} - \vec{q} = 12\hat{x} + 4\hat{y}$$

The tail to tip method carries over directly for two-dimensional vectors, as we show in Figure 2.6. Example 2.4 gives some additional practice with this and with the process of

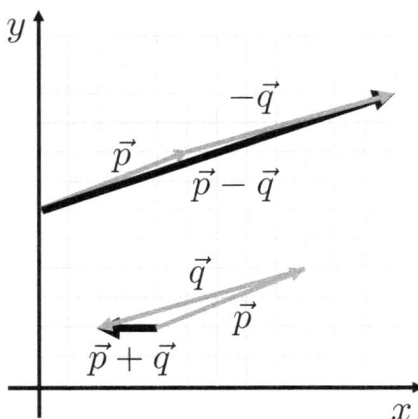

FIGURE 2.6

The tail to tip method of visually combining vectors in two dimensions. Note that a vector is not defined by *where* it is on the coordinate system; only the length and orientation of each component matters.

multiplying a vector by a positive scalar (which can modify the magnitude of a vector but doesn't modify the direction in which it points) or a negative scalar (which reverses the direction of the vector in addition to modifying the magnitude). We then consider a series of examples that apply these ideas to motion diagrams and vectors in two dimensions.

Example 2.4 Two Dimensional Vectors $\star$

Consider the vectors $\vec{p} = 2\hat{x} + 3\hat{y}$ and $\vec{q} = -5\hat{x} + \hat{y}$. Consider (a) $-\vec{p}$, (b) $\vec{p} + \vec{q}$, (c) $\vec{q} - \vec{p}$, and (d) $3\vec{p}$. In each case express the vector mathematically (in terms of its components) and visually (on a Cartesian coordinate system).

For (a), the negative of a vector reverses the direction of each of its components; visually, the vector simply points in the opposite direction.

$$-\vec{p} = -(2\hat{x} + 3\hat{y})$$
$$= -2\hat{x} - 3\hat{y}$$

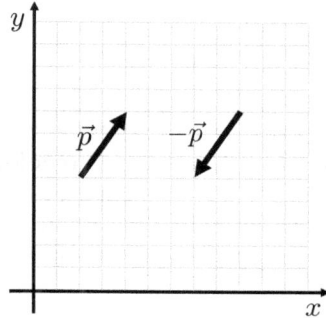

FIGURE 2.7

For (b), adding the vectors involves adding their components individually; visually, we apply the tail to tip method.

$$\vec{p} + \vec{q} = (2\hat{x} + 3\hat{y}) + (-5\hat{x} + \hat{y})$$
$$= (2 - 5)\hat{x} + (3 + 1)\hat{y}$$
$$= -3\hat{x} + 4\hat{y}$$

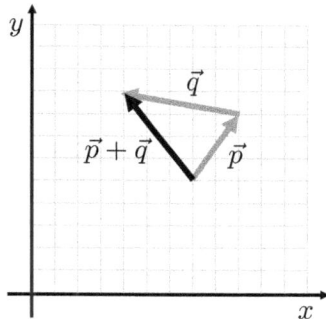

FIGURE 2.8

For (c), we need to take care to distribute the negative sign into the components of $\vec{p}$; visually we combine the previous two steps.

$$\vec{q} - \vec{p} = (-5\hat{x} + \hat{y}) - (2\hat{x} + 3\hat{y})$$
$$= (-5 - 2)\hat{x} + (1 - 3)\hat{y}$$
$$= -7\hat{x} - 2\hat{y}$$

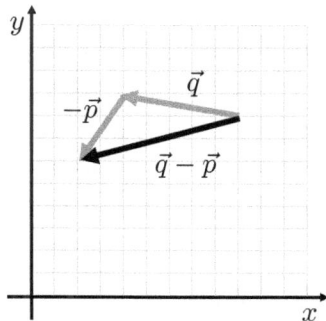

FIGURE 2.9

For (d), Because each component of the vector increases by the same factor, the direction does not change; only the length.

$$3\vec{p} = 3\left(2\hat{x} + 3\hat{y}\right)$$
$$= 6\hat{x} + 9\hat{y}$$

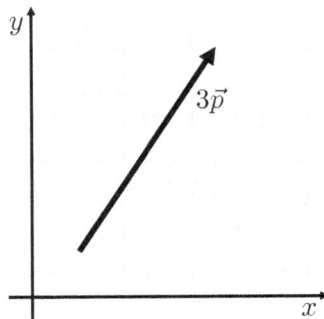

FIGURE 2.10

Example 2.5 Average Acceleration in Two Dimensions   ⋆
An object starts at point A, as shown in the left side of Figure 2.11, at $t = 0$ s, then moves to point B at $t = 1.0$ s, then finally to point C at $t = 2.0$ s. What is the object's average acceleration from $t = 0$ s to $t = 2.0$ s?

The velocity vector $\vec{v}_{AB}$ points from A to B, and the velocity vector $\vec{v}_{BC}$ points from B to C. Thus the average acceleration is

$$\vec{a}_{ABC} = \frac{\Delta \vec{v}}{\Delta t}$$
$$\vec{a}_{ABC} = \frac{\vec{v}_{BC} - \vec{v}_{AB}}{\Delta t}$$

This is shown graphically on the right side of Figure 2.11; we see

$$\vec{a}_{ABC} = (2.0\hat{x} - 2.0\hat{y})\ \text{m/s}^2$$

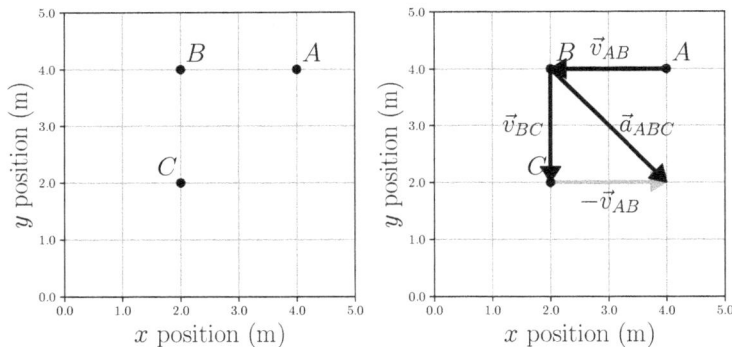

FIGURE 2.11
The fact that the acceleration points *right* and *down* corresponds to the fact that in the horizontal direction the velocity goes from a value to the left (specifically, -2.0 m/s) to 0, and in the vertical direction the velocity goes from 0 to a value (specifically, -2.0 m/s) in the downward direction.

Example 2.6 A Hiker's Journey ★★

A hiker leaves their car and travels 3.0 mi 45° North of East, then travels 2.0 mi North, then finally travels 4.0 mi West. How far and in what direction should they travel to return to the car?

We sketch the three displacement vectors in Figure 2.12. To mathematically determine the hiker's location, we add the three vectors, which we'll label $\vec{d}_1$, $\vec{d}_2$ and $\vec{d}_3$. Let's first determine the components of $\vec{d}_1$:

$$d_{1,x} = 3.0 \cos\left(45°\right) \text{ mi} = 2.1 \text{ mi}$$

and

$$d_{1,y} = 3.0 \sin\left(45°\right) \text{ mi} = 2.1 \text{ mi}$$

Here I am using the x axis to run East-West and the y axis to run North-South. (Our results demonstrate the fact that $\sin\left(45°\right) = \cos\left(45°\right)$.) Now we can add the vectors in component form:

$$\vec{d}_1 + \vec{d}_2 + \vec{d}_3 = (2.1 + 0 - 4.0) \text{ mi}\hat{x} + (2.1 + 2.0 + 0) \text{ mi}\hat{y}$$
$$\vec{d}_1 + \vec{d}_2 + \vec{d}_3 = (-1.9\hat{x} + 4.1\hat{y}) \text{ mi}$$

Graphically, the sum of these vectors runs from the tail of the first vector to the tip of the third vector; in other words, it points from the car (the origin of the coordinate system) to where the hiker is standing after the first three legs of their journey. Now if we want a fourth vector $\vec{d}_4$ that brings the hiker back to their car, then the total sum must be 0:

$$\vec{d}_1 + \vec{d}_2 + \vec{d}_3 + \vec{d}_4 = 0$$
$$\vec{d}_4 = -\left(\vec{d}_1 + \vec{d}_2 + \vec{d}_3\right)$$
$$\vec{d}_4 = -\left(-1.9\hat{x} + 4.1\hat{y}\right) \text{ mi}$$
$$\vec{d}_4 = (1.9\hat{x} - 4.1\hat{y}) \text{ mi}$$

We can determine the magnitude of $\vec{d}_4$ with the Pythagorean Theorem:

$$|\vec{d}_4| = \left(1.9^2 + 4.1^2\right)^{1/2} \text{ mi} = 4.5 \text{ mi}$$

And the angle follows from the trigonometry:

$$\theta = \tan^{-1}\left(\frac{4.1}{1.9}\right) = 65°$$

So the hiker needs to walk 4.5 miles 65° South of East to return to the car. Figure 2.12 shows the four vectors added tail to tip, demonstrating a complete loop that brings the hiker back to the origin.

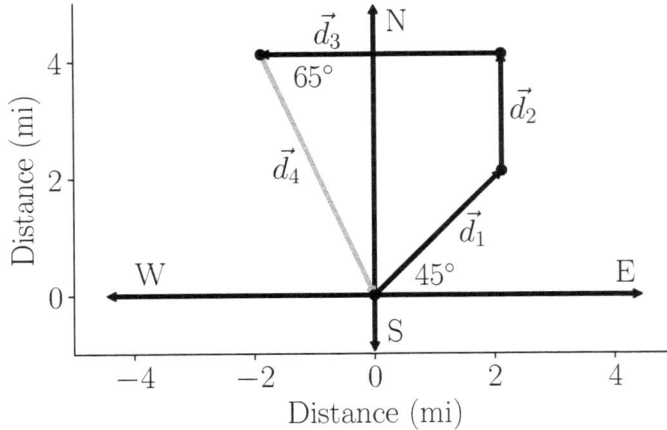

FIGURE 2.12

Example 2.7 Motion Diagram, Velocity, And Acceleration of a Puck ⋆⋆
The table below shows the position of a hockey puck as it is pushed across ice. Draw
the corresponding motion diagram and include both velocity and acceleration vectors.

time (s)	x position (m)	y position (m)
0.00	0.00	0.00
0.25	0.00	0.50
0.50	0.00	1.00
0.75	0.50	1.50
1.00	1.50	2.00
1.25	3.00	2.50

Figure 2.13 shows the motion diagram with velocity vectors drawn in gray. Recall
that a vector from one position to the next is, formally, the *displacement* vector
$\Delta \vec{r}$; however, because $\vec{v} = \Delta \vec{r}/\Delta t$, the velocity points in the same direction as the
displacement and their magnitudes are proportional. (They differ by a factor of Δt,
which has the same value–here, 0.25 s–for every pair of successive positions.) Thus
the standard approach is to connect successive positions with a vector labeled as the
average velocity as the object moves between those positions.

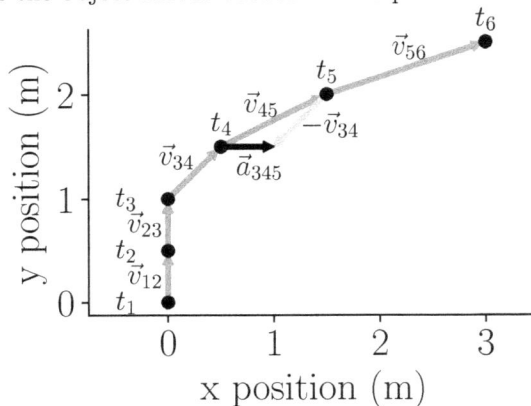

FIGURE 2.13

Similarly, to calculate an average *acceleration*, we need two successive *velocity* vectors: $\vec{a} = (\vec{v}_f - \vec{v}_i)/\Delta t$. Using the tail to tip method, we flip the direction of the initial velocity vector and place its tail on the tip of the final velocity vector. Then $\Delta \vec{v}$ runs from the tail of $\vec{v}_f$ to the tip of $-\vec{v}_i$. Just as we map $\Delta \vec{x}$ to $\vec{v}$, as we discussed above, so too we can map $\Delta \vec{v}$ to $\vec{a}$. In Figure 2.13 we demonstrate this for $\vec{v}_{34}$ as the first (initial) vector and $\vec{v}_{45}$ as the second (final) vector; we use a lighter gray color for $-\vec{v}_{34}$ and show the acceleration $\vec{a}_{345}$ in black.

Figure 2.14 repeats this process for every pair of velocity vectors; it gets a bit cluttered but if you check each acceleration vector you will see that it is drawn with the same logic discussed above. Note in particular that there is no $\vec{a}_{123}$: the velocity is *the same* for $\vec{v}_{12}$ and $\vec{v}_{34}$ and so $\Delta\vec{v}_{123}$ (and therefore $\vec{a}_{123}$) equals 0.

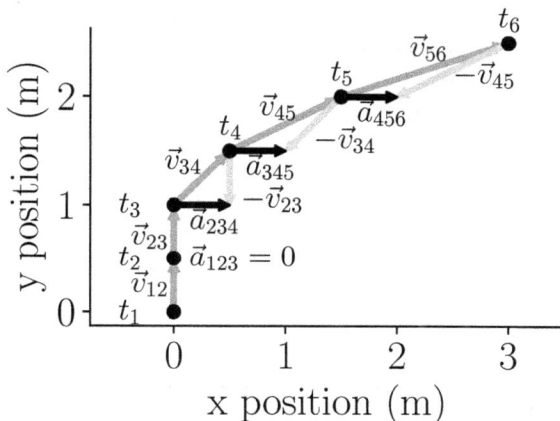

FIGURE 2.14

Example 2.8 Analyzing a Trajectory ★★

Consider the motion diagram of Figure 2.1. Suppose someone analyzed a video of such a toss and found $\vec{r}_6 = 2.60 \text{ m}\hat{x} + 1.63 \text{ m}\hat{y}$, $\vec{r}_7 = 3.11 \text{ m}\hat{x} + 1.36 \text{ m}\hat{y}$, and $\vec{r}_8 = 3.62 \text{ m}\hat{x} + 0.95 \text{ m}\hat{y}$. If these positions are separated in time by 0.12 s, then what are the average velocities $\vec{v}_{67}$ and $\vec{v}_{78}$? What is the average acceleration $\vec{a}_{678}$?

Recall that the average velocity is given by $\vec{v} = \Delta\vec{r}/\Delta t$. So we have

$$\vec{v}_{67} = \frac{\vec{r}_7 - \vec{r}_6}{\Delta t}$$

$$\vec{v}_{67} = \frac{(3.11 \text{ m}\hat{x} + 1.36 \text{ m}\hat{y}) - (2.60 \text{ m}\hat{x} + 1.63 \text{ m}\hat{y})}{0.12 \text{ s}}$$

$$\vec{v}_{67} = \frac{(3.11 - 2.60) \text{ m}\hat{x} - (1.36 - 1.63) \text{ m}\hat{y}}{0.12 \text{ s}}$$

$$\vec{v}_{67} = (4.3\hat{x} - 2.3\hat{y}) \text{ m/s}$$

Going through the same procedure yields $\vec{v}_{78} = (4.3\hat{x} - 3.4\hat{y})$ m/s, as you can check for yourself.

Then to find the acceleration, we'll make use of $\vec{a} = \Delta\vec{v}/\Delta t$:

$$\vec{a}_{678} = \frac{\vec{v}_{78} - \vec{v}_{67}}{\Delta t}$$

$$\vec{a}_{678} = \frac{(4.25 \text{ m/s}\hat{x} - 2.25 \text{ m/s}\hat{y}) - (4.25 \text{ m/s}\hat{x} - 3.42 \text{ m/s}\hat{y})}{0.12 \text{ s}}$$

$$\vec{a}_{678} = (0.0\hat{x} - 9.8\hat{y}) \text{ m/s}^2 = -9.8\hat{y} \text{ m/s}^2$$

The acceleration matches what we would expect due to gravity. Notice that in calculating the acceleration, I used numerical values for the velocities to greater precision than I reported above–this is a good example where rounding values to the appropriate precision and then using rounded values in subsequent calculations can throw off your result.

Figure 2.15 shows a zoomed-in view of the motion and demonstrates the tail top tip method to graphically find the acceleration vector. As we would expect, the acceleration vector points straight down.

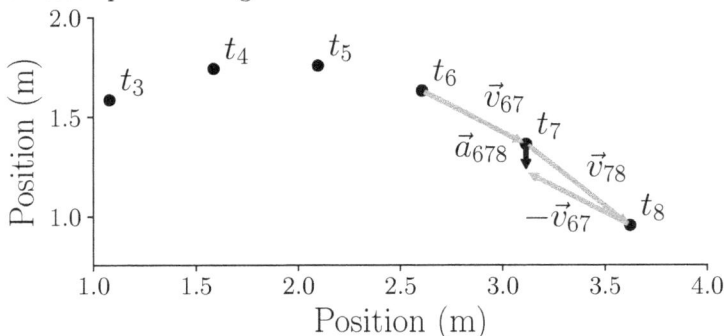

FIGURE 2.15

Example 2.9 Two-Dimensional Motion Diagram   ⋆⋆

Draw a complete motion diagram for an object that begins at $\vec{r}_1 = 1.0 \text{ m}\hat{x}$ and then moves in a counterclockwise circle, 30° every second, until it returns to the starting position.

Recall that in a motion diagram, the velocity vectors simply connect each position to the next. Figure 2.16 shows a dot for each position (at 1-second intervals) and the corresponding velocity vectors.

To be complete, we also need to determine the acceleration vectors. For instance, we can use the first two velocity vectors to determine the average acceleration across the first three time points: $\vec{a}_{012} = (\vec{v}_{12} - \vec{v}_{01})/\Delta t$. In Figure 2.17, we zoom in on the first quadrant of the motion and apply the tail-to-tip method to identify $\vec{a}_{012}$ (recall that $\vec{a}$ is proportional to $\Delta\vec{v}$, so determining the direction of $\Delta\vec{v}$ suffices to orient the direction of $\vec{a}$).

It is no coincidence that the acceleration vector points toward the center of the object's motion. As we shall see later, objects moving at a constant speed around the circumference of a circle *always* have an acceleration vector pointing toward the center of the circle; this acceleration is called the **centripetal acceleration**. You are asked to verify that the acceleration consistently points in toward the center of the circle in Problem 2.22.

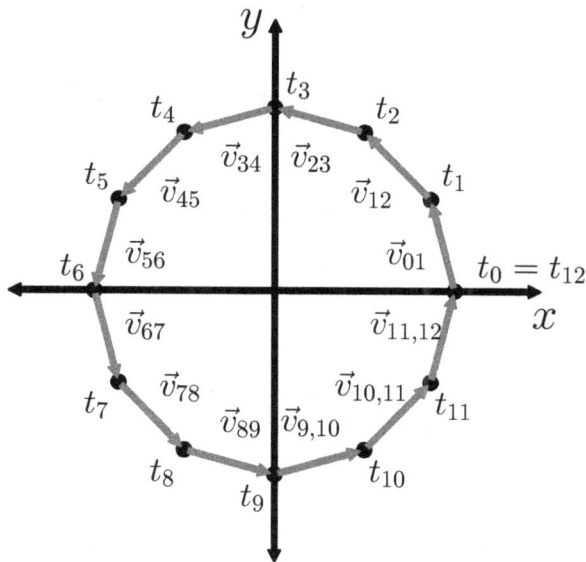

FIGURE 2.16

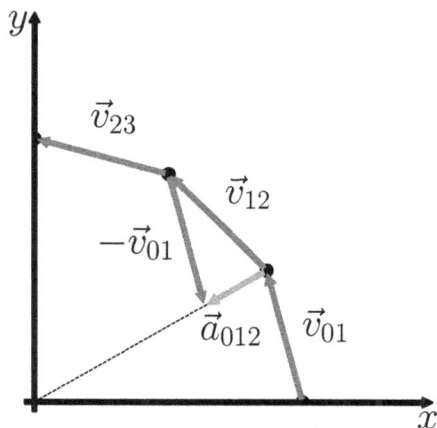

FIGURE 2.17

2.2 Two-Dimensional Kinematics

In the last section we saw how we can describe a two-dimensional vector with two one-dimensional vectors (via the component method). It turns out that there is a similar statement for two-dimensional kinematics: a two-dimensional kinematics problem can be analyzed as two one-dimensional kinematics problems. Let's take as an example a ball that is rolled horizontally off the edge of a table, as shown in Figure 2.18. Suppose that we want to know how far from the edge of the table the ball strikes the ground, given its initial velocity and the height of the table. We set our origin on the ground, below the edge of the table, and call *to the right* the positive x direction and *up* the positive y direction.

Let's analyze the vertical motion independently from the horizontal motion. If we follow our usual procedure from Chapter 1, we can make a list of the kinematic variables separately for each dimension:

vertical motion: horizontal motion:

$$\vec{a}_y = -g \qquad\qquad\qquad\qquad \vec{a}_x = 0\,\frac{m}{s^2}$$

$$\vec{y}_i = h \qquad\qquad\qquad\qquad \vec{x}_i = 0\text{ m}$$

$$\vec{y}_f = 0\text{ m} \qquad\qquad\qquad\qquad \vec{x}_f = ?$$

$$(\vec{v}_y)_i = 0\,\frac{m}{s} \qquad\qquad\qquad (\vec{v}_x)_i = v$$

$$(\vec{v}_y)_f = ? \qquad\qquad\qquad\qquad (\vec{v}_x)_f = v$$

$$\Delta t = ? \qquad\qquad\qquad\qquad \Delta t = ?$$

Note in particular that because there is no acceleration in the horizontal direction (gravity only acts vertically), the horizontal velocity does not change from its initial value. A key insight is that the *time in flight* is the same for both the vertical motion and the horizontal motion.[3] This means that if we determine Δt by analyzing one component of the motion, we can then use that same value when analyzing the other component of the motion. Here, the time in flight is determined by the vertical motion (the ball falls down until it strikes the ground; meanwhile, there is no barrier restricting the horizontal motion of the ball). Thus the kinematic equations (1.12) can be applied to the vertical motion to determine Δt (we choose the equation that omits $\vec{v}_f$ since we do not know it, and swap y for x):

$$\vec{y}_f = \vec{y}_i + (\vec{v}_y)_i\,\Delta t + \frac{1}{2}\vec{a}_y\,(\Delta t)^2$$

$$0\text{ m} = h + \left(0\,\frac{m}{s}\right)\Delta t - \frac{1}{2}g\,(\Delta t)^2$$

$$\frac{1}{2}g\,(\Delta t)^2 = h$$

Or, solving for Δt,

$$\Delta t = \left(\frac{2h}{g}\right)^{\frac{1}{2}} \tag{2.6}$$

This is how long the ball is in flight.[4] How far does it travel horizontally in this time? Again, we can turn to the kinematic equations (1.12):

$$\vec{x}_f = \vec{x}_i + (\vec{v}_x)_i\,\Delta t + \frac{1}{2}\vec{a}_x\,(\Delta t)^2$$

$$\vec{x}_f = 0\text{ m} + \vec{v}\Delta t + \frac{1}{2}\left(0\,\frac{m}{s^2}\right)(\Delta t)^2$$

$$\vec{x}_f = \vec{v}\Delta t$$

Or, finally,

$$\vec{x}_f = \vec{v}\left(\frac{2h}{g}\right)^{\frac{1}{2}} \tag{2.7}$$

---

[3] The ball might bounce or roll once it hits the ground, but for now we're only concerned with the path it takes from when it leaves the table until it first hits the ground.

[4] As always, it is worth your time to consider what this equation says: if the table is higher, the ball takes longer to fall; if the acceleration due to gravity is greater, it takes less time.

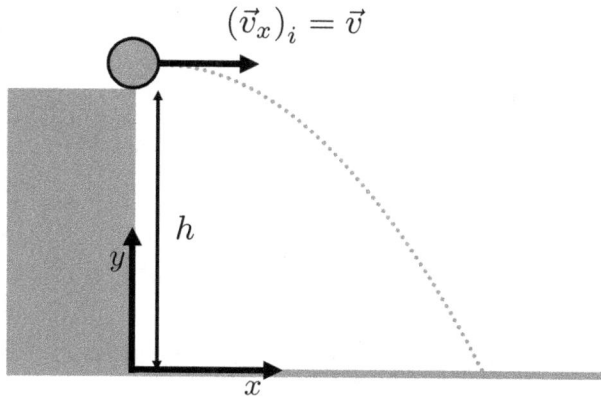

FIGURE 2.18
A ball rolled horizontally off of a table.

Equation (2.7) tells us how far the projectile will travel before striking the ground (this is sometimes called its **range**). This is just one example of a wide variety of problems in **free fall** motion, which encompasses every kind of projectile that only accelerates because of gravity (meaning we neglect things like air resistance and propulsion via rockets). The realization that two-dimensional motion can be analyzed in terms of two one-dimensional problems *linked by time* is the key takeaway from this example.

The next set of examples provide additional practice with two dimensional kinematics. Example 2.10 provides a simple example outside the context of free fall, then we return to free fall in Examples 2.11 and 2.12. We close the section with Example 2.13, which returns to the idea of *relative motion*, this time in a (you guessed it!) two-dimensional context.

Example 2.10 Two-Dimensional Acceleration of a Boat ⋆

A boat is moving at 4.0 m/s due east when a strong gust of wind gives it an acceleration of 2.0 m/s^2 directed 60.° north of west for 0.75 s. What is the boat's velocity after the gust ends?

We set up a coordinate system according to the coordinates of the compass, with $+\hat{x}$ pointing east and $+\hat{y}$ pointing north. We can decompose the acceleration into its components:

$$\vec{a}_x = -\left(2.0\cos\left(60.°\right)\hat{x}\right)\ \text{m/s}^2 = -1.0\ \text{m/s}^2$$

$$\vec{a}_y = \left(2.0\sin\left(60.°\right)\hat{y}\right)\ \text{m/s}^2 = 1.7\ \text{m/s}^2$$

where the negative sign for $\vec{a}_x$ accounts for the orientation of the acceleration (Fig. 2.19).

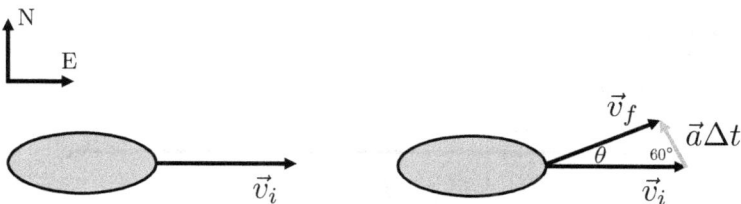

FIGURE 2.19

Then in the x direction we can determine the final velocity using the kinematic equation that omits Δx:

$$(v_x)_f = (v_x)_i + a_x (\Delta t)$$

$$(v_x)_f = \left(4.0\frac{\text{m}}{\text{s}}\right) + \left(-1.0\frac{\text{m}}{\text{s}^2} (0.75 \text{ s})\right)$$

$$(v_x)_f = 3.3\frac{\text{m}}{\text{s}}$$

Similarly in y direction:

$$(v_y)_f = (v_y)_i + a_y (\Delta t)$$

$$(v_y)_f = \left(0\frac{\text{m}}{\text{s}}\right) + \left(1.7\frac{\text{m}}{\text{s}^2} (0.75 \text{ s})\right)$$

$$(v_y)_f = 1.29\frac{\text{m}}{\text{s}}$$

If we'd prefer to express the final velocity in magnitude angle form, we can do so with a bit of trigonometry:

$$v_f = \left((v_x)_f^2 + (v_y)_f^2\right)^{1/2} = 3.5\frac{\text{m}}{\text{s}}$$

$$\theta = \tan^{-1}\left(\frac{(v_y)_f}{(v_x)_f}\right) = 22°$$

Figure 2.19 shows the initial and final velocities and also graphically demonstrates the kinematic equation $\vec{v}_f = \vec{v}_i + \vec{a}\Delta t$ in two dimensions.

Example 2.11 A Half Trajectory   ★★
Equation (2.7) gives the range of a projectile that is rolled horizontally off of a ledge. What is the range of a projectile on a flat field that is given an initially velocity $\vec{v}_i$ at an angle θ above the ground (such as a kicked soccer ball)? What is its maximum height?

As usual, we begin with a diagram (Figure 2.20).

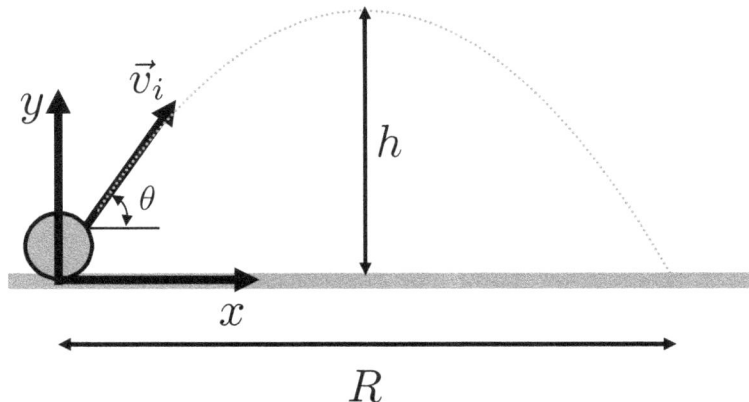

FIGURE 2.20
A ball launched over level ground.

Our coordinate system is such that the ball is at the origin, and its initial velocity has a positive x component and a positive y component. Because the acceleration is only in the y direction (due to gravity), the horizontal velocity is *constant* and the vertical velocity will decrease to 0 m/s, then increase in magnitude, this time pointing down (Fig. 2.21).

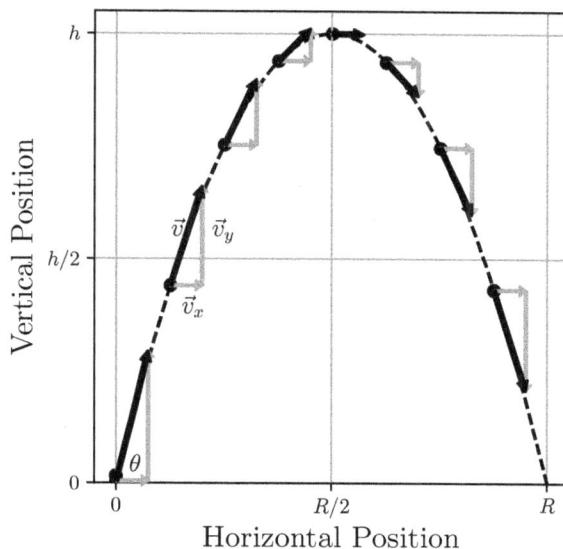

FIGURE 2.21

A full trajectory with velocity vectors marked.

Let's begin first by considering the vertical motion. Because we're curious about how high the ball goes, let's consider just the motion from when it is kicked until it reaches its maximum height (remember that this corresponds to the instant when the ball's vertical velocity is 0 m/s). The kinematic variables are:

vertical motion, to top:

$$\vec{a}_y = -g$$

$$\vec{y}_i = 0 \text{ m}$$

$$\vec{y}_f = h$$

$$(\vec{v}_y)_i = v_i \sin\theta$$

$$(\vec{v}_y)_f = 0\frac{\text{m}}{\text{s}}$$

$$\Delta t = ?$$

If the expressions for $(\vec{v}_y)_i$ is unclear, refer back to Example 2.2; you'll do this kind of analysis frequently enough that eventually it should become nearly automatic. We see that we know all of the variables except time and the final height. We'd like to solve for height, so we can use the kinematic equation that does not involve time:

$$(\vec{v}_y)_f^2 = (\vec{v}_y)_i^2 + 2\vec{a}\Delta\vec{y}$$

$$0\frac{\text{m}}{\text{s}} = (v_i \sin\theta)^2 - 2gh$$

$$h = \frac{(v_i \sin\theta)^2}{2g}$$

We also want to know how far the ball travels horizontally, but to know this we need to know how long it travels. Like in our original example, this is determined by the vertical motion (the ball stops once it strikes the ground). Thus let's stick with the vertical motion for now, and write down the kinematic variables for the entire motion:

vertical motion, complete:

$$\vec{a}_y = -g$$
$$\vec{y}_i = 0 \text{ m}$$
$$\vec{y}_f = 0 \text{ m}$$
$$(\vec{v}_y)_i = v_i \sin \theta$$
$$(\vec{v}_y)_f = ?$$
$$\Delta t = ?$$

The final height is now the same as the initial height, and we've marked the final velocity as unknown. With this in mind we can use the kinematic equation that omits final velocity:

$$\vec{y}_f = \vec{y}_i + (\vec{v}_y)_i \Delta t + \frac{1}{2} \vec{a}_y (\Delta t)^2$$
$$0 \text{ m} = 0 \text{ m} + v_i \sin \theta \Delta t - \frac{1}{2} g (\Delta t)^2$$
$$0 \text{ m} = v_i \sin \theta \Delta t - \frac{1}{2} g (\Delta t)^2$$

Clearly $\Delta t = 0$ s is a solution to this equation; this just corresponds to the instant the ball is kicked. We're interested in when the ball *returns* to the ground, though, so we can divide both sides by Δt before proceeding:

$$0 \text{ m} = v_i \sin \theta - \frac{1}{2} g \Delta t$$
$$\frac{1}{2} g \Delta t = v_i \sin \theta$$
$$\Delta t = \frac{2 v_i \sin \theta}{g}$$

This is the time in flight, and we can now turn to the horizontal motion to see how far the ball goes in this time:

horizontal motion, complete:

$$\vec{a}_x = 0 \frac{\text{m}}{\text{s}^2}$$
$$\vec{x}_i = 0 \text{ m}$$
$$\vec{x}_f = R = ?$$
$$(\vec{v}_x)_i = v_i \cos \theta$$
$$(\vec{v}_x)_f = v_i \cos \theta$$
$$\Delta t = \frac{2 v_i \sin \theta}{g}$$

We proceed as before:

$$\vec{x}_f = \vec{x}_i + (\vec{v}_x)_i \Delta t + \frac{1}{2} \vec{a}_x (\Delta t)^2$$
$$R = (0 \text{ m}) + v_i \cos \theta \Delta t + \frac{1}{2} \left(0 \frac{\text{m}}{\text{s}^2}\right) (\Delta t)^2$$
$$R = v_i \cos \theta \left(\frac{2 v_i \sin \theta}{g}\right)$$
$$R = \frac{2 v_i^2 \sin \theta \cos \theta}{g}$$

When algebraically handling multiple trigonometric arguments, it is sometimes useful to use **trigonometric identities** to simplify our work (some common trig identities are listed in Appendix B). In this case, note

$$2 \sin \theta \cos \theta = \sin (2\theta)$$

We can use this to simplify our expression, which leaves us with

$$R = \frac{v_i^2 \sin (2\theta)}{g} \tag{2.8}$$

Before moving on, note that (as if often the case) we could have proceeded through this problem differently. For instance, the motion of the ball is symmetric: it takes just as long for it to go up to height h as it does to fall from height h back to the ground, and so the ball's horizontal position when it is at height h is $R/2$ (Problem

2.36 asks you to verify these statements). Thus, we could have analyzed just the first half of the motion to determine $R/2$, then doubled the answer directly to arrive at the same equation for R that we found here by explicitly analyzing the entire motion. Recognizing relationships such as these will help you see multiple approaches to solving problems in kinematics.

Example 2.12 A Basketball Shot ★★

An athlete is standing 4.5 m in front of a 3.0 m-tall basketball hoop. A basketball is thrown from a height of 2.0 m and with an initial speed of 8.0 m/s. We define θ as the angle at which the ball should be thrown for it to fall directly through the hoop. Determine an equation involving θ and known quantities (you needn't solve it for θ).

We begin, as usual, with a diagram (Figure 2.12). We'll choose our coordinate system such that the origin is directly below the ball, level with the ground, with the usual horizontal x axis and vertical y axis.

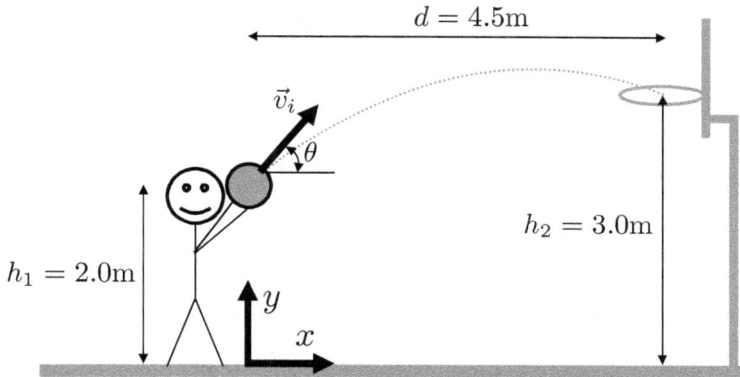

FIGURE 2.22
A basketball shot.

Let's proceed by listing our kinematic variables in each dimension:

vertical motion:

$$\vec{a}_y = -g$$
$$\vec{y}_i = h_1$$
$$\vec{y}_f = h_2$$
$$(\vec{v}_y)_i = v_i \sin \theta$$
$$(\vec{v}_y)_f = ?$$
$$\Delta t = ?$$

horizontal motion:

$$\vec{a}_x = 0\frac{\text{m}}{\text{s}^2}$$
$$\vec{x}_i = 0 \text{ m}$$
$$\vec{x}_f = d$$
$$(\vec{v}_x)_i = v_i \cos \theta$$
$$(\vec{v}_x)_f = v_i \cos \theta$$
$$\Delta t = ?$$

Note that, as is always the case in free fall, the horizontal velocity is constant. If you assess these tables of variables you'll note that we know everything except Δt and θ in the horizontal direction. This means that we can analyze the horizontal motion to relate these variables in terms of other things that we do know:

$$\vec{x}_f = \vec{x}_i + (\vec{v}_x)_i \, \Delta t + \frac{1}{2} \vec{a}_x \, (\Delta t)^2$$

$$d = 0 \text{ m} + v_i \cos\theta \, (\Delta t) + \frac{1}{2} \left(0 \frac{\text{m}}{\text{s}^2} \right) (\Delta t)^2$$

$$d = v_i \cos\theta \, (\Delta t)$$

$$\Delta t = \frac{d}{v_i \cos\theta}$$

This is one equation with two variables, which we can't solve directly (if we have two unknown variables, then we need at least two independent equations involving them). Let's turn to the vertical motion to see if we can find another equation. We're not particularly concerned with the final velocity, so we can turn to the same kinematic equation that we just used for the horizontal motion:

$$\vec{y}_f = \vec{y}_i + (\vec{v}_y)_i \, \Delta t + \frac{1}{2} \vec{a}_y \, (\Delta t)^2$$

$$h_2 = h_1 + v_i \sin\theta \, (\Delta t) - \frac{1}{2} g \, (\Delta t)^2$$

If we now plug in our above expression for Δt, we arrive at an equation involving θ and things we know:

$$h_2 = h_1 + v_i \sin\theta \left(\frac{d}{v_i \cos\theta} \right) - \frac{1}{2} g \left(\frac{d}{v_i \cos\theta} \right)^2$$

This is the requested relation, and we can stop here. We're not proceeding further because isolating θ in this equation is quite messy. In such cases it is often useful to rely instead on a computer; see Problem 2.53.

Example 2.13 Crossing a River   ★★

A swimmer seeks to cross a 50. m-wide river that has a current of 0.80 m/s. The swimmer's speed relative to the water is 1.2 m/s. (a) What is the swimmer's velocity relative to the ground? (b) If she swims directly toward the opposite bank, how long will she spend crossing the river? (c) How far downstream will the river push her as she crosses? (d) Is it possible for her to make the trip such that she doesn't actually move downstream? If so, how can she do it and how long will it take her?

The first thing to note is that this is an example of relative motion in two dimensions (we are given the speed of the swimmer "relative to the water"). We discussed relative motion in Section 1.5; the relevant equation is (1.14) on page 40, which gives the general relationship

$$\vec{v}_{BA} = \vec{v}_{BT} + \vec{v}_{TA}$$

This equation applies to two-dimensional motion, as well; we're just adding two-dimensional vectors. Let's consider the situation described in (a) and draw a diagram (Figure 2.13).

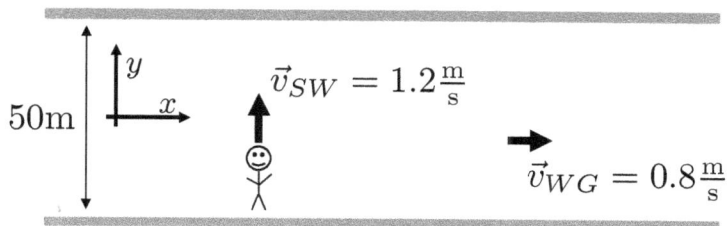

FIGURE 2.23
A swimmer crossing a river with a current.

We'll place the origin at the swimmer's initial position; in the diagram I've shown the orientation of the coordinate system off to the side to avoid having too many things overlapping on the diagram. Note also that I'm using a subscript S, W, and G to denote the swimmer, water, and ground, respectively. For (a), we need to know the velocity of the swimmer relative to the ground (denoted $\vec{v}_{SG}$). According to equation (1.14), this can be written as

$$\vec{v}_{SG} = \vec{v}_{SW} + \vec{v}_{WG}$$
$$\vec{v}_{SG} = \left(0.80\frac{m}{s}\right)\hat{x} + \left(1.2\frac{m}{s}\right)\hat{y}$$

Only the velocity in the y dimension actually contributes to crossing the river, so for (b) we have

y dimension:

$$\vec{a}_y = 0\frac{m}{s^2}$$
$$\vec{y}_i = 0 \text{ m}$$
$$\vec{y}_f = 50. \text{ m}$$
$$(\vec{v}_y)_i = 1.2\frac{m}{s}$$
$$(\vec{v}_y)_f = 1.2\frac{m}{s}$$
$$\Delta t = ?$$

Because we have no acceleration, the algebra simplifies quickly:

$$\vec{y}_f = \vec{y}_i + (\vec{v}_y)_i\,\Delta t + \frac{1}{2}\vec{a}_y\,(\Delta t)^2$$
$$\vec{y}_f = (0 \text{ m}) + (\vec{v}_y)_i\,\Delta t + \frac{1}{2}\left(0\frac{m}{s^2}\right)(42 \text{ s})^2$$
$$\vec{y}_f = (\vec{v}_y)_i\,\Delta t$$
$$\Delta t = \frac{\vec{y}_f}{(\vec{v}_y)_i}$$
$$\Delta t = \frac{50. \text{ m}}{1.2\frac{m}{s}} = 42 \text{ s}$$

For (c), we need to determine how far the swimmer goes in the x direction in this time. We've done this a few times now so I'll skip listing the variables and go straight to the equation; see if you can follow along:

$$\vec{x}_f = \vec{x}_i + (\vec{v}_x)_i\,\Delta t + \frac{1}{2}\vec{a}_x\,(\Delta t)^2$$
$$\vec{x}_f = (0 \text{ m}) + \left(0.80\frac{m}{s}\right)(42 \text{ s}) + \frac{1}{2}\left(0\frac{m}{s^2}\right)(42 \text{ s})^2$$
$$\vec{x}_f = 33 \text{ m}$$

Finally, for (d) we want to negate the effect of the river's current so $\vec{v}_{SG}$ has no x component. In other words, the swimmer needs to angle partially upstream so the effect of the current is cancelled by the horizontal component of the swimmer's velocity. Visually, the vectors need to be arranged as in Figure 2.24.

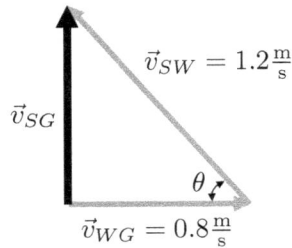

FIGURE 2.24

When thinking this through, note that if $v_{SW} = v_{WG}$, then the cancellation would only be possible if the swimmer directed all of her efforts to swimming upstream, in which case she would simply stay in place (until, of course, they were too tired to keep swimming at that speed). The angle formed by the swimmer's velocity axis and the negative x axis is marked with a θ; its value is given by

$$\cos\theta = \frac{0.80\frac{m}{s}}{1.2\frac{m}{s}}$$

$$\theta = \cos^{-1}\left(\frac{0.80\frac{m}{s}}{1.2\frac{m}{s}}\right)$$

$$\theta = 48°$$

We can determine the magnitude of $\vec{v}_{SG}$ with the Pythagorean Theorem:

$$v_{SW}^2 = v_{SG}^2 + v_{WG}^2$$

$$v_{SG}^2 = v_{SW}^2 - v_{WG}^2$$

$$v_{SG}^2 = \left(1.2\frac{m}{s}\right)^2 - \left(0.80\frac{m}{s}\right)^2$$

$$v_{SG} = 0.89\frac{m}{s}$$

Then finally, using the same procedure as before, we can determine the time it takes to cross the 50 m river:

$$\Delta t = \frac{\vec{y}_f}{(\vec{v}_y)_i}$$

$$\Delta t = \frac{50.\ m}{0.89\frac{m}{s}}$$

$$\Delta t = 56\ s$$

We summarize by saying that it is possible for the swimmer to cross the river by swimming at an angle of 48° relative to the negative x axis, which will allow her to cross in 56 seconds.

2.3   Angular Kinematics

When describing vectors we have made use of *component* form and *magnitude-angle* form. While we have not invoked this term so far in this book, when we are working with magnitudes and angles, we are making use of a **polar coordinate system**. While a Cartesian coordinate system defines a two-dimensional vector with components $\hat{x}$ and $\hat{y}$, a polar coordinate system uses components $\hat{r}$ (what *radius*?) and $\hat{\theta}$ (what *angle*?). Figure 2.25 shows the orientation of these unit vectors. Notice that the $+\hat{r}$ direction is *radially away* from the origin and the $+\hat{\theta}$ direction points counterclockwise, tangent to the circle.

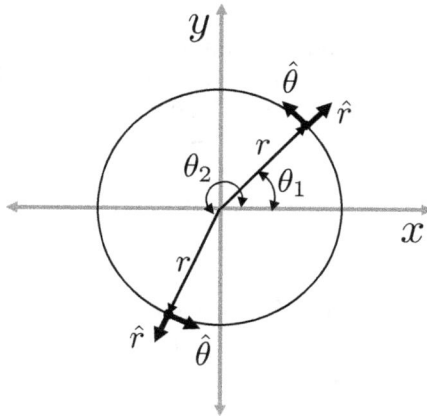

FIGURE 2.25

Two vectors described using polar coordinates. Both vectors point to a position on the edge of a circle of radius r, and so have a $r\hat{r}$ component: in polar coordinates, the $\hat{r}$ unit vector points *radially away* from the origin. In any coordinate system, the axes must be perpendicular, so the other unit vector, $\hat{\theta}$, is tangent to the circle (and by convention, is directed counter-clockwise from the $+x$ axis). Thus the two vectors can be written $r\hat{r} + \theta_1\hat{\theta}$ and $r\hat{r} + \theta_2\hat{\theta}$.

I introduce this notation now because we can make a powerful comparison between these coordinate systems. So far when we have described motion in two dimensions, we used the kinematic equations separately in the $\hat{x}$ dimension and the $\hat{y}$ dimensions (and made use of the fact that the motion in each dimension shared the same time). If, rather, we are considering *circular* motion, we can describe the object's position with a constant $\hat{r}$ component (equal to the radius of the circle) and a varying $\hat{\theta}$ component. Because all of the motion occurs in the angular dimension, circular motion is, in a sense, one-dimensional.[5]

Consider again the positions shown on Figure 2.25; to be specific, let's suppose you're sitting on the edge of a merry-go-round of radius r.[6] At time t_1 you're at position θ_1, then at some later time t_2 you're at a position θ_2. Thus you have experienced an angular displacement $\Delta\vec{\theta}$ in a time Δt, and we can define **angular velocity**, written with the Greek

---

[5]Of course, we need a plane to draw a circle; not a line. The point here is that only one quantity – the angle – *changes* when an object is experiencing circular motion.

[6]Weee!

lowercase omega:

$$\vec{\omega} = \frac{\Delta\vec{\theta}}{\Delta t} \tag{2.9}$$

Following the orientation of $\hat{\theta}$, we see that the positive direction for a angular velocity is counter-clockwise (and a negative direction is clockwise). We also see that angular velocity has units of "angle per time". We shall see shortly that the SI unit of angle is the radian (Equation (2.14)), so the SI units for angular velocity are rad/s. In a similar way, if the angular speed is changing (as the merry-go-round speeds up from rest, for instance), you experience an **angular acceleration**, written with the Greek lowercase alpha:

$$\vec{\alpha} = \frac{\Delta\vec{\omega}}{\Delta t} \tag{2.10}$$

with SI units of rad/s^2.[7] These are average quantities over the time interval we are considering; if the interval becomes infinitesimally small, then (just as in the linear case) these equations become exact and are expressed with derivatives:

$$\vec{\omega} = \frac{d\vec{\theta}}{dt} \tag{2.11}$$

$$\vec{\alpha} = \frac{d\vec{\omega}}{dt} \tag{2.12}$$

The quantities $\vec{\theta}$, $\vec{\omega}$, and $\vec{\alpha}$ are directly analogous to their linear counterparts ($\vec{x}$, $\vec{v}$, and $\vec{a}$), and the **angular kinematic equations** are likewise directly analogous to the linear kinematic equations (Equations (1.12) on page 28):

$$
\begin{aligned}
\vec{\omega}_f &= \vec{\omega}_i + \vec{\alpha}\Delta t & \left(\text{no } \Delta\vec{\theta}\right) \\
\vec{\theta}_f &= \vec{\theta}_i + \vec{\omega}_i \Delta t + \frac{1}{2}\vec{\alpha}\left(\Delta t\right)^2 & \left(\text{no } \vec{\omega}_f\right) \\
\vec{\omega}_f^2 &= \vec{\omega}_i^2 + 2\vec{\alpha}\Delta\vec{\theta} & \left(\text{no } \Delta t\right)
\end{aligned}
\tag{2.13}
$$

While this new context can take some getting used to, the mathematics and problem-solving strategies for angular kinematics carry over directly from one-dimensional kinematics in Chapter 1.

Before moving on to some examples, a comment on the *units* of these angular quantities is in order. Thus far in this book, we have referred to angles in degrees; we know there are 360° in a circle. However, consider the well-known equation for the circumference C of a circle of radius r:

$$C = 2\pi r$$

Thus "all the way around the circle" is 2π multiplied by the radius; "half way around the circle" is π multiplied by the radius, and so on. Evidently it is the quantity 2π, not 360, that is inherently useful when discussing angles. Thus we define the **radian** as the SI unit of angular displacement (Figure 2.26),[8] and

$$2\pi \text{ rad} = 360° = 1 \text{ rotation} = 1 \text{ revolution} \tag{2.14}$$

---

[7]The sign conventions are the same for α as for ω. Here's an example: If a traditional clock is running out of batteries, then the clockwise angular velocity is *negative*, but because that velocity is decreasing toward 0, the angular acceleration α points in the opposite (here, positive or counter-clockwise) direction.

[8]This means that when working with Equations (2.13), be sure to work exclusively with angles measured in radians.

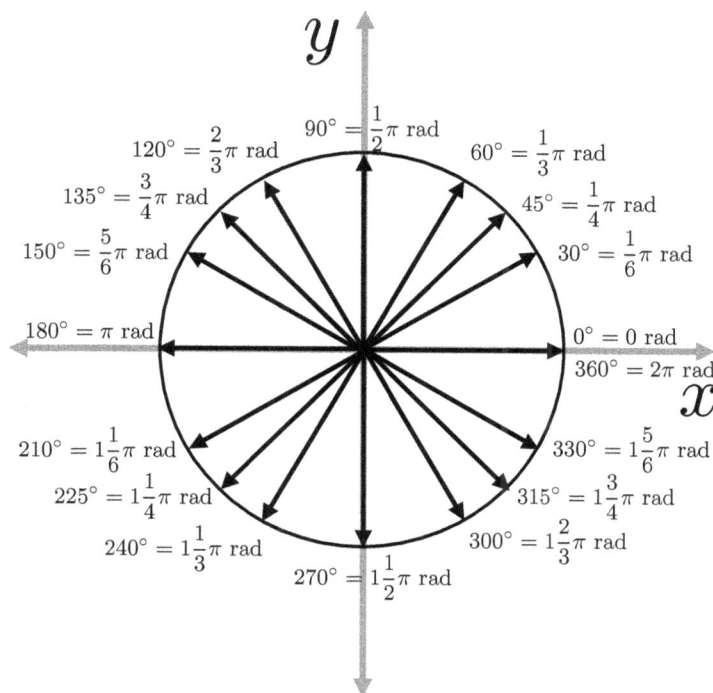

FIGURE 2.26
Some standard angles in both degrees and radians.

This means that we can generalize the equation for circumference for an arbitrary **arc length** s,

$$s = \theta r$$

If this is unclear, substitute $\theta = 2\pi$, in which case the arc length is the entire circumference, and we obtain $C = 2\pi r$, as noted above. Note also that the radian is a *unitless quantity*, since from the above we see $\theta = \frac{s}{r}$ and the units on the right hand side cancel. Some quantity of radians, therefore, serve as a scaling factor that relates the radius of a circle to a distance along the circle's circumference.[9]

We can extend this argument to relate the angular velocity and acceleration to their linear counterparts:

$$s = \theta r \quad v = \omega r \quad a = \alpha r \tag{2.15}$$

In this context a and v are referred to as the **tangential acceleration** and **tangential velocity** because these quantities point *tangent* to the circle. Equations (2.15) allow us to convert from angular quantities to linear (tangential) quantities and vice versa. We demonstrate some applications of angular kinematics in the following examples.

---

[9]The reason we introduce radians here is in part to help keep track of what we're talking about: an angular velocity talks about how quickly an *angle* is changing, even if an angle has no units.

Example 2.14 Angular Speed in Planetary Motion   ⋆

What is the angular speed of (a) the Earth and (b) Jupiter, assuming they are moving with constant angular speeds in circular orbits around the Sun? (c) What are the corresponding tangential speeds? (Relevant data is in Appendix B.)

If the planets have *constant* angular velocity, then $\alpha = 0$ and the angular kinematic equations boil down to $\omega = \Delta\theta/\Delta t$. A complete orbit means the objects are cutting out a full rotation of 2π radians (Equation (2.14)), so if we generically denote the length of time needed to complete an orbit as T, we have

$$\omega = \frac{2\pi}{T}$$

Referring to the back inside cover, for Earth we have

$$\omega = \frac{2\pi}{T}$$
$$\omega_E = \frac{2\pi}{365.2 \text{ d}} \times \frac{1 \text{ d}}{8.64 \times 10^4 \text{ s}}$$
$$\omega_E = 1.99 \times 10^{-7} \text{ rad/s}$$

In the second line I have used the conversion from days to seconds given in Appendix B. (You could also convert days to hours to minutes to seconds, as you likely have those conversions in your head already!) Repeating this process for Jupiter yields

$$\omega_J = 1.68 \times 10^{-8} \text{ rad/s}$$

These are exceedingly small because the planets take quite a long time to make a complete orbit! If we convert to tangential speeds using

$$v = \omega r$$

we find

$$v_E = \omega_E r_E = \left(1.99 \times 10^{-7} \text{ rad/s}\right)\left(1.496 \times 10^{11} \text{ m}\right) = 2.98 \times 10^4 \text{ m/s}$$
$$v_J = \omega_J r_J = \left(1.68 \times 10^{-8} \text{ rad/s}\right)\left(7.785 \times 10^{11} \text{ m}\right) = 1.31 \times 10^4 \text{ m/s}$$

Thus the Earth is hurtling through space (relative to the Sun) at a speed of nearly 30 kilometers every second (around 67,000 miles per hour!), and Jupiter is moving at more then 13 kilometers every second. The astonishingly massive difference between ω and v in SI units can be attributed to the hard-to-fathom radii of the planets' orbits around the Sun.

Example 2.15 Accelerating a Merry-Go-Round   ⋆⋆

A student sits 2.95 m from the center of a merry-go-round ("MGR") which is initially at rest and then accelerates at a rate of $0.050 \frac{\text{rad}}{\text{s}^2}$ for 8.3 s. It then travels at a constant angular velocity for 180 s before slowing down to rest with an angular acceleration of $-0.019 \frac{\text{rad}}{\text{s}^2}$.

a. What angular displacement does the student experience while the MGR is accelerating?

b. What (constant) angular speed will the MGR have before beginning to slow down? Provide your answer in both $\frac{\text{rad}}{\text{s}}$ and revolutions per minute (rpm).

c. What is the student's corresponding linear speed?

d. How long does it take for the MGR to slow down to rest? How long does the entire journey take?

e. Draw the ω-t motion graph for the entire motion.

We can split the motion into three phases. During the initial phase, the MGR is accelerating from rest. We proceed, as usual, by listing the kinematic variables:

Phase 1:

$\vec{\alpha} = 0.050 \dfrac{\text{rad}}{\text{s}^2}$

$\vec{\theta}_i = 0 \text{ rad}$

$\vec{\theta}_f = ?$

$\vec{\omega}_i = 0 \dfrac{\text{rad}}{\text{s}}$

$\vec{\omega}_f = ?$

$\Delta t = 8.3 \text{ s}$

For (a) we are interested in θ; because we do not know ω_f we choose to use the equation that omits it:

$$\vec{\theta}_f = \vec{\theta}_i + \vec{\omega}_i \Delta t + \frac{1}{2}\vec{\alpha}\left(\Delta t\right)^2$$

$$\Delta\theta = \left(0\frac{\text{rad}}{\text{s}}\right)(8.3 \text{ s}) + \frac{1}{2}\left(0.050\frac{\text{rad}}{\text{s}^2}\right)(8.3 \text{ s})^2$$

$$\Delta\theta = 1.7 \text{ rad}$$

For (b) we are asked to determine ω_f:

$$\vec{\omega}_f = \vec{\omega}_i + \vec{\alpha}\Delta t$$

$$\omega_f = \left(0\frac{\text{rad}}{\text{s}}\right) + \left(0.050\frac{\text{rad}}{\text{s}^2}\right)(8.3 \text{ s})$$

$$\omega_f = 0.42\frac{\text{rad}}{\text{s}}$$

We also need to convert this to rpm:

$$\omega_f = 0.42\frac{\text{rad}}{\text{s}} \times \frac{1 \text{ rotation}}{2\pi \text{ rad}} \times \frac{60 \text{ s}}{1 \text{ min}} = 4.0\frac{\text{rotation}}{\text{min}}$$

For (c) we note from Equations (2.15) that $v = \omega r$, so

$$v = \left(0.42\frac{\text{rad}}{\text{s}}\right)(2.95 \text{ m}) = 1.2\frac{\text{m}}{\text{s}}$$

Note that because the radian is a unitless quantity, we can simply drop it once we have moved from an inherently angular quantity – angular speed – to a linear quantity.

For (d), we need to consider the third phase of motion:

Phase 3:

$$\vec{\alpha} = -0.019\frac{\text{rad}}{\text{s}^2}$$

$$\vec{\theta}_i = ?$$

$$\vec{\theta}_f = ?$$

$$\vec{\omega}_i = 0.42\frac{\text{rad}}{\text{s}}$$

$$\vec{\omega}_f = 0\frac{\text{rad}}{\text{s}}$$

$$\Delta t = ?$$

We are interested in how long the journey takes, but we don't know anything about the angular distance traveled in this time, so we use the equation that omits $\Delta\vec{\theta}$:

$$\vec{\omega}_f = \vec{\omega}_i + \vec{\alpha}\Delta t$$

$$\frac{\omega_f - \omega_i}{\alpha} = \Delta t$$

$$\frac{(0 - 0.42)\frac{\text{rad}}{\text{s}}}{-0.019\frac{\text{rad}}{\text{s}^2}} = \Delta t$$

$$22\text{ s} = \Delta t$$

Thus the entire journey takes $(8.3 + 180 + 22)$ s $= 210$ s. Finally, in Figure 2.27 we sketch the motion as requested in (e).

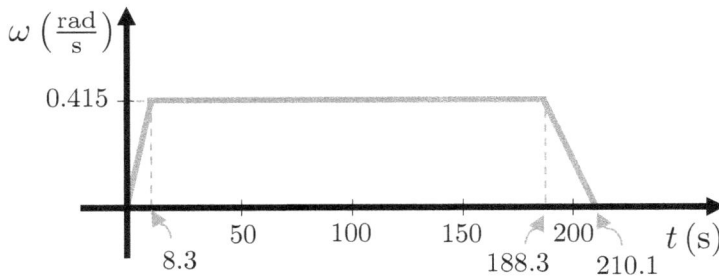

FIGURE 2.27
Angular velocity vs. time motion diagram for a merry-go-round.
Let me emphasize one more time that this should all seem very familiar to problems we solved earlier in this chapter, for 2-dimensional motion *not* constrained to a circle!

2.4   Tangential and Centripetal Acceleration

In Example 2.9 we introduced the idea of *centripetal* acceleration: if an object is moving in a circle at a constant speed, then the change in the velocity vector (as the object turns) is reflected in an acceleration that points in toward the center of the circle. We typically denote the magnitude of this acceleration a_c.

However, in Section 2.3 we saw that an object moving in a circular track can experience an *angular* acceleration α, and we correspondingly discussed the *tangential* acceleration $a_t = \alpha r$ where r is the radius of motion.[10] Figure 2.28 depicts all three of these vectors *and* the **net linear acceleration**, $\vec{a}_{\text{NET}}$:

$$\vec{a}_{\text{NET}} = \vec{a}_c + \vec{a}_t = -a_c\hat{r} + a_t\hat{\theta} \tag{2.16}$$

This can get to be confusing at first, so a pithy example may help: suppose a car starts from rest and then begins driving in a circle:

---

[10]I previously called this quantity just a, to draw a parallel between the angular and linear kinematic properties. We're now juggling multiple different quantities that are all different varieties of acceleration, so some added notation will be helpful to draw a distinction.

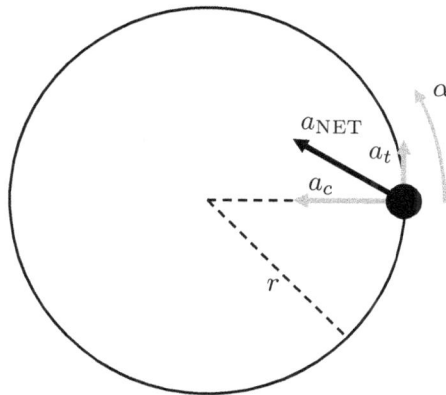

FIGURE 2.28

Accelerations in circular motion. Here we suppose the tangential and angular accelerations are counter clockwise. Notice that the angular speed ω is not shown; it could be clockwise or counter clockwise. If ω is counter clockwise, then ω will increase in magnitude (α and ω would be pointing in the same direction); if ω is clockwise, it will decrease in magnitude (α and ω would be pointing in opposite directions).

- The **tangential** acceleration accounts for the change in the *magnitude* of the velocity (what the speedometer on the car's dashboard reads). The tangential acceleration always points in the direction of the car's instantaneous velocity, tangent to the circle.

- The **centripetal** acceleration accounts for the change in the *direction* of the velocity and always points in toward the center of the circular motion.

- The **angular** acceleration maps the tangential acceleration to angular coordinates which, as we've seen, can be useful for analyzing motion via the angular kinematic equations.

We've already quantified the tangential and angular acceleration, but what can we say about the centripetal acceleration? We'll proceed by focusing our attention on the case where the angular speed ω is a constant (so $\alpha = 0$ and $a_t = 0$); this is so-called **uniform circular motion**. If ω is a constant, then the linear speed $v = \omega r$ is also a constant.

This means, for instance, that it takes just as long to travel around the circumference of the circle the first time as the second time. It is worth noting that time to go all the way around the circle (or, if you prefer, the time to complete one *cycle*) is referred to as the **period**, T:

$$T = \frac{2\pi}{\omega} = \frac{2\pi r}{v} \tag{2.17}$$

While the *speed* is constant, the *velocity* is not because its *direction* is always changing (Figure 2.29).

Now if $\vec{v}$ has a constant magnitude and always points in the $\hat{\theta}$ direction, then the total acceleration *can't* have a component in the $\hat{\theta}$ direction: if it *did*, then the velocity would be changing in magnitude, which is precisely what we're saying *isn't* the case.[11] Thus the only option is for the total acceleration to point in the $\pm\hat{r}$ direction. This acceleration is

---

[11]Of course there isn't anything fundamentally wrong with having an acceleration in the $\hat{\theta}$ direction: this is simply the tangential acceleration, which we are for the moment assuming is 0.

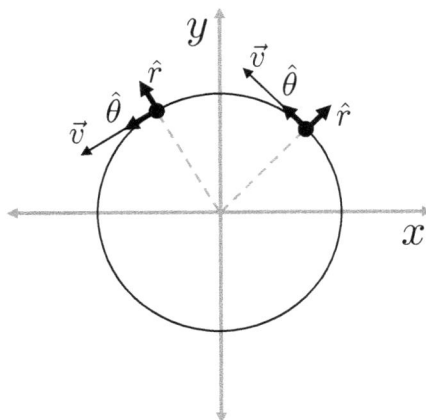

FIGURE 2.29
In uniform circular motion, the velocity vector is always tangent to the object's position vector. The velocity vector therefore changes along with the position vector.

constantly "dragging" the velocity vector so it points in a different direction without ever changing its magnitude. In other words, the acceleration points along a line connecting the object and the center of its circular path. Indeed, as we sketch in Figure 2.30, to achieve uniform circular motion, the acceleration always points radially in, toward the center of the circle.

To define these concepts mathematically, consider the arrangement of $\vec{v}_i$ and $\vec{v}_f$ in Figure 2.30. We see

$$a_c \Delta t = \Delta v = v_f \sin \theta$$

Now, if an angle θ is very small, then $\sin \theta \approx \theta$ (see Problem 2.85). Thus if we consider our two velocity vectors to be arbitrarily close together and with the same magnitude, we see $\Delta v = v\theta$. Now consider the definition of acceleration:

$$a_c = \frac{\Delta v}{\Delta t} = \frac{v\theta}{\Delta t}$$

But now recall $s = r\theta \Rightarrow \theta = s/r$ and $\Delta t = s/v$ where s is the arc length, i.e. the distance along the circumference over which the object is moving. Substituting these expressions in to the above equation, we have

$$a_c = \frac{v\frac{s}{r}}{\frac{s}{v}} = v\frac{s}{r}\frac{v}{s}$$

Canceling the factors of s, we arrive at the definition for centripetal acceleration in terms of the linear speed v and the radius of motion r:

$$a_c = \frac{v^2}{r} \tag{2.18}$$

If the magnitude of the linear speed v is changing because there is a nonzero *tangential* acceleration, then the centripetal acceleration changes along with v according to Equation (2.18).

The arguments we've made in this section certainly aren't simple, but here is the key point: when an object is experiencing uniform circular motion, the acceleration points toward the center of the circle with magnitude given by Equation (2.18). We explore these ideas further in the following example.

$$\vec{a}_c \Delta t = \Delta \vec{v}$$

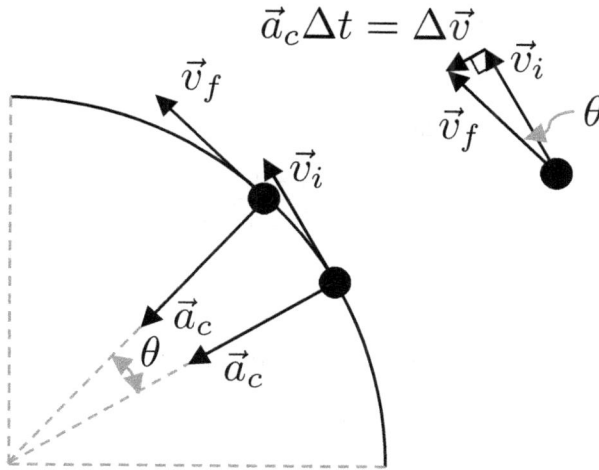

FIGURE 2.30
An object undergoing uniform circular motion experiences a net acceleration directed toward the center of the circle, $\vec{a}_c$. Thus for some small interval of time Δt, the object experiences an inward change in velocity $\vec{a}_c \Delta t$ that changes the initial velocity $\vec{v}_i$ to a final velocity $\vec{v}_f$, as shown to the side. (If it seems odd to you that the two angles marked θ are the same, imagine sliding $\vec{v}_f$ along the circle toward $\vec{v}_i$: both angles approach $0°$.) If we consider Δt to be infinitely small, then the momentum vector only changes direction, not magnitude, as the object moves, resulting in circular motion.

Example 2.16 Tangential Acceleration of a Turning Car ★★
A car is at rest on a circular track with a circumference of 2.50 km. It experiences a uniform tangential acceleration of 20.0 m/s^2 until it reaches a top linear speed of 50.0 m/s.

a. How long does the car take to reach its top linear speed?

b. What linear and angular distance does the car cover in the time you found in (a)?

c. What is the car's total acceleration just as it reaches its top linear speed?

d. What is the car's total acceleration after it has reached its top linear speed?

e. How long does it take for the car to complete one lap around the track?

First, note the radius of the track is given by

$$C = 2\pi r \implies r = \frac{C}{2\pi} = \frac{2.50 \times 10^3 \text{ m}}{2\pi} = 4.00 \times 10^2 \text{ m}$$

It follows that the angular acceleration while the car's linear speed is increasing is

$$\alpha = \frac{a_t}{r} = \frac{20.0 \text{ m/s}^2}{4.00 \times 10^2 \text{ m}} = 0.0503 \text{ rad/s}^2$$

and the car's final angular speed is

$$\omega_f = \frac{v_f}{r} = \frac{50.0 \text{ m/s}^2}{4.00 \times 10^2 \text{ m}} = 0.126 \text{ rad/s}$$

How long does this take? Well,

$$\omega_f = \omega_i + \alpha \Delta t$$

$$\Delta t = \frac{\omega_f - \omega_i}{\alpha}$$

$$\Delta t = \frac{(0.126 - 0) \,\text{rad/s}^2}{0.0503 \,\text{rad/s}^2}$$

$$\Delta t = 2.50 \text{ s}$$

This answers (a). We can continue the analysis for (b):

$$\theta_f = \theta_i + \omega_i \Delta t + \frac{1}{2}\alpha \left(\Delta t\right)^2$$

$$\theta_f = \frac{1}{2}\left(0.0503 \,\text{rad/s}^2\right)(2.50 \text{ s})^2$$

$$\theta_f = 0.157 \text{ rad} = 9.00°$$

where in the second line I've noted that $\theta_i = 0$ rad and $\omega_i = 0$ rad/s. Converting to the arc length (and keeping some extra digits in our calculator),

$$s = \theta r = (0.157 \text{ rad})\left(4.00 \times 10^2 \text{ m}\right) = 62.8 \text{ m}$$

For (c), we need both the tangential acceleration (which was provided in the problem statement) and the centripetal acceleration:

$$a_c = \frac{v^2}{r} = \frac{(50.0 \text{ m/s})^2}{4.00 \times 10^2 \text{ m}} = 6.25 \text{ m/s}^2$$

With this, we can find that net acceleration through the Pythagorean Theorem (keep in mind that a_θ is the tangential acceleration and a_r is the centripetal acceleration; these quantities are the perpendicular components of $\vec{a}_{\text{NET}}$, as shown in Figure 2.28):

$$a_{\text{NET}} = \left(a_c^2 + a_t^2\right)^{1/2}$$

$$a_{\text{NET}} = \left(\left(6.25 \text{ m/s}^2\right)^2 + \left(20.0 \text{ m/s}^2\right)^2\right)^{1/2}$$

$$a_{\text{NET}} = 21.0 \text{ m/s}^2$$

(d) is straightforward: once the car has reached its top speed, $a_t = 0$ and the total acceleration is just the centripetal acceleration, which we've already calculated to be $a_c = 6.25\text{m/s}^2$. (The centripetal acceleration has a *smaller* value when the car hasn't yet reached its top speed because $a_c = v^2/r$.)

Finally, for (e), we need to consider the rest of the motion around the track. We've already seen that the car travels 0.157 rad as its linear speed increases, so the remaining angular distance it must cover to complete a lap is

$$\Delta\theta = (2\pi - 0.157)\,\text{rad} = 6.126 \text{ rad}$$

At a constant angular speed, angular kinematics simplifies to $\omega = \Delta\theta/\Delta t$; here

$$\Delta t = \frac{\Delta\theta}{\omega}$$

$$= \frac{6.126 \text{ rad}}{0.126 \text{ rad/s}^2}$$

$$= 48.8 \text{ s}$$

This is the time required to finish the lap once the car has reached its top speed. We already found that the time for the initial speed up is 2.50 s, so the total time is the sum of these values, or 51.3 s. You're asked to draw motion graphs for this motion in Problem 2.78. Figure 2.31 sketches the motion along the perimeter of the circle and indicates the total linear acceleration during the initial "speed up" and again after the car reaches its top speed. Note that the nonzero $\vec{a}_t$ during the initial speed up means that the net acceleration points ahead of the center of the circular track (the angular displacement during the initial linear acceleration, shown in black, is intentionally exaggerated to emphasize this effect).

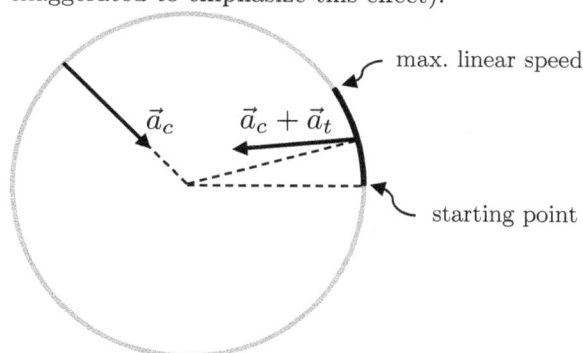

FIGURE 2.31

2.5 Problems for Chapter 2

2.1 Two-Dimensional Vectors and Motion Diagrams

★ **Problem 2.1.** Convert the following vectors from component form to magnitude-angle form. Include a sketch of each vector and label the components, magnitude, and angle.

a. $\vec{A} = 10.0\hat{x} + 5.0\hat{y}$

b. $\vec{B} = 5.0\hat{x} - 5.0\hat{y}$

c. $\vec{C} = -8.0\hat{x} + 2.0\hat{y}$

d. $\vec{D} = -3.0\hat{x} - 7.0\hat{y}$

★ **Problem 2.2.** Convert the following vectors from magnitude-angle form to component form. Include a sketch of each vector and label the components, magnitude, and angle.

a. $\vec{A} = 8.0$ at $30.°$ above the $+x$ axis

b. $\vec{B} = 3.5$ at $60.°$ below the $+x$ axis

c. $\vec{C} = 11.0$ at $50.°$ left of the $+y$ axis

d. $\vec{C} = 5.0$ at $225°$ counter-clockwise from the $+x$ axis

★ **Problem 2.3.** A ball is thrown at an angle of $35.0°$ above the horizontal, and $|\vec{v}| = 22.0$ m/s. Express this velocity in component form; include a sketch.

★ **Problem 2.4.** The components of a velocity vector are given as $v_x = -3.0$ m/s and $v_y = 4.0$ m/s. Express the velocity in magnitude-angle form.

★ **Problem 2.5.** You drive 0.60 miles east in 2.0 minutes, then turn and drive 0.70 miles south in 1.5 minutes. Draw your displacement vector for both legs of the journey as well as your overall displacement vector. What is your average velocity during the entire trip?

★ **Problem 2.6.** You follow a path that runs west for 20.0 mi and then heads 10.0 mi north, before heading $(10.0\hat{x} + 20.0\hat{y})$ mi where $\hat{x}$ points east and $\hat{y}$ points north. What is your total displacement?

★ **Problem 2.7.** An object moves west in a straight line, then south in a straight line. In what general direction (north, south, east, west, northeast, southeast, northwest, or southwest) is the object's average acceleration vector?

★ **Problem 2.8.** Determine the components of the vector in Figure 2.32 according to the shown coordinate system.

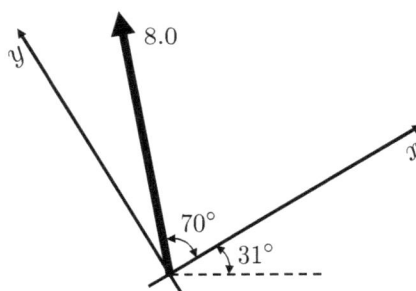

FIGURE 2.32
Problem 2.8

★ **Problem 2.9.** Express each of the vectors shown in Figure 2.33 in both component form and magnitude-angle form.

★★ **Problem 2.10.** Refer to the vectors in Figure 2.33 and calculate the following. Express your answers in component form, magnitude-angle form, and visually using the tail to tip method.

    a. $\vec{p} + \vec{q}$

    b. $\vec{p} - \vec{r}$

    c. $\vec{s} + 2\vec{q}$

    d. $2\vec{p} + \vec{q} - \vec{r}$

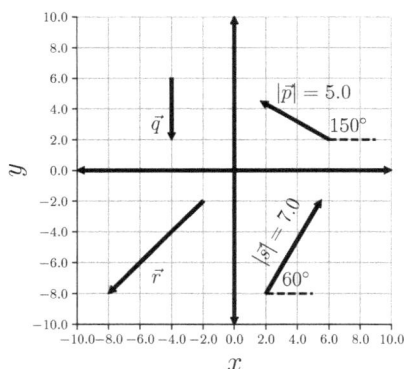

FIGURE 2.33
Problems 2.9 and 2.10

★★   **Problem 2.11.** Consider the situation shown in Figure 2.34, which depicts one force acting on an object perpendicular to the surface it is resting on (we shall see later that additional forces are present in this system; here we're just focusing on one). Express $\vec{F}$ in component form using the coordinate system shown. Your answer will be an algebraic expression in terms of the magnitude of the vector, F, and θ.

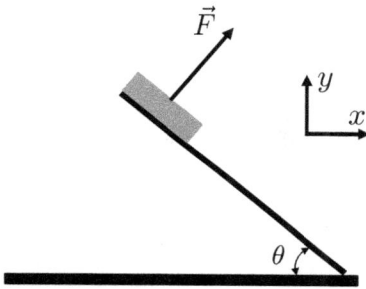

★★   **Problem 2.12.** Consider the situation shown in Figure 2.35, which depicts two forces acting on a ball suspending from a support rod by a string. Two coordinate systems are shown to the side: the first is labeled with components x and y and has the usual horizontal and vertical orientations. The second is labeled with components x' and y' and is oriented such that $\vec{F_1}$ is aligned with $\hat{y}'$.

    a. Express $\vec{F_1}$ in component form using the x, y coordinate system.

    b. Express $\vec{F_2}$ in component form using the x, y coordinate system.

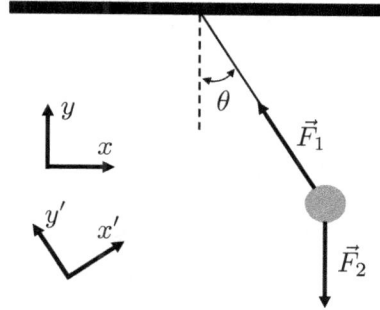

    c. Express $\vec{F_1}$ in component form using the x', y' coordinate system.

    d. Express $\vec{F_2}$ in component form using the x', y' coordinate system.

FIGURE 2.34
Problem 2.11

FIGURE 2.35
Problem 2.12

★   **Problem 2.13.** Consider the situation shown in Figure 2.36. At the instant shown, will the object's speed increase or decrease? Will it begin curving upward, downward, or maintain its trajectory?

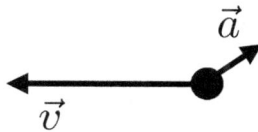

FIGURE 2.36
Problem 2.13

★★   **Problem 2.14.** Consider the vectors $\vec{p} = 3.0\hat{x} + 4.0\hat{y}$, $\vec{q} = -5.0\hat{x} + 5.0\hat{y}$, and $\vec{r} = 3.0\hat{x} - 9.0\hat{y}$.

    a. Sketch each vector on a Cartesian coordinate system.

    b. Express each in magnitude-angle form.

    c. Calculate $\vec{p} + \vec{q}$ and sketch the result using the tail-to-tip method. Express your answer in both component form and magnitude-angle form.

    d. Repeat (c) for the vector $\vec{q} + \vec{r}$.

    e. Repeat (c) for the vector $2\vec{p} - \vec{r}$.

★★   **Problem 2.15.** Consider the vectors shown in Figure 2.37. Respond to the following visually (using the tail top tip method) and mathematically (by adding components).

    a. Calculate $\vec{A} + \vec{B}$

b. Calculate $\vec{B} - \vec{A}$

c. Calculate $2\vec{A} - \vec{C}$

d. For what vector $\vec{D}$ will $\vec{A} + \vec{B} + \vec{C} + \vec{D} = 0$?

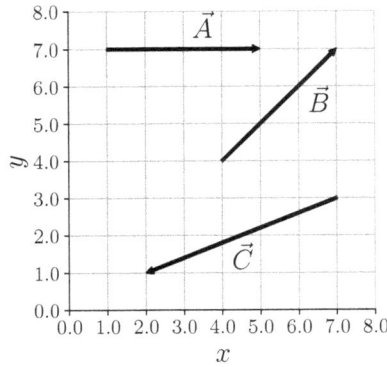

FIGURE 2.37
Problem 2.15

★ **Problem 2.16.** Consider the vectors $\vec{p} = 10.0$ m at $\theta = 30.0°$ above the $+x$ axis and $\vec{q} = 10.0$ m at $\theta = 30.0°$ above the $-x$ axis.

a. Sketch each vector on a Cartesian coordinate system.

b. Determine $\vec{p} + \vec{q}$. Express your answer in both component form and magnitude-angle form and show the result using the tail-to-tip method.

c. Repeat part (b) for the vector $\vec{p} - \vec{q}$.

★★ **Problem 2.17.** Consider the vectors shown in Figure 2.38. Calculate the following visually (using the tail top tip method) and mathematically (by adding components).

a. $\vec{A} + \vec{B}$

b. $\vec{A} - \vec{B}$

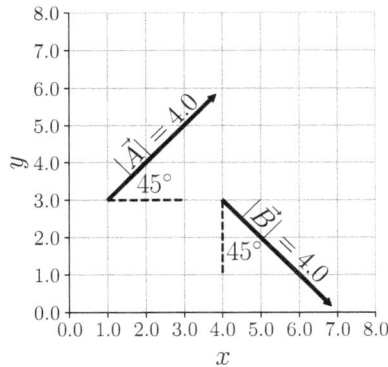

FIGURE 2.38
Problem 2.17

★ **Problem 2.18.** Suppose you're on your way to a picnic. To reach your destination, you could walk on a trail 0.60 km East, then 0.80 km South on another trail. Alternatively, you could walk directly across a field to reach your destination. Whichever route you take, you'll walk at a steady 1.2 m/s.

a. How long will it take you to reach your destination if you follow the trails?

b. How long will it take you to reach your destination if you cut across the field?

c. What angle will you walk at if you cut across the field? Express your angle relative to the compass points, e.g. "10 degrees S of E".

★★ **Problem 2.19.** Consider the vectors shown in Figure 2.39. Calculate the following visually (using the tail top tip method) and mathematically (by adding components) to two significant figures.

a. $\vec{B} - \vec{C}$

b. $\vec{A} + \vec{B} + \vec{C}$

c. $\frac{1}{2}\vec{B} + 3\vec{C}$

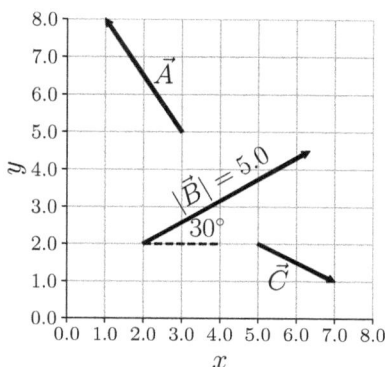

FIGURE 2.39
Problem 2.19

★★ **Problem 2.20.** Suppose you are flying a remote-controlled drone. From your location, it flies 1.2 km East, then 0.50 km North, then 0.30 km Southwest. Suppose it maintains a steady speed of 13 m/s for the entire trip.

a. Sketch the drone's path.

b. What was the drone's displacement from the beginning of its motion to the end? Express your answer in component form and in magnitude-angle form.

c. What was the drone's average velocity for the entire trip?

★★ **Problem 2.21.** You go hiking through some woods but are quickly distracted by how delicious your trail mix is. You end up traveling 3.0 mi E, then 1.5 mi S, and finally 1.0 mi E. At this point you run out of trail mix and realize you haven't been particularly efficient in reaching your destination, which was 6.0 mi E and 8.0 mi N of where you started.

a. Make a neat sketch of your path and ultimate destination.

b. How far are you from your destination?

c. What direction should you walk in to get to your destination? Express your angle relative to the compass points, e.g. "10 degrees S of W".

★★ **Problem 2.22.** Refer to Example 2.9 on page 66. Carefully re-draw the positions and velocity vectors, then graphically determine each of the following vectors. Show that each acceleration vector points toward the center of the object's circular path.

a. $\vec{a}_{345}$

b. $\vec{a}_{456}$

c. $\vec{a}_{9,10,11}$

★★   **Problem 2.23.** A heavy stone block is being pulled into place. When looking down from above and using the usual coordinate system where $+x$ is directed to the right and $+y$ is directed up, the forces being applied to the block are $\vec{F}_1 = -25\mathrm{N}\hat{y}$, $\vec{F}_2 = 20\mathrm{N}$ at $45°$ above the $+x$ axis, and $\vec{F}_3 = 55\mathrm{N}$ at $30°$ to the left of the $+y$ axis. (We haven't talked about forces yet, but don't worry: for our present purposes they're just vectors!)

   a. Draw a diagram: a dot to represent the box and a vector for each of the three forces.

   b. What is the total (or *net*) force, $\vec{F}_{\text{NET}} = \vec{F}_1 + \vec{F}_2 + \vec{F}_3$? Express your answer in component form.

★★   **Problem 2.24.** Consider the motion diagram shown in Figure 2.24. The coordinates are (in meters) $(x_1, y_1) = (0.0, 0.0)$, $(x_2, y_2) = (4.0, 3.0)$, $(x_3, y_3) = (12.0, 9.0)$, and $(x_4, y_4) = (24.0, 18.0)$ and successive positions are separated in time by 0.50 s. Make your own copy of this diagram and annotate it with average velocity and average acceleration vectors. Mathematically determine the velocities and accelerations and express each in component form.

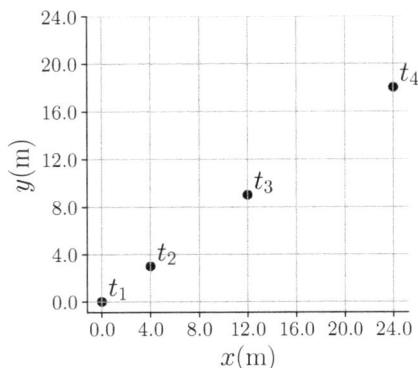

FIGURE 2.40
Problem 2.24

★★★   **Problem 2.25.** Consider the motion diagram shown in Figure 2.41. Make your own copy of this diagram and annotate it with average velocity and average acceleration vectors (the corresponding data is shown in Table 2.1). Mathematically determine the velocities and accelerations and express each in both component and magnitude angle form.

TABLE 2.1
Problem 2.25

Time (s)	x (m)	y (m)
0.500	0.000	-0.500
1.000	0.625	-0.375
1.500	1.000	-0.250
2.000	1.125	-0.125
2.500	1.000	0.000
3.000	0.625	0.125
3.500	0.000	0.250
4.000	-0.875	0.375

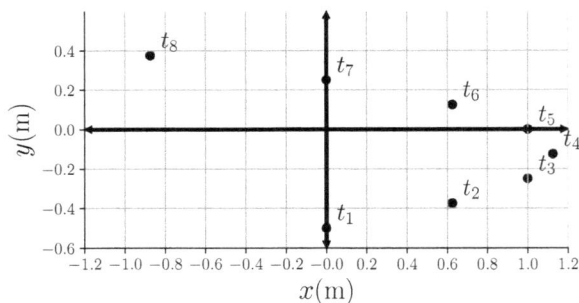

FIGURE 2.41
Problem 2.25

★★   **Problem 2.26.** Figure 2.42 shows the position of a hiker in 1.0 hour time intervals. Make your own copy of this figure and draw in the average velocity and acceleration vectors using the tail to tip method.

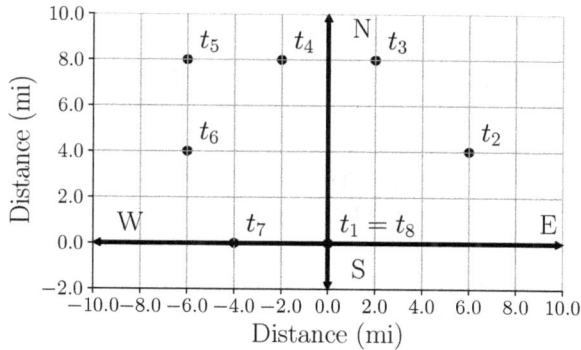

FIGURE 2.42
Problems 2.26 and 2.27

★★ **Problem 2.27.** Mathematically work out the average velocity and acceleration vectors for the hiker's motion described in Figure 2.42. Confirm that your mathematical results agree with the visual results you worked out in Problem 2.26.

2.2 Two-Dimensional Kinematics

★ **Problem 2.28.** A tumbleweed is tumbling east at 5.0 m/s. A sudden gust of wind gives the tumbleweed an acceleration $\vec{a} = (0.80 \text{ m/s}^2, 40.°$ north of east). What is the tumbleweed's (a) x-component and (b) y-component of the velocity 6.0 s later when the gust subsides?

★ **Problem 2.29.** A cart moving on a level table fires a ball straight up. If the cart is constantly accelerating in its direction of motion, will the ball land in the launch tube, in front of the cart, or behind the cart? Neglect air resistance.

★ **Problem 2.30.** Two balls are thrown from the same location and follow the trajectories shown in Figure 2.43. Which lands first, or do they land at the same time?

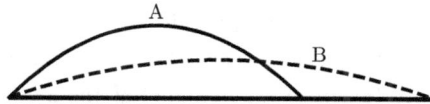

FIGURE 2.43
Problem 2.30

★ **Problem 2.31.** A bullet is dropped from rest from a height of 1.8 m. Simultaneously, a bullet is fired from the same height, but with a horizontal velocity of 150 m/s.

a. How long does it take for each bullet to strike the ground?

b. What is the range of the bullet that was fired?

c. Calculate the x and y coordinates of the bullet that was fired for 0.10 second intervals. Make a table, then make a motion diagram. Draw the velocity and acceleration vectors. In words, describe how the velocity and acceleration changes.

★ **Problem 2.32.** A ball is kicked at an initial speed of 20. m/s and an angle of 20.° above level ground.

a. What is the ball's range?

b. What is the ball's maximum height?

   c. How long is the ball in the air?

   d. Repeat the entire problem for an initial angle of 50.°.

★   **Problem 2.33.** A ball is thrown and caught (at the same height) 3.0 s later by a person a horizontal distance of 25.0 m from the person who threw the ball. At what velocity (magnitude and angle) was the ball thrown?

★   **Problem 2.34.** A ball is thrown at $t = 0.0$ s. At $t = 1.0$ s, the velocity is $v = (15.0\hat{x} + 9.8\hat{y})$ m/s and the ball reaches its maximum height at $t = 2.0$ s. At what velocity (magnitude and angle) was the ball thrown?

★   **Problem 2.35.** A basketball player shoots a basketball toward the center of the backboard. The ball and hoop are positioned so its displacements vector (from being launched to hitting the backboard) is $(\Delta\vec{x}, \Delta\vec{y}) = (3.00 \text{ m}, 1.60 \text{ m})$. The ball hits the backboard at the instant when $v_y = 0.0$ m/s. What is the ball's speed as it leaves her hands?

★★ **Problem 2.36.** Consider a "full trajectory" problem where an object is launched with a speed v at an angle θ above the horizontal and lands at the same height from which it was launched. Use the kinematic equations to show that (a) the object spends as much time *going up* as it does *going down* and (b) if the range is R, then at its highest point it has traveled a horizontal distance $R/2$.

★★ **Problem 2.37.** On the Apollo 15 mission to the Moon, astronaut Alan Shepard hit a golf ball with a 6 iron. The free-fall acceleration on the moon is $1/6$ of its value on Earth. Suppose he hit the ball on a large flat surface with a speed of 25.0 m/s at an angle of 30.0° above the horizontal.

   a. What would be the total flight time of the ball?

   b. At the ball's maximum height, explain whether the value of velocity AND acceleration are (+, 0 or -). Be sure to justify each with theory and words as needed. No calculations are required.

   c. How far did the ball travel horizontally?

   d. What is the ball's impact velocity (magnitude and direction)?

★★ **Problem 2.38.** During a lecture, one of my colleagues threw a marker over his shoulder toward the whiteboard, and it landed in the board rail. (I am told there was thunderous applause.) Assume the instructor was 3.5 m horizontally from the rail when the marker was released, the release was 2.0 m above the floor, and that the velocity of the marker at launch was 5.0 m/s at an angle of 60.° above horizontal.

   a. How long did it take the marker to reach the rail from release?

   b. What is the y-component of the velocity when the marker impacts the rail?

   c. What is the impact velocity (magnitude and angle) as the marker impacts the rail?

   d. At the top of the marker's motion, explain whether the velocity AND acceleration are zero or not.

★   **Problem 2.39.** A package is to be dropped from a plane that is cruising at speed v a distance h above level ground. The plane's horizontal position is $x = 0$ m the instant the object is released.

   a. Sketch the plane, ground, and the subsequent path of the package as it travels to the ground. If you like, replace the plane with a flying superhero.

   b. Determine an algebraic value for x_{imp}, the point of impact for the package.

   c. What is x_{plane} (or x_{hero}), the horizontal position of the plane (or hero) at the instant the package impacts the ground?

★   **Problem 2.40.** A plane is cruising at 115 m/s when a package is released from a hatch. What is its velocity after 7.0 s, neglecting air resistance and assuming it hasn't struck the ground?

★   **Problem 2.41.** A supply plane needs to drop a package of supplies (you decide the context. Equipment for soldiers? Food to scientists? The possibilities are endless). The plane flies horizontally 1.00×10^2 m above the ground at a speed of 150 m/s. How far short of the target should it drop the package?

Problem 2.42. The object of table shuffleboard is to slide pucks across a table to a point near the end of the table without falling off. In order to decrease friction, the table is sprinkled liberally with shuffleboard wax, but it isn't quite friction free.

a. You launch a puck with an initial velocity of 4.0 m/s and it comes to rest after 2.0 seconds while experiencing a constant -2.0 m/s^2 acceleration. What distance did the puck travel?

b. You launch a second puck and it travels down the level table and flies off the table with a velocity of 1.0 m/s and it lands on the floor below after falling 1.4 m vertically. How far did the second puck travel horizontally from the edge of the table?

c. If a third puck leaves the table with twice the velocity as the second puck, explain how the flight time of the third puck compares to the flight time of the second puck.

Problem 2.43. A rifle is aimed horizontally at a target 45.0 m away. The bullet hits the target 2.3 cm below the aim point. What was the bullet's speed as it left the barrel?

Problem 2.44. You'd like to measure the speed at which you throw a ball, but you don't have any fancy electronic sensors. Instead, you throw the ball horizontally from a height of 2.0 m and measure the point of impact on the ground to be 15 m away.

a. What was the speed of the ball as it left your hand?

b. We are, as usual, ignoring air resistance. Given that air resistance does, in fact, exist, do you expect that the *actual* speed is lower or higher than what you found in (a)? Explain.

Problem 2.45. In a game show, contestants sit upon a large cannon that spins and bucks. As this is happening, they fire balls from a cannon in an attempt to hit a target with the ball. We will pretend there is no air resistance and ignore any impact of spinning. Suppose the total flight time for a particular ball that hit the target is 0.82 s and the time to the max height to be 0.70 s.

a. What is the vertical velocity at launch?

b. How high is the target from the launch point?

c. If the launch angle is 20.°, what horizontal distance did the ball travel?

d. A second ball is fired that with the same initial velocity, but at a larger launch angle. If the ball still hits the target, is the flight time larger, smaller or the same? No calculations needed, but explain your logic clearly!

Problem 2.46. Person A is riding in a car that is driving down a straight road at 34 mph. They toss a small ball straight up and catch it as it falls back into their hand.

a. If the ball reaches a maximum height of 0.25 m above their hand, what was the velocity of the ball as it left person A's hand?

b. Person B is riding in a second car traveling at 35 mph parallel to Person A. Person B looks out the window to observe person A toss the ball. Person B says "the ball is at the origin at the instant it is thrown, and both cars are traveling in the $+x$ direction". What is the position of the ball (according to Person B) when person A catches it?

Problem 2.47. A 1.0-m-long board is resting on the top of a table that is 1.5 m above the ground. You lift one end of the board so it is 0.30 m above the table.

a. What is the angle between the board and the table?

b. An object is placed on the elevated side of the board and slides down to the lower edge of the board. If you ignore friction, what is the velocity of the object once it reaches the end of the board?

c. The object continues by falling off of the edge of the board and impacting the ground. How far from the edge of the table does it strike the ground?

d. Just as it hits the ground, what is the object's velocity?

Problem 2.48. Suppose you are a consultant for an action movie. In one scene, the hero is riding a motorcycle at 50. m/s and rides off a very tall cliff that is inclined at 35° above the horizontal.

a. What is the hero's maximum elevation above the edge of the cliff?

b. Suppose the director wants a static shot of the stunt taken from the side (so we can use the usual xy coordinate system) and wants the hero to be in frame for 8.0 seconds after she leaves the edge of the cliff. How far will she fall below her launch point in this time?

c. How far in the horizontal direction will the hero travel in this time? (Knowing the answers to all three parts of this question determines where the camera will be positioned to ensure she doesn't leave the frame too quickly.)

★   **Problem 2.49.** A ball is thrown toward a bridge of height h with a speed of 21 m/s and an angle of 63° above the horizontal. It lands on the bridge 3.0 s later. How high is the bridge relative to the height from which the ball is thrown?

★★   **Problem 2.50.** Suppose a small bottle rocket is placed on a level table that is 1.50 m above the ground. When launched, the rocket accelerates horizontally at 4.0 m/s². The table is 2.0 m long.

a. How fast will the rocket be moving when it reaches the edge of the table?

b. Once the rocket reaches the edge of the table, it continues to experience the same horizontal acceleration but also begins to experience a vertical acceleration g. What horizontal distance from the edge of the table will the rocket be when it has fallen halfway to the ground?

c. You place a circular hoop at the location found in (b), so the rocket flies through it. How long does it take from the instant the rocket is launched to the instant it passes through the hoop?

★★★   **Problem 2.51.** Suppose one picturesque winter's eve I am standing on my roof putting up decorations when I slip and slide (from rest) a distance of 3.0 m along the icy (frictionless) roof, which is inclined at an angle of 30° above the horizontal.

a. What speed will I have as I reach the bottom edge of the roof?

b. I enter into beautiful parabolic motion as I slide off the roof. A big pile of snow is 2.5 m below the edge of the roof. How long will I be in the air (from leaving the roof to hitting the snow bank)?

c. With what speed will I impact the snow bank?

d. Suppose the vertical compaction of the snow is 1.0 m as I come to rest. What was my average vertical acceleration as I slowed to rest?

★★★   **Problem 2.52.** A 1.0 kg puck is launched up a 30.° frictionless ramp: it travels 2.0 m along the ramp and attains a speed of 4.0 m/s as it leaves the top of the ramp.

a. What was the launch speed of the puck?

b. The puck now undergoes beautiful projectile motion as it leaves the ramp at 4.0 m/s, 30° above the horizontal. How long after leaving the ramp does the puck reach its maximum height?

c. What is the puck's maximum height, as measured from the top of the ramp?

d. If the top of the ramp is 0.50 m above the ground (which is horizontal), then how long from the instant the puck leaves the edge of the ramp will it be in the air?

e. What horizontal distance, measured from the edge of the ramp, will the puck travel before hitting the ground?

★★   **Problem 2.53.** We concluded Example 2.12 with a messy equation that relates the kinematic properties of a basketball shot to get the ball through the hoop: solving for the launch angle θ is problematic in part because it came up in multiple locations in the equation. Your task here is to determine a value for θ using the numerical constants in the Example and a computer. Perhaps the simplest way is to use spreadsheet software (like Microsoft Excel or Google Sheets), set up a column of possible values for θ (in small steps of, say, 0.1°) and a column for the height h_2 when the ball has traveled horizontally to the hoop. Then, identify the value of θ among all the possibilities that gets closest to $h_2 = 3.0$ m. (You can do this efficiently with a simple loop in a full programming language such as Python, and you can do even better if you use numerical optimization libraries.)

a. Using your preferred method (or the method requested by your instructor), determine *two* values of θ that get the ball to $h_2 = 3.0$ m. Which of the two do you think an actual basketball player would opt for?

b. Include a graph of h_2 vs. θ. Your graph should demonstrate that there are two possibilities; explain how.

2.3 Angular Kinematics

⋆ **Problem 2.54.** A fan is spinning counter clockwise and slowing down. What are the signs (positive or negative) of ω and α?

⋆ **Problem 2.55.** A standard clock, like the one hanging on the wall in many classrooms, is running out of batteries. As the second hand slows to a stop, what is the sign of ω and α?

⋆ **Problem 2.56.** For each rotating wheel described below, identify the sign (positive or negative) of ω and α.

a. Rotating clockwise and speeding up

b. Rotating clockwise and slowing down

c. Rotating counter-clockwise and speeding up

d. Rotating counter-clockwise and slowing down

⋆ **Problem 2.57.** A fan experiences a counter-clockwise angular acceleration of 2.0 rad/s^2 for 2.0 s. If it is initially rotating with a clockwise angular velocity of 6.0 rad/s, what is the final angular velocity of the fan?

⋆ **Problem 2.58.** A merry-go-round (MGR) has a radius of 3.0 m. A passenger at the edge is supposed to have a linear velocity of 1.5 m/s when the MGR is at "full speed". The MGR begins from rest and can accelerate at 0.025 rad/s^2.

a. How many rotations will the rider experience before coming to full speed?

b. How long will it take for them to achieve top speed?

⋆⋆ **Problem 2.59.** A 5.0-m-diameter merry-go-round (MGR) is initially rotating such that a complete rotation occurs every 3.5 s. It slows down and comes to a halt in 20.0 s.

a. What was the initial speed (in m/s) of my son Joey, who was riding at the edge of the MGR?

b. Suppose I was standing on the ground and gave Joey a high-five every time he rotated by. If the first high-five occurred at the instant the MGR started to slow down, then how many high-fives were delivered in total before the MGR came to a halt?

⋆ **Problem 2.60.** A drill is spinning counter-clockwise at 20.0 rad/s slows to 10.0 rad/s and makes 20.0 complete revolutions. What is the angular acceleration of the drill?

⋆ **Problem 2.61.** The 0.40 m diameter wheels on your car makes 40.0 revolutions as it rolls with constant velocity along the road for 60.0 seconds. How far did the car travel?

⋆⋆ **Problem 2.62.** A space station is constructed as a rotating cylinder that rotates about its axis. The space station spins up for 7.20×10^2 seconds from rest until the outer deck is rotating at 85 m/s. If (once it is up to speed) it takes the space station 55 s to complete one rotation, then what is the radial distance to the outer deck from the axis of rotation?

⋆⋆ **Problem 2.63.** A spool has a thin, negligible-mass cable attached to it which is pulled with an acceleration of 1.5 m/s^2. The diameter of the spool is 6.0 cm and it rotates on a frictionless bearing from rest. After 1.0 m of the cable has been unwound, what is the angular speed of the spool in revolutions per minute? Assume the cable is pulled at an angle that allows for the maximum speed to be obtained.

★★ **Problem 2.64.** A bicycle wheel is set spinning and students measure the velocity of a point on the edge by allowing a 13 cm card to pass through a photogate; the speed is recorded as 5.5 m/s. 15 revolutions later, they do it again and find that the speed is 5.0 m/s. The card was located 15 cm from the center of the wheel.

    a. What was the angular acceleration of the wheel during this time?

    b. How much time passed between the two measurements?

★★ **Problem 2.65.** Consider the motion described in Problem 2.64. Draw:

    a. Draw the α vs time graph.

    b. Draw the ω vs time graph.

    c. Draw the θ vs time graph.

    d. In a few sentences, describe how each of the graphs in (a)–(c) would change if the first measured velocity was the same but the magnitude of the ensuing acceleration was twice as large.

★ **Problem 2.66.** A tire is rolling without slipping over level ground at 350 rotations per minute (rpm).

    a. What is this angular velocity in rad/s?

    b. If the tire is to come to rest in 5.0 s, what angular acceleration must it experience?

    c. How many rotations will the tire experience during the motion described in (b)?

    d. If the tire has a 75 cm diameter, the how far (in m) will the tire have traveled during the motion described in (b)?

★★ **Problem 2.67.** A 5.0-cm-diameter hard disk starts at rest at $t = 0$ s, then accelerates for 0.50 s at 250 rad/s^2.

    a What is the angular speed of a point on the rim of the disk at $t = 0.50$ s?

    b What is the linear speed of a point on the disk at $t = 0.50$ s?

    c If the disk continues to rotate at the angular speed you found in (b), then how many rotations will it have turned through at $t = 2.0$ s?

★ **Problem 2.68.** An object is traveling in the $+x$ direction at a speed v. A short time later, it is traveling at the same speed in the $+y$ direction. In between, it travels in a *circular arc*.

    a. Invent a physics problem that uses this as a setup: give some additional context and specify what the student is to solve for/explain/etc.

    b. Solve the problem you invented in (a).

★★ **Problem 2.69.** Consider the Earth as a sphere rotating about an axis through its poles. What are the magnitudes of the linear velocity (v) and angular velocity (ω) of a person standing on Earth's surface at a latitude θ? (The latitude varies from 0° at the equator to 90° at the North and South Poles.)

2.4 Tangential and Centripetal Acceleration

★ **Problem 2.70.** What is the (a) angular velocity, (b) tangential velocity, (c) angular acceleration, (d) tangential acceleration, and (e) centripetal acceleration of the Earth orbiting about the Sun? (f-j) Repeat for the Moon orbiting about the Earth. Relevant data is in Appendix B.

★ **Problem 2.71.** Figure 2.44 shows the centripetal acceleration of an object at four positions as it travels counter clockwise around a circular track, starting from the top of the figure. What is the direction of the object's angular acceleration? Explain.

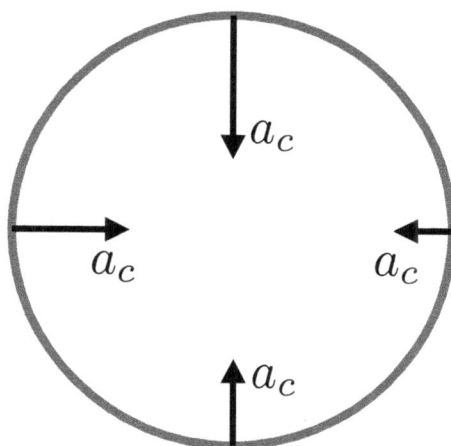

FIGURE 2.44
Problem 2.71

★ **Problem 2.72.** A ball on a string spins with a constant angular speed with centripetal acceleration a_c at a radius r. If the ball spins with half the centripetal acceleration and half the angular speed, what is the new radius of the system (in terms of r)?

★ **Problem 2.73.** A ball is spun in a clockwise circle at radius of 0.50 m with an acceleration of 1.00×10^2 m/s^2. What is the angular velocity of the motion, assuming it is traveling at constant speed?

★ **Problem 2.74.** A top is spinning counter-clockwise at 10.0 rev/s slows as it experiences a 2.0 rad/s^2 magnitude acceleration. After 10.0 seconds, through what angle has the top turned?

★ **Problem 2.75.** You spin a ball on the end of a 1.2 m long string from rest with an angular acceleration of 0.90 rad/s^2. What arc length has the ball traveled after the ball has traveled for 1.0 second?

★ **Problem 2.76.** A particle has a horizontal velocity at a particular instant and experiences a downward acceleration that (a) remains downward as the object moves or (b) remains perpendicular to the object's velocity. In each case, describe and sketch the object's subsequent motion.

★★ **Problem 2.77.** A car starts at rest and begins to accelerate around a circular curve with a radius of 150 m. The tangential acceleration is 1.5 m/s^2. How long does it take for the car's total acceleration to reach 3.0 m/s^2?

★★ **Problem 2.78.** Consider the situation described in Example 2.16. For the first 3.0 seconds of the motion:

a. Draw a θ vs. time graph

b. Draw a ω vs. time graph

c. Draw a α vs. time graph

d. Draw a a_t vs. time graph

e. Draw a a_c vs. time graph

f. Draw a a_{NET} vs. time graph

★★ **Problem 2.79.** Assuming that the Moon's orbital motion is a uniform circular motion, what is the Moon's acceleration as it "falls" toward the Earth? Relevant data is in Appendix B.

★★ **Problem 2.80.** A high-speed drill is rotating counter clockwise at 2400 rpm (251.3 rad/s).

a. If a point on the edge of drill chuck experiences a centripetal acceleration of 1.9×10^3 m/s^2, what is the radius of the drill chuck?

b. Suppose the drill comes to a halt in 2.5 s. As it slows, identify all accelerations acting on the edge of the drill. Include a properly labeled diagram.

c. What is the drill's angular acceleration as it comes to rest?

d. How many revolutions does the drill make as it stops?

★★ **Problem 2.81.** A gear (that we'll treat as a disk) is 8.0 cm in diameter and initially at rest. The gear accelerates at 3.75×10^2 rad/s^2 for 0.50 s to achieve its steady operational speed.

a. What is the speed of a point on the rim when at operational speed?

b. Through how many revolutions has the disk turned during the startup?

c. Explain whether (or not) all points on the disk experience the same tangential acceleration while at operational speed.

d. What is the acceleration of a point on the rim once at operational speed? Provide a numerical answer along with a statement about the direction of the acceleration.

★★ **Problem 2.82.** The end sheets of a 12.0 cm diameter roll of toilet paper are pulled with an angular acceleration of 10.0 rad/s^2 from rest unwinding 1.00 m of paper. Assume the diameter of the roll is constant during this unwinding.[12]

a. After 0.30 m of the toilet paper has been unwound, how many revolutions has the roll completed?

b. What is its angular speed in rad/s when 0.30 m of paper has been unwound?

c. What is the centripetal acceleration of the spool at this instant?

d. Explain/show the direction of the total acceleration vector at this instant. What acceleration(s) determine the total?

★ **Problem 2.83.** A 3.0 cm diameter drill rotates from rest to an operational angular speed of 1.00×10^3 rad/s, while it experiences an angular acceleration of 100.0 rad/s^2. What is the total angular distance traveled by the drill during this process?

★★ **Problem 2.84.** In a game show, two contestants sit back-to-back in chairs on a platform that spins, making them dizzy. At operational speed the chairs move with uniform circular motion completing one rotation every 1.5 seconds counter clockwise with a contestant's head located 0.20 m from the center axis of rotation.

a. What centripetal acceleration does the contestant's head experience when spinning at operational speed? Provide a direction as well as a magnitude.

b. The second contestant's head is at a radius slightly bigger than the first contestant's head. Explain which contestant's head experiences a larger tangential acceleration while at operational speed.

c. The platform slows uniformly and stops in half a revolution. What is the angular acceleration of the platform as it slows?

d. How long did it take to stop?

Additional Problems for Chapter 2

★★★ **Problem 2.85.** We claimed in Section 2.4 that if θ is small, $\sin\theta \approx \theta$. Indeed, this is the leading order term in the series expansion for $\sin\theta$ provided in Appendix B. The **Taylor series** formula provided in Appendix B allows you to expand *any* well-behaved function[13] in terms of a power series of its derivatives.

---

[12]This is clearly unrealistic. *Don't worry*, we'll look at this situation in more detail in Problem 3.153.

[13]Essentially, functions with continuous derivatives.

a. Use the Taylor series formula in Appendix B to verify the series expansion for $\sin\theta$ provided in Appendix B. (Formally using the notation in Appendix B, we are expanding $\sin\theta$ around $\theta = 0$ by looking for $\sin(\theta + \delta)$ where δ is small.)

b. Repeat (a) for $\cos\theta$

c. Repeat (a) for $\tan\theta$.

d. Use a computer or a calculator to determine the smallest value for θ for which the percent difference between $\sin\theta$ and θ is greater than 5%.

e. Use a computer or a calculator to determine the smallest value for θ for which the percent difference between $\tan\theta$ and θ is greater than 5%.

★★ **Problem 2.86.** On another planet, you launch a ball at an angle θ. After 2.0 seconds, the ball is traveling 1.0 m/s vertically and 3.0 m/s horizontally. At 4.0 seconds, the ball has reached its maximum vertical height.

a. What is the acceleration due to gravity on this planet?

b. What is the velocity of the ball at 0.0 seconds and 6.0 seconds? Be sure to justify both the horizontal and vertical components.

c. What is the launch angle of the ball?

★★ **Problem 2.87.** Three forces act on an object,[14] and the net force is 2.0 N in the positive y direction. If the first force has components of 3.0 N $\hat{x}$ and -2.0 N $\hat{y}$ and the second force is -2.0 N $\hat{x}$, then...

a. what are the x and y components of the third force?

b. what is the magnitude and angle of the third force?

c. draw and label the three forces to scale on a x/y axes centered at the origin. Also include the net force.

---

[14]We'll get to forces shortly; for now it is just another kind of vector!

3

Forces

In Chapters 1 and 2 we discussed the process of *describing* motion: we considered how a coordinate system defines a mathematical framework that we can then use to talk about the position, velocity, and acceleration of some object(s). We visualized how these quantities change over time with motion diagrams and motion graphs, and we related them mathematically with the kinematic equations.

However, there is a deeper question: why do any of these quantities change in the first place? Answering this question is the topic of this chapter; as the title suggests, the answer revolves around understanding *force*. As we will see, if we know the forces acting on an object, then (with the help of the kinematic equations) we are typically well-equipped to describe its subsequent motion.

Learning Objectives

After reading this chapter, you should be able to:

- Define and explain the idea of a force in the context of Newton's Three Laws.

- Analyze the forces in a given situation to determine, for instance, the acceleration of a mass.

- Recognize and analyze the specific forces discussed in Sections 3.4–3.10:

  - Newton's Law of Gravitation determines the gravitational attraction between objects with mass (and, as a result, the acceleration due to gravity on a planet).

  - The normal force is directed outward from (and perpendicular to) a surface and balances the total inward force.

  - Friction is directed parallel to a surface and is directed to decrease speed to $0\frac{\text{m}}{\text{s}}$ (or keep it there).

  - Drag always points opposite the velocity of an object, and increases with increasing velocity.

  - Tension exists in any taut rope; the tension is applied to both ends of the rope and points in toward the center of the rope. A pulley effectively redirects a tension force.

  - Hooke's law applies to a stretchable/compressible material. The force increases with an increasing stretch amount.

  - Every object immersed in a fluid experiences a buoyant force equal to the weight of the displaced fluid (Archimedes' Principle).

- Use both forces and the kinematic equations to analyze the motion of an object.

- Recognize that the *moment of inertia* plays the role of *mass* in analyzing rotational motion.

DOI: 10.1201/9781003571568-4

- Calculate the moment of inertia for a series of point masses and use the moment of inertia for continuous objects.

- Recognize that *torque* plays the role of *force* in analyzing rotational motion, and apply Newton's Second Law for rotational motion (Equation (3.27)).

- Calculate the center of mass of an extended body, and recognize that it is the point about which there is no net gravitational torque.

3.1   Introduction

If you were asked to define a force, you might say something like, "a force is what an object experiences if you push or pull on it". This is in fact a good starting point, because it immediately raises an interesting question: what happens to an object that experiences a force? Or, conversely, what can we say about an object that experiences no force?

We'll address these questions and define the basic properties of forces in Sections 3.2-3.3. We'll then discuss the most common types of forces that arise in the study of mechanics and provide some cumulative examples. We conclude the chapter by considering the causes of *circular motion* that we first considered at the end of Chapter 2. Because there are a variety of types of forces, this chapter is rather lengthy. If you start to feel overwhelmed, keep in mind that the central ideas are presented in the (rather short) Sections 3.2-3.3.

3.2   Newton's 1st and 2nd Laws: Forces Can Cause Acceleration

Suppose you have a small box or hockey puck, and you give it a gentle kick so it slides across level ground. Eventually, of course, the object will slow to a stop. More formally, if we define the direction of its initial velocity to be positive, then the puck has a negative acceleration (Figure 3.1). But how far will it travel before it stops? (In other words: what will its displacement be?) This depends in part on what kind of material the ground is made of: the puck won't go very far at all over concrete compared to ice!

This is because of a force called **friction**. We'll define friction more precisely later, but for now you can think of it intuitively as an interaction between a surface (like the ground) and an object (like the puck) that tends to make the object slow down. Evidently, there is something about the ground and the object – you can think of it loosely like *roughness* – that determines how severe the effect of friction will be. Thus there is more friction on the grassy field than on concrete, as evidenced by the fact that the puck goes farther on the concrete than on the grass.

Can you think of a surface that would have even less friction? One answer is ice: if you've ever seen an ice skater or a hockey player, you know that an athlete can move very quickly and then glide for a long time before even appearing to slow down, much less coming to a stop. In fact, if you push this idea even further and try to imagine a situation where you kick a ball over a surface with no friction at all, then it should continue on with the same velocity forever! This concept is famously known as **Newton's First Law**[1]:

---

[1] After Isaac Newton (1642-1726), one of the most influential scientists of all time. (Of all time!)

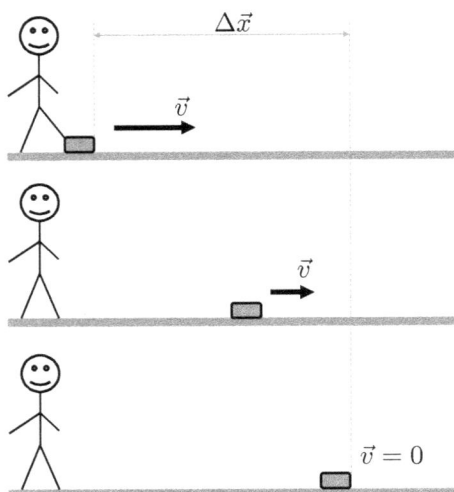

FIGURE 3.1
A kicked puck slows to a stop because of friction.

An object will move with constant velocity unless there is a net force acting on it.

This statement deserves some reflection, because it runs counter to our experiences: no surface really has no friction, so in practice we need to overcome it (by pedaling our bicycle, kicking the ball again, etc.). This is why the language "net force" is included in Newton's First Law: a cart, for instance, can move at a constant velocity if someone pushes it from behind so as to balance the effect of friction. Note that any object that is moving at constant nonzero velocity (that is, it is moving but with no net force) is said to be in **translational equilibrium**.[2] If the object is experiencing no net force and is at rest, then it is said to be in **static equilibrium**.

Example 3.1 Translational Equilibrium   ⋆
Which of the following people/objects are in translational or static equilibrium?

(a) A bicyclist picking up speed as she coasts down a hill.

(b) A person riding up an elevator at a constant velocity.

(c) A book sitting on a desk.

(d) A book falling after it is knocked over a desk.

(e) A car making a circular turn at a constant speed.

(a) The bicyclist is accelerating (picking up speed) and therefore cannot be in translational or static equilibrium.

---

[2]A related term is **dynamic equilibrium**, which states that the object is moving at a constant velocity (as with translational equilibrium) *and* at a constant *rotational* speed. If you spin a basketball on your fingertip, then as you're spinning it and increasing its rotational speed the basketball is in translational but not dynamic equilibrium.

(b) The person is not accelerating and the velocity is nonzero, so they are in translational equilibrium.

(c) The book is not accelerating and the velocity is 0 m/s; it is therefore in static equilibrium.

(d) The book is accelerating at a rate g toward the ground and is not in translational or static equilibrium.

(e) The car has a constant *speed*, but the velocity is a vector defined by both a magnitude and a direction. Because the direction is changing as the car turns, the velocity also changes and the car is not in translational or static equilibrium (We'll discuss more about circular motion starting in Section 3.12).

If you carefully read Newton's First Law, you may realize that it implicitly makes another statement: if there is a net force acting on an object, then the velocity will change (that is, it will *accelerate*). **Newton's Second Law** formalizes this relationship by relating the *net force*, the *mass*, and the *acceleration* of an object:

$$\vec{F}_{\text{NET}} \equiv \sum \vec{F} = m\vec{a} \tag{3.1}$$

Let's unpack what this says. We will always use $\vec{F}$ to represent a force, and in general we'll use a subscript to differentiate between forces. In equation (3.1) we use the subscript "NET" to indicate that we're talking about the *net force*, which is to say the combination of all of the forces acting on an object, which is equivalent to the middle term $\sum \vec{F}$: the symbol $\sum$, an uppercase Greek sigma, represents a *summation* ("add up whatever follows") when used in a mathematical equation like equation (3.1).[3] Since what follows is $\vec{F}$, it says "add up all of the forces":

$$\vec{F}_{\text{NET}} = \sum \vec{F} = \vec{F}_1 + \vec{F}_2 + \vec{F}_3 + \cdots$$

Note that the mass is not a vector. It is measured in kilograms (kg) and is a measure of how much matter there is in an object. Because the units on both sides of an equation must be equal, we see immediately that 1 unit of force is equal to $1\text{kg}\frac{\text{m}}{\text{s}^2}$ ("one kilogram meter per second squared"). This gets to be a mouthful, and so we typically refer to this quantity instead as a **Newton**, which is the SI unit of force. Specifically, 1 Newton is the force that accelerates a 1 kg object at a rate of $1\frac{\text{m}}{\text{s}^2}$.[4]

Before we move on, it is a good idea to pause and acquire some familiarity with Newton's 2nd Law; in particular, it can be challenging to combine the forces on an object to determine the net force. However, if you've read Chapters 1 and 2 you already know how to combine vectors in one and two dimensions, so don't be thrown off by the new context. A key takeaway here is that you now have a powerful new tool for your physics toolbox: while Chapters 1 and 2 focused on how the kinematic variables are related, you can now relate those same variables (via acceleration) to an object's mass and the forces acting on it. The following examples will give you some practice with these ideas. In particular, we introduce the **force diagram**: a drawing that indicates all of the forces acting on an object. Carefully representing the

---

[3]The $\equiv$ symbol means *equivalent to*; we use it here to indicate that the net force F_{NET} is equivalent to the vector sum of all forces $\sum \vec{F}$ *by definition*.

[4]This is a good point to formally define a **force** as an interaction that obeys Newton's Second Law. You can loosely think of a force with magnitude F as a push or a pull that (in the absence of other forces) makes an object of mass m accelerate at a rate $a = F/m$ in the direction of the force.

magnitude and orientation of each force in a force diagram helps you to determine the net force acting on the object.

Example 3.2 Newton's 2nd Law Graphed   ⋆

The acceleration of two objects are recorded as the net force is varied, as shown in Figure 3.2. What is the mass of each object?

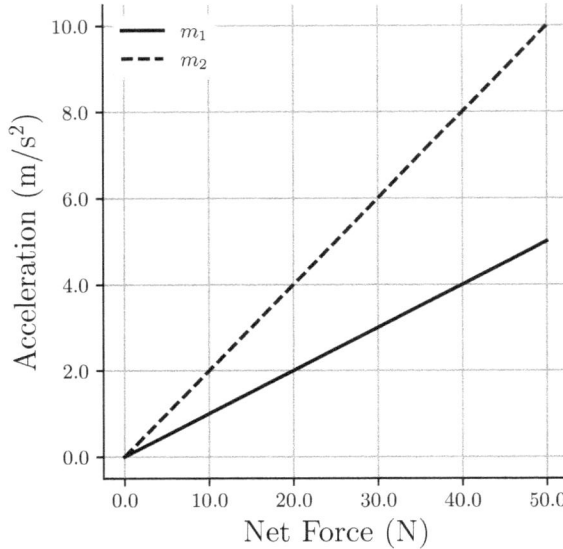

FIGURE 3.2

Acceleration of two objects are recorded as the net force is varied.

Newton's Second Law tells us that $\vec{F}_{\text{NET}} = m\vec{a}$, which we can equivalently write as

$$\vec{a} = \frac{1}{m}\left(\vec{F}_{\text{NET}}\right)$$

I've rewritten it this way because our data is presented with the acceleration on the vertical axis and the net force on the horizontal axis; we see that the data follows a straight line of the form $y = (\text{slope})\,x$, where y (the vertical coordinate) is acceleration and x (the horizontal coordinate) is the net force. Thus the slopes of the straight lines in Figure 3.2 are the inverses of the masses.

For mass 1, we see the slope is $\frac{10.0 \text{ m/s}^2}{50.0 \text{ N}} = 0.200 \text{ kg}^{-1}$ (check the units by substituting in $1 \text{ N} = 1 \text{ kg m /s}^2$), which indicates that

$$m_1 = \frac{1}{0.200 \text{ kg}^{-1}} = 5.00 \text{ kg}$$

Similarly, for mass 2 the slope is $\frac{4.0 \text{ m/s}^2}{40.0 \text{ N}} = 0.10 \text{ kg}^{-1}$, which indicates that

$$m_1 = \frac{1}{0.10 \text{ kg}^{-1}} = 1.0 \times 10^1 \text{ kg}$$

Example 3.3 Pushing a Box (with Friction?) ★
You push on a 20.0 kg box over level ground with a force of 100.0 N. The box is initially at rest and it takes 2.7 s for the box to travel 4.0 m. Is there a frictional force acting horizontally on the box while it is moving? If so, how large is it and in what direction does it point? Draw a picture of the box and include vectors representing all forces acting horizontally on the box.

We'll set the direction of the acceleration (and the pushing force) to be in the $\hat{x}$ direction. We are given kinematic information sufficient to solve for the acceleration of the box:

$$x_f = x_i + v_i \Delta t + \frac{1}{2} a \left(\Delta t \right)^2$$

$$2 \left(\frac{\Delta x - v_i \Delta t}{\left(\Delta t \right)^2} \right) = a$$

$$2 \left(\frac{4.0 \text{ m}}{\left(2.7 \text{ s} \right)^2} \right) = a = 1.1 \text{ m/s}^2$$

According to Newton's 2nd Law:

$$\sum \vec{F} = m\vec{a}$$

$$\sum \vec{F} = (20.0 \text{ kg}) \left(1.1 \hat{x} \text{ m/s}^2 \right) = 22 \hat{x} \text{ N}$$

The net force is clearly not equal to the pushing force: the net force is smaller than the pushing force, so there must be another force that opposes the push. As we've argued (and will see in more detail later), this opposing force is frictional, so we'll call if $\vec{F}_f$. Turning to Newton's 2nd Law, we see:

$$\sum \vec{F} = 22 \hat{x} \text{ N}$$

$$F_P \hat{x} + \vec{F}_f = 22 \hat{x} \text{ N}$$

$$F_f = \left(22\hat{x} - 100.0\hat{x} \right) \text{ N} = -78\hat{x} \text{ N}$$

The negative sign confirms that our frictional force points opposite the push, leading to a lower acceleration than if the box really did only experience the pushing force. Figure 3.3 shows the force diagram (notice that the length of the vectors are proportional to their magnitudes; we can see that the net force, given by adding $\vec{F}_p$ and $\vec{F}_f$ tail-to-tip will yield a net force to the right).

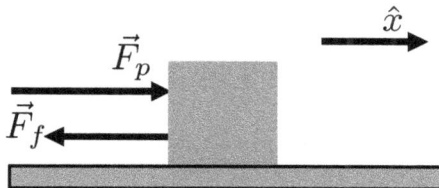

FIGURE 3.3
Force diagram.

Example 3.4 Adding Forces ★★

Figure 3.4 shows a top-down view of a 5.0 kg box. Determine the acceleration of the box.

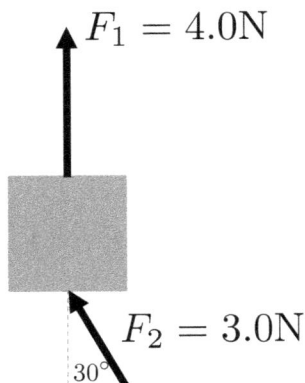

FIGURE 3.4

Top-down view.

Before we begin, note that it doesn't matter that one force has the *tip* on the box and the other has the *tail* on the box. This suggests that $\vec{F}_1$ is a pulling force and $\vec{F}_2$ is a pushing force, but this does not affect our analysis. Our first task, as usual, is to set up a coordinate system. We'll make the usual choice of calling "up the page" the $+y$ direction and "to the right" the $+x$ direction. Once we've done this we have a framework for describing $\vec{F}_1 + \vec{F}_2 = \vec{F}_{\text{NET}}$, which (just like in Chapter 2!) we can describe as "some amount along the x axis and some amount along the y axis":

$$\vec{F}_{\text{NET}} = \left(\vec{F}_{\text{NET}}\right)_x \hat{x} + \left(\vec{F}_{\text{NET}}\right)_y \hat{y}$$

Because we have two forces vectors here, we can represent the total x component as just the sum of the two individual x components and likewise for the y components:

$$\vec{F}_{\text{NET}} = \left((F_1)_x + (F_2)_x\right) \hat{x} + \left((F_1)_y + (F_2)_y\right) \hat{y}$$

Now $\vec{F}_1$ points directly along the y axis, so it has no x component and $(F_1)_y = 4.0$ N. $\vec{F}_2$, however, does not point directly along either axis and therefore has both an x component and a y component. Moreover, the x component of $\vec{F}_2$ is *negative*. We shall make this explicit in our net force equation by including a negative sign:

$$\vec{F}_{\text{NET}} = \left(-(F_2)_x\right) \hat{x} + \left((F_1)_y + (F_2)_y\right) \hat{y}$$

We can do a bit of trigonometry to describe these components, as shown in Figure 3.4.

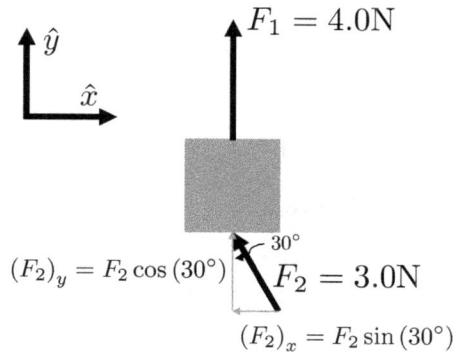

$F_1 = 4.0\text{N}$

$(F_2)_y = F_2 \cos(30°)$

$30°$

$F_2 = 3.0\text{N}$

$(F_2)_x = F_2 \sin(30°)$

FIGURE 3.5

If the trigonometry is unclear to you, try writing it out explicitly, like so:

$$\cos(30°) = \frac{(F_2)_y}{F_2} \implies F_2 \cos(30°) = (F_2)_y$$

Decomposing vectors into their components is a common enough task that you should have the goal of recognizing the situation and directly saying to yourself, "Ah, the non-hypotenuse side adjacent to the 30° angle is the y component, so $(F_2)_y = F_2 \cos(30°)$". You can make a similar statement for the non-hypotenuse side *opposite* the 30° angle and the *sine* component of $\vec{F}_2$.

Now we can express $\vec{F}_{\text{NET}}$ in terms of its components:

$$\vec{F}_{\text{NET}} = \left(-(F_2)_x\right)\hat{x} + \left((F_1)_y + (F_2)_y\right)\hat{y}$$
$$= \left(-3.0\sin(30°)\,\hat{x}\right)\text{ N} + \left(4.0\text{N} + 3.0\cos(30°)\,\hat{x}\right)\text{ N}$$
$$= \left(-1.5\hat{x} + 6.6\hat{x}\right)\text{ N}$$

This is the net force in component form, and we can invoke Newton's 2nd Law to determine the acceleration, also in component form:

$$\vec{F}_{\text{NET}} = m\vec{a}$$
$$\vec{a} = \frac{\vec{F}_{\text{NET}}}{m}$$
$$\vec{a} = \frac{\left(-1.5\hat{x} + 6.6\hat{x}\right)\text{ N}}{5.0\text{ kg}}$$
$$\vec{a} = \left(-0.30\hat{x} + 1.3\hat{y}\right)\text{ m/s}^2$$

We could stop here, but for the sake of completeness (and practice) let me point out that we can also express the net force in terms of the overall magnitude and a direction (Fig. 3.6). From the Pythagorean Theorem we have

$$F_{\text{NET}} = \left((F_{\text{NET}})_x^2 + (F_{\text{NET}})_y^2\right)^{1/2}$$
$$F_{\text{NET}} = \left((-1.5\text{ N})^2 + (6.6\text{ N})^2\right)^{1/2}$$
$$F_{\text{NET}} = 6.8\text{ N}$$

and from the geometry,

$$\theta = \tan^{-1}\left(\frac{(F_{\text{NET}})_y}{(F_{\text{NET}})_x}\right)$$

$$\theta = \tan^{-1}\left(\frac{6.6 \text{ N}}{1.5 \text{ N}}\right)$$

$$\theta = 78°$$

Thus we can equivalently state that the net force is 6.8 N at 78° above the $-x$ axis. To determine the *acceleration* from this point we can invoke Newton's 2nd Law:

$$\vec{F}_{\text{NET}} = m\vec{a}$$

$$\vec{a} = \frac{\vec{F}_{\text{NET}}}{m}$$

The *magnitude* of the acceleration is therefore

$$a = \frac{6.8 \text{ N}}{5.0 \text{ kg}} = 1.4 \text{ m/s}^2$$

and the *direction* of the acceleration is the same as the direction of the net force, because the two are linked by a positive scalar that can only change the magnitude (and units) of a vector, not the direction. Figure 3.6 shows both the net force and the net acceleration in all of their detail: the vectors are labeled with an overall magnitude and direction *and* in terms of their components. (Keep in mind that because the units of force are different from the units of acceleration, comparing the overall lengths of the vectors isn't meaningful; I've chosen to draw them with the same length just to address and emphasize this exact point.)

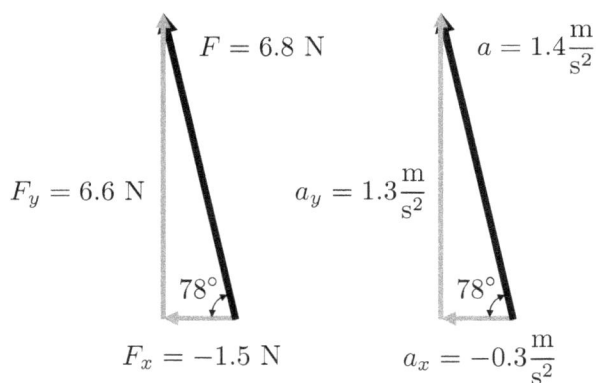

FIGURE 3.6
Displaying both the net force and the net acceleration.

However we express the acceleration, it is important to recognize that we could now, in principle, go on and solve any number of related problems in kinematics. For instance, if the object starts from rest at the origin, how far will it have traveled in 3.0 seconds, and what will its speed be? This question is in fact asked in Problem 3.10.

3.3   Newton's 3rd Law: Forces Come in Pairs

We've now seen several examples of how we can analyze the forces *on an isolated object*. This is a good start, but in general objects interact with one another, and we would naturally like to know how to deal with them. Consider as an example the situations shown in Figure 3.7, where a person is standing on a frictionless surface (ice is a reasonable approximation to this for our present purposes). If the person pushes on the box with some force $\vec{F}_{\mathrm{BP}}$ (the force on the box, B, from the person, P), then the two will separate and move apart.

$$\vec{v}_P = \vec{v}_B = 0\tfrac{\mathrm{m}}{\mathrm{s}}$$

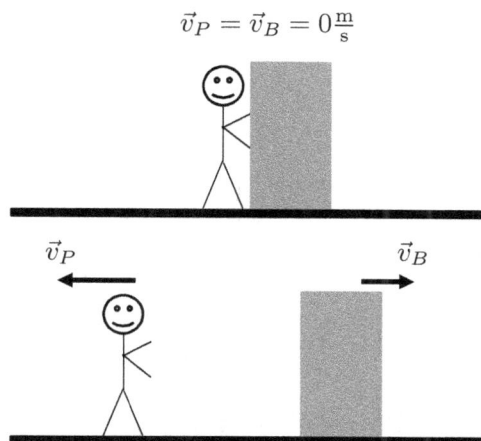

FIGURE 3.7
While on a frictionless surface, a person pushes on a box. Later, the person and box move away from one another at constant velocities.

In other words, the velocity of the box has changed *and* the velocity of the person has changed (initially they were at rest, which is to say with a velocity of 0 m/s; after the push the velocities are not zero and point in opposite directions). But, according to Newton's 1st Law, this can only happen if there was a force acting on the box and a force acting on the person during the push. How can this be, if the only thing that occurred was that the person pushed on the box? We explain this by saying that *the box pushes back on the person*, with a force that is *equal in magnitude but in the opposite direction* of the force of the person on the box (Figure 3.8).[5]
Mathematically,

$$\vec{F}_{\mathrm{BP}} = -\vec{F}_{\mathrm{PB}} \tag{3.2}$$

This is **Newton's Third Law**, which we can rewrite with more general notation as

$$\vec{F}_{\mathrm{AB}} = -\vec{F}_{\mathrm{BA}} \tag{3.3}$$

You may have heard of this law being informally phrased as, "for every action there is an equal and opposite reaction". However, this is somewhat sloppy language for our purposes

---

[5]Importantly, if the *masses* are different, the accelerations will have different magnitudes even if the only horizontal force on each object is the pushing force ($\vec{F}_{\mathrm{NET}} = m\vec{a}$). In general, then, the objects will separate away from one another at *different speeds*. Note also that in the absence of other forces (like friction) that result in a nonzero net force on the objects, they will then move in opposite directions at constant velocities.

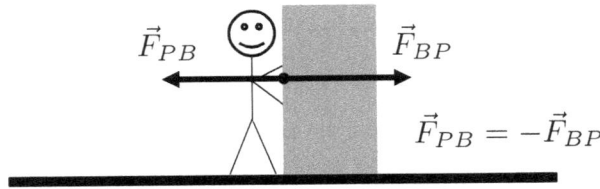

FIGURE 3.8
While the person of Figure 3.7 is pushing on the box, the box is pushing back with a force of equal magnitude but opposite direction.

since it is unclear what an "action" is. It is more precise to say, "If A applies a force on B, then B applies an equal and opposite force on A". As the next example shows, Newton's Third Law is a powerful tool for analyzing systems of interacting objects.

Example 3.5 Newton's Third Law with Two Boxes ★★
A 15.0 kg box and a 10.0 kg box sit side by side on a flat, frictionless surface. A 30.0 N force is applied horizontally to the 10.0 kg box, as shown. At what rate do the boxes accelerate? What is the net force on each box?

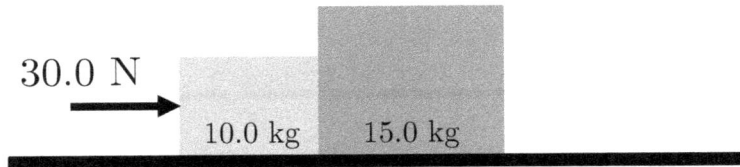

FIGURE 3.9

Your intuition may be to treat the two boxes as one object with a mass of $(10.0 + 15.0)$ kg $= 25.0$ kg, in which case (if we set our coordinate system such that the force aligns with the positive x axis)

$$\vec{F}_{\text{NET}} = m\vec{a}$$
$$30.0 \text{ N}\hat{x} = (25.0 \text{ kg})\,\vec{a}$$
$$1.20\frac{\text{m}}{\text{s}^2} = \vec{a}$$

This turns out to be the correct answer, but it is important to understand why. The applied 30.0 N force pushes on the 10.0 kg box; as a result the 10.0 kg box pushes on the 15.0 kg box. (The 10.0 kg box pushes back on whatever is applying the 30.0 N force, as well, according to Newton's Third Law. However, in this problem we're ignoring whatever is applying the 30.0 N force so this doesn't affect our analysis.) By Newton's Third Law, this means that the 15.0 kg box pushes back on the 10.0 kg box with a force of equal magnitude. To keep our notation clear we'll label the 10.0 kg box as box 1 and the 15.0 kg box as box 2 (Figure 3.10).

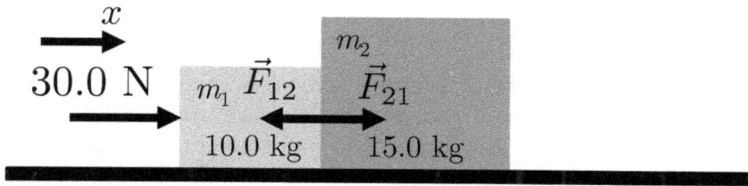

FIGURE 3.10

With this in mind we can explain the principle of **selective grouping**: you can consider objects that are moving together (with the same acceleration) as a single object, as long as you consider their *combined mass* and consider only *external forces*. (This is because the internal forces cancel, which means they may as well not be there at all as far as Newton's Second Law is concerned.) In this case the only external force acting on the boxes is the 30.0 N force, and their combined mass is 25.0 kg. As we showed above, this yields an acceleration for both boxes of $1.20\frac{m}{s^2}\hat{x}$.

The second part of the question asks us to determine the net force on each object. From the above diagram we see that the only force acting on the 15.0 kg box is $\vec{F}_{21}$, which clearly points in the positive x direction. Because we have already determined the acceleration, we can determine the magnitude of $\vec{F}_{21}$ with Newton's Second Law:

$$\sum \vec{F}_{\text{on } 2} = m_2 \vec{a}$$
$$\vec{F}_{21} = (15.0 \text{ kg}) \left(1.20\frac{m}{s^2}\hat{x}\right)$$
$$\vec{F}_{21} = 18.0 \text{ N}\hat{x}$$

Then, by Newton's Third Law,

$$\vec{F}_{12} = -\vec{F}_{21}$$
$$\vec{F}_{12} = -18.0 \text{ N}\hat{x}$$

We can double check this result by looking at the net force on box 1:

$$\sum \vec{F}_{\text{on } 1} = m_1 \vec{a}$$
$$(30.0 - 18.0) \text{ N}\hat{x} = (10.0 \text{ kg}) \vec{a}$$
$$1.20\frac{m}{s^2}\hat{x} = \vec{a}$$

We find the same acceleration, as expected.

This kind of analysis can be rather confusing at first, so let's take a moment to summarize what we've done. The two boxes are moving together, so we are allowed to treat them as a single object with a combined mass. This allows us to ignore the internal forces (box 1 pushing on box 2 and vice versa), which in turn allows us to determine the acceleration of the boxes. Once we know this we can go back and consider the boxes individually in order to analyze the properties of the internal forces.

3.4   Gravity

Up until now we've mostly thought of forces as some sort of generic push or pull. However, there are many kinds of forces (or, if you prefer, there are many different ways in which a force can be generated), and it is now time to discuss some of the most common examples.

Very broadly, a force is either a **contact force** or a **non-contact force**. As the name suggests, a contact force requires two things to be touching: you can only kick a ball if your foot touches it. All but one force in our study of mechanics are contact forces. The sole exception, as you may have guessed, is **gravity**. The fact that an object doesn't need to be in contact with anything to experience the force of gravity should be obvious if you reflect on the fact that an object accelerates from rest when you drop it. In fact, *every object that has mass is attracted to every other object that has mass*. This relationship is defined mathematically by **Newton's Law of Gravitation**:

$$\vec{F}_G = G\frac{m_1 m_2}{r^2} \text{ (attractive)} \tag{3.4}$$

In equation (3.4), G is a fundamental constant of nature called the *Gravitational Constant*:[6]

$$G = 6.67 \times 10^{-11} \text{m}^3/(\text{kg s}^2) \tag{3.5}$$

The masses of the two objects are denoted by m_1 and m_2, and the distance between them is denoted r.[7] The direction of the vector is denoted in equation (3.4) simply as "attractive": object 1 is attracted toward object 2, and object 2 is attracted toward object 1.[8] A more mathematically precise way of saying the same thing is

$$\vec{F}_{G12} = G\frac{m_1 m_2}{r^2}\hat{r} \tag{3.6}$$

Where $\hat{r}$ is a unit vector pointing from object 1 to object 2 (Figure 3.11) and the subscript indicates that we are talking about the force on object 1 due to object 2.

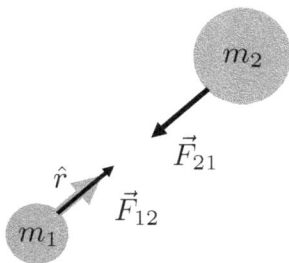

FIGURE 3.11
Gravitational attraction between two objects is directed along the unit vector $\hat{r}$.

---

[6]The odd units of G are necessary if the right side of the equation is to have units of force (check it for yourself!).

[7]Specifically, r is the distance between the center of one object and the center of the other (assuming, as we always will, that the mass is evenly distributed through the object).

[8]This pair of forces is an example of Newton's Third Law.

As we shall see in Example 3.7, to two significant figures $GM_E/r_E^2 = 9.8$ m/s^2, i.e. near the surface of the Earth (with mass M_E and radius R_E) an object with mass m experiences a gravitational force of magnitude

$$F_G = mg \qquad (3.7)$$

where g is the acceleration due to gravity originally defined in Equation (1.6). We *define g* as 9.8 m/s^2 in part because factors like the rotation of the Earth (influenced also by your latitude) and your elevation relative to sea level can affect the local value of g.

The idea that everything with mass is attracted to everything else with mass is striking: it means, for example, that there is a literal force of attraction between you and the people walking past you on the street. However, as the next example shows, for most practical purposes the only gravitational force that matters for everyday situations is due to the Earth. As we will see in Section 3.12, the force of gravity plays a major role in explaining planetary motion, such as that of the Earth around the Sun.[9]

Example 3.6 Three Body Problem (a Special Case) ★★

Suppose a small meteoroid of mass m is out in space along a line that connects two stars that are separated by a distance R, as shown in Figure 3.12. If $M_2 = 2M_1$ and the net force on the meteoroid is 0 N, how far from M_1 is the meteoroid? (Your answer will be a multiple of the star separation R).

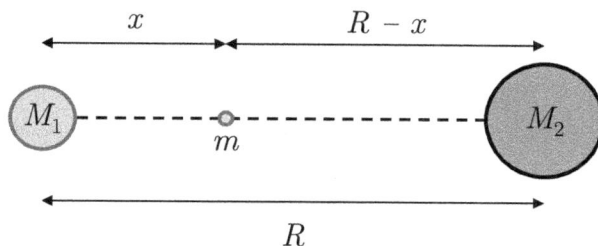

FIGURE 3.12

If the meteoroid is a distance x from M_1, then it is a distance $R - x$ from M_2, as indicated in the figure. We'll call "to the right" the $\hat{x}$ direction, as usual. Turning to Newton's Second Law and invoking Newton's Law of Gravitation, we have

$$\sum \vec{F} = m\vec{a}$$

$$-G\frac{mM_1}{x^2} + G\frac{mM_2}{(R-x)^2} = 0$$

$$-\frac{M_1}{x^2} + \frac{M_2}{(R-x)^2} = 0$$

where in the last line we canceled the common terms. Now we will insert the condition

---

[9]It is worth noting that explaining *how* a force is felt at a distance requires some advanced topics that would take us far from the goal of this chapter, which is to get you to the position where you feel comfortable working with forces and Newton's Laws. For our purposes we will simply assert that Newton's Law of Gravitation is *observationally true*. If this bothers you, you should consider a career in physics!

$M_2 = 2M_1$ and solve for x:

$$\frac{2M_1}{(R-x)^2} = \frac{M_1}{x^2}$$

$$\frac{2^{1/2}}{(R-x)} = \frac{1}{x}$$

$$2^{1/2}x = R - x$$

$$x = \frac{R}{1 + 2^{1/2}} \approx 0.41R$$

Thus we see that the object is *closer* to the star with less mass: a smaller numerator in Newton's Law of Gravitation is paired with a smaller denominator to ensure that the overall magnitudes of the forces balance.

Example 3.7 The Force of Gravity   ⋆⋆
Use Newton's Law of Gravitation to respond to the following:

a. Show that the acceleration due to gravity of some object near the surface of the Earth is given by $g = 9.81\frac{\text{m}}{\text{s}^2}$.

b. What is the gravitational force of attraction between an 80.0 kg person and the Earth (this force is the student's **weight**)? What about between the person and the Sun? How do the values compare?

c. What is the gravitational force of attraction between an 80.0 kg person and a 65.0 kg person if they are standing 2.00 meters from one another?

Note that you can find relevant physical values, such as the mass of the Earth, in Appendix B.

a. We'll begin by denoting the mass of the person as m_p and the mass of the Earth as m_E. If the person is near the surface of the Earth, then the distance between the person and the center of the Earth is equal to the radius of the earth, which we'll denote r_E. Then Newton's Law of Gravitation says

$$\vec{F}_{\text{pE}} = G\frac{m_p m_E}{r_E^2}\hat{r}$$

$$m_p\vec{a} = G\frac{m_p m_E}{r_E^2}\hat{r}$$

$$\vec{a} = G\frac{m_E}{r_E^2}\hat{r}$$

$$\vec{a} = \left(6.67 \times 10^{-11}\frac{\text{m}^3}{\text{kg s}^2}\right)\frac{5.972 \times 10^{24}\text{ kg}}{(6.371 \times 10^6\text{ m})^2}\hat{r}$$

$$\vec{a} = 9.81\frac{\text{m}}{\text{s}^2}\hat{r}$$

where $\hat{r}$ points from the person to the center of the Earth. This is equivalent to the value of g given above (we typically take g to two significant figures instead of three, as given here).

b. The gravitational force of attraction is given by

$$\vec{F}_{\text{pE}} = G\frac{m_p m_E}{r_E^2}\hat{r}$$

We've just shown that on the Earth $g = G\frac{m_E}{r_E^2}\hat{r}$, so we can avoid repeating the algebra and note directly that

$$\vec{F}_{\text{pE}} = m_p g\hat{r}$$

$$\vec{F}_{\text{pE}} = (80.0 \text{ kg})\left(9.81\frac{\text{m}}{\text{s}^2}\right)\hat{r}$$

$$\vec{F}_{\text{pE}} = 785 \text{ N}\hat{r}$$

In the United States most people are more used to thinking of weight in pounds than in Newtons. If you perform the unit conversion (see Appendix B) you find that the 80 kg student weighs about 180 pounds. Importantly, while the person's mass is a constant, their weight depends on g. This is why, for instance, astronauts on the Moon weigh less than on the Earth (see Problem 3.36).

If we're concerned instead with the attraction between the person and the Sun, we need to work with the mass of the Sun (which we'll denote m_S) and the distance between the Earth and the Sun (a good approximation for the distance between a person on Earth and the Sun; we'll denote this as r_{pS}). Appealing again to Newton's Law of Gravitation,

$$\vec{F}_{\text{pS}} = G\frac{m_p m_S}{r_{pS}^2}\hat{r}$$

$$\vec{F}_{\text{pS}} = \left(6.67 \times 10^{-11}\frac{\text{m}^3}{\text{kg s}^2}\right)\frac{(80.0 \text{ kg})\left(1.990 \times 10^{30} \text{ kg}\right)}{\left(1.496 \times 10^{11} \text{ m}\right)^2}\hat{r}$$

$$\vec{F}_{\text{pS}} = 0.474 \text{ N}\hat{r}$$

Comparing our answers, we see that the attraction toward the Sun is tiny compared to the attraction toward the Earth. We should expect a large difference – we are, after all, stuck to the Earth and not the Sun! One way to compare these forces quantitatively is to take their ratio (for instance, if the forces were equivalent, their ratio would be 1):

$$\frac{F_{\text{pE}}}{F_{\text{pS}}} = \frac{785 \text{ N}}{0.474 \text{ N}} \approx 1.7 \times 10^4$$

c. Once again we appeal to Newton's Law of Gravitation:

$$\vec{F}_{12} = G\frac{m_1 m_2}{r^2}\hat{r}$$

$$\vec{F}_{12} = \left(6.67 \times 10^{-11}\frac{\text{m}^3}{\text{kg s}^2}\right)\frac{(80.0 \text{ kg})(65.0 \text{ kg})}{(2.00 \text{ m})^2}\hat{r}$$

$$\vec{F}_{12} = 8.67 \times 10^{-8} \text{ N}\hat{r}$$

And $\vec{F}_{21}$ has the same magnitude (and points in the opposite direction). This is much too small to be detected.[a]

---

[a]If you are ever the recipient of an unwanted statement of affection that invokes Newton's Law of Gravitation, you can correctly state that you are more attracted to the Earth than you are to the person in question.

3.5   Normal Force

If the Earth exerts a gravitational force on you, then why aren't you accelerating to the center of the Earth right now? The obvious answer is that the ground holds you up; to accelerate to the center of the Earth you'd need to somehow phase through the ground, and that simply doesn't happen. How can we formalize this in terms of Newton's Laws? According to Newton's First Law, the fact that you're in equilibrium (that is, you're not accelerating) as you stand on the ground means that there must be no net force acting on you:

$$\vec{F}_{\text{NET}} = 0 \text{ N}$$

But we also know that there is a gravitational force acting on you. Therefore there must be one or more other forces that combine to perfectly balance the force of gravity. In this case there is just one force, which we call the **normal force**:

$$\vec{F}_G + \vec{F}_N = 0 \text{ N}$$
$$\vec{F}_G = -\vec{F}_N$$

The situation is depicted in Figure 3.13 (a). This may strike you as an odd name for a force. The use of the term *normal* is because of its geometrical meaning, which is "perpendicular to" – the normal force is always *perpendicular to a surface*. When you have some fixed surface like the ground, a table, or a ramp, the normal force is what prevents an object from accelerating through the surface: it perfectly balances whatever force is applied straight in to the surface. Thus, if you place a box on an incline plane Figure 3.13(b)), the normal force only balances the *component* of gravity perpendicular to the surface (i.e. $\left(\vec{F}_G\right)_\perp = \vec{F}_N$). We analyze this scenario in more detail in Example 3.9, after first considering a common misconception regarding the normal force in Example 3.8.

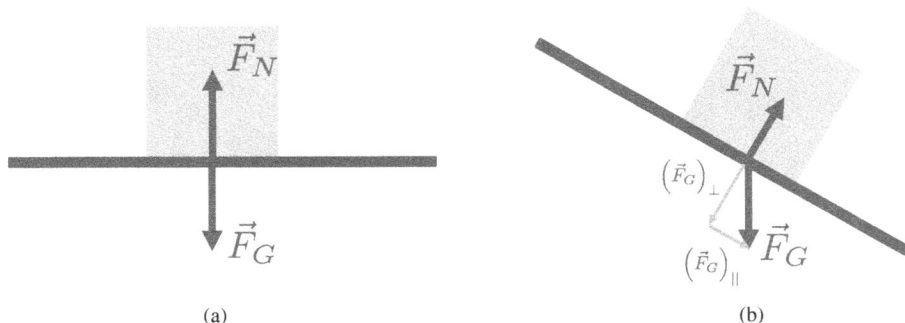

(a)   (b)

FIGURE 3.13
The gravitational and normal forces on a box on (a) a flat surface and (b) an inclined surface.

Example 3.8 Newton's Third Law Pairs ★★

Consider your favorite food sitting on a table (I will assume it has a square profile). Students commonly think of the normal force and the gravitational force as a Newton's Third Law pair, as they're drawn in Figure 3.13(a), but they aren't. (a) Why not? (b) What *are* the forces that form Newton's Third Law pairs with the normal and gravitational forces acting on the food?

Both of these forces act *on the object*, but a Newton's Third Law pair involves *different* objects (e.g. I push on the wall and the wall pushes back on me). Thus the gravitational force *on the object* and the normal force *on the object* cannot possibly be a Newton's Third Law pair, even if they have equal magnitudes and point in opposite directions.

That settles (a). What, then, are the pair forces? Well, the Newton's Third Law partner for the gravitational force of the Earth on the object is the gravitational force of the object on the Earth (Equation (3.4)). Similarly, if the table is pushing up on the object, the object is pushing down on the table. All of these forces have the same magnitude, but the paired forces point in opposite directions. Figure 3.14 shows the forces; I've kept the forces drawn in Figure 3.13 black for ease of comparison, with the newly drawn Third Law pairs drawn in gray.

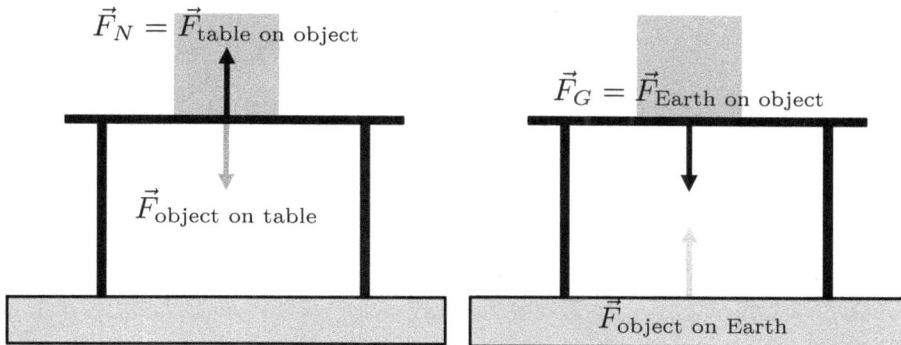

$$\vec{F}_N = \vec{F}_{\text{table on object}}$$

$$\vec{F}_{\text{object on table}}$$

$$\vec{F}_G = \vec{F}_{\text{Earth on object}}$$

$$\vec{F}_{\text{object on Earth}}$$

FIGURE 3.14
Forces.

Example 3.9 Normal Force on a Ramp ★★

Consider a 23 kg box resting on a frictionless ramp inclined at an angle of $\theta = 30°$ above the horizontal (in this example I will assume the angle is exactly 30°). What is the magnitude of the normal force acting on the box? At what rate will it accelerate down the ramp?

This is similar to the situation depicted in Figure 3.13 (b), but now we need to carefully describe the details. As usual we need to first choose a coordinate system. While we've typically chosen the x axis as horizontal and the y axis as vertical, this is an excellent example where our analysis is easier if we make a different choice and set *one axis parallel to the surface and the other perpendicular to the surface*. (We can make *any* choice that we like for our coordinate system, so long as the axes are perpendicular to one another.) Referring to Figure 3.13, we see that

$\vec{F}_N$ is perpendicular to the surface (as it must be), and $\vec{F}_G$ can be split in to a component perpendicular to the surface (and as we argued above, $\left(\vec{F}_G\right)_\perp = -\vec{F}_N$) and a component parallel to the surface.

We know that the box will slide along the surface of the ramp, so directing one of our axes along the surface means we can describe the displacement, velocity, acceleration, and net force on the box in terms of one vector component. We depict the situation in Figure 3.9 (as we've done in the past, we're showing the coordinate system off to the side just for visual clarity).

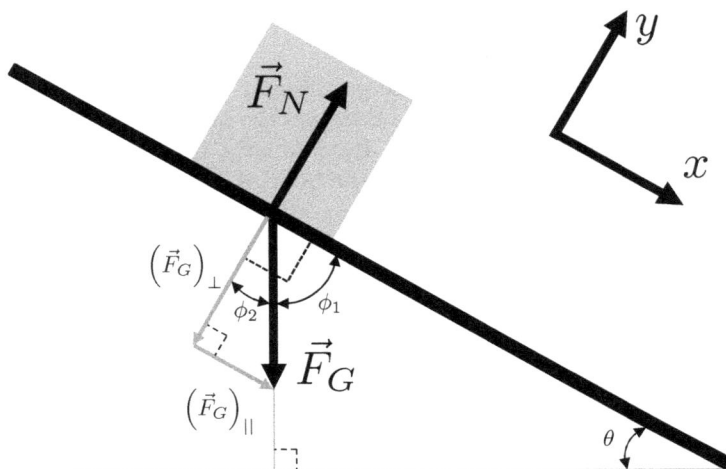

FIGURE 3.15

Note that we've labeled several angles to facilitate our analysis. Considering just the y dimension, we have already noted that the normal force will balance the component of gravity perpendicular to the surface. To be thorough, let's show it methodically:

$$\left(\vec{F}_{\text{NET}}\right)_y = m\vec{a}_y$$

$$\vec{F}_N + \left(\vec{F}_G\right)_\perp = 0 \text{ N}$$

$$\vec{F}_N = -\left(\vec{F}_G\right)_\perp$$

Furthermore, we know from Example 3.7 that the magnitude of $\vec{F}_G$ is

$$mg = (230 \text{ kg})\left(9.8\frac{\text{m}}{\text{s}^2}\right) = 230 \text{ N}$$

Analyzing the geometry of $\vec{F}_G$ and its components, we see

$$\vec{F}_G \cos\phi_2 = \left(\vec{F}_G\right)_\perp$$

If we combine all this, we find

$$F_G \cos\phi_2 = (F_G)_\perp = F_N$$

$$F_G \cos\phi_2 = F_N$$

$$(230 \text{ N}) \cos\phi_2 = F_N$$

We're almost there: if we knew ϕ_2 we could plug it into this last equation and determine a value for F_N. We can determine a value for ϕ_2 by noting that $\theta = 30°$ and analyzing the geometry of the diagram. First, observe the large right triangle formed by θ and ϕ_1. Because the angles inside a triangle must add up to 180°, we know

$$180° = 90° + \theta + \phi_1$$
$$\phi_1 = 90° - \theta$$

Then, note that $\phi_1 + \phi_2 = 90°$, which means $\phi_2 = 90° - \phi_1$. Combining with the above, we find

$$\phi_2 = 90° - (90° - \theta)$$
$$\phi_2 = \theta = 30°$$

This relationship is summarized visually in Figure 3.9. It will come up frequently, and it is a good idea to memorize it. (In fact, we already saw this geometry in Example 2.3 on page 59, where we were talking about the acceleration due to gravity being broken in to components parallel and perpendicular to the surface.)

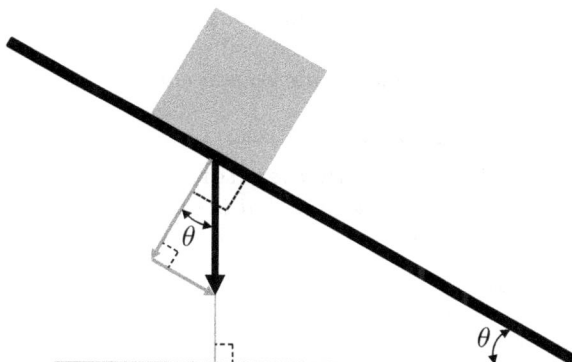

FIGURE 3.16

Now that we know a value for ϕ_2, we can solve for $\vec{F}_N$:

$$(230 \text{ N}) \cos \phi_2 = F_N$$
$$(230 \text{ N}) \cos (30°) = F_N$$
$$2.0 \times 10^2 \text{ N} = F_N$$

where the direction of the vector is in the positive y direction.

The second part of the question asks us to determine the acceleration of the box. Because $\vec{F}_N$ perfectly balances the component of the gravitational force perpendicular to the surface, the only force that remains to actually accelerate the box is the component of the gravitational force parallel to the surface. If we look back to the geometry of Figure 3.9, we find

$$\vec{F}_G \sin \phi_2 = (F_G)_{\|}$$
$$(230 \text{ N}) \sin (30°) = (F_G)_{\|}$$
$$110 \text{ N} = (F_G)_{\|}$$

We can now use Newton's Second Law along the surface of the ramp:

$$\left(\vec{F}_{\text{NET}}\right)_x = m\vec{a}_x$$

$$\left(\vec{F}_G\right)_\parallel = m\vec{a}_x$$

$$110 \text{ N}\hat{x} = (23 \text{ kg})\,\vec{a}_x$$

$$4.9\frac{\text{m}}{\text{s}^2} = \vec{a}_x$$

As we discussed in Example 3.7, the weight of an object of mass m is given by mg, where g is the acceleration due to gravity. However, our *sensation* of weight is actually determined by the normal force, and the normal force determines what a scale reads (the normal force and the force on the scale form a Third Law pair, as in Example 3.8). A skydiver or anyone else falling a long distance (by jumping off of a tall diving board into a pool, for example) feels weightless, though of course gravity is still acting on them. This is why the normal force is sometimes referred to as the **apparent weight**. Conversely, if you bend your knees and then jump into the air, during the push off you temporarily feel heavier than when you are standing still. We analyze the related phenomenon of an elevator ride in the following example.

Example 3.10 Normal Force in an Elevator   ⋆⋆
A person of mass m stands on an elevator that is initially at rest on the ground floor. Consider the following phases of an elevator ride:

a. The elevator is at rest.

b. The elevator accelerates up at a rate a until it reaches a cruising speed v.

c. The elevator maintains this speed for most of the trip.

d. As the elevator approaches the top floor of the building, it comes to rest; the magnitude of the acceleration is a.

What is the apparent weight of the passenger for each phase of the trip? When are they in static or translational equilibrium?

Let's choose a coordinate system where up is the positive direction (we'll call it the y direction, though we could just as well call it x or anything else we please) and write down Newton's Second Law. Initially, the person is standing at rest and

$$\sum \vec{F} = m\vec{a}$$
$$(F_N - mg)\,\hat{y} = 0$$
$$F_N = mg$$

Thus for (a) the passenger's apparent weight is just their weight, mg. The situation is more complicated in (b)–(d); we sketch these phases of the journey in Figure 3.17.

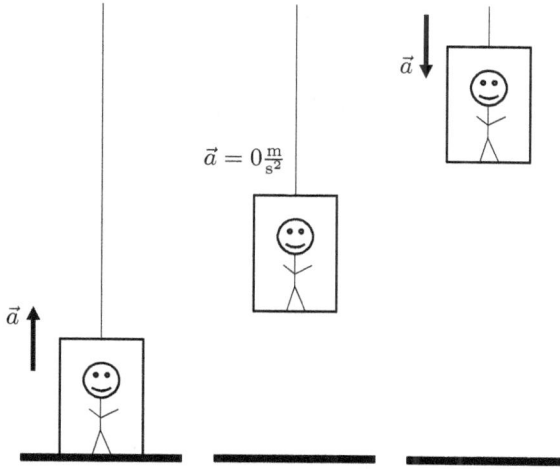

FIGURE 3.17

If the reasoning for the directions of the acceleration vectors is unclear, try drawing the velocity vectors. During part (b), for instance, the velocity is initially 0 m/s and ends up with some upward value v. Thus the change in the velocity $\Delta \vec{v} = \vec{v}_f - \vec{v}_i$ also points up (and because $\vec{a} = \Delta \vec{v}/\Delta t$ and time is a scalar, acceleration points in the same direction as $\Delta \vec{v}$). During each stage of the journey, there are only two forces acting on the passenger: the force of gravity (down) and the normal force (up). Again, we turn to Newton's Second Law:

$$\sum \vec{F} = m\vec{a}$$
$$\vec{F}_N + \vec{F}_G = m\vec{a}$$
$$(F_N - F_G)\,\hat{y} = m\vec{a}$$
$$(F_N - mg)\,\hat{y} = m\vec{a}$$

This equation holds throughout the journey. When we consider a particular phase of the journey, we can insert what we know of the acceleration. For (b), we have a positive upward acceleration, so

$$(F_N - mg) = ma$$
$$F_N = m\,(a + g)$$

Thus the apparent weight, which is mg if we're not accelerating at all, is now $m\,(g + a)$. Apparently the effect of accelerating upwards is to increase our sensation of the acceleration due to gravity, and we feel heavier.

During the middle portion of the journey, the acceleration is $0\frac{\text{m}}{\text{s}^2}$ and so

$$(F_N - mg) = ma\hat{y}$$
$$(F_N - mg) = 0\frac{\text{m}}{\text{s}^2}$$
$$F_N = mg$$

In other words, if we're moving up at a constant velocity, we feel just as heavy as if we're not moving at all.

Finally, during the last portion of the journey we have a negative acceleration, so

$$(F_N - mg) = m(-a)$$
$$F_N = m(g - a)$$

Thus the apparent weight *decreases* as if the acceleration due to gravity has decreased by a.

Finally, let's consider the question regarding equilibrium. The passenger will be in static equilibrium in (a): the net force is 0 N and their velocity is 0 m/s. They'll be in translational equilibrium (or if you prefer, dynamic equilibrium since there is no rotation anywhere in this situation) in (c): the net force is 0 N but $v \neq 0$ m/s. (Both of these statements boil down to a Newton's First Law argument: there is no net force and therefore the velocity doesn't change.) In (b) and (d) the net force is not 0 N and therefore the passenger is not in equilibrium.

It is worth emphasizing that if the person were standing on a scale in the elevator, the reading would change during the ride according to these equations for the normal force! Indeed, if you've ever taken an elevator ride up a building, you've likely experienced the sensation of being pushed against the floor as you accelerate up, then feeling lighter than normal as the elevator slows to a stop. You may also have some intuition about what occurs if you take an elevator ride down, instead of up, which you're asked to analyze in Problem 3.45.

3.6   Friction

Now that we've discussed the normal force we're in a position to discuss friction, which we briefly described at the beginning of this chapter. While the normal force is *perpendicular* to a surface (so, for instance, objects don't somehow fall through a table due to the force of gravity), friction is *parallel* to a surface. Consider the situation shown in Figure 3.18, where someone is pushing on a heavy box with an increasing amount of force.

Based on your own experiences you probably recognize that for a very light applied force $\vec{F}_{\mathrm{BP}}$, the box won't accelerate at all. This is because of **static friction**, denoted by $\vec{F}_s$, which will exactly match the applied force (parallel to the surface) up to a maximum value with a magnitude given by

$$(F_s)_{\mathrm{max}} = \mu_s F_N \tag{3.8}$$

In Equation (3.8), μ_s is referred to as the **coefficient of static friction**; it accounts for the "roughness" between the two surfaces (this is how we describe the different behavior of grass, packed dirt, and so on in the example that opened the chapter).[10] Now, if you apply a force greater than $(F_s)_{\mathrm{max}}$, the box will begin to accelerate and experience **kinetic friction**, which has a magnitude of

$$F_k = \mu_k F_N \tag{3.9}$$

---

[10] Note that the coefficient of static friction must be unitless if both sides of the equation are to have the same units.

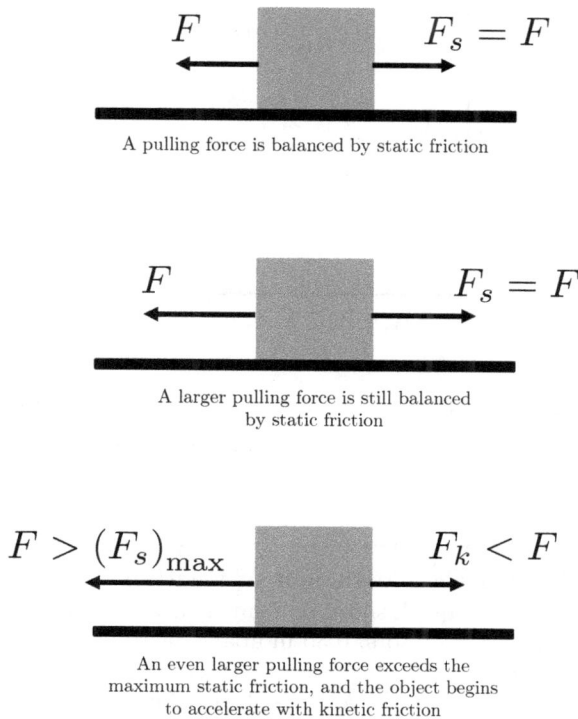

A pulling force is balanced by static friction

$$F \qquad F_s = F$$

A larger pulling force is still balanced
by static friction

$$F > (F_s)_{max} \qquad F_k < F$$

An even larger pulling force exceeds the
maximum static friction, and the object begins
to accelerate with kinetic friction

FIGURE 3.18

On a frictional surface, static friction increases to match the applied horizontal force up until a breaking point given by Equation (3.8).

where μ_k is the **coefficient of kinetic friction** (and generally $\mu_k <$ μ_s; typically it is easier to keep something moving than it is to get it moving in the first place). The relationship between the applied force and the frictional force is shown in Figure 3.19. In both cases we see that friction depends only on the normal force and a quantity that describes how the two surfaces interact: for in-

TABLE 3.1

Coefficients of Friction

Materials	μ_s	μ_k
Rubber and concrete (dry)	0.90	0.70
Rubber and concrete (wet)	0.70	0.50
Wood and wood	0.35	0.25
Metal and metal (dry)	0.70	0.50
Metal and metal (lubricated)	0.15	0.050
Waxed wood and snow	0.15	0.10

stance, a tire on a wet road has a smaller coefficient of friction than on a dry road. Table 3.1 provides typical coefficients of friction.[11] The fact that friction (which is parallel to a surface) depends on the normal force (which is perpendicular to a surface) links our analysis of forces along perpendicular axes, as the next set of examples demonstrates.

---

[11]The values in Table 3.1 are safe for use in the examples and problems in this book, but you should be aware that these are representative values only. When we talk about wood rubbing against wood, for instance, it matters what *type* of wood we're talking about, if they've been sanded (and to what grit), if the grains are aligned or not, the ambient temperature, how humid it is, etc.

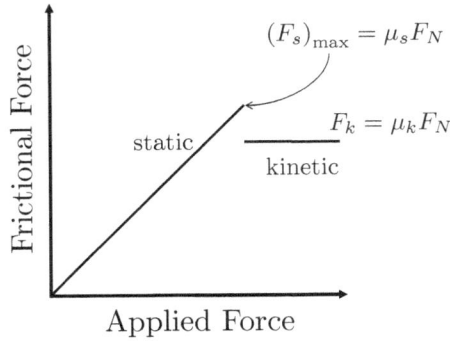

FIGURE 3.19
The relationship between the applied force parallel to a surface and the frictional force. We see here that once the applied force exceeds the maximum static frictional force, the object will begin to accelerate.

Example 3.11 Finding the Coefficient of Friction   ⋆

In a physics lab, you give a quick shove to a small cart and observe as it travels along a level track. You have two sensors, 0.30 m apart; each measures the velocity of the cart as it passes the sensor. The first sensor measures the cart moving at 0.80 m/s and the second measures 0.70 m/s. What is the coefficient of kinetic friction between the cart and the track?

The force diagram is shown in Figure 3.20. We set *up* to be the $\hat{y}$ direction and *right* to be the $\hat{x}$ direction, as usual. In the y direction, Newton's Second Law says

$$\sum \vec{F}_y = m\vec{a}_y$$
$$F_N - F_G = 0$$
$$F_N = mg$$

Meanwhile, in the x direction:

$$\sum \vec{F}_x = m\vec{a}_x$$
$$-F_f = ma_x$$
$$-\mu_k F_N = ma_x$$

where in the last line we've substituted in the kinetic frictional force (Equation (3.9)). We'll next eliminate F_N by using the result of our analysis of the forces in the y direction:

$$-\mu_k mg = ma_x$$
$$-\mu_k g = a_x$$
$$\mu_k = -\frac{a_x}{g}$$

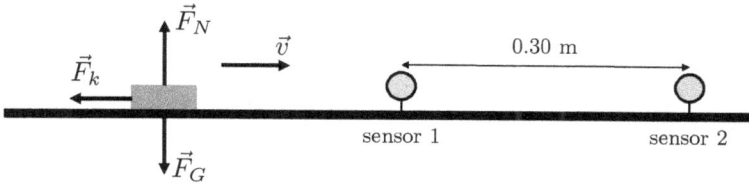

FIGURE 3.20

This gives us the kinetic coefficient of friction in terms of the acceleration and g. To find the acceleration, we can apply kinematics:

$$v_f^2 = v_i^2 + 2a_x\Delta x$$

$$\frac{v_f^2 - v_i^2}{2\Delta x} = a_x$$

$$\frac{(0.70 \text{ m/s})^2 - (0.80 \text{ m/s})^2}{2\,(0.30 \text{ m})} = a_x = -0.25 \text{ m/s}^2$$

Therefore, the coefficient of kinetic friction is

$$\mu_k = -\frac{-0.25 \text{ m/s}^2}{9.8 \text{ m/s}^2} = 0.026$$

Example 3.12 Motion with Friction ★★

An 8.0 kg box rests on level ground where $\mu_s = 0.35$ and $\mu_k = 0.20$. A force of 5.0 N is directed downward on the box, at an angle of 45° below the horizontal. Choose the x coordinate to be directed to the right, parallel to the ground. Describe the acceleration of the box if the box's velocity is (a) $\vec{v} = 3.4\hat{x}$ m/s, (b) $-\vec{v} = 3.4\hat{x}$ m/s, and (c) $\vec{v} = 0$ m/s.

We choose the y component to be vertically up. We'll begin drawing the force diagram for each of the situations; it is useful to compare them. For (a) and (b) we have kinetic friction that points in the opposite direction of the velocity. In (c) we have static friction that opposes the direction of the net force along the surface.

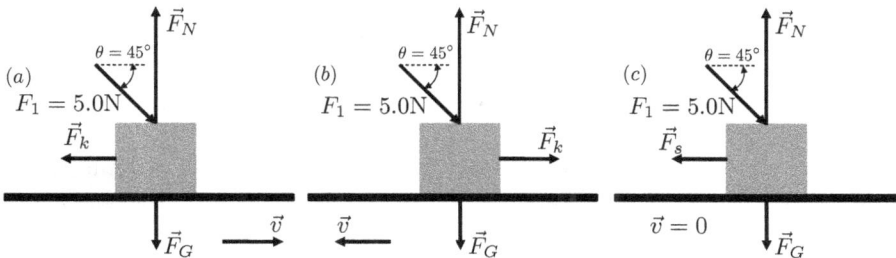

FIGURE 3.21

Refer back to equations (3.8–3.9) and note that friction (a force that is parallel to the surface) depends on the normal force (a force that is perpendicular to the surface).

Thus in this case we need to analyze the forces in the y dimension to determine the normal force before we can analyze the forces in the x dimension (this is an important point and worth some reflection).

In fact, the vertical forces are the same for each of parts (a)–(c). Writing out Newton's Second Law, we find

$$\left(\vec{F}_{\text{NET}}\right)_y = m\vec{a}_y$$

$$F_N - F_1 \sin\theta - mg = m\left(0\frac{\text{m}}{\text{s}^2}\right)$$

$$F_N = F_1 \sin\theta + mg$$

$$F_N = (5.0 \text{ N})\sin(45°) + (8.0 \text{ kg})\left(9.8\frac{\text{m}}{\text{s}^2}\right)$$

$$F_N = 82 \text{ N}$$

Now let us consider the horizontal forces for each part. In (a) we have

$$\left(\vec{F}_{\text{NET}}\right)_x = m\vec{a}_x$$

$$F_1 \cos(45°) - F_k = ma_x$$

$$F_1 \cos(45°) - \mu_k F_N = ma_x$$

$$\frac{1}{m}\left(F_1 \cos(45°) - \mu_k F_N\right) = a_x$$

$$\frac{1}{8.0 \text{ kg}}\left((5.0 \text{ N})\cos(45°) - (0.20)(82 \text{ N})\right) = a_x$$

$$-1.6\frac{\text{m}}{\text{s}^2} = a_x$$

We find a negative acceleration, indicating that the force of friction (to the left, in the negative x direction) is larger than the horizontal component of the applied force (which is to the right).

In (b), we have

$$\left(\vec{F}_{\text{NET}}\right)_x = m\vec{a}_x$$

$$F_1 \cos(45°) + F_k = ma_x$$

$$F_1 \cos(45°) + \mu_k F_N = ma_x$$

$$\frac{1}{m}\left(F_1 \cos(45°) + \mu_k F_N\right) = a_x$$

$$\frac{1}{8.0 \text{ kg}}\left((5.0 \text{ N})\cos(45°) + (0.2)(82\text{N})\right) = a_x$$

$$2.5\frac{\text{m}}{\text{s}^2} = a_x$$

The positive acceleration should make sense: looking at the force diagram, it is clear that all of the horizontal forces (friction and the horizontal piece of the applied force) are in the positive direction.

Finally, in (c) the situation is slightly different because we don't explicitly know if the applied force is larger than the maximum value of static friction. This is crucial because if the applied force is less than the maximum possible value of static friction, then friction perfectly matches the applied force and the net force (in the x dimension) is 0 N. To make the comparison, let's first compute the horizontal component of the applied force explicitly:

$$(F_1)_x = F_1 \cos\theta$$
$$(F_1)_x = (5.0 \text{ N}) \cos 45°$$
$$(F_1)_x = 3.5 \text{ N} = F_s$$

In the last line I've noted that the x component of the applied force has the same magnitude as the static friction force: we're still in the "static" regime where static friction can perfectly match the applied horizontal forces to keep the object in static equilibrium.

We can show that this is the case formally by considering the maximum value of static friction given by Equation (3.8):

$$(F_s)_{\max} = \mu_s F_N$$
$$(F_s)_{\max} = (0.35)(82 \text{ N})$$
$$(F_s)_{\max} = 29 \text{ N}$$

Clearly $(F_s)_{\max} > F_s = (F_1)_x$, so static friction can match the applied force and the acceleration in this case is $0\frac{\text{m}}{\text{s}^2}$.

Example 3.13 Critical Angle of a Ramp   ★★
A book of mass m rests on a horizontal plank of wood. One end of the wood is slowly raised such that it forms an angle θ with the horizontal. If the coefficients of static and kinetic friction between the book and plank are given by μ_s and μ_k, respectively, then at what angle will the book begin to slide?

This is similar to the situation shown in Figure 3.13 and discussed in Example 3.9, though now we have friction involved. The force diagram is shown in Figure 3.22.

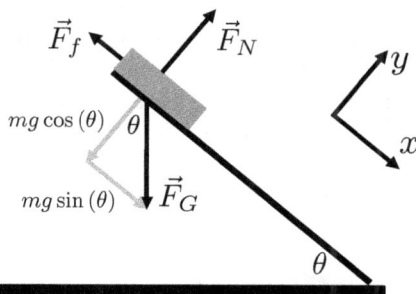

FIGURE 3.22

Note that as before we've chosen our coordinate system so the x coordinate is parallel to the surface. At some values of θ the friction will be static (the book won't move), while for others it will be kinetic (the book will be sliding). Note also that I've jumped straight to the conclusion that the angle between the board and the horizontal, θ, is the same as between mg and the component of mg perpendicular to the surface (refer back to Example 3.9 if this is unclear).

Newton's Second Law in the y dimension states $F_N = mg \cos \theta$, which you may be able to see directly from the force diagram (there are two opposing forces and we know the normal force will balance any forces that point perpendicular into the surface of the board). In the x dimension,

$$\left(\vec{F}_{\text{NET}} \right)_x = m\vec{a}_x$$

$$-\vec{F}_f + mg \sin \theta = m\vec{a}_x$$

The maximum value of static friction is $\mu_s F_N$, and by definition if the friction is static then the acceleration is $0 \frac{m}{s^2}$. Importantly, once static friction is at a maximum, the corresponding value of θ is the maximum angle before static friction gives way to kinetic friction and the book starts to slide. Thus we can plug $\mu_s F_N$ in for friction and $0 \frac{m}{s^2}$ in for the acceleration in the above equation to determine an expression for $\theta_{\max}$:

$$-\mu_s F_N + mg \sin \theta_{\max} = 0 \text{ N}$$

$$-\mu_s \left(mg \cos \theta_{\max} \right) + mg \sin \theta_{\max} = 0 \text{ N}$$

$$-\mu_s \left(\cos \theta_{\max} \right) + \sin \theta_{\max} = 0 \text{ N}$$

$$\sin \theta_{\max} = \mu_s \left(\cos \theta_{\max} \right)$$

$$\frac{\sin \theta_{\max}}{\cos \theta_{\max}} = \mu_s$$

$$\tan \theta_{\max} = \mu_s$$

$$\theta_{\max} = \tan^{-1} \left(\mu_s \right)$$

Note that we've made use of the relationship $\tan \theta = \frac{\sin \theta}{\cos \theta}$ in the above. Evidently the critical angle depends only on the value of μ_s. You may be inclined to think that mg should appear in our answer, but (as we see in the second line of the above equations) the fact that an increased weight increases the normal force (and therefore the frictional force) is perfectly balanced by the fact that an increased weight *also* increases the component of the gravitational force that is parallel to the surface.

At $\theta_{\max}$, the book will begin to accelerate down the board with the slightest tap; above $\theta_{\max}$ it will accelerate on its own. You can test this for yourself, if you like, by getting two similar objects (books with the same texture, for example) with different weights. Place them on a board and slowly lift one end: they will start to slide at the same time.

3.7   Air Resistance and Drag

We are intuitively familiar with air resistance; bicyclists experience the feeling of the wind against their face, and anyone who has ridden in a car at high speeds recognizes the sound of the air whistling by. As we discussed in Chapter 1, air resistance comes into play in cases of simple free fall, though it is complicated enough that (as we first became familiar with ideas and methods of kinematics) we typically neglected it entirely.

Now, however, we have the tools necessary to discuss air resistance in greater detail. Air resistance is an example of a **drag force**, which encompasses resistive forces that arise from moving through fluids (such as air and water). For this reason we'll use the symbol $\vec{F}_d$ to represent **air resistance**, which, like all drag forces, *depends on the velocity of the object.* Specifically, for everyday objects it is given by

$$\vec{F}_d = -bv^2\hat{r} \tag{3.10}$$

where $\hat{r}$ is a unit vector pointing in the direction of the object's motion and b is a constant that depends on the object and the properties of the fluid. We see from Equation (3.10) that an object that is completely motionless experiences no drag, but as its speed increases, the drag force increases rapidly.

As an example, consider a ball dropped from a tower (Figure 3.23). Initially the velocity is 0 m/s and therefore the only force acting on the ball is the gravitational force; its initial acceleration is g. As it accelerates, however, the drag force increases, and eventually the drag forces completely balances the gravitational force:

$$\vec{F}_{\text{NET}} = m\vec{a}$$
$$\vec{F}_g + \vec{F}_d = 0 \text{ N}$$
$$mg - bv_t^2 = 0 \text{ N}$$
$$mg = bv_t^2$$

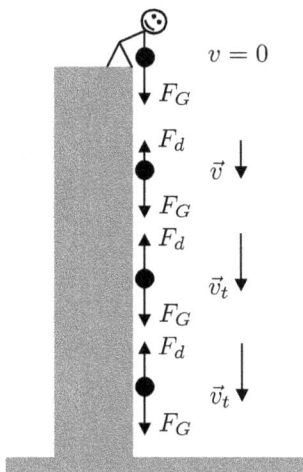

FIGURE 3.23
A ball dropped under the influence of air resistance experiences an increasing drag force as the velocity increases, until the drag force balances the gravitational force.

Or, finally,

$$v_t = \left(\frac{mg}{b}\right)^{1/2} \tag{3.11}$$

We've labeled the speed here as v_t because it is the **terminal velocity**; when the forces balance in this way the object stops accelerating and so the drag force stops increasing (thus once the ball reaches its terminal velocity, it is in translational equilibrium). It is worth reflecting on the fact that from the instant the ball is released up until it reaches its terminal velocity, the acceleration is *changing* (from g down to the ground to $0\frac{m}{s^2}$). This means that until the ball reaches its terminal velocity, we cannot analyze the motion with the kinematic equations, because they were derived under the assumption of *constant* acceleration. Figure 3.24 shows the height, velocity, and acceleration of an object falling both *with* and *without* drag.

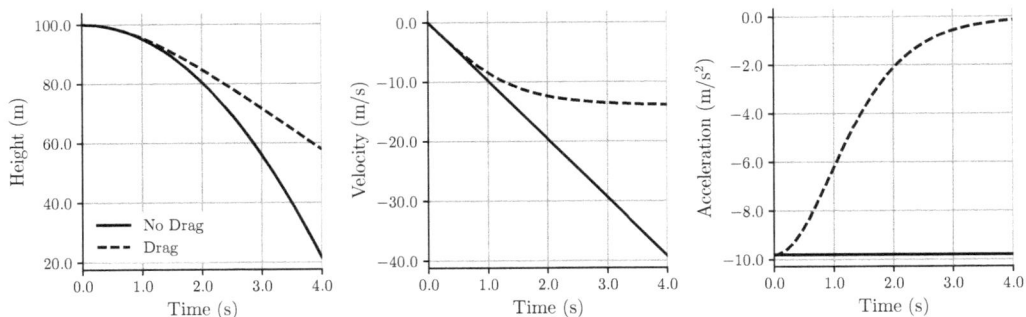

FIGURE 3.24
Motion graphs for an object dropped from rest from an initial height of 100.0 m. If we neglect air resistance, the ball drops with an acceleration g directed down to the ground: the position graph is a downward (half) parabola, the velocity graph is a straight line with a slope $-g$, and the acceleration is a constant $-g$. If we include air resistance, the resistive effect of air resistance means that for any specific time, the object won't have fallen as far as it would have if air resistance was neglected. Similarly, the velocity approaches a terminal velocity, which corresponds to the acceleration becoming 0, as the drag force eventually grows to balance the gravitational force.

Example 3.14 Skydiver's Terminal Velocity   ⋆
The terminal velocity of a particular 80.0 kg skydiver (with their body in a specific orientation) is 50.0 m/s. What is the corresponding value of b (Equation 3.11) for this skydiver? How long will it take a skydiver moving at this speed to fall 1450 feet, the height of Willis Tower in Chicago, IL?

Directly from Equation (3.11) we find

$$v_t = \left(\frac{mg}{b}\right)^{1/2}$$

$$v_t^2 = \frac{mg}{b}$$

$$b = \frac{mg}{v_t^2}$$

$$b = \frac{(80.0\text{kg})\left(9.8\frac{m}{s^2}\right)}{\left(50.0\frac{m}{s^2}\right)^2}$$

$$b = 0.314\left(\text{kg}\frac{m}{s^2}\frac{s^2}{m^2}\right) = 0.314\frac{\text{kg}}{m}$$

where in the last line we've explicitly shown the algebra involving unit simplification (since we didn't explicitly mention the units on b above). Note that

$$1450\text{ft}\left(\frac{0.3048\text{m}}{1\text{ft}}\right) = 442 \text{ m}$$

and recall that if there is no acceleration, the speed, time, and displacement are related by

$$v = \frac{\Delta x}{\Delta t}$$

$$\Delta t = \frac{\Delta x}{v}$$

$$\Delta t = \frac{442 \text{ m}}{50.0\frac{\text{m}}{\text{s}^2}}$$

$$\Delta t = 8.84 \text{ s}$$

3.8   Tension and Pulleys

We considered the normal force in terms of an object being supported by whatever it is resting upon (such as the ground or a table). What if an object is not supported from beneath, but instead from above? A sign hanging from the ceiling in a store is a common example of this. In this case the force that balances the gravitational force is a **tension force**, which exists in any taut rope, string, chain, etc. We show the forces in this scenario in Figure 3.25. If the sign (of mass m) is hanging in equilibrium, then by Newton's Second Law $F_T = mg$. Importantly, a taut rope exerts an equal and opposite force on both ends: the ceiling feels a downward force equal to the upward force that suspends the sign in the first place.[12] Thus the ceiling experiences a downward force equal to mg, which you may have suspected was the case intuitively. Example 3.15 considers a related situation involving multiple ropes and nonzero acceleration.

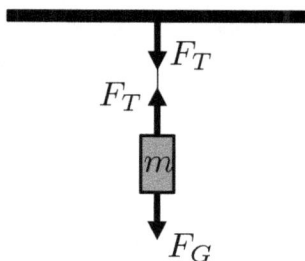

FIGURE 3.25
An object suspended from a ceiling by a rope. The gravitational force is balanced by a tension force in the rope, which is also felt by the ceiling.

---

[12]We are here making the common assumption of a *massless* rope; things get a bit more complicated when accounting for mass.

Example 3.15 Elevator Spring Scale   ★★

A **spring scale** is attached to the ceiling of an elevator (a spring scale measures the tension in the rope connected to it). Attached to the scale's rope is a block with a mass $m_1 = 15.0$ kg. A second block with a mass $m_2 = 5.0$ kg is hanging from the bottom of the first mass by way of a second rope. The elevator begins to accelerate up at a rate of $a = 1.5$ m/s^2. What does the spring scale read (a) before and (b) during the acceleration?

As always, a clear diagram is crucially important; we show the force diagram Figure 3.26 (a). Before we get to writing down Newton's Second Law for this situation, let's make a few general observations:

- Because we have two ropes, we need to consider two different tensions.

- The tension in each rope will be the same (and point inward from either end of the rope) so in the force diagram we draw the two $(F_T)_1$ vectors with the same length. Similarly, we draw the two $(F_T)_2$ vectors with the same length.

- When the elevator is accelerating up, the net force on each mass must be up. Thus we ensure that the net upward force vector on each mass is longer than the net downward force.

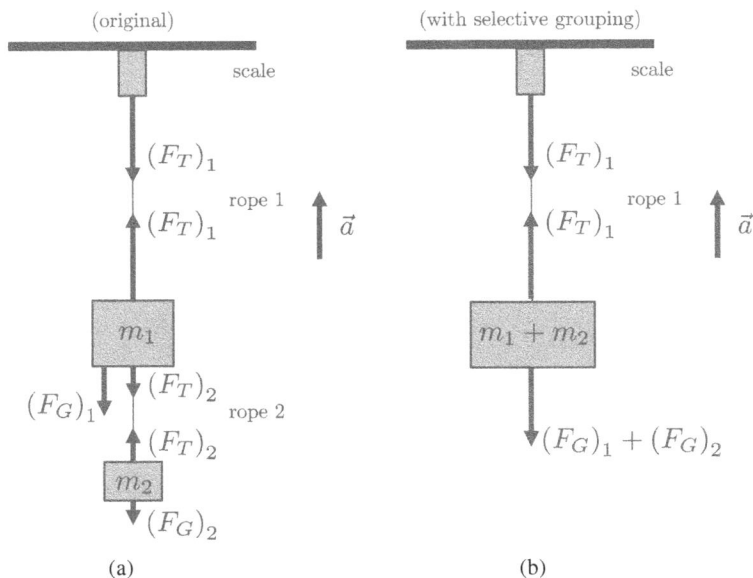

FIGURE 3.26

Now to set up the details. The lower mass is simpler, so let's start there:

$$\sum \vec{F}_2 = m_2 \vec{a}$$
$$(F_T)_2 - (F_G)_2 = m_2 a$$

For the upper mass we have an additional tension force:

$$\sum \vec{F}_1 = m_1 \vec{a}$$
$$(F_T)_1 - (F_G)_1 - (F_T)_2 = m_1 a$$

Ultimately we're after $(F_T)_1$, but we don't know $(F_T)_2$. We can eliminate the second tension from our second result by adding it to our first result:

$$((F_T)_2 - (F_G)_2) + ((F_T)_1 - (F_G)_1 - (F_T)_2) = m_2 a + m_1 a$$
$$- (F_G)_2 + (F_T)_1 - (F_G)_1 = m_2 a + m_1 a$$
$$(F_T)_1 = (m_2 + m_1) a + (F_G)_1 + (F_G)_2$$
$$(F_T)_1 = (m_2 + m_1)(a + g)$$

where in the last step I noted $F_G = mg$ for each mass, as usual. Thus if $a = 0$, as in (a), we have

$$(F_T)_1 = (20.0 \text{ kg})\left(9.8 \text{ m/s}^2\right) = 196 \text{ N}$$

and if $a = 1.5 \text{ m/s}^2$, as in (b), we have

$$(F_T)_1 = (20.0 \text{ kg})\left((9.8 + 1.5) \text{ m/s}^2\right) = 226 \text{ N}$$

I've gone through the detailed mathematics here, but it is worth mentioning that this is an excellent opportunity for use of selective grouping (Example 3.5): if we combine the two masses, then the internal forces from rope 2 ($(F_T)_2$) add to zero and we're left with a single mass $m = m_1 + m_2 = 20.0$ kg and the tension force $(F_T)_1$. (We see this explicitly above: when we added the Newton's Second Law equation separately for mass m_1 and mass m_2, the tension in the second rope canceled.) This gives

$$\sum \vec{F} = m\vec{a}$$
$$(F_T)_1 - (F_G) = ma$$
$$(F_T)_1 = (m_1 + m_2) g + (m_1 + m_2) a$$
$$(F_T)_1 = (m_1 + m_2)(a + g)$$

just as we found more laboriously above. The corresponding force diagram shown in Figure 3.26 (b).

Ropes are commonly used in conjunction with pulleys, which effectively serve to redirect forces.[13] Consider the **Atwood's machine**[14] shown in Figure 3.27. The heavier mass will fall while the lighter will rise, but the value of the acceleration depends on both masses, as we explore in the following example.

---

[13]For now we will assume that pulleys are *massless* and *frictionless* (or in practical language, we will assume they are light enough that it is reasonable to treat them as massless, and we will assume that friction is small enough to be negligible). To properly analyze pulleys with mass, we need to understand *torque*, which we'll get to in Section 3.13.

[14]After George Atwood (1745-1807).

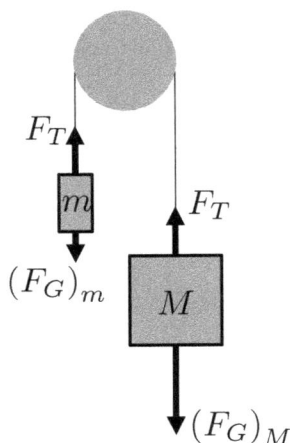

FIGURE 3.27
In an Atwood's machine, a light rope is wrapped over a pulley that is free to rotate. The rope is attached to a hanging mass on either end.

Example 3.16 Acceleration of an Atwood's Machine ★★
Determine an expression for the acceleration of the blocks shown in Figure 3.27 involving only m, M, and g.

We can write out Newton's Second Law for both masses; both equations will include F_T, and we can use the two equations to arrive at one combined equation that does not include F_T. However, there is a subtlety here that requires careful attention. Thus far we have always used one coordinate system for any problem. But because we have two objects here, we are free to use a *different coordinate system for each object*. This might seem needlessly complicated, but consider that the two blocks are locked together, so either "m goes up and M goes down" or "m goes down and M goes up". If we choose "m up, M down" to be positive, then our diagram and coordinate systems are shown in Figure 3.28.

If we apply Newton's Second Law to each block, we find

$$\vec{F}_{\text{NET}} = m\vec{a} \qquad \qquad \vec{F}_{\text{NET}} = M\vec{a}$$
$$F_T - mg = ma \qquad \qquad -F_T + Mg = Ma$$

We could solve one equation for F_T and plug it into the other equation, then simplify. However, here again we can add the equations. Like in Example 3.15, we simply add the left hand side of the equations separately from the right hand side. (This is similar to saying $3 - 2 = 1$ and $2 + 3 = 5$, so $(3 - 2) + (2 + 3) = (1 + 5) = 6$.) Applied to our equations, we find

$$(F_T - mg) + (-F_T + Mg) = (ma) + (Ma)$$
$$-mg + Mg = (m + M)a$$
$$\frac{M - m}{M + m}g = a$$

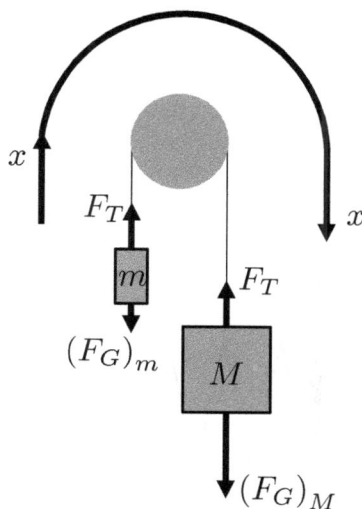

FIGURE 3.28
Coordinate systems.

Note that adding the equations directly leads to the F_T terms canceling, since it is positive in one equation and negative in the other. Let us take a moment to analyze our result. If $m = 0$, we find $a = g$, which is to say mass M would be in free fall. (Indeed, if you are analyzing an Atwood's machine and predict $a > g$, you've done something wrong; the acceleration is gravity limited.) Conversely, if $M = 0$, $a = -g$, meaning mass m is in free fall (the negative sign arises by virtue of the fact that on the left side of the pulley, up is positive). Finally, if $m = M$, $a = 0\frac{m}{s^2}$, indicating the masses are in balance.

Canceling F_T in this way is more than just a neat mathematical trick: the fact that we can do so says something interesting about the physics. If two objects are moving together, as in this example, their interaction forces cancel. This is the principle behind selective grouping, which we introduced in Example 3.5 on page 112. If we follow that example and treat the two masses as one combined object, we have to carefully note that the two external forces (with magnitudes mg and Mg) pull the system in opposite directions, even though they both point down (mg is in the "m goes down, M goes up" direction, while Mg is in the "m goes up, M goes down" direction). Thus if we call the "m goes up, M goes down" direction positive, we could jump immediately to

$$Mg - mg = (m + M)\,a \Rightarrow \frac{M - m}{M + m}g = a$$

as we determined above.

If, on the other hand, we always called up the positive direction, then things would become confused when we combined the two objects. Thus, while it is worth stressing that you *could* solve this problem by always calling, say, *up the page* the positive direction,[a] when objects are connected by pulleys it is often convenient to call one direction of their combined motion positive and the other negative, to straightforwardly apply selective grouping. We will sometimes refer to this framework as a *curved coordinate system.*

---

[a]You are asked to do so in Problem 3.84.

3.9   Hooke's Law and Young's Modulus

Stretchable materials like rubber bands and rigid springs have a preferred length, and when the object is stretched and/or compressed, they display a restoring force that attempts to return the object to its preferred length (Figure 3.29).[15]

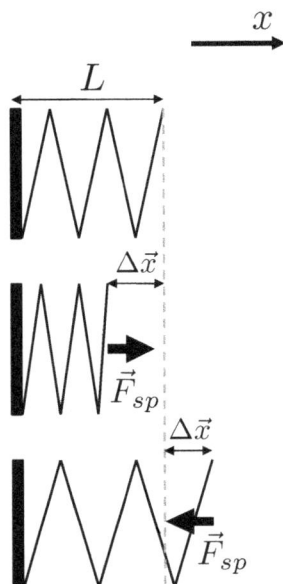

FIGURE 3.29
A stiff spring displays a restoring force if it is compressed or stretched.

If we define the unit vector $\hat{x}$ to point in the direction of the object's displacement from equilibrium, then Hooke's Law[16] describes the restoring force as

$$\vec{F}_{\text{sp}} = -k\left(\Delta x\right)\hat{x} \tag{3.12}$$

where k is the spring constant of the material measured in N/m (a stiffer spring has a larger value of k: a larger force is required to stretch it or compress it a given amount). Importantly Δx is the spring's *displacement from equilibrium*, not the total length of the spring or the distance of the spring's end from some other reference point. Fig 3.30 shows this relationship graphically.

It is important to note that not all objects obey Hooke's Law at all displacements; clearly if you stretch something far enough the material will warp and/or break! Thus, despite the name, Hooke's Law isn't a *law* in the same sense as, say, Newton's Law of Gravitation. That being said, Hooke's Law accurately describes the behavior of many materials for reasonably small displacements, and in this book we will typically assume that it is valid.

---

[15]Some objects, like a rubber band, will only exhibit a restoring force if it is stretched; it you compress it, it will tend to crumple into a ball. This is important to keep in mind but here we're talking about the general case where the restoring force comes up when the material is compressed just as when it is stretched (like with a thick rubber rod).

[16]After Robert Hooke (1635–1703), a contemporary and rival of Newton.

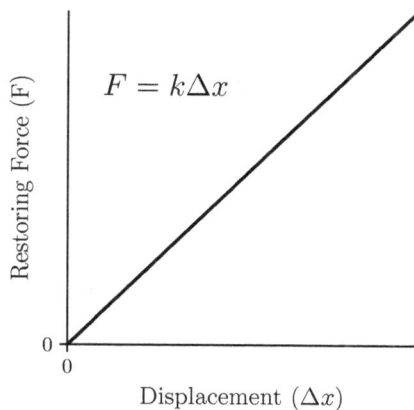

FIGURE 3.30
The magnitude of a Hooke's Law force is linear with the displacement; the slope of the curve is given by the spring constant k.

Example 3.17 Stretching a Spring ⋆
An object is hung vertically from a spring with $k = 75$ N/m. At equilibrium, the spring is stretched by 15 cm. What is the object's mass?

We'll choose down to be the positive direction. The "before and after" is sketched in Figure 3.31.

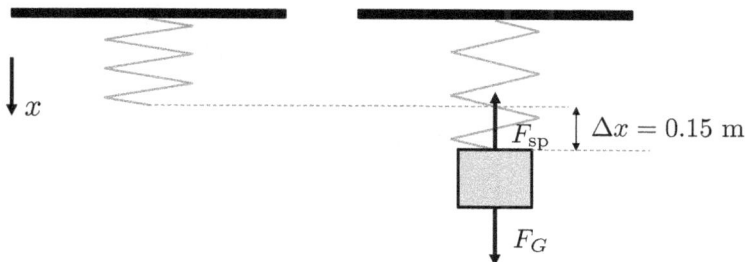

FIGURE 3.31

From the force diagram on the right, we can write out Newton's Second Law and determine an expression for the mass (remember that "at equilibrium" means the acceleration is $0\frac{\text{m}}{\text{s}^2}$):

$$\vec{F}_{\text{NET}} = m\vec{a}$$
$$-k\Delta x + mg = m\left(0\frac{\text{m}}{\text{s}^2}\right)$$
$$k\Delta x = mg$$
$$\frac{k\Delta x}{g} = m$$
$$\frac{\left(75\frac{\text{N}}{\text{m}}\right)(0.15 \text{ m})}{9.8\frac{\text{m}}{\text{s}^2}} = m = 1.1 \text{ kg}$$

As an extension of Hooke's Law, we can define **Young's modulus,**[17] which defines the spring constant for a solid chunk of some material in terms of its physical properties:

$$Y = k\frac{L}{A} \tag{3.13}$$

In Equation (3.13), Y is the material's Young's modulus, k is the spring constant, A is the cross-sectional area of the material perpendicular to the force and L is its length parallel to the force. We show the relevant geometry for a cylinder in Figure 3.32.

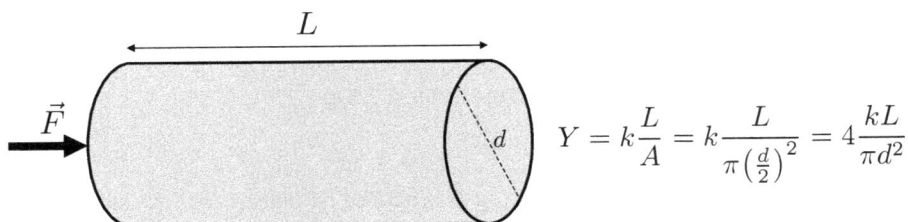

$$Y = k\frac{L}{A} = k\frac{L}{\pi\left(\frac{d}{2}\right)^2} = 4\frac{kL}{\pi d^2}$$

FIGURE 3.32
The Young's modulus of a cylinder of length L and diameter d.

Example 3.18 Compressing Rubber ⋆
Consider a block of rubber with the dimensions shown below. If the Young's Modulus is $5.0 \times 10^7 \frac{\text{N}}{\text{m}}$, by how much will the block compress if a 8.0×10^2 N force is applied to each face of the block?

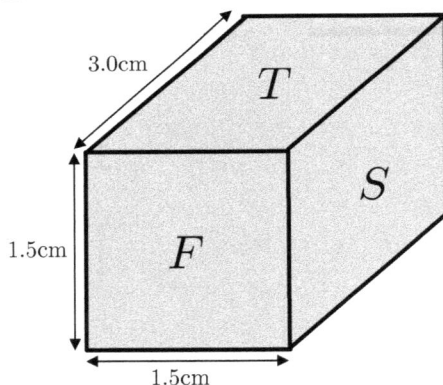

FIGURE 3.33

Let us first compute the area of each face:

$$A_F = (0.15 \text{ m})^2 = 0.023 \text{ m}^2$$
$$A_T = A_S = (0.15 \text{ m})(0.30 \text{ m}) = 0.045 \text{ m}^2$$

Note that we've only labeled 3 of the 6 faces of the block, but the block will compress just the same if we apply the force to, say, the back instead of the front.

---

[17] After Thomas Young (1773-1829).

We ultimately wish to know by how much the block will compress, which will require Hooke's Law. Concerning ourselves with only the magnitude of the force for now, we can write

$$F_{\text{sp}} = k\left(\Delta x\right)$$

We do not know the spring constant k, but we do know the Young's modulus and the dimensions of the object, so we can determine an expression for k:

$$Y = k\frac{L}{A}$$
$$k = \frac{YA}{L}$$

Before we proceed, it is worth our time to reflect on what this equation says: the spring constant decreases if the cross-sectional area decreases and/or if the length increases. Looking at our diagram, we see that the cross sectional area is smaller for face F than for faces T and S; similarly, the length is larger for F. Thus the spring constant for face F is smaller and, according to Hooke's Law, we should expect the compression to be larger for face F than for either of the other two faces.

Let us confirm this analysis. Inserting into our expression for force and solving for Δx, we find

$$F_{\text{sp}} = \frac{YA}{L}\left(\Delta x\right)$$
$$\frac{F_{\text{sp}}L}{YA} = \Delta x$$

This is a general expression that we can use to determine Δx for the force being applied to any of the faces of the block:

$$\Delta x_F = \frac{\left(8.0 \times 10^2 \text{ N}\right)\left(0.30 \text{ m}\right)}{\left(5.0 \times 10^7 \frac{\text{N}}{\text{m}}\right)\left(0.023 \text{ m}^2\right)} = 2.1 \times 10^{-4} \text{ m} = 0.21 \text{ mm}$$

$$\Delta x_T = x_S = \frac{\left(8.0 \times 10^2 \text{ N}\right)\left(0.15 \text{ m}\right)}{\left(5.0 \times 10^7 \frac{\text{N}}{\text{m}}\right)\left(0.045 \text{ m}^2\right)} = 5.3 \times 10^{-5} \text{ m} = 0.53 \text{ mm}$$

As we predicted, $\Delta x_F > \Delta x_T = \Delta x_S$.

3.10   Pressure and Buoyancy

Thus far we've been considering forces as applied at a point. For instance, if a chandelier is suspended from the ceiling by a rope, then the tension force is applied at the location where the rope is tied to the chandelier. This view isn't always helpful, however. For instance, if I try to hammer a nail into a block of wood, it matters quite a bit if the nail comes to a narrow point or if there is some defect that has resulted in a blunt tip. To quantify this we define the **pressure** as the force divided by the area:

$$P = \frac{F}{A} \tag{3.14}$$

The unit of pressure is $\frac{\text{N}}{\text{m}^2}$; the SI name for this quantity is the Pascal (abbreviated "Pa").[18] Pressure increases if the force increases and/or the area over which it is distributed decreases. As another example, someone wearing stiletto heels will focus most of their weight on a very small area, which can damage the floor they are standing on. The same floor, meanwhile, can happily support a much heavier object (a piano, for example) without issue because its weight is distributed over a broad area.

Example 3.19 The Bed of Nails ⋆

A common example of pressure involves the "bed of nails", which consists of many nails that have been hammered through a large board and then laid on the ground, points side up. The physics instructor (or a brave volunteer) lays down on the bed, amazingly (?) unharmed.

a. What is the average pressure on the instructor/volunteer's feet if their mass is 85 kg and total surface area of the bottom of their shoes is 6.00×10^2 cm^2?

b. Suppose the instructor/volunteer has a mass of 85 kg. If all of their weight was focused on a single nail, with a surface area of 0.20 cm^2, what is the pressure of the nail on the foot?

c. How many nails would you need to ensure that the average pressure on any one nail is the same as what you found in (a)?

For (a), we can invoke the definition of pressure directly, keeping in mind that the force from the nail is due to the force of gravity:

$$P_{\text{shoes}} = \frac{F}{A} = \frac{mg}{A} = \frac{(85 \text{ kg}) \left(9.8 \text{ m/s}^2\right)}{6.00 \times 10^{-2} \text{ m}^2} = 1.4 \times 10^4 \text{ Pa}$$

Notice that I've converted from square centimeters to square meters, which involves a factor of 100 *twice*.

For (b), we repeat the process but use the given surface area for the nail. This yields

$$P_{\text{nail}} = 4.2 \times 10^7 \text{ Pa}$$

Ouch! To bring the pressure back down to what we found in (a), the total surface area of the nails needs to match the surface area of the shoes. For N nails the total surface area is

$$A_{\text{total}} = N A_{1 \text{ nail}}$$

Therefore

$$N = \frac{A_{\text{total}}}{A_{1 \text{ nail}}} = \frac{6.00 \times 10^2 \text{ cm}^2}{0.20 \text{ cm}^2} = 3.0 \times 10^3$$

So we should plan for 3,000 nails; increasing the number of nails decreases the pressure further. Some extensions of this demonstration involve putting another board on the volunteer's chest, stacking concrete blocks on top, then smashing them with a sledgehammer (all while the volunteer is on the bed of nails)! Another option is to *not* do that.

One topic where pressure is particularly important is in fluids: if an object is immersed in a fluid then there will be a force, and the force will be spread over the surface of the object.

---

[18]After Blaise Pacal (1623-1662)

Thus it is often convenient to discuss the pressure associated with objects in fluids. Let's begin our discussion of fluids with an obvious statement: a heavy boat will float in water, but a small marble will sink. This phenomenon is explained by **Archimedes' principle:**[19]

> **An object immersed in a fluid will experience a buoyant force equal to the weight of the fluid it displaces.**

To explore Archimedes' principle we first need to introduce the idea of **density**, which is simply the ratio of an object's mass and volume:

$$\rho = \frac{m}{V} \tag{3.15}$$

Equation (3.15) allows us to determine an object's mass given its density and volume. For instance, steel has a density of about $7.80\frac{g}{cm^3}$, or $7800\frac{kg}{m^3}$.[20] If we have a square cube of steel measuring 1.5 m to a side ("1.5 cubic meters"), then its weight is

$$m_s g = \rho_s V_s g = \left(7800\,\frac{kg}{m^3}\right)(1.5\text{ m})^3\left(9.8\frac{m}{s^2}\right) = 2.6 \times 10^5 \text{ N}$$

where I am using the s subscript to specify that we're talking about the solid. Now, if we submerge this entire block in water, Archimedes' principle says that it experiences a buoyant force equal to the weight of 1.5 cubic meters of water. Water's density is $1.00\frac{g}{cm^3}$, or $1.00 \times 10^3 \frac{kg}{m^3}$, so the weight of the displaced water is

$$m_w g = \rho_w V_w g = \left(1.00 \times 10^3\,\frac{kg}{m^3}\right)(1.5\text{ m})^3\left(9.8\frac{m}{s^2}\right) = 3.3 \times 10^4 \text{ N}$$

where the w subscript indicates that we're talking about the water. We see that if we hold the steel under water and release it, the downward force due to its weight is larger than the upward buoyant force, and the cube sinks.

The reason a boat can float even though it is constructed from dense material (like steel) is because its volume is not *entirely* composed of dense materials: the interior of the boat has a significant amount of open space to decrease the average density of the entire boat to be less than that of water (air has a density of just $1.3\frac{kg}{m^3}$). As we see in the next example, in cases where an object's density is less than the fluid in which it is immersed, the object will *partially* submerge (that is, it floats) until the buoyant force matches the gravitational force on the object.

Example 3.20 Floating Ice ⋆
A cylinder of ice floats upright in a glass of water. The cylinder has a radius of 2.5 cm and a length of 0.10 m. Ice has a density of $0.92\frac{g}{cm^3}$. What fraction of the cylinder is under water?

The situation is sketched in Figure 3.34.

---

[19]Archimedes lived in the 3rd century BC; his many contributions to physics, mathematics, and astronomy make him one of the most important scientists of all time.
[20]Check the conversion yourself!

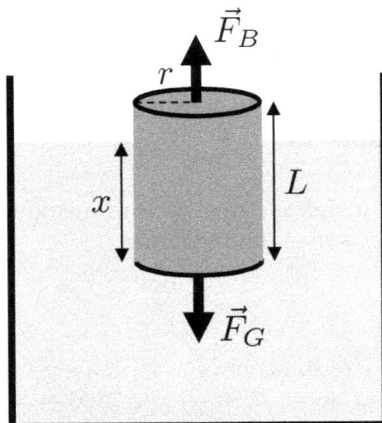

FIGURE 3.34

We've labeled the length of the portion of the cylinder that is submerged as x. This means that the volume of displaced water is $V_{\text{water}} = \pi r^2 x$ and the volume of the entire block of ice is $V_{\text{ice}} = \pi r^2 L$. We wish to know x as a fraction of L, which is to say we want an equation that looks like "$\frac{x}{L} = \cdots$". If the ice is floating, then there is no net force on it. We'll call up the positive direction, in which case Newton's Second Law says

$$\vec{F}_{\text{NET}} = m\vec{a}$$
$$F_B - F_G = 0\text{N}$$
$$F_B = F_G$$
$$m_{\text{water}}g = m_{\text{ice}}g$$
$$m_{\text{water}} = m_{\text{ice}}$$
$$\rho_{\text{water}}V_{\text{water}} = \rho_{\text{ice}}V_{\text{ice}}$$
$$\rho_{\text{water}}\pi r^2 x = \rho_{\text{ice}}\pi r^2 L$$
$$\frac{x}{L} = \frac{\rho_{\text{ice}}}{\rho_{\text{water}}}$$

Note that our solution says that if the density of ice and the density of water were the same, then $x = L$ and the object would be totally submerged at equilibrium (such an object is said to be **neutrally buoyant**; if it floats without being completely submerged then it is said to be **buoyant**). Note also that the radius of the ice plays no role: it contributes equally to the weight of the object and the buoyant force, and the term cancels out. If we plug in our values for the densities, we find $\frac{x}{L} = 0.92$, meaning the ice is 92% submerged.

The fact that fluids have weight raises another important point: if an object is submerged in a fluid, it experiences the weight of the fluid pushing down on it from above (much like the box on the bottom of a stack feels the boxes above it pressing down). As an example, suppose a square box with a surface area A is sitting on the bottom of a tub of water such that the top face of the box is a distance d below the surface (Figure 3.35). A square column of water with volume $V = Ad$ is pushing down on the box, and from our discussion above

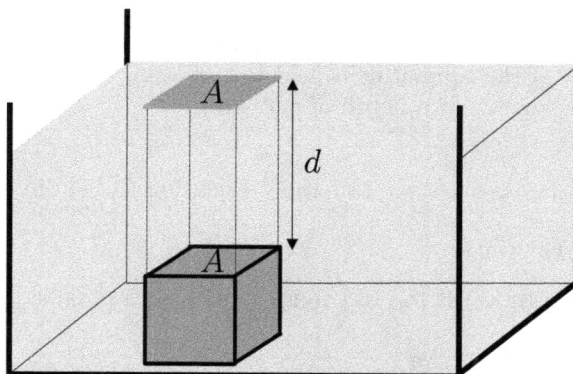

FIGURE 3.35
The weight of a rectangular column of water of volume Ad pushes down on the top of a box with surface area A.

we can determine its weight:

$$m_{\text{water}}g = \rho_{\text{water}}Vg$$
$$= \rho_{\text{water}}Adg$$

It is often advantageous to think of this situation in terms of the pressure, which we defined in Equation (3.14) as $P = F/A$. Thus we see

$$P_{\text{water}} = \frac{m_{\text{water}}g}{A}$$
$$P_{\text{water}} = \rho_{\text{water}}dg$$

The usefulness of doing this comes from the fact that the pressure does not depend on the properties of the box (or submarine or whatever else we're talking about): it reflects what is constant at a given depth in the fluid.

While it is natural to think of water as exerting pressure as you descend farther down into some body of water, the air in the atmosphere exerts a pressure on us as well. **Atmospheric pressure** varies with elevation above sea level, temperature, and a host of other factors. That being said, we define "1 atmosphere" (abbreviated "1 atm") as a good approximate value, and denote it with the symbol P_0:

$$P_0 = 1 \text{ atm} = 1.01 \times 10^5 \text{ Pa} \tag{3.16}$$

Thus the total pressure at some depth below the surface of a fluid depends both on the pressure exerted *by the fluid* and the pressure exerted *by the atmosphere*. (You can think of it like two books stacked on your stomach as you lay on the ground: you certainly feel the effect of both!) We summarize by saying that if we start at atmospheric pressure and then descend into some fluid of density ρ, then at a depth d the pressure P_d is given by

$$P_d = P_0 + \rho g d \tag{3.17}$$

Example 3.21 Water Pressure   ★★

Consider the effects of pressure below the surface of water:

(a) What is the pressure in sea water, which has a density of $1030\frac{\text{kg}}{\text{m}^3}$, at a depth of 8.00×10^3 m? Give your answer in atm.

(b) Suppose a submarine has a circular window with a 30.0 cm diameter. If the pressure inside of the submarine is held to 1.00 atm, then what net force will be applied to the window at a depth of 8.00 km?

To determine the pressure, we can apply Equation (3.17) directly:

$$P_d = P_0 + \rho g d$$
$$= \left(1.01 \times 10^5 \text{ Pa}\right) + \left(1030 \frac{\text{kg}}{\text{m}^3}\right) \left(9.8 \frac{\text{m}}{\text{s}^2}\right) \left(8.00 \times 10^3 \text{ m}\right)$$
$$= 8.07 \times 10^7 \text{ Pa}$$

Note that we've used SI units throughout to ensure we don't end up doing something along the lines of trying to add inches and meters (or, in this case, atmospheres and Pascals). Now that we have our answer in SI units, we can convert to the desired units:

$$P_d = 8.07 \times 10^7 \text{ Pa} \left(\frac{1.00 \text{ atm}}{1.01 \times 10^5 \text{ Pa}}\right) = 801 \text{ atm}$$

Apparently the pressure at this depth is about 800 times what it is at the surface!

For (b), note that the submarine window feels an *outward* force due to the pressure *inside* the submarine, and it feels an *inward* force due to the pressure *outside* the submarine. We'll call the inward force positive and the outward force negative, so the net force is given by

$$F_{\text{NET}} = F_{\text{inward}} - F_{\text{outward}}$$
$$= P_{\text{inward}}A - P_{\text{outward}}A$$
$$= (P_0 + \rho g d - P_0) A$$
$$= \rho g d A$$
$$= \left(1030 \frac{\text{kg}}{\text{m}^3}\right) \left(9.8 \frac{\text{m}}{\text{s}^2}\right) \left(8.00 \times 10^3 \text{ m}\right) \left(\pi (0.150 \text{ m})^2\right)$$
$$= 5.71 \times 10^6 \text{ N}$$

Unsurprisingly, the net force is huge; the window must be designed to withstand it (and the submarine must take care to avoid descending beyond the window's maximum safe depth).

3.11 Cumulative Examples

Now that you've been exposed to Newton's Laws and the most common types of forces, we are in a good position to go through some more complicated examples. Example 3.22 considers the identification of forces (and force pairs) now that we are in a position to consider the different types of forces at play in some typical scenarios. Examples 3.23 and 3.24 bring together many of the different types of forces and problem-solving techniques that we've discussed in this chapter (and kinematics from Chapters 1 and 2). I encourage you to read through these carefully; as always, it is a good idea to try some problems for yourself soon after reading about a new concept!

Example 3.22 Newton's Wagon ★★

A girl pulls a cart over level ground. Identify all of the forces acting on the wagon and girl and determine the forces with which they are paired according to Newton's Third Law. Determine an expression for the net horizontal force on the girl.

First let us consider the horizontal dimension. The girl is walking forward, so the ground is pushing on her (this is a friction force; if she were walking on ice she'd be prone to slip and fall). However, by Newton's Third Law, she must also be pushing on the ground. Similarly, the girl is pulling the wagon, and so the wagon is pulling back on her. The situation is sketched in Figure 3.36.

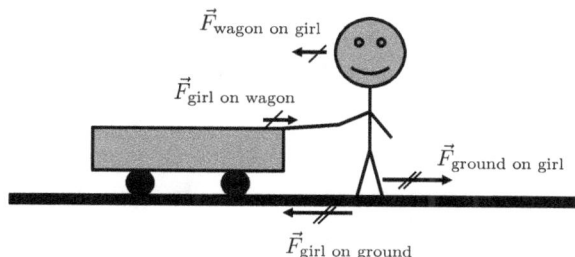

FIGURE 3.36

The horizontal forces on a girl pulling a wagon. Forces with equal magnitudes are indicated with slashes.

Looking at only the forces acting on the girl, we see

$$\left(\vec{F}_{\text{NET}}\right)_{\text{girl}} = \vec{F}_{\text{ground on girl}} + \vec{F}_{\text{wagon on girl}}$$

We see from the diagram that $\vec{F}_{\text{ground on girl}}$ and $\vec{F}_{\text{wagon on girl}}$ are in opposite directions, but they are not necessarily of the same magnitude (if they were, the net force would be 0N and the girl could never accelerate!).

Meanwhile, in the vertical dimension, both the wagon and the girl are gravitationally attracted down to the Earth, and experience an equal and opposite normal force directed up (Figure 3.37.)

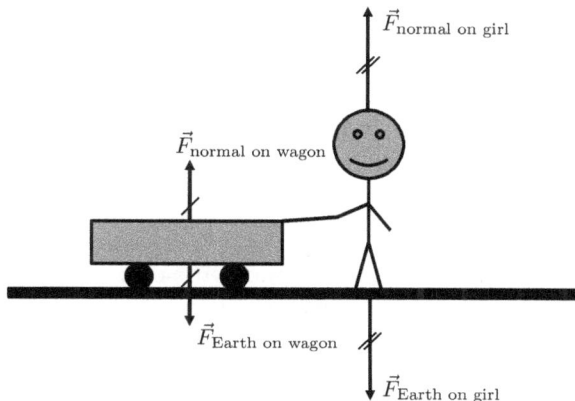

FIGURE 3.37

The vertical forces on a girl pulling a wagon. Forces with equal magnitudes are indicated with slashes.

You may be tempted to say that the gravitational and normal forces are another example of a Newton's Third Law pair of forces; they are of equal magnitude and point in opposite directions, after all! However, this is incorrect. For instance, we're saying that a gravitational force acts *on the girl*, and a normal force acts *on the girl*. This cannot be a Newton's Third Law pair of forces because they're both acting *on the girl*, whereas Newton's Third Law refers to a pair of forces acting on two different things. For instance, $\vec{F}_{\text{Earth on girl}}$ (the attraction felt by the girl, toward the Earth) is paired with $\vec{F}_{\text{girl on Earth}}$, the gravitational attraction felt by the Earth toward the girl.[a] Similarly, $\vec{F}_{\text{normal on girl}}$ is paired with $\vec{F}_{\text{girl on normal}}$, which is to say there is a downward normal force felt by the ground equal to the upward normal force felt by the girl. The analysis of the vertical forces on the wagon are exactly analogous.

---

[a]The incredible mass of the Earth means its acceleration due to this force would essentially be $0\frac{m}{s^2}$, even if the force wasn't balanced by the Earth being attracted to people and objects on the other side of the Earth from the girl in question.

Example 3.23 A Pulley and Ramp ★★

Two blocks of masses m and M are attached by a rope that is strung over a massless pulley, as shown in Figure 3.38. Block m and the surface of the ramp have a coefficient of static friction μ_s. The blocks are released from rest. What must be true if the blocks are to remain stationary?

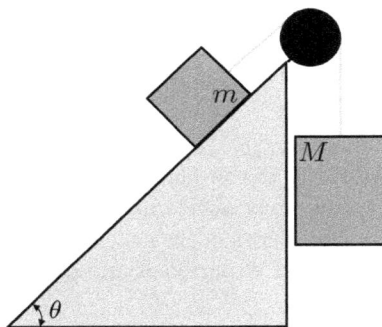

FIGURE 3.38
Two blocks of masses m and M are attached by a rope that is strung over a massless pulley.

The short answer is to say, "The net force on each block is 0 N", or equivalently "Both blocks must be in static equilibrium". However, we can go further than this and write out Newton's Second Law to show how the forces must be related if the net force is indeed to be 0 N. Our goal in such problems is to express our answer in terms of the quantities given (m, M, μ_s, and θ) and standard constants (such as g).

To proceed, we should recognize that we have a pulley and so we can make use of a curved coordinate system (Example 3.16 on page 136). We'll choose the positive direction to be "M goes vertically down while m goes up the ramp". Let's carefully draw out all of the forces and note our choice of coordinate system (Figure 3.39).

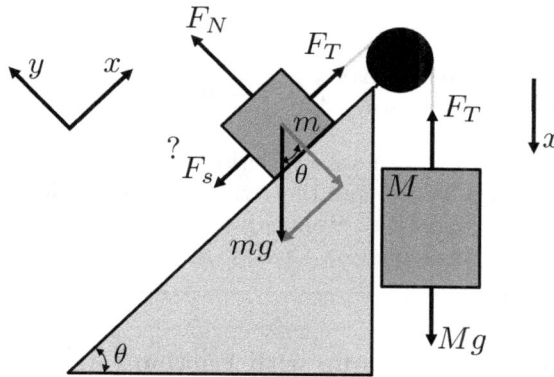

FIGURE 3.39

We've placed a "?" by $\vec{F}_s$ because we actually don't know if it points down the ramp or up the ramp: friction always opposes the net effect of all of the other forces parallel to the surface, and it is certainly possible that in the absence of friction, block m will prefer to slide in the positive x direction (if M is very large, for instance) or in the negative x direction (if m is very large, for instance).

Now we can take advantage of *selective grouping* (as in Examples 3.5 and 3.16) to treat both blocks as a combined mass: the internal (tension) forces cancel and Newton's Second Law in the x direction says:

$$\left(\vec{F}_{\text{NET}}\right)_x = m\vec{a}_x$$

$$Mg - mg\sin\theta + \vec{F}_s = (m + M)\,a_x$$

Note that F_s could be positive or negative, as we argued above. Now if we insist that the net force is 0 N, then the acceleration is $0\frac{m}{s^2}$ and

$$Mg - mg\sin\theta + \vec{F}_s = 0\text{N}$$

$$\vec{F}_s = mg\sin\theta - Mg$$

This says precisely what we reasoned above: the direction of the frictional force depends on the relative sizes of m and M (and g and θ). We aren't done, however, since F_s is not a given/known value. To proceed, recall that there is a maximal value of static friction given by

$$(F_s)_{\text{max}} = \mu_s F_N$$

By inspecting the force diagram and recognizing that the acceleration of mass m perpendicular to the surface must be $0\frac{m}{s^2}$, we see that Newton's Second Law in the y direction gives

$$\left(\vec{F}_{\text{NET}}\right)_y = m\vec{a}_y$$

$$F_N - mg\cos\theta = 0\text{N}$$

$$F_N = mg\cos\theta$$

Inserting into the equation for $(F_s)_{\text{max}}$, we see

$$(F_s)_{\text{max}} = \mu_s mg\cos\theta$$

Because this is the maximum value for static friction, it follows that the condition for the blocks to remain at rest is

$$\left.\begin{array}{rcl}(F_s)_{\max} & = & \mu_s mg \cos\theta \\ F_s & = & mg\sin\theta - Mg\end{array}\right\} \Rightarrow \mu_s m\cos\theta \geq |m\sin\theta - M|$$

If this inequality *isn't* true, then static friction cannot overcome the net effect of gravity acting on the two blocks, and they will begin to accelerate. We see that our answer involves only the quantities given, so we are content to stop here (if we were asked to, we could solve this equation for, say, μ_s in terms of the other constants).

Example 3.24 Sliding on a Ramp with Friction (and Projectile Motion?)
★ ★ ★
A 3.00 kg block is placed at the bottom of an elevated ramp, as shown in Figure 3.40. On the ramp, $\mu_s = 0.200$ and $\mu_k = 0.100$. At time $t = 0.00$ seconds, the block is given a velocity of 1.5 m/s directed up the ramp. Determine the location of the block a long time later, assuming it does not bounce or roll if it strikes the ground.

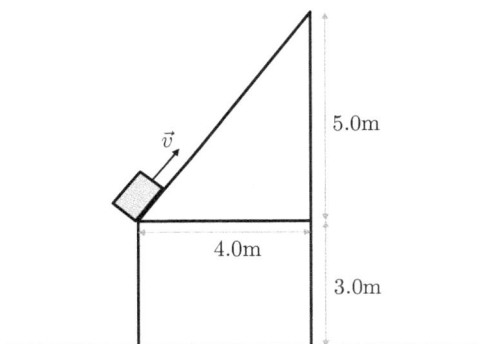

FIGURE 3.40
A 3.00 kg block is placed at the bottom of an elevated ramp.

Before we dig in to the details of the problem, it is worth thinking through the possible outcomes of the motion. The block will slow down as it moves up the ramp; either it will reach the top of the ramp and topple over the right edge or it will come to a stop somewhere along the ramp. If it comes to a stop, it will either stay there (because of static friction), or it will reverse course and slide back down the ramp, ultimately toppling over the left edge.

    Let me also note that this is quite the long example, but it is perhaps less intimidating if you think of it as a series of linked problems: we will analyze the first leg of the journey, then use our results to inform our analysis of the second leg, and so on.

    The first order of business is to determine the block's acceleration right after it is given a shove: knowing this will allow us to determine how far up the ramp it travels. We'll set up our coordinate system so the positive x coordinate points up the ramp, and carefully draw out the force diagram (Figure 3.41).

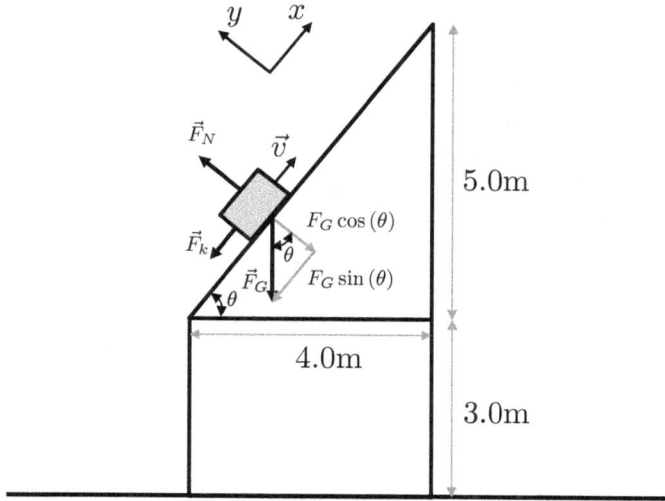

FIGURE 3.41

Refer back to Examples 3.9 (page 119) and 2.2 (page 58) if this force diagram is confusing to you. Once we have the force diagram, we can turn to Newton's Second Law, but as always we should have our plan in mind: we want to determine the acceleration in the x direction, but because this involves friction we first need to know what the normal force is. For this reason we'll start with Newton's Second Law in the y direction:

$$\left(\vec{F}_{\text{NET}}\right)_y = m\vec{a}_y$$
$$F_N - F_G \cos\theta = 0\text{N}$$
$$F_N = F_G \cos\theta$$

In the above we've made use of the fact that (1) the block is not accelerating perpendicular to the surface, and (2) the gravitational force for an object near the surface of the earth is just mg. We were given the mass but not the angle θ; however, we can solve for it with a bit of trigonometry:

$$\tan\theta = \frac{5.0 \text{ m}}{4.0 \text{ m}}$$
$$\theta = \tan^{-1}\left(\frac{5.0}{4.0}\right) = 51°$$

We now know everything on the right hand side of the equation for the normal force. We could plug in to get a number, but instead we'll just use the expression $mg\cos\theta$ for the normal force when evaluating Newton's Second Law in the x direction:

$$\left(\vec{F}_{\text{NET}}\right)_x = m\vec{a}_x$$
$$-F_k - mg\sin\theta = ma_x$$
$$-\mu_k F_N - mg\sin\theta = ma_x$$
$$-\mu_k\left(mg\cos\theta\right) - mg\sin\theta = ma_x$$
$$-g\left(\mu_k\cos\theta + \sin\theta\right) = a_x$$

Using a general expression for the normal force has given us a result that holds regardless of the specific values of μ_k and θ (and we see that mass plays no role, having canceled as it did in Example 3.13). If we plug in the known values in the left hand side of the equation for a_x, we find (check for yourself!) $-8.3\frac{m}{s^2}$.

Now that we know the acceleration, we can determine how far the box would like to go before it comes to rest. If the ramp is shorter than this distance (whatever it is), then the box will tumble over the edge. This is a one-dimensional kinematics problem and we will proceed, as we did in Chapter 1, by listing our kinematic variables:

variables:

$$\vec{a} = -8.3\frac{m}{s^2}$$

$$\vec{x}_i = 0 \text{ m}$$

$$\vec{x}_f = ?$$

$$\vec{v}_i = 1.5\frac{m}{s}$$

$$\vec{v}_f = 0\frac{m}{s}$$

$$\Delta t = ?$$

In this case we don't care about the time it takes for the box to come to rest, so we use the "no Δt" equation:

$$v_f^2 = v_i^2 + 2a\Delta x$$

$$\frac{v_f^2 - v_i^2}{2a} = \Delta x$$

$$\frac{\left(0\frac{m}{s}\right)^2 - \left(1.5\frac{m}{s}\right)^2}{2\left(-8.3\frac{m}{s^2}\right)} = \Delta x = 0.14 \text{ m}$$

Compare this to the length of the ramp, which (by the Pythagorean Theorem) is $h = \left((4.0 \text{ m})^2 + (5.0 \text{ m})^2\right)^{1/2} = 6.4$ m. Clearly the box comes to a stop well before reaching the top! We now know that the box will either stay there (if static friction is large enough), or otherwise will slide back down the ramp. Let us consider static friction, which comes into play the instant the box comes to rest. The force diagram for this instant is shown in Figure 3.42.

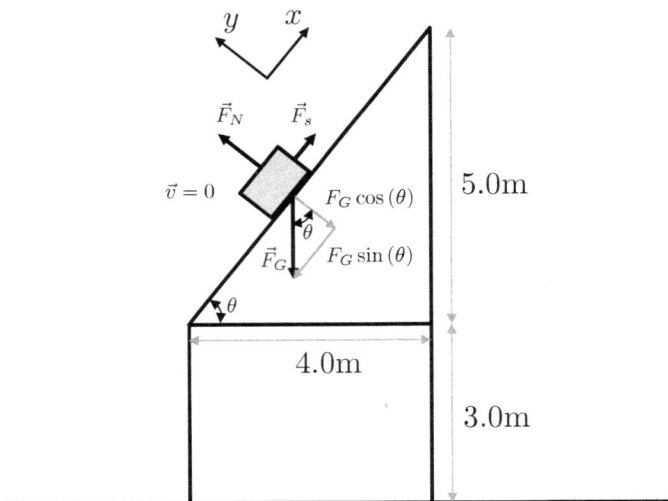

FIGURE 3.42

We know that friction now points *up* the plane because it opposes the effect of all other forces in the x direction, which points *down* the plane (due to $F_G \sin \theta$). Now the maximum value of static friction is given by

$$(F_s)_{max} = \mu_s F_N$$
$$(F_s)_{max} = \mu_s mg \cos\theta$$
$$(F_s)_{max} = (0.200)\,(3.00\text{ kg})\left(9.8\frac{\text{m}}{\text{s}^2}\right)\cos 51° = 3.7\text{ N}$$

Compare this to $F_G \sin\theta = mg\sin\theta = (3.00\text{ kg})\left(9.8\frac{\text{m}}{\text{s}^2}\right)\sin 51° = 23\text{ N}$, and we see that static friction cannot match the force of gravity directed down the ramp. Thus the block begins to slide back the way it came. Eventually it will leave the ramp, and we need to know its velocity at this instant if we are to analyze the subsequent free fall motion (Figure 3.43).

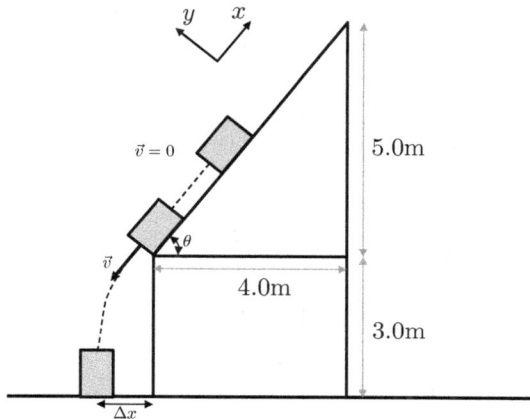

FIGURE 3.43

To determine its velocity as it leaves the ramp, we need to consider the one-dimensional kinematics problem where the box is initially at its highest point, at rest, and then it begins to accelerate down the ramp. To analyze the kinematics, we'll need the acceleration, which we can obtain from force analysis. The forces when the box is accelerating down the ramp are summarized in Figure 3.44.

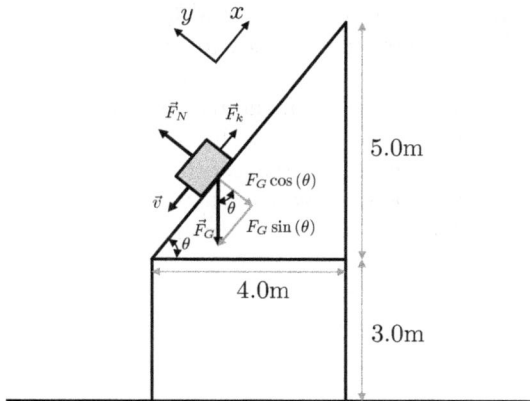

FIGURE 3.44

Newton's Second Law in the y dimension hasn't changed, so we can turn immediately to the x dimension:

$$\left(\vec{F}_{\text{NET}}\right)_x = m\vec{a}_x$$

$$F_k - mg\sin\theta = ma_x$$

$$\mu_k F_N - mg\sin\theta = ma_x$$

$$\mu_k\left(mg\cos\theta\right) - mg\sin\theta = ma_x$$

$$g\left(\mu_k\cos\theta - \sin\theta\right) = a_x$$

This is similar to our expression for the initial acceleration up the ramp, except now friction is pointing in the opposite direction (because the box's velocity is in the opposite direction). If we plug in to this equation, we find $a_x = -7.0\frac{\text{m}}{\text{s}^2}$.
Having analyzed the forces, we can turn to kinematics:

variables:

$$\vec{a} = -7.0\frac{\text{m}}{\text{s}^2}$$

$$\vec{x}_i = 0.14 \text{ m}$$

$$\vec{x}_f = 0 \text{ m}$$

$$\vec{v}_i = 0\frac{\text{m}}{\text{s}}$$

$$\vec{v}_f = ?\frac{\text{m}}{\text{s}}$$

$$\Delta t = ?$$

We still don't care about Δt, so we can use the same equation as before:

$$v_f^2 = v_i^2 + 2a\Delta x$$

$$v_f = \left(v_i^2 + 2a\Delta x\right)^{\frac{1}{2}}$$

$$v_f = \left(\left(0\frac{\text{m}}{\text{s}}\right)^2 + 2\left(-7.0\frac{\text{m}}{\text{s}^2}\right)(0 \text{ m} - 0.14 \text{ m})\right)^{\frac{1}{2}}$$

$$v_f = -1.4\frac{\text{m}}{\text{s}}$$

Remember that if we have an equation involving a square root, like $(4)^{\frac{1}{2}}$, then there are two solutions, like ± 2 (because $(2)^2 = (-2)^2 = 4$). It is up to us to interpret the situation and determine which solution is physically appropriate; here, we know the object is sliding in the negative x direction, so we choose the negative root in the final line above.

We're almost there! We know the velocity of the object as it departs the ramp and enters free fall. This is a question that we could have asked in Chapter 2. Because gravity acts straight down, it now makes sense for us to adjust our coordinate system into the usual vertical/horizontal configuration. However, we'll choose "to the left" to be the positive x direction because all of the motion will be in that direction. We'll set the origin on the ground, directly below the point of departure (Figure 3.45).
Note that as a consequence of changing our coordinate system, the initial velocity is no longer one-dimensional. Thus we need to carefully assess the geometry to describe the components of the velocity in our new coordinate system.

As in all two-dimensional kinematics problems, we can analyze the motion as two one-dimensional kinematics problems that are linked by time. The motion will stop when the object hits the ground, which is a question of the vertical motion. Thus we analyze the y dimension first to determine time:

variables:

$$\vec{a}_y = -9.8 \frac{\text{m}}{\text{s}^2}$$

$$\vec{y}_i = 3.0 \text{ m}$$

$$\vec{y}_f = 0 \text{ m}$$

$$(\vec{v}_y)_i = -1.4 \sin 51° \frac{\text{m}}{\text{s}}$$

$$= -1.1 \frac{\text{m}}{\text{s}}$$

$$(\vec{v}_y)_f = ? \frac{\text{m}}{\text{s}}$$

$$\Delta t = ?$$

We neither know nor care to know $(\vec{v}_y)_f$, so we might be tempted to use

$$y_f = y_i + (\vec{v}_y)_i \Delta t + \frac{1}{2} a_y (\Delta t)^2$$

However, because $(\vec{v}_y)_i$ is not 0 m/s, this gives us a quadratic equation.

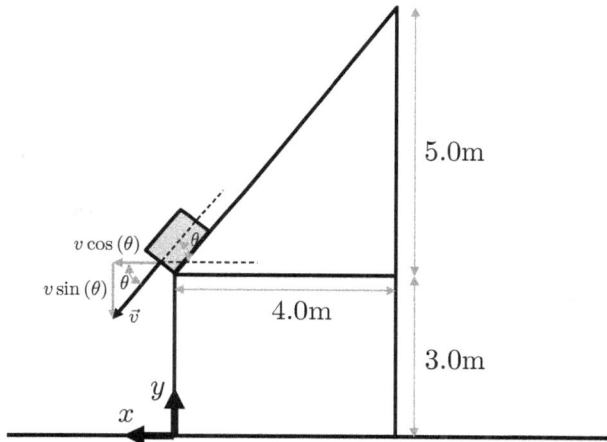

FIGURE 3.45

We could solve the quadratic equation, but instead let's use the other two kinematic equations. First we solve for $(\vec{v}_y)_i$:

$$(\vec{v}_y)_f^2 = (\vec{v}_y)_i^2 + 2a_y \Delta y$$

$$(\vec{v}_y)_f = \left((\vec{v}_y)_i^2 + 2a_y \Delta y \right)^{\frac{1}{2}}$$

$$(\vec{v}_y)_f = \left(\left(-1.1 \frac{\text{m}}{\text{s}}\right)^2 + 2\left(-9.8 \frac{\text{m}}{\text{s}^2}\right)(0 \text{ m} - 3.0 \text{ m}) \right)^{\frac{1}{2}}$$

$$(\vec{v}_y)_f = -7.7 \frac{\text{m}}{\text{s}}$$

Then we can use this to solve for Δt:

$$(\vec{v}_y)_f = (\vec{v}_y)_i + a_y \Delta t$$

$$\frac{(\vec{v}_y)_f - (\vec{v}_y)_i}{a_y} = \Delta t$$

$$\frac{\left(-7.7 \frac{\text{m}}{\text{s}}\right) - \left(-1.1 \frac{\text{m}}{\text{s}}\right)}{-9.8 \frac{\text{m}}{\text{s}^2}} = \Delta t = 0.68 \text{ s}$$

Now, finally, we can analyze the horizontal motion during this time:

$$\vec{a}_x = 0\frac{\text{m}}{\text{s}^2}$$

$$\vec{x}_i = 0 \text{ m}$$

$$\vec{x}_f = ?$$

$$(\vec{v}_x)_i = (\vec{v}_x)_f = 1.4\cos 51°\frac{\text{m}}{\text{s}}$$

$$= 0.86\frac{\text{m}}{\text{s}}$$

$$\Delta t = 0.68 \text{ s}$$

$$x_f = x_i + (\vec{v}_x)_i\,\Delta t + \frac{1}{2}a_x\,(\Delta t)^2$$

$$x_f = (0 \text{ m}) + \left(0.86\frac{\text{m}}{\text{s}}\right)(0.68 \text{ s})$$

$$+ \frac{1}{2}\left(0\frac{\text{m}}{\text{s}^2}\right)(0.68 \text{ s})^2$$

$$x_f = 0.59 \text{ m}$$

This is the answer we've been after this whole time: the box lands 0.59 meters from the left end of the ramp.

This one example, which we've spent quite a bit of time going through, has really been several linked problems: we analyzed the kinematics of the box sliding up the ramp, but that required analyzing the forces (so we could apply Newton's Second Law to determine the acceleration). Doing this allowed us to establish that the box comes to rest before tipping over the right edge of the ramp. Once we knew that it comes to rest on the ramp, we checked to see if static friction pinned it in place. It didn't, so then we re-analyzed the forces as it slid back down. This allowed us to determine the velocity of the box right before it left the left edge of the ramp. Then, finally, we could analyze the box flying through the air, a standard problem from two-dimensional kinematics. Thus in the course of solving this one problem, we've applied much of what we've covered in the first three chapters of this book.

3.12   Tangential and Centripetal Force

We concluded Chapter 2 by taking up the case of *circular motion*, which is to say motion where the polar coordinate θ can change but the *radius* is fixed. We found that the one-dimensional kinematic equations map to the case where it is θ that changes, and we saw that the velocity $\vec{v}$ can be changed via the tangential acceleration $\vec{a}_t$ (in the $\hat{\theta}$ direction, affecting the magnitude of $\vec{v}$) and the centripetal acceleration $\vec{a}_c$ (in the $-\hat{r}$ direction, affecting the direction of $\vec{v}$) according to Equation (2.16):

$$\vec{a}_{\text{NET}} = -\vec{a}_c\hat{r} + \vec{a}_t\hat{\theta}$$

where the centripetal acceleration obeys (Equation (2.18))

$$a_c = \frac{v^2}{r}$$

Equipped with our study of Newton's Laws, we are now ready to consider *forces* in the context of circular motion. Indeed, the basic idea maps directly from our analysis of accelerations. For an object moving on a circular path, it is convenient to express the net force in polar coordinates, in which case we have centripetal and tangential components (refer back to Figure 2.29):

$$\vec{F}_{\text{NET}} = -F_c\hat{r} + F_t\hat{\theta} \tag{3.18}$$

Similarly, we define the centripetal force (refer back to Figure 2.30) as

$$F_c = ma_c = m\frac{v^2}{r} \tag{3.19}$$

for an object of mass m moving at speed v in a circle of radius r. As the next series of examples shows, this is a very useful result.

Example 3.25 Tension in Circular Motion ⋆
A 375 g ball is tied to the end of a string of length 48 cm. The ball is then swung about in a horizontal circle. If the string will break if the tension force exceeds 35 N, what is the maximum angular speed of the ball? Provide your answer in rpm.

We sketch the force diagram in Figure 3.46.

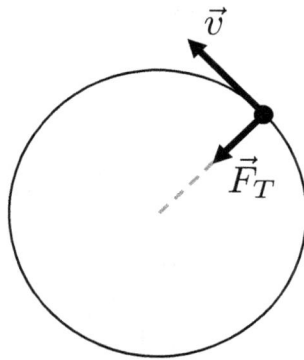

FIGURE 3.46

The only force is the tension force, so

$$\vec{F}_{\text{NET}} = m\vec{a}$$
$$F_T = ma_c$$
$$F_T = m\frac{v^2}{r}$$
$$F_T = m\frac{(\omega r)^2}{r}$$
$$F_T = m\omega^2 r$$
$$\left(\frac{F_T}{mr}\right)^{\frac{1}{2}} = \omega$$

Thus we see that the greatest angular velocity corresponds to the greatest tension force, and

$$\left(\frac{35 \text{ N}}{(0.375 \text{ kg})(0.48 \text{ m})}\right)^{\frac{1}{2}} = \omega_{\text{max}}$$
$$14\frac{\text{rad}}{\text{s}} = \omega_{\text{max}}$$

Converting to rpm, we find

$$\omega_{\text{max}} = 14\frac{\text{rad}}{\text{s}} \times \frac{1 \text{ rotation}}{2\pi \text{ rad}} \times \frac{60 \text{ s}}{1 \text{ min}} = 130 \text{ rpm}$$

Example 3.26 Gravity in Planetary Motion ★★

Mercury orbits the Sun once every 88 days at an average distance of 58 million kilometers. Estimate the mass of the Sun, and compare to the known value of 1.99×10^{30} kg. (Your answer will not be exact because Mercury's orbit is actually quite different from a perfect circle.)

First, note that the force diagram for Mercury is identical to the one sketched in the previous example (Figure 3.46), except here instead of a *tension* force, we have the *gravitational* force from Newton's Law of Gravitation ($F_G = G\frac{m_1 m_2}{r^2}$, Equation (3.6) on page 114). Thus according to Newton's Second Law,

$$\vec{F}_{\text{NET}} = m_M \vec{a}$$
$$F_G = m_M a_c$$
$$G\frac{m_M m_S}{r^2} = m_M \frac{v^2}{r}$$
$$m_S = \frac{v^2 r}{G}$$

We don't directly know v, but we do know the period, $T = \frac{2\pi r}{v} \Rightarrow v = \frac{2\pi r}{T}$ (Equation (2.17)). Thus we can substitute this in to the above equation:

$$m_S = \frac{v^2 r}{G}$$
$$m_S = \frac{r}{G}\left(\frac{2\pi r}{T}\right)^2$$
$$m_S = \frac{r^3}{G}\left(\frac{2\pi}{T}\right)^2$$

Before substituting in our values, we need to convert the period into the SI unit of time (seconds):

$$T = 88 \text{ days} \times \frac{24 \text{ hr}}{1 \text{ day}} \times \frac{60 \text{ min}}{1 \text{ hr}} \times \frac{60 \text{ sec}}{1 \text{ min}} = 7.6 \times 10^6 \text{ s}$$

Thus

$$m_S = \frac{\left(5.8 \times 10^{10} \text{ m}\right)^3}{6.67 \times 10^{-11} \text{ m}^3/(\text{kg s}^2)}\left(\frac{2\pi}{7.60 \times 10^6 \text{ s}}\right)^2$$
$$m_S = 2.0 \times 10^{30} \text{ kg}$$

This agrees with the value given in the question when rounded to two significant figures!

Example 3.27 A Banked Turn ★★★

Consider a car of mass m driving in a large circle of radius r on a road with a coefficient of static friction μ_s. If the ground is (a) flat or (b) inclined at an angle θ, as shown, what is the maximum speed at which the car can travel without slipping?

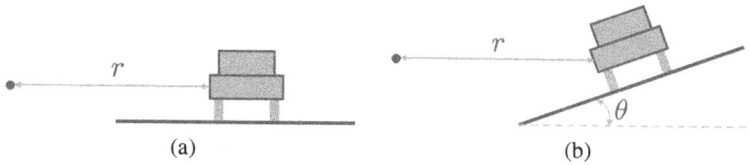

(a)   (b)

FIGURE 3.47
A car traveling in a circle of radius r, viewed from behind, (a) a flat road and (b) a road banked at an angle θ.

If the car is traveling in a circle, it must experience a net force directed toward the center of the circle; if it is traveling on flat ground (as in (a)), the only force acting along the plane of the ground is friction; so it *must* point to the center of the circle. (We're arguing that friction points this way because it must. This is correct but it isn't terribly satisfying; if you'd like to think through the details, check out Problem 3.111.)

To determine the maximum safe speed, we will assume that friction has its maximum value: $F_s = \mu_s F_N$ (beyond this value we shift to kinetic friction – where the wheels are slipping and skidding, rather than rolling smoothly). Then

$$\vec{F}_{\text{NET}} = m\vec{a}$$
$$F_s = ma_c$$
$$\mu_s F_N = m\frac{v^2}{r}$$
$$\mu_s mg = m\frac{v^2}{r}$$
$$\mu_s g = \frac{v^2}{r}$$
$$(\mu_s g r)^{\frac{1}{2}} = v$$

Thus we see, for instance, that the maximum safe speed increases with a larger radius. In case (b) the situation is more complicated and it is worth our time to carefully draw the force diagram (Figure 3.48). Because the net force must point to the center of the circle – which is to say, horizontally to the left – it seems appropriate to set up our coordinate system with one coordinate aligned horizontally (and the other vertically, where we know there is no net force). Thus we have broken the normal force and the frictional force into their horizontal and vertical components, and marked (from analysis of the geometry) where the angle θ appears.

Now, you may be thinking that $F_N = mg\cos\theta$, as for an object sliding along an incline (Example 3.9). However, this relationship comes from the fact that the net force is *along the incline*. Here the object is moving around the banked turn such that the net force is horizontal and toward the center of the circular path. Thus it is decidedly *not* the case that $F_N = mg\cos\theta$.

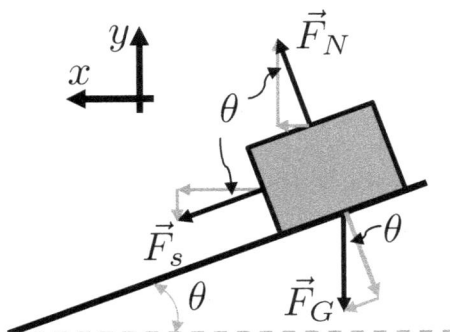

FIGURE 3.48

We will apply Newton's Second Law separately in each dimension, noting first that the net force is the centripetal force, which points along the positive x dimension. Thus in the x dimension

$$(F_{\text{NET}})_x = F_N \sin\theta + F_s \cos\theta$$

$$m\frac{v^2}{r} = F_N \sin\theta + \mu_s F_N \cos\theta$$

Meanwhile, in the y dimension we have

$$(F_{\text{NET}})_y = F_N \cos\theta - F_s \sin\theta - mg$$

$$0\text{ N} = F_N \cos\theta - \mu_s F_N \sin\theta - mg$$

Both of these equations involve v and F_N, the two parameters that are not given (symbolically) in the problem. Because we are interested in v, we will solve the y equation for F_N:

$$F_N = \frac{mg}{\cos\theta - \mu_s \sin\theta}$$

and substitute it into the x equation:

$$m\frac{v^2}{r} = \frac{mg}{\cos\theta - \mu_s \sin\theta}(\sin\theta + \mu_s \cos\theta)$$

$$\frac{v^2}{r} = g\frac{\sin\theta + \mu_s \cos\theta}{\cos\theta - \mu_s \sin\theta}$$

$$v = \left(gr\frac{\sin\theta + \mu_s \cos\theta}{\cos\theta - \mu_s \sin\theta}\right)^{\frac{1}{2}}$$

For reasonable values of μ_s and appropriate values of θ, it turns out that v for a banked turn is higher than for a flat road (Problem 3.112). This is why racetracks and turns on highways are banked: drivers are in less danger of slipping at high speeds (especially if the road becomes slippery from rain). The flipside of this is that a car is liable to slip down the surface if they are traveling too slow (and the definition of *too slow* depends on the value of θ and μ_s). Problem 3.113 explores some of these details.

3.13 Torque

In the previous section we saw that a net force acting in the $-\hat{r}$ direction allows an object to move with a constant angular speed ω (and therefore constant linear speed v because $v = \omega r$) in so-called uniform circular motion. We reasoned this in part because any component of the net force along the $\pm\hat{\theta}$ direction would change the magnitude of v and ω, which by definition does not occur in *uniform* circular motion.

Let us now proceed to consider an object experiencing a centripetal force (i.e. moving in a circle) *and* a force that *is* directed in the $\pm\hat{\theta}$ direction: its angular velocity $\vec{\omega}$ will no longer be constant (Figure 3.49). We will write down Newton's Second Law in the $\hat{\theta}$ direction for the object shown in the figure, then re-arrange some terms to arrive at a particularly useful expression:[21]

$$\left(\vec{F}_{\text{NET}}\right)_\theta = m\vec{a}_\theta$$
$$F_\theta = ma_\theta$$
$$F_\theta = m\alpha r$$
$$F_\theta r = mr^2\alpha \tag{3.20}$$

Before we re-write Equation (3.20) in its general form, we need to make a few observations. First, the left hand side of the equation is, in words, the product of the *magnitude of the vector pointing to where the force is applied* (which is by definition in the $\hat{r}$ direction) and the *magnitude of the force perpendicular to that vector* (i.e. the $\hat{\theta}$ component). This is an important quantity that we refer to as the magnitude of the **torque**, written with the Greek lowercase tau: τ. Mathematically,[22]

$$\tau = F_\theta r \tag{3.21}$$

We can now write Equation (3.20) as $\tau = mr^2\alpha$. The mr^2 term comes from the fact that we are considering an object of mass m a distance r from the point of rotation. We see that, mathematically, this term represents a resistance of sorts to the torque: for a fixed value of τ, a larger value of mr^2 leads to a smaller value of α. (This should seem awfully familiar; we are drawing parallels to $\vec{F} = m\vec{a}$ for linear motion.)

Indeed, if we doubled the value of m or, equivalently, if we added a second identical mass to our system (e.g. two masses stuck to the ends of a thin rod, then rotated it about its midpoint, as sketched in Figure 3.50), then our equation would simply become $\tau = 2mr^2\alpha$. To generalize this, we refer to the quantity mr^2 as the **moment of inertia** of a point mass:

$$I_{\text{point mass}} = mr^2 \tag{3.22}$$

---

[21]We will also use the fact that the linear acceleration, a, is in the $\hat{\theta}$ direction, so $a = a_\theta = \alpha r$.

[22]I should mention that the full story is three-dimensional: if $\vec{r}$ and $\vec{F}$ lie on the plane of the page, $\vec{\tau}$ points either in to or out of the page (for clockwise and counter-clockwise rotations, respectively). The mathematical notation that relates these quantities is referred to as a **cross product**: $\vec{\tau} = \vec{r} \times \vec{F}$. The vector orientation of $\vec{\tau}$ is important for three-dimensional problems that we *won't* consider here. We will tackle the cross product in greater detail when we discuss magnetic fields. For now, if you're curious, check out Problem 3.134.

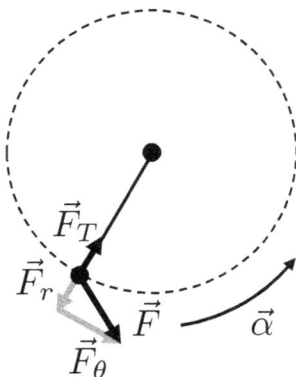

FIGURE 3.49

A top-down view of a light, rigid rod that pivots around one end such that a mass attached to the other end is constrained to move in a circle. An applied force $\vec{F}$ can be broken down into components in the $\hat{r}$ direction and the $\hat{\theta}$ direction. While $\vec{F}_r$ and a tension force from the rod account for the centripetal acceleration of the mass, $\vec{F}_\theta$ accounts for an angular acceleration $\vec{\alpha}$.

and if the object that we're rotating consists of a series of point masses, we have

$$I_{\text{point masses}} = \sum m_i r_i^2 \tag{3.23}$$

This is useful for situations like the light rod and masses shown in Figures 3.49 and 3.50, but it is also useful when we have a continuous object, like a heavy rod, disk, or sphere. In these cases we can extend Equation (3.23) to represent a sum over an infinite number of infinitely tiny point masses m_i, each with their own position r_i relative to the rotation point (Figure 3.51). We are, of course, referring to an integral. We will discuss the calculus involved in computing these moments of inertia, but if you'd like to skip to the results, note that a list of the moments of inertia for some common objects is provided in Figure 3.52.

To think through the integration involved in calculating the moment of inertia, suppose we have a one-dimensional object, like a rod pivoting about its center. According to Equation

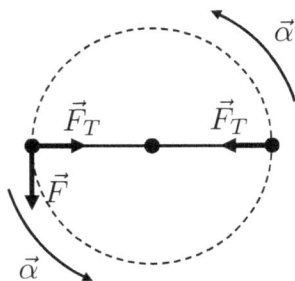

FIGURE 3.50

A top-down view of a light, rigid rod that pivots around its midpoint such that two masses attached to either end are constrained to move in a circle. An applied force $\vec{F}$ in the $\hat{\theta}$ direction (or, more generally, a component in the $\hat{\theta}$ direction) causes the rod and both masses to experience an angular acceleration $\vec{\alpha}$.

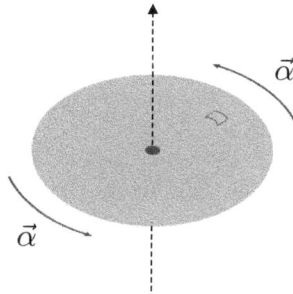

FIGURE 3.51
A disk can be viewed as a series of point masses stuck together, much like the simpler example shown in Figure 3.50. Here only one such "point mass" is outlined in black (the piece is drawn as a tiny component of a thin ring; you can imagine other pieces arrayed to complete the ring, then other rings with different radii to cover the entire disk).

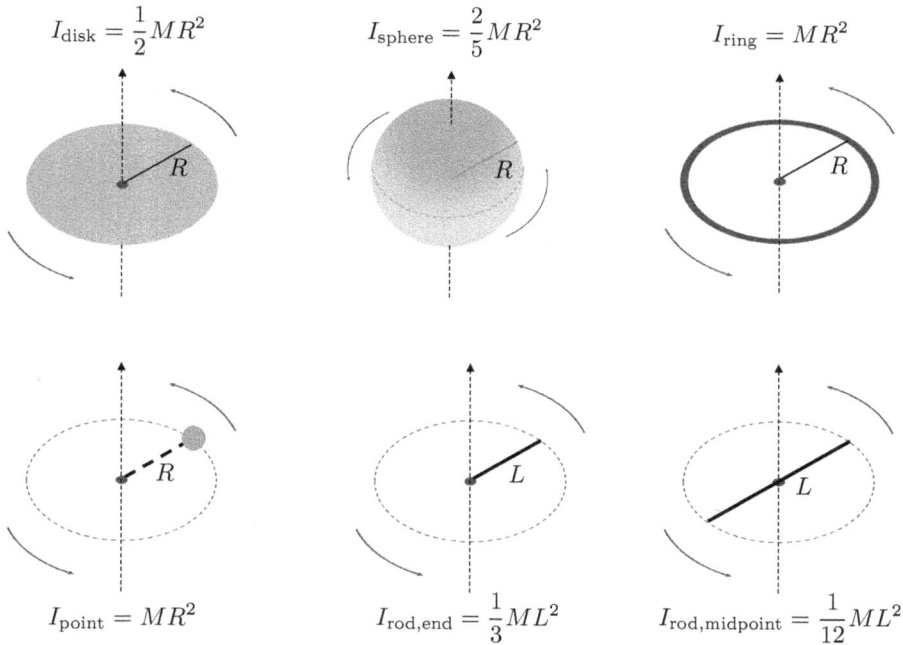

$$I_{\text{disk}} = \frac{1}{2}MR^2 \qquad I_{\text{sphere}} = \frac{2}{5}MR^2 \qquad I_{\text{ring}} = MR^2$$

$$I_{\text{point}} = MR^2 \qquad I_{\text{rod,end}} = \frac{1}{3}ML^2 \qquad I_{\text{rod,midpoint}} = \frac{1}{12}ML^2$$

FIGURE 3.52
The moments of inertia for some common objects. Each object has a mass M, while circular or spherical objects have a radius R and rods have a length L.

(3.23), our integral is

$$I = \int x^2 dm$$

which is to say, we integrate over every differential chunk of mass multiplied by the square of its distance from our rotation point. This isn't very useful, though; when we perform an integral over some geometric object (here, a rod) we tend to think of the limits of integration corresponding to a *position*, not a mass.

To move to this more intuitive framework, we introduce the **linear mass density**, i.e. the "mass per unit length": $\lambda = dm/dx$. Solving for dm and inserting into the integral above, we find

$$I = \int \lambda x^2 dx \tag{3.24}$$

This equation can be generalized to two and three dimensions,[23] though the ensuing double and triple integrals are beyond the scope of this book. In some cases, though, it isn't terribly challenging to break these multivariable integrals into a sequence of one-dimensional integrals; see Problem 3.118 for an example.

Example 3.28 Calculating Moment of Inertia $\qquad\qquad \star\star \int dx$

Use Equation (3.24) to determine the moment of inertia of (a) a rod of mass m and length L rotating about its midpoint and (b) a circular loop of mass m and radius r rotating about an axis through the middle of the loop and perpendicular to its face.

For (a) we have $\lambda = m/L$ (the mass per unit length is the same everywhere and so the ratio of a tiny chunk dm/dx is the same as the overall ratio). We will integrate from the center of the rod to one end and double our result to account for the other half:

$$I = \int \lambda x^2 dx$$

$$I = 2 \int_0^{L/2} \frac{m}{L} x^2 dx$$

$$I = \frac{2m}{L} \left(\frac{x^3}{3}\right)\Big|_0^{L/2}$$

$$I = \frac{mL^2}{12}$$

This agrees with the result shown in Figure 3.52.

Turning to (b), you might think that this is really a two-dimensional problem since the loop lies on a plane rather than a line. But as we walk along the loop, we only need one parameter, θ, to indicate where we are. Thus this is a one-dimensional situation in the same sense as the angular kinematic equations are one-dimensional.

Note first that $\lambda = m/(2\pi r)$. The quantity x^2 inside Equation (3.24) refers to the square of the distance of a piece of material from the axis of rotation: here this is simply the radius of the wire, so $x^2 \to r^2$. Meanwhile, dx refers to a step along the material; if an arc length is given by $x = r\theta$, then a differential step along the arc is

---

[23]Specifically, $I = \iint \sigma r^2 dA$ and $I = \iiint \rho dV$ for "mass per unit area" σ and "mass per unit volume" (i.e. density) ρ.

given by $dx = rd\theta$. Putting this together, the integral becomes

$$I = \int_0^{2\pi} \lambda r^2 \left(rd\theta \right)$$

$$I = \frac{m}{2\pi r} r^3 \int_0^{2\pi} d\theta$$

$$I = mr^2$$

Again, this agrees with the result shown in Figure 3.52. Note that this is the same moment of inertia as a point mass a distance r from the axis of rotation, which shouldn't be surprising – *every* chunk of mass in this situation is the same distance from the axis!

As a final comment on the moment of inertia, note that the total moment of inertia for multiple objects that are stuck together and rotating about some common axis (such as two rods crossed at their midpoint, or two concentric rings connected by light wire) is simply the sum of each individual moment of inertia:

$$I_{\text{TOT}} = \sum I \tag{3.25}$$

and if we have an extended body rotating about an axis parallel to the center of mass a distance d from the center of mass (like a meter stick pivoting around the 20 cm mark rather than the 50 cm mark), we can use the **parallel axis theorem**:

$$I = I_{\text{CM}} + Md^2 \tag{3.26}$$

Once we recognize that an object can experience multiple torques (and that the net torque is simply the sum of all torques, as with forces) we're finally in a position to rewrite Equation (3.21) in its general form:

$$\vec{\tau}_{\text{NET}} = I\vec{\alpha} \tag{3.27}$$

where the **net torque** is either counter-clockwise (for a positive torque) or clockwise (for a negative torque) and can be written as a vector sum over individual torques:

$$\vec{\tau}_{\text{NET}} = \sum \vec{\tau} = \vec{\tau}_1 + \vec{\tau}_2 + \dots \tag{3.28}$$

You'll note that I've introduced vector notation here now that we are accounting for the *direction* of the torque as well as the magnitude. There are only two choices (clockwise or counter-clockwise, i.e. negative or positive), so adding torques is analogous to adding one-dimensional vectors.[24]

Equation (3.27), with the net torque defined in Equation (3.28), is **Newton's Second Law for rotational motion**. Torque plays the role of force: it is the quantity that tends to make an object experience rotational acceleration, $\vec{\alpha}$; the moment of inertia I is the resistive quantity that takes the role of mass. We will see that these analogous relationships are very useful as we move on to discuss conservation of energy and conservation of momentum for rotational motion. Before we do so, however, it is well worth our time to carefully go through some examples demonstrating the ideas we have developed in this section.

---

[24]This is only because the lever arms and forces we will consider lie in the plane of the page. Things get to be more complicated in the general three-dimensional case.

Example 3.29 Torque on a Merry-Go-Round   ★★

The left side of Figure 3.53 shows a top-down view of a merry-go-round (MGR) that has a mass $m = 1.50 \times 10^3$ kg and a radius $R = 7.5$ m. Three ropes are tied to the MGR and pulled, as shown: $\vec{F}_1$ is applied at the edge, $\vec{F}_2$ halfway to the edge, and $\vec{F}_3$ at the center of the MGR.

a. What is the net torque applied to the MGR?

b. What is the moment of inertia of the MGR?

c. What is the angular acceleration of the MGR?

d. If the MGR begins from rest, how long will it take for a point on the edge to reach a linear speed of 5.0 m/s? (Assume that the orientations between the forces and the MGR remain constant as it rotates.)

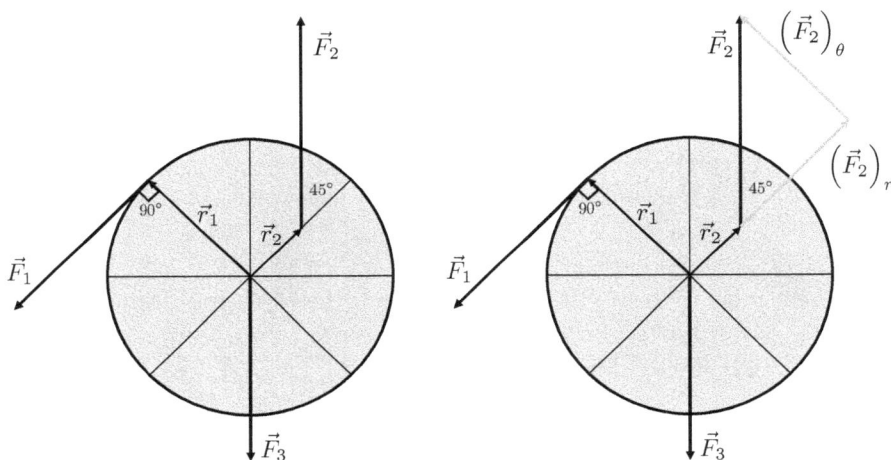

FIGURE 3.53

(a) The force labeled F_1 is applied tangent to the rim of the circle, i.e. it points entirely in the $\hat{\theta}$ direction. Thus

$$\tau_1 = F_1 R = (350 \text{ N})(7.5 \text{ m}) = 2.6 \times 10^3 \text{ Nm}$$

and the direction is counterclockwise, i.e. positive. Meanwhile, the second force is applied at a distance $r/2$ from the center and the $\hat{\theta}$ component is $F_2 \sin(45°)$ (as shown on the right side of Figure 3.53), so

$$\tau_2 = F_2 \sin(45°) \frac{R}{2}$$

$$= (4.0 \times 10^2 \text{ N}) \sin(45°) \frac{7.5 \text{ m}}{2}$$

$$= 1.1 \times 10^3 \text{ Nm}$$

and the direction is also counterclockwise and therefore positive.

Finally, for the third force the torque is 0 Nm. Why? There is no lever arm because the force is being applied directly at the point of rotation, so $\tau_3 = F_3 \, (0 \text{ m}) = 0$ Nm. We conclude that the net torque is

$$\sum \tau = \tau_1 + \tau_2 = 3.7 \times 10^3 \text{ Nm}$$

(b) The MGR is (approximately) a disk, so

$$I = \frac{1}{2}mR^2 = \frac{1}{2}\left(1.50 \times 10^3 \text{ kg}\right)(7.5 \text{ m})^2 = 4.2 \times 10^4 \text{ kg m}^2$$

(c) From Newton's Second Law,

$$\sum \vec{\tau} = I\vec{\alpha}$$
$$\alpha = \frac{\sum \vec{\tau}}{I}$$
$$\alpha = \frac{3.7 \times 10^3 \text{ Nm}}{4.2 \times 10^4 \text{ kg m}^2}$$
$$\alpha = 8.7 \times 10^{-2} \text{ rad/s}^2$$

(d) Here we can apply angular kinematics. Recall first that the tangential velocity and angular velocity are related by

$$\omega = \frac{v}{r} = \frac{5.0 \text{ m/s}}{7.5 \text{ m}} = 0.67 \text{ rad/s}$$

Then to find the time to accelerate to this final velocity:

$$\omega_f = \omega_i + \alpha \Delta t$$
$$\Delta t = \frac{\omega_f}{\alpha}$$
$$\Delta t = \frac{0.67 \text{ rad/s}}{8.7 \times 10^{-2} \text{ rad/s}^2}$$
$$\Delta t = 7.6 \text{ s}$$

where in the second like we've noted $\omega_i = 0$ rad/s.

Example 3.30 Moment of Inertia of a Loaded See-Saw ⋆

Consider a see-saw that consists of a 10.7 kg board that is 2.30 m in length and pivots about its midpoint. If a 20.0 kg child sits on one end and a 23.5 kg child sits on the other end, what is the moment of inertia of the see-saw and children?

We sketch the situation in Figure 3.54. We have three objects to consider: the board (a rod of length L rotating about its midpoint) and each child (whom we will consider to be point masses each a distance $L/2$ from the rotation point). According to Equation (3.25), the total moment of inertia is then

$$I_{\text{TOT}} = \sum I$$

$$I_{\text{TOT}} = I_{\text{board}} + I_{\text{child 1}} + I_{\text{child 2}}$$

$$I_{\text{TOT}} = \frac{1}{12} m_b L^2 + m_{\text{c1}} \left(\frac{L}{2}\right)^2 + m_{\text{c2}} \left(\frac{L}{2}\right)^2$$

$$I_{\text{TOT}} = \left(\frac{1}{12} m_b + \frac{1}{4} (m_{\text{c1}} + m_{\text{c2}})\right) L^2$$

$$I_{\text{TOT}} = \left(\frac{1}{12} (10.7 \text{ kg}) + \frac{1}{4} (20.0 \text{ kg}) + (23.5 \text{ kg})\right) (2.30 \text{ m})^2$$

$$I_{\text{TOT}} = 62.2 \text{ kg m}^2$$

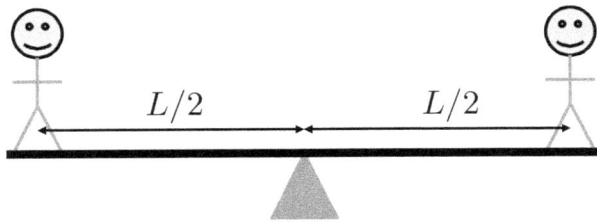

FIGURE 3.54

Example 3.31 Torque and Angular Acceleration on a Door   ★★

Figure 3.55 shows several top-down views of a door of mass M; in each the door is experiencing a force of magnitude F applied in a different location/direction. Determine an expression for the magnitude of the door's angular acceleration in each case, and identify the largest. Note that a door has the same moment of inertia as a rod rotating about its end (the "depth" of the rod – i.e. the height of the door – does not affect its moment of inertia).

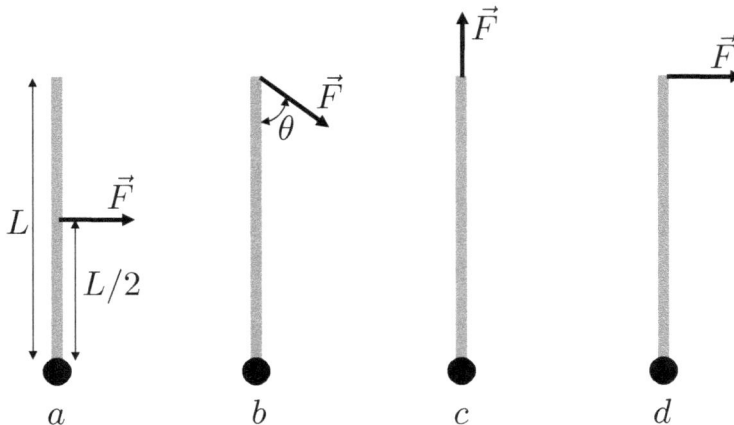

FIGURE 3.55

It is always a good idea to arrive at an intuitive answer before you think through the physics in detail (to then confirm your intuition or otherwise see precisely how and why your intuition is incorrect). Here, your intuition is probably that (d) is the correct choice: when you want to push open a door you tend to push on the end farthest from the hinge, and you push straight on rather at an angle. This is the correct choice; in the language of physics this is because it *maximizes the torque*. In Figure 3.56 we re-draw each situation with $\vec{r}$ drawn explicitly and with $\vec{F}$ broken into its radial ($\hat{r}$) and tangential ($\hat{\theta}$) components (in panel (a) we also show part of the circular path of the tip of the door, in case the circular motion isn't clear from the context).

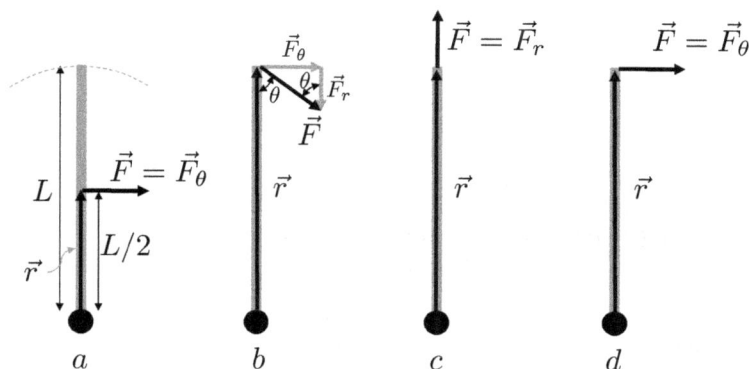

FIGURE 3.56

The moment of inertia of the door is $I = \frac{1}{3}ML^2$; because the moment of inertia is a property of the door, it does not change based on what torque(s) we apply (much as mass does not depend on the force(s) applied to an object). The torques (Equation (3.21)), however, do change:

(a) The force is entirely in the $\hat{\theta}$ direction, but it is being applied only halfway down the length of the door, so $\tau_a = FL/2$.

(b) Here the force has only a component in the $\hat{\theta}$ direction, $F_\theta = F\sin\theta$, and it is applied at a distance L from the end of the door, so $\tau_b = FL\sin\theta$.

(c) Here the entire force is in the $\hat{r}$ direction so $F_\theta = 0$ N and $\tau_c = 0$ Nm.

(d) Here the entire force is in the $\hat{\theta}$ direction and we have the maximum "lever arm", L, so $\tau_d = FL$.

Now we can insert these expressions for τ, along with the moment of inertia for the door, into Newton's Second Law for rotational motion (Equation (3.27)). We care only about magnitudes so we will drop the vector notation,[a] and we'll start by simplifying the expression while referring to a generic τ before plugging in the specific expressions for each part:

$$\tau = I\alpha$$

$$\alpha = \frac{\tau}{I}$$

$$\alpha = \frac{\tau}{\frac{1}{3}ML^2}$$

$$\alpha = \frac{3\tau}{ML^2}$$

Now we will insert the expressions for τ:

(a) $\alpha_a = \frac{3}{ML^2}\tau_a = \frac{3}{ML^2}\frac{FL}{2} = \frac{3}{2}\frac{F}{ML}$

(b) $\alpha_b = \frac{3}{ML^2}\tau_b = \frac{3}{ML^2}FL\sin\theta = 3\sin\theta\frac{F}{ML}$

(c) $\alpha_c = \frac{3}{ML^2}\tau_c = \frac{3}{ML^2}(0\text{ Nm}) = 0\text{ Nm}$

(d) $\alpha_d = \frac{3}{ML^2}\tau_d = \frac{3}{ML^2}FL = 3\frac{F}{ML}$

Thus we see that (d) does indeed provide the largest value of α (remember $\sin\theta$ ranges between -1 and 1, depending on the value of θ).

---

[a]That being said, by inspecting the figure you should be able to convince yourself that all of the motion is either 0 (for part (c)) or clockwise, i.e. in the negative $\hat{\theta}$ direction.

Example 3.32 Torque on a Sphere   ⋆

A 0.65 N force is applied tangent to the edge of a 3.5 kg sphere with an 8.0 cm diameter. If it begins at rest, what will its angular speed be after 0.52 seconds experiencing the force?

We can apply Newton's Second Law to determine the angular acceleration α, then use angular kinematics to determine the sphere's angular velocity ω. First, we'll consider Newton's Second Law. There is only one force acting on the sphere, and because it is tangential to its surface the net torque is just $\tau = FR$ where R is the radius. Thus

$$\vec{\tau}_{\text{NET}} = I\vec{\alpha}$$

$$FR = \frac{2}{5}MR^2\alpha$$

$$\frac{5}{2}\frac{F}{MR} = \alpha$$

$$\frac{5}{2}\frac{0.65\text{ N}}{(3.5\text{ kg})(0.04\text{ m})} = \alpha$$

$$1.2\frac{\text{rad}}{\text{s}^2} = \alpha$$

We can now turn to angular kinematics:

$$\vec{\alpha} = 1.2 \frac{\text{rad}}{\text{s}^2}$$

$$\vec{\theta}_i = 0 \text{ rad}$$

$$\vec{\theta}_f = ?$$

$$\vec{\omega}_i = 0 \frac{\text{rad}}{\text{s}}$$

$$\vec{\omega}_f = ?$$

$$\Delta t = 0.52 \text{ s}$$

We would like to know $\vec{\omega}_f$ but we don't know $\Delta\vec{\theta}$, so we choose the equation that omits it:

$$\vec{\omega}_f = \vec{\omega}_i + \vec{\alpha}\Delta t$$

$$\omega_f = 0 \frac{\text{rad}}{\text{s}} + \left(1.2 \frac{\text{rad}}{\text{s}^2}\right)(0.52 \text{ s})$$

$$\omega_f = 0.60 \frac{\text{rad}}{\text{s}}$$

Example 3.33 The Center of Mass   ★★

A light, 45-cm-long rod has a small 3kg mass attached to one end and a small 5.0 kg mass attached to the other end. You intend to impress someone by balancing the entire rod on your finger. Where should you place it so the contraption will experience no net torque due to gravity? (This location is referred to as the object's **center of mass**.)

We sketch the rod and masses in Figure 3.57. If your finger is anywhere between them, the rod will experience opposing torques (due to the gravitational force on each of the masses). If the torques are to balance so the net torque is 0 Nm, the larger mass must have a smaller lever arm:

$$\vec{\tau}_{\text{NET}} = I\vec{\alpha}$$

$$m_1 g r_1 - m_2 g r_2 = 0$$

$$m_1 r_1 = m_2 r_2$$

There are many cases where we want the torques to cancel so the object is in so-called **rotational equilibrium**; some other examples are included in the problems at the end of the chapter.

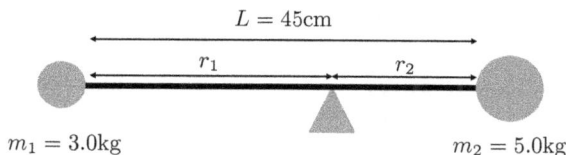

FIGURE 3.57

Back to the matter at hand, we see that we have two unknowns, r_1 and r_2, but only one equation. The other part of this is that we know that their sum is equal to the length of the rod:

$$r_1 + r_2 = L$$

Thus we can solve this expression for, say, r_2 and then re-insert it into our equation from Newton's Second Law to determine r_1:

$$m_1 r_1 = m_2 r_2$$
$$m_1 r_1 = m_2 \left(L - r_1 \right)$$
$$\left(m_1 + m_2 \right) r_1 = m_2 L$$
$$r_1 = \frac{m_2 L}{m_1 + m_2}$$

If we insert the values provided in the problem, we find $r_1 = 0.28$ m, so you need to place your finger 28 cm from the 3.0 kg mass for the object to balance. (Though I won't promise that anyone will be impressed by the display.)

The equation for r_1 is an example of the general equation for finding the position of the center of mass of any object consisting of masses spread along a straight line:

$$x_{\mathrm{COM}} = \frac{\sum m_i \vec{x}_i}{\sum m_i} \tag{3.29}$$

In words, this says that you need to choose an origin for your coordinate system, then multiply each mass by its distance from that origin (the position is a vector; because we are assuming a one-dimensional situation, this means it can be either positive or negative). You do this for every mass and add up your results, then divide by the total mass.[a] Note also that if we were considering the situation where the rod itself had some mass, then we could consider all of its mass to be centered at its midpoint (you are invited to investigate this in Problem 3.120).

Equation (3.29) is a useful result that is useful in its own right (e.g. Problems 3.125 and 3.126) and can be extended straightforwardly to two-dimensional and three-dimensional objects as well (simply calculate the center of mass independently along each Cartesian axis; see e.g. Problems 3.127 and 3.128). Example 3.34 provides some additional practice with these ideas.

---

[a]You should be able to see that our equation for r_1 above agrees with this general equation if we place the origin of our coordinate system on top of m_1.

Example 3.34 Building a Mobile   ★ ★ ★

You are competing in a physics competition and have been asked to construct a *mobile* consisting of rods, objects, and string. You are expected to hang every object from a rod, and to connect one more string to each rod that connects either to a rod above it or (for the top rod) to the ceiling. The objective is to connect the strings at appropriate positions so each rod hangs completely horizontal.

You have a somewhat simple mobile consisting of just two rods (each 1.00 m long and light enough to consider massless) and three objects of masses 2.40 kg, 3.20 kg, and 6.90 kg. Unfortunately, your arch-nemesis has used superglue to attach the objects to the rods as shown in Figure 3.58.

(a) Where should you attach a string to the bottom rod so it will hang horizontally? Is there a range of viable choices, only one, or none?

(b) Where should you attach the bottom rod to the top rod, and where should you attach the support string to the top rod, so the top rod will hang horizontally? Is there a range of viable choices, only one, or none?

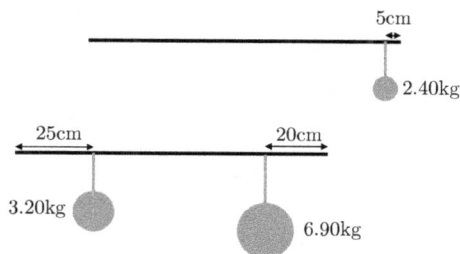

FIGURE 3.58

In Figure 3.59 we re-draw the schematic with the additional strings; forces and positions are also labeled. Now the support point needs to be at the center of mass of the rod, so we can jump straight to Equation (3.29) (as long as we note that the effect of the tension force in the strings connected to the masses is to apply a downward force on the rod equal to the weight of each mass). Setting the origin on the left end of the lower rod, we find

$$x_{\text{COM}} = \frac{\sum m_i \vec{x}_i}{\sum m_i}$$

$$x_{\text{COM}} = \frac{(3.20 \text{ kg}) (0.25 \text{ m}) + (6.90 \text{ kg}) (0.80 \text{ m})}{3.20 \text{ kg} + 6.90 \text{ kg}}$$

$$x_{\text{COM}} = 0.63 \text{ m}$$

Thus the support point should be 63 cm from the left end of the rod; there is no other way to balance the lower rod.

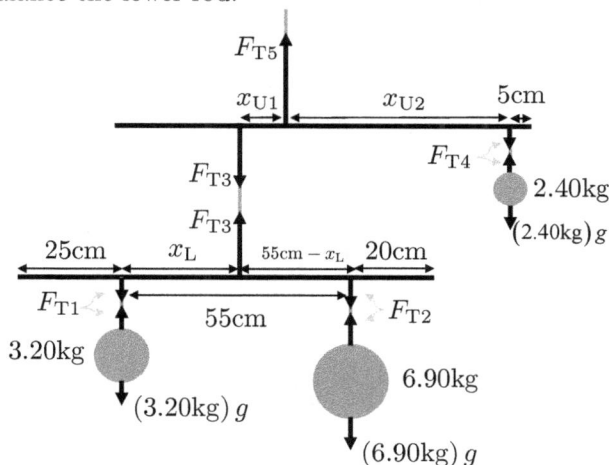

FIGURE 3.59
The mobile with additional labeling. The downward tension applied to the ceiling from the top string is not shown.

We can proceed in the same way for (b), but we now have some additional flexibility because we can choose where to attach the lower rod to the upper rod *and* where to attach the support string for the upper rod. To think this through, let's set up the center of mass equation for the top rod, setting the origin at the center of mass (so the desired position of the center of mass is 0 m, i.e. at the origin) and calling right the positive direction. While we have a string on the left side, its effect is to apply the weight of both lower masses to the top rod, so we consider both masses to be a distance x_{U1} to the left of the origin:

$$x_{COM} = \frac{\sum m_i \vec{x}_i}{\sum m_i}$$

$$0 \text{ m} = \frac{(3.20 \text{ kg} + 6.90 \text{ kg})(-x_{U1}) + (2.40 \text{ kg})x_{U2}}{3.20 \text{ kg} + 6.90 \text{ kg} + 2.40 \text{ kg}}$$

$$0 \text{ m} = 10.1(-x_{U1}) + (2.40)x_{U2}$$

$$10.1 x_{U1} = 2.40 x_{U2}$$

Now we have no other constraints beyond the fact that $x_{U1} + x_{U2} \leq 0.95$ m. Thus

$$x_{U1} \leq 0.95 \text{ m} - x_{U2}$$

If we insert this into our center of mass equation, we find

$$10.1 x_{U1} = 2.40 x_{U2}$$

$$x_{U1} = \frac{2.40}{10.1} x_{U2}$$

$$\frac{2.40}{10.1} x_{U2} \leq 0.95 \text{ m} - x_{U2}$$

$$\left(\frac{2.40}{10.1} + 1\right) x_{U2} \leq 0.95 \text{ m}$$

$$x_{U2} \leq \frac{0.95 \text{ m}}{\frac{2.40}{10.1} + 1}$$

$$x_{U2} \leq 0.77 \text{ m}$$

This specifies a range of possible values for x_{U2}, then our center of mass equation determines x_{U1}:

$$10.1 x_{U1} = 2.40 x_{U2}$$

$$x_{U1} = \frac{2.40}{10.1} x_{U2}$$

As two examples, let's consider the extreme values of $x_{U2} = 0.77$ m and $x_{U2} = 0$ m. Inserting in to the above equation, we find

$$x_{U2} = 0.77 \text{ m} \Rightarrow x_{U1} = 0.18 \text{ m}$$

$$x_{U2} = 0 \text{ m} \Rightarrow x_{U1} = 0 \text{ m}$$

Thus in the first case, we are able to balance the rods by attaching the lower rod to the extreme left edge of the upper rod (since $0.18 + 0.77 + 0.05 = 1.00$) and attaching the top rod's support string 77 cm to the left of the right hanging mass. In the second case, we tie both the bottom rod and the support string directly on top of the right hanging mass (this is the rather uninteresting case where everything tied to the top rod is stacked on top of one another).

Example 3.35 Atwood's with a Massive Pulley   ⋆ ⋆ ⋆

In Example 3.16 on page 136 we considered an Atwood's machine, which consists of a rope slung over a pulley and connected to a mass on either end (Figure 3.60). While in Example 3.16 we treated the pulley as massless, here we will suppose that we cannot ignore the mass of the pulley: the rope grips the edge of the pulley due to friction, and it experiences an angular acceleration as a result. Specifically, we will assume that the masses are initially at rest and the heavier mass m_2 is initially hanging a distance h above the floor, then determine:

(a) the magnitude of the linear acceleration a of the blocks

(b) the magnitude of the angular acceleration α of the pulley

(c) the speed of the heavier block when it strikes the ground

We will also confirm that in the case where the mass of the pulley is very small, the linear acceleration of the blocks agrees with the result found in Example 3.16.

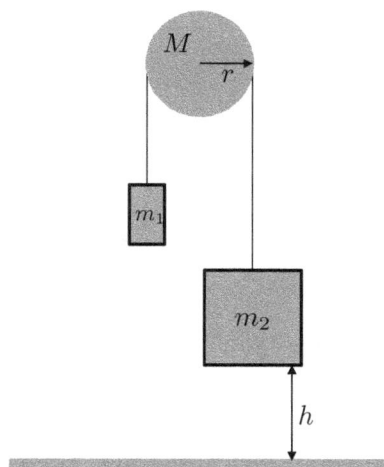

FIGURE 3.60

When the pulley was massless, we noted that the rope applies an equal tension force to each mass. The first thing to recognize here is that this is no longer the case: the pulley rotates because the rope is pulling on it (i.e. exerting a force on it). Thus there must be a downward tension force exerted by the rope on either side of the pulley, and they can't be the same (if they were, the torques due to those forces would cancel and the pulley would never rotate). Meanwhile, the masses still experience upward tension forces. We sketch the situation in Figure 3.61.

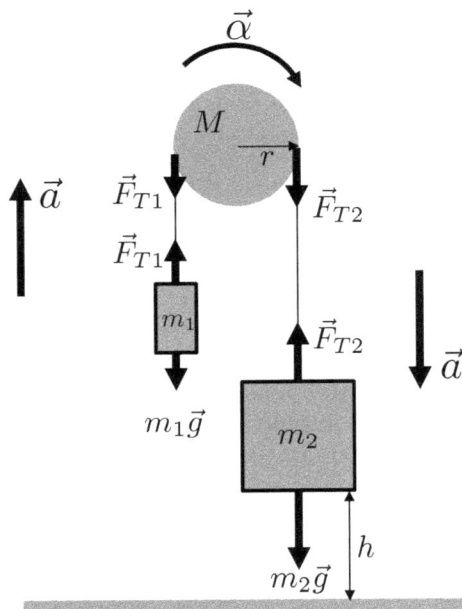

FIGURE 3.61

For the blocks, we will consider the direction of the acceleration to be the positive direction ("m_1 goes up and m_2 goes down"). In Example 3.16 this allowed us to treat the tension on either end of the rope as a pair of (unknown) internal forces that cancel when we considered both blocks as one combined mass. We can do something similar here by combining both masses *and the pulley* into one combined mass. To see how, let's first write out Newton's Second Law for each object.

For m_1:	For m_2:	For the pulley:
$\vec{F}_{NET} = m\vec{a}$	$\vec{F}_{NET} = m\vec{a}$	$\vec{\tau}_{NET} = I\vec{\alpha}$
$F_{T1} - m_1 g = m_1 a$	$-F_{T2} + m_2 g = m_2 a$	$F_{T2}r - F_{T1}r = \dfrac{1}{2}Mr^2\alpha$

The idea behind canceling internal forces is that we can add equations for each mass and find, for instance, $-F$ on the left hand side of one equation and F on the left hand side of another, so when the equations are added the terms cancel. We see that we have F_{T1} in the equation for m_1 and $-F_{T1}r$ in the equation for the pulley (and similarly for the other tension force). To make the tension forces cancel when we add the equations, we will therefore divide both sides of the equation for the pulley by r. While we are at it, we will also note that $a = \alpha r \Rightarrow \alpha = a/r$, since a point on the rim of the pulley must accelerate at the same linear rate as either block. Then the equation for the pulley becomes:

$$F_{T2}r - F_{T1}r = \frac{1}{2}Mr^2\alpha$$

$$F_{T2} - F_{T1} = \frac{1}{2}Mr\frac{a}{r}$$

$$F_{T2} - F_{T1} = \frac{1}{2}Ma$$

Now we can add this equation with the equations for each mass (the sum of the left hand sides equals the sum of the right hand sides):

$$(F_{T1} - m_1 g) + (-F_{T2} + m_2 g) + (F_{T2} - F_{T1}) = m_1 a + m_2 a + \frac{1}{2} M a$$

$$(m_2 - m_1) g = \left(m_2 + m_1 + \frac{1}{2} M \right) a$$

$$\frac{m_2 - m_1}{m_2 + m_1 + \frac{1}{2} M} g = a$$

The unknown tension forces have cancelled, which was the whole point. We have our answer to (a), and we see that if M is small enough to be ignored (i.e. if $M \approx 0$ kg or, more formally, if $M \ll m_2 + m_1$), the equation reduces to

$$\frac{m_2 - m_1}{m_2 + m_1} g = a$$

If you compare this to Example 3.16 (and note that we have re-named some objects), you will see that the answers agree, as expected.

For (b), we simply need to recall $\alpha = \frac{a}{r}$, so

$$\alpha = \frac{m_2 - m_1}{m_2 + m_1 + \frac{1}{2} M} \frac{g}{r}$$

Finally, for (c), we have standard one-dimensional kinematics. We already have the acceleration; the other kinematic variables are listed below. (Note that because we're calling *down* the positive direction, $x_f > x_i$. This is why we set $x_f = h$ and $x_i = 0$ m.)

We use the kinematic equation without t:

$$\vec{x}_i = 0 \text{ m}$$
$$\vec{x}_f = h$$
$$\vec{v}_i = 0 \frac{\text{m}}{\text{s}}$$
$$\vec{v}_f = ?$$
$$\Delta t = ?$$

$$\vec{v}_f^2 = \vec{v}_i^2 + 2 \vec{a} \Delta \vec{x}$$

$$v = \left(\left(0 \frac{\text{m}}{\text{s}} \right)^2 + 2 \left(\frac{m_2 - m_1}{m_2 + m_1 + \frac{1}{2} M} g \right) (h - 0 \text{ m}) \right)^{\frac{1}{2}}$$

$$v_f = \left(2gh \frac{m_2 - m_1}{m_2 + m_1 + \frac{1}{2} M} \right)^{\frac{1}{2}}$$

There we have it. Before we move on, it is worth noting that we *could* have determined the linear acceleration a by solving the equation for m_1 for F_{T1} and the equation for m_2 for F_{T2}, then plugging these expressions into the equation for the pulley (if we have three equations and three unknowns, we can eliminate any two of those unknowns to arrive at an equation for the third).

3.14   Problems for Chapter 3

3.2 Newton's 1st and 2nd Laws: Forces Can Cause Acceleration

⋆   **Problem 3.1.** Are the following objects in equilibrium? If *yes*, is it static equilibrium or translational equilibrium? Briefly explain.

a. An apple hanging from a tree branch.

b. An apple as it falls to the ground.

⋆   **Problem 3.2.** Are the following objects in equilibrium? If *yes*, is it static equilibrium or translational equilibrium? Briefly explain.

a. A passenger in a car that is driving at a constant speed over a road that is straight (no turns) and flat (level ground).

b. A passenger in a car that is driving at a constant speed over a road that is straight and has an upward incline.

c. A passenger in a car that is driving at a constant speed around a circular turn (flat, but not straight).

⋆   **Problem 3.3.** What is the weight, in Newtons, of the following objects?

a. A person with a weight of 150 pounds.

b. A 1.0 kg box of pasta.

c. A passenger airplane that weighs 330 metric tons. (1 metric ton is 1000 kg.)

⋆   **Problem 3.4.** Suppose I hold a coffee mug with mass m in my hand. Determine an expression for the force I need to apply to the mug in each of the following situations:

a. I hold the mug motionless.

b. I move the mug at a constant horizontal speed v.

c. I move the mug at a constant vertical speed v.

d. I move the mug straight down so it accelerates at a rate g. (Hopefully there isn't anything in it!)

⋆   **Problem 3.5.** Figure 3.62 shows the acceleration versus force graphs for two objects. What is the mass of each object?

⋆   **Problem 3.6.** Two people are fighting over a 12 kg trunk (Fig. 3.63). The person on the left is exerting a force $F_L = 50$. N to the left. If the box is accelerating at 1.0 m/s^2 to the left, what force is the person on the right exerting? (Neglect the effect of friction acting on the trunk).

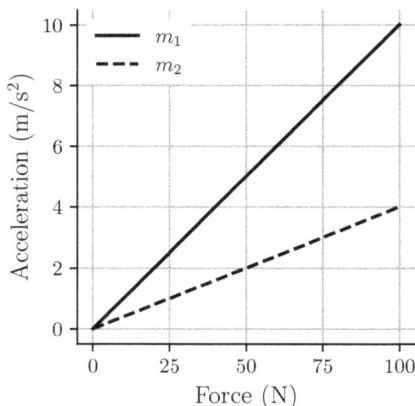

FIGURE 3.62
Problem 3.5

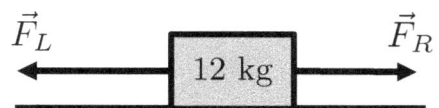

FIGURE 3.63
Problem 3.6

★   **Problem 3.7.** A 150 kg object is subjected to forces $\vec{F}_1 = 125\hat{x}$ N and $\vec{F}_2 = 250\hat{y}$ N. What third force would place the object into equilibrium? Include a force diagram.

★   **Problem 3.8.** A car is advertised as being able to accelerate from rest to a speed of 60.0 mph in 1.9 seconds.

   a. What average net force (in Newtons!) is needed to achieve this if the car weighs 5.0×10^3 pounds?

   b. How far does the car travel in this time?

★   **Problem 3.9.** A musket with a barrel length of 1.0 m fires a 15g bullet with a horizontal speed of 2.70×10^2 m/s. The bullet strikes a thick target and penetrates to a depth of 8.0 cm. What average force does the target apply to the bullet?

★   **Problem 3.10.** Consider the situation described in Example 3.4. If the object starts from rest at the origin, how far will it have traveled in 3.0 seconds, and what will its speed be?

★   **Problem 3.11.** A 0.16 kg billiard ball is shot east at 2.0 m/s by striking it with a pool cue with a 1.0 kN force. Through what distance was the force applied from the cue to the ball?

★★ **Problem 3.12.** A 0.50 kg ice hockey puck is traveling in the $+\hat{y}$ direction (up the page, as usual) at a speed of 2.5 m/s when a hockey stick exerts a force of 4.0 N directed $30.^\circ$ left of the $+\hat{y}$ direction. Once the puck has traveled 0.30 m in the $\hat{y}$ direction, the force stops. What is the final velocity of the puck?

★★ **Problem 3.13.** A person bends their knees and then jumps straight into the air. As they're pushing off the ground, their head moves up 0.30 m. Then, once they're airborne, their head moves up another 0.20 m. If the person's mass is 55 kg, what average force did they experience as they were pushing off the ground?

★★ **Problem 3.14.** A bottle rocket has a mass of 1.0 kg and a thrust of 3.0 N. It is fired horizontally 2.0 m from the edge of a frictionless table that is 1.5 m above the ground. How far from the base of the table does the rocket land? (Assume the rocket maintains its horizontal orientation even after leaving the table.)

★★ **Problem 3.15.** A 5.0 kg object experiences the forces $\vec{F}_1 = (3.0\hat{x} - 3.0\hat{y})$ N, $\vec{F}_2 = (-3.0\hat{x} + 4.0\hat{y})$ N, and a third force pushing the box in the $-\hat{x}$ and $-\hat{y}$ direction: $F_3 = 8.0$ N oriented 45° below the $-x$ axis.

   a. What net force is being applied to the box?

   b. If the box starts from rest at $t = 0.0$ s, where will it be at $t = 2.0$ s? What velocity will it have?

   c. If the box has a velocity of $\vec{v}_i = 1.2\hat{y}$ m/s at $t = 0.0$ s, where will it be at $t = 2.0$ s? What velocity will it have?

★★ **Problem 3.16.** A heavy stone block is being pulled into place. When looking down from above and using the usual coordinate system where $+x$ is directed to the right and $+y$ is directed up, the forces being applied to the block are $\vec{F}_1 = -25N\hat{y}$, $\vec{F}_2 = 20N$ at 45° above the $+x$ axis, and $\vec{F}_3 = 55N$ at $30.^\circ$ to the left of the $+y$ axis.

   a Draw a diagram: a dot to represent the box and a vector for each of the three forces.

   b What is the total (or *net*) force, $\vec{F}_{\text{NET}} = \vec{F}_1 + \vec{F}_2 + \vec{F}_3$? Express your answer in component form *and* in magnitude-angle form.

★★ **Problem 3.17.** Figure 3.64 shows the force diagram for a point particle. If $\sum \vec{F} = (10.1\hat{x} - 3.1\hat{y})$ N with $\hat{x}$ directed to the right and $\hat{y}$ directed up, as usual, then what is $\vec{F}_4$?

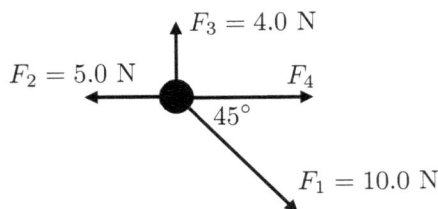

FIGURE 3.64
Problem 3.17

★★   **Problem 3.18.** Figure 3.65 shows three forces being applied to a ring. If the ring is to experience no net force, then what numerical values (to two significant figures) should θ and F_3 have?

★★   **Problem 3.19.** Figure 3.66 shows a ring. If the ring is to experience no net force, then what third force (magnitude and direction!) should be applied to the ring? Provide two significant figures in your response.

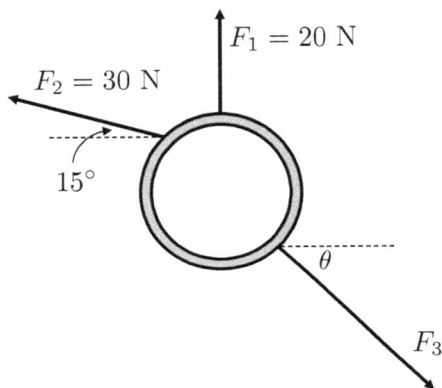

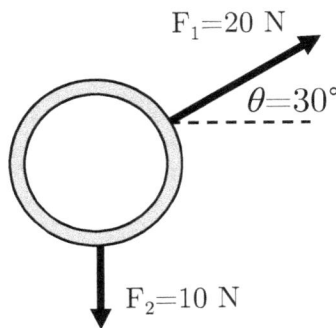

FIGURE 3.65
Problem 3.20

FIGURE 3.66
Problem 3.19

★★   **Problem 3.20.** Suppose a person with a mass of 62.0 kg is driving at 20.0 m/s when they are in a head-on collision.

a. Cars are designed to crumple over a distance of about 1.0 m. What average force does the person experience if this is the case?

b. If the person isn't wearing a seatbelt, they'll instead come to rest in about 5.0 mm as they collide with the windshield of their car. What is the average force on the person in this situation?

3.3 Newton's 3rd Law: Forces Come in Pairs

★   **Problem 3.21.** A friend complains that Newton's 3rd Law can't possibly be true: "If every force has an equal and opposite partner, then nothing can ever accelerate!" In a few sentences (and perhaps a diagram), how might you correct your friend's confusion?

★   **Problem 3.22.** A pickup truck is pulling a trailer; they are slowing down. Is the force of the truck on the trailer larger than, equal to, or smaller than the force of the trailer on the truck? Explain.

★   **Problem 3.23.** A coconut filled with a firecracker explodes into two pieces one that has a larger mass than the other. During the explosion which piece experiences a bigger magnitude of acceleration?

★★ **Problem 3.24.** Two people ($m_A = 75$ kg and $m_B = 150$ kg) are wrestling on an icy pond, which we can assume to be friction free. They are at a standstill when person A pushes person B with a force of 1.5 kN. If the push takes 1.00×10^{-3} s, what is the speed of person B post push?

★★ **Problem 3.25.** Consider the situation described in Problem 3.6. Sketch the situation including the box, both people, and the ground (make your life a bit easier and ignore the rope; assume instead the person pulls directly on the box). Identify all of the forces and identify which form Newton's 3rd Law pairs. (*Hint:* The Earth is gravitationally attracted toward every object that experiences a gravitational force toward the Earth; the magnitudes are the same!)

★ **Problem 3.26.** A pushing force is applied to a box of mass M that is touching a box of mass m (Fig. 3.67). The ground is level and frictionless. Write out Newton's Second Law for (a) mass M, (b) mass m, and (c) the combined system. (d) If $M = 30.0$ kg, $m = 15.0$ kg, and $F_{push} = 50.0$ N, what is the acceleration of the system?

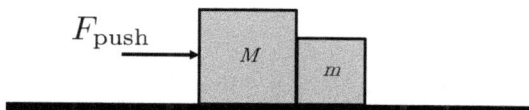

FIGURE 3.67
Problem 3.26

★★ **Problem 3.27.** Three blocks of mass $m_A = 3.0$ kg, $m_B = 2.0$ kg, and a third block C with an unknown mass are lined up in a row, left to right, on a frictionless table. Block A is pushed by a 12 N force directed to the right.

a. Draw a force diagram clearly identifying all forces acting on each block.

b. If the acceleration of the system is 2.0 m/s^2, determine the force that block B exerts on block A.

c. What is the net force on block B?

d. What is the mass of block C?

★ **Problem 3.28.** A train car has a mass of 3.0×10^5 kg and is sitting on level, frictionless ground.

a. What net force is required to make it accelerate at 0.20 m/s^2?

b. Suppose the force you found in (a) is being applied to the car when it runs into a second train of equal mass. The cars lock together. At what rate will they accelerate if the applied force does not change?

c. What force is each cart applying to the other? Draw a clear force diagram.

d. Repeat parts (b) and (c) in the scenario where the second car has a mass of 4.0×10^5 kg.

★★ **Problem 3.29.** Three blocks, of masses 15 kg, 20. kg, and 25 kg are sitting in a row on a frictionless table. All three masses are pushed forward by a 50. N force applied horizontally to the 15 kg block.

a. At what rate do the blocks accelerate?

b. How much force does the 15 kg block exert on the 20. kg block?

c. How much force does the 20. kg block exert on the 15 kg block?

d. How much force does the 20. kg block exert on the 25 kg block?

e. How much force does the 25 kg block exert on the 20. kg block?

★★ **Problem 3.30.** Three boxes of equal mass m are stacked, one on top of the other, and are in static equilibrium.

a. Draw a clear force diagram for each of the three boxes. Label the forces.

b. Which (if any) of the forces you drew in (a) are pairs according to Newton's 3rd Law?

c. Rank the forces according to their magnitudes from greatest to least. Explain.

3.4 Gravity

★★ **Problem 3.31.** What is the gravitational force between the Earth and the Sun? What about between Earth and Jupiter?

★★ **Problem 3.32.** Use the information in the appendices and Newton's Law of Gravitation to determine the acceleration due to gravity on:

a. The Moon.

b. Mercury.

c. Mars.

★ **Problem 3.33.** What is the radius of a planet that has the same mass as the Earth but has a free-fall acceleration of 20.0 m/s^2?

★ **Problem 3.34.** Planet Z is 1.00×10^4 km in diameter. The free-fall acceleration on Planet Z is 8.0 m/s^2. What is the mass of planet Z?

★ **Problem 3.35.** A space station orbits 3.0×10^2 km above the surface of the earth. What is the gravitational force on a 2.0 kg sphere inside the space station? Compare to the gravitational force of the sphere when on the surface of the Earth.

★ **Problem 3.36.** What is the weight of a 75 kg astronaut on (a) the Moon and (b) Mars?

★ **Problem 3.37.** A rocket has a motor than can produce 3.0×10^5 N of thrust vertically causing it to accelerate straight up at $a = 5.2$ m/s^2. Ignoring air resistance, what is the mass of the rocket?

★★ **Problem 3.38.** A device catches objects as they fall and slows their descent to prevent damage (for example, bales of hay thrown from the loft of a barn).

a. Draw a force diagram for the object as it falls through the air. Draw a separate force diagram for the object while it is being brought to rest by the catcher.

b. If the object impacts the catcher with a velocity of 10.0 m/s directed down toward the Earth and it was dropped from rest, how far did it fall?

c. If the catcher slows the object to rest over a distance of 2.00 m and a time of 2.00 s, then what is the object's acceleration? In which direction?

d. What is the object's mass if the average force registered by the catcher is 9.00×10^2 N?

★★ **Problem 3.39.** Fun with gravity near the surface of the Earth:

a. What is the acceleration due to gravity for a weather balloon 3.0×10^4 m above the surface of the Earth?

b. Repeat (a) for a satellite orbiting at 7.0×10^5 m above the surface of the Earth.

c. At what distance above the surface of the Earth is the acceleration due to gravity half of the value found *at* the surface?

d. Make a sketch of the acceleration due to gravity (on the vertical axis) as a function of distance above the surface of the Earth. Mark the values you found in (a)–(c).

★★ **Problem 3.40.** Consider an object of mass $M_1 = 4M$ fixed the origin of a one-dimensional coordinate system. A second object, of mass $M_2 = M$, is fixed at $x = R$. A third object of a mass $M_3 = M$ is free to move anywhere on the line connecting the two larger masses.

a. Make a graph of $\vec{F}_{G13}$ (the gravitational force of mass 3 acting on mass 1) vs. x for $0 < x < R$.

b. On the same graph, show $\vec{F}_{G23}$ vs. x for $0 < x < R$.

c. On the same graph, show the net gravitational force acting on mass 3 for $0 < x < R$.

d. Mathematically determine the value(s) of x in the range $0 < x < R$ where the net force acting on mass 3 is 0N.

3.5 Normal Force

★★ **Problem 3.41.** You are pulling your friend on a sled across a level bit of ground that is covered with snow with a rope at an angle above the ground. You have been walking at a steady 1.50 m/s and the mass of your friend and the sled is 60.0 kg. The rope pulls the sled with components of 100.0 N upward and 50.0 N to the right. What is the normal force acting on the sled?

★★ **Problem 3.42.** A 52.0 kg student stands on a 5.0 kg scale in an elevator. If the elevator is accelerating up at 1.0 m/s^2, what magnitude force will the student exert on the scale?

★★ **Problem 3.43.** A 25.0 kg child wants to go down a frictionless hill on a sled. The hill is 12.0 m long and is inclined at 25° relative to the horizontal. If the child begins at rest, how fast will they be moving when they reach the bottom of the hill? What normal force acts on the child while on the slope?

★★ **Problem 3.44.** A puck is sliding across a flat, frictionless surface at a speed of 0.50 m/s. It then begins to move up a gentle 10.° incline.

  a. Make a force diagram. What is the component of the gravitational force perpendicular to the ramp? Parallel to the ramp?

  b. How far along the incline will it travel before coming instantaneously to rest?

  c. What will the object's velocity be once it returns to the bottom of the incline?

  d. Make motion graphs for the object's position, velocity, and acceleration vs. time from the instant it begins moving up the incline to the instant that it returns to the bottom of the incline.

★★ **Problem 3.45.** A person with mass m stands on an elevator at the top floor of a building. The elevator begins at rest, then (a) accelerates at a constant rate a until it has a downward speed v, (b) continues with speed v until it is close to the bottom of the building, and finally (c) comes to rest with an acceleration with magnitude a. For each phase of the motion, determine an expression for the person's apparent weight.

★ **Problem 3.46.** Consider the *bed of nails* demonstration: a large rectangular board has thousands of nails driven through it, and a brave volunteer carefully lies down. The crowd goes wild as the person gets back up without injury. Suppose the person in question has a mass of 75 kg and normally stands on their feet, which have a surface area of 640 cm^2.

  a. What is the pressure exerted on the shoes by the floor when the person stands on their feet?

  b. Suppose each nail has a surface area of 0.060 cm^2 and when the person is laying down, 3000 nails support their weight. What is the pressure exerted on the person by the nails?

3.6 Friction

★ **Problem 3.47.** You have a 10.0 kg box sitting in the bed of your truck. The truck is initially traveling at 10.0 m/s and comes to rest in 5.0 s, and the box does not slide. What is the magnitude of the friction force between the bed of the truck and box?

★ **Problem 3.48.** A pickup truck with a steel bed is carrying a steel file cabinet. If the truck's speed is 15 m/s, the shortest distance the truck can stop without the file cabinet sliding is 14.3 m. What is the coefficient of friction between the truck and file cabinet?

★ **Problem 3.49.** A martial artist kicks a falling target pad and pins it against the wall with her foot. Is there a frictional force acting on the pad? Why or why not? If there *is*, which way does it point? Explain with words and a diagram.

★ **Problem 3.50.** A large 1.5 kg book is pushed horizontally into a wall with a force of magnitude 45 N. The book is slipping down the wall with an acceleration of −1.0 m/s^2.

  a Draw a force diagram for the book.

b What is the normal force acting on the book? Explicitly state a direction for the force.

c What is the friction force on the book? Explicitly state a direction for the force.

d What is the coefficient of friction between the book and wall?

e For a moment you are able to stop the book and you again apply the same 45 N push force; the book does not move. Why?

Problem 3.51. A worker is pushing an 82 kg crate to the right across a level barn floor such that the crate is accelerating at $+1.00$ m/s^2 where the positive direction is to the right, as usual. The worker is pushing the crate with 200.0 N downward and 250.0 N to the right.

a. What is the normal force acting upon the crate?

b. What is the friction force? Explicitly state what direction it is acting.

c. What is the coefficient of friction between the crate and floor?

d. If the crate stops and the worker pushes again with the same force, is it possible that the crate doesn't move? Explain.

Problem 3.52. A block is pushed along the floor with a force with magnitude F. When the block has a speed v, the pushing force is removed and the block slides a distance d before coming to rest.

a. What velocity is necessary when the push ends if you want the block to travel a distance of $2d$ before stopping?

b. Can the block be given the velocity you found in (a) with the same force F? If *yes*, how? If *no*, why not?

Problem 3.53. A block of mass m is pushed along the floor with velocity, v and slides a distance, d, after the pushing force is removed. What mass with the same initial velocity would slide a distance of $2d$ before stopping with the same surfaces in contact?

Problem 3.54. A drinking glass has mass m and is sliding along a table with an initial velocity v_0 and comes to rest after traveling a distance d.

a. If the mass is doubled (i.e. $m \to 2m$), what distance will the glass cover before coming to rest? Express your answer in terms of d.

b. If instead the velocity is doubled (i.e. $v_0 \to 2v_0$), what distance will the glass cover before coming to rest? Express your answer in terms of d.

c. If the coefficient of kinetic friction is doubled (i.e. $\mu_k \to 2\mu_k$), what distance will the glass cover before coming to rest? Express your answer in terms of d.

Problem 3.55. A 52 kg box is sitting on a level surface where the coefficients of friction are $\mu_s = 0.30$ and $\mu_k = 0.15$.

a. What horizontal force will just barely suffice to make the box accelerate?

b. What will the box's acceleration be if the applied force is twice what you found in (a)?

c. Repeat (a) and (b) in the case where the applied force is no longer horizontal but is instead applied with a slight downward angle, 15° below the horizontal.

d. Repeat (a) and (b) in the case where the applied force is applied at 15° *above* the horizontal.

Problem 3.56. In your quest for the perfect room arrangement, you decide to move your desk from one side of the room to another. The 75 kg desk slides with constant velocity, and you push with a 1.00×10^2 N horizontal force. What is the coefficient of friction between the desk and the floor?

Problem 3.57. In a physics lab, a 250 g cart is rolling on a horizontal track. Initially it has a speed of 1.5 m/s, and it comes to rest after traveling 0.70 m. What is the coefficient of kinetic friction between the cart and the track if (a) the track is horizontal and (b) if the track is inclined at 5.0°?

★ **Problem 3.58.** A block is sliding up and down a ramp and there is friction. When is the net force parallel to the ramp the greatest, or is it always the same? Explain.

★ **Problem 3.59.** A 0.200 kg wooden block is set on a wooden ramp that has an angle of 20.0° with respect to the horizontal. The block remains at rest. What is the magnitude of the friction force acting on the block?

★★ **Problem 3.60.** A wooden block of 2.0 kg slides up a wooden ramp angled at 30.° above the horizontal. As the block slides back down the ramp, what is the magnitude of the net force acting on the block? Please include friction.

★★ **Problem 3.61.** A 2.0 kg block is sliding up a ramp that has an angle of 20.0° with respect to the horizontal. The block experiences an acceleration of 6.0 m/s² down the ramp.

   a. What is the magnitude of the normal force acting on the block?

   b. What is the magnitude and direction of the friction force acting on the block?

   c. What is the coefficient of (kinetic) friction between the block and ramp?

   d. If the block was instead sliding *down* the ramp, would the magnitude of the net force on the block be the same or different? If different, would it be larger or smaller? Explain.

★★ **Problem 3.62.** A 2.0×10^3 kg car with rubber tires is parked on a concrete hill with a 40.° slope.

   a What is the magnitude and direction of the normal force acting on the car? Describe in words and with a diagram.

   b What is the magnitude and direction of the friction force acting on the car? Describe in words and with a diagram.

   c The car would still be stationary if the angle was somewhat larger (say, 45°). Explain how this is possible in terms of your answer to (b) and the concept(s) of friction.

   d After a winter storm, the owner of the car attempted to park on the original 40.° sloped hill, but unfortunately the car begins to slide. If the coefficient of kinetic friction between the tires and the road are 0.10, what is the acceleration (magnitude and direction) of the car?

★★ **Problem 3.63.** You hold a book (mass m) against a vertical wall by applying a force with magnitude F at an angle θ: $\theta = 0°$ means you're pushing perpendicular to the wall, and $\theta = 90°$ means you're not pushing the book into the wall at all and are instead supporting the book from below. The coefficient of static friction between the book and the wall is μ_s.

   a. Draw a force diagram for an arbitrary angle θ.

   b. Determine an expression for the magnitude of the force you need to apply if the book is to remain in static equilibrium. Check that your answer makes sense in the limits that $\theta = 0°$ and $\theta = 90°$.

★★ **Problem 3.64.** You press your 1.0 kg physics book against a vertical wall with a force that has components of $(10.0\hat{x} + 2.5\hat{y})$ N. You may not ignore friction ($\mu_k = 0.25$, $\mu_s = 0.60$). What magnitude of static friction force is necessary to have the book remain motionless along the wall? Is it possible to have this much frictional force with this applied force?

★★ **Problem 3.65.** You are hanging a picture of mass m when the support cable snaps and the picture begins to slide vertically down the wall. With exceptional reflexes, you put a hand at the bottom of the picture and push "in and up" at an angle of 45° at an instant when the picture's vertical speed is v. Determine an algebraic expression for the magnitude of the force you should apply, F, to keep the painting moving at a steady speed. Include a clear force diagram and denote the coefficient of sliding friction between the painting and wall as μ_k.

★ *dx* **Problem 3.66.** Complete Problem 3.63 if you haven't already, then determine the value of θ that minimizes the magnitude of the applied force.

★★ **Problem 3.67.** A car of mass m uses front wheel drive, meaning only the front wheels are rotated by the engine. A fraction γ (the Greek letter *gamma*; here $0 < \gamma < 1$) of the car's mass rests on the front wheels. The coefficient of friction between the wheels and the concrete is μ.

    a. Determine an algebraic expression for the car's maximum possible acceleration, a.

    b. Suppose $\gamma = 0.55$. Use your answer to (a), suppose $m = 1.5 \times 10^3$ kg, and use an appropriate value for μ from Table 3.1.

3.7 Air Resistance and Drag

★★ **Problem 3.68.** A bicyclist coasts down a 5.0° incline and finds that their terminal velocity is 11.0 m/s. The combined mass of the bike and bicyclist is 80.0 kg and the frictional force is 25.0 N. Draw a free body diagram and then determine the drag coefficient b.

★★ $\int dx$ **Problem 3.69.** In the text we discussed the case of quadratic drag forces, where the drag force has magnitude bv^2. For small, slow objects the drag force is instead *linear*: $\vec{F}_d = -cv$ where c is again a positive constant. If this is the only force on an object, Newton's 2nd Law can be written $\vec{F}_{NET} = m\vec{a} = -c\vec{v}$. Because acceleration is the derivative of velocity, this can also be written as $m\frac{dv}{dt} = -cv$. Apply separation of variable to solve this differential equation and obtain velocity as a function of time, $v(t)$.

3.8 Tension and Pulleys

★ **Problem 3.70.** A 2.5 kg mass is attached to a string and held in equilibrium with the horizontal force as shown in Figure 3.70. If the hold force is 12 N, what is the tension in the string? Include a sketch and provide both the magnitude and the angle of the tension force acting on the mass.

★ **Problem 3.71.** Figure 3.69 shows a top-down view of a ring tied to a wall with two ropes. A force $F = 250N$ is applied down and right, at an angle $\theta = 25°$ below the horizontal. What is the tension in each rope? What tension is being applied to each wall?

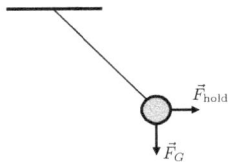

FIGURE 3.68
Problem 3.70

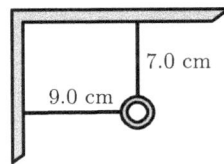

FIGURE 3.69
Problem 3.71

★ **Problem 3.72.** A crane is lowering its payload, which is connected to the crane by a single cable, at a constant speed. Draw a force diagram that includes all forces acting on the payload. Is the net force zero or nonzero? Explain. If it is nonzero, which way does the net force point?

★★ **Problem 3.73.** In a game show, a contestant hangs from a cable and is yanked up to a higher height when they get a question wrong. Suppose the contestant has a mass of 75 kg, starts from rest, and (when they get a question wrong) is moved upward 1.0 m in 0.50 s.

    a. Draw a force diagram when the contestant is hanging at rest.

    b. Draw a force diagram when the contestant is accelerating up (right after they answered a question wrong).

    c. What is the acceleration of the contestant when they're being pulled up?

    d. What is the tension in the cable *before*, *during*, and *after* they're pulled upward? How do each of your answers compare to the gravitational force on the contestant?

★   **Problem 3.74.** A 1,500 kg elevator car is accelerating upward at a rate of 1.0 m/s² due to a cable attached to the top of the car. What is the tension in the cable?

★   **Problem 3.75.** A 0.050 kg mass is attached to a string that you are holding. Initially the mass is hanging in static equilibrium, then you decrease the tension in the string by lowering your hand toward the ground. If it takes 2.0 s to travel 1.0 m, what is the net force on the mass?

★   **Problem 3.76.** A cable is pulling upwards on a beam with 350 N, as the beam is lowered into place. If the upward acceleration of the beam is 3.6 m/s², what is the beam's mass?

★   **Problem 3.77.** A 50.0 kg box hangs from a rope and is being lowered with a speed of 5.0 m/s that is slowing to 2.5 m/s over a distance of 2.0 m.

   a. Draw a force diagram showing the tension and force of gravity acting on the box drawn to scale.

   b. Determine the acceleration of the box as it slows.

   c. What is the tension in the rope as the box slows?

   d. The rope breaks and the box hits the ground traveling at 2.0 m/s. If the net force it experiences is 2.00 × 10² N upwards from the floor, what is the acceleration of the box as it comes to rest?

★   **Problem 3.78.** Two people are playing tug-of-war, and are presently at a stand-still. If each person is pulling horizontally away from the center of the rope with a force of 1.00 × 10² N, what is the tension in the rope? Ignore the mass of the rope, and assume it is equally stretched.

★   **Problem 3.79.** Figure 3.70 shows two boxes connected by a massless string on a frictionless surface. You pull on the right box with a force of 25.0 N.

   a. Draw a force diagram for both boxes.

   b. What is the tension force in the string?

   c. At what rate will the boxes accelerate?

★★ **Problem 3.80.** Repeat Problem 3.79 for the situation where the boxes are on an incline, as shown in Figure 3.71.

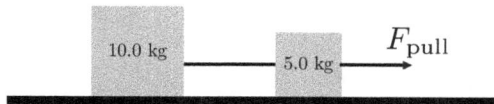

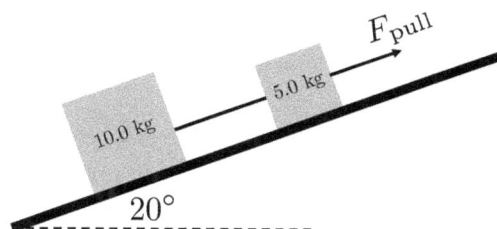

FIGURE 3.70   **FIGURE 3.71**
Problem 3.79   Problem 3.80

★   **Problem 3.81.** A 15.0 kg box is suspended from the ceiling by two cords. One cord makes an angle $\theta_1 = 20.0°$ to the left of the vertical and the other cord makes an angle $\theta_2 = 20.0°$ to the right of the vertical (so the cords make a "V" with an angle of 40.0° between them). What is the tension in each cord?

★★ **Problem 3.82.** Repeat Problem 3.81 for the case where the cords are *not* symmetric: $\theta_1 = 30.0°$ to the left of the vertical and $\theta_2 = 20.0°$ to the right of the vertical.

★★ **Problem 3.83.** Two blocks $M = 3.0$ kg and $m = 2.0$ kg are released from rest. Block m rests on a frictionless table, the string is massless and equally stretched, and the massless pulley has no friction (Fig. 3.72).

   a. What is the net acceleration for the system?

b. If block m is given a gentle shove such that is has an initial velocity of 0.20 m/s to the right, how far will it travel before stopping? Assume the strings are long enough that the hanging mass won't bump into the pulley.

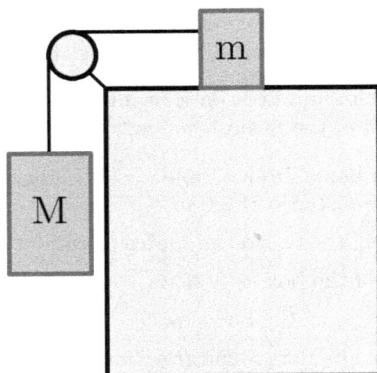

FIGURE 3.72
Problem 3.83

★★ **Problem 3.84.** Refer to the Atwood's machine discussed in Example 3.16. Show that you obtain the same result for the acceleration of the masses if you choose a coordinate system where *up* is consistently chosen as the positive direction for your coordinate system.

★★ **Problem 3.85.** Consider an Atwood's machine where $m_A = 2.0$ kg and $m_B = 1.0$ kg.

   a. What is the acceleration of the system when released? Include a diagram and clearly indicate the direction of acceleration for each mass.

   b. If the heavier mass is 1.5 m from the ground when the objects are released from rest, then how long will it take for the heavier mass to reach the ground?

   c. What speed will the masses have when the heavier mass strikes the ground?

★★ **Problem 3.86.** In a movie, a hero grabs a rope tied around a light pulley. The hero has a mass of 75 kg and the other end of the rope is attached to a 200. kg object initially suspended 3.0 m above the ground. The hero dramatically kicks a lever that released the pulley and allows the box to fall and pull the hero into the air.

   a. Draw a force diagram for the suspended object and the hero.

   b. What is the hero's acceleration while in the air?

   c. What is the tension in the cable?

   d. How long from the instant the lever is kicked will it take for the object to crash into the ground?

   e. How fast will the hero be moving when the object hits the ground?

3.9 Hooke's Law and Young's Modulus

★ **Problem 3.87.** A **spring scale** is a simple device for determining the mass of an object: a cord at the top of the scale attaches the scale to the ceiling (or some other support), then the object to be weighed is attached to a spring at the bottom of the scale. As the spring stretches, a dial on the scale twists to indicate the force being applied.

   a. Draw a force diagram for a spring scale weighing an object of mass m.

   b. Is the tension in the support cord equal to, greater than, or less than the weight of the object?

   c. If a certain spring scale uses a spring with a spring constant of 5.00×10^2 N/m, then how far will the string stretch when weighing a 0.300 kg object?

★ **Problem 3.88.** A bungee jumper is connected to a bungee cable (of negligible mass) that has a spring constant of 55.0 N/m. If the cord is 20.0 m long when it isn't stretched, then how long will it be when a 60.0 kg bungee jumper is hanging vertically (in equilibrium) from the cord?

★★ **Problem 3.89.** Consider the following simple model of the shocks in a car: 60% of the car's mass is supported by the front wheels, and 40% is supported by the back wheels. Each of the 4 wheels is supported by an identical spring with a spring constant k. Suppose a car originally has a mass of 1,500 kg and then it increases by 3.00×10^2 kg (when passengers get in). If the front springs compress by 2.0 cm, (a) what is k and (b) by how much do the back springs compress?

★ **Problem 3.90.** A circular beam has length L_1 and cross-sectional radius r_1. A second beam is made from the same material, but has length $L_2 = 1.1L_1$ and $r_2 = 1.5r_1$. What is the ratio of the spring constants, k_1/k_2?

★ **Problem 3.91.** A typical value for Young's modulus for steel is 2.00×10^{11} Pa. Consider a beam that is 3.0 m long with a square 0.50 m x 0.50 m cross section. The beam is oriented so the long edge is vertical. What mass can the beam support if it is to compress by no more than 50.0 mm? How might you adjust the dimensions of the beam if you'd like to support more mass without increasing the compression?

3.10 Pressure and Buoyancy

★ **Problem 3.92.** A wooden block has an edge length of 2.0 cm. When placed in a bucket of water, 1.2 cm of the vertical edge is visible above the waterline. What is the density of the cube if the density of the water is 1.00×10^3 kg/m^3?

★ **Problem 3.93.** The density of water is 1.0×10^3 kg/m^3. How far below the surface of a lake do you need to swim for the pressure to be twice atmospheric pressure?

★ **Problem 3.94.** Exciting story: once, I dropped a spherical knob into a bucket of paint. If the wood has a density of 250 kg/m^3 and the paint has a density of 1.3 g/cm^3, will the knob sink or float? If it will sink, what percentage of its total volume will be submerged?

★ **Problem 3.95.** Two identical cylinders are capped at one end and open at the other. The open sides are placed together, then the air inside is removed with a vacuum pump attached to a small nozzle in one of the cylinders. The cylinders have a 8.0 cm diameter.

a. How much pressure is applied to one of the capped ends?

b. If one end of the joined cylinder is firmly attached to the floor and you lift vertically on the other end, how much force (neglecting the mass of the cylinders) must you apply to separate the cylinders?

★★ **Problem 3.96.** A flotation device is in the form of a large rectangular board measuring 0.300 m by 1.000 m by 0.040 m. The board's density is 50.0 kg/m^3 and it is laying flat in a large pool of water so a person can stand on it. Suppose the density of the water is 1.00×10^3 kg/m^3.

a. How far below the surface of the water will the board submerge?

b. If a child with a mass of 25 kg climbs onto the board, how far will the board submerge?

c. What maximum mass can the board support before it will be completely submerged?

★ **Problem 3.97.** Consider a circular window with a 32 cm diameter in a submarine. The inside of the submarine is kept at 1.00 atm, while the pressure outside depends on the submarine's depth d. The density of salt water is $\rho = 1.02 \times 10^3$ kg/m^3. What is the total force on the window at:

a. $d = 0.00$ m (not submerged at all)

b. $d = 1.00 \times 10^3$ m

c. $d = 3.80 \times 10^3$ m (the approximate depth of the Titanic wreckage)

★★ **Problem 3.98.** A 1,400 kg car's weight is supported evenly by each of its 4 tires. The tires are inflated to a **gauge pressure** of 35 psi (pounds per square inch). The gauge pressure is what a pressure gauge actually measures; it is simply the pressure inside the tires minus the exterior atmospheric pressure: $P_G = P_{\text{inside}} - P_0$.

a. What is the magnitude of the normal force applied to each tire from the ground?

b. What surface area of each tire is in contact with the road?

c. If the gauge pressure drops, will the surface area increase, decrease, or remain the same? Explain.

★★ **Problem 3.99.** A popular (in physics circles) demonstration of the effects of air pressure involves a hollow metal sphere that is cut into two equal halves. The two halves are placed together, then a vacuum pumped is attached to a nozzle in one of the hemispheres so the air inside can be drawn out. The resulting low pressure inside the sphere is called a **vacuum**. To be specific, suppose the inner radius of the sphere is 9.0 cm, the outer radius is 10.0 cm, and the vacuum pressure is 1.0×10^{-5} atm.

a. What is the total force acting on the sphere due to the air trapped inside the sphere?

b. What is the total force acting on the sphere due to the atmospheric pressure in the room outside the sphere?

c. What is the net force acting on the sphere?

d. Convert your answer to (c) to pounds.[25]

3.12 Tangential and Centripetal Force

★ **Problem 3.100.** Which of the following configurations results in the largest centripetal force?

a. m, v, r

b. $2m, v, \frac{1}{4}r$

c. $\frac{1}{2}m, 2v, \frac{1}{2}r$

d. $2m, 2v, 2r$

★ **Problem 3.101.** You're riding a rollercoaster. Draw a force diagram for the instant you are at the top of a hill and moving with a velocity directed to the right. As always, label each force and make sure the vector lengths convey their relative magnitudes.

★ **Problem 3.102.** A 2.0 kg ball is spun vertically in uniform circular motion at 4.0 rad/s, and the tension in the string is 40. N at the top of the motion. What is the net force acting on the ball at the top of the loop?

★ **Problem 3.103.** A 1.0 kg ball is spinning on a 1.0 m long string in uniform circular motion. What is the tension in the string when the ball is at the top of its motion if the centripetal force is 25 N?

★ **Problem 3.104.** One method for simulating gravity in space is to set a long cylindrical tube rotating about a central axis. Suppose one such cylinder has a radius of 5.00×10^3 m. What angular speed is necessary for people standing on the inside surface of the cylinder to experience a centripetal acceleration equal to g?

★★ **Problem 3.105.** A Ferris wheel is an amusement park ride that is essentially a large, upright wheel that spins in place. Suppose a Ferris wheel has radius R and is spinning at a constant angular speed ω.

---

[25]This gives a sense of the force needed to pull apart the hemispheres, though there isn't an exact equivalency because the atmospheric force is spread over the surface of the sphere and is directed "radially in": formally, in the $-\hat{r}$ direction in spherical polar coordinates. Meanwhile, if you pull on a handle attached to one of the hemispheres, you'll be applying a force aligned with a single Cartesian axis ($\hat{x}$, say). So you'd need to overcome the component of the crushing force aligned with that axis. Working this out is a nice application of multivariable calculus, which is beyond the scope of this book. Continue your education and eventually you'll circle back to problems like this!

a. Consider a rider at the *lowest* point on the wheel. Draw a force diagram for the rider and determine an expression for their *net* force. What is their apparent weight?

b. Repeat (a) for a rider at the *highest* point on the wheel.

c. If a second Ferris wheel has a radius that is 20% larger than the first but has the same angular speed, by what factor will the rider's apparent weight change at the (a) bottom and (b) top of the ride?

Problem 3.106. Consider a large Ferris wheel with a diameter of 35.0 m and a rotational period of 50.0 s. Suppose a 60.0 kg passenger is riding the wheel.

a. What is the passenger's angular speed ω and linear speed v?

b. What is the passenger's apparent weight at the top of the ride?

c. What is the passenger's apparent weight at the bottom of the ride?

Problem 3.107. A coin sits on the edge of a 1.00 m diameter solid disc that steadily speeds up, starting from rest, with a counter clockwise rotation. At operational speed, the centripetal acceleration on the coin is 8.0 m/s^2.

a. What is angular velocity of the coin at operational speed?

b. If it takes 4.0 seconds to reach the operational speed, what is the angular acceleration of the coin starting from rest to the operational speed?

c. How many revolutions has the coin undergone while it started from rest to the operational speed?

d. If a second coin is placed at half the radius of the first coin, would the centripetal acceleration on the second coin be bigger, smaller or the same? Explain.

Problem 3.108. A 10.0 kg block is placed 3.0 m from the center of a merry-go-round (MGR). The static and kinetic coefficients of friction between the block and the MGR are $\mu_s = 0.70$ and $\mu_k = 0.50$. The MGR rotates with a period of 25.0 s.

a. Does the block slide off?

b. If your answer to (a) is *yes*, then what mass would keep it in place? If your answer to (a) is *no*, what rotational period would make it slide?

Problem 3.109. A conveyer belt carries packages in a shipping factory. One section of the belt is circular with a radius of curvature $r = 12.0$ m. The coefficient of static friction between the belt and the packages is $\mu_s = 0.20$. What is the maximum speed at which the belt can move without packages slipping?

Problem 3.110. In an amusement park ride, passengers stand on a level floor and place their backs against the inside of a large cylinder of radius R. The cylinder and riders begin to rotate about a central axis. Eventually the floor falls away and the riders are held in place against the wall. Suppose a rider has mass m and the coefficient of static friction between the rider and the wall is μ_s.

a. Draw a force diagram for the rider.

b. What is the minimum angular speed at which the rider can be held against the wall by friction?

Problem 3.111. In Example 3.27 we considered a car driving at a constant speed in a circle of radius r. Suppose the car is on a flat (i.e. unbanked) turn. Draw a force diagram for the car's tires and explain how the car rolling along the ground yields a net force that points in toward the center of the circular path.

Problem 3.112. Consider the result of Example 3.27. Suppose $r = 120$ m and $\mu_s = 0.60$.

a. What is v for (a) $\theta = 0°$, (b) $\theta = 10.°$, and (c) $\theta = 15°$?

b. Use a computer to make a graph of v as a function of θ. What happens as θ increases toward 90°? Can you give a physical explanation?

★★ **Problem 3.113.** Consider the result of Example 3.27 and now suppose that the car is at risk of slipping down a banked turn.

    a. Draw the force diagram for this situation and write out Newton's 2nd Law for the car. What is different about the force diagram compared to the one that was worked out in Example 3.27?

    b. Determine an equation for the minimum velocity that will prevent the car from slipping down the incline. (Your answer will be a function of some or all of the parameters given in the example.)

★★ **Problem 3.114.** A centrifuge is used to analyze proteins in biological samples. It essentially consists of a disk that spins at incredibly high speeds; samples are placed along the perimeter. Suppose the disk has a radius r and spins at an angular speed ω.

    a. Two samples of equal masses m are placed opposite one another. Draw a force diagram and demonstrate that the net force acting on the shaft (which runs through the middle of the centrifuge and is responsible for making the entire apparatus rotate) due to the samples is 0 N.

    b. Now suppose that one sample has mass m but the other sample has mass $1.1\,m$. What is the net force on the shaft now?

    c. Use your answer to (b) to calculate a numerical value for the net force when $m = 1.0$ g, $r = 0.10$ m, and the centrifuge rotates at 75,000 rpm.[26]

3.13 Torque

★★ **Problem 3.115.** An equilateral triangle with an edge length of 0.30 m has a 5.0 kg mass fixed to each corner. If the rods connecting the masses are light enough to be ignored, then what is the moment of inertia if the object pivots around an axis perpendicular to the plane of the object and through (a) one of the masses or (b) a point at the exact center of the triangle?

★ **Problem 3.116.** A 15 kg rod is 2.0 m long. On each end is a small 30. kg weight. What is the moment of inertia of the object if it will spin about its midpoint?

★ **Problem 3.117.** Calculate the moment of inertia of a 1.0 kg meter stick oscillating about its 0.30 m mark.

★★★∫ **Problem 3.118.** Derive the equation for the moment of inertia of a flat disk of mass M and radius R. *Hint:* Example 3.28 has demonstrated how to find a ring. Walk through this again on your own (show your work!), then integrate a series of thin rings with radii varying from 0 to R.

★★ **Problem 3.119.** A thin rod has a length L and mass m. It is lying flat on a table such that a distance $x < L$ is sticking over the edge, unsupported. (Recall that for the purposes of assessing torque, you can treat all of the mass of the rod as if it were concentrated at its midpoint.)

    a. How large can x be (as a fraction of L) before the rod begins to rotate off of the edge?

    b. Suppose x is larger than the value you found in (a). Determine an expression for the magnitude of the rod's angular acceleration α when it is still horizontal (i.e. at the instant it begins to rotate).

    c. Consider the situation described in (b), but now suppose the rod has rotated from the horizontal position by some angle θ. What is the angular acceleration now? Show that your answer reduces to what you found in (b) if $\theta = 0°$.

★★★∫ **Problem 3.120.** In the text we asserted that we can consider the mass of a uniform object to be concentrated at its center. Let's *show* it for the situation described in part (b) of Problem 3.119.

    a. Draw the situation and set the origin of your coordinate system to be the edge of the table.

    b. Consider an infinitesimally small chunk of mass, dm, and draw the corresponding force due to gravity on your diagram.

---

[26]Your answer should demonstrate the fact that it is important to use equal mass samples in a centrifuge!

c. If this chunk of mass is a distance x from the origin, the torque due to the chunk has a magnitude of $d\tau = xgdm$. But the ratio of the entire mass to the entire length is the same as the ratio of a chunk of mass to a chunk of length:

$$\frac{M}{L} = \frac{dm}{dx} \implies dm = \frac{M}{L}dx$$

Use this fact to set up an integral for the torque on the rod in terms of dx.

d. Solve the integral you set up in (c).

e. Compare your result in (d) to the case where we assume all of the mass is concentrated at the midpoint of the rod. Show that the torques are the same.

★★ **Problem 3.121.** A 20.0 kg cylinder with a diameter of 5.0 cm is free to pivot without friction around a central axis. Two forces are applied to the edge of the cylinder, as shown in Figure 3.73.

a. What is the net torque about the axle?

b. What is the moment of inertia of the cylinder?

c. What is the angular acceleration of the cylinder?

★★ **Problem 3.122.** Two forces with $F_1 = 1.0$ N and $F_2 = 2.0$ N are applied to a 2.0 kg solid disc with radius 2.0 m, as shown in Figure 3.74. Force F_1 is applied 1.0 m from the axle and F_2 is applied on the rim.

a. What is the net torque on the disc?

b. What is the angular acceleration of the disc?

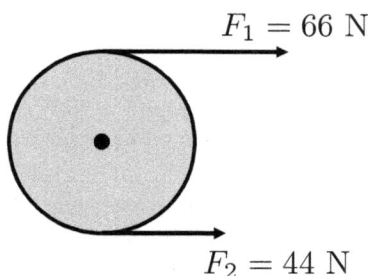

FIGURE 3.73
Problem 3.121

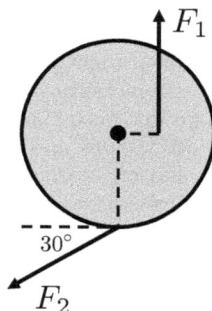

FIGURE 3.74
Problem 3.122

★★ **Problem 3.123.** A 250 kg cylinder has a radius of 0.75 m and a length of 1.0 m. It lies flat and is free to rotate about an axis through its center. A cable is wrapped around the cylinder and a free edge is pulled straight down with a constant force of 5.00×10^2 N.

a. What is the angular acceleration of the cylinder?

b. What is the angular velocity of the cylinder after 6.00×10^2 seconds?

c. What length of the cable will have been unwound after 6.00×10^2 seconds? (Assume the mass of the cylinder does not change, i.e. the cable has negligible mass.)

★★ **Problem 3.124.** A 40.0 kg bag of sand is placed on the right end of a board that is 2.2 m long and has a mass of 10.0 kg; the center of mass of the bag is 2.0 m from the left edge of the board. The board is placed near the edge of a flat roof such that 0.80 m of the board is over the edge (Fig. 3.75). How much mass could be placed on the left end of the board without it tipping over?

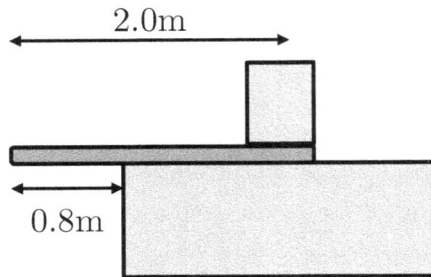

FIGURE 3.75
Problem 3.124

★   **Problem 3.125.** A see-saw is 3.0 m long and is supported at its midpoint. If a 40.0 kg child is seated 1.5 m from the pivot (i.e. on the edge of the see-saw), where should a 60.0 kg parent sit so the center of mass is centered over the pivot?

★   **Problem 3.126.** A 2.0 m stiff board has a mass of 10.0 kg and is lying flat on a roof with the left half of its length projecting over the edge. The right edge is weighed down with a 30.0 kg bag of concrete. A 75.0 kg adult stands on the board just next to the concrete and slowly begins walking toward the left edge of the board.

   a. Determine an expression for the position of the center of mass of the system as a function of the adult's position on the board.

   b. How far from the right edge of the board can the adult stand before the board begins to tip over the edge?

★★ **Problem 3.127.** When considering objects that span more than one dimension, Equation (3.29) can be applied independently to each Cartesian axis. Consider a children's toy consisting of a light square plastic plate measuring 8.0 cm × 8.0 cm and a round ball on each corner. The plate is held parallel to the ground such that the balls are at positions $(x_A, y_A) = (0, 0)$ cm, $(x_B, y_B) = (8.0, 0)$ cm, $(x_C, y_C) = (0, 8.0)$ cm, and $(x_D, y_D) = (8.0, 8.0)$ cm and their masses are $m_A = 150$ g, $m_B = 75$ g, $m_C = 125$ g, and $m_D = 225$ g. If the child would like to balance the toy on her fingertip, where should she place her finger?

★★ **Problem 3.128.** Two light rods are 0.75 cm in length. They are attached to the same 2.5 kg ball such that the rods form a "V" with an interior angle of 55°. On the opposite ends of the rods (the tips of the "V") the rods are attached to separate 1.0 kg balls. Where is the center of mass of the contraption? Provide a clear sketch.

★★ **Problem 3.129.** Figure 3.76 shows a simple model of a wheelbarrow. If the load (the material in the wheelbarrow) is 25.0 kg and it is located 15.0 cm from the fulcrum (the wheel), how much lifting force must be applied to the handles to keep the front of the wheelbarrow off of the ground? Suppose the force is applied 90.0 cm from the wheel.

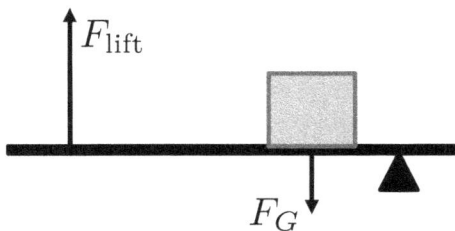

FIGURE 3.76
Problem 3.129

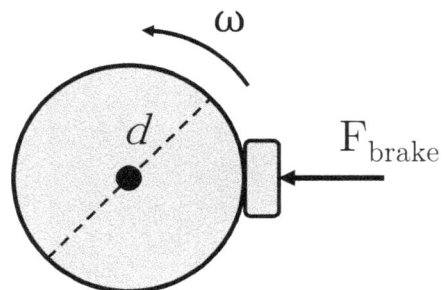

FIGURE 3.77
Problem 3.130

** **Problem 3.130.** A wheel with a diameter of $d = 0.40$ m and a moment of inertia of 0.45 kg m^2 is rotating at 75.0 rpm when a 55.0 N braking force is applied, as shown in Figure 3.77. The coefficient of friction between the brake pad and the wheel is 0.20. How much time passes before the wheel comes to rest?

* **Problem 3.131.** A 3.50 kg, 25.0 cm diameter disk is spinning at 40.0 rad/s. How much friction force must a brake apply to the rim to bring the disk to a halt in 2.50 s?

** **Problem 3.132.** A 1.50 m long crowbar is positioned so its right edge is on the ground, wedged under a heavy box. A point 0.30 m from the right edge is resting on a narrow 0.15-m-tall rock. If a 75 kg person on the left edge of the bar pushes down with all of their weight but the box doesn't move, what force is being applied to the box from the bar?

** **Problem 3.133.** The left side of a 1.0-m-long rod is attached horizontally to a building. A cable is attached to the right end of the rod and fixed to the wall, 0.50 m above the left end of the rod. The rod has a mass of 15.0 kg. Write a Newton's 2nd Law equation for the rod, both for translational motion ($\sum \vec{F} = m\vec{a}$) and for rotational motion ($\sum \vec{\tau} = I\vec{\alpha}$). What is the tension in the cable?

** **Problem 3.134.** Another way of defining torque is with a **cross product**:

$$\vec{\tau} = \vec{r} \times \vec{F} \implies \tau = rF\sin\theta \qquad (3.30)$$

where $\vec{r}$ points from the origin of your coordinate system (the hinge of a door, say) to the location where the force $\vec{F}$ is being applied, and θ is the angle between $\vec{r}$ and $\vec{F}$.

    a. Consider each of the doors (a-d) in Example 3.31. Draw your own copy of each door and force. Show that the torque defined as $rF\sin\theta$ is the same as rF_θ, i.e. show that the $\sin\theta$ argument serves to "pick out" the piece of the force perpendicular to the lever arm $\vec{r}$.

    b. Work Problem 3.133 using Equation (3.30) as your definition of torque. (Your answer, of course, should be the same!)

Additional Problems for Chapter 3

* **Problem 3.135.** You are driving downhill when you see an obstacle ahead and slam on the brakes. Draw a force diagram that includes all forces acting on the car. Is the net force zero or nonzero? Explain. If it is nonzero, which way does the net force point?

* **Problem 3.136.** A car runs out of gas just before the top of a hill. It rolls over the top of the hill and starts down the other side. Draw a force diagram for the car at the very top of the hill, assuming friction and drag forces are negligible.

* **Problem 3.137.** A block sits on a scale on the floor of an elevator. The elevator is moving downward at an increasing speed. is the magnitude of reading of the scale smaller than, larger than, or equal to is the magnitude of the force of the Earth on the block? Explain.

** **Problem 3.138.** You're pulling a 70.0 kg sled over level ground at a steady 1.20 m/s; the rope forms an angle of 30.0° with the horizontal and the tension force is 80.0 N. What is the coefficient of friction between the sled and the ground?

** **Problem 3.139.** Mr. Scribbles (a heroic pet hamster) has been sent to Mars. He is attempting to land and is approaching the surface. The 50.0 kg lander is slowing its descent by firing its rocket motor, which exerts an upward thrust force of 236 N on the lander. Ignore "air" resistance.

    a. If the lander's acceleration is 1.0 m/s^2, what is the magnitude of the acceleration due to Martian gravity?

    b. If Mr. Scribbles' mass is 0.070 kg, what normal force does he experience?

    c. The rocket runs out of fuel, and it continues to travel downwards. Explain whether the normal force experienced by Mr. Scribbles is greater than or less than the force of gravity acting upon him. Explain with words and a diagram.

   d. Luckily, Mr. Scribbles deploys the *Anti-Fall-Device*, causing the lander to safely come to rest on the surface of the planet over 10.0 m in 10.0 seconds. What is the net force (magnitude and direction) on the lander?

★★ **Problem 3.140.** A suitcase of mass m is dropped onto a conveyer belt that is moving at speed v. The static and kinetic coefficients of friction between the suitcase and conveyer belt are μ_s and μ_k, respectively. How far will the suitcase travel before it begins to move smoothly with the belt?

★★ **Problem 3.141.** In a game show, contestants climb up a slippery ramp with an adjustable angle.

   a. Suppose the incline is at $30.0°$ when a 80.0 kg contestant slides down the ramp with an acceleration of magnitude 2.50 m/s^2. What is the force of friction on the contestant?

   b. What is the normal force acting on the contestant as they slide down the ramp?

   c. What is the coefficient of friction between the contestant and the ramp?

   d. If the coefficient of static friction is twice what you found in (c), what is the steepest angle at which the contestant could remain stationary on the slope?

★★ **Problem 3.142.** Consider the situation described in Problem 3.79. This time, suppose the boxes are initially moving to the right with a velocity of 2.5 m/s and the coefficient of kinetic friction is 0.15.

   a. Re-draw the force diagram.

   b. What is the acceleration of the boxes?

   c. Determine the tension in the string.

   d. If the acceleration is to the right, then determine how long it takes the boxes to travel 3.0 m. If the acceleration is to the left, determine how much time passes until the boxes come to rest.

★★ **Problem 3.143.** You hold in your hand a book of mass m_a. On top of that book is a second book of mass m_b. You lift the books with a uniform acceleration a.

   a. Draw a force diagram including your hand, the books, and the Earth. Identify the Newton's 3rd Law pairs.

   b. Write a Newton's 2nd Law equation for book b.

   c. Write a Newton's 2nd Law equation for book a.

   d. Write a Newton's 2nd Law equation for the combination of book a and book b.

★★ **Problem 3.144.** A block of mass $m_A = 5.0$ kg is next to (and touching) a block of mass $m_B = 10.0$ kg on a flat, frictionless surface.

   a. Suppose a horizontal force F is applied to mass A in the direction of mass B. If the force of A on B is 20. N, what is the acceleration of the system?

   b. What is the magnitude of F?

   c. What is the net force on A? on B?

   d. Suppose the blocks are traveling at 2.0 m/s when they move onto a section of the surface that *isn't* frictionless. If it takes 1.0 seconds to come to rest, what is the friction force (magnitude and direction) between the block and the surface?

★★★ **Problem 3.145.** A block of mass $M = 10.0$ kg is sitting on a roof and attached to a hanging block of mass $m = 5.0$ kg by way of a massless rope and pulley, as shown in Figure 3.78. The masses are in stable equilibrium.

   a. If the masses are in equilibrium because of friction, what is the magnitude of the frictional force acting on mass M?

   b. Now suppose the surface is frictionless and the masses are released from rest. What is the acceleration of the objects? Explain the direction of the acceleration for each object as well as the magnitude.

   c. What is the tension in the rope as the objects accelerate?

** **Problem 3.146.** A 5.0 kg object is connected by a string via a massless, frictionless pulley to a 2.5 kg object that is hanging freely. The 5.0 kg object is given a push along the frictionless horizontal surface and released with a speed of 2.0 m/s (Fig. 3.79).

a. Make a force diagram for both objects.

b. What is the acceleration of the objects after being released from the push?

c. How far does the 5.0 kg object move before coming to rest?

d. Sketch a velocity vs. time graph for the 5.0 kg object.

e. What is the tension in the string?

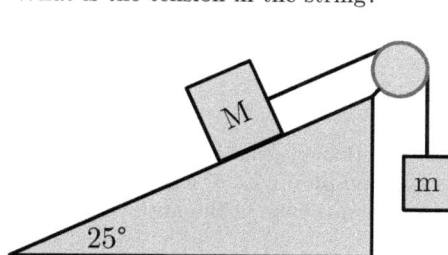

FIGURE 3.78
Problem 3.145

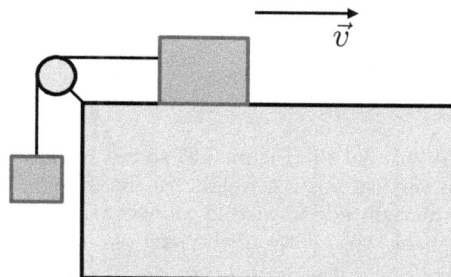

FIGURE 3.79
Problem 3.146

** **Problem 3.147.** You are unloading a moving van and are using a ramp to make your life easier. You have a box connected to a rope that you are pulling parallel to the ramp (Fig. 3.80); the ramp is frictionless (which is nice as long as you don't try to walk on it) and forms an angle of $\theta = 25°$ with the horizontal. The 47 kg box begins from rest and accelerates at 1.0 m/s² down the ramp.

a. What is the net force acting upon the box?

b. What is the magnitude of the force you are applying to the rope?

c. After 0.20 s, what is the velocity of the box?

d. Suppose that after 0.20 s, you let go of the rope and it slides down the ramp, destroying a mirror. Does the box or the mirror experience a larger magnitude force as they collide, or are they equal? Explain.

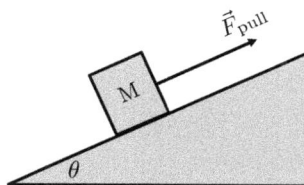

FIGURE 3.80
Problem 3.147

** **Problem 3.148.** An 8.0 kg box and a 20.0 kg box are attached with a light string and placed on an incline, then the 8.0 kg box is attached to a 40.0 kg box by way of a massless, frictionless pulley, as shown in Figure 3.81. The 20.0 kg box is gripped and given a brief shove down the ramp such that at $t = 0.0$ s the boxes are moving with a speed of 0.75 m/s. Suppose $\mu_k = 0.20$, $\mu_s = 0.40$, and that the strings and ramp are long enough that the blocks won't impact the pulley, floor, etc.

a. Draw a force diagram.

b. What is the acceleration of the boxes?

c. Determine the tension in the strings.

d. Will the boxes continue moving down the ramp, come to a halt and stay there, or reverse direction such that the 8.0 kg box and the 20.0 kg box move up the ramp?

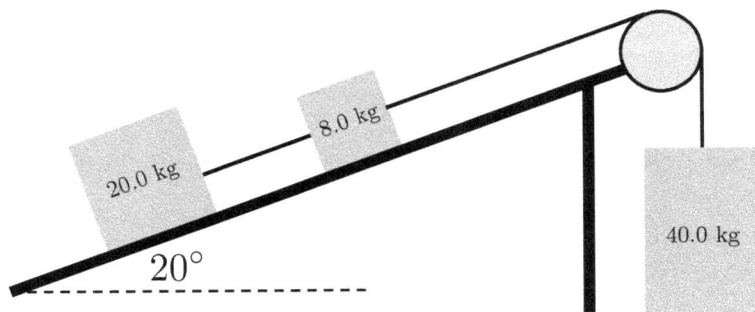

FIGURE 3.81
Problem 3.148

★★★   **Problem 3.149.** Figure 3.82 shows a diagram of a crane: a horizontal bar of mass m_{bar} is supported by a vertical post. A weight on the left edge has mass m_w and an object of mass m_{ball} is hanging on the right edge. A cable connects the support post to the right edge of the horizontal post. The "x" marks the center of the bar.

   a. Draw and label all of the forces acting on the horizontal bar.

   b. Determine an equation for the moment of inertia of system comprising the bar, ball, and weight.

   c. Write an equation for the net torque acting on the bar. Use the location of the black circle as the origin of your coordinate system.

   d. Determine an equation for the tension in the cable. Assume the crane is in static equilibrium. Does increasing m_w increase or decrease the tension in the cable?

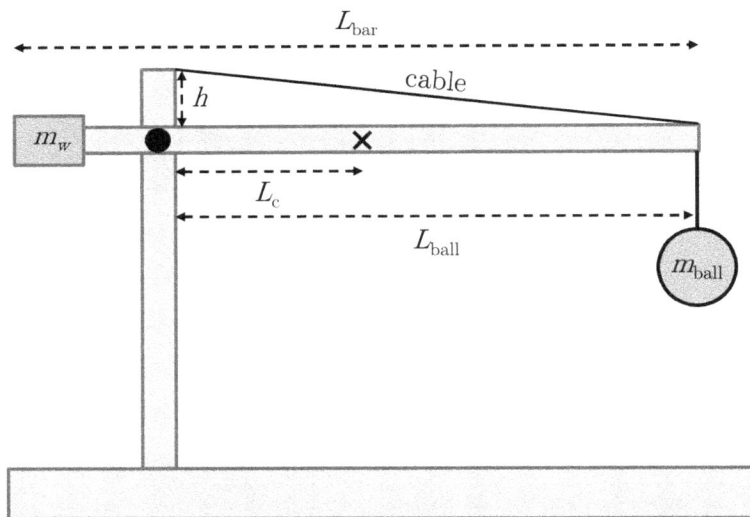

FIGURE 3.82
Problem 3.149

★   **Problem 3.150.** Two wheels with fixed hubs, each having a mass of 2.0 kg, start from rest, and have a force applied tangent to their rims. Assume the hubs and spokes are massless, so that the rotational inertia is $I = mR^2$. In order to impart identical angular accelerations, how large must F_2 be if $F_1 = 1.0$ N and $R_2 = 2R_1$?

★★   **Problem 3.151.** Three 0.50 kg blocks are connected by ideal strings (massless and unstretchable) and are being pulled to the right by a horizontal force F over level, frictionless ground. Label the blocks A, B, and C from left to right, so the pulling force is applied to block C.

a. If the system travels 62.5 cm in 0.500 s from rest, what is the acceleration of the system?

b. What is the magnitude of F?

c. What are the tensions in the strings?

d. Are the net forces on block B and C the same or different? Explain. (No detailed calculations needed!)

** **Problem 3.152.** In an amusement park ride, passengers are strapped to the inside of a large cylindrical ring of radius R that rotates around a vertical axis that runs straight through the middle of the ring. Suppose there are n passengers of equal mass m and the ring itself has a mass M. The ring is initially rotating at an angular speed ω. If an external torque τ is applied to slow down the ring, determine an equation for the time Δt needed to bring the ring (and passengers!) to rest.

** **Problem 3.153.** In Problem 2.82 on page 100 you considered a roll of toilet paper unwinding. The point there was to consider the angular kinematics and accelerations involved with the roll as it rotates. (If the context strikes you as silly, consider instead a large roll of paper or sheet metal in an industrial facility.) However, for simplicity we assumed that the thickness of the roll didn't change, and we weren't yet in a position to consider torque and moment of inertia. Use the data in Problem 2.82 and additionally suppose that the diameter of the inner tube for the toilet paper roll is 4.0 cm, the density of toilet paper is 0.10 g/cm^3, and the height of the roll is 12.0 cm.

a. What is the mass of the original toilet paper roll?

b. The moment of inertia of an annular disc with an inner radius r_i and an outer radius r_2 is

$$I = \frac{1}{2}M\left(r_1^2 + r_2^2\right) \tag{3.31}$$

Use this equation to determine the moment of inertia of the original roll of toilet paper.

c. Once 1.00 m of paper has been unrolled, the outer radius has been reduced to 11.8 cm. What is the new moment of inertia of the roll? (Keep in mind the mass has changed as well!)

d. What torque was applied to the roll to generate an initial angular acceleration of 10.0 rad/s^2?

e. If the same torque is applied during the unwinding process, what is the roll's angular acceleration after 1.00 m of paper has been pulled?

f. You can't use the angular kinematic equations to analyze a situation with a non-constant acceleration. As an *approximation*, then, set α equal to the *average* of 10.0 rad/s^2 and your answer to (e) and re-solve Problem 2.82 part (b). Compare to the answer when you just use 10.0 rad/s^2.[27]

** **Problem 3.154.** Hooke's Law is an example of a force that varies *linearly* with displacement. This turns out to be a very useful example because many nonlinear forces are *approximately linear* for *small displacements*. In this problem you'll verify this claim for some of the forces we've considered in this chapter. You'll want to make use of the binomial approximation, which states that $(1 + x)^n \approx 1 + nx$ if $x \ll 1$.

a. Consider Newton's Law of Gravitation (Equation (3.4)) for variations in distances near the surface of the Earth: $F_G = G\frac{m_1 m_2}{(R_E + h)^2}$ where R_E is the radius of the Earth and h is the distance above the surface of the Earth. Show that if $h \ll R_E$, then F_G varies linearly with h.

b. Consider a cube of edge length L and density ρ_c floating in a fluid of density ρ_f. You grab the top of the cube and push it up (or pull it down) a small distance x, then release it. Determine an expression for the net force acting on the cube. Is it linear in x?

---

[27]A better solution involves setting $\alpha = \tau/I$ for a constant τ and I given by Equation (3.31), noting that $\alpha = \frac{d\omega}{dt}$, and solving for ω via integration: $\omega = \int \frac{\tau}{I} dt$. The premise of the problem is that τ is a constant, but I is not because r_2 changes as the paper is unwound. This is a nice example where computational techniques are handy: a computer can use a variety of sophisticated techniques to simulate the motion in a sequence of tiny advancements of time, updating both α and r_2 along the way. If you continue your education in a math-intensive STEM field you'll likely have the opportunity to study these techniques intensively at some point.

c. Hooke's Law refers to a **restoring force** because, for small displacements away from equilibrium, the force pulls the object back toward equilibrium. Thus a **Hookean force** is both *linear* and *restoring*. In (a) and (b) you demonstrated that the net force is *linear* for small displacements. But are the net forces Hookean? In other words, does gravity point back to the surface of the Earth for small displacements (positive or negative) away from the surface? Does the net force on a buoyant object point back to an equilibrium depth for small displacements into or out of the water?

⋆⋆ **Problem 3.155.** A spring scale is used to measure a 2.0 kg disco ball. The mass stretches the spring a vertical distance of 0.10 m from equilibrium.

a. What is the spring constant?

b. You apply an upward force of 10. N. By how much has the spring stretched if the disco ball is still motionless?

⋆⋆ **Problem 3.156.** Consider an object undergoing circular motion in a vertical loop (such as a rollercoaster car or – this is a classic physics classroom demonstration – a bucket filled with water spun in a circle). The **critical speed** is the lowest speed that allows the object to continue in its circular path at the top of the trajectory.

a. For an object of mass m traveling in a circular loop of radius R, what is the critical speed v_c? (You're after an algebraic expression here. Set up Newton's Second Law for the object and consider which of the forces can and cannot change.)

b. For a rollercoaster cart with a mass of 750 kg going through a vertical loop with a 15 m radius, what is the critical speed?

c. Suppose you are on such a rollercoaster traveling at the critical speed. At the highest point, you dramatically pull out a spring scale attached to a disco ball with mass $m = 2.0$ kg and hang it vertically, so the scale is above the mass from the perspective of a ground-based observer. What mass would the spring scale read?

4

Conservation Laws

I encourage you to think of your mastery of concepts in physics as tools in your intellectual toolbox: you now have an array of tools from kinematics and force analysis at your disposal. In this chapter, we will make two important additions to your toolbox. We will first tackle the concept of **energy**, which is immensely useful in large part because the total energy in a closed system never changes (i.e. it is *conserved*). Knowing this allows us to quickly analyze some situations that would otherwise be very complex.

In the second part of this chapter we will take up the study of **momentum**. We will see that we can make a similar statement about momentum conservation, which is very useful when we consider a new topic: collisions and explosions.

Learning Objectives

After reading this chapter, you should be able to:

- Describe the relationship between forces, work, and energy (specifically using the work-energy theorem).

- Articulate the relationship between force and potential energy.

- Define gravitational potential energy and spring potential energy.

- Use the principle of conservation of energy to analyze the motion of objects.

- Apply the concept of impulse to describe the effect of collisions.

- Apply the principle of conservation of momentum to analyze collisions and explosions.

- Differentiate elastic and inelastic collisions.

- Use the concept of rotational kinetic energy when applying the law of conservation of energy.

- Use the concept of conservation of angular momentum when analyzing collisions and explosions involving angular motion.

4.1   Work and Kinetic Energy

To begin our study of energy, let's analyze a situation with which we are already familiar: a force acting on an object. In particular, suppose we have a block of mass m moving on a frictionless, horizontal surface at some initial velocity $\vec{v}_i$. If this was the whole story then the object would continue on at this speed forever (Newton's First Law!), but now a pushing

DOI: 10.1201/9781003571568-5

force, $\vec{F}_P$, is applied in the same direction as $\vec{v}_i$. If we check back in on the block after it has traveled some distance Δx, it will have a new velocity $\vec{v}_f$. The situation is sketched in Figure 4.1.

Because only one force is acting on the block in the horizontal direction, $\vec{F} = \left(\vec{F}_{\text{NET}}\right)_x = m\vec{a}_x$. Then from standard kinematics,

$$v_f^2 = v_i^2 + 2a\Delta x$$
$$v_f^2 = v_i^2 + 2\left(\frac{F}{m}\right)\Delta x$$
$$\frac{1}{2}m\left(v_f^2 - v_i^2\right) = F\Delta x \tag{4.1}$$

Let's write Equation (4.1) in a slightly different way. First we will re-write the left hand side of the equation by defining **kinetic energy**, K, as[1]

$$K = \frac{1}{2}mv^2 \tag{4.2}$$

If you analyze the units in this equation you will see that they correspond to $\frac{\text{kg m}^2}{\text{s}^2}$ or a "Newton-meter". This quantity is so important that we give it a new name, the *joule*, which we abbreviate as J.[2] This is the SI unit of energy.

Equation (4.1) can be written in terms of kinetic energy:

$$\Delta K = F\Delta x \tag{4.3}$$

The right hand side of Equation (4.3) supposes that the force and the displacement point in the same direction. In general, however, this need not be the case. In general, only of the *component* of the force vector that lies on the same axis as the displacement vector affects its velocity (and therefore its kinetic energy). Consider, for example, the angled pulling force shown in Figure 4.2: the vertical component of the force will reduce the normal force (which affects the frictional force), but only the horizontal component directly contributes to the object's acceleration.

In mathematical language, taking two vectors and determining the product of their parallel components is referred to as the **dot product**. If you visually place the "tails" of the vectors at the same position and call the angle between them θ (Figure 4.3), then the dot product is given by:[3]

$$\vec{A} \cdot \vec{B} = AB\cos\theta \tag{4.4}$$

If we prefer to work with the vectors in component form, then

$$\vec{A} \cdot \vec{B} = A_x B_x + A_y B_y \tag{4.5}$$

You are asked to show that these equations are in fact identical in Problem 4.5.

---

[1]We're presenting this in a one-dimensional context, but as we shall see, this definition extends directly to three-dimensional motion.

[2]After James Prescott Joule (1818-1889)

[3]Notice that the dot product takes two vectors and returns a scalar.

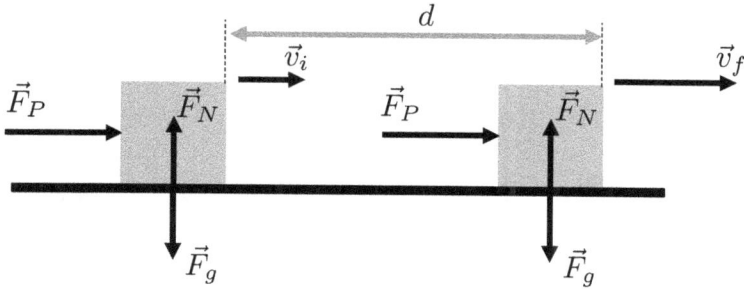

FIGURE 4.1
A force applied to a block traveling over a frictionless surface accelerates it from a sped v_i to a speed v_f over a distance $d = \Delta x$.

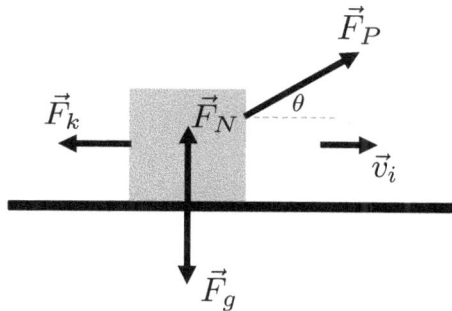

FIGURE 4.2
A block sliding to the right on a level surface. The displacement will be horizontal to the right, but none of the four forces point (entirely) in this direction.

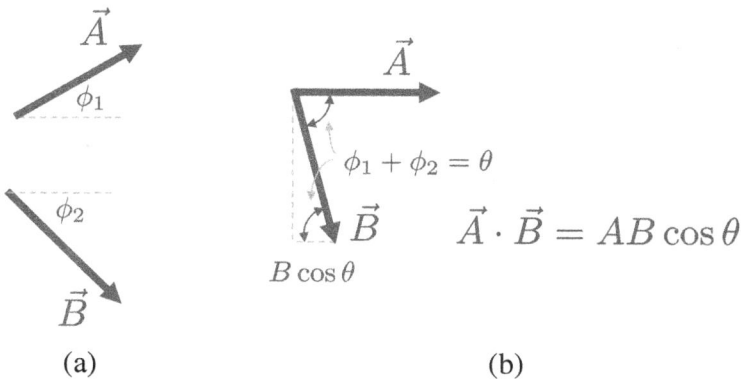

(a)   (b)

FIGURE 4.3
(a) Two arbitrary vectors. (b) Their dot product is the product of their parallel components; the arrangement of the vectors on the right shows the geometry (to make the geometry easier to analyze, both vectors have been rotated so $\vec{A}$ is horizontal).

Example 4.1 The Dot Product   ⋆
Consider the vectors $\vec{q} = 5.0\hat{x} + 3.0\hat{y}$, $\vec{r} = 6.0$ at $30.°$ below the $+x$ axis, and $\vec{s} = 4.0$ at $20.°$ above the $-x$ axis. Determine (a) $\vec{q} \cdot \vec{q}$, (b) $\vec{q} \cdot \vec{r}$, (c) $\vec{q} \cdot \vec{s}$, and (d) $\vec{r} \cdot \vec{s}$.

Let's begin by drawing the vectors. We'll place all of their tails at the origin so we can more readily identify the angles between them according to Equation (4.4) (Fig. 4.4).

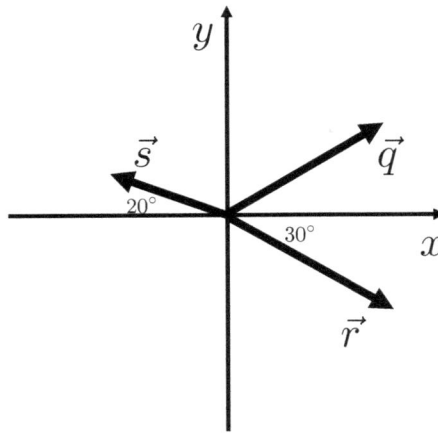

FIGURE 4.4

(a) Every vector is by definition *parallel with itself*, so $\vec{q} \cdot \vec{q} = q^2 \cos 0° = \left(\left(5.0^2 + 3.0^2 \right)^{\frac{1}{2}} \right)^2 = 34$.

(b) $\vec{q}$ is given in component form, but $\vec{r}$ is given as a magnitude and angle. Thus we must convert one vector to the other form to use either Equation (4.4) or (4.5). Let's put $\vec{r}$ into component form: $\vec{r}_x = r \cos 30.°\hat{x} = 5.2\hat{x}$ and $\vec{r}_y = -r \sin 30.°\hat{y} = -3.0\hat{y}$. Thus

$$\vec{q} \cdot \vec{r} = q_x r_x + q_y r_y$$
$$\vec{q} \cdot \vec{r} = (5.0)(5.2) + (3.0)(-3.0)$$
$$\vec{q} \cdot \vec{r} = 17$$

(c) We will proceed as with the previous question and determine the components of $\vec{s}$: $\vec{s}_x = -s \cos 20.°\hat{x} = -3.8\hat{x}$ and $\vec{s}_y = s \sin 20.°\hat{y} = 1.4\hat{y}$. Thus

$$\vec{q} \cdot \vec{s} = q_x s_x + q_y s_y$$
$$\vec{q} \cdot \vec{s} = (5.0)(-3.8) + (3.0)(1.4)$$
$$\vec{q} \cdot \vec{s} = -15$$

While $\vec{q}$ and $\vec{r}$ have colinear components that point in the same direction and so have a positive dot product, $\vec{q}$ and $\vec{s}$ has colinear components that point in opposite directions and so they have a negative dot product.

(d) We already have $\vec{r}$ and $\vec{s}$ in magnitude-angle form, so we will appeal to Equation (4.4). We need to know the angle between them, which (working counter-clockwise from $\vec{r}$ toward $\vec{s}$) is $\theta = (30° + 90° + (90° - 20°)) = 190°$.[a] Then:

$$\vec{r} \cdot \vec{s} = rs \cos \theta$$
$$\vec{r} \cdot \vec{s} = (6.0)(4.0) \cos 190°$$
$$\vec{r} \cdot \vec{s} = -24$$

---

[a]If you work clockwise instead of counter-clockwise, you will arrive at 170°. The final answer you get for the dot product does not depend on your choice – as it shouldn't!

In this context our vectors are a force and a displacement, and the dot product is referred to as **work**:

$$W = \vec{F} \cdot \Delta \vec{x} \tag{4.6}$$

This equation assumes that we have a constant force acting on an object that moves in a straight line. In general, though, the force acting on an object can vary as it moves, and the object's path need not be a straight line. In this case we need to express Equation (4.6) in integral form:

$$W = \int \vec{F} \cdot d\vec{r} \tag{4.7}$$

Like any integral, this equation represents a *cumulative effect* of many calculations: in this case, the instant-by-instant result of some component of a force causing an object to change velocity by being aligned (or anti-aligned) with the object's displacement. Graphically, the work is given by the area bounded by a force-position curve (Problem 4.6).

If we express Equation (4.3) in terms of work and keep in mind that in general there can be multiple forces doing work on an object (such as a pushing force, friction, tension in an attached cable, gravity...), we arrive at the **work-energy theorem**:

$$W_{\text{TOT}} = \sum W = \Delta K \tag{4.8}$$

In words, this says that *the total work done to an object is equal to its change in kinetic energy*.[4] We've jumped through a few hoops to arrive at this point, but keep in mind that all we've done is generalize Equation (4.1), which we obtained directly from force analysis and kinematics. As the next example demonstrates, the work-energy theorem is very useful for analyzing situations where many forces act on an object.

---

[4]In fact, a good definition of energy is *the capacity to do work*. A good example here is a beam being lifted with a cable: if the tension force is equal in magnitude to the gravitational force, then from force analysis we'd argue $\vec{F}_{\text{NET}} = m\vec{a} = 0$ and the velocity (and therefore kinetic energy) is constant. In the context of total work, we see that the tension and gravitational forces do equal and opposite work and so $W_{\text{TOT}} = 0$ and the kinetic energy is constant. Thus we come to the same conclusion either way. In contrast, if the tension force is larger than the gravitational force then $\vec{a} \neq 0$ and $W_{\text{TOT}} \neq 0$, leading to a nonzero acceleration and a change in kinetic energy.

Example 4.2 Working on Work   ⋆
Refer back to Figure 4.2 and suppose the block has a mass of 3.2 kg, $\mu_k = 0.15$,
$F_P = 12$ N, $\theta = 25°$, and $v_i = 0.50\frac{m}{s}$. Consider the block after it has moved 1.5m.
How much work will each of the forces shown on the figure have done? What will the
final speed of the block be, and how much kinetic energy will it have?

We can note straightaway that $\vec{F}_G$ and $\vec{F}_N$ will do no work because they are
perpendicular to the displacement (the forces are vertical while the displacement is
horizontal). Mathematically, this is because $\cos 90° = 0$, so $W = Fd\cos\theta = 0$. For
$\vec{F}_P$ we find

$$\vec{F}_P \cdot \vec{d} = F_P d \cos\theta$$
$$= (12 \text{ N})(1.5 \text{ m})\cos 25° = 16 \text{ J}$$

Meanwhile, friction points directly opposite the displacement (they are *anti-parallel*),
so $\theta = 180°$, which means $\cos\theta = -1$, and

$$\vec{F}_k \cdot \vec{d} = F_k d \cos\theta$$
$$= \mu_k F_N d \cos\theta$$
$$= \mu_k mgd \cos\theta$$
$$= (0.15)(3.2 \text{ kg})\left(9.8\frac{m}{s^2}\right)\cos 180° = -4.7 \text{ J}$$

We see that friction does negative work: it is trying to slow the block down, or, in
other words, decrease its kinetic energy.

Now that we have the work due to each of the forces, let's turn to the work-energy
theorem:

$$W = \Delta K$$

$$16 \text{ J} - 4.7 \text{ J} = \frac{1}{2}(3.2 \text{ kg})\left(v_f^2 - \left(0.50\frac{m}{s}\right)^2\right)$$

$$\left(\frac{2(11.6 \text{ J})}{3.2 \text{ kg}} + \left(0.50\frac{m}{s}\right)^2\right)^{\frac{1}{2}} = v_f = 2.7\frac{m}{s}$$

And finally we see that the final kinetic energy is

$$K_f = \frac{1}{2}mv_f^2 = \frac{1}{2}(3.2 \text{ kg})\left(2.74\frac{m}{s}\right)^2 = 12 \text{ J}$$

Before moving on, it is worth emphasizing that before opening this chapter, we were
perfectly capable of analyzing this situation by considering the forces and applying
the kinematic equations because the acceleration is *constant* (recall that this was an
assumption for the derivation of the kinematic equations). As we shall see, the work-
energy theorem allows us to characterize situations with non-constant acceleration,
as well!

Example 4.3 Work on a Roller Coaster   ⋆
An 8.00×10^2 kg roller coaster cart is at rest at the top of a 15.0 m hill. If we neglect
friction and air resistance, (a) how much work is done by gravity during the descent
and (b) how fast will the cart be moving at the bottom of the hill?

The work done by gravity is given by

$$W = \int \vec{F}_g \cdot d\vec{r}$$

At first glance it can seem like this will be rather complicated and depend on the exact curvature of the hill. However, the force in question, gravity, is constant and always points down toward the Earth; if up is the $\hat{y}$ direction then $F_g = -mg\hat{y}$. Thus, when evaluating the dot product we are only ever concerned with the vertical part of the displacement, which amounts to a total change $\Delta y = -15.0$ m. The work is therefore

$$W = mg\Delta y$$
$$= \left(8.00 \times 10^2 \text{ kg}\right)\left(9.8\frac{\text{m}}{\text{s}^2}\right)(15.0 \text{ m})$$
$$= 1.18 \times 10^5 \text{ J}$$

This, according to the work-energy theorem, is equal to the change in kinetic energy:

$$W = \Delta K$$
$$W = \frac{1}{2}mv^2$$
$$\left(\frac{2W}{m}\right)^{1/2} = v$$
$$\left(\frac{2\left(1.18 \times 10^5 \text{ J}\right)}{8.00 \times 10^2 \text{ kg}}\right)^{1/2} = v$$
$$17.1 \text{ m/s} = v$$

Example 4.4 Work on a Spring $\qquad \star\star \int dx$

An elastic cord obeys Hooke's Law with spring constant k when it is stretched beyond its relaxed length L. One end of the cord is tied to a cart on a frictionless air track; the other end is tied to the adjacent end of the track. The cart is given a swift push to give it an initial speed v_i along the track. How far will it travel before stopping? Neglect friction and assume, as usual, that the "spring" (in this case, the cord) is massless.

The cart will move at its initial speed v_i until the cord becomes taut; then it will begin to slow down as the cord stretches beyond its relaxed length. Because the force *changes* as the cart moves, we must rely on the *integral* definition of work. We may as well set the origin of the coordinate system at the position of the cart when the cord has been stretched to a length L. Then the work is

$$W = \int \vec{F}(x) \cdot d\vec{x} = \int_0^{x_f} (-k\vec{x}) \cdot d\vec{x}$$

The negative sign falls out of Hooke's Law: if the object is moving in the positive direction (to the right, say), then the force is pointing in the negative direction (to the left). Now $\vec{x}$ and $d\vec{x}$ point in the same direction, so the angle between them is $0°$ and $\vec{x} \cdot d\vec{x} = x\,dx$. Thus we have

$$W = \int_0^{x_f} (-kx)\,dx = -\frac{1}{2}kx_f^2$$

This equation relates the work done to how far it moves once the cord begins to stretch beyond its equilibrium length. To relate this to the block's initial velocity, we invoke the work-energy theorem (Equation (4.8)) and note that $v_f = 0$:

$$W = \Delta K$$
$$-\frac{1}{2}kx_f^2 = \frac{1}{2}m\left(v_f^2 - v_i^2\right)$$
$$kx_f^2 = mv_i^2$$
$$x_f = \left(\frac{m}{k}\right)^{\frac{1}{2}} v_i$$

This expresses the distance the block travels once the cord became taut. The *total* distance traveled from the end of the air track is $L + x_f$.

4.2   Conservative Forces and Potential Energy

Suppose you hold a ball of mass m a distance h above the ground and release it. What speed will it have when it strikes the ground? According to the work-energy theorem and the result of Example 4.3,

$$W = \Delta K$$
$$mgh = \frac{1}{2}mv_f^2$$
$$(2gh)^{\frac{1}{2}} = v_f$$

Note that mgh has units of energy. This energy is due to the force of gravity and indicates that the object in question has the *potential* to gain kinetic energy (if it falls). We therefore refer to it as **gravitational potential energy**. We refer to potential energy generically with the symbol U, and so the gravitational potential energy is written[5]

$$U_g = mgh \tag{4.9}$$

Note in particular here that h refers to a height measured *from the ground*.[6] Thus *when the ball is released*, $U_i = mgh$ and *when it strikes the ground*, $U_f = 0$. Conversely, the *initial* kinetic energy is $K_i = 0$ and the *final* kinetic energy is $K_f = \frac{1}{2}mv_f^2$. It follows from our analysis of the work done that

$$K_i + U_i = K_f + U_f \implies \Delta K = -\Delta U \tag{4.10}$$

---

[5] Remember that $F_G = mg$ holds near the surface of the Earth only. We explore the gravitational potential energy for cases far from the surface of the Earth in Problem 4.58.

[6] We could, however, choose any other position as the position of no gravitational potential energy: all that matters is how the energy changes from one position to another, and that doesn't depend on our choice of origin. This is the same as saying that 2 m − 0 m = 2 m just as 4 m − 2 m = 2 m.

This conversion of energy (from gravitational potential to kinetic) is an example of **conservation of energy:**[7]

> Energy can neither be created nor destroyed; it can only be transformed.

If we generically refer to *any* kind of energy as E, then we can write the law of conservation of energy mathematically like so:

$$\sum E_i = \sum E_f \tag{4.11}$$

Now, it is important to keep in mind that all of this energy business comes from a *force* acting over a *distance*. A good example of conservation of energy is a block sliding along a table and slowing to rest because of friction: initially all of the energy is associated with the block's kinetic energy, and later it has been transformed to thermal energy because of the work done by friction acting as the block slides (Problem 4.12).

As a second example, let's look at the relationship between the gravitational force and gravitational potential energy (Figure 4.5). The "flat $\to$ linear" relationship between the magnitude of the force and the potential energy should seem familiar; we discussed this already in the context of kinematics! The other part of that argument was "linear $\to$ quadratic"; in this context we say that *a potential energy that varies quadratically with position corresponds to a force that varies linearly with position.*

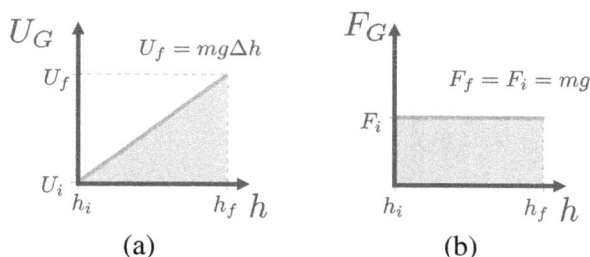

(a)   (b)

FIGURE 4.5
(a) The gravitational potential energy, mgh, varies linearly with height h with a slope equal to the magnitude of the force (b).

You may recall from Chapter 3 that one such force is given by Hooke's Law: $\vec{F}_{\text{sp}} = -k\Delta\vec{x}$. Indeed, the **spring potential energy** is given by

$$U_{\text{sp}} = \frac{1}{2}k\left(\Delta x\right)^2 \tag{4.12}$$

where Δx is the displacement *from the equilibrium length of the spring*. The relationship between Hooke's Law and the spring potential energy is shown in Figure 4.6 (the area under the force-position graph corresponds to the energy: the further we stretch the spring, the more energy we add).[8]

---

[7]It may be useful to think of this like dumping water from one bucket to another: the amount of water (i.e. energy) never changes, though it can move between buckets (i.e. change form, like from kinetic to gravitational potential and vice versa).

[8]Of course, you already saw this result if you looked at the calculus-based Example 4.4. This should not be surprising; all we've done here is "manually" take an integral by inspecting the geometry of the force-displacement graph.

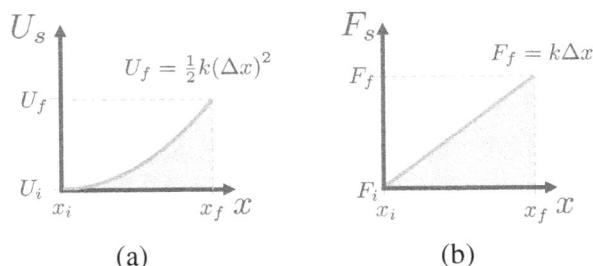

(a)   (b)

FIGURE 4.6
(a) The spring potential energy, $\frac{1}{2}k\left(\Delta x\right)^2$, varies quadratically with the magnitude of the displacement Δx, corresponding to a force that varies linearly with displacement (b).

Note that the above considers the energy (a scalar) and the *magnitude* of the corresponding force (a vector). What can we say about the direction of the force as it relates to its corresponding potential energy? The answer is that *the force points in the direction of decreasing potential energy*. In the case of gravity, the potential energy *increases up from the Earth*, while the force points *down toward the Earth*. In the case of a spring, the potential energy *increases as the spring is stretched/compressed from equilibrium*; the spring force points *back toward equilibrium*. We summarize this in Figure 4.7.

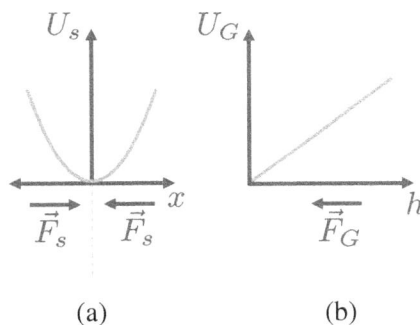

(a)   (b)

FIGURE 4.7
The relationship between potential energy and the direction of the corresponding force. The spring potential energy (a) increases away from equilibrium, and the spring force points back toward equilibrium. The gravitational potential energy (b) increases away from the ground, and the gravitational force points back toward the ground.

A characteristic of both the gravitational force and the spring force is that they are **conservative forces**: the work done by the force only depends on where the object starts and where it ends, but not the path it takes in between (e.g. see Example 4.3). In contrast, consider a block sliding across a surface with friction: the act of the block and the surface rubbing is what slows it down, so it doesn't matter *where* the object starts and ends; rather, *how far it goes* matters. The work done by gravity, on the other hand, only depends on your starting height and final height ($W = mg\Delta y$).[9] Thus friction is a **non-conservative**

---

[9]Another way of thinking about this is that friction makes the block lose energy – to thermal energy, as the surface over which it moves is warmed – that it cannot easily regain. In contrast, energy can easily move back and forth between kinetic energy and spring/gravitational potential energy. The fact that no energy is "lost" to the environment in this case is precisely why we refer to these forces as *conservative*.

force. This is an important distinction because *only conservative forces have a potential energy*: a potential energy tells us a corresponding force based only on the position of some object, and this simply isn't possible if you also need to know other information (like the past motion of the object).

Now if $W = \Delta K$ and $\Delta K = -\Delta U$, it follows that

$$W = \Delta K = -\Delta U \qquad (4.13)$$

Combining this with Equation (4.7) yields

$$U(x) = -\int F(x)\,dx \qquad (4.14)$$

and we can call out the associated conservative force by taking the derivative of the potential energy with respect to position:

$$F(x) = -\frac{d}{dx}U(x) \qquad (4.15)$$

Thus, for instance, $U_g = mgh \implies F_g = -mg$ and $U_{\text{sp}} = \frac{1}{2}kx^2 \implies F_{\text{sp}} = -kx$, and we see from Figures 4.5 and 4.6 that the force is equivalent to the negative of the slope of the potential energy graph. This is also an important result because it says that if a conservative force has 0 magnitude, the corresponding energy is at a local minimum or maximum (Fig. 4.8).

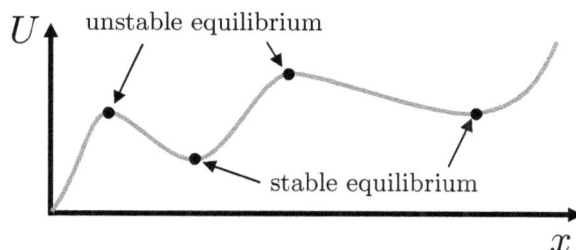

FIGURE 4.8
An energy vs. position diagram for the potential energy of some object; one intuitive example is gravitational potential energy of a rollercoaster cart. Positions where the derivative of the energy function is 0 are equilibrium positions. Because the corresponding force points in the direction of decreasing energy, local *minima* in the energy are *stable* equilibrium positions: objects tend to return to these positions if displaced slightly. In contrast, local *maxima* are positions of *unstable* equilibrium: nudge them slightly and they will continue to move away.

This argument generalizes to the situation where an object is experiencing multiple conservative forces (and no non-conservative forces): if the *net* force is 0, then the derivative of the *total* energy is also 0. This allows us to analyze the equilibrium positions of complicated systems strictly by analyzing the energy. Example 4.8 examines this in a familiar context.

We've introduced a few nuanced ideas so far, so before we move on to some examples, let's recap:

- A force acting over a distance *does work* and can change the kinetic energy of an object. If the work done by a particular force depends only on its initial and final positions, it is referred to as a *conservative force*.

- Every conservative force has a corresponding *potential energy*. We discussed *gravitational potential energy*, which depends on height h ($U_G = mgh$), and *spring potential energy*, which depends on the spring's displacement from equilibrium Δx ($U_{sp} = \frac{1}{2}k(\Delta x)^2$).

- The law of conservation of energy, which states that *energy cannot be created or destroyed*, allows us to analyze some complicated-looking problems fairly easily (Examples 4.5-4.7).

- If an object experiences only conservative forces, then the positions of equilibrium can be determined strictly by analyzing the energy of the system (Example 4.8).

Example 4.5 Energy of a Spring   ★★

A spring with a spring constant $k = 220\frac{\text{N}}{\text{m}}$ is attached to a block of mass $m = 2.2$ kg and stretched over a horizontal frictionless surface until it has a potential energy of 12 J. (a) How far has the spring been stretched? (b) What force is required to keep the mass in place? (c) If the mass is released, how fast will it be moving once the spring has contracted back to its equilibrium length?

We sketch the situation in Figure 4.9.

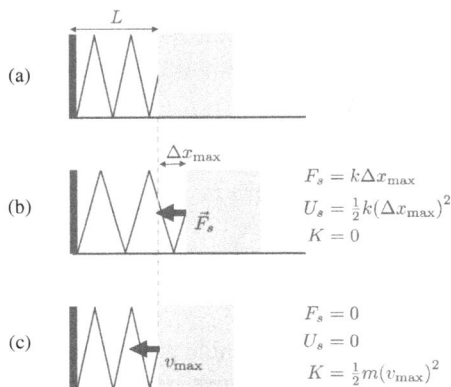

FIGURE 4.9

(a) For reference, the block is shown at equilibrium. (b) The initial configuration, where the mass has been pulled to some maximum extension Δx_{max}. (c) After being released, the block will slide to the left, reaching its maximum speed at the equilibrium position.

(a) From Equation (4.12),

$$U_{sp} = \frac{1}{2}k(\Delta x_{\text{max}})^2$$

$$\left(\frac{2U_{sp}}{k}\right)^{\frac{1}{2}} = \Delta x_{\text{max}}$$

$$\left(\frac{2(12\text{ J})}{220\frac{\text{N}}{\text{m}}}\right)^{\frac{1}{2}} = \Delta x_{\text{max}}$$

$$0.33\text{ m} = \Delta x_{\text{max}}$$

(b) Using the result from (a) with Hooke's Law, we find

$$F_s = k\Delta x_{\max}$$

$$= \left(220\frac{N}{m}\right)(0.33 \text{ m})$$

$$= 73 \text{ N}$$

(c) When the block is released, the spring force pulls the block toward equilibrium. In the language of energy, the spring potential energy decreases while the kinetic energy increases. When the block is at the equilibrium point, there is no more spring potential energy; the energy has completely transformed to kinetic energy:

$$(U_{\text{sp}})_i = K_f$$

$$12 \text{ J} = \frac{1}{2}mv_{\max}^2$$

$$\left(\frac{2\,(12 \text{ J})}{2.2 \text{ kg}}\right)^{\frac{1}{2}} = v_{\max}$$

$$3.3\frac{m}{s} = v_{\max}$$

This answers the question, but let's briefly describe the subsequent motion of the mass. The block is back at its equilibrium position, meaning there is no force acting on it, but it has kinetic energy and will therefore continue to move to the left. As it does, the kinetic energy will transform back into potential energy (the spring force pulls the mass back toward equilibrium, this time toward the right). Eventually the mass will stop; then, by the same logic we applied above, it will move back toward the equilibrium position. This cyclic motion is an example of *simple harmonic motion*, which we will study in more depth in Chapter 6.

It is also worth noting that we couldn't solve this problem with the tools we built up before this chapter: the spring force *changes* as the block moves, and therefore so too does the acceleration. The kinematic equations, however, assume a constant acceleration, and therefore can't be correctly applied here.

Example 4.6 Bouncing Between Springs ★★
A block of mass m is placed between two springs with spring constants k_1 and k_2, as shown. The space between the springs is frictionless except for a rough patch of width R, which has a coefficient of friction μ_k. The mass is pushed against the left spring, compressing it a distance x_1, and then released. How many times will it strike a spring before coming to rest?

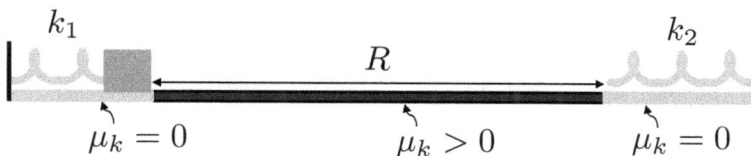

FIGURE 4.10

First note that this is an algebraic problem, so our answer will also be algebraic. The initial energy stored in the system is $U_{sp} = \frac{1}{2}k_1 x_1^2$; this energy will slowly transfer to thermal energy due to the work done by friction. The block will stop moving when this transfer is complete, meaning

$$\frac{1}{2}k_1 x_1^2 = F_k d = \mu_k F_N d = \mu_k mgd$$

where we have noted that because there is no net force in the vertical direction, $F_N = mg$. Thus the total distance traveled is

$$d = \frac{k_1 x_1^2}{2\mu_k mg}$$

Now we can compute the ratio $n = \frac{d}{R}$ as the number of times the mass strikes a spring. Thus, for example, if $n = 3.75$, it strikes a spring three times (right, left, right), then stops 75% of the way to the next spring.

Example 4.7 Spring on a Ramp   ★★
A 0.20 kg mass is pushed against a spring with a spring constant $k = 250\frac{\text{N}}{\text{m}}$ until it compresses by 8.5 cm. It is then released and allowed to slide across a frictionless surface and then up a ramp, as shown. Suppose (a) the ramp is also frictionless and (b) the ramp has a coefficient of kinetic friction $\mu_k = 0.32$. In each case, determine the speed of the mass when it reaches the top of the ramp or, if it doesn't reach the top, determine how far up the ramp it travels before stopping.

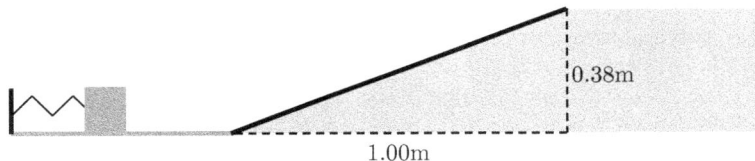

FIGURE 4.11

The initial energy is

$$U_{sp} = \frac{1}{2}k\left(\Delta x\right)^2 = \frac{1}{2}\left(250\frac{\text{N}}{\text{m}}\right)(.085\text{ m})^2 = 0.90\text{ J}$$

The spring will push the block to the right; as it does so the spring potential energy will decrease to 0 J and the kinetic energy will increase to 0.90 J. Now let us consider case (a), where there is no friction on the ramp. Moving up a distance of 0.38 m requires gravitational potential energy:

$$mgh = (0.20\text{ kg})\left(9.8\frac{\text{m}}{\text{s}^2}\right)(0.38\text{ m}) = 0.74\text{ J}$$

Now, the block has more than the required amount of energy, so it will make it to the top, transferring kinetic energy to gravitational potential energy along the way, and reach the top while still having some kinetic energy to spare. Specifically,

$$(U_{\text{sp}})_i = (U_G)_f + K_f$$

$$0.90\text{J} = 0.74 \text{ J} + \frac{1}{2}(0.20 \text{ kg}) v_f^2$$

$$\left(2\frac{0.90 \text{ J} - 0.74 \text{ J}}{0.20 \text{ kg}}\right)^{\frac{1}{2}} = v_f$$

$$1.3\frac{\text{m}}{\text{s}} = v_f$$

If there *is* friction, then to make it to the top the block's kinetic energy will be transferred to both gravitational potential energy and lost to the environment due to the work of kinetic friction. We already know how much gravitational potential energy is required to make it to the top, so let's consider friction. First, note that the angle of inclination of the ramp is $\theta = \tan^{-1}\left(\frac{0.38}{1.00}\right) = 21°$ and the length of the ramp is $d = \left((1.00 \text{ m})^2 + (0.38 \text{ m})^2\right)^{\frac{1}{2}} = 1.07 \text{ m}$. Then

$$W = F_k d$$

$$= \mu_k F_N d$$

$$= \mu_k mgd \cos \theta$$

$$= (0.32)(0.20 \text{ kg})\left(9.8\frac{\text{m}}{\text{s}^2}\right)(1.07 \text{ m}) \cos(21°)$$

$$= 0.63 \text{ J}$$

Thus the total energy required to make it to the top is $U_G + W = 0.74 \text{ J} + 0.63 \text{ J} = 1.4 \text{ J}$, more than the 0.90 J given to the block by the spring. The question is then *how far does it go?* Let's call its final elevation y_f and its final distance along the ramp d_f, so $y_f = d_f \sin 21°$, and invoke conservation of energy:

$$(U_{\text{sp}})_i = F_k d_f + mgy_f$$

$$(U_{\text{sp}})_i = \mu_k mgd_f \cos\theta + mgd_f \sin\theta$$

$$(U_{\text{sp}})_i = (\mu_k \cos\theta + \sin\theta) mgd_f$$

$$d_f = \frac{(U_{\text{sp}})_i}{mg(\mu_k \cos\theta + \sin\theta)}$$

$$d_f = \frac{0.90 \text{ J}}{(0.20 \text{ kg})\left(9.8\frac{\text{m}}{\text{s}^2}\right)((0.32)\cos 21° + \sin 21°)}$$

$$d_f = 0.70 \text{ m}$$

We can compare this to the total length of the ramp: $\frac{.70}{1.07} = 0.66$, meaning the block makes it 66% of the way up the ramp before stopping.

Example 4.8 Energy with a Spring and Gravity $\star \star dx$

A rubber band that obeys Hooke's Law (with spring constant k) when stretched beyond its equilibrium length L is attached to the ceiling on one end and to a marble of mass m on the other. The marble is slowly lowered from near the ceiling until the rubber band fully supports its weight. How far below the ceiling is the marble when this occurs?

First note that this is a straightforward problem to analyze via forces. If we call *up* the positive direction, then when the marble is at equilibrium we have

$$F_{\text{NET}} = k\Delta x - mg = 0$$

$$\Delta x = \frac{mg}{k}$$

Here Δx describes how far beyond its equilibrium length L the rubber band has stretched, so the marble's total distance below the ceiling is

$$y = L + \frac{mg}{k}$$

We can obtain the same result from energy analysis by noting that the total energy is the sum of the spring potential energy and the gravitational potential energy. We'll call the ceiling the position of 0 gravitational energy and so once the rubber band is taut we have

$$U_{\text{TOT}} = U_G + U_{\text{sp}}$$

$$U_{\text{TOT}} = -mg\left(L + x\right) + \frac{1}{2}k\left(\Delta x\right)^2$$

The gravitational potential energy is negative because we're talking about a position *below* the zero point: the ceiling. (You will obtain the same final answer with a different choice; see Problem 4.47 for one example.)

As suggested by Equation (4.15), we proceed by taking the negative derivative of the energy

$$-\frac{d}{dx}\left(U_{\text{TOT}}\right) = mg - k\Delta x$$

and setting it equal to 0:

$$0 = mg - k\Delta x$$

$$\Delta x = \frac{mg}{k}$$

Once again, we conclude that the marble's equilibrium position is a distance

$$y = L + \frac{mg}{k}$$

below the ceiling.

Energy is not restricted to the types we have discussed so far; we will encounter more as we proceed through the rest of this book (for instance, electrical circuits make use of the energy associated with interacting *charged* particles). For now, let's make just one final comment. We are often concerned with *how quickly work is being done* (or, if you prefer, how quickly energy is being transformed). This is referred to as **power**, which is defined (for cases of constant force) as:

$$P = \frac{W}{\Delta t} = \frac{\vec{F} \cdot \vec{d}}{\Delta t} = \vec{F} \cdot \vec{v} \qquad (4.16)$$

The units of power are the Joule/second, which has the SI unit *watt* (1 W = 1 J/s).[10] Example 4.9 demonstrates these ideas.

Example 4.9 Energy and Power in an Elevator ⋆

A particular skyscraper has 50 floors and is 2.00×10^2 m tall. Consider a 1.00×10^3 kg elevator, and suppose it travels from the ground floor to the top floor in (a) 30. seconds or (b) 1.0 minute. In each case, determine the energy and the power required for the trip. Assume the elevator's acceleration at the beginning and end of the trip occurs rapidly enough to be neglected; it travels the bulk of the trip at a constant speed.

In each case the energy required to get the elevator to the top floor is

$$U_G = mgh$$
$$= \left(1.00 \times 10^3 \text{ kg}\right) \left(9.8 \frac{\text{m}}{\text{s}^2}\right) \left(2.00 \times 10^2 \text{ m}\right)$$
$$= 1.96 \times 10^6 \text{ J}$$

Then the power involved is just this energy divided by the time it takes to get to the top. Thus for (a) we have $P = 1.96 \times 10^6$ J/30. s $= 6.5 \times 10^4$ W and for (b) we have $P = 1.96 \times 10^6$ J/60. s $= 3.3 \times 10^4$ W.

Equivalently, we could use $P = \vec{F} \cdot \vec{v}$ and note $\vec{F} = mg$ directed straight up (if the elevator travels at a constant speed, the upward force must balance the downward force of gravity) and the velocity has magnitude $v = h/t$ and is also directed straight up. This results in the same values for power we determined above.

4.3 Momentum, Collisions, and Explosions

We introduced the ideas of work and energy by thinking about a force acting over a *distance*. Let us now think about a force acting over some amount of *time*. As before, let's consider a block on a horizontal frictionless surface experiencing a single horizontal force $\vec{F}_P$ (Figure 4.1). While before we considered it traveling some distance Δx, it obviously takes some amount of time Δt to do this, and we can note

$$\vec{F}_P = m\vec{a} = m\frac{\Delta \vec{v}}{\Delta t} \tag{4.17}$$

If we rearrange some terms and talk generically about an average force $\vec{F}_{\text{avg}}$ over some time interval Δt, we find

$$\vec{F}_{\text{avg}} \Delta t = m\Delta \vec{v} \tag{4.18}$$

We call the term on the left hand side of this equation (a force acting for some duration of time) an **impulse**, $\vec{J}$:

$$\vec{J} = \vec{F}_{\text{avg}} \Delta t \tag{4.19}$$

---

[10]After James Watt (1736-1819)

We also define an object's **momentum**, $\vec{p}$:

$$\vec{p} = m\vec{v} \tag{4.20}$$

While not immediately important, it is worth mentioning that from this definition it is straightforward to express an object's kinetic energy in terms of the magnitude of its momentum:

$$K = \frac{mv^2}{2} = \frac{(mv)^2}{2m} = \frac{p^2}{2m} \tag{4.21}$$

Now, if we combine Equations (4.19) and (4.20) with Equation (4.18), we can write

$$\vec{J} = \vec{F}_{\text{avg}}\Delta t = m\Delta\vec{v} = \Delta\vec{p} \tag{4.22}$$

We see from this equation that the units of impulse and momentum are the Newton-second (Ns), or equivalently $\frac{\text{kg m}}{\text{s}}$. Notice that this equation refers to the impulse in terms of a time interval Δt. It is possible to recast these results in terms of the *instantaneous* change in momentum in the usual way:[11]

$$\vec{F} = d\vec{p}/dt \tag{4.23}$$

and

$$\vec{J} = \int \vec{F}dt \tag{4.24}$$

Graphically, then, the impulse is the area bounded by a force vs. time curve (Problem 4.60).

Equation 4.22 is useful in and of itself, especially when analyzing objects bouncing off of rigid surfaces such as a table or wall (as we demonstrate in Example 4.10). However, there is more to the story of momentum than this. Consider two balls that have been thrown and collide mid-air (Figure 4.12). They each have some initial velocity (just prior to the collision), and after they are done colliding they will each have some final velocity. *During* the collision, they will each be experiencing a force: ball 1 pushes on ball 2, and ball 2 pushes back on ball 1. Importantly, the forces are equal in magnitude but opposite in direction (Newton's Third Law!):

$$\vec{F}_{12} = -\vec{F}_{21}$$
$$m_1\vec{a}_1 = -m_2\vec{a}_2$$
$$m_1\frac{\Delta\vec{v}_1}{\Delta t} = -m_2\frac{\Delta\vec{v}_2}{\Delta t}$$
$$\Delta\vec{p}_1 = -\Delta\vec{p}_2$$
$$(\vec{p}_1)_f - (\vec{p}_1)_i = -\left((\vec{p}_2)_f - (\vec{p}_2)_i\right)$$
$$(\vec{p}_1)_i + (\vec{p}_2)_i = (\vec{p}_1)_f + (\vec{p}_2)_f$$

What the final line above says is *the combined initial momentum of both objects is the same as the combined final momentum of both objects*.[12] While we here considered only two

---

[11]In fact, Equation (4.23) is a more fundamental definition of Newton's second law. If you use the product rule, $F = d(mv)/dt = m(dv/dt) + (dm/dt)v$. If m is constant, this reduces to Newton's second law as we normally consider it, $F = ma$. If the mass is changing, though (like a rocket burning fuel), the second term is needed to accurately describe the physics.

[12]It is worth remembering that when we say $\vec{a} = \Delta\vec{v}/\Delta t$, we are talking about the *average* acceleration, which only becomes exact if Δt becomes infinitely small. This is important when we don't have a constant force – and the force between the two balls during the collision generally won't be constant.

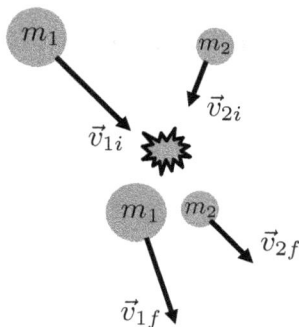

FIGURE 4.12
Two balls collide and then separate.

objects, the argument generalizes for any number of objects, so long as there are no external forces (i.e. all forces have an equal and opposite partner according to Newton's Third Law). Thus we have the law of **conservation of momentum**:

The total momentum of a closed system is constant.

As with the law of conservation of energy, we can write this mathematically:

$$\sum \vec{p}_i = \sum \vec{p}_f \qquad (4.25)$$

Examples 4.11 and 4.12 demonstrate the law of conservation of momentum, which can be used to describe both collisions (objects coming together) and explosions (objects separating).

Example 4.10 Impulse from a Ball Bounce ★★
A 0.030 kg rubber ball is dropped from a height of 1.70 m onto a metal scale that is resting on the floor. The ball bounces back to a maximum height of 1.39 m, and the scale reports an average force of 4.5 N. (a) What impulse did the ground impart to the ball? (b) How long was the ball in contact with the floor?

We could draw a diagram for this, but see if you can paint a clear mental picture (and by all means sketch a diagram if you like!). Initially the ball is at a height h_1 and has potential energy mgh_1; when it hits the ground this energy will have been converted to kinetic energy $\frac{1}{2}mv^2$. If we refer to this speed as the *impact speed* v_{impact}, then

$$mgh_1 = \frac{1}{2}mv_{\text{impact}}^2$$

$$(2gh_1)^{\frac{1}{2}} = v_{\text{impact}}$$

After the impact (at the instant the ball leaves the floor), the ball has some upward speed, which we'll call the *rebound speed*, v_{rebound}. As the ball moves up, it loses kinetic energy and gains gravitational potential energy, until it reaches a maximum height h_2. Thus

$$\frac{1}{2}mv_{\text{rebound}}^2 = mgh_2$$

$$v_{\text{rebound}} = (2gh_2)^{\frac{1}{2}}$$

We need to be careful about the direction of these velocities. If we call *up* the positive y direction, then $\vec{v}_{\text{impact}} = -(2gh_1)^{\frac{1}{2}}\,\hat{y}$ and $\vec{v}_{\text{rebound}} = (2gh_2)^{\frac{1}{2}}\,\hat{y}$. Thus

$$\vec{J} = m\Delta\vec{v}$$

$$\vec{J} = m\left(\vec{v}_{\text{rebound}} - \vec{v}_{\text{impact}}\right)$$

$$\vec{J} = m\left((2gh_2)^{\frac{1}{2}}\,\hat{y} - \left(-(2gh_1)^{\frac{1}{2}}\,\hat{y}\right)\right)$$

$$\vec{J} = m\left((2gh_2)^{\frac{1}{2}} + (2gh_1)^{\frac{1}{2}}\right)\hat{y}$$

$$\vec{J} = m\,(2g)^{\frac{1}{2}}\left(h_2^{\frac{1}{2}} + h_1^{\frac{1}{2}}\right)\hat{y}$$

$$\vec{J} = (0.030 \text{ kg})\left(2\left(9.8\frac{\text{m}}{\text{s}^2}\right)\right)^{\frac{1}{2}}\left((1.39 \text{ m})^{\frac{1}{2}} + (1.70 \text{ m})^{\frac{1}{2}}\right)\hat{y}$$

$$\vec{J} = 0.33\frac{\text{kg m}}{\text{s}}$$

Now we know the impulse points in the same direction as the force on the ball (up), so

$$J = F\Delta t$$

$$\frac{J}{F} = \Delta t$$

$$\frac{0.33\frac{\text{kg m}}{\text{s}}}{4.5 \text{ N}} = \Delta t = 0.073 \text{ s}$$

Example 4.11 Momentum Conservation in a Cannon Shot   ⋆

The *1841 Model Gun* was an 810 kg artillery cannon that fired a 6.0 pound (2.7 kg) projectile. Suppose one of these cannons is sitting on a frictionless horizontal surface when it fires a projectile horizontally such that someone standing nearby measures its speed to be 110 m/s. What is the recoil velocity of the cannon?

We will set the $+x$ direction to be the direction of the projectile. The situation is sketched in Figure 4.13.

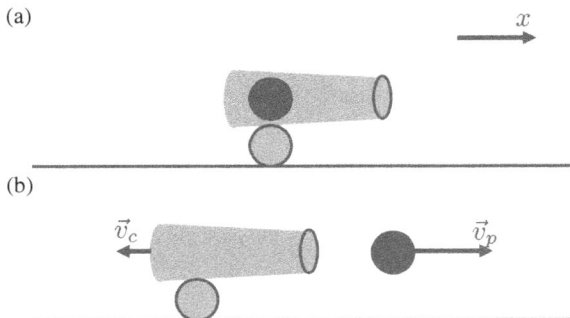

FIGURE 4.13
(a) A cannon sitting at rest on a frictionless surface with a projectile resting inside.
(b) After the projectile has been fired with some velocity $\vec{v}_p$, the cannon recoils with a velocity $\vec{v}_c$.

We can invoke conservation of momentum, noting that initially there is *no momentum* because both the projectile and the cannon are at rest:

$$\sum \vec{p}_i = \sum \vec{p}_f$$

$$0 \text{ Ns} = m_c\left(\vec{v}_c\right) + m_p\left(\vec{v}_p\right)$$

$$0 \text{ Ns} = m_c\left(-v_c\right) + m_p\left(v_p\right)$$

$$\frac{m_p v_p}{m_c} = v_c$$

$$\frac{(2.7 \text{ kg})\left(110\frac{\text{m}}{\text{s}}\right)}{810 \text{ kg}} = v_c = 0.37\frac{\text{m}}{\text{s}}$$

Before moving on, it is worth pointing out that this isn't really a collision: both objects start at rest and then separate; we typically refer to this as an *explosion*. The law of conservation of momentum applies, even so! (An explosion is essentially a collision run in reverse and all of the arguments we made about equal and opposite interaction forces apply in precisely the same way.)

Example 4.12 Momentum in a Two-Dimensional Collision ★★
In a game of pool, two billiard balls (each of mass 160 g) have initial velocities shown in the top of Figure 4.14. Ball 1's velocity after the collision is shown in the bottom of the figure. What is the final velocity of ball 2, calculated to two significant figures?

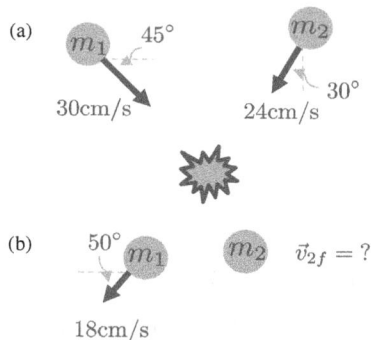

FIGURE 4.14
A top down view of the velocities of two billiard balls before (a) and after (b) colliding.

Unlike the previous example, here we have motion occurring in two dimensions. Just as we considered Newton's Second Law independently in two different dimensions, so too can we apply the law of conservation of momentum. Let's call "to the right" the $+x$ direction and "up" the $+y$ direction. Then in the x dimension we have

$$\sum \left(\vec{p}_i\right)_x = \sum \left(\vec{p}_f\right)_x$$

$$m\left(\vec{v}_{1i}\right)_x + m\left(\vec{v}_{2i}\right)_x = m\left(\vec{v}_{1f}\right)_x + m\left(\vec{v}_{2f}\right)_x$$

$$\left(0.30\frac{\text{m}}{\text{s}}\right)\cos 45° + \left(-0.24\frac{\text{m}}{\text{s}}\right)\sin 30° = \left(-0.18\frac{\text{m}}{\text{s}}\right)\cos 50° + \left(v_{2x}\right)_f$$

$$0.21\frac{\text{m}}{\text{s}} = \left(v_{2x}\right)_f$$

Similarly, in the y dimension we have

$$\sum (\vec{p}_i)_y = \sum (\vec{p}_f)_y$$

$$m (\vec{v}_{1i})_y + m (\vec{v}_{2i})_y = m (\vec{v}_{1f})_y + m (\vec{v}_{2f})_y$$

$$\left(-0.30\frac{m}{s}\right) \sin 45° + \left(-0.24\frac{m}{s}\right) \cos 30° = \left(-0.18\frac{m}{s}\right) \sin 50° + (v_{2f})_y$$

$$-0.28\frac{m}{s} = (v_{2y})_f$$

Thus the final velocity of the second ball is

$$(\vec{v}_2)_y = 0.21\frac{m}{s}\hat{x} - 0.28\frac{m}{s}\hat{y}$$

If we would prefer to have our answer in magnitude-angle form, some trigonometry yields (check it yourself!)

$$(\vec{v}_2)_y = 0.35\frac{m}{s} \text{ at } 54° \text{ below the } +x \text{ axis.}$$

as shown in Figure 4.15.

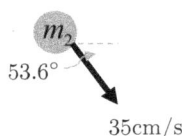

FIGURE 4.15
The final velocity of ball 2 from Figure 4.12.

4.4   Elastic and Inelastic Collisions

As the examples in the previous section demonstrate, conservation of momentum is a very useful law. It does not, however, provide as much information as we might always like. Consider a situation where two objects with known masses and velocities are about to experience a head-on collision (Figure 4.16).[13] We have six variables (a mass, initial velocity, and final velocity for each object), and we don't know two of them (the final velocities). Thus in general we *cannot* determine both final velocities using only conservation of momentum (we require two independent equations to solve for two unknown variables).

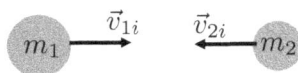

FIGURE 4.16
Two masses about to experience a head-on collision.

---

[13]and, as usual, assuming no other forces – like friction from the ground – are at play.

I say "in general" because there is one situation where conservation of momentum tells us everything we want to know: if the objects *stick together*, they have the same final velocity and therefore we have one less unknown variable. Such a collision is referred to as an **inelastic collision**. Many car crashes are inelastic collisions; below we consider a different example.

Example 4.13 A Ballistic Pendulum ⋆
A 5.3 g bullet traveling at 460 m/s strikes a 4.5 kg target hanging from a rope, as shown. The bullet becomes lodged in the target. What is their speed immediately after the collision?

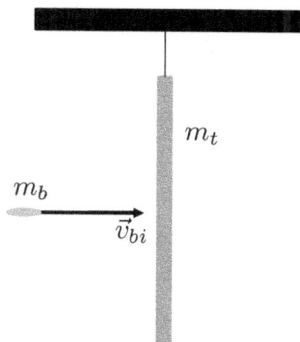

FIGURE 4.17

Here we can apply conservation of momentum directly. We'll call "to the right" the positive direction, so

$$\sum \vec{p}_i = \sum \vec{p}_f$$

$$m_b \left(v_b\right)_i + m_t \left(0\frac{m}{s}\right) = (m_b + m_t)\, v_f$$

$$\frac{(.0053 \text{ kg}) \left(460\frac{m}{s}\right)}{(0.0053 + 4.5)\, \text{kg}} = v_f = 0.54\frac{m}{s}$$

In Example 4.13, the final kinetic energy is much smaller than the initial kinetic energy (you are asked to verify this is Problem 4.105). Evidently most of the initial kinetic energy becomes some other type of energy. Indeed, in any collision some kinetic energy is transformed to other types of energy (sound and thermal energy, for example). In an inelastic collision, energy is also required to deform the objects so they stick together. In fact, it turns out that an inelastic collision involves losing the maximum amount of kinetic energy possible without violating the law of conservation of momentum.[14]

The other end of the spectrum is an **elastic collision**, where all of the energy remains kinetic energy (we always lose *some* energy to the environment, but in some cases – firm rubber balls colliding, for example – this is an excellent approximation). In an elastic collision we can appeal to conservation of momentum:

$$m_1 \left(\vec{v}_1\right)_i + m_2 \left(\vec{v}_2\right)_i = m_1 \left(\vec{v}_1\right)_f + m_2 \left(\vec{v}_2\right)_f$$

---

[14]You are asked to investigate this in Problem 4.81.

and conservation of energy:

$$\frac{1}{2}m_1\left(v_1\right)_i^2 + \frac{1}{2}m_2\left(v_2\right)_i^2 = \frac{1}{2}m_1\left(v_1\right)_f^2 + \frac{1}{2}m_2\left(v_2\right)_f^2$$

Consider again the standard example where we have two masses colliding with known initial velocities (Figure 4.16). We now have everything we need to determine both final velocities: we can algebraically combine the two equations above to obtain equations for $(\vec{v}_1)_f$ and $(\vec{v}_2)_f$ involving just the (known) masses and initial velocities. The algebra is somewhat tedious (see Problem 4.106), so we'll jump right to the end:

$$\left(v_1\right)_f = \left(\frac{m_1 - m_2}{m_1 + m_2}\right)\left(v_1\right)_i + \left(\frac{2m_2}{m_1 + m_2}\right)\left(v_2\right)_i \qquad (4.26)$$

$$\left(v_2\right)_f = \left(\frac{2m_1}{m_1 + m_2}\right)\left(v_1\right)_i + \left(\frac{m_2 - m_1}{m_1 + m_2}\right)\left(v_2\right)_i \qquad (4.27)$$

These equations allow us to analyze many interesting situations, as we'll see in the problems at the end of this chapter and again later, when we consider interactions between charged particles.

Example 4.14 Elastic and Inelastic Collisions   ⋆
A 2.50 kg ball traveling at $(v_1)_i = 1.00\hat{x}$ m/s collides with a 1.50 kg ball traveling at $(v_2)_i = -2.00\hat{x}$ m/s. What will their velocities be if the collision is (a) elastic (if the balls are made from hard rubber, say) and (b) inelastic (for instance, if the balls are wrapped in Velcro).

For (a), we can invoke Equation (4.26):

$$\left(v_1\right)_f = \left(\frac{m_1 - m_2}{m_1 + m_2}\right)\left(v_1\right)_i + \left(\frac{2m_2}{m_1 + m_2}\right)\left(v_2\right)_i$$
$$= \left(\frac{1.00 \text{ kg}}{4.00 \text{ kg}}\right)(1.00 \text{ m/s}) + \left(\frac{3.00 \text{ kg}}{4.00 \text{ kg}}\right)(-2.00 \text{ m/s})$$
$$= -1.25 \text{ m/s}$$

and Equation (4.27):

$$\left(v_2\right)_f = \left(\frac{2m_1}{m_1 + m_2}\right)\left(v_1\right)_i + \left(\frac{m_2 - m_1}{m_1 + m_2}\right)\left(v_2\right)_i$$
$$= \left(\frac{5.00 \text{ kg}}{4.00 \text{ kg}}\right)(1.00 \text{ m/s}) + \left(\frac{-1.00 \text{ kg}}{4.00 \text{ kg}}\right)(-2.00 \text{ m/s})$$
$$= 1.75 \text{ m/s}$$

We see that both balls bounce away in the opposite direction compared to their initial direction of motion. You can confirm that momentum is indeed conserved for this collision (Problem 4.79). For (b), conservation of momentum allows us to evaluate the common final velocity for both balls:

$$\sum \vec{p}_i = \sum \vec{p}_f$$

$$m_1 \left(v_1\right)_i + m_2 \left(v_2\right)_i = \left(m_1 + m_2\right) v_f$$

$$\frac{m_1 \left(v_1\right)_i + m_2 \left(v_2\right)_i}{\left(m_1 + m_2\right)} = v_f$$

$$\frac{(2.50 \text{ kg}) (1.00 \text{ m/s}) + (1.50 \text{ kg}) (-2.00 \text{ m/s})}{(4.00 \text{ kg})} = v_f$$

$$-0.125 \text{ m/s} = v_f$$

4.5   Conservation Laws in Angular Motion

As we've proceeded through the book so far, we've encountered the angular version of kinematics (i.e. angular kinematics) and force (i.e. torque). To complete our comparison of linear and rotational motion, in this section we will take up conservation of energy and conservation of momentum in the context of rotational motion.

In the case of energy, we have only one new quantity to consider, the **rotational kinetic energy**. Recall that the moment of inertia I is the angular analog of the mass m, and the angular velocity ω is the angular analog of the speed v. If we apply these substitutions to the equation for linear kinetic energy $K_{\text{lin}} = \frac{1}{2}mv^2$, we find[15]

$$K_{\text{rot}} = \frac{1}{2}I\omega^2 \tag{4.28}$$

Thus, if we have an object that *isn't* a point particle, we can think of it as having linear kinetic energy due to its *translational* motion, rotational kinetic energy due to its *rotational* motion, or *both* kinds of kinetic energy due to *both* kinds of motion (Fig. 4.18):

$$K = \frac{1}{2}mv^2 + \frac{1}{2}I\omega^2 \tag{4.29}$$

A prime example of this is a ball or wheel rolling over level ground. If the object is rolling smoothly (i.e. the object is not *slipping* like a car might when the driver slams on the brakes), then the velocity of the center of mass is given by $v = \omega r$ and Equation (4.29) can be written[16]

$$K = \frac{v^2}{2} \left(m + \frac{I}{r^2}\right) \tag{4.30}$$

$$K = \frac{\omega^2}{2} \left(mr^2 + I\right) \tag{4.31}$$

---

[15]In Chapter 4 we referred to this simply as the kinetic energy. Now that we have two different types of kinetic energy, we'll be careful to refer to them as "linear" and "rotational".

[16]If this is unclear, imagine dabbing some paint on the edge of the wheel and then sending it rolling. The distance between adjacent paint marks on the ground will be equal to the circumference of the wheel, but the center of the wheel will be directly above the paint when both marks are made. Thus a point on the rim covers the same distance ($2\pi r$ around the center) as the center of the wheel ($2\pi r$ parallel to the ground) in the same amount of time. We explore this in more detail in Example 4.17.

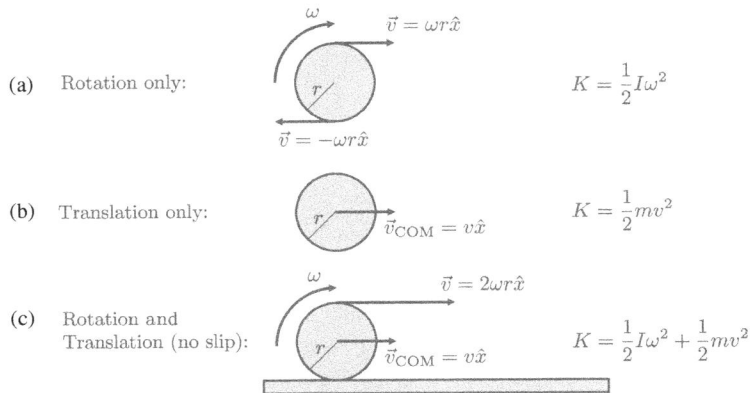

FIGURE 4.18
(a) A ball or wheel rotating about its center with an angular speed ω has rotational kinetic energy. The tangential velocity of two points are drawn. (b) A ball or wheel that is not rotating but is *translating* has linear kinetic energy. The velocity of the center of mass is drawn. (c) a ball or wheel that is both translating and rotating has both forms of kinetic energy. Here we suppose the object is not slipping, so $v = \omega r$. The velocities at the top and bottom points are given by the vector sum of the tangential velocities from the top and middle diagrams: at the top of the wheel the velocities add (they both point right, in the $+\hat{x}$ direction), but at the bottom they cancel: the point in contact with the ground is instantaneously *at rest* (hence the "no slip" designation).

The principle of conservation of energy states that energy can only be transformed (never created or destroyed); when we work with energy to describe some situation in mechanics we are saying "the energy is initially split up *this* way" (gravitational potential and spring potential, say) and later "the energy is split up *that* way" (kinetic, say). We now have a new kind of energy, but the approach hasn't changed at all. As one particular example, note that we defined power as the rate at which energy is delivered (or consumed). In the linear case we are concerned with the power involved with a *force* (Equation (4.16); $P = \vec{F} \cdot \vec{v}$), while in the rotational case we are concerned with a *torque* (see e.g. Problem 4.135):

$$P = \tau \omega \qquad\qquad (4.32)$$

The next set of examples demonstrate how to integrate rotational kinetic energy into our analysis.

Example 4.15 Kinetic Energies of a Rolling Ball   ⋆
A small ball has a mass of 150 g and a radius of 2.0 cm. If it is rolling without slipping over level ground with a translational speed of 2.0 m/s, then what is its (a) linear, (b) rotational, and (c) total kinetic energy?

The rotational kinetic energy is

$$K_{\text{lin}} = \frac{1}{2}mv^2 = \frac{1}{2}\left(0.150 \text{ kg}\right)\left(2.0 \text{ m/s}\right)^2 = 0.30 \text{ J}$$

Meanwhile, the rotational kinetic energy is

$$K_{\text{lrot}} = \frac{1}{2}I\omega^2 = \frac{1}{2}\left(\frac{2}{5}mr^2\right)\left(\frac{v}{r}\right)^2 = \frac{mv^2}{5}$$

$$= \frac{(0.150 \text{ kg})(2.0 \text{ m/s})^2}{5} = 0.12 \text{ J}$$

For (c), we combine our results:

$$K_{\text{tot}} = K_{\text{lin}} + K_{\text{rot}} = 0.42 \text{ J}$$

Example 4.16 Unwinding a Pulley ★

A light string has been coiled around a pulley of mass M and radius R; one end is attached to a block of mass m which hangs below the pulley, as shown. The system is initially at rest, then the block is allowed to fall, unwinding the string in the process. Consider a short time later, when the block has fallen a distance h.

(a) Determine an expression for the speed of the block.

(b) Confirm that if $M = 0$ kg, your answer for (a) reduces to the simple case where an object of mass m is dropped from rest and allowed to fall a distance h.

We sketch the situation in Figure 4.19. We set the origin of our coordinate system to be level with the final position of the block, so we go from gravitational potential energy mgh initially to no gravitational potential energy. This energy is converted into the falling (i.e. linear kinetic) energy of the mass and the spinning (i.e. rotational kinetic) energy of the pulley:

$$(U_G)_i = (K_{\text{lin}})_f + (K_{\text{rot}})_f$$
$$mgh = \frac{1}{2}mv^2 + \frac{1}{2}I\omega^2$$

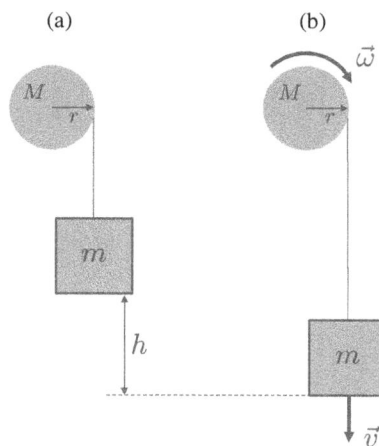

FIGURE 4.19

A mass suspended from a pulley that unwinds as the mass falls. (a) The mass and pulley are initially at rest. (b) a short time later, the mass has fallen a distance h and has a downward velocity v, while the pulley has an angular speed ω.

Now the linear speed of the block is the same as the linear speed on the rim of the pulley, so $v = wR$ and

$$mgh = \frac{1}{2}mv^2 + \frac{1}{2}\left(\frac{1}{2}MR^2\right)\left(\frac{v}{R}\right)^2$$

$$mgh = \left(\frac{1}{2}m + \frac{1}{4}M\right)v^2$$

$$\left(\frac{mgh}{\frac{1}{2}m + \frac{1}{4}M}\right)^{\frac{1}{2}} = v$$

This is our answer to (a). For (b), we set $M = 0$ kg and simplify:

$$v = \left(\frac{mgh}{\frac{1}{2}m}\right)^{\frac{1}{2}}$$

$$v = (2gh)^{\frac{1}{2}}$$

This is indeed the same result as if we had a mass falling from rest through a height h:

$$(U_G)_i = (K_{\text{lin}})_f$$

$$mgh = \frac{1}{2}mv^2$$

$$v = (2gh)^{\frac{1}{2}}$$

You may have noticed that this problem is rather similar to Example 3.35; all that is missing is a second mass. Indeed, analyzing that example using energy can be considerably quicker than using torque/force analysis.

Example 4.17 The Rotation Race   ★★

A classic demonstration of rotational energy involves placing a ring, disk, and sphere of equal masses M and equal radii R at the top of an inclined plane with a sufficiently high coefficient of friction that all of the objects roll smoothly (rather than slip) as they move to the bottom of the ramp (Figure 4.20). Determine an expression for the speed of each object at the bottom of the ramp, and rank them. This ranking corresponds to the order in which the objects will reach the bottom of the ramp (the fastest object reaches the bottom first and so on); why? (You can readily find video demonstrations of this "race" online.)

FIGURE 4.20

A circular object at the top of an inclined plane of height h. Here the object is drawn as a disk, though in this example we are considering a disk, ring, and sphere lined up side by side at the top of the inclined plane. (Use your imagination!).

Initially the objects are at rest and at a height h above the bottom of the ramp. If we set the position of 0 gravitational potential energy to be at the bottom of the ramp, then the initial energy of each object is simply Mgh. The final energy, meanwhile, will be a combination of linear kinetic energy (from the fact that the object is *translating* down the ramp) and rotational kinetic energy (because the object is rotating). If the distinction is not clear, consider the fact that the object could be translating without rotating–if you threw it, say–or rotating in place like a pulley. Here the object is doing both. Mathematically, we have

$$(U_G)_i = (K_{\text{lin}})_f + (K_{\text{rot}})_f$$

$$Mgh = \frac{1}{2}Mv^2 + \frac{1}{2}I\omega^2$$

Now, like we've done in other cases, we'll note that $v = \omega R \Rightarrow \omega = v/R$ to eliminate ω in our energy equation. You may object that when we say $v = \omega R$, the variable v refers to the linear velocity of a point at a distance R from the axis of rotation, and *not* the translational velocity of the entire object. This argument is correct, but it turns out that the two velocities are the same: imagine a point on the object touching the ramp at some instant. When the object has made a complete rotation, that same point will have traveled a distance down the ramp equal to the circumference of the object. However, the center of mass of the object must have traveled the same distance in the same time, meaning it has the same speed (Figure 4.21).

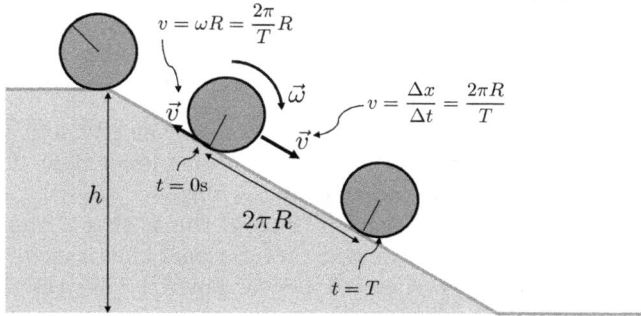

FIGURE 4.21

The object is shown at two positions along the ramp. At time $t = 0$ s, a radius to the point touching the ramp is drawn. One period T later, the object has made a complete rotation and moved a distance down the ramp equal to its circumference. Thus the linear speed of a point on the rim equals the linear speed of the center of the object, as shown mathematically on the diagram.

Making the substitution to eliminate ω, we find

$$Mgh = \frac{1}{2}Mv^2 + \frac{1}{2}I\omega^2$$

$$Mgh = \frac{1}{2}Mv^2 + \frac{1}{2}I\left(\frac{v}{R}\right)^2$$

$$2Mgh = \left(M + \frac{I}{R^2}\right)v^2$$

$$\left(\frac{2Mgh}{M + \frac{I}{R^2}}\right)^{\frac{1}{2}} = v$$

We have purposefully done all of this in terms of a generic moment of inertia I because we have arrived at an equation that applies to each of the objects that we are considering. All that remains is to insert the moment of inertia for each object and simplify:

$$v_{\text{sphere}} = \left(\frac{2Mgh}{M + \frac{\frac{2}{5}MR^2}{R^2}}\right)^{\frac{1}{2}} = \left(\frac{2gh}{1 + \frac{2}{5}}\right)^{\frac{1}{2}}$$

$$v_{\text{disk}} = \left(\frac{2Mgh}{M + \frac{\frac{1}{2}MR^2}{R^2}}\right)^{\frac{1}{2}} = \left(\frac{2gh}{1 + \frac{1}{2}}\right)^{\frac{1}{2}}$$

$$v_{\text{ring}} = \left(\frac{2Mgh}{M + \frac{MR^2}{R^2}}\right)^{\frac{1}{2}} = \left(\frac{2gh}{2}\right)^{\frac{1}{2}}$$

We haven't simplified these equations as far as we could so they each have the same numerator; we can therefore rank them by examining the denominators. The ring has the largest denominator (and therefore the lowest speed), followed by the disk, followed by the sphere (which will therefore have the highest speed).

Now the objects have all traveled the same distance (the length of the ramp), which we'll call Δx. From basic kinematics we have $\Delta x = v_{\text{avg}}\Delta t$, so it follows that

$$(v_{\text{avg}})_{\text{sphere}} (\Delta t)_{\text{sphere}} = (v_{\text{avg}})_{\text{ring}} (\Delta t)_{\text{ring}} = (v_{\text{avg}})_{\text{disk}} (\Delta t)_{\text{disk}}$$

A higher *final* velocity clearly implies a higher *average* velocity, and so if the above equality is to be met, a higher final velocity must mean a lower time. Thus the sphere finishes first, followed by the disk, followed by the ring.

Strictly in terms of energy, a good way to think of this is that a higher moment of inertia means that more energy is "invested" in rotational kinetic energy as the object rolls down the ramp, leaving less energy for linear kinetic. Thus the ring, which has the highest moment of inertia of these objects, has the lowest speed and finishes last (and so on for the other two objects).

The other conservation law we have considered in this chapter is conservation of *momentum*. Recall that the linear momentum of an object is defined as the product of its mass and velocity: $\vec{p} = m\vec{v}$. We use the symbol $\vec{L}$ to represent **angular momentum**, and if (as we argued in our discussion of forces vs. torques) we substitute I for m and $\vec{\omega}$ for $\vec{v}$, we have

$$\vec{L} = I\vec{\omega} \tag{4.33}$$

For a particle of mass m moving in a circle of radius r at speed v, the angular momentum has magnitude (Problem 4.126)

$$L = mvr \tag{4.34}$$

I asserted Equation (4.33) by analogy to linear momentum, but we can also construct this result from the ground up; see Problem 4.137.[17]

---

[17]The complete story of angular momentum, like torque, is three dimensional. If the object in question is rotating about on the plane of the page (as objects have been drawn), $\vec{L}$ points in to or out of the plane of the page. For our purposes we can think of angular momentum as a quantity that corresponds to an angular velocity, which is either positive (counterclockwise) or negative (clockwise).

Just as a force acting on an object for some period of time gives the object an impulse $\vec{J}$, a torque acting on an object for some period of time gives the object an **angular impulse** $\vec{J}_L$:[18]

$$\vec{J}_L = \vec{\tau}\Delta t = I\Delta\vec{\omega} = \Delta\vec{L} \tag{4.35}$$

As in the linear case, the concept of an angular impulse is primarily useful when some external torque acts on some rotating object (such as friction acting to slow a rolling wheel; Example 4.18).

The law of conservation of momentum states that the "total momentum of a closed system is conserved". When we were talking strictly about linear momentum we defined this mathematically as $\sum\vec{p}_i = \sum\vec{p}_f$ (page 219). Now we can generalize this to say that *both* linear momentum *and* angular momentum must be conserved, so

$$\sum\vec{p}_i = \sum\vec{p}_f$$

and

$$\sum\vec{L}_i = \sum\vec{L}_f \tag{4.36}$$

In the linear case we saw that conservation of momentum is useful when isolated objects collide (or explode). This is still true, but now we are able to consider rotational motion, as well. Examples 4.19 and 4.20 demonstrate these ideas.

Example 4.18 Rolling with Friction ★★

A wheel of mass M and radius R is rolling on a horizontal surface with translational speed v; the surface has a coefficient of static friction μ_s. The situation is sketched in Figure 4.22.

(a) In what direction is the force of friction acting on the wheel? What is the force that is paired with the frictional force according to Newton's Third Law?

(b) Determine an expression for the amount of time it takes for the wheel to roll to a stop.

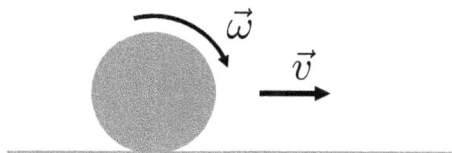

FIGURE 4.22
A disk rolling to the right over a rough horizontal surface has a translational speed v and a rotational speed ω.

As drawn, the wheel is rolling to the right, meaning $\vec{\omega}$ is clockwise. Thus if friction is going to slow it down, there must be a *counterclockwise* torque acting on it. Now friction must point parallel to the surface, meaning left or right at the bottom of

---

[18]The linear analog of this equation is Equation (4.22) on page 218.

the wheel. Consider a force pointing to the *left*: it would create a counterclockwise torque and increase ω. Evidently the frictional force must instead point to the *right* in order to slow down the wheel (Figure 4.23).

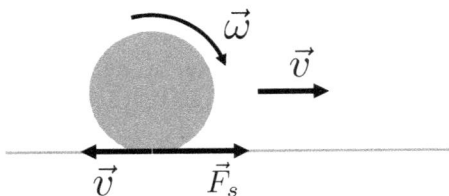

FIGURE 4.23
At the point where the wheel touches the surface, its tangential velocity is to the left. Friction therefore points to the right.

This is perhaps counterintuitive; in Chapter 3 we thought of a number of cases where an object slides across the floor and kinetic friction points opposite the velocity. In contrast, here it appears that friction points in the same direction as the object's velocity!

To make sense of this, recall that friction comes in to play when two surfaces are in contact. Here it is the *rim* of the wheel touching the floor, so it is the *tangential* velocity at the bottom of the wheel that matters when considering friction. The tangential velocity at the bottom of a disk with a clockwise angular velocity is to the left (Figure 4.23). Friction opposes this velocity and so points to the right, as we reasoned above from torque analysis. (Note also that, as usual, for the disk to be rotating smoothly we must be talking about static, rather than kinetic, friction.)

Part (a) asks us to also identify the "equal and opposite" force paired with the frictional force. The answer is that *the wheel pushes on the ground with a force directed to the left* to balance the frictional force on the wheel, which can be described in words as *the ground pushes on the wheel with a force directed to the right.*[a]

For (b) we are asked to make this reasoning quantitative and determine how long it takes for the wheel to come to rest. From Equation (4.35) we find[b]

$$\vec{\tau}\Delta t = I\Delta\vec{\omega}$$

$$F_s R\Delta t = \frac{1}{2}MR^2\omega$$

$$\mu_s F_N \Delta t = \frac{1}{2}MR\omega$$

$$\mu_s Mg\Delta t = \frac{1}{2}MR\omega$$

$$\Delta t = \frac{\omega R}{2\mu_s g}$$

---

[a]If this seems odd, consider that the same pairing occurs when you walk on the ground: it is friction that propels you forward. If you try to walk on a (nearly) frictionless surface such as ice, your foot just slips backward.

[b]Here we note, as usual, that $v = \omega r$ and recall from force analysis that $F_s = \mu_s F_N = -\mu_s Mg$.

Example 4.19 Angular Momentum in a Collision ⋆

An old-fashioned LP record measures 0.30 mm in diameter, has a mass of 130 g, and rotates at a rate of $33\frac{1}{3}$ rpm. Suppose such an LP is rotating at this rate on a frictionless axis when your friend drops a wad of sticky gum 7.5 cm from the center of the disk. If the gum weighs 8.0 g, what will the new rotational speed of the LP (and gum) be? Provide your answer in rpm.

We sketch the situation in Figure 4.19. Initially we have a disk rotating, so $L_i = I_{\text{disk}}\omega_i$. After the gum has been attached to the disk we need to invoke Equation (3.25), $I_{\text{TOT}} = \sum I$, so $L_f = (I_{\text{disk}} + I_{\text{gum}})\omega_f$. We will treat the gum as a point mass and insert these equations into a statement of conservation of angular momentum:

$$\vec{L}_i = \vec{L}_f$$
$$I_{\text{disk}}\omega_i = (I_{\text{disk}} + I_{\text{gum}})\omega_f$$
$$\frac{\frac{1}{2}MR^2}{\left(\frac{1}{2}MR^2 + mr^2\right)}\omega_i = \omega_f$$

(a)   (b)

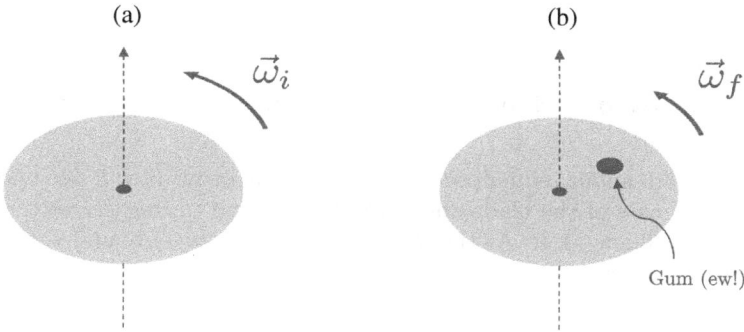

FIGURE 4.24

(a) A disk is initially rotating at a constant angular velocity ω_i. (b) when a small object sticks to the disk, the total moment of inertia increases and, by conservation of angular momentum, the angular velocity decreases.

In the above we have respectively used M and R for the mass and radius of the disk and m and r for the mass and radial position of the gum. We could convert ω_i to $\frac{\text{rad}}{\text{s}}$, solve for ω_f, then convert that back to rpm, but if you look at the above equation you'll see that all of the units *other* than those for ω cancel: we have kg m^2 in the numerator and the denominator. Thus we don't need to worry about combining the units of ω with any other quantity, and we're free to stick with rpm.[a] Inserting our values, we find

$$\frac{\frac{1}{2}\left(0.130 \text{ kg}\right)\left(0.15 \text{ m}\right)^2\left(33\frac{1}{3} \text{ rpm}\right)}{\frac{1}{2}\left(0.130 \text{ kg}\right)\left(0.15 \text{ m}\right)^2 + \left(0.0080 \text{ kg}\right)\left(0.075 \text{ m}\right)^2} = \omega_f = 32 \text{ rpm}$$

---

[a]As a counterexample, if we were working with $\Delta\theta = \omega\Delta t$, we couldn't safely multiply ω measured in rpm by time measured in seconds, because the units on time don't agree.

Example 4.20 Angular Velocity in a Collision ★★

A helicopter blade is 8.00 m long and weighs 125 kg. In an action movie, the heroine (who has a mass of 60.0 kg) leaps and grabs on to the edge of the blade; her impact velocity is perpendicular to the blade with a magnitude of $2.0\frac{m}{s}$ (Figure 4.25). The heroine rotates with the blade, then lets go to collide with an unsuspecting villain. What speed will she have when she lets go? Treat the blade as a rigid bar and the heroine as a point mass.

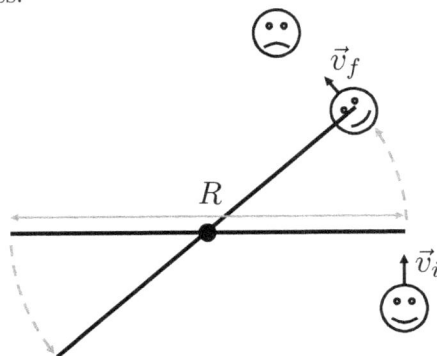

FIGURE 4.25

A heroine jumps on to a helicopter blade, rotates with it, then releases to surprise a villain.

The bar will rotate about its midpoint and angular momentum will be conserved. We will refer to the mass of the blade and heroine as M and m, respectively, and denote the length of the blade as R. At the instant of impact, the only angular momentum is that of the heroine:

$$L_i = I_{\text{heroine}}\omega_i = m\left(\frac{R}{2}\right)^2\omega_i$$

Once the two are attached, we have

$$L_f = (I_{\text{heroine}} + I_{\text{blade}})\,\omega_f = \left(m\left(\frac{R}{2}\right)^2 + \frac{1}{12}MR^2\right)\omega_f$$

Setting the two equal, we find

$$\vec{L}_i = \vec{L}_f$$

$$m\left(\frac{R}{2}\right)^2\omega_i = \left(m\left(\frac{R}{2}\right)^2 + \frac{1}{12}MR^2\right)\omega_f$$

$$\frac{m\left(\frac{R}{2}\right)^2}{m\left(\frac{R}{2}\right)^2 + \frac{1}{12}MR^2}\omega_i = \omega_f$$

When the heroine lets go, she'll maintain her linear velocity, v_f, and move off tangent to the circular path of the tip of the blade. If we note that $v = \omega\frac{R}{2} \Rightarrow \omega = \frac{2v}{R}$ (because the radius of motion is only half the length of the rotor) for both the initial and final velocities, we find

$$\frac{m\left(\frac{R}{2}\right)^2}{m\left(\frac{R}{2}\right)^2 + \frac{1}{12}MR^2}\frac{2v_i}{R} = \frac{2v_f}{R}$$

$$\frac{m\left(\frac{R}{2}\right)^2}{m\left(\frac{R}{2}\right)^2 + \frac{1}{12}MR^2}v_i = v_f$$

$$\frac{(60.0 \text{ kg})\left(\frac{8.00 \text{ m}}{2}\right)^2}{(60.0 \text{ kg})\left(\frac{8.00 \text{ m}}{2}\right)^2 + \frac{1}{12}(125 \text{ kg})(8.00 \text{ m})^2}\left(2.0\frac{\text{m}}{\text{s}}\right) = v_f$$

$$1.2\frac{\text{m}}{\text{s}} = v_f$$

We see that the heroine gives up some of her momentum to get the blade rotating, so has a reduced speed when she collides with the villain.

Before closing this chapter, it is worth mentioning that *circular* motion is something of a special case: the equation $F_{\text{NET}} = m\frac{v^2}{R}$ places very specific constraints on how the net force acting on an object is related to its mass, velocity, and radius of motion. What happens, for example, if (at some instant) a planet is moving tangent to a star with the *speed* required for circular motion, but the wrong *velocity* (i.e. the velocity is not perpendicular to the gravitational force)? The answer is that it moves in an *ellipse*, rather than a circle, and the ideas we have developed in this chapter can be applied to understand the details of this motion.[19] While the mathematics required to treat this fully are beyond the scope of this book, we address some of the details in Problem 4.138.

4.6 Problems for Chapter 4

4.1 Work and Kinetic Energy

★ **Problem 4.1.** Refer to Figure 4.26. What is $\vec{A} \cdot \vec{B}$?

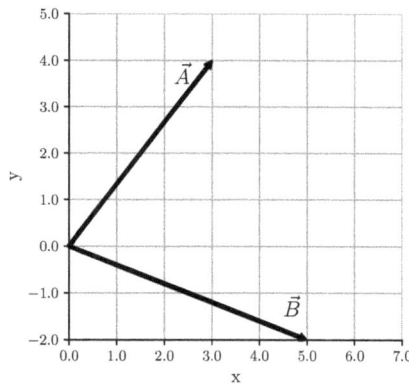

FIGURE 4.26
Problem 4.1

---

[19] *Kepler's Laws*, after Johannes Kepler (1571-1630), summarize some key results. They were originally developed from very detailed observations of the movement of heavenly bodies. The fact that Newton was later able to explain these laws mathematically contributed to his great fame.

★★ **Problem 4.2.** Consider the vectors $\vec{p} = 3.0\hat{x} + 4.0\hat{y}$, $\vec{q} = -5.0\hat{x} - 5.0\hat{y}$, and $\vec{r} = 8.0$ at $45°$ above the $-x$ axis. You can calculate three distinct dot products from among these vectors (e.g. there is no need to compute both $\vec{p} \cdot \vec{q}$ and $\vec{q} \cdot \vec{p}$ since they are identical). List and calculate them. Support your answers by sketching the vectors and identifying their parallel components.

★★ **Problem 4.3.** Show that Equation (4.5) is valid by starting with the left side and showing that it is equal to the right side:

a. Express both vectors on the left side in component form.

b. Distribute terms: you'll end up with a sum of four terms and each will have a dot product.

c. Use Equation (4.4) to evaluate the dot products from (b) and show that your expression simplifies to the right side of Equation (4.5).

★★ **Problem 4.4.** Show that Equation (4.4) and Equation (4.5) are equivalent definitions of the dot product in the *special case* where $\vec{B}$ lies entirely on the x axis, i.e. $\vec{B} = B\hat{x}$ and $\vec{A}$ has magnitude A and forms an angle θ with the $+x$ axis.

a. With the help of a sketch, show that $AB\cos\theta$ (the right side of Equation (4.4)) reduces to $A_x B_x$.

b. Show that the right side of Equation (4.5) also reduces to $A_x B_x$, which is to say that two definitions of the dot product are, in this case at least, equivalent.[20]

★★ **Problem 4.5.** Show that Equation (4.4) and Equation (4.5) are equivalent definitions of the dot product in the *general case*. Suppose $\vec{A} = A\cos\theta_1 \hat{x} + A\sin\theta_1 \hat{y}$ and $\vec{B} = B\cos\theta_2 \hat{x} + B\sin\theta_2 \hat{y}$.

a. Make a sketch of the vectors (assume both $\vec{A}$ and $\vec{B}$ are in the first quadrant). If θ is the angle between the two vectors, then how is θ related to θ_1 and θ_2?

b. Use Equation (4.5) to evaluate $\vec{A} \cdot \vec{B}$ in component form.

c. Use appropriate trigonometric identities (refer to the Appendix) to show that your result from (b) reduces to what you would expect from Equation (4.4).

★ **Problem 4.6.** A 5.0 kg box is sitting on a flat, frictionless surface. A pushing force increases linearly from 0 N to 10.0 N when the box has traveled 2.0 m. A steady 10.0 N force is then applied for the next 3.0 m. How much work was done by the pushing force? Include a graph of the force and refer to it in your solution.

★ **Problem 4.7.** A force F is used to horizontally push a cart of mass m a distance d over a horizontal frictionless surface. Then, the same force is applied over the same distance to a larger mass M. Both masses are initially at rest. Is the kinetic energy of mass M larger than, less than, or equal to the kinetic energy of mass m? Explain.

★ **Problem 4.8.** A 2.0 kg block is pulled from rest over a wooden floor with a force of 20.0 N over a distance of 2.0 m. The net work applied to the block is 10.0 J. What is the magnitude of the work done by friction?

★★ **Problem 4.9.** A 20.0 kg child slides down a 3.00 m high playground slide. She starts from rest, and her speed at the bottom is 3.0 m/s. How much work is done by friction during the slide?

★ **Problem 4.10.** A 0.50 kg ice cube is sitting on a friction free table top. Figure 4.27 shows the force you apply and the cube's displacement. What is the ice cube's kinetic energy after the push?

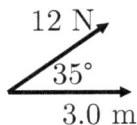

FIGURE 4.27
Problem 4.10

---

[20]You can always consider a coordinate system such that one of your vectors is parallel to the x axis, so the result here neatly demonstrates that the dot product says "multiply the parallel components of the vectors".

★ **Problem 4.11.** Consider the kinetic energy of each of the following objects:

a. A 2.00 kg ball dropped from a height of 10.0 m.

b. A 4.00 kg ball dropped from a height of 10.0 m.

c. A 2.00 kg ball dropped from a height of 20.0 m.

First, *rank* them from lowest kinetic energy to greatest without doing any calculations. Explain your logic. Then, calculate numerical values and check if your predictions were correct. If not, identify the error in your thinking and explain.

★ **Problem 4.12.** A 5.00 kg box is sliding across a flat surface with a coefficient of kinetic friction $\mu_k = 0.25$. Initially the box has a velocity of 0.70 m/s.

a. What is the box's initial kinetic energy?

b. How much work does friction do to the box in bringing it to rest?

c. How far does the box go before coming to rest? (Answer this with the work-energy theorem.)

d. Confirm your answer to (c) with force analysis and kinematics.

★ **Problem 4.13.** You push horizontally on a box across a level floor with a constant force of 5.0 N. The box travels for 10.0 m.

a. How much work have you done?

b. If you push the box twice as far with the same force, what work will you have done?

★★ **Problem 4.14.** A 50.0 kg box is sliding across a flat floor with a constant velocity. You are pulling with a 30.0 N force at an angle 30.0° above the horizontal.

a. What is the work done by the pulling force after the box has slid 10.0 m?

b. What is the total work done on the box?

c. What is the work done by friction?

d. What is the force of friction acting on the box?

e. What is the normal force acting on the box?

f. What is the coefficient of kinetic friction?

★ **Problem 4.15.** A crane lowers a steel girder into place. The girder moves with constant speed.

a. What is the sign (positive or negative) of the work done by gravity?

b. What is the sign (positive or negative) of the work done by the tension in the cable supporting the girder?

c. What would your answers to (a) and (b) be if the girder was being *lifted* (rather than *lowered*) at constant speed?

★ **Problem 4.16.** You attach a rope to a 20.0 kg box and pull it (from rest) 2.50 m across a frictionless surface with a force of magnitude 75.0 N. Calculate the work done to the box and use the work-energy theorem to find the box's final velocity if:

a. You pull horizontally on the box.

b. You pull at an angle of 10.0° above the horizontal.

c. You pull at an angle of 45.0° above the horizontal.

★★ **Problem 4.17.** Consider Problem 4.16 for the situation where the box is being pulled across a surface with a coefficient of kinetic friction $\mu_k = 0.300$. For each part (a)–(c), determine the work done by *you*, the work done by *friction*, and use the work-energy theorem to determine the final velocity of the box.

★ **Problem 4.18.** A 4.0 kg mass is moving along the x axis at 1.0 m/s. Initially it is located at $x = 0$ m and then it is subjected to the force shown in Figure 4.28.

a. What is the object's initial kinetic energy?

b. What is the work done by the force after 4.0 m?

c. What is the object's kinetic energy when it reaches 4.0 m?

d. What is the object's speed when it reaches 4.0 m?

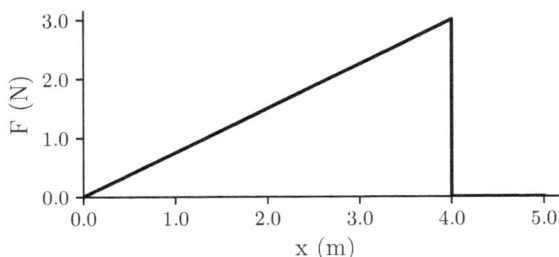

FIGURE 4.28
Problem 4.19

★★ **Problem 4.19.** A 2.0 kg pumpkin is fired straight up at 25 m/s.

a. Draw a force diagram for the pumpkin.

b. What is the pumpkin's initial kinetic energy?

c. Calculate the work done by gravity after the pumpkin has traveled 20. m. Is it positive or negative?

d. What is the *change in* kinetic energy after the pumpkin has traveled 20. m?

e. What is the kinetic energy when the pumpkin has traveled 20. m? What is its speed?

★ **Problem 4.20.** You are lifting blocks from the floor to a 2.0 m high shelf.

a. If you lift a heavy block with $m = 100.0$ kg at a constant velocity, what is the total work done on the block?

b. If you lift a light block with $m = 1.0$ kg with a constant acceleration of 1.0 m/s^2, what is the work done by gravity?

c. What is the total work done to the light block?

d. What work do you do to the light block?

★★ **Problem 4.21.** A sledder is traveling down a long hill inclined at 12.0° relative to the horizontal; when it is 10.0 m from the bottom it is traveling at 2.00 m/s. The sled and rider have a combined mass of 50.0 kg and the coefficient of sliding friction is 0.100. Use the definition of work and the work-energy theorem to answer the following questions:

a. How much work does gravity do to the sledder over the last 10.0 m?

b. How much work does friction do to the sledder over the last 10.0 m?

c. What is the sledder's kinetic energy at the bottom of the hill?

d. What is the sledder's speed at the bottom of the hill?

★★ **Problem 4.22.** Consider an Atwood's Machine where one mass is $M = 50.0$ kg and the other is $m = 10.0$ kg. Both masses are 1.50 m above the floor when they are released from rest. Assume the pulley is far above the masses so neither mass will strike the pulley in the subsequent motion.

a. Use force analysis and kinematics to determine the speed of the masses when the heavier mass strikes the ground. What is the corresponding kinetic energy of the masses?

b. What is the work done by gravity on each of the masses?

c. What is the net work done on the two mass system? (What other force(s) act on the masses besides gravity?)

d. Use the work-energy theorem to determine the total kinetic energy of the masses just before the heavier mass strikes the floor. Verify that you obtain the same result as what you found in (a).

4.2 Conservative Forces and Potential Energy

★★   **Problem 4.23.** A 1.0 kg rock is thrown vertically with a speed of 25 m/s.

a. Using energy conservation, determine the maximum height of the rock.

b. Set the launch point to be the point of 0 gravitational potential energy. What is the gravitational potential energy when the rock reaches its maximum height?

c. Set the launch point to be the point of 0 gravitational potential energy. What is the gravitational potential energy when the rock returns to its launch point?

d. Repeat (b) and (c) where the point of 0 gravitational potential energy is set to the rock's maximum height.

e. For both "zero points" considered here, what is the total change in potential energy and kinetic energy from launch to the rock's highest location?

f. Does the choice of "zero point" affect the underlying physics?

★   **Problem 4.24.** Consider the three equal-height slides shown in Figure 4.29. Ignore friction and suppose the child starts from rest at the top of the slide. Rank the child's speed (v_A, v_B, and v_C) at the bottom of each slide.

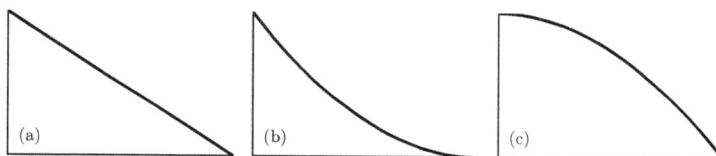

FIGURE 4.29
Problem 4.24

★   **Problem 4.25.** Two marbles are dropped from the roof of a building. One marble has twice the mass of the other: $m_A = 2m_B$. Just before hitting the ground, how do the kinetic energies compare? Report your answer in the form $K_A = \alpha K_B$ where your job is to determine a numerical value for α.

★★   **Problem 4.26.** A ball of mass m rolls down a frictionless ramp oh height H and then up another ramp of height $h < H$. It flies off the second ramp at an angle θ relative to the horizontal.

a. What is the ball's potential energy and kinetic energy at the top of the first ramp?

b. What is the ball's potential energy and kinetic energy at the bottom of the first ramp?

c. What is the ball's potential energy and kinetic energy at the top of the second ramp?

d. What maximum height (relative to the bottom of the ramps) does the mass reach after it leaves the second ramp? Does your answer make sense in the cases where $\theta = 0°$ and $\theta = 90°$?

e. With what speed will the ball impact the ground?

★★   **Problem 4.27.** A 5.00×10^2 kg rollercoaster cart is traveling at $v = 7.0$ m/s at the bottom of a hill with height $h = 2.0$ m. A spring with a spring constant of $k = 1500$ N/m is at the top of the hill (Fig. 4.30). If the track is frictionless, by how much does the spring compress as it slows the cart to rest?

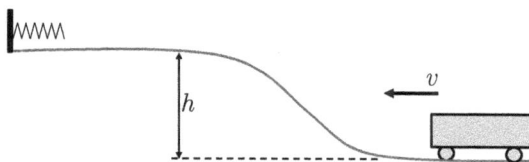

FIGURE 4.30
Problem 4.27

★★ **Problem 4.28.** You are designing a roller coaster that involves horizontally launching a car with a compressed spring.[21] Each car will have a mass of 350 kg, and after launching the cars will go up a 15 m high hill and then descend 25 m. The springs can be compressed 1.5 m.

    a What spring constant will just barely get the cart to the top of the hill?

    b Assume you use a spring constant 115% the value you found in (a). What speed will the cart have (i) just after launch, (ii) at the top of the hill, and (iii) at the bottom of the hill?

★★ **Problem 4.29.** A particle moves along the x axis has total and potential energies shown in Figure 4.29. Suppose it is just barely moving to the left at $x = 5.0$ m.

    a. At what value(s) of x is the particle's speed a maximum?

    b. At what value(s) of x is the particle's speed a minimum?

    c. Does the particle have any turning points in the range shown in the graph? If so, where?

    d. What location(s) in the region shown are stable equilibria?

    e. What location(s) in the region shown are unstable equilibria?

★★ **Problem 4.30.** Two blocks of masses m and M ($m < M$) are connected by a massless string over a massless, frictionless pulley (Fig. 4.32). The blocks are released from rest such that both blocks are a distance h above the ground.

    a. Use conservation of energy to determine an expression for the speed of the blocks, v, when block M strikes the ground.

    b. What maximum height will block m reach? (Assume the pulley is sufficiently high that the mass won't impact it.)

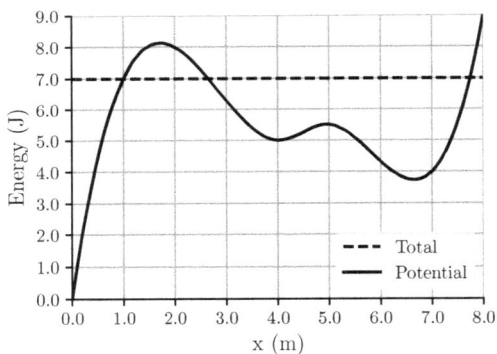

FIGURE 4.31
Problem 4.29

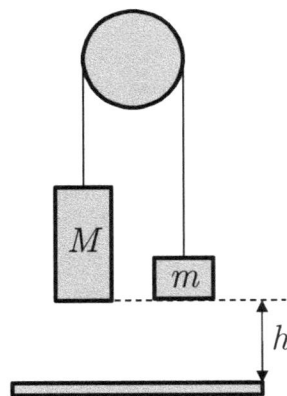

FIGURE 4.32
Problem 4.30

★★ **Problem 4.31.** A block of mass $M = 10.0$ kg is sitting on a roof and attached to a hanging block of mass $m = 5.0$ kg by way of a massless rope and pulley (Fig. 4.33). The blocks are at rest when mass m is given a quick tap so it begins moving down with a velocity $v = 0.20$ m/s. A short time later, the blocks are back at rest with mass m $h = 12$ cm lower than it was. Using work and energy concepts, explain what has happened and determine the coefficient of kinetic friction between M and the ramp.

---

[21] Magnetic launchers are used in practice.

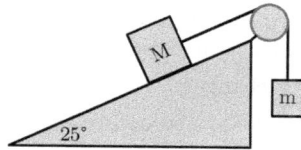

FIGURE 4.33
Problem 4.31

★★ **Problem 4.32.** Consider the situation shown in Figure 4.34, where two blocks are connected by a thin rope strung over a light, frictionless pulley. Use the concepts of work and energy to determine an expression for the speed of block m just before it hits the floor if (a) if the table is frictionless and (b) the coefficient of friction is μ_k (assuming that static friction is insufficient to prevent any movement).

★★ **Problem 4.33.** Consider the situation shown in Figure 4.35, where two blocks are connected by a thin rope strung over a light, frictionless pulley.

  a. If the incline plane is frictionless, what values of θ will result in the hanging mass moving down rather than up? (Assume θ is in this range for the rest of the problem.)

  b. If the incline plane is frictionless, with what speed will mass m strike the ground?

  c. What coefficient of static friction μ_s will prevent the masses from moving?

  d. If μ_s is insufficient to prevent motion and there is a coefficient of kinetic friction μ_k, then with what speed will mass m strike the ground?

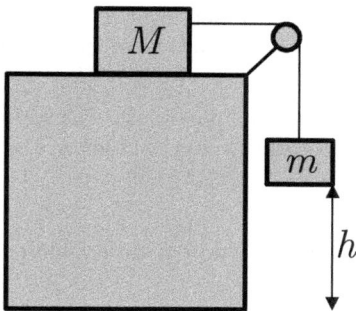

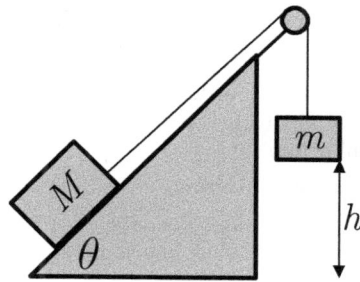

FIGURE 4.34   **FIGURE 4.35**
Problem 4.32   Problem 4.33

★★ **Problem 4.34.** A ball with mass m is fired with a speed v out of a spring-loaded ball launcher that can be adjusted to fire at an angle θ above the horizontal. Consider the ball at the instant of its launch and a short time later, when it is a small distance h above its launch point.[22]

  a. Write down a conservation of energy equation comparing these two instances.

  b. How does the value of θ affect the velocity of the ball when it is at height h?

★ **Problem 4.35.** A diver steps off of a cliff and splashes into the water below. Determine an expression for the diver's energy at the top of the cliff and again at the bottom of the cliff, just as they impact the water. Use the principle of conservation of energy (Equation (4.11)) to determine an equation for their speed on impact with the water, v, in terms of the height of the cliff, h. If the cliff is 12.0 m tall, what is their velocity as they impact the water?

★ **Problem 4.36.** A particle is moving in a region with potential energy as shown in Figure 4.36. It begins at $x = 1.0$ m and is moving in the $+\hat{x}$ direction with 1.0 J of kinetic energy.

---

[22]If $\theta = 0°$ or is sufficiently close to $0°$ then it won't reach h, but here we're supposing θ is large enough to ensure that doesn't happen.

a. In the region 1.0 m $< x <$ 2.0 m, is the particle accelerating or moving at constant velocity? If it is accelerating, in which direction is the acceleration? Explain.

b. What is the total energy? Does it change?

c. Make a sketch of the figure and include a second curve for the total energy.

d. Would the particle ever stop moving to the right? If *yes*, where? If *no*, why not?

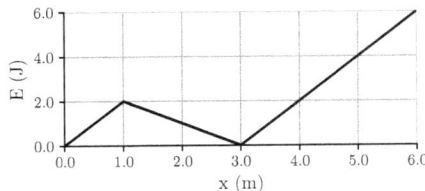

FIGURE 4.36
Problem 4.36

★   **Problem 4.37.** A spring-loaded gun shoots a plastic ball with a speed of 4.0 m/s. What will the launch speed be if the spring is compressed twice as far?

★   **Problem 4.38.** A spring-loaded gun shoots a plastic ball with a speed v when a spring with spring constant k is compressed a distance Δx. What will the launch speed be if the gun is modified so two identical springs are used? (Assume that the springs are side by side and each compress the same distance Δx.)

★   **Problem 4.39.** A spring has a normal equilibrium length of 12.0 cm and a spring constant of 350 N/m. You apply a 20.0 N force to the end of the spring. What is the spring's displacement from equilibrium? What potential energy is stored in the spring?

★★ dx **Problem 4.40.** A vertical spring scale is used to measure the mass of an object. The spring has a constant of $k_1 = 1.0 \times 10^3$ N/m, and the mass stretches the spring a vertical distance of 0.10 m from equilibrium.

a. Draw a force diagram.

b. What is the mass of the object? Answer *both* from force analysis and from energy analysis (show that your results are the same!)

c. Suppose the object is lowered onto a second vertical spring that is on the floor. The second spring has a constant $k_2 = 5.0 \times 10^2$ N/m. When the second spring has been compressed by 10.0 cm, what is the reading on the scale?

★   **Problem 4.41.** A toy gun fires a small ball of mass $m = 50.0$ g by compressing a spring with spring constant k by a distance $\Delta x = 12$ cm. When the gun is fired, the spring is released and the ball is launched. If the ball is supposed to have a velocity of 5.0 m/s when fired horizontally, then what should the spring constant be?

★   **Problem 4.42.** A 2.00×10^2 N/m spring launches a 2.0 kg block. Due to a frictional force of 2.0×10^1 N, it comes to rest after traveling 2.0 m. How far was the spring compressed?

★★   **Problem 4.43.** A 5.5 kg box is placed on a vertical 2.0 m long spring with a constant $k = 5.0 \times 10^4$ N/m. Take the zero point for potential energy to be the ground.

a. How much force is needed to compress the spring 0.80 m?

b. What is the total energy of the system when the spring is compressed as in (a)?

c. If the spring is released from (a), how high into the air will the object go (assuming g is constant)?

★★   **Problem 4.44.** A horizontal spring in a toy gun is loaded with a 0.50 kg mass. The spring constant is 100.0 N/m and the spring is compressed from equilibrium by 0.20 m. Assume that friction is negligible.

a. What is the work done by the spring while it is being compressed?

b. What is the kinetic energy of the mass as it leaves the cannon?

c. What is the velocity of the mass as it leaves the cannon?

d. If you double the compression, by what factor does the kinetic energy change?

⋆ **Problem 4.45.** A light string of length L is tied to a post on one end and attached to a ball of mass m on the other. The ball is originally hanging vertically, but then you pull it to the side by 90°. You hold the ball motionless, then release it from this position. What kinetic energy will the ball have when it passes through its original "hanging vertically" position?

⋆⋆ **Problem 4.46.** A 3500 N/m spring has an equilibrium length of 0.50 m. It is sitting vertically on the floor when a 20.0 kg object is placed on the spring.

a. How far does the spring compress?

b. How much work is required to compress the spring by an additional 8.0 cm?

c. Suppose the spring is compressed as described in (b), then released so the object flies into the air. What is the object's maximum height above the ground?

dx **Problem 4.47.** In Example 4.8 we obtained an expression for the equilibrium position of an object hanging from a spring (or rubber band) by calculating the energy and finding the minimum: $y = L + mg/k$ where y is measured down from the ceiling, L is the equilibrium length of the spring, m is the mass of the object, and k is the spring constant. Repeat this analysis, but set the gravitational potential energy to be 0 *on the floor* (rather than the ceiling). Suppose the total height of the room is h. Show that the equilibrium position of the hanging object is consistent with what we found in Example 4.8. What should h be to ensure the object doesn't hit the ground?

⋆ **Problem 4.48.** An object is initially at rest and then slides down a frictionless ramp. The object's speed is v at the bottom of the ramp. If the ramp is made to be twice as high, what speed (in terms of v) will the object have at the bottom?

⋆ **Problem 4.49.** You throw a 5.5 g coin straight down at 5.0 m/s from a 20.-m-high bridge.

a. How much work does gravity do as the coin falls to the water below?

b. What is the speed of the coin just as it hits the water?

⋆ **Problem 4.50.** A car is rolling from rest at the top of a hill that is inclined at 30.0° relative to the horizontal. The car will reach the bottom of the hill once it has traveled 10.0 m.

a. What is the velocity at the bottom of the ramp? Answer using kinematics.

b. What height has the car moved through as it moves to the bottom of the hill?

c. Use energy conservation to find the velocity. Confirm that your answer agrees with (a).

d. What is the car's velocity when it is halfway down the hill?

⋆⋆ **Problem 4.51.** A 71.0 kg student is skiing down a hill that is 50.0 m high and inclined at 25.0° relative to the horizontal. As he goes down the slope, he is skiing into a wind that applies a 2.00×10^2 N horizontal force.

a. Sketch the situation. What is Δr, the student's displacement?

b. What is the work done by the wind?

c. What is the work done by gravity?

d. What is the student's final kinetic energy?

⋆⋆ **Problem 4.52.** An 85 kg student is sledding (Fig. 4.37). A large spring with a constant of 7.5×10^4 N/m is compressed 0.50 m. Suppose the launch area and first slope is without friction, but once the student starts up the far incline there is a coefficient of friction $\mu_k = 0.10$.

a. What is the total energy of the system when the spring is compressed but the student has yet to be launched? (Specify your zero point for gravitational potential energy!)

b. How far up the incline does the student slide?

c. What work is done by non-conservative forces from the moment of launch to the moment when the student is instantaneously at rest?

d. If the student slides back down the incline after coming to rest, how fast will they be moving when they get back to the bottom of the incline?

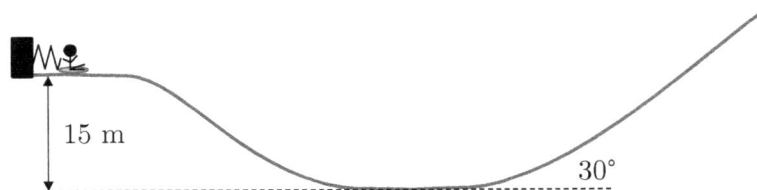

FIGURE 4.37
Problem 4.52

★   **Problem 4.53.** You lift a 100.0 kg box from the ground to the table at constant velocity, covering a vertical height of 1.0 m in 15.0 s. Your friend lifts another box with the same mass, also with constant velocity, to the same table in 10.0 s. (a) What work have each of you done? (b) How much power have each of you used/generated?

★   **Problem 4.54.** How much work does an elevator do to lift a 1.0×10^3 kg elevator at constant velocity up a height of 80. m? What power is needed to do this at a constant speed over the course of 50. s?

★★ **Problem 4.55.** Three students run up some stairs. Student A has a mass of 60.0 kg and climbs a vertical distance of 10.0 m in 7.5 s. Student B has a mass of 80.0 kg and climbs the same distance in 10.0 s. Student C has a mass of 80.0 kg and climbs a total of 15.0 m in 19.0 s. Rank the power outputs from largest to smallest.

★   **Problem 4.56.** Consider the potential energy shown in Problem 4.36. Draw a corresponding "force vs. position" graph.

★   **Problem 4.57.** Consider the potential energy graph shown in Figure 4.38.

a. What is the x component of the force acting on a particle at $x = 3.0$ m?

b. Draw a graph of F_x vs x.

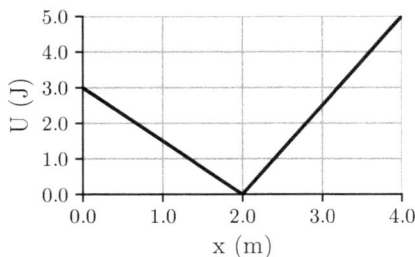

FIGURE 4.38
Problem 4.57

★★ ∫ **Problem 4.58.** Newton's Law of Universal Gravitation states that for masses m and M a distance r from one another, each mass is attracted toward the other with magnitude $F_G = GmM/r^2$.

a. What is the corresponding gravitational potential energy of two interacting objects with masses m and M at a separation r?

b. Make a graph of $U(r)$.

4.3 Momentum, Collisions, and Explosions

★ **Problem 4.59.** If you jump off of a ledge, you naturally bend your knees as you land to avoid damaging your knees. Explain this in the context of Equation (4.22).

★ **Problem 4.60.** An object experiences an impulse of 1.38 Ns. If the force it experiences is shown in Figure 4.39, then for how long did the object experience the force?

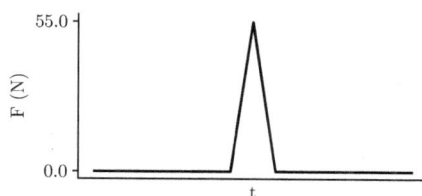

FIGURE 4.39
Problem 4.60

★ **Problem 4.61.** An object experiences a force that increases linearly from 0 N to 5.0 N in 12.0 ms. What impulse does it experience? Support your answer with a graph of force vs. time.

★ **Problem 4.62.** A 5.0 kg object is moving at a velocity $\vec{v} = 2.0$ m/s $\hat{x}$ when it suddenly experiences a 4.0 N $\hat{x}$ force that steadily decreases to 0 N in 0.50 s. (a) What is the object's new velocity? (b) If the object is then subjected to a force that increases from 0 N to -4.0 N $\hat{x}$ in another 0.50 s, what will the object's final velocity be? Support your answer with a graph of force vs. time.

★ **Problem 4.63.** In a physics lab, a 0.50 kg cart is placed 1.00 m from the bottom of a track that is elevated at 30.° above the horizontal. The cart bounces off of a force sensor at the bottom and stops 0.80 m above the bottom. The sensor indicates that it was in contact with the cart for 0.030 s. What was (a) the impulse and (b) the average force exerted on the cart by the sensor?

★ **Problem 4.64.** Consider a force that increases linearly from 0 to 12,000 N in 7.5 s.

  a What is the average force?

  b What impulse is imparted by this force?

  c What is the change in momentum of a 1300 kg car that experiences this force?

  d If the car was initially at rest, what is its speed after 7.5 s?

★ **Problem 4.65.** Two rubber balls of unequal masses experience a head-on collision. Just before the collision, their velocities were equal and opposite. Are the impulses experienced by the balls the same, or is one larger than the other? Explain.

★ **Problem 4.66.** Two boxes, one heavier than the other, are traveling with the same initial momentum. They experience the same constant force from a wind that causes them to slow over the same amount of time. Which box has less momentum after experiencing the wind, or are they the same? Explain.

★ **Problem 4.67.** A car of mass m and a truck of mass $2m$ are driving in opposite directions.

  a. The car has twice the velocity of the truck. Algebraically compare their momentums.

  b. If the truck is moving at 16 m/s and has a mass of 3,200 kg, what is the momentum of the car and the truck?

  c. What is the kinetic energy of each vehicle?

★ **Problem 4.68.** An astronaut is floating motionless in space and wants to throw a tool to begin moving toward their colleague. Clearly explain (with words, a diagram, and some mathematics) the direction in which the tool should be thrown.

★ **Problem 4.69.** You have a choice of punching a wall or a pillow with your fist. Compare (a) the impulse and (b) the force in each case.

★ **Problem 4.70.** You dramatically throw a hard rubber ball (with a mass of 150 g) against a wall: it impacts at a speed of 30.0 m/s and rebounds in the opposite direction at the same speed. What impulse did the wall exert on the ball? If the ball was in contact with the wall for 0.20 s, then what average force did the wall exert on the ball?

★★ **Problem 4.71.** A hard ball is thrown against a wall at an angle θ. When it rebounds it has the same magnitude momentum but the orientation has changed such that the incident angle and the reflected angle are the same (Fig. 4.40). What is the change in momentum $\Delta \vec{p}$? Show the result (a) graphically and (b) mathematically.

★★ **Problem 4.72.** You see me one day, scream "I'll teach *you* some physics!" and throw a book (this book?) at me. As it impacts my face, the book experiences the force shown in Fig. 4.41. Suppose the book has a mass of 1.0 kg.

    a. What impulse does the book experience?

    b. What is the change in momentum of the book?

    c. If the book came to rest because of this force, what was the initial velocity?

    d. What average force was applied to the book?

    e. What average force was applied to my face?

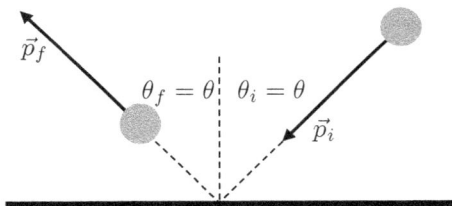

FIGURE 4.40
Problem 4.71

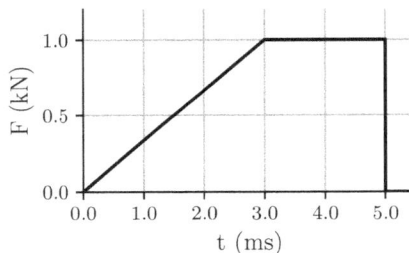

FIGURE 4.41
Problem 4.72

★ **Problem 4.73.** A 0.500 kg cart is rolling along a track in a physics lab when it rebounds off of a force sensor. The sensor detects an impulse of 0.200 Ns. If the cart was moving to the left at 0.30 m/s before the impact, what is its velocity after the impact?

4.4 Elastic and Inelastic Collisions

★ **Problem 4.74.** A car and a truck experience a head on collision and stick together. Which has the larger change in momentum, or are they the same?

★ **Problem 4.75.** Two trucks with the same mass of 2.00×10^3 kg are traveling in opposite directions. They are each traveling with a speed of 35.0 m/s, collide, and stick together.

    a. What is the total momentum of the system before the collision?

    b. What is the total momentum of the system after the collision?

    c. What is the final velocity of the trucks?

    d. If the collision time for the trucks is 10.0 ms, what is the magnitude of the average force exerted on either truck?

    e. Modern vehicles are designed to crumple in a collision such as this. Why?

★ **Problem 4.76.** Paintball guns shoot marble-sized balls of paint (enthusiasts use these to play games like capture the flag). A paintball has a mass of about 1.25 g and has a velocity of about 85 m/s. A paintball is shot at a 0.300 kg target that is hanging by a thin rope. What is the combined speed of the target-paintball system just after impact?

★ **Problem 4.77.** A train car is traveling at a speed v when it rolls into a stationary train car of mass m. The two cars couple together. What is their final velocity if the first car has a mass of (a) m or (b) $2m$?

★★ **Problem 4.78.** Three train cars, each of mass m, are coupled together and are traveling in the $+x$ direction at speed v_0. First, a fourth car of mass m traveling at a speed $2v_0$ catches up to the original three and couples with them. Then, the four cars strike a fifth car, also mass m, which was originally stationary. What is the final speed of the five-car train?

★ **Problem 4.79.** Show that momentum is conserved in the elastic collision of Example 4.14.

★ **Problem 4.80.** An empty 5.0 kg sled is sliding on ice with a speed of 5.0 m/s when a 72 kg student drops vertically from a tree above the sled.

a. When the student lands in the sled, does it speed up, slow down, or continue at the same speed?

b. What is the momentum before the collision?

c. What is the momentum after the collision?

d. What is the final velocity? (Does your answer here support your answer to (a)?)

★ **Problem 4.81.** A friend angrily claims that all collisions in a closed system should be elastic. Their reasoning goes something like this: "Momentum in a closed system is conserved because of Newton's Third Law: objects apply equal and opposite forces for the same time intervals. But if colliding objects are applying the forces over the same time intervals, they must be applying them over the same distances, too. So if momentum is conserved, energy is conserved, and all collisions are elastic!" Your friend is so angry they flip a table. Aside from the needless violence, where have they gone wrong? (You don't need any complex mathematics here!)

★★ **Problem 4.82.** An unstable atom is initially at rest, then splits into two unequal pieces that move away from one another due to equal and opposite repulsive forces. One piece is measured to have a mass of 6.6×10^{-27} kg and a speed of 1.5×10^6 m/s. If the other piece moves in the opposite direction with a speed of 2.5×10^5 m/s, what was the mass of the original unstable atom?

★ **Problem 4.83.** A 10.0 g bullet moving at 1.00×10^3 m/s strikes and passes through a 5.0 kg block that is initially at rest. The bullet emerges from the block with a speed of 3.00×10^2 m/s. What is the velocity of the block post-collision?

★★ **Problem 4.84.** A rocket accelerates straight up at 10.0 m/s² for 2.00 s, then explodes into two pieces of masses m and $2m$. The lighter piece travels straight up and reaches a maximum height of 0.600 km. What was the velocity (magnitude and direction) of the other piece immediately after the explosion?

★ **Problem 4.85.** A squid is floating around with some water inside its body, minding its own business, when it senses danger nearby. It starts with a combined mass of 1.8 kg and is moving with a speed of 0.30 m/s. It ejects 0.15 kg of water to increase its speed to 3.2 m/s. What was (a) the speed and (b) the momentum of the liquid as it was ejected?

★★ **Problem 4.86.** Two particles collide. Before the collision one was at rest.

a. Is it possible for *both* particles to be at rest after the collision?

b. Is it possible for *one* particle to be at rest after the collision?

In both cases, give an example or explain why it isn't possible.

★★ **Problem 4.87.** Two billiard balls have equal masses and experience a glancing collision. Suppose $\vec{v}_{1i} = 2.0\hat{x}$ m/s, $\vec{v}_{2i} = -1.0\hat{x}$ m/s, and $\vec{v}_{2f} = 1.5\hat{y}$ m/s. What is the final velocity of ball 1?

★   **Problem 4.88.** You observe as a 35 kg child rides past you on a 3.0 kg skateboard. The child and skateboard are coasting at 1.5 m/s when the child leaps forward off of the skateboard; you observe her moving through the air at a speed of 2.0 m/s. What is the speed of the skateboard just after the child leaps off?

★   **Problem 4.89.** You're playing tee-ball and smack a great hit that goes straight to the pitcher. The velocity of the ball is 20.0 m/s after the bat makes contact with the 0.15 kg ball for 2.0 ms. The 1.0 kg bat had a velocity of 10.0 m/s before the impact.

     a. What is the velocity of the bat immediately after the collision?

     b. What is the average net force on the ball from the bat?

★   **Problem 4.90.** A 1.0 kg ball is traveling at 6.0 m/s when it strikes a stationary 2.0 kg ball. The balls stick together.

     a. What is the momentum of the system before the collision?

     b. What is the momentum of the system after the collision?

     c. What is their velocity after the collision?

     d. How far do the balls travel in the first 2.0 s after the collision, assuming friction is negligible?

★★   **Problem 4.91.** A block ($m = 1.0$ kg) slides down a frictionless ramp of height $h = 1.3$ m and collides with a second block ($M = 3.0$ kg).

     a. Right before the collision, what is the speed of mass m?

     b. If the blocks bounce elastically, what is the velocity of the larger block right after the collision?

     c. Explain whether or not the magnitude of the larger block's change in momentum is smaller, larger or the same as the momentum change of the smaller block.

     d. Post collision, the larger block slides over a rough area with $\mu_k = 0.50$. If the block slides to rest, how far did it slide over the rough area?

★★★ **Problem 4.92.** A car of mass m_1 is traveling East at a speed v_1 when it collides with a car of mass m_2 traveling North at a speed v_2. The cars become locked together and skid a distance d at an angle θ (measured counter clockwise from East) before coming to rest. Suppose the coefficient of kinetic friction between the tires and the road is μ_k. Determine expressions for v_1 and v_2 in terms of the other parameters and any other constants needed.

★★   **Problem 4.93.** A 0.300 kg puck is sliding with a velocity $\vec{v}_{1i} = (10.0\hat{x} + 15.0\hat{y})$ cm/s when it impacts a second puck that has a mass of 0.500 kg and a velocity $\vec{v}_{2i} = (-12.0\hat{x} - 8.0\hat{y})$ cm/s. After the collision, the first puck has a velocity $\vec{v}_{1f} = (3.0\hat{x} - 5.0\hat{y})$ cm/s. What is the final velocity of the second puck?

★★   **Problem 4.94.** Consider the collision described in Problem 4.93. Is this collision inelastic, elastic, or neither? Explain. What fraction of the initial energy is lost to the environment?

★★   **Problem 4.95.** Student A rides a sled (combined mass 3.0×10^1 kg) down a steep hill and achieves a speed of 1.0×10^1 m/s as she travels over level ground. Student B is traveling at 2.0×10^1 m/s and catches up to Student A. The pair collide, stick together, and move off with a velocity of 15 m/s.

     a. What is the mass of student B and his sled?

     b. Student A and Student B continue on after colliding with each other and slam into Student C (mass 3.0×10^1 kg), who was initially stationary. The three stick together: what is their final speed?

★   **Problem 4.96.** Two balls approach each other. Ball A has mass $m_A = 2.0$ kg and initial velocity $v_{Ai} = 3.0\hat{x}$ m/s. Ball B has mass $m_B = 5.0$ kg and initial velocity $v_{Bi} = 5.0\hat{y}$ m/s. The balls collide and stick together.

     a. What are the components of the balls' velocity just after the collision?

b. What are the components of the final momentum?

c. What is the final momentum in magnitude angle form?

★★ **Problem 4.97.** One end of a (massless) spring is attached to a 0.250 kg air track glider; the other is fixed in place to the end of the track. A 0.500 kg glider hits and sticks to the first glider, and the spring compresses from its original length of 0.35 m to a minimum length of 0.25 m. If the spring constant is 25 N/m, what was the speed of the 0.500 kg glider just before impact?

★★ **Problem 4.98.** One end of a (massless) spring is attached to a 0.750 kg air track glider; the other is fixed in place to the end of the track. A 0.250 kg glider hits the first glider with an impact velocity of 0.50 m/s. If the collision is elastic and the spring constant is 35 N/m, then (a) how far will the spring compress and (b) what velocity will the 0.250 kg glider have immediately after the impact?

★ **Problem 4.99.** Two skaters are initially at rest and push off of one another. If $m_A > m_B$, (a) which skater will have the greater speed after they separate? (b) Which exerts a greater force on the other? (c) Which skater has greater momentum after they separate?

★ **Problem 4.100.** Your friend, who has a mass of 75 kg, is stuck on slippery ice. Your friend is holding a 0.500 kg rock and can throw it with a speed of 30.0 m/s. The distance through which his had moves as he accelerates the rock is 1.0 m.

a. What is the total momentum of the system prior to the throw?

b. What is the velocity of your friend after the throw?

★ **Problem 4.101.** An object initially at rest explodes into three pieces. One flies off west and another flies off south.

a. In which direction, generally speaking, will the third object move?

b. Suppose the object that flies off west has a mass of 0.50 kg and a speed of 2.8 m/s, the object that flies off south has a mass of 1.3 kg and a speed of 1.5 m/s, and that the original object had a mass of 3.0 kg. What is the speed and direction of the third fragment just after the explosion?

★★ **Problem 4.102.** A 3.00×10^2 kg rocket is traveling straight up at a speed of 120 m/s when it explodes into two pieces. One piece has a mass of 1.20×10^2 kg that is traveling up at a speed of 75 m/s immediately after the collision. What is the speed of the other piece?

★★ **Problem 4.103.** Football Player A is diving toward the goal line when a defender, Player B, tackles him midair. Consider a top down view where the $+\hat{y}$ direction points toward the goal line and the $+\hat{x}$ direction is parallel to it. Player A has a mass of 90.0 kg and has an initial velocity $\vec{v}_{Ai} = 1.1\hat{y}$ m/s. Player B has a mass of 100.0 kg and has an initial velocity $\vec{v}_{Bi} = (0.70\hat{x} - 0.30\hat{y})$ m/s. The players are effectively stuck together immediately after the collision.

a. What is the velocity of the players immediately after the collision?

b. If the collision happens 1.1 m above the ground and there is no vertical velocity at the instant after the collision, where will they land? Set the origin of your coordinate system to be on top of the players just as they collide. If the goal line is at $y = 0.40$ m, does Player A cross the line and score before landing?

★★ **Problem 4.104.** Derive Equations (4.26)–(4.27).

★ **Problem 4.105.** Determine the total kinetic energy both before and after the collision of Example 4.13. What is the ratio K_f/K_i?

★ **Problem 4.106.** Analyze Example 4.13 in the case where the collision is elastic: what is the velocity of (a) the bullet and (b) the target immediately after the collision?

★★ **Problem 4.107.** An object of mass m_1 is traveling at a speed v_{1i} when it collides elastically with a stationary mass m_2. What is the final velocity of each mass if (a) $m_2 = m_1$, (b) $m_2 = 2m_1$, (c) $m_2 = m_1/2$, or (d) $m_2 \gg m_1$?

★★ **Problem 4.108.** Two billiard balls of equal mass m experience a head on collision; the balls have the same initial speeds v_i but are traveling in opposite directions. What are the final velocities of the balls if the collision is (a) elastic and (b) inelastic?

★ **Problem 4.109.** Two balls have equal masses and equal magnitude velocities in opposite directions. They collide elastically. After, will they (a) move off together, (b) be at rest, (c) move in the same direction, or (d) recoil in opposite directions? Support your answer mathematically.

★★ **Problem 4.110.** Two balls have equal masses m and approach each other with different velocities v_1 and v_2. What will their final velocities be if (a) they experience a head on elastic collision or (b) they just barely miss one another? (c) Compare the answers to (a) and (b). If the balls look identical and you can't observe them just before or after the collision (suppose they roll into a box on the ground, where holes have been cut for the balls to enter/exit), could you tell if they collide or miss one another?

★★ **Problem 4.111.** A 1.0 kg puck is traveling in one direction with a velocity of 5.0 m/s when it elastically strikes a second stationary puck with a mass of 2.0 kg. Assume the motion occurs entirely in one dimension.

    a. What are the velocities of the pucks after the collision?

    b. Show that linear momentum is conserved.

★★ **Problem 4.112.** Consider the pucks of Problem 4.111. This time, suppose you send the 1.0 kg puck with a velocity of $5.0\hat{x}$ m/s and your friend sends the 2.0 kg puck with a velocity of $-5.0\hat{x}$ m/s. The pucks collide elastically and all motion occurs in one dimension.

    a. What are the velocities of the pucks after the collision?

    b. Show that linear momentum is conserved.

4.5 Conservation Laws in Angular Motion

★ **Problem 4.113.** A 2.0 kg ball is spun on a 1.0 m long massless string at a speed of 4.0 m/s.

    a. What is the moment of inertia?

    b. What is the rotational kinetic energy?

    c. By what factor would your answers to (a) and (b) change if you double the radius but kept the same mass and angular speed?

    d. By what factor would your answers to (a) and (b) change if you double the mass but kept radius and angular speed constant?

★ **Problem 4.114.** What is the rotational kinetic energy of the Earth due to (a) its daily rotation about its axis and (b) its yearly revolution around the sun?

★ **Problem 4.115.** A 270 kg barrel has a diameter of 0.45 m and is filled with a thickened, tar-like substance (ew!). It rolls across the floor at 1.5 m/s. What is its kinetic energy?

★ **Problem 4.116.** A 65 kg break dancer is spinning on his head. Initially he is rotating at 4.0 rad/s. If you crudely model the dancer as a solid cylinder with a radius of 0.30 m, then what will his rotational velocity be if he pulls his arms and legs in so his new radius is 0.20 m?

★★ **Problem 4.117.** A disk of mass m_d and a ring of mass m_r have equal radii r and are rolling at equal speeds v when they encounter an incline at an angle θ above the horizontal. How far along the slope does each travel before coming to rest?

★★ **Problem 4.118.** A wheel with a diameter of $d = 0.40$ m and a moment of inertia of 0.30 kg m^2 is rotating at 65 rpm when a 35 N braking force is applied (Fig. 4.42). The coefficient of friction between the brake pad and the wheel is 0.20. How much work does the braking force do in bringing the wheel to rest? How many rotations does it go through as it comes to rest?

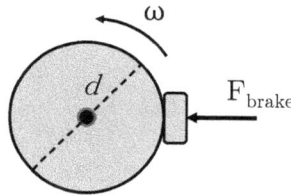

FIGURE 4.42
Problem 4.118

★★ **Problem 4.119.** Consider the situation described in Example 3.35. Solve part (c) using energy conservation.

★ **Problem 4.120.** An ice skater spins in place with her arms extended. As she pulls her arms closer to her torso, will her angular velocity increase or decrease? Why?

★ **Problem 4.121.** A 0.50 kg meter stick is held vertically against a wall. It then falls, pivoting about its base, until it hits the ground. Just as it strikes the ground, what is (a) the angular velocity and (b) the tangential velocity of the tip?

★★ **Problem 4.122.** A hoop and a solid disk have the same radius and mass. They roll down an incline plane. Which reaches the bottom first?

★★ **Problem 4.123.** A piece of homemade playground equipment consist of a vertical post that runs through the middle of a horizontal board, which is attached to the top of the post by rope (Fig. 4.43).

    a. Suppose the board has length L, mass M, and that two children of equal masses m and m are seated at opposite ends of the seesaw. What is the moment of inertia of the board+children system?

    b. An adult pushes on the side of the board so it slowly rotates around the post. In the process the board lifts a distance h as the cords wrap around the post. How much work has the adult done?

    c. The adult releases the board and it begins to unwind. What rotational velocity will the children have when the board is back to its original height?

    d. What centripetal acceleration will the children experience when the board is back to its original height?

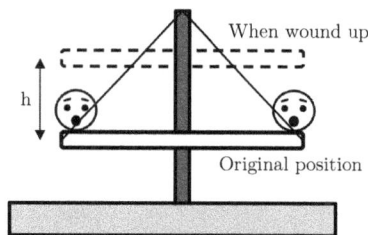

FIGURE 4.43
Problem 4.123

★★ **Problem 4.124.** A hoop has a mass of 0.10 kg and a radius of 0.15 m. It rolls down an incline that is 0.50 m tall.

    a. What is the total mechanical energy at the top of the hill? At the bottom of the hill?

    b. What is the velocity of the center of mass of the hoop as it reaches the bottom of the hill?

★★ **Problem 4.125.** A particular fly wheel consists of a 250 kg solid cylinder with a diameter of 1.50 m. It is spinning at 126 rad/s about a central axis and is connected to a machine to which it will deliver energy. If half the energy stored in the flywheel is delivered in 2.00 seconds, what is the power delivered to the machine?

⋆ **Problem 4.126.** Starting with Equation (4.33), show that the angular momentum of a point particle of mass m moving in a circle of radius r at a speed v has angular momentum given by equation (4.34).

⋆⋆ **Problem 4.127.** On a game show, you spin a large wheel. Suppose it has a radius of 1.50 m, mass of 2.00×10^2 kg and you apply a constant torque of 50.0 Nm for 1.00 s.

  a. What is the change in angular momentum of the wheel?

  b. If it started from rest, what is the final angular velocity of the wheel?

  c. If the wheel makes 2.60 rotations before coming rest (after you stop spinning it), what is the magnitude of the torque acting on the wheel due to friction?

⋆ **Problem 4.128.** Consider a bowling ball with a 22 cm diameter and a mass of 5.0 kg. If it has an angular momentum of 0.23 kg m^2/s, what is its angular speed in rpm?

⋆⋆ **Problem 4.129.** Conservation of angular momentum is an important principle in celestial mechanics. Suppose a star with a core radius of 7.0×10^5 km rotated once every 2.5×10^6 seconds. During a supernova, the core collapses to a radius of 10.0 km. If the mass of the core doesn't change, what is the new rotational period of the star?

⋆⋆ **Problem 4.130.** A 0.100 kg bullet is traveling at 1.00×10^3 m/s when it strikes tangent to the edge of a 1.0 m diameter disk spinning about its center. If the moment of inertia of the disk is 10.0 kg m^2 and after the collision the bullet and disk are at rest, then what was the angular speed of the disk prior to the collision?

⋆⋆ **Problem 4.131.** A 2.5-m-long see-saw has a mass of 16.0 kg. It is sitting horizontally when a 2.0 kg rubber ball elastically bounces off of the right edge of the see saw. If the ball was dropped from rest from a height of 2.0 m above the see-saw and bounces back to a height of 1.5 m above the see-saw, then what is the angular velocity of the see-saw immediately after the bounce?

⋆⋆ **Problem 4.132.** A 2.0 kg turntable has a radius of 0.20 m and is spinning at $33\frac{1}{3}$ rpm (about 3.5 rad/s). What is the angular velocity of the system if you drop a 1.0 kg record with a diameter of 0.15 m onto the turntable?

⋆⋆ **Problem 4.133.** A merry-go-round (MGR) has a 2.5m diameter, a mass of 2.00×10^2 kg, and is rotating at 25 rpm. I am on the ground running tangent to the MGR in the same direction that it is turning, and I jump on to the outer edge with negligible tangential velocity. If my mass is 72 kg, what is the MGR's new rotational speed, in rpm, after I jump on?

⋆⋆ **Problem 4.134.** A 50. kg passenger is standing on the edge of a 120. kg merry-go-round (MGR) with a 4.0 m diameter. The MGR is initially rotating counter clockwise (when looked down from above) with a rotational period of 5.0 seconds. The passenger suddenly jumps off of the MGR, tangent to the circle and in the direction of rotation, with a horizontal velocity of 3.5 m/s relative to the ground. What is the new angular velocity of the MGR?

⋆ **Problem 4.135.** A boat engine consumes 8.5×10^4 W while the propeller shaft rotates at 4.00×10^2 rpm. What is the torque on the shaft?

⋆ **Problem 4.136.** A motor is advertised as having a torque of 15 Nm and rotates at 3.5 rad/s. What is the power output?

⋆⋆⋆ **Problem 4.137.** The complete description of the angular momentum of a particle of mass m and momentum $\vec{p}$ is given by

$$\vec{L} = \vec{r} \times \vec{p} \qquad (4.37)$$

where $\vec{r}$ is the particle's position vector. The $\times$ symbol here indicates a **cross product**. We shall discuss the cross product in some detail when discussing magnetism, but the basic idea is that the cross product of two vectors will also be *perpendicular to the plane formed by the vectors*. So if $\vec{r}$

and $\vec{p}$ are on the plane of a table (formed by the x and y axes, say), $\vec{L}$ points either straight up or straight down (along the z axis). Mathematically,

$$\vec{A} \times \vec{B} = (A_y B_z - A_z B_y)\,\hat{x} + (A_z B_x - A_x B_z)\,\hat{y} + (A_x B_y - A_y B_x)\,\hat{z} \qquad (4.38)$$

where, if $\hat{x}$ points right along the page and $\hat{y}$ points up along the page (as usual), then $\hat{z}$ points up out of the page, toward you.

a. A particle has momentum $\vec{p} = (5\hat{x} + 5\hat{y})$ kg m/s when it is located at $\vec{r} = -10$ m $\hat{y}$. Use Equation (4.38) to determine the angular momentum.

b. Repeat (a), this time using Equation (4.34) and show that you obtain the same result.

c. A ball of mass m is spinning in a horizontal circle. At a certain instant it is at a position $\vec{r} = r_x \hat{x}$ with a velocity $\vec{v} = v_y \hat{y}$. If, at this instant, the ball is given a sharp blow in the $+\hat{x}$ direction (so v_x increases but r does not (or at least the change in r can be neglected), then in what direction is the change in angular momentum, $\Delta\vec{L}$?

$\star\star$ **Problem 4.138.** For planetary bodies in circular orbits, Kepler's third law can be stated as $T^2/r^3 = \sigma$ where T is the period of the orbit, r is the radius of the orbit, and σ is a value that is the same for every planet in a solar system. Determine an expression for σ in terms of the mass of the star at the center of a solar system and any other constants needed.

Part II

Applying the Fundamentals

In Part I we introduced the fundamental tools of introductory physics: kinematics, forces, and conservation laws. Now it is time to extend these ideas to more complicated systems. First recall that up until now we mostly thought of thermal energy as wasteful: if friction causes a sliding block's kinetic energy to become thermal energy, that energy is "lost" to the environment (unlike, say, an object bouncing on an ideal spring, where the energy smoothly and cyclically transfers between spring potential energy, gravitational potential energy, and kinetic energy). However, if you have ever used a refrigerator, water cooler, or air conditioner, then you likely recognize that thermal energy can be harnessed for useful purposes.

In Chapter 5 we shall see that thermal energy is a measure of the energy inside a system: for example, the gas molecules in a "hot" container tend to have more kinetic energy than the molecules in a "cold" container. By analyzing this system with the tools we developed in Part I, we will derive the ideal gas law–an incredibly important law in both physics and chemistry. From there, we will take up the study of nature's rules concerning thermal energy–the so-called laws of thermodynamics–and their applications to practical devices.

Then, in Chapter 6 we will consider motion when we have a *non-constant* force, in contrast with what we normally considered in Part I. We will find that a particular type of non-constant force gives rise to motion that describes vastly diverse phenomena including the sound of musical instruments, radio waves used in telecommunications, and even the behavior of subatomic particles.

5

Thermodynamics

Learning Objectives

After reading this chapter, you should be able to:

- Use the ideal gas law to relate changes in the pressure, volume, and temperature of a gas.

- Explain the relationship between the temperature of an ideal gas and the average kinetic energy of the gas particles.

- Explain how a version of the ideal gas law can be obtained by thinking through the motion of the gas particles.

- Define and relate essential concepts from thermodynamics, including thermal energy, temperature, thermal equilibrium, heat, thermal contact, and work.

- Explain the first law of thermodynamics and use it to mathematically describe situations involving heat transfer.

- Explain the second law of thermodynamics and use it to characterize the fundamental limitations of heat engines and heat pumps.

- Analyze situations where liquids and/or solids exchange heat in terms of the specific heat(s) of the materials involved.

5.1 Introduction

Suppose you have two identical blocks of metal. One has been in a freezer for quite some time, while the other has been sitting on a hot stove. Which block has the higher temperature? The answer, of course, is the one that has been on the stove. But what exactly does this mean – what *is* temperature?

As you consider this question, you might think something along the lines of, "Well, if I touch something with a high temperature, I become warmer; if I touch something with a low temperature, I become colder". This is the language of *energy transfer*: always from the object with the higher temperature to the objet with the lower temperature. This is in fact a good starting definition of **temperature**: the tendency for an object to give up energy.[1]

While you are probably most familiar with measuring temperature in degrees Fahrenheit or degrees Celsius, The SI unit of temperature is the kelvin (abbreviated "K").[2] Probably

---

[1]We will refine this definition later in this chapter, once we've introduced some important ideas.

[2]After William Thomson (1824-1907), also known as Lord Kelvin. It is worth your time to refer to Appendix B to get a handle on converting between these units. Importantly, a 1°C change in temperature is the same as a 1 K change in temperature. Consequently, any equation involving ΔT can be equivalently expressed in degrees Celsius and Kelvin.

DOI: 10.1201/9781003571568-7

the most famous equation involving temperature is the **ideal gas law**, which relates the pressure, P, volume, V, and temperature T of a sealed container holding N particles of a gas:

$$PV = Nk_BT \qquad (5.1)$$

k_B is a constant of nature called **Boltzmann's constant**:[3]

$$k_B = 1.381 \times 10^{-23} \text{ J K}^{-1} \qquad (5.2)$$

We often refer to "standard temperature and pressure" (STP), which refers to a pressure of 1 atmosphere (or in SI units, 1.01×10^5 Pa) and a temperature of 273 K.

You may have seen the ideal gas law written in the form most often used by chemists:

$$PV = nRT \qquad (5.3)$$

Here R is the **gas constant**,

$$R = 8.314 \text{ J K}^{-1} \text{ mol}^{-1} \qquad (5.4)$$

and n is the number of particles measured not as the raw count (which is N) but rather as the number of **moles**. If you compare Equations (5.1) and (5.3) you will see

$$Nk_B = nR$$

which indicates that the number of particles in 1 mole (denoted N_A) is

$$N_A = (1 \text{ mol}) \frac{R}{k_B} = (1 \text{ mol}) \frac{8.314 \text{ J K}^{-1} \text{ mol}^{-1}}{1.381 \times 10^{-23} \text{ J K}^{-1}} = 6.022 \times 10^{23} \qquad (5.5)$$

This unitless quantity is called **Avogadro's number**.[4] When dealing with ideal gases it is often more convenient to deal with quantities measured in moles–for instance, Table 5.1 shows the mass of common gases in grams *per mole* of the gas–and it is important to become comfortable working with them mathematically (see e.g. Problem 5.4).

TABLE 5.1

Molar masses of some gases.

Name	Symbol	Molar mass (g/mol)
Hydrogen	H_2	2.01
Helium	He	4.00
Nitrogen	N_2	28.0
Air	–	29.0
Oxygen	O_2	32.0
Carbon Dioxide	CO_2	44.0

The ideal gas law was originally developed to characterize experimentally observed properties of gases. However, in the next section we will *derive* a version of the ideal gas law very similar to Equation (5.1); you may be surprised to see that the derivation requires no new fundamental physics. The version of the ideal gas law that we derive will serve to underscore the relationship between temperature and energy. (If you find yourself getting bogged down in the details, the key result is presented and summarized in the last two paragraphs of the section – but it is an excellent idea to take the time to obtain intellectual ownership of the derivation at some point!)

---

[3]After Ludwig Boltzmann (1844-1906)
[4]After Amedeo Avogadro (1776-1856). N_A is equal to the number of atoms in 12 g of carbon-12.

5.2 The Ideal Gas Law

Suppose we have a cubical box resting on the floor; we'll label the length of an edge L. Now suppose we put a single gas molecule (we'll treat it like a small ball) of mass m in the box and give it a velocity $\vec{v}$ parallel to the floor such that it bounces back and forth between two opposite walls. We will ignore gravity and assume that energy is conserved when the molecule strikes a wall, so the motion can proceed indefinitely (Fig. 5.1).[5] We are interested in determining an expression for the *pressure* exerted on the box by the molecule.

To begin, consider a single "bounce": the molecule impacts the wall with velocity $\vec{v}$ and bounces back with velocity $-\vec{v}$. (The rebound velocity must have the same magnitude as the impact velocity if energy is to be conserved.) In other words, the molecule experiences an impulse with magnitude $J = \Delta p = 2mv$ (the direction of the vector is, of course, in the direction of the final velocity, toward the inside of the box). Now, this change in momentum occurs because the molecule experiences a force from the wall $\vec{F}_{MW}$, and by Newton's third law we know that the wall experiences an equal and opposite force due to the ball $\vec{F}_{WM}$. The direction is toward the outside of the box, and we can express its magnitude in the following equation:

$$F_{WM}\Delta t = \Delta p = 2mv \tag{5.6}$$

In this equation, Δt corresponds to the time during which the molecule is in contact with the wall. This is not a known quantity and we would like to eliminate it, leaving us with an equation for the force felt by the wall in terms of known quantities (from there we can move quickly to an expression for the pressure, since from Equation (3.14) we have $P = F/A = F/L^2$).

To eliminate Δt from Equation (5.6) we will make use of the fact that the impact we are thinking about occurs at regular intervals: after an impact, the molecule travels down the length of the box in a time $t_1 = L/v$ and then comes back to the original wall in an equal amount of time. Thus, the wall experiences a force F_{WM} for a time Δt and then no force for a time $T = 2L/v$; this cycle repeats, in principle, forever.

With this in mind, what is the average force experienced by the wall during one such interval? To answer this we need to compute a *weighted average*: if an object experiences a series of forces F_1, F_2,... for times Δt_1, Δt_2,..., then the average force is given by

$$F_{avg} = \frac{F_1\Delta t_1 + F_2\Delta t_2 + ...}{\Delta t_1 + \Delta t_2 + ...} \tag{5.7}$$

The denominator is the total time we are considering and the numerator is the sum of each force multiplied by the amount of time during which that force is felt.[6] In the situation we are considering, the total time interval is $T + \Delta t$, during which time the wall experiences just one force, F_{WM}, for a period Δt. Thus,

$$F_{avg} = \frac{F_{WM}\Delta t}{T + \Delta t}$$

$$F_{avg} \approx \frac{F_{WM}\Delta t}{T}$$

$$F_{avg} = \frac{F_{WM}\left(\Delta t\right)v}{2L}$$

---

[5]You might object that it is ridiculous to ignore gravity, but for gas particles this is actually a reasonable assumption. See Problem 5.11.

[6]For instance, suppose there are just two forces $F_1 = 10$ N and $F_2 = 20$ N. If the object experiences these forces for equal times, then the average force is just 15 N. If the times are unequal, then the average force will shift toward the force experienced for the longer period of time. See Problems (5.12)–(5.13) for some practice with this.

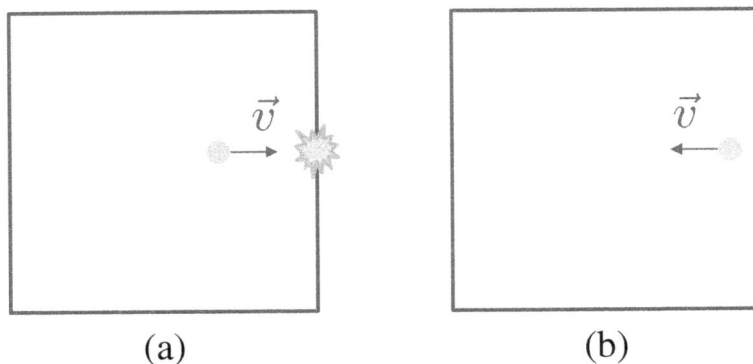

(a)   (b)

FIGURE 5.1
A gas particle in a sealed container. (a) The particle approaches the right wall with a velocity $\vec{v}$ directed to the right. (b) The particle rebounds with the same speed; the process repeats on the left wall and some time later the particle returns to the configuration shown in the left panel.

In the second line we have made the assumption $T + \Delta t \approx T$, meaning that it takes much longer for the molecule to travel the length of the box, T, than it does in the process of bouncing off of the wall, Δt (this is reasonable for any sort of macroscopic box). Solving for F_{WM} yields

$$F_{WM} = \frac{2LF_{avg}}{(\Delta t)\,v}$$

We can substitute this result into Equation (5.6):

$$F_{WM}\Delta t = 2mv$$
$$\frac{2LF_{avg}}{(\Delta t)\,v}\Delta t = 2mv$$
$$\frac{LF_{avg}}{v} = mv$$
$$F_{avg} = \frac{mv^2}{L}$$

This is what we were looking for: Δt algebraically cancels on the second line, and we are left with an equation for the force experienced by the wall in terms of known quantities. Note, however, that we have shifted from the force experienced during a *single* impact by the molecule to the average force felt over *many* impacts. The corresponding pressure is therefore also an average quantity:

$$P_{avg} = \frac{F_{avg}}{A}$$
$$P_{avg} = \frac{mv^2}{L^3}$$
$$P_{avg} = \frac{mv^2}{V} \tag{5.8}$$

In the above we have made use of the fact that the surface area of the wall is $A = L^2$ and the volume of the entire box is $V = L^3$.

We've been analyzing the special case of our molecule moving back and forth in a straight line, but now that we've completed our analysis it is just a short step to go to the much more realistic case of *many* molecules, each moving with a three-dimensional velocity. Note that *on average*, the component of a particle's velocity is the same in each dimension (there is nothing to differentiate any one dimension from the others), so we can express the magnitude of the overall velocity vector in terms of its components:

$$(v_{avg})^2 = (v_x)^2_{avg} + (v_y)^2_{avg} + (v_z)^2_{avg}$$
$$(v_{avg})^2 = 3 (v_i)^2_{avg}$$

In the second line, the subscript i refers to any of the three Cartesian components.

This equation relates the one-dimensional component of the velocity (which is what we have been dealing with so far) to the *total* velocity. Thus, to update Equation (5.8), we need to change v^2 to $v^2/3$ and introduce a factor of N on the right hand side to account for the cumulative effect of our collection of particles:[7]

$$P_{avg} = N \frac{m (v_{avg})^2}{3V}$$

This is the average pressure experienced by each of the six walls of our container. If we rearrange this equation, drop the "avg" subscripts,[8] and characterize the particles in terms of the average kinetic energy $K = \frac{1}{2}mv^2$, we find

$$PV = N \frac{mv^2}{3}$$
$$PV = N \frac{2}{3}K \tag{5.9}$$

This equation relates the pressure and volume of our container to the number of gas molecules contained therein and their average kinetic energy. While our derivation has assumed that our container is a cubical box, a more detailed treatment can show that the shape of the container does not affect this result.

If we compare Equation (5.9) to Equation (5.1), we find

$$k_B T = \frac{2}{3}K \tag{5.10}$$

We see that, aside from some constants, **the temperature of an ideal gas is proportional to the average kinetic energy of a particle in the gas.** This is the key relationship we set out to show in this section, though I hope you appreciate the derivation for its own sake, too![9]

---

[7]We can only insert a factor of N on the right hand side of the equation if we assume that the particles don't interact with each other; our analysis of a single particle assumed that it could travel across the center of the box without interruption. If the box is too crowded, this is not a good assumption (see Problem (5.14)). The fact that we assume that the molecules do not interact with one another as they move about the container is why the law we are deriving is called the *ideal* gas law.

[8]In a typical gas, the bombardment of molecules against the surface of the container is essentially constant over any reasonable time scale because N tends to be so large (see Problem (5.9)). Thus, we generally refer to what is, strictly speaking, an average quantity as simply "the pressure".

[9]I should mention that this holds for a *monatomic* gas, i.e. cases where we can consider the gas as a tiny ball with no internal dynamics. For instance, diatomic molecules can also vibrate because of their internal molecular bonds, which complicates the analysis but simply changes the 3 in Equation (5.10) to a 5; we say that the diatomic gas has two more *degrees of freedom* than the monatomic gas, which has 3 due to the usual three Cartesian axes along which it can independently translate.

At the beginning of this chapter, we claimed that temperature is a measure of the tendency for an object to *give up* thermal energy. Now that we have directly related temperature to energy, we see that we might paraphrase this to say "objects with high thermal energy tend to lose it". To think this through, in the next section we will consider what happens when two initially isolated ideal gases are allowed to interact.

Example 5.1 Air Particles in a Closet $\star$

A particular closet measures 1.0 m × 2.0 m × 2.5 m. Assuming a typical temperature of $T = 295$ K (about 70 degrees Fahrenheit) and atmospheric pressure, about how many particles are in the closet?

We need to solve the ideal gas law for N:

$$PV = Nk_BT$$

$$N = \frac{PV}{k_BT}$$

From here we can substitute $P = 1\,\text{atm} = 1.01 \times 10^5$ Pa, $V = (1.0 \times 2.0 \times 2.5)\,\text{m}^3 = 5.0$ m³, the given temperature and the numerical value for k_B:

$$N = \frac{\left(1.01 \times 10^5 \frac{\text{N}}{\text{m}^2}\right)\left(5.0\,\text{m}^3\right)}{\left(1.38 \times 10^{-23}\,\text{m}^2\,\text{kg s}^{-2}\,\text{K}^{-1}\right)295\,\text{K}}$$

$$N = 1.2 \times 10^{26}$$

Because N is a number, it has no units – it is good practice to go through and check that the units in the above equation do indeed cancel (I have taken one helpful step by replacing "Pa" with the equivalent N/m²; remember 1 N = 1 kg m/s²).

In a mathematical sense, for typical values of P, V, and T, it follows that N is so large because Boltzmann's constant is so small. Physically, this result serves to demonstrate how many particles typically exist even in relatively small containers: even if a container had just one tenth of the volume of the closet considered here, there would still be on the order of 10^{25} particles inside!

5.3   Thermal Equilibrium

Consider a box cut into two equal volumes by a rigid and impermeable barrier (Fig. 5.2); each half is filled with an equal number of ideal gas particles (each with the same mass m). Mathematically, then, $V_L = V_R = V$ and $N_L = N_R = N$. Suppose, however, that the temperature in the left side is twice that of the right: $T_L = 2T_R$. It follows from the ideal gas law that $P_L = 2P_R$; from Equation (5.10) it follows that $K_L = 2K_R$.

Now suppose that the barrier is suddenly removed. What do you suppose will happen? The answer is that the particles will *mix*: particles that were on a collision course with the barrier will instead encounter no resistance and move into the other side of the container. If you wait long enough, the particles that were initially on the left side of the container will be evenly spread throughout the container; the same is true for the particles initially on the right side. (All we're describing is the natural tendency for a gas to distribute itself evenly in a container – here we have two gases doing so simultaneously.)

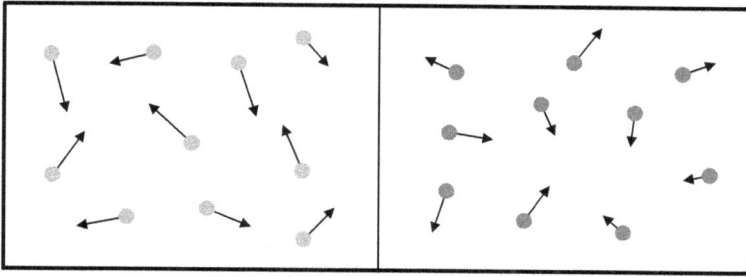

FIGURE 5.2
A sealed container with a removable partition that separates the interior into two sides, each with an equal number of gas particles. The particles on the left are drawn with a light color and tend to have longer velocity vectors (indicating higher temperature) than the darker particles on the right.

As the last step of our experiment, suppose now that the barrier is replaced a long time after it was initially removed. The volume of the two halves will be the same, but what about the other properties we've been thinking about: the number of particles, the temperature, and the pressure? Let's consider each in turn:

- The number of particles in each half of the container ought to be the same as at the beginning: we had a total of $2N$ particles that were allowed to mix evenly, then the volume they occupied was cut in half.[10] It follows that each half of the container should hold N particles, just as in the beginning.

- The temperature, remember, corresponds to the average kinetic energy of the particles (Equation (5.10)). If each container now has an equal mix of the particles that were originally in the left side and in the right side, then the new average energy for each half of the container should be the average of the original values: $(K_L)_{\text{new}} = (K_R)_{\text{new}} = (K_L + K_R)/2$ and so $(T_L)_{\text{new}} = (T_R)_{\text{new}} = (T_L + T_R)/2 = 1.5 T_R$.

- The pressure will behave the same as the temperature, as we can see from the ideal gas law: $PV = N k_B T$. V and N are unchanged, while T has changed (as we just argued). To balance the equation, the pressure must change by the same factor as the temperature: $(T_R)_{\text{new}} = 1.5 T_R$, so $(P_R)_{\text{new}} = 1.5 P_R$.

We can summarize all of this by saying that allowing the gases to mix allows the two halves of the box to be in **thermal equilibrium**: the average kinetic energy of the particles on the left side is the same as the average kinetic energy of the particles on the right side. The fact that the temperature of the left side of the container drops while the temperature on the right side increases follows from the fact that the particles – with their different kinetic energies – are mixing. We sketch this final configuration in Figure 5.3.

However, it is very important to note that we can achieve the same result *without* allowing the particles to mix. Suppose the barrier between the two halves is a very thin membrane, and we *don't* remove it. A particle from the left chamber can collide with a particle from the right chamber as they simultaneously strike the barrier from opposite sides. Using Equations (4.26) and (4.27), it is not terribly difficult to show that the velocities of the two particles simply exchange: the final velocity of the particle on the left is the initial velocity of the

---

[10]It is worth mentioning that when you have something like 10^{26} particles, you won't necessarily get *exactly* half when you go through a procedure like this. But any deviations will be small enough to be ignored.

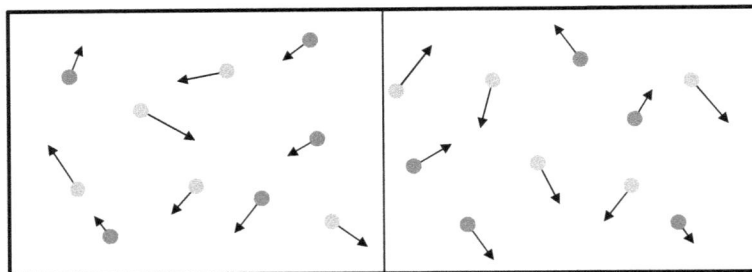

FIGURE 5.3
If the partition in the box shown in Figure 5.2 is removed, the gases will mix. Reinserting the partition a long time later results in an even mixture of the gases on both sides of the partition.

particle on the right, and vice versa (see Problem (5.17)). In this scenario, no *particles* are exchanged, but *energy* is (we refer to energy that can be exchanged in this way as **thermal energy**). If we allow this process to go on long enough, the energies will become evenly mixed between the two chambers. Just as when we allowed the particles to mix, the eventual temperatures of the two halves of the container are the same (i.e. they are in thermal equilibrium).

We have analyzed the scenario where the initial values of N and V are the same for each half of the container because it helps to simplify the analysis. The following example demonstrates how to handle a similar, but somewhat more complicated, instance of ideal gases mixing.

Example 5.2 Temperature and Mixing Gases   ★★
Consider a box with cross-sectional area A and length L. A barrier is inserted at $x = 0.3L$, cutting the box into two separate chambers with unequal volumes. The smaller chamber is filled with an ideal gas with pressure P and temperature T; the larger volume is filled with an ideal gas with pressure $1.2P$ and temperature $1.1T$. If the barrier is then removed, what will the temperature of the box be a long time later? Assume all values are exact.

Before we delve into the mathematical details, let's try to get a rough sense for what we should *expect*. If the chambers had an identical number of particles, we'd expect to find the exact mean of the two temperatures, $1.05T$ (as in the scenario we just analyzed in the main text). It does not seem likely that the number of particles are the same here, and so we should expect a shift from $1.05T$ toward the temperature corresponding to the chamber with more particles.

Let's start with the smaller chamber. We're given a value for every quantity in the idea gas law except the number of particles, which we'll denote N_S, so

$$PV = N_S k_b T$$
$$N_S = \frac{PV}{k_B T}$$
$$N_S = 0.3 \frac{P(LA)}{k_B T}$$

Meanwhile, the larger chamber has N_L particles:

$$N_L = \frac{(1.2P)\,A\,(0.7L)}{k_B\,(1.1T)}$$

$$N_L = 0.924 \frac{P\,(LA)}{k_B T}$$

We see that the larger chamber has more particles, and we therefore expect to obtain a final temperature between $1.05T$ and $1.1T$.

We can determine the appropriate value by computing a weighted average, in the same way as in Equation (5.7). While there we were thinking about a series of forces and corresponding times during which the forces were felt, here we have a series of temperatures and corresponding numbers of particles. To simplify the algebra, we'll set $\alpha = P\,(LA)\,/k_B T$, so:

$$T_{\text{final}} = \frac{N_L T_L + N_S T_S}{N_L + N_S}$$

$$T_{\text{final}} = \frac{(0.924\alpha)\,(1.1T) + (0.3\alpha)\,(T)}{0.924\alpha + 0.3\alpha}$$

$$T_{\text{final}} = 1.08T$$

All of the α terms cancel, leaving us with units of temperature, and we see that the value does indeed fall within the expected range.

5.4   Some Definitions

We have introduced a number of ideas so far in this chapter, and it would be good to pause now and formally define the key concepts and laws of thermodynamics. We will begin with some basic definitions:

- **Thermal energy** is a measure of an object's internal energy.[11] In an ideal gas, as we have seen, this corresponds to the total kinetic energy of the molecules. Molecules in a *solid* do not move about freely like in an ideal gas, but they can *vibrate*; higher thermal energy corresponds to stronger vibrations. (We will not worry about the details – the point is that the idea of thermal energy is not restricted to ideal gases only!)

- **Temperature** is a measure of the average thermal energy of a particle in an object (like a molecule of an ideal gas).[12]

- Two objects are in **thermal equilibrium** if they have the same temperature.

- **Heat**, which we will symbolize Q, refers to some energy that is transferred from one object to another due to a difference in temperature.

---

[11]There are other kinds of internal energy than thermal – in chemical bonds that bind molecules, for instance. *Thermal* energy refers to the sort of internal energy that can flow between objects if they are in "thermal contact", like two blocks of metal touching.

[12]Keep in mind that thermal energy is a *total* quantity, while temperature is an *average* quantity. If I take a container of ideal gas and partition it into two uneven sections, they'll have the same temperature (the average kinetic energy of a molecule on the left is the same as on the right) but different thermal energies because the total number of particles is different.

- Two items are in **thermal contact** if heat can flow between them. Most simply, this means the two objects are touching or are connected by something that can conduct heat, like a metal bar.

- **Work** means the same thing that it always has: a force acting over some distance. Here we are interested in work that changes the thermal energy of some object. For instance, if I quickly compress a gas-filled piston (which involves doing work), I can increase the thermal energy of the gas.

If you think that the definitions of these terms are somewhat circular, with each definition invoking several other terms, then you're right. Perhaps this summary will help: **When two objects of different temperatures are placed in thermal contact, heat flows from the hotter to the cooler object until they reach thermal equilibrium.** The mechanism for how this occurs, even for the case of solids, is very similar to the "molecules bouncing off of a membrane" example discussed in the context of an ideal gas on page 263.

5.5   The First Law of Thermodynamics

It is worth pausing for a moment and contemplating the fact that we are now thinking of objects as *reservoirs* of thermal energy (it would be fair to argue that this is the baseline idea that drives all of the physics that we are developing in this part of the book). While thermal energy can flow between objects in the form of heat, it is important to keep in mind that the total energy of a closed system is still conserved: if the hot object loses some energy Q, then the cool object gains that same energy Q.

Moreover, if a hot object loses some energy Q, then that loss is reflected by a decrease in its thermal energy E_{th} (we have removed the energy Q from the reservoir, so to speak). If we instead *add* the energy Q, the thermal energy increases. We summarize this mathematically as

$$\Delta E_{th} = Q$$

where we have defined the quantity Q to be positive if the energy is being added to the object and negative if it leaves the object.

As mentioned above, we can change an object's thermal energy by doing work on it, as well. Thus in general we can say

$$\Delta E_{th} = Q + W \qquad (5.11)$$

where both Q and W are defined to be positive if they are adding energy to the object and negative if they are removing energy from the object. This is the **first law of thermodynamics**. In words, it says that energy flowing in to or out of a system in the form of work or heat changes the system's internal energy in accordance with the law of conservation of energy. (Put simply: you can neither create nor destroy thermal energy out of nothing.)

What does this have to do with temperature? Well, the tendency for an object to change its temperature based on a change in thermal energy depends on material specific properties that we summarize in the **heat capacity** C such that $\Delta E_{th} = C\Delta T$, or, inserting Equation (5.11):

$$\Delta E_{th} = Q + W = C\Delta T \qquad (5.12)$$

The heat capacity has units of J/K and depends on *what kind* of material we have and also *how much* material we have. For an ideal gas, the heat capacity is different if the process occurs at a constant pressure:

$$C_P = \frac{5}{2}R \qquad \text{(monatomic)} \qquad (5.13)$$

$$C_P = \frac{7}{2}R \qquad \text{(diatomic)} \qquad (5.14)$$

or at constant volume:

$$C_V = \frac{3}{2}R \qquad \text{(monatomic)} \qquad (5.15)$$

$$C_V = \frac{5}{2}R \qquad \text{(diatomic)} \qquad (5.16)$$

Notice that these quantities have the same units as R, the gas constant (J K^{-1} mol^{-1}). Thus they describe the heat capacity (J/K) *per mole* and are so-called *molar heat capacities*. We will consider heat capacity in greater depth in Section 5.7; for now the key point is that thermal energy is related to the temperature of an object. As a specific example, for an ideal gas we have[13]

$$\Delta E_{th} = nC_V\Delta T \qquad (5.17)$$

where the heat needed to drive a particular change in temperature ΔT is

$$Q = nC_V\Delta T \qquad (5.18)$$

if the gas is at constant volume and

$$Q = nC_P\Delta T \qquad (5.19)$$

if the gas is at constant pressure (Problem 5.49).

Example 5.3 Cooking with Gravity ★
The heat capacity of 1.0 kg of beef is about 3.4×10^3 J/K. From what height should you drop the meat if all of the potential energy will be converted to thermal energy and you wish to increase its temperature from 0°C to 65°C (from frozen to cooked)?

First, take a moment to appreciate that this is a ridiculous question; you can't cook food this way. Much of the potential energy would be wasted in deforming the meat as it dramatically splats against the ground. Still, the problem gives a sense of scale and can be used to impress all the right kind of people in idle conversation, so let's see it through. If we assume the gravitational potential energy really will be converted to thermal energy, then Equation (5.12) becomes

$$mgh = C\Delta T$$

$$h = \frac{C\Delta T}{mg}$$

$$h = \frac{\left(3.4 \times 10^3 \text{ J/K}\right)(65 \text{ K})}{(1.0 \text{ kg})\left(9.8\text{m/s}^2\right)}$$

$$h = 2.3 \times 10^4 \text{ m}$$

---

[13] C_V is the appropriate heat capacity because holding volume constant directly relates the internal energy to the temperature.

Note that we've used the fact that a temperature difference in degrees Celsius is the same as in Kelvin. This is quite the height; most commercial airliners cruise at less than half this altitude. In fact, this is high enough that the approximation that gravitational potential energy is given by mgh begins to break down (see Problem (5.22)).

We will have more to say about Equation (5.12) as we proceed, but for now let's focus our attention on the First Law of Thermodynamics (Equation (5.11)). As you probably surmised from the impressive title, it is very important. In the next pair of examples, we will use it to begin thinking about two important types of devices that use thermal energy: heat engines (e.g. power plants) and heat pumps (e.g. air conditioners).

Example 5.4 Thermodynamic Cycles: Heat Engine   ⋆
A **heat engine** operates between a *hot reservoir*, such as a coal furnace, and a *cold reservoir*, such as a river. Every second, the engine takes in some amount of heat from the hot reservoir, Q_H, uses it to do work, W (for instance, by turning a turbine that generates electricity), and dumps some amount of heat Q_C into the cold reservoir (Figure 5.4).

    a. From the perspective of the engine, which of these quantities are positive? Which are negative?

    b. How are they related according to the first law of thermodynamics? Assume that the engine's internal thermal energy is constant.

    c. The **efficiency** of a heat engine is defined to be W/Q_H, i.e. the net work output divided by the input heat energy. If $Q_H = 100$ J and $Q_C = 30$ J, what is the efficiency of the heat engine?

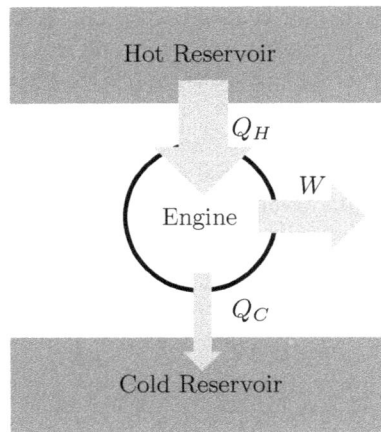

FIGURE 5.4
Schematic diagram of a heat engine.

We define energies as positive if they are added to the system and negative if they are removed. Here we are adding energy Q_H (from the hot reservoir) and losing both Q_C (to the cold reservoir) and W (e.g. to the turbine). So, Q_H is positive and both Q_C and W are negative.

Turning to (b), we can insert our results into the right hand side of the first law of thermodynamics. We will assign the positive and negative signs explicitly in the equation, so the algebraic symbols Q_H, Q_C, and W will refer to the corresponding magnitudes:

$$\Delta E_{th} = Q + W$$
$$\Delta E_{th} = Q_H - Q_C - W$$

The left side of this equation is ΔE_{th}, the change in thermal energy of the heat engine. We are told in the problem that we are to assume that this is 0, so after a bit of algebra we have

$$Q_H = Q_C + W$$

In other words, the energy added to the system from the hot reservoir splits into useful energy (in the form of work generated by the engine) and "waste" energy that is dumped into the cold reservoir.

Finally, for (c) we are asked to determine the efficiency, which we will symbolize η (Greek lowercase *eta*):

$$\eta = \frac{W}{Q_H}$$
$$= \frac{Q_H - Q_C}{Q_H}$$
$$= 1 - \frac{Q_C}{Q_H}$$
$$= 1 - \frac{30\,\text{J}}{100\,\text{J}} = 0.7$$

We can see from the algebra here that if $Q_C = 0$, the efficiency would be 1, meaning that energy would be transferred entirely from the hot reservoir to useful work output by the engine. As we will see, a result of the *second* law of thermodynamics is that this is not possible. In the meantime, we have seen how the *first* law can be used to describe the flow of energy through the heat engine.

Example 5.5 Thermodynamic Cycles: Heat Pump ⋆

A **heat pump** uses some energy W to draw some energy Q_C out of a cold reservoir (such as the inside of an apartment or refrigerator) and dumps some energy Q_H into a hot reservoir (outside the apartment or outside the refrigerator); see Figure 5.5.

(a) If energy flowing into the pump is defined to be positive, then which of these quantities are positive? Which are negative?

(b) How are they related according to the first law of thermodynamics? Assume that the pump's internal thermal energy is constant.

(c) The **coefficient of performance** of a heat pump when used as a refrigerator ($\text{COP}_{\text{cooling}}$) is defined to be Q_C/W, i.e. the heat drawn out of the cool reservoir divided by the input work. If $Q_H = 50.$ J and $Q_C = 10.$ J, what is the $\text{COP}_{\text{cooling}}$? (Incidentally, $\text{COP}_{\text{heating}} = \frac{Q_H}{W}$.)

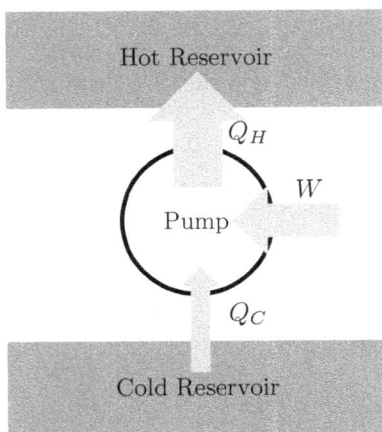

FIGURE 5.5
Schematic diagram of a heat pump.

This example deliberately parallels the preceding example to (hopefully) make the comparison between heat engines and heat pumps clear. Here we are *adding* energy in the form of work, W; it is being used to *draw* some energy Q_C from the cold reservoir and to *deposit* some energy Q_H in the hot reservoir.[a] Thus, W and Q_C are positive and Q_H is negative.

Turning to the mathematical form of the first law of thermodynamics for (b), we will again write the positive and negative signs explicitly so algebraic variables are magnitudes:

$$\Delta E_{th} = Q + W$$
$$0 = Q_C - Q_H + W$$
$$Q_H = W + Q_C$$

Finally, for (c) we have

$$\text{COP}_{\text{cooling}} = \frac{Q_C}{W}$$
$$= \frac{Q_C}{Q_H - Q_C}$$
$$= \frac{10.\ \text{J}}{(50. - 10.)\ \text{J}}$$
$$= 0.25$$

Here we might wonder if we could set $W = 0$, meaning that energy moves from the cold reservoir to the hot reservoir with no energy input to the engine. In this case the $\text{COP}_{\text{cooling}}$ would be undefined (we would be dividing by 0), and it should come as no surprise that an object will not spontaneously cool to a temperature *below* its surroundings. This, too, is a consequence of the second law of thermodynamics, as we shall see.

---

[a]If we're talking about an air conditioner, then you are sitting in the cool reservoir and want to remain comfortable, so the removal of Q_C is your principle concern.

5.6 Thermodynamic Cycles

In Examples 5.4 and 5.5 we considered heat engines and heat pumps, but we did so without discussing the underlying mechanisms: for example, *how* is heat added to a heat engine, and *how* is some of the energy converted to work? In this section we will take a closer look at the thermodynamic processes that underly these devices. Consider as our model "engine" a sealed cylinder and piston holding an ideal gas (Fig. 5.6).

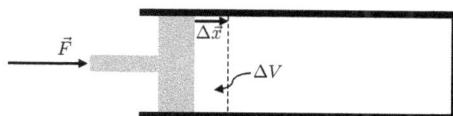

FIGURE 5.6
A sealed piston contains an ideal gas. Here, a force causes the piston to compress a distance $\Delta \vec{x}$, which causes the volume of the piston to decrease by an amount $\Delta V = A \Delta x$ where A is the cross-sectional area of the piston.

Four Standard Thermodynamic Processes

Note first that if the container is sealed, the number of gas particles is fixed. Thus from the ideal gas law

$$\frac{PV}{T} = Nk_B$$

is a constant. In other words, whatever happens to our piston, we can compare any two instances and note that

$$\frac{P_1 V_1}{T_1} = \frac{P_2 V_2}{T_2} \tag{5.20}$$

Equation (5.20) will be an important component of our toolkit. As special cases, we often consider processes where one of these three variables is held constant. Figure 5.7 shows these processes on a pressure vs. volume ("PV") diagram:

- An **isobaric process** is one in which the pressure does not change. On a PV diagram, it is simply a horizontal line. As an example, if our piston is allowed to move while the container is heated, then the temperature and volume will both increase while the pressure remains constant.

- A **isochoric process** is one in which the volume does not change. On a *PV* diagram, it is a vertical line. If our cylinder is heated but the piston is *not* allowed to move, then the pressure increases along with the temperature.

- An **isothermal process** is one in which the temperature does not change. Equation (5.20) here simplifies to $P_2 = P_1 V_1 / V_2$, meaning that the pressure at some arbitrary point (P_2) is inversely proportional to the volume (V_2). As an example, consider slowly compressing a piston: if the piston is in good thermal contact with its surroundings (heat can flow easily), it will maintain a constant temperature with its surroundings. As the pressure increases, the volume decreases.

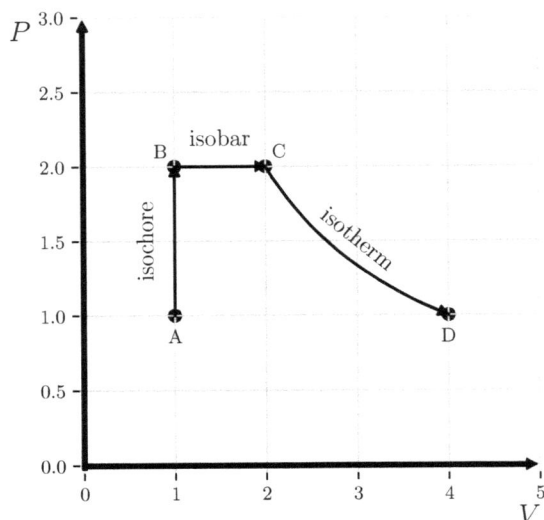

FIGURE 5.7

A *PV* diagram showing three typical processes on a sealed container of an ideal gas. The axes are shown with arbitrary units. The system begins at state *A*, with $P = V = 1$, then moves to state *B* through an *isochoric* process (constant volume; the curve is called an *isochore*). It then proceeds to state *C* through an *isobaric* process (constant pressure, the curve is called an *isobar*). Finally, it moves to state *D* through an *isothermal* process (constant temperature; the curve is called an *isotherm*). Note that if we included an additional isobaric process $D \to A$, the system would be back where it started and the diagram would depict a complete *thermodynamic cycle*.

A fourth process that we will consider is one in which all three of these quantities (the pressure, volume, and temperature) change but no *heat* flows, i.e. $Q = 0$. This is called an **adiabatic process**; an example is compressing a piston so rapidly that the compression outpaces the ability for the cylinder to exchange heat with its surroundings. It turns out (see Problem 5.51) that for adiabatic processes,

$$P_1 V_1^\gamma = P_2 V_2^\gamma \qquad (5.21)$$

and

$$T_1 V_1^{(\gamma-1)} = T_2 V_2^{(\gamma-1)} \qquad (5.22)$$

where γ is appropriately referred to as the **adiabatic exponent**. For diatomic gasses, $\gamma = 7/5 = 1.4$ and for monatomic gasses, $\gamma = 5/3 \approx 1.67$. Figure 5.8 shows a complete *PV* diagram cycle including two adiabatic processes.

Thermal Energy in the Four Processes

We've now described the four main processes used to analyze thermodynamic cycles, but to be complete we need to characterize how the thermal energy of our sealed gas *changes* in each process. Recall the First Law of Thermodynamics (Equation (5.11)), which states

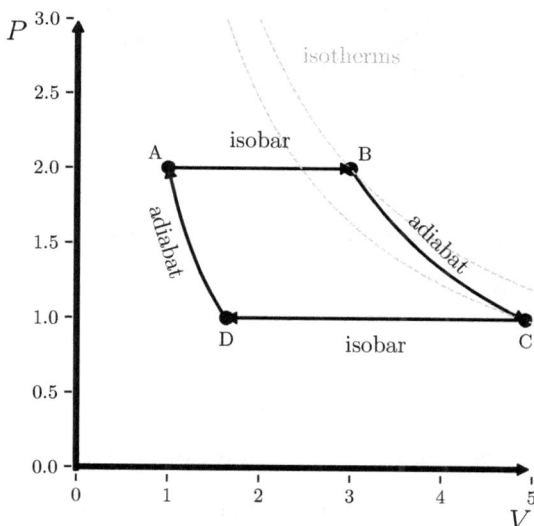

FIGURE 5.8
A PV diagram showing a cycle consisting of two isobars and two adiabats. The light gray dashed curve show isotherms connected to states B and C; this demonstrates that an adiabat is *steeper* than an isotherm.

$\Delta E_{th} = W + Q$, i.e. the change in thermal energy depends both on the work done on (or by) the gas and the heat transferred in to (or out of) the gas. We therefore need to think about both the work and the heat involved in each of our four processes.

We'll first consider the work. If you grip the piston and push on it to compress the gas, then you are doing work on the gas:

$$W = \int \vec{F} \cdot d\vec{x} = \int F \, dx = \int PA \, dx$$

In the second step I have carried out the dot product (you are applying a force in the direction of the piston's displacement) and in the third I have expressed the force in terms of the pressure in the cylinder and the cross-sectional area of the piston (recall that $P = F/A$).

Now the quantity $A \, dx$ is the area multiplied by a tiny displacement of the piston. This is a tiny change in the *volume* of the cylinder, but we have to be careful: when the piston compresses, as we're considering here, the volume *decreases*, so $A \, dx = -dV$ and

$$W = -\int P \, dV \tag{5.23}$$

In other words, the work done on the gas is the negative of the area bounded by the pressure curve and the volume axis in a PV graph.[14] The sign convention here can be a little tricky at first, so to reiterate: the work is a positive quantity if work is being done *on*

---

[14]In our example of compressing the piston, the volume decreases and the work will be overall positive.

the gas (i.e. energy is added to the gas) and is negative if work is being done *by* the gas (i.e. energy is being removed). Equation (5.23) is a general result that can be applied to each of the four processes we've been considering:

- In an isobaric process, the pressure is constant and so

$$W = -P\Delta V \tag{5.24}$$

  Graphically, the PV curve forms a simple rectangle with dimensions P and ΔV.

- In an isochoric process, the volume does not change and so

$$W = 0 \text{ J} \tag{5.25}$$

  Graphically, the PV curve is a vertical line and so there is no bounded area.

- In an isothermal process, the integral evaluates to (Problem (5.52))

$$W = -Nk_B T \ln (V_2/V_1) = -nRT \ln (V_2/V_1) \tag{5.26}$$

- In an adiabatic process, the integral evaluates to (Problem (5.53))

$$W = \frac{P_2 V_2 - P_1 V_1}{\gamma - 1} \tag{5.27}$$

Finally, what about heat? Here we refer back to the heat capacity defined in Equation (5.12): $Q = C\Delta T$ and the molar heat capacities described in Equations (5.13)–(5.16). For an ideal gas we have:

- In an isobaric process,

$$Q = nC_P \Delta T \tag{5.28}$$

- In an isochoric process,

$$Q = nC_V \Delta T \tag{5.29}$$

- In an isothermal process, $\Delta T = 0$ by definition and because temperature is a measure of the thermal energy, it follows that $\Delta E_{th} = C\Delta T = 0$. Thus from the first law we find

$$Q = -W \tag{5.30}$$

- In an adiabatic process, $Q = 0$ by definition.

Phew! Table 5.2 summarizes these relationships. We've gone through quite a bit of fundamental theory here; let's apply these ideas in some examples. Example 5.6 considers a single thermodynamic process and Example 5.7 considers a complete **thermodynamic cycle**, where a piston goes through a series of steps (always involving work and/or the flow of heat) that eventually returns to its original state. This framework allows us to analyze the functioning of practical thermodynamic cycles such as the **Otto cycle**,[15] which describes most automobile engines (see Problem 5.54).

---

[15]After Nicolaus Otto (1832-1891)

TABLE 5.2
Summary of four main thermodynamic processes

Process	Defining property	Work	Heat
Isobaric	$\Delta P = 0$	$-P\Delta V$	$nC_P\Delta T$
Isochoric	$\Delta V = 0$	0	$nC_V\Delta T$
Isothermal	$\Delta T = 0$	$-nRT \ln(V_f/V_i)$	$-W$
Adiabatic	$Q = 0$	$(P_2V_2 - P_1V_1)/(\gamma - 1)$	0

Example 5.6 Compression in a Piston ★★
A cylinder and piston have a cross-sectional radius of 2.00 cm. An object of mass
m sits on the piston, which is initially held in place 8.00 cm above the bottom
of the cylinder. The initial pressure both inside and outside the cylinder is 1.00
atm. The piston is then released and is observed to sink until it is 4.00 cm above
the bottom of the cylinder. Assume that the temperature of the gas inside the
cylinder did not change from its initial value of 3.00×10^2 K. (a) What is the
final pressure inside the cylinder? (b) What is m? (c) Sketch the pressure as a
function of volume as the piston settles into the cylinder. (d) How much energy is
added to (or removed from) the piston via work? (e) How much energy is added
to (or removed from) the cylinder via heat? (f) How do your results from (d)
and (e) compare to the change in gravitational potential energy from the sinking
piston?

For (a), note that if the temperature is fixed, then the right side of the
ideal gas law ($PV = Nk_BT$) is *constant* (we're dealing with an isothermal process). From the description of the problem we see that the volume decreases by
a factor of two, so the pressure must double, from 1.00 atm to 2.00 atm. Formally:

$$P_iV_i = P_fV_f$$

$$P_f = P_i\frac{V_i}{V_f}$$

$$P_f = P_i\frac{\pi r^2 h_i}{\pi r^2 h_f}$$

$$P_f = (1.00 \text{ atm})\left(\frac{0.0800 \text{ m}}{0.0400 \text{ m}}\right)$$

$$P_f = 2.00 \text{ atm}$$

For (b), note that when the piston has sunk to its new position, it is in equilibrium.
The forces on it include the gravitational force, a force from the atmospheric pressure
due to the air in the room (this accounts for a *downward* force) and a force from
the pressure we just identified inside the cylinder (this accounts for an *upward*
force).

With this in mind, Newton's Second Law says

$$\vec{F}_{\text{NET}} = m\vec{a}$$

$$P_{\text{in}}A - P_{\text{out}}A - mg = 0$$

$$\frac{(P_{\text{in}} - P_{\text{out}})A}{g} = m$$

$$\frac{\left(1.00 \text{ atm}\left(\frac{1.01\times10^5\,\text{Pa}}{1\,\text{atm}}\right)\right)\left(\pi\,(0.0200 \text{ m})^2\right)}{9.8\text{m/s}^2} = m$$

$$13.0 \text{ kg} = m$$

How can we sketch the pressure in the cylinder as the piston sinks (part (c))? Well, we've already reasoned that $P_f = P_iV_i/V_f$, so here

$$P = \frac{P_iV_i}{V} = (1.0 \text{ atm})\left(\frac{1.01 \times 10^2 \text{ cm}^3}{V_f}\right)$$

where V varies from $V_i = \pi r^2 h_i = 1.01 \times 10^2$ cm^3 (when $h = h_i = 8.00$ cm) to $V = 5.03 \times 10^1$ cm^3 (when $h = h_f = 4.00$ cm). The graph of this relationship is shown in Figure 5.9.

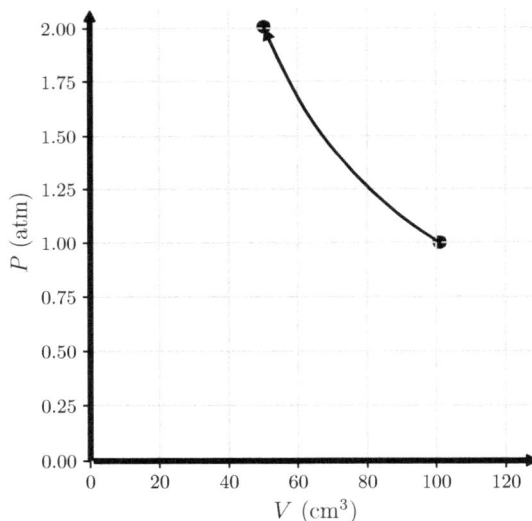

FIGURE 5.9

For (d), we'd like to apply Equation (5.26):

$$W = -nRT\ln(V_2/V_1)$$

but we don't have n, the number of moles of gas in the cylinder. Never fear, we can apply the Ideal Gas Law:

$$PV = nRT$$

$$\frac{PV}{RT} = n$$

$$\frac{\left(1.01 \times 10^5 \text{ Pa}\right) \left(1.01 \times 10^{-4} \text{ m}^3\right)}{\left(8.314 \text{ J K}^{-1} \text{ mol}^{-1}\right) \left(3.00 \times 10^2 \text{ K}\right)} = 4.07 \times 10^{-3} \text{ mol}$$

Thus armed, we can determine the work done:

$$W = -nRT \ln\left(V_2/V_1\right)$$
$$W = -\left(4.07 \times 10^{-3} \text{ mol}\right) \left(8.314 \text{ J K}^{-1} \text{ mol}^{-1}\right) \left(3.00 \times 10^2 \text{ K}\right) \ln\left(0.5\right)$$
$$W = 7.04 \text{ J}$$

We see that the work is positive (energy is *added to the cylinder*), as it should be whenever the volume is decreasing.

For (e), we turn to Equation (5.30), which tells us $Q = -W$ in an isothermal process. Therefore we have $Q = -7.04$ J; heat *leaves* the system.

Finally, for (f) we note that the gravitational potential energy of the piston decreased as it sunk:

$$\Delta U = mg\Delta y = (13.0 \text{ kg}) \left(9.8 \text{m/s}^2\right) (-0.04 \text{ m}) = -5.08 \text{ J}$$

Is this all self-consistent? Let's summarize all of the work done as the piston descends. Gravity does 5.08 J of work on mass m. Meanwhile, the air outside the piston does work because it is applying downward pressure as the piston moves down ($PA\Delta h = 5.08$ J, as you can check), and the air inside does *negative* work because it is applying an upward pressure as the piston moves down (7.04 J, as we calculated above).

This gives a discrepancy of 3.11 J: this must either be kinetic energy (so the mass moves down past equilibrium), work due to friction (as the piston rubs against the cylinder to keep a tight seal and prevent gas from escaping), or a mixture of both.

Example 5.7 Thermodynamic Cycles: PV Diagram ★★

Figure 5.10 shows a complete thermodynamic cycle for a sealed container holding an ideal gas with a moveable piston. Suppose that the gas is N_2, for which $C_V = \frac{5}{2}R$ and $C_P = \frac{7}{2}R$, and that the temperature of the gas at point A is 5.00×10^2 K. For each of the four "strokes" ($A \rightarrow B, B \rightarrow C, C \rightarrow D, D \rightarrow A$) determine the energy and heat that is added to or removed from the system. Show that energy is conserved for the cycle, i.e. $\Delta E_{\text{cycle}} = 0$.

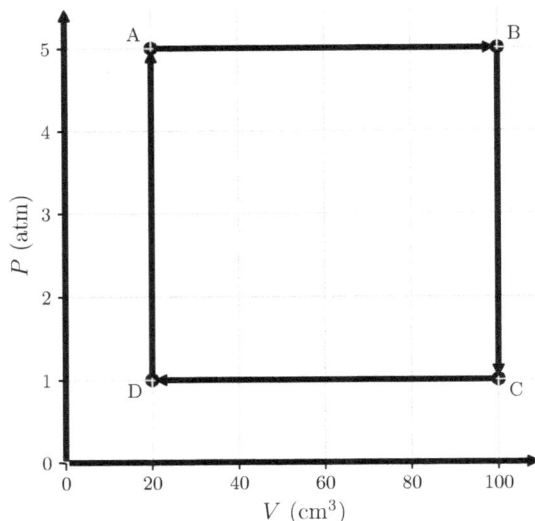

FIGURE 5.10

To begin, note that we know the pressure, volume, and temperature at point A, so we can readily apply the Ideal Gas Law to determine *how much* gas we have:

$$PV = nRT$$
$$n = \frac{PV}{RT}$$
$$n = \frac{(5.0)\left(1.01 \times 10^5 \text{ Pa}\right)\left(20.0 \times 10^{-6} \text{ m}^3\right)}{\left(8.314 \text{ J K}^{-1} \text{ mol}^{-1}\right)\left(5.00 \times 10^2 \text{ K}\right)}$$
$$n = 2.4 \times 10^{-3} \text{ mol}$$

We'll tuck this away for a moment and consider the work. Two of the strokes ($D \rightarrow A$ and $B \rightarrow C$) are isochores and therefore involve no work (Equation (5.25)). The other two strokes are isobars, so from Equation (5.24) we have $W = -P\Delta V$. Applied to this system:

$$W_{AB} = -\left(5.0 \text{ atm}\right)\left(100.0 - 20.0\right) \text{ cm}^3$$
$$= -\left(5.0\left(1.01 \times 10^5 \text{ Pa}\right)\right)\left(80.0 \times 10^{-6} \text{ m}^3\right)$$
$$= -40. \text{ J}$$

Notice that the work here is negative (the gas is doing work; energy is *leaving* the system) because the volume is increasing. For the other isobar:

$$W_{CD} = -\left(1.0 \text{ atm}\right)\left(20.0 - 100.0\right) \text{ cm}^3$$
$$= -\left(1.0\left(1.01 \times 10^5 \text{ Pa}\right)\right)\left(-80.0 \times 10^{-6} \text{ m}^3\right)$$
$$= 8.1 \text{ J}$$

The magnitude of the total work is the area bounded within the cycle. We see here an example of the general trend that a clockwise cycle yields an overall negative work done (the system does net work); for counter-clockwise cycles the net work is positive.

We now turn to the heat, where we will need to apply Equations (5.29) and (5.28). To do so, though, we need to know the temperature of the gas at each state; we can apply Equation (5.20). For an isobaric process this reduces to

$$\frac{V_1}{T_1} = \frac{V_2}{T_2} \implies T_2 = \frac{V_2}{V_1}T_1$$

Similarly, for an isochoric process we have

$$T_2 = \frac{P_2}{P_1}T_1$$

Thus for the stroke $A \to B$ we have

$$T_B = \frac{100.0 \text{ cm}^3}{20.0 \text{ cm}^3}\left(5.00 \times 10^2 \text{ K}\right) = 2.50 \times 10^3 \text{ K}$$

Following the same procedure for points C and D, we find (check it!) $T_C = 5.00 \times 10^2$ K and $T_D = 1.00 \times 10^2$ K.

Now we are equipped to apply (5.29) and (5.28):

$$Q_{AB} = nC_P\Delta T_{AB} = n\left(\frac{7}{2}R\right)(T_B - T_A) = 1.4 \times 10^2 \text{ J}$$

$$Q_{BC} = nC_V\Delta T_{AB} = n\left(\frac{5}{2}R\right)(T_C - T_B) = -1.0 \times 10^2 \text{ J}$$

$$Q_{CD} = nC_P\Delta T_{AB} = n\left(\frac{7}{2}R\right)(T_D - T_C) = -28 \text{ J}$$

$$Q_{DA} = nC_V\Delta T_{AB} = n\left(\frac{5}{2}R\right)(T_A - T_D) = 20. \text{ J}$$

If you add up the two works (W_{AB} and W_{CD}) we identified above you will find $W_{\text{cycle}} = -32$ J; if you add the heats we found above you will find $Q_{\text{cycle}} = +32$ J. Thus this cycle absorbs heat to do work, and the total energy $\Delta E_{\text{cycle}} = W_{\text{cycle}} + Q_{\text{cycle}} = 0$, as it must for any thermodynamic cycle: we're back where we started.

5.7   Specific Heat and Calorimetry

When we considered the ideal gas law in Section 5.2, we found (Equation (5.10))

$$k_B T = \frac{2}{3}K$$

This equation says that the temperature of an ideal gas is a measure of the average kinetic energy of the gas particles. We also asserted that this concept of some *internal thermal energy* is not limited to gases alone; for instance, molecules in a solid can store thermal energy in the form of *vibrations*. In fact, we've already seen in Equation (5.12) that thermal energy is related to the temperature of an object according to its heat capacity C:

$$\Delta E = Q + W = C\Delta T$$

where C depends on what kind of material we have and *how much* of it we have. In this section, we will explore these ideas in a bit more detail.

Suppose that we would like to characterize just the type of material (e.g. aluminum or lead) without regard to how much of it we have: we can use the material-dependent quantity called the **specific heat capacity** measured in J/(kg K)

$$c = C/m \qquad (5.31)$$

If no work is being done, so $\Delta E_{th} = Q$, then

TABLE 5.3

Specific heat capacities

Material	Specific heat $\left(\frac{\text{J}}{\text{kg K}}\right)$
Aluminum	9.00×10^2
Copper	3.87×10^2
Glass	8.37×10^2
Iron	4.52×10^2
Marble	8.58×10^2
Silver	2.36×10^2
Water (liquid)	4.19×10^3
Water (ice)	2.09×10^3

$$Q = mc\Delta T \qquad (5.32)$$

You can check the units in this equation to see that specific heat is measured in $\frac{\text{J}}{\text{kg K}}$. In words, the specific heat is the amount of heat required to raise the temperature of 1 kilogram of the material by 1 Kelvin. The specific heat for a material is sensitive to a number of factors, including the temperature, pressure, and volume of the material both before and after the heat is added. For this reason, you will often see some variation among specific heats reported in tables of standard values. For the examples and problems in this book, refer to Table 5.3. An interesting application of Equation (5.32) involves the exchange of heat between objects: if we consider a closed system (i.e. we consider everything that gains/loses heat in a process) and no mechanical work is being done, then conservation of energy tells us

$$\sum Q = 0 \qquad (5.33)$$

Analyzing systems in this way is an example of **calorimetry**. The next set of examples demonstrates the idea. Here we are not concerned with *phase changes*, such as ice melting into water, though we address these situations in Problem 5.59.

Example 5.8 Cooling Tea   ⋆

A cup of tea has a mass of 0.250 kg and is initially at a very hot temperature of 70.0°C. Later, the tea is found to have cooled to a more manageable temperature of 50.5°C. Assuming the tea has the same specific heat as water, how much thermal energy did the tea lose to the environment?

This involves a fairly straightforward invocation of Equation (5.32). We will make use of the convenient fact that because any ΔT has the same value when measured in °C and in K–refer to Appendix B if you haven't already!–we can change our units directly from °C to K without actually converting the individual temperatures (check it if you don't believe me!):

$$Q = mc\Delta T$$

$$Q = (0.250 \text{ kg}) \left(4.19 \times 10^3 \frac{\text{J}}{\text{kg K}}\right)(50.5 - 70.0)\text{ K}$$

$$Q = -2.04 \times 10^4 \text{ J}$$

The fact that this value is negative indicates that the tea has *lost* this energy, as indicated in the problem statement.

Example 5.9 Calorimetry: Metal in a Liquid ★★

A 0.300 kg block of some unknown metal is dunked into a pot of boiling water, for which the temperature is held to a constant 100.0°C. The block is then quickly transferred to a 0.300 L container of water at room temperature (20.0°C, say). You measure the temperature of the tap water for the next few minutes and see that it increases to a constant value of 24.2°C. Assuming that no heat is lost to the environment (such as the cup holding the water), what is the specific heat of the metal? Does it seem reasonable to think that the metal is one of those listed in Table 5.3? (Hint: 1 mL of water has a volume of 1 cm^3.)

Let's think through what the problem is describing and add some labels to relevant quantities. When the block is lifted out of the boiling water, it will have a temperature $T_{b,i} = 100.0$°C. When it is dumped into the cup of water, which has temperature $T_{w,i} = 20.0$°C, heat will flow from the block (the hotter object) to the water (the cooler object). Eventually the two equilibrate to the same final temperature $T_{b,f} = T_{w,f} = 24.2°.$[a]

Now the first law of thermodynamics states that energy must be conserved when heat is exchanged between objects, so the heat lost by the block must be equal to the heat gained by the water, for a net change of 0J:

$$Q_b + Q_w = 0$$

Substituting in Equation (5.32) for both heats, we find

$$m_b c_b \Delta T_b + m_w c_w \Delta T_w = 0$$

We can algebraically solve this equation for c_b, but we don't have a numerical value for the mass of the water. We *do* know its volume, though, so we can calculate the mass from its volume and the standard density of water:

$$m_w = \rho_w V_w$$

$$m_w = \left(1000 \frac{\text{kg}}{\text{m}^3}\right) \left(3.00 \times 10^2 \text{ mL} \times \frac{1 \text{ cm}^3}{1 \text{ mL}} \times \left(\frac{1 \text{ m}}{100 \text{ cm}}\right)^3\right)$$

$$m_w = 0.300 \text{ kg}$$

In the second term on the right side of the second line, I have explicitly written out the conversion from mL to cm^3 to m^3.

With this value in hand, we can return to our energy conservation equation and solve it for c_b. As in the preceding example, we don't actually need to convert the temperatures to K because we only invoke them when computing a *difference* in temperatures:

$$m_b c_b \Delta T_b + m_w c_w \Delta T_w = 0$$
$$m_b c_b \Delta T_b = -m_w c_w \Delta T_w$$
$$c_b = \frac{-m_w c_w \Delta T_w}{m_b \Delta T_b}$$
$$c_b = \frac{-(0.300 \text{ kg}) \left(4.19 \times 10^3 \frac{\text{J}}{\text{kg K}}\right) (24.2 - 20.0) \text{ K}}{(0.300 \text{ kg}) (24.2 - 100.0) \text{ K}}$$
$$c_b = 232 \frac{\text{J}}{\text{kg K}}$$

If we refer back to our table of specific heats, we see that we are awfully close to the value for silver!

---

^aRealistically we would expect the objects to slowly cool; if you've ever held your hand over a hot cup of tea, you know heat escapes into the air. Here, though, we're assuming this doesn't happen, which is a reasonable assumption as long as the equilibration happens fairly quickly.

5.8   The Second Law of Thermodynamics

We've already seen that the Second Law of Thermodynamics has much to say about how devices such as refrigerators work. To introduce the law, let's return to the example of two gases, each consisting of N particles, inside a partitioned container (Figure 5.2 on page 263). We made the intuitive argument that the gases would be evenly mixed if the partition was removed and later reinserted. Let me now make the same argument in a slightly different way: if we count up *all* the possible configurations of the gas particles (*"This* set on the left and *that* set on the right" or "*this other* set on the left and *that other* set on the right"...), we will see that the *overwhelming majority* correspond to the even mixing we described before. Thus, if the final state of the gases is chosen randomly from all possibilities, we are almost certain to find the gases have mixed evenly.

This probabilistic statement *is* the second law of thermodynamics. We will define it in more precise language, but let's first justify the argument we just made – that nearly all of the gas configurations correspond to even mixing. To begin, consider this question: if you were free to individually assign each particle to sit in either the left side or the right side of the box when we reinsert the partition, then how many ways could you do it? Well, the first particle you consider can be assigned to the left side or the right side (L or R for short), and so too can the second particle, for a total of $2 \times 2 = 4$ combinations for those two particles (LL, LR, RL, and RR). If we have a total of $2N = M$ particles, then there are 2^M distinct assignments you may make.[16] If you decide which assignment to make for each particle by flipping a coin, then we're equivalently talking about all the possible sequences of M heads and tails that you can get (with, say, all heads corresponding to putting all of the particles in the left side of the box).

We refer to a particular configuration as a **microstate**. If $M = 6$, We might write down the "every particle ends up on the left side" microstate as "LLLLLL" and one possible "even split" microstate as "LLLRRR". How many microstates correspond to the case where *one* particle ends up on the right side, and the remaining five end up on the left? The answer, of course, is *six*: any of the particles could be the one chosen to be on the right side. We refer to this collection of "five on the left, one on the right" microstates as a **macrostate**. What about the macrostate where we choose *two* of the six particles to end up on the right? This is not as obvious, but we should expect to find more than six, since there are multiple combinations of positions we might consider. Indeed, if you count them up, you'll see that the answer is 15. The "3 macrostate" has even more, at 20, but then the number decreases; the final three macrostates – for 4, 5, and 6 of the particles being assigned to the right side of the box – have 15, 6, and 1 microstates, respectively.[17] Evidently, once you move beyond

---

[16]When you consider that a typical container of gas can have something like $M = 10^{25}$ particles, then it should be apparent that 2^M is, technically speaking, *unimaginably big*.

[17]You can add all of these up to confirm that we have a total of $2^6 = 64$ microstates.

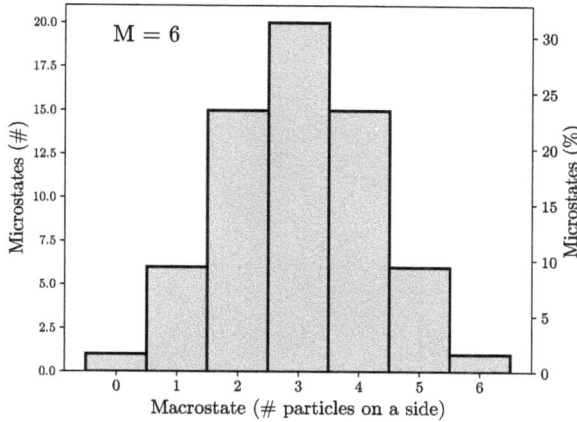

FIGURE 5.11

The distribution of microstates in each macrostate when $M = 6$. The horizontal axis labels the macrostates according to how many particles are in one side (the right side, say – though the distribution is symmetric so the image would be the same if we thought about the left side instead). The left vertical axis identifies the raw number of microstates in the corresponding macrostate; the right vertical axis indicates what percentage this is out of the total number of macrostates ($2^6 = 64$).

an even split of 3 to a side, the number of microstates begins to decrease. We show this distribution graphically in Figure 5.11.

By the way, the general equation for determining how many ways you can choose k spots out of M possibilities is

$$\binom{M}{k} = \frac{M!}{k!\,(M-k)!} \tag{5.34}$$

The left side of the equation is read "M choose k", and the exclamation point indicates a factorial.[18]

We have seen that when $M = 6$, the macrostate corresponding to an even split between left and right is the most likely if you choose randomly among all possible microstates: it corresponds to 20 of the 64 microstates, more than any other macrostate. The flip side of that argument is that there is a $\frac{44}{64} \approx 69\%$ chance that you will pick *any other* macrostate. If M increases, however, this changes; *most* of the microstates become concentrated in and near the "even split" macrostate (see Figure 5.12). In other words, if M is an absurdly large number, as it is in the case of mixing gases, then randomly choosing a microstate is essentially guaranteed to correspond to the "even split" macrostate or one very close to it.

To sum up, we have shown that when choosing randomly among all possible configurations (or *microstates*) of the gas particles being in the left or right side of the box when the partition is reinserted, it is probabilistically likely that we will find very close to half of them on one side and half on the other (the "even split" *macrostate* or a macrostate very close to it). We could go through a similar procedure to formally show that you're also almost certain find both sides will contain an even mixture of particles that were originally on the left and the right – but instead I'll just point out that this should seem logical enough: we're talking about assigning positions based on coin flips, and there is no reason to expect favoritism.

---

[18]To carry out a factorial, multiply the number by every smaller integer, down to 1: $N! = N \times (N-1) \times (N-2)\ldots 2 \times 1$. So for instance $4! = 4 \times 3 \times 2 \times 1 = 24$.

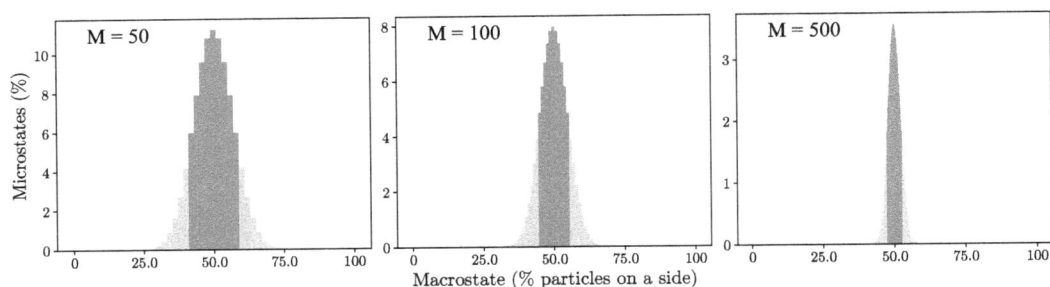

FIGURE 5.12
The distribution of microstates in each macrostate for $M = 50$, $M = 100$, and $M = 500$. The horizontal axes indicate the macrostates according to the percentage of all particles that exist in one side of the container. The vertical axes identify the percentage of the microstates that exist in the corresponding macrostate. The darker shading of the bars near the center of the distributions represent the "90% confidence interval": there is a 90% chance that a randomly selected microstate will exist in one of the shaded macrostates. As M increases, this region becomes more tightly clustered around the "even split" macrostate, and the probability of getting a macrostate on the extreme ends of the ranges becomes vanishingly small.

Thus, the intuitive fact that the gases mix evenly can be framed as a statement of probability: left to its own devices, the gas goes to the *most probable configuration*. With this in mind, we can now define the **second law of thermodynamics** formally as follows:

> Isolated systems move toward the macrostate with the most microstates.

The second law of thermodynamics is sometimes called the *law of entropy*. We won't deal with entropy in a mathematical sense; for our purposes it suffices to point out that it is a measure of how many microstates correspond to a given macrostate. As we have seen, the macrostate describing a well-mixed system has a large number of microstates, so it is common (although ultimately imprecise) to think of entropy as a measure of how *disordered* (i.e. mixed) the system is. With this in mind, we can equivalently write the second law of thermodynamics as

> The entropy of an isolated system will never decrease.

I want to stress the *isolated system* component of these definitions. It is emphatically *not* true that entropy never decreases under any circumstances. For instance, in a refrigerator, the cold reservoir experiences a *decrease* in entropy, but the hot reservoir experiences an *increase* in entropy. The second law of thermodynamics says that the gain from the hot reservoir cannot be smaller than the loss from the cold reservoir, so the *total* effect is an increase in entropy, or, in the limiting case, no change at all.[19]

This bears directly on a common argument, which you may have heard, involving the second law of thermodynamics and the natural process of evolution. We will take this up in the next example before applying the second law more thoroughly to heat engines and heat pumps.

---

[19]It would be fair to describe heat pumps and heat engines as devices that make entropy decrease in some region of interest – but always at the cost of a net increase when considering the larger system.

Example 5.10 Evolution and the Second Law ⋆

Some skeptics of evolution raise an argument that can be summarized as follows:

- According to evolution, life becomes increasingly complex.

- According to the second law of thermodynamics, everything moves from order to disorder.

- Therefore, if science has proven the second law of thermodynamics (which it has), then evolution cannot be real.

This argument is flawed. Why?

The second point says that "everything" tends to disorder (i.e. increasing entropy), but the second law applies only to an *isolated system*. As we just argued in the main text, the entropy of the cold reservoir in a refrigerator can decrease, but the entropy of the hot reservoir (the room in which the refrigerator is located) increases *more*, so the net effect is an increase in entropy.

This happens in a biological context, even setting aside evolution. For instance, a protein can fold into a particular three-dimensional shape that involves a *decrease* in the protein's entropy. The net effect of this procedure, however, involves a comparatively large *increase* in the entropy of the protein's environment, so there is no violation of the second law of thermodynamics.

Thus, it is not necessarily the case that there is a fundamental violation of physics if a cell generates a lower entropy structure, such as a folded or entirely new (and evolutionarily favorable) protein.

Let us now turn to the topic that brought us into this section: the constraints that the second law places on heat engines and heat pumps. We will skip some of the mathematical details and simply assert the following relationship between the energies involved in these devices, Q_H and Q_C, and the temperatures of the reservoirs, T_H and T_C:

$$\frac{Q_C}{T_C} \geq \frac{Q_H}{T_H} \tag{5.35}$$

The equality is only met in the ideal case where operating the engine involves *no increase in entropy*. This simply can't be done in practice: even if the inner workings of the engine proceed with no increase in entropy, the act of exchanging heat with the reservoirs must involve a net increase in entropy.[20] We can therefore use this inequality to characterize the limits of real heat engines and heat pumps.

Consider first the heat pump we described in Example 5.4, where we described the efficiency of the engine as

$$\eta = \frac{W}{Q_H} \tag{5.36}$$

or equivalently

$$\eta = 1 - \frac{Q_C}{Q_H} \tag{5.37}$$

---

[20]I apologize that I am simply asking you to take my word on this and asserting Equation (5.35) seemingly out of the blue, but doing justice to the mathematical details would take more time and effort than I want to devote to this topic. As with any topic that piques your interest, I invite you to do some additional reading elsewhere... and if you continue your formal education in physics you will likely take at least one course (or the equivalent) in *thermodynamics*.

Rearranging Equation (5.35), we see

$$\frac{Q_C}{Q_H} \geq \frac{T_C}{T_H}$$

and so it follows from Equation (5.37) that the efficiency of a heat engine obeys

$$\eta \leq 1 - \frac{T_C}{T_H} \tag{5.38}$$

where the equality is only ever met in the ideal case; it can never be achieved in practice.

We can perform a similar analysis for the coefficient of performance of a heat pump from the perspective of the reservoir being cooled (Example 5.5), which can be written

$$\mathrm{COP}_{\mathrm{cooling}} = \frac{Q_C}{W} \tag{5.39}$$

or equivalently

$$\mathrm{COP}_{\mathrm{cooling}} = \frac{Q_C}{Q_H - Q_C} \tag{5.40}$$

$$= \left(\frac{Q_H - Q_C}{Q_C}\right)^{-1}$$

$$= \left(\frac{Q_H}{Q_C} - 1\right)^{-1}$$

Again invoking Equation (5.35), it can be shown (check it yourself) that

$$\mathrm{COP}_{\mathrm{cooling}} \leq \frac{T_C}{T_H - T_C} \tag{5.41}$$

where, again, the equality is only met in the ideal scenario.

Similarly, if the reservoir of interest is the one that is warmed, we have

$$\mathrm{COP}_{\mathrm{heating}} = \frac{Q_H}{W} \tag{5.42}$$

$$\mathrm{COP}_{\mathrm{heating}} = \frac{Q_H}{Q_H - Q_C} \tag{5.43}$$

and the same approach yields

$$\mathrm{COP}_{\mathrm{heating}} \leq \frac{T_H}{T_H - T_C} \tag{5.44}$$

The following examples demonstrate how these ideas can be used to further characterize how these devices work.

Example 5.11 Efficiency of a Power Plant   ⋆

You are researching generators powered by utility boilers for your employer. One in particular uses coal to generate steam at a temperature of 5.40×10^2 degrees Celsius and a water supply at 20.0 degrees Celsius for cooling. The manufacturer claims that the generator can operate at up to 70% efficiency. Can this be true?

If we convert to Kelvin and invoke Equation (5.38), we see

$$\eta_{\text{max}} = 1 - \frac{T_C}{T_H}$$

$$= 1 - \frac{293 \text{ K}}{813 \text{ K}}$$

$$= 0.640$$

Thus the theoretical maximum efficiency of this generator is just 64%, and the manufacturer's claim is in violation of physical (and perhaps civil) law.

Example 5.12 Coefficient of Performance of a Heat Pump ★

A geothermal heat pump can be used during the winter months to transfer heat from deep underground into a house. Suppose the underground temperature is 53 degrees Fahrenheit and the desired temperature of the house is 65 degrees Fahrenheit. Because the heat pump is being used as a heater rather than a refrigerator, here we are concerned with $\text{COP}_{\text{heating}} = \frac{Q_H}{W}$. What is the maximum COP of a heat pump operating in these conditions?

Here it is important to note that a heat pump draws energy out of the cold reservoir and dumps it into the hot reservoir (see Example 5.5), so in this case the ground is the cold reservoir and the house is the hot reservoir.

With this in mind, we convert to Kelvin use Equation (5.44) to find

$$\text{COP}_{\text{max}} = \frac{T_H}{T_H - T_C}$$

$$= \frac{291 \text{ K}}{(291 - 285) \text{ K}}$$

$$= 48.5$$

Keep in mind that the COP is equivalent to $\frac{Q_H}{W}$, which is to say the ratio between the heat pumped into the house and the energy you provide to the pump (in the form of electricity). So, the argument here is that we can drive 48.5 Joules of heat into the house for every 1 Joule of electricity used by the pump.

For a host of reasons, however, COP values achieved in practice are generally much less than their theoretical limits; a typical value is 3.5.

Example 5.13 Violations of the 1st and 2nd Laws ★

Which, if any, of the heat engines proposed below violate the first law of thermodynamics? Which, if any, violate the second law of thermodynamics? Assume all values are exact.

(a) $T_C = 300$ K, $T_H = 800$ K, $Q_C = 50$ J, $Q_H = 100$ J, $W = 70$ J

(b) $T_C = 290$ K, $T_H = 750$ K, $Q_C = 35$ J, $Q_H = 115$ J, $W = 80$ J

(c) $T_C = 285$ K, $T_H = 650$ K, $Q_C = 74$ J, $Q_H = 116$ J, $W = 42$ J

The first law deals with energy conservation: the total energy put in to the engine must equal the total energy put out of the engine. In a heat engine, Q_H is the input and Q_C and W are output, so we expect $Q_H = Q_C + W$ (see Example 5.4 for a refresher). Checking the given values, we see that this equality is only met in (c), so both (a) and (b) are in violation of the first law of thermodynamics.

The second law deals with limitations that arise from the reservoir temperatures. For a heat engine, we refer to Equation (5.38):

$$\eta_{max} = 1 - \frac{T_C}{T_H}$$

Inserting our values, we find that for (a), $\eta_{max} = 0.625$; for (b), $\eta_{max} = 0.613$; and for (c), $\eta_{max} = 0.562$. Now, what are the *actual* efficiencies of these engines? Well, from Equation (5.36) we have

$$\eta = \frac{W}{Q_H}$$

We find that for (a), $\eta = 0.70$; for (b), $\eta = 0.70$; and for (c), $\eta = 0.36$. Comparing to the theoretical maximum values, we see that engines (a) and (b) violate the second law and that engine (c) does not.

Taken together, only engine (c) is in agreement with both laws and could correspond to a real heat engine. (You could also calculate η with $\eta = (Q_H - Q_C)/Q_H$. Because (a) and (b) violate the *first* law, you'll get different efficiencies here, which underscores how nonsensical it is for an *actual* heat engine to violate the first law.)

5.9   Problems for Chapter 5

5.2 The Ideal Gas Law

★   **Problem 5.1.** "Standard Temperature" is 273.15 K. What is this temperature in (a) degrees Celsius and (b) degrees Fahrenheit?

★   **Problem 5.2.** The Celsius scale is defined such that 0°C is the temperature at which water freezes and 100°C is the temperature at which water boils. What are both of these temperatures in (a) ° F and (b) K?

★   **Problem 5.3.** A student uses a bicycle pump to inflate a bike tire. Describe how each term in the ideal gas law changes (thinking of inside the tire as the container) during this process. If any terms are ambiguous, explain why.

★   **Problem 5.4.** How many moles is 10.0 g of (a) oxygen and (b) helium gas? (c) What volume would each gas occupy at STP?

★   **Problem 5.5.** Pure helium fills a leak-proof cylinder. The volume, pressure, and temperature of the gas are 15.0 L, 2.00 atm, and 300.0 K.

  a. How many helium atoms are there?

  b. What number of moles of helium are there?

  c. What is the number density, N/V?

  d. What is the total mass of the gas?

★ **Problem 5.6.** Consider a small room that is 8.0 feet tall and is both 10.0 feet long and wide. If the room is filled with an ideal gas at STP, how many particles are there? How many moles is this?

★ **Problem 5.7.** Two identical cylinders contain the same type of gas. The pressure is the same in both cylinders, but cylinder A has twice as many molecules of gas as cylinder B. Are the temperatures the same? If not, which cylinder has the higher temperature? Can you explain your result in terms of what is happening in the cylinders?

★★ **Problem 5.8.** A sealed piston is filled with air and set outside. The circular piston has a mass of 15 g and a cross-sectional area of 2.0×10^{-3} m^2. In the morning, when the air temperature is 10.°C, the height of the air column in the piston is 5.0 cm. What height should you measure in the afternoon if the air temperature is 20.°C?

★ **Problem 5.9.** A thin spherical balloon is initially inflated to a radius of 8.0 cm. It (and the surrounding air) is at 1.0 atm and the temperature is 293 K. For each process described below, determine P, V, n, and T inside the balloon.

a. The balloon is then slowly inflated until its radius is 12.0 cm.

b. Then, the balloon is placed in a freezer at $T = 0°C$ for a long time.

★ **Problem 5.10.** A vertical tube is sealed with a cap on the bottom and a movable piston at the top. It is filled with an ideal gas at STP and has a length $y = h$.

a. If the piston is compressed slowly (so the temperature of the gas doesn't change), what will the pressure be when $y = h/2$?

b. Generalize your answer to (a) by sketching the pressure as a function of the length for $0 < y < h$.

★★ **Problem 5.11.**

a. Use Equation (5.9) to determine the kinetic energy of an ideal gas particle at 3.00×10^2 K.

b. Suppose diatomic nitrogen (consisting of two nitrogen atoms, each with 7 neutrons and 7 protons) has this kinetic energy. What is its speed?

c. If the nitrogen molecule is traveling horizontally with the initial speed you found in (a), how far will it drop due to gravity as it crosses a room of width 12 m?

★ **Problem 5.12.** You tap your finger against a table in a steady rhythm: you apply 5.0 N of force for 0.20 s, pause for 1.0 s, then repeat the process. What is the average force applied to the table?

★★ **Problem 5.13.** You have a bucket of a dozen 0.250 kg rubber balls. You throw them, one at a time, against a wall. If you throw one ball every 3.0 s, they collide elastically with the wall at an impact velocity of 19 m/s, and each impact lasts 0.30 s, then what average force does the wall experience from impact to impact?

★★ **Problem 5.14.** To get a sense for how tightly packed ideal gas particles are in a typical gas, suppose you have a cubical box that measures 1.0 m to a side. The box is filled with an ideal gas at STP.

a. What is N, the number of gas particles in the box?

b. What is the volume *per gas particle*, V/N?

c. If you could place each particle in its own little container with the volume you found in (b), you'd fill up the entire box. If this tiny container is also in the shape of a cube, what should the edge length be?

d. Compare the edge length you found in (c) to the radius of a typical gas molecule, which is about 3.0×10^{-10} m (The precise value depends on how you define the radius and what gas you're talking about). Which is larger? By what factor?

5.3 Thermal Equilibrium

★ **Problem 5.15.** A sealed and insulated container has a total volume of 3.50×10^2 cm^3. It contains a gas at 293 K and 1.20 atm. If the gas was previously sealed in one side of the container with a volume equal to 1/3 the total volume of the container, then what was (a) the temperature and (b) the pressure of the gas at that time?

★★ **Problem 5.16.** A sealed and insulated container has a total volume of 2.00×10^2 cm^3. The container is split into two equal volume halves by a removable membrane. Initially, the left side contains 0.020 mol of nitrogen gas at 25°C and the right side contains 0.020 mol of nitrogen gas at 35°C.

   a. What is the pressure on each side of the membrane?

   b. The membrane is removed and the gas particles mix. What is the final temperature?

   c. What is the final pressure?

★★ **Problem 5.17.** Consider a gas molecule of mass m traveling with a velocity $v_1\hat{x}$ when it experiences a head-on collision with a gas molecule, also of mass m, with initial velocity $-v_2\hat{x}$. Use Equations (4.26) and (4.27) to determine the final velocity of each particle.

5.5 The First Law of Thermodynamics

★ **Problem 5.18.** 0.500 kJ of work is done on a system in a process that decreases the system's thermal energy by 0.200 kJ. How much heat energy is transferred to or from the system?

★ **Problem 5.19.** 0.050 mol of oxygen gas is initially at 290 K. If the gas absorbs 240 J of heat at (a) constant volume or (b) constant pressure, what will its final temperature be?

★ **Problem 5.20.** A 1.20 kg block of wood has a heat capacity of 1.80×10^3 J/K and the coefficient of friction between the block and the floor is 0.400. If the wood is initially at a temperature of 22.0°C, what temperature will it have after it is pushed 2.50×10^2 m over level ground? (In practice the thermal energy will be concentrated at the bottom of the block – here you will be determining the *average* temperature.)

★ **Problem 5.21.** With what speed do have to slap a 180 g chicken breast to raise its temperature from 2.0°C to 74°C (refrigerated temperature to cooked)? Assume all of your hand's kinetic energy is transferred to the chicken as thermal energy,[21] that your hand and arm have a mass of 4.0 kg, and that the heat capacity of the chicken is 605 J/K.

★★ ∫ **Problem 5.22.** In Example 5.3 we considered the (silly) case of converting gravitational potential energy to cook 1.0 kg of frozen beef. Repeat the exercise, but this time consider the gravitational potential energy using Newton's Law of Universal Gravitation rather than the "constant field" assumption $U_g = mgh$. (The result of Problem 4.58 will be helpful.)

★ **Problem 5.23.** What is the efficiency of a heat engine that draws 0.100 kJ out of the hot reservoir, converts 0.040 kJ to work and deposits 0.060 kJ into the cold reservoir?

★ **Problem 5.24.** What is the coefficient of performance of a refrigerator that uses 0.040 kJ of work to pull 0.060 kJ out of the cold reservoir and deposit 0.100 kJ into the hot reservoir?

★ **Problem 5.25.** What is the efficiency of a heat engine that draws 50.0 J of heat from the hot reservoir and dumps 25.0 J of heat into the cold reservoir every cycle? How much work is done every cycle for this engine?

★ **Problem 5.26.** What is the coefficient of performance for a refrigerator that draws 25.0 J of heat from the cold reservoir and dumps 50.0 J of heat into the hot reservoir every cycle? How much work is required every cycle to accomplish this task?

---

[21] Just to be clear, this is a terrible assumption.

5.6 Thermodynamic Cycles

★ **Problem 5.27.** A gas is at standard temperature is in a sealed container. The pressure increases isochorically from 2.0 atm to 3.0 atm. What is the final temperature in °C?

★ **Problem 5.28.** A gas is in a sealed container with a pressure of 1.0 atm and temperature of 293 K. The volume decreases isobarically from 3.50×10^2 cm^3 to 2.75×10^2 cm^3. What is the final temperatures in °C?

★ **Problem 5.29.** A gas is in a sealed container with an initial pressure of 3.0 atm, volume of 150 cm^3, and temperature 293 K. The volume expands isothermally to 450 cm^3. What is the final pressure?

★★ **Problem 5.30.** A cylinder contains 25 g of helium. How much work is required to compress the gas at a constant temperature of 75°C until the volume is half of its original value?

★★ **Problem 5.31.** A piston contains 0.050 mol of oxygen gas. It expands isobarically from a volume $V_A = 5.0 \times 10^2$ cm^3 to $V_B = 8.0 \times 10^2$ cm^3. The initial temperature is 20°C.

 a. What is P_A? Express your answer in atm.

 b. What is T_B? Express your answer in °C.

 c. How much heat flows during this process? Does it flow into or out of the piston?

 d. How much work was done in this process? Is it being done on the gas or by the gas?

★★ **Problem 5.32.** A container of gas initially has a volume of 200.0 cm^3 and a pressure of 100.0 kPa. The pressure increases linearly to 300.0 kPa while the volume decreases to 100.0 cm^3.

 a. Sketch this process on a PV diagram.

 b. How much work is done on the gas during this process?

 c. If the initial temperature is 25°C, how many moles of gas are in the container?

 d. What is the final temperature of the gas?

 e. Is the heat added in this process positive, negative, or zero? Explain.

★★ **Problem 5.33.** A container holds 0.015 mol of helium. Initially the container has a volume of 100.0 cm^3 and a pressure of 5.0 atm. The pressure decreases linearly to 1.0 atm while the volume increases to 300.0 cm^3. The final temperature is 255 K.

 a. Sketch this process on a PV diagram.

 b. What is the initial temperature of the gas?

 c. What work is done on the gas during this process?

 d. How much heat is added to or removed from the container during this process?

 e. What is the change in thermal energy of the gas?

★★ **Problem 5.34.** A thermal vent at the bottom of a lake emits a small gas bubble filled with nitrogen gas (N_2). The pressure at the bottom of the lake is 3.0 atm and the initial radius of the bubble is R_1. What is the radius (as a multiple of R_1) when the bubble has risen to a depth where the pressure is 1.5 atm? Assume the bubble moves adiabatically, i.e. so rapidly it doesn't have time to exchange heat with the environment.

★★ **Problem 5.35.** A gas cylinder holds O_2 at 150°C, a pressure of 3.25 atm and a volume of 1.50×10^{-3} m^3. The gas expands adiabatically until the pressure is 2.00 atm. What is (a) the final volume and (b) the final temperature?

★ **Problem 5.36.** A cylinder expands isobarically from 1.00×10^2 cm^3 to 3.00×10^2 cm^3 while at 3.00×10^2 kPa. How much work was done on the gas during this process?

★★ **Problem 5.37.** A cylinder contains 0.010 g of helium gas. How much work is required to compress the gas at a constant temperature of 80°C until the volume is one third of its original value?

★★ **Problem 5.38.** 15.0 g of He gas is initially at temperature $T_A = 40°C$ in a container with volume V_A and pressure $P_A = 2.0$ atm. The pressure increases isochorically to $T_B = 620°$ and pressure P_B. Then, the gas expands isothermally to a temperature T_C and pressure $P_C = P_A$.

   a. Sketch the process $A \rightarrow B \rightarrow C$ on a *PV* diagram.

   b. What is V_A?

   c. What is P_B?

   d. What is V_C?

   e. What is T_C?

★★ **Problem 5.39.** A container of gas has an initial pressure of 2.5 atm and an initial temperature of 175°C. It is compressed isothermally to half the original volume. Then the gas is compressed isobarically until the volume is one quarter the original volume.

   a Sketch the process on a *PV* diagram.

   b What is the final pressure of the gas?

   c What is the final temperature of the gas?

★★ **Problem 5.40.** For the process shown in Figure 5.13, what is the ratio $\frac{T_C}{T_A}$?

★★ **Problem 5.41.** A sealed glass bottle at 27.0°C contains a gas at atmospheric pressure and has a volume of 25.0 cm^3. It is tossed into a fire and the temperature inside eventually reaches 200°C.

   a. What type of process is this? Why?

   b. What is the final pressure inside the bottle?

   c. Graph this process on a *PV* diagram.

★★ **Problem 5.42.** Figure 5.14 shows a three-state thermodynamic cycle. At point A, the gas is at 2.0 atm and 200°C. $V_A = 125$ cm^3.

   a. Identify the type of each process $A \rightarrow B$, $B \rightarrow C$, and $C \rightarrow A$.

   b. Will the temperature at C be higher, lower, or the same as the temperature at A?

   c. Determine T_B, V_B, and P_B.

   d. Determine T_C, V_C, and P_C.

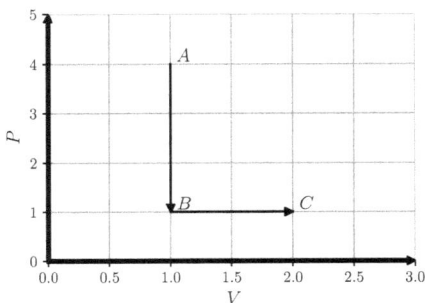

FIGURE 5.13
Problem 5.40

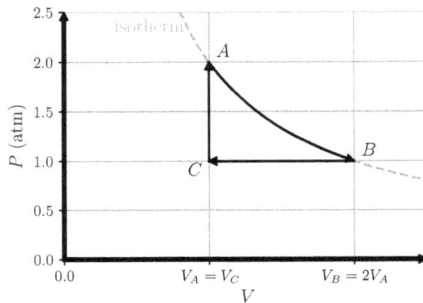

FIGURE 5.14
Problem 5.42

★★ **Problem 5.43.** Figure 5.16 shows a thermodynamic process.

   a. What type of process is this?

   b. What is the work done by the gas?

   c. What would the answer to (b) be if the process was run in reverse ($B \rightarrow A$ instead of $A \rightarrow B$)?

d. Explain the sign (+ or −) of your answers to (b) and (c).

★ **Problem 5.44.** Two identical containers are at equal temperatures and contain the same mass of nitrogen. You supply 25 J of heat to container A while holding the *volume* constant and you supply 25 J of heat to container B while holding the *pressure* constant. After, is the temperature greater in container A, in container B, or are the temperatures the same? Explain.

★★ **Problem 5.45.** A sealed container is initially at a state (V_A, P_A). Later, it is at a state (V_B, P_B) where $V_B > V_A$ and $P_B > P_A$. Consider three ways to move from $A \rightarrow B$: (1) the pressure increases linearly with the volume, (2) the pressure expands isochorically and then the volume expands isobarically, and (3) the volume expands isobarically and then the pressure expands isochorically. (Option 3 is the same as option 2 but with the steps reversed.)

a. Sketch all three routes on a PV diagram.

b. Which route results in the greatest change in temperature ΔT? Explain.

c. Which route results in the greatest amount of heat added to the gas? Explain.

★ **Problem 5.46.** A cylinder contains 10.0 g of Helium, which has a mass of 4.0 g/mol. How much work must be done to compress the gas at a constant temperature of 80°C until the volume is half its original value?

★ **Problem 5.47.** A gas cylinder holds 0.10 mol of O_2 at 150°C and a pressure of 3.0 atm. The gas expands adiabatically until the pressure is halved.

a. What is the initial volume?

b. What are the final volume and temperature?

c. How much work is done?

★★ **Problem 5.48.** Figure 5.15 shows a thermodynamic process for 250 mg of nitrogen gas. $P_A = 1.0$ atm and $P_B = 8.0$ atm.

a. Determine the pressure (in atm), temperature (in °C), and volume (in cm³) for each point A, B, and C. Provide your answers in a table.

b. How much work is done for each stroke $A \rightarrow B$, $B \rightarrow C$, and $C \rightarrow A$?

c. How much heat energy is transferred into or out of the gas during each stroke?

d. Show that the cycle obeys the first law of thermodynamics.

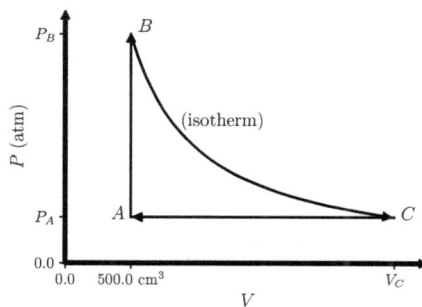

FIGURE 5.15
Problem 5.48

★★ **Problem 5.49.** 3.0 mol of O_2 gas are at 20°C.

a. If 0.600 kJ of heat energy are transferred to the gas at constant pressure, what is the temperature of the gas?

b. Following (a), 0.600 kJ of heat energy are removed at constant volume. What is the final temperature?

c. Sketch a PV diagram for the processes described above.

★   **Problem 5.50.** A gas is brought from state A to state C either *directly* or *indirectly*, via intermediate state B, as shown in Figure 5.17.

   a. Which route has the larger change in thermal energy, or are they the same?

   b. Which route involves more work done to the gas, or are they the same?

   c. Which route involves larger heat transfer, or are they the same?

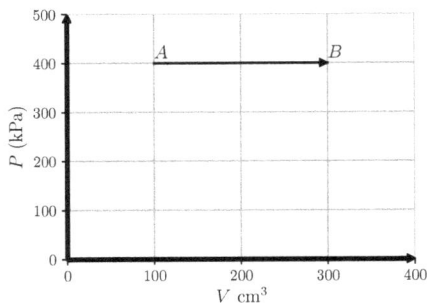

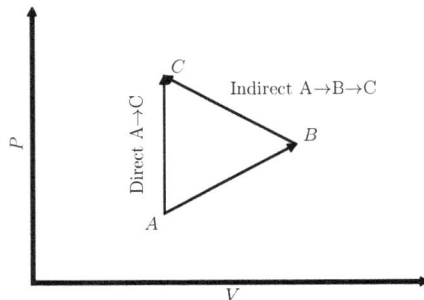

FIGURE 5.16
Problem 5.43

FIGURE 5.17
Problem 5.50

★★★ $\int dx$ **Problem 5.51.** In this problem we will work through the derivation of Equations (5.21) and (5.22) for an adiabatic process.

   a. Demonstrate that for an adiabatic process, the first law of thermodynamics reduces to $\Delta E_{th} = W$, or, in the more precise language of calculus, $dE_{th} = dW$.

   b. For an ideal gas, the thermal energy is stored as kinetic energy. Thus, use Equation (5.10) to show that for a gas of N particles, $dE_{th} = \frac{3}{2} N k_B dT$.

   c. Combine (a) and (b) with the definition of work (Equation (5.23)) to show $\frac{3}{2} N k_B dT = -P dV$.

   d. Use your result from (c) and the ideal gas law to show $\frac{3}{2} \frac{dT}{T} = -\frac{dV}{V}$.

   e. Integrate both sides of the equation from (d); you start from an initial temperature and volume and end at a final temperature and volume. After some algebra and appropriate use of the ideal gas law, show that you've arrived at Equations (5.21) and (5.22) for a monatomic gas. (See the footnote surrounding Equation (5.10) for a comparison to the diatomic case.)

★★ $\int$ **Problem 5.52.** Integrate Equation (5.23) in the isothermal case and show that Equation (5.26) follows.

★★★ $\int$ **Problem 5.53.** Integrate Equation (5.23) in the adiabatic case and show that Equation (5.27) follows.

★★★   **Problem 5.54.** The Otto cycle models the thermodynamic cycle in most internal combustion engines. From an initial state (P_1, V_1) the gas undergoes the following processes:

   • an adiabatic compression to a state (P_2, V_2) where $P_2 > P_1$ and $V_2 < V_1$,

   • an isochoric increase in pressure to state $(P_3, V_3 = V_2)$,

   • an adiabatic expansion to a state $(P_4, V_4 = V_1)$, and

   • an isochoric decrease in pressure to return to the initial state.

We are after the efficiency of the Otto cycle. To obtain it, consider the following procedure:

   a. Sketch the cycle on a PV diagram.

   b. Determine the total work of the cycle.

   c. What is the heat absorbed during the first isochoric process? (This corresponds to the heat added to the system).

   d. Show that the efficiency is given by

$$\eta = 1 - \left(\frac{V_2}{V_1}\right)^{\gamma - 1}$$

5.7 Specific Heat and Calorimetry

⋆ **Problem 5.55.** The same amount of heat Q is added to two objects with the same mass. if $\Delta T_1 > \Delta T_2$, which specific heat (c_1 or c_2) is larger?

⋆ **Problem 5.56.** The specific heat of a substance depends on the *phase* of the substance. For instance, for water

$$c_{\text{ice}} = 2090 \frac{\text{J}}{\text{kg K}}$$

$$c_{\text{liquid}} = 4190 \frac{\text{J}}{\text{kg K}}$$

$$c_{\text{steam}} = 2009 \frac{\text{J}}{\text{kg K}}$$

For a fixed mass of water and a fixed heat Q to be added to the water, which phase of liquid will have the largest change in temperature?

⋆ **Problem 5.57.** You mix 0.50 kg of 100°C water with 1.0 kg of 5.0°C water in a well-insulated cup. Assuming no heat is lost to the cup or surrounding air, what is the water's equilibrium temperature? (The specific heat of water is provided in Problem 5.56.)

⋆ **Problem 5.58.** A 2.0 kg aluminum sphere has been heated to 300°C and is dropped into a pool filled with 35 kg of water that is initially at 20°C. Assuming that no heat is lost to the environment, what will the equilibrium temperature of the water and lead sphere be?

⋆⋆ **Problem 5.59.** Our discussion of the heat necessary to change the temperature of an object did not encompass phase changes. If you have water at 0°C, for instance, and would like to extract enough heat to make it freeze, then you need to know the **latent heat** of fusion, L_f. Given a mass m, the heat needed to melt or freeze the substance is

$$Q = \pm m L_f \tag{5.45}$$

where the $+$ is used for heat flowing in to the system (to melt the substance) and the $-$ is used for heat flowing out of the system (to freeze the substance). Similarly, for a substance at the boiling/condensation point one needs to know the latent heat of vaporization, L_v:

$$Q = \pm m L_v \tag{5.46}$$

For water,

$$L_f = 3.33 \times 10^5 \text{ J/kg} \tag{5.47}$$

$$L_v = 2.26 \times 10^6 \text{ J/kg} \tag{5.48}$$

With all this in mind, respond to the following:

a. How much ice at 0°C must be added to 1.0 kg of water at 100°C to end up with all of the liquid at 20°?

b. You have 1.0 kg of ice at 0.0°C. The ice is heated until it melts, raised to 100.0°C, and evaporates into water vapor. Assuming the system is isolated, which step requires the most energy?

⋆ **Problem 5.60.** How much heat energy must be added to 0.25 kg of ice at 0.0°C to melt it and then raise the temperature to 20°C? (Refer to Problem 5.59).

⋆⋆ **Problem 5.61.** Iced coffee is generally made by brewing hot coffee and then mixing it with ice. Suppose 240 cm³ (about 1 cup) of hot coffee is brewed at a temperature of 70°C and mixed with ice at a temperature of 0°C. Treat the coffee like water (so for instance the density is 1000 kg/m³) and note that the density of ice is 920 kg/m³.

a. What is the mass of the coffee?

b. How much heat must be extracted from the coffee to cool it to a temperature of 4.0°C?

   c. What mass of ice must be provided if no energy is lost to the environment and the final temperature of the melted ice and coffee is to be 4.0°C? (Refer to Problem 5.59)

   d. Use your result from (c) to find the initial volume of the ice in cm^3. Compare to the volume of the coffee by reporting V_{ice}/V_{coffee}. Interpret your answer: should your iced coffee have more coffee or more ice, measured by volume? (Though note here we're assuming you wait to drink the coffee until the ice has completely melted!)

★★   **Problem 5.62.** A cylindrical pond is 5.0 m in diameter and 1.50 m deep. Energy from the sun is added at an average rate of 4.00×10^2 W/m^2. How many hours will it take to warm the water from 52°F to 68°F if the water absorbs all of the solar energy and doesn't exchange any energy with its surroundings?

5.8 The Second Law of Thermodynamics

★   **Problem 5.63.** You flip a fair coin 5 times.

   a. How many outcomes (microstates) are there?

   b. How many ways can you get 2 heads?

   c. How many ways can you get 2 tails?

   d. What number(s) of heads is/are most likely?

★★   **Problem 5.64.** You roll a 6 sided die two times (in the lingo of games that use dice rolls, this is a "2d6".

   a. How many outcomes (microstates) are there?

   b. If all you're concerned with is the sum of the two rolls, then what are all of the possibilities? What is the probability of achieving each?

   c. Repeat (a) and (b) for one roll of a 12 sided die (A "1d12"). Which configuration is more likely to yield a 12?

★   **Problem 5.65.** A power plant uses steam at 280°C to generate electricity. It uses a nearby lake as the cold reservoir.

   a. What is the maximum possible efficiency of the power plant if the lake's temperature is 1.0°C?

   b. What is the maximum possible efficiency of the power plant if the lake's temperature is 20.0°C?

★   **Problem 5.66.** A heat engine is operating between reservoirs at 20.0°C and 620°C. If it has 25% of its maximum possible efficiency, how much energy does this engine extract from the hot reservoir to do 1.0 kJ of work?

★   **Problem 5.67.** What is the maximum possible coefficient of performance of a freezer operating at 0°C in a 20°C room?

★   **Problem 5.68.** Is a heat pump's coefficient of performance larger when the temperature difference between the reservoirs is large or small? Support your answer with a sketch of the COP's dependence on temperature (fix one reservoir temperature and consider different temperatures for the other reservoir).

★   **Problem 5.69.** A heat engine is proposed to operate between reservoirs at 7.00×10^2 K and 3.00×10^2 K. A designer argues that every cycle the engine will draw 0.500 kJ from the hot reservoir, do 0.250 kJ of useful work, and deposit 0.250 kJ in the cold reservoir. Does this engine violate (a) the first and/or (b) the second law of thermodynamics?

★★   **Problem 5.70.** A heat engine operates between reservoirs at 6.00×10^2 K and 3.00×10^2 K. For each theoretical heat engine cycle described below, describe if the engine violates the first law of thermodynamics, the second law of thermodynamics, both, or neither. Assume the given heats are exact.

a. 50 J are drawn from the hot reservoir, 30 J are used as useful work, and 20 J are deposited into the cold reservoir.

b. 50 J are drawn from the hot reservoir, 25 J are used as useful work, and 15 J are deposited into the cold reservoir.

c. 30 J are drawn from the hot reservoir, 10 J are used as useful work, and 20 J are deposited into the cold reservoir.

⋆⋆ **Problem 5.71.** A power plant generates steam at 425°C. The cold reservoir is a nearby river at 20°C. If the heat generated by burning coal is 85 MJ, what is the minimum amount of heat that must be transferred to the river?

⋆ **Problem 5.72.** Two containers of water are connected by a heat pump such that one is used as the hot reservoir and the other is used as the cold reservoir. If the hot reservoir is to be maintained at a piping-hot 80.°C and the cold reservoir is to be maintained at a cool 10.°C, then what is the maximum coefficient of performance for the process of heating the hot reservoir?

⋆ **Problem 5.73.** A tropical ocean's surface temperature is about 30.°C. Deep below the surface, the temperature is about 5°C. What is the maximum possible efficiency for an engine that uses the ocean at these temperatures as the reservoirs for a heat engine?

⋆ **Problem 5.74.** The temperature a few meters below the surface of the earth is fairly constant year round. The exact temperature depends on where you are on the Earth, but suppose it is about 55°F near where you live. Consider a geothermal heat pump that heats your home during winter by using the earth as the cold reservoir and the inside of your house as the hot reservoir. What is the maximum possible coefficient of performance of such a heat pump if the inside of your house is to be warmed to 68°F?

6

Simple Harmonic Motion and Waves

Learning Objectives

After reading this chapter, you should be able to:

- Recognize the relationship between the "sine wave" and the right-triangle definition of the sine function.

- Determine if an object will experience simple harmonic motion.

- Characterize the motion of an object experiencing simple harmonic motion in terms of:

  - forces

  - energy conservation

  - The equation of motion (Equation (6.4))

  - a position-time motion graph

- Analyze the behavior of traveling wave pulses (e.g. down a taut string) when interacting with other pulses or a hard barrier (e.g. wall).

- Describe the behavior of traveling waves with Equations (6.12) and (6.13).

- Analyze the wave speed of a traveling wave.

- Explain how two traveling waves can interfere so as to generate a standing wave.

- Characterize standing waves in terms of the "boundary conditions" of a string or tube (e.g. two fixed ends as with a guitar string).

6.1 Introduction

In our study of kinematics we found that, given enough information about the kinematic variables (for instance, the acceleration along with the initial position and velocity), we could determine an equation for the position of an object at every subsequent instant in time. We sometimes refer to such an equation as *the position as a function of time*, written $x(t)$. As you may recall, the kinematic equations we developed only apply to cases where the acceleration is constant, so we can't use them to determine $x(t)$ for cases where the net force is *not* constant, like when an object is connected to a spring (where the force depends on how extended/compressed the spring is from its equilibrium length).

While you hopefully agree that what we've done since studying the kinematic equations has been interesting and useful, we haven't actually discussed how to obtain $x(t)$ for a

DOI: 10.1201/9781003571568-8

non-constant net force; this is (in part) because doing so is generally quite difficult. In this chapter, however, we will discuss one particular case where we can determine $x(t)$, called *simple harmonic motion*. While simple harmonic motion is well worth our attention on its own merits, a major motivation for studying it is that the language of simple harmonic motion is the language of *waves*, which come up time and time again in the study of physics. In particular, a major theme of Parts IV and V of this book is the study of *light*; we will see that in many ways *light is a wave*. Thus this chapter will set the stage for much of the rest of this book.

6.2 Two Seemingly Different Versions of the Sine Function

So far in this book, when we have used $\sin\theta$ it has been in the context of *right triangle trigonometry* (as shown in the left panel of Figure 6.1). However, you may have some experience with the sine function depicted as in the right panel of Figure 6.1. We will be using both versions of the sine function in this chapter, and before we delve into the physics it is worth our time to explain how these seemingly different definitions of the sine function are related.

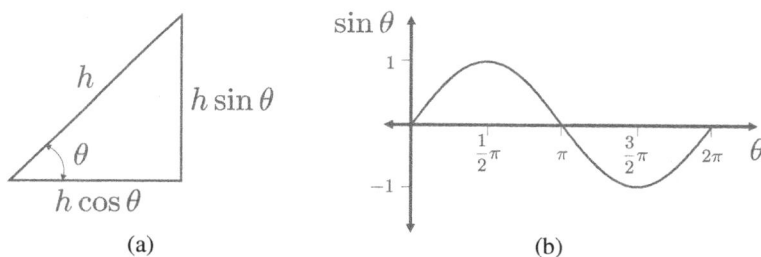

(a) (b)

FIGURE 6.1
Two common ways of representing the sine function. (a) In a right triangle, $\sin\theta$ is equal to the ratio of the side of the triangle opposite θ and the hypotenuse (here shown with a magnitude h). In this book we have often thought of the hypotenuse as a vector, so the "opposite" component can be expressed as $h\sin\theta$ ($h\cos\theta$ is the "adjacent" component). (b) Represented graphically as a function of θ, $\sin\theta$ oscillates between -1 and 1 with a period of 2π. In Figure 6.2 the equivalency of these two representations is depicted.

To begin, recall that we've spent quite a bit of time discussing objects moving in circular paths. Consider such an object moving at an angular speed ω on the unit circle (i.e. with a radius of motion equal to 1). If, every so often, we compute its vertical displacement and plot this against the total time elapsed, we obtain a sine curve (Figure 6.2). As we shall see, this mathematical relationship is very useful in discussing simple harmonic motion.

6.3 Simple Harmonic Motion

As we argued in the introduction to this chapter, **simple harmonic motion** (sometimes abbreviated *SHM*) occurs when an object experiences a particular kind of non-constant net force. Let us begin by describing the net force that leads to SHM:

An object will experience simple harmonic motion if the net force

1. is one-dimensional

2. points toward a position of stable equilibrium

3. grows in magnitude linearly with the distance from equilibrium

This may sound familiar: Hooke's Law ($\vec{F}_{\text{sp}} = -k\left(\Delta x\right)\hat{x}$) fits this description.[1] When we initially introduced Hooke's Law in Chapter 3, we were primarily concerned with questions of equilibrium, like "if a block of mass m is hung from a spring with spring constant k, how far will it stretch from its original length?" Now, however, we will consider *motion* that arises from Hooke's Law: suppose we have a mass attached to a spring on a flat, frictionless surface, pull it to the side, then release it.

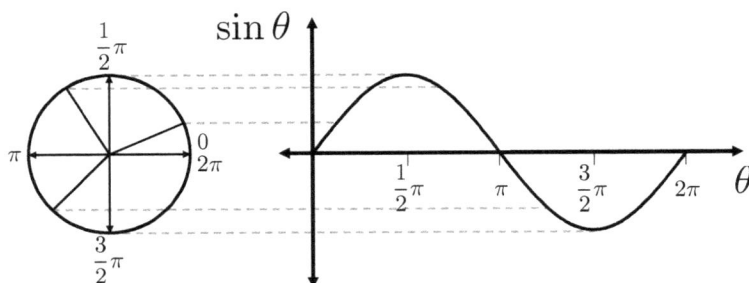

FIGURE 6.2
If we think of the $h\sin\theta$ component of a right triangle as the vertical coordinate of a point on the edge of the unit circle (where $h = 1$), then plotting that vertical coordinate as a function of θ generates the sine curve shown on the right panel of Figure 6.1. The gray dashed horizontal lines connect a few points on the circle and graph. (Recall that on the unit circle, the angle θ is measured counter-clockwise from the right side of the horizontal axis, and a complete circle contains 2π radians.)

In fact, we've already considered this situation in terms of *energy*, in Example 4.5 on page 212. As we argued in that example, if we pull the mass to the side we store spring potential energy in the mass-spring system. When the mass is released, that energy is converted to kinetic energy as the spring force pulls the mass back toward equilibrium. *At* equilibrium, all of the energy has become kinetic energy, so the mass moves on past the equilibrium position until all the energy has converted to spring potential energy again (this time with the mass an equal distance on the other side of the equilibrium position from where it began). It then begins to move back toward equilibrium in a cycle that, in the absence of friction (or other resistive forces like air resistance), will repeat forever. This process is summarized in Figure 6.3.

While we have a firm grasp on what is happening in this situation in terms of the cyclic energy transfer, we do not have a complete description of exactly where the object will be in its cycle at any given instant. To move toward obtaining the function $x(t)$ for the mass, we

---

[1]We first defined Hooke's Law in Equation (3.12) on page 138.

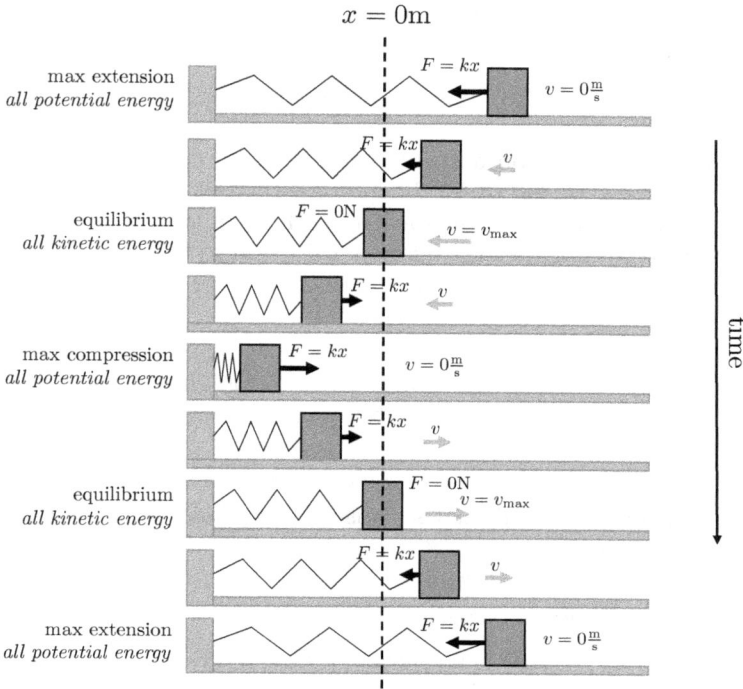

FIGURE 6.3

The motion of a mass attached to a spring on a flat, frictionless surface. The mass is originally pulled to the right from equilibrium ($x = 0$ m, marked with the vertical dashed line), then released. It picks up speed as the spring force pulls it back toward equilibrium, then passes through the equilibrium point and compresses the spring to the left a distance equal to its original extension to the right. The motion then repeats in reverse, and it ends up just where it began (and then the entire process begins again). This cyclic motion is an example of simple harmonic motion.

will turn to Newton's Second Law. When the block is at an arbitrary position x, we have

$$\vec{F}_{\text{NET}} = m\vec{a}$$
$$F_{\text{sp}} = ma$$
$$-kx = ma$$

where we have set $x_{eq} = 0$ m. It is possible to determine $x(t)$ from this equation, but first we will re-write the above equation in a general way that applies to *all* SHM rather than just to a mass on a spring:

$$a(t) = -\omega^2 x(t) \tag{6.1}$$

Here we are noting explicitly that both a and x are changing with time. ω is some constant that is often called the **angular frequency**, though we shall see that it is really the same thing as the angular speed from Section 2.3 (which is why we use the same symbol in both cases). The value of ω is context-dependent; evidently for a mass on a spring

$$\omega_{\text{spring}} = \left(\frac{k}{m}\right)^{\frac{1}{2}} \tag{6.2}$$

Now, if we recall that the acceleration is the second derivative of position with respect to time, we could just as well write Equation (6.1) as

$$\frac{d^2 x(t)}{dt^2} = -\omega^2 x(t) \tag{6.3}$$

An equation involving a function and its own derivative is referred to as a differential equation; here we have a *second order* differential equation because we have the second derivative of our function. There is a mathematical theorem that states that if we can come up with two different solutions to this equation, call them $p(t)$ and $q(t)$, then *any* function $x(t)$ that satisfies the equation can be expressed as a sum of the independent solutions:

$$x(t) = Ap(t) + Bq(t)$$

Here, A and B are constants that vary from solution to solution. So the question is this: can we think of two different functions that satisfy Equation (6.3)? Well, we came across one of them back in Example 1.10:

$$p(t) = C_1 \sin(\omega t)$$

If you've studied calculus, it should not surprise you if I point out that a second equation that meets our requirements is

$$q(t) = C_2 \cos(\omega t)$$

Putting these solutions together, our general solution is

$$x(t) = C_1 \sin(\omega t) + C_2 \cos(\omega t)$$

If we use some trig identifies and relabel some of the constants (see Problem 6.72), this equation becomes

$$x(t) = A\sin(\omega t + \phi) + x_0 \tag{6.4}$$

Moreover, it is a straightforward application of differential calculus (see Problem 6.27) to show that

$$v(t) = A\omega \cos(\omega t + \phi) \tag{6.5}$$

and

$$a(t) = -A\omega^2 \sin(\omega t + \phi) \tag{6.6}$$

Equation (6.4) is the general solution to Equation (6.1) that we've been after.[2] Before we discuss the meaning of the parameters in this equation, it is worth pointing out that this is why we introduced the graphical depiction of the sine function at the beginning of this chapter: instant by instant, an object experiencing SHM traces out a sine curve (Figure 6.4).

Let us now describe the variables in Equation (6.4). Here A is the **amplitude**: the maximum distance the object travels from equilibrium (for a mass on a spring, this is the maximum compression/extension, as in Figure 6.3). For a mass on a spring the value of ω is given by Equation (6.2), but for all types of SHM we can also relate it to the **period of oscillation** (or simply the **period**) T, i.e. the time for a complete cycle:[3]

$$\omega = \frac{2\pi}{T} \tag{6.7}$$

---

[2]We could use a *cosine* in place of a *sine* if we wished; the consequence is just that the value of ϕ would be different. See Problem 6.26.

[3]You may recall that we already discussed the period of cyclic motion in the context of circular motion; the meaning is the same here.

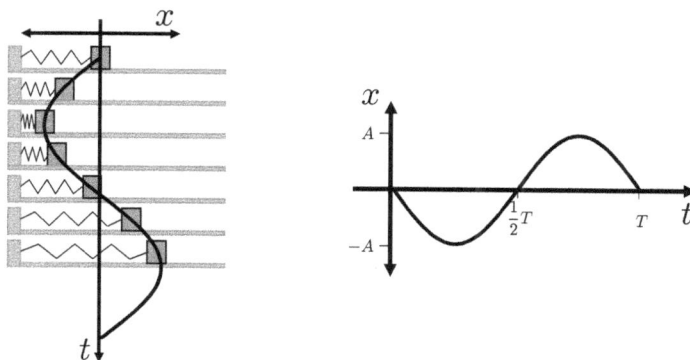

FIGURE 6.4
As a mass on a spring undergoes simple harmonic motion, its position vs. time motion graph creates a sine curve. We show the (a) overlap directly, then (b) rotate the sine curve to the usual orientation (with time on the horizontal axis and position on the vertical axis).

The meaning of ω is the same as angular velocity, as shown in Figure 6.2: given ω, after some amount of time t the object will have traveled an angular distance $\omega t = \theta$, which we can visualize as either an angular distance around a circle or a horizontal distance on a motion graph.[4]

The variable ϕ in Equation (6.4) is called the **phase shift**. This variable simply accounts for *when we start observing the motion*. In Figure 6.3 the first instant drawn is the instant of maximum extension, but in some cases we might begin looking at some other position, such as when the object is passing through equilibrium and traveling in the negative direction. (Coincidentally, this is how Figure 6.4 is drawn![5]) Graphically, ϕ acts as a horizontal shift of the sine function relative to the instant when $t = 0$.

Now, if there can be a *horizontal* shift, there can also be a *vertical* shift. Physically, this means that the oscillation is around some point other than $x = 0$. For instance, if a mass is hanging vertically from a spring and experiencing simple harmonic motion, we might record the oscillation with a sensor on the ground, below the mass. Reasonably enough, the sensor likely sets the $x = 0$ point as on the sensor itself, so if the mass oscillated between, say, 0.9 m and 1.1 m above the sensor, we'd say it is oscillating about the point $x_0 = 1.0$ m with an amplitude of $A = 0.1$ m. Figure 6.5 shows some sine curves with key parameters labelled.

Before moving on to some examples demonstrating applications of these ideas, let's introduce one more quantity of interest. While the period answers, "how long does it take for a complete cycle?", the **frequency** answers, "how many cycles do we observe in every second?" Mathematically the frequency f is simply the inverse of the period:

$$f = \frac{1}{T} \tag{6.8}$$

Combining this with Equation (6.7), we have

$$\omega = \frac{2\pi}{T} = 2\pi f \tag{6.9}$$

Frequency is measured in inverse seconds, or Hertz ($1\frac{1}{s} = 1$ Hz).[6]

---

[4]Of course it is important to keep in mind that in SHM, nothing is *actually* moving in a circle! The point is that because an object undergoing SHM moves sinusoidally, we can make use of the mathematics we used to describe circular motion, as summarized in Figure 6.2 and Equation (6.7). This can be confusing to digest at first, but recognizing the connection between circular motion and SHM will help you to improve your mastery of both topics.

[5]It isn't really a coincidence; I did this on purpose.

[6]After Heinrich Hertz (1857-1894).

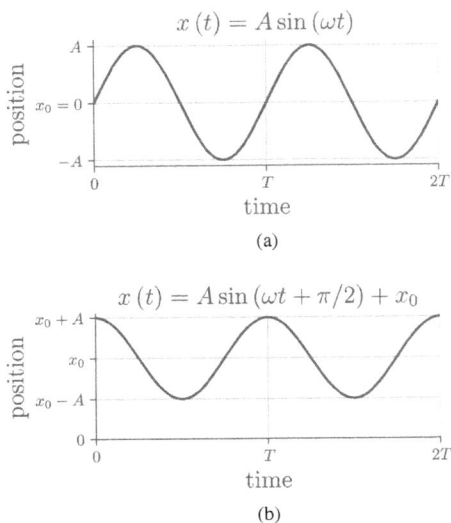

$$x(t) = A \sin(\omega t)$$

(a)

$$x(t) = A \sin(\omega t + \pi/2) + x_0$$

(b)

FIGURE 6.5
(a) A sine curve where $\phi = 0$ and $x_0 = 0$ is shown with the amplitude A (the maximum distance from equilibrium) and period T (the time required for a complete cycle) labelled. (b) A sine curve with $x_0 > 0$ and $\phi = \pi/2$. Note that at $t = 0$ the function evaluates to $x(0) = \sin(\pi/2) + x_0$, which effectively shifts the top curve to the left by $\frac{1}{4}T$ (a complete oscillation corresponds to a phase shift of 2π) and up by x_0.

Example 6.1 Oscillation Frequency ★
A 5.00 kg mass is attached to a spring with a spring constant $k = 35.0\frac{\text{N}}{\text{m}}$ and set on a flat, frictionless surface (as we sketched in Figure 6.3). It is then pulled to the side a distance A and released. How long will it take for a complete cycle to occur if (a) $A = 3.00$ cm or (b) $A = 5.00$ cm? Conversely, how many cycles occur every second in each case?

The time to complete one cycle is the period, T. If we note $\omega = \left(\frac{k}{m}\right)^{\frac{1}{2}}$ from Equation (6.2) and insert into Equation (6.7), we can solve for T:

$$\omega = \frac{2\pi}{T}$$

$$\left(\frac{k}{m}\right)^{\frac{1}{2}} = \frac{2\pi}{T}$$

$$T = \frac{2\pi}{\left(\frac{k}{m}\right)^{\frac{1}{2}}}$$

$$T = \frac{2\pi}{\left(\frac{35.0\frac{\text{N}}{\text{m}}}{5.00 \text{ kg}}\right)^{\frac{1}{2}}}$$

$$T = 2.38 \text{ s}$$

Note that this does not depend on the amplitude (a somewhat surprising result, perhaps!), so this is our answer for both (a) and (b). If we want instead the *frequency*, we can simply note

$$f = \frac{1}{T} = \frac{1}{2.38 \text{ s}} = 0.421 \text{ Hz}$$

Alternatively, we could have solved for the frequency first by noting $\omega = 2\pi f$, then determined the period by noting $T = \frac{1}{f}$. The answer is the same either way, as you can check for yourself!

Example 6.2 Describing Simple Harmonic Motion ★★
A 0.40 kg mass and spring system obeys the equation

$$x(t) = (6.4 \text{ cm}) \sin\left(\left(5.9\frac{\text{rad}}{\text{s}}\right)t - 0.45 \text{ rad}\right)$$

(a) What is the amplitude of oscillation for this mass?

(b) What is the period of oscillation for this mass?

(c) How much energy is stored in this system?

(d) When the position of the mass is 30% of its amplitude, what is its speed?

Equation (6.4) provides the general form of $x(t)$ for SHM:

$$x(t) = A \sin(\omega t + \phi)$$

By comparing to the given equation, we see that the amplitude is $A = 6.4$ cm. The period is not explicitly given in this equation, but we see $\omega = 5.9\frac{\text{rad}}{\text{s}}$, so

$$\omega = \frac{2\pi}{T}$$
$$T = \frac{2\pi}{\omega}$$
$$T = \frac{2\pi}{5.9\frac{\text{rad}}{\text{s}}} = 1.1 \text{ s}$$

This takes care of parts (a) and (b). For (c) we need to determine the total energy in the system. In general there is a mix of spring potential and kinetic energy, so if the total energy is denoted E_{TOT}, we can say

$$E_{\text{TOT}} = K + U_{\text{sp}}$$

We know that when the spring is maximally extended (i.e. when $x(t) = A$) all of the energy is spring potential energy,[a] and when it is passing through equilibrium (i.e. when $x(t) = 0$ m) all of the energy is kinetic energy.[b] Thus when we are concerned with determining the total energy it is often most convenient to consider one of these two cases. Here we have already determined the amplitude, so let's consider when the energy is all spring potential:

$$E_{\text{TOT}} = \frac{1}{2}kA^2$$

To proceed we need to know k, which is not given in the problem. However, we have m and ω, so we can invoke Equation (6.2):

$$\omega_{\text{spring}} = \left(\frac{k}{m}\right)^{\frac{1}{2}}$$

$$m\omega_{\text{spring}}^2 = k$$

$$0.40 \text{ kg} \left(5.9\frac{\text{rad}}{\text{s}}\right)^2 = k = 14\frac{\text{N}}{\text{m}}$$

Inserting into our energy equation (and making sure we convert our units to SI), we have

$$E_{\text{TOT}} = \frac{1}{2}kA^2$$

$$E_{\text{TOT}} = \frac{1}{2}\left(13.9\frac{\text{N}}{\text{m}}\right)(0.064 \text{ m})^2$$

$$E_{\text{TOT}} = 2.9 \times 10^{-2} \text{ J}$$

Finally, for part (d) we will return to the general expression for the energy in the system, and insert the condition that $x = 0.3A$:

$$E_{\text{TOT}} = K + U_{\text{sp}}$$

$$E_{\text{TOT}} - U_{\text{sp}} = K$$

$$E_{\text{TOT}} - \frac{1}{2}kx^2 = \frac{1}{2}mv^2$$

$$\left(\frac{2}{m}\left(E_{\text{TOT}} - \frac{1}{2}k\left(0.3A\right)^2\right)\right)^{\frac{1}{2}} = v$$

$$0.36\frac{\text{m}}{\text{s}} = v$$

---

[a]This is analogous to throwing a ball up and considering the instant of greatest height: at the instant between when it is moving up and when it begins falling, $v = 0\frac{\text{m}}{\text{s}}$ and therefore it has no kinetic energy.

[b]This is analogous to the ball initially leaving your hand (or the instant when the ball falls back to it).

So far we have considered only one example of SHM (a mass connected to a spring), but SHM occurs in completely different circumstances, as well. Here we will consider just one other example: consider a small mass suspended from a horizontal support by a light string, as shown in Figure 6.6. This setup is referred to as a **simple pendulum**. Clearly the mass will remain hanging vertically if undisturbed (it is a position of *stable equilibrium*); if it is instead pushed a small distance to either side, it will sway back and forth in simple harmonic motion.

To prove this, we will analyze Newton's Second Law for the mass (sometimes called a *bob* in this context) in much the same way that we analyzed the motion of the mass and spring. First, note that the best choice of coordinate system here is polar coordinates: if we place the origin at the support point, we see that the tension force is always in the $-\hat{r}$ direction, and we can split the force of gravity in to a component in the $\hat{r}$ direction and a component in the $\hat{\theta}$ direction that always points toward the position of stable equilibrium. (Note also that while we normally refer to "horizontally to the right" as the position where

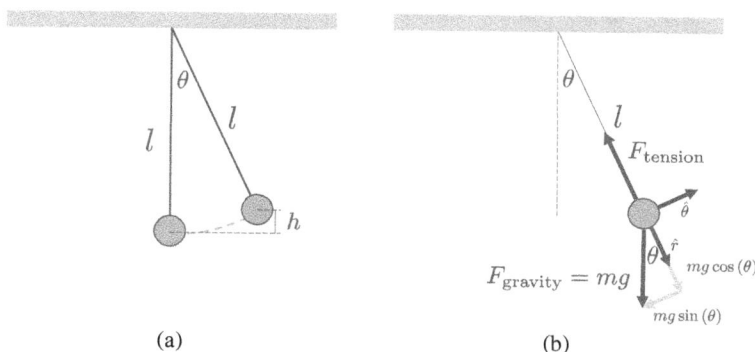

(a)   (b)

FIGURE 6.6

A simple pendulum consists of a string fixed at one end and attached to a small mass that hangs from the other end. (a) If the mass is moved to the side so the string forms an angle θ with the vertical, then the mass will move along the arc of a circle and end up a distance h above the point of stable equilibrium. (b) The same situation is shown with the forces drawn and broken into their (polar) components.

$\theta = 0$ rad, here it makes sense to call "straight down" this position, so the bob will oscillate around $\theta = 0$ rad.)

Let's consider each requirement for SHM in turn:

1. Is the net force one-dimensional? We see from the force diagram that the net force in the $\hat{r}$ direction is 0 N,[7] so the net force only has a component in the $\hat{\theta}$ dimension.

2. Does the force always point toward a position of stable equilibrium? The position of stable equilibrium is $\theta = 0$ rad; if the bob is at some other position, the $\hat{\theta}$ component of gravity will always point back toward equilibrium.

3. Does the force grow linearly with position? Here we must assume that θ is very small, so $\sin\theta \approx \theta$ and $F_\theta \approx mg\theta$ (see Problem 6.40). In this case the net force does grow linearly with the (angular) position. (If we don't make this assumption we have the so-called **large angle pendulum**, which is much harder to analyze mathematically.)

All that remains is to determine how the angular frequency ω relates to the parameters of the simple pendulum (recall that $\omega_{\text{spring}} = \left(\frac{k}{m}\right)^{\frac{1}{2}}$). To be explicit, let's write out Newton's Second Law; because we are dealing with polar coordinates we will work with the rotational version:

$$\vec{\tau}_{\text{NET}} = I\vec{\alpha}$$
$$-(mg\theta)\, l\hat{\theta} = \left(ml^2\right)\alpha\hat{\theta}$$
$$-\frac{g}{l}\theta = \alpha$$

To more directly compare this to the mass on a spring example, let's consider the linear variables x and a by recalling $x = r\theta$ and $a = r\alpha$ (where r is the radius of motion that we are here calling l, the length of the string). Thus we have

$$-\frac{g}{l}x = a$$

---

[7] This is analogous to a block on a ramp, where the net force perpendicular to the surface is 0 N because the normal force perfectly balances the force(s) applied directly into to the surface. Here the tension force plays the role of the normal force by balancing the component of the gravitational force that is directed along the length of the string (i.e. in the $\hat{r}$ direction).

Compare this to Equation (6.1), $a(t) = -\omega^2 x(t)$, and we see that for a pendulum,

$$\omega_{\text{pendulum}} = \left(\frac{g}{l}\right)^{\frac{1}{2}} \tag{6.10}$$

As the next example demonstrates, much of what we considered when analyzing the mass on a spring carries over directly to a simple pendulum, so long as you note the different definition for ω. Indeed, this is part of the beauty of simple harmonic motion: the context may change, but the underlying mathematics does not.

Example 6.3 The Simple Pendulum   ★★

A grandfather clock measures time by the swaying of a pendulum.

(a) How long must a simple pendulum be so its period of oscillation is 1.00 seconds?

(b) If this simple pendulum was taken to the moon, how long would the period of oscillation be? (Hint: it may be helpful to refer back to Example 3.7 on page 116.)

(c) Back on Earth, suppose the pendulum is pulled $+10.0°$ from equilibrium and, at $t = 0$ s, it is released. Write down the equation of motion $x(t)$ for the pendulum.

(d) Continuing from part (c): how fast will the pendulum be moving (in m/s) as it passes back through equilibrium?

For (a), we know that in general $\omega = \frac{2\pi}{T}$. If we combine this with Equation (6.10), we find

$$\left(\frac{g}{l}\right)^{\frac{1}{2}} = \frac{2\pi}{T}$$

$$\frac{g}{l} = \left(\frac{2\pi}{T}\right)^2$$

$$g\left(\frac{T}{2\pi}\right)^2 = l$$

$$9.8\frac{\text{m}}{\text{s}^2}\left(\frac{1.0 \text{ s}}{2\pi}\right)^2 = l = 0.248 \text{ m}$$

For (b), we can start with the same equation as (a), but now solve for T:

$$\left(\frac{g}{l}\right)^{\frac{1}{2}} = \frac{2\pi}{T}$$

$$T = 2\pi\left(\frac{l}{g}\right)^{\frac{1}{2}}$$

Now, we also need to recall that $g = 9.8\frac{\text{m}}{\text{s}^2}$ *on Earth*. Thus to proceed we need to calculate g for the Moon. If, as the problem suggests, we refer back to Example 3.7 (and part (a) in particular), we see that from Newton's Law of Universal Gravitation, the acceleration due to gravity for an object near the surface of some planetary body of mass m and radius r is

$$g = G\frac{m}{r^2}$$

If we insert the mass and radius of the Moon (see the Appendix), we find

$$g_{\text{moon}} = \left(6.67 \times 10^{-11} \frac{\text{m}^3}{\text{kgs}^2}\right) \frac{7.35 \times 10^{22} \text{ kg}}{(1.74 \times 10^6 \text{ m})^2} = 1.62 \frac{\text{m}}{\text{s}^2}$$

Returning to the equation for the period of oscillation for the pendulum, we find

$$T = 2\pi \left(\frac{l}{g}\right)^{\frac{1}{2}}$$

$$T = 2\pi \left(\frac{0.248\text{m}}{1.62\frac{\text{m}}{\text{s}}}\right)^{\frac{1}{2}}$$

$$T = 2.46 \text{ s}$$

Thus we see that when the acceleration due to gravity is *smaller*, the period of oscillation becomes *larger*.

Moving on to part (c), we need to determine A, ω, and ϕ for the general equation of motion for SHM: $x(t) = A\sin(\omega t + \phi)$. The *angular* amplitude is

$$\theta_{\text{max}} = 10.0° \left(\frac{\pi \text{ rad}}{180°}\right) = 0.174 \text{ rad}$$

And so the *linear* amplitude (the maximum arc length along the circular path of the bob) is

$$A = l\theta_{\text{max}} = (0.248 \text{ m})\, 0.174 \text{ rad} = 0.043 \text{ m}$$

The premise of the problem is that $T = 1.00$ s, so

$$\omega = \frac{2\pi}{T} = 6.28 \frac{\text{rad}}{\text{s}}$$

To determine ϕ, consider that at $t = 0$ s, we have

$$x(0 \text{ s}) = A\sin(\omega(0 \text{ s}) + \phi) = A\sin(\phi)$$

Here, however, we know that $x(0 \text{ s}) = A$, which means

$$x(0 \text{ s}) = A = A\sin(\phi)$$

$$1 = \sin(\phi)$$

$$\sin^{-1}(1) = \phi$$

$$\frac{\pi}{2} \text{ rad} = \phi$$

Putting this all together, we see

$$x(t) = (0.043 \text{ m})\sin\left(6.28\frac{\text{rad}}{\text{s}}t + \frac{\pi}{2} \text{ rad}\right)$$

Finally, for part (d), we can invoke conservation of energy. If we set the position of no gravitational potential energy to be the lowest position of the bob, then initially the energy is $U_G = mgh$, and as it passes through equilibrium we have $K = \frac{1}{2}mv^2$. Setting these equal, we find

$$mgh = \frac{1}{2}mv^2$$

$$(2gh)^{\frac{1}{2}} = v$$

The problem is that we don't have an expression for h. However, we can determine this with l and θ, as shown in Figure 6.7. Inserting the expression for h, we find

$$(2gh)^{\frac{1}{2}} = v$$

$$(2gl\,(1 - \cos\theta))^{\frac{1}{2}} = v$$

$$\left(2\left(9.8\frac{\text{m}}{\text{s}}\right)(0.248\text{ m})(1 - \cos(0.174\text{ rad}))\right)^{\frac{1}{2}} = v$$

$$0.272\frac{\text{m}}{\text{s}} = v$$

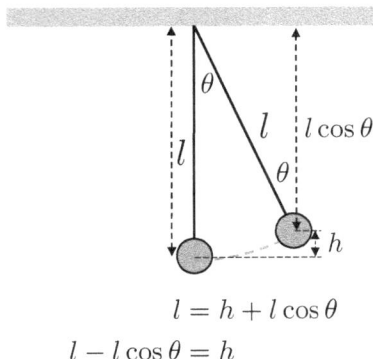

$$l = h + l\cos\theta$$
$$l - l\cos\theta = h$$

FIGURE 6.7

When the bob hangs vertically, its distance below the support is equal to the length of the string, l. If the bob is instead swung through some angle θ, the distance becomes $l\cos\theta$, meaning it has been elevated a distance $h = l - l\cos\theta = l(1 - \cos\theta)$.

6.4   Traveling Waves

Suppose you tie one end of a string to a wall and pull it taut. If you then jerk your hand up and down, you will send a *pulse* down the string. In Figure 6.8 we show this from the perspective of your friend such that the pulse travels to the right, toward the wall. It is important to note that no piece of the string is actually moving to the right, and yet the energy associated with the pulse *is* being transported to the right.[8] This is in fact the key characteristic of a **wave**, which refers to any sort of oscillation (like the pulse in the rope) that transports energy through some medium.

The pulse is spread out in space: we can describe the vertical displacement of the rope (we'll call it y) in terms of the horizontal position (call it x). But the pulse is moving, so

---

[8]If it is unclear that there is energy associated with the pulse, consider that the pulse involves a piece of the rope (which has mass) having some speed – which means the pulse contains kinetic energy. Incidentally, this is an example of a **transverse** wave: the oscillation is up-and-down while the propagation of energy is side-to-side. This is in contrast to a **longitudinal** wave, where both the oscillation and the direction of energy are in the same direction. An example of this is sound: air molecules jostle into one another, creating differences in pressure that propagate through the air. The mathematics for describing each kind of wave is the same.

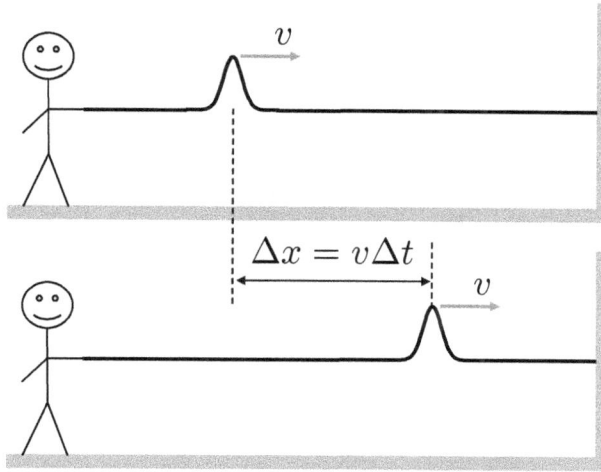

FIGURE 6.8
If a taut rope is quickly jerked up and down, a pulse will propagate down the rope with some speed v.

the displacement is fundamentally a multivariable function: $y(x, t)$. To visualize a pulse, we often fix one variable or the other. For instance, if we "freeze time" to some instant t_0 we can consider what the entire wave looks like as a function of position: $y(x, t_0)$. We call such a graph a **snapshot graph** because it represents what we'd see in a high-speed photograph of the rope taken at time t_0. Alternatively, we can "freeze the position" to some value x_0 and consider what has happened to the vertical displacement of that bit of rope over time. We call such a graph a **history graph**. Example 6.4 demonstrates the idea.

Example 6.4 Snapshot of a Traveling Pulse ⋆
Figure 6.9(a) shows a snapshot graph of a traveling pulse taken at $t = 0.00$ s. Figure 6.9(b) shows the history graph at $x = 0.50$ cm. (a) In which direction is the pulse moving? (b) What is its speed? (c) Explain how the *shape* of the history graph can be predicted from your answer to (b) and the shape of the snapshot graph.

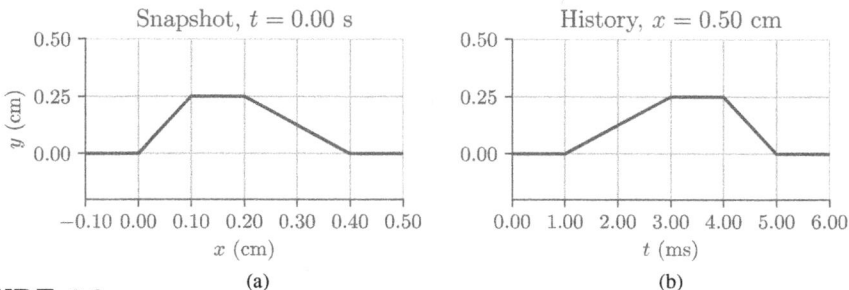

FIGURE 6.9
Graph of a traveling pulse.

(a) The snapshot graph shows that at $t = 0.00$ s, the pulse spans $x = 0.00$ cm to $x = 0.40$ cm. The history graph shows that the disturbance moves through $x = 0.50$

cm (a position to the right of the pulse at $t = 0.00$ s) at *later times*. If the pulse is to the right of its original position at a later time, it is moving to the right.

(b) Now we need to get numerical. The rightmost edge of the pulse is, at $t = 0.00$ s, at $x = 0.40$ cm. This is the first part of the pulse that will reach the position considered in the history graph for $x = 0.50$ cm. The history graph shows the disturbance begins (for $x = 0.50$ m) at $t = 1.00$ ms. Thus the pulse travels $\Delta x = (0.50 - 0.40)$ cm $= 0.10$ cm in a time $\Delta t = (1.00 - 0.00)$ s $= 1.00$ ms, so

$$v = \frac{\Delta x}{\Delta t} = \frac{1.0 \times 10^{-3} \text{ m}}{1.00 \times 10^{-3} \text{ s}} = 1.0 \text{ m/s}$$

Alternatively, if we look at the left edge of the pulse, we see it is as $x = 0.00$ cm at $t = 0.00$ s and the position $x = 0.50$ cm finished the disturbance (when the left edge finishes passing that point) at $t = 5.00$ ms. Calculating the speed with these values yields the same result, as it must.

(c) From the snapshot graph we know the width of the pulse is $\Delta x = 0.40$ cm, and because the speed of the pulse if $v = 1.0$ m/s, the time it takes to completely pass through any given point on the string is

$$t = \frac{\Delta x}{v} = \frac{4.0 \times 10^{-3} \text{ m}}{1.0 \text{ m/s}} = 4.0 \text{ ms}$$

This is indeed the duration of the distortion observed in the history graph. By similar logic, the time it takes to go from no displacement to maximum displacement is $t = 2.0$ ms (using the distance from the leading edge of the pulse to the peak, 0.40 cm to 0.20 cm, in $t = (\Delta x)/v$), the time it takes to finish experiencing the "flat top" of the wave is $t = 1.0$ ms (using the distance 0.20 cm to 0.10 cm) and the time it takes to return to equilibrium as the pulse finishes passing past the position in question is also 1.0 ms (using the distance from the end of the flat top to the trailing edge, 0.10 cm to 0.00 cm).

Put differently: the leading edge (on the right, since the pulse is traveling right) of the pulse reaches $x = 0.50$ cm *first*, so in the history graph we see that disturbance *earlier* (further left on the time axis). This is why the curves look to be mirror images of one another.

Of course, you don't need to create just one pulse. Suppose instead that you rhythmically move your hand so the end of the string experiences SHM. Just like with the pulse, the motion of the end of the string will propagate along the length of the string, and every little chunk of rope will experience simple harmonic motion according to Equation (6.4). While every chunk of rope is oscillating *in time*, the effect of the propagation along the length of the rope leads to a *spatial* sine wave that travels along the length of the string (this is a so-called **traveling sine wave**). To describe it mathematically, we need to introduce some new terminology (Figure 6.10).

- The **wavelength** λ refers to the linear distance occupied by a complete sine wave: it tells us how far apart two chunks of rope are that are at the exact same phase of their motion (at their largest positive displacement from equilibrium, say). Thus the wavelength, which measures the *distance* needed for a complete cycle, is analogous to the period, which measures the *time* needed for a complete cycle.

- While the angular frequency ω $\left(= \frac{2\pi}{T}\right)$ quantifies the rate with which any chunk of rope oscillates, the **wave number** k makes a parallel measurement in terms of the wavelength

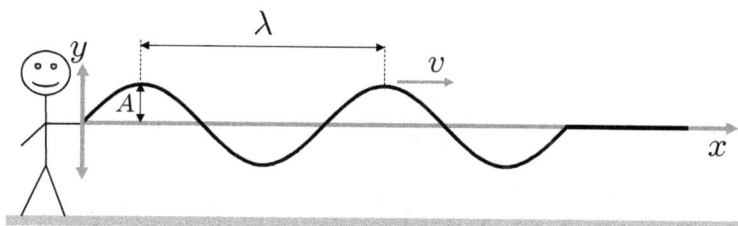

FIGURE 6.10
If the end of a taut rope is oscillated sinusoidally, the oscillation will propagate down the length of the rope, much like the pulse of Figure 6.8. Here we are supposing the oscillation has not yet propagated to the end of the rope. The horizontal distance occupied by a complete oscillation is referred to as the wavelength, λ. Note also that because the horizontal position is (as usual) labeled x, the amplitude of oscillation here occurs in the y dimension.

(as we argue above, we have simply swapped λ for T):

$$k = \frac{2\pi}{\lambda} \tag{6.11}$$

The amplitude A and phase shift ϕ carry the same meaning in both cases, so the same symbols are used. Thus this snapshot of a traveling sine wave can be described by

$$y(x) = A\sin(kx + \phi) \tag{6.12}$$

We can in fact combine Equations (6.4) and (6.12) to arrive at one equation that gives us *the vertical displacement y* of any chunk of rope with a horizontal position x at any time t. This is the **equation for a traveling sine wave**:

$$y(x, t) = A\sin(kx \pm \omega t + \phi) \tag{6.13}$$

We use the symbol $\pm$ to draw out the fact that a *negative* value of ω refers to a wave moving to the *right*; a *positive* value means the wave is moving to the *left* (see Problem 6.72).

So far we've considered a snapshot of the traveling wave, but (just like a pulse) the wave is moving down the string. With the terminology we have just introduced, we can define the speed of the sinusoidal wave as it travels down the string (this is unsurprisingly referred to as the **wave speed**). The definition for average velocity in one dimension is $v = \frac{\Delta x}{\Delta t}$. If we use the wavelength λ as Δx and the period T as Δt, we have[9]

$$v = \frac{\lambda}{T} = \lambda f \tag{6.14}$$

This is a general relationship for all traveling waves (including, for example, sound and light), but we can also characterize the wave speed in terms of the material through which the wave is propagating. For a wave moving down a rope, the wave speed depends on both

---

[9]For the traveling wave to move a complete wavelength down the string, a tiny chunk of the rope needs to undergo a complete cycle of its simple harmonic motion. This is by definition one period, so T is the appropriate amount of time to match with λ.

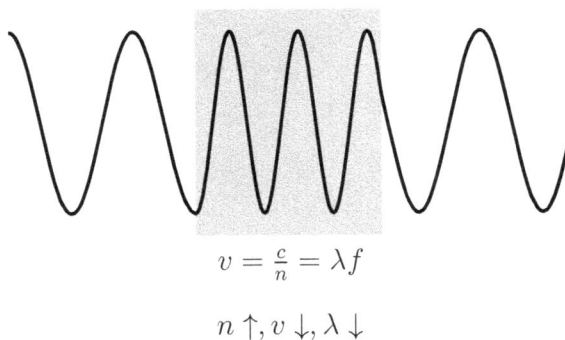

$$v = \frac{c}{n} = \lambda f$$

$$n \uparrow, v \downarrow, \lambda \downarrow$$

FIGURE 6.11

A light wave passing from air into some medium (e.g. glass or water), represented by the shaded box. Because the medium has an index of refraction $n > 1$, the speed v of the wave decreases. Consequently, the wavelength λ also decreases. If the wave emerges on the other side of the medium, the speed and wavelength revert to their original values.

the tension in the rope (F_T) and the rope's **linear mass density**, μ.[10] The linear mass density is simply the rope's mass, m, divided by its length, L:

$$\mu = \frac{m}{L} \tag{6.15}$$

With this definition in mind, it turns out that the wave speed for a taut rope is given by

$$v_{\text{string}} = \left(\frac{F_T}{\mu}\right)^{\frac{1}{2}} \tag{6.16}$$

In contrast, the speed of sound depends on a host of factors, such as the ambient temperature. For our purposes, we shall assume that the speed of sound takes the standard value

$$v_{\text{sound}} = 343\frac{\text{m}}{\text{s}} \tag{6.17}$$

Light, meanwhile, travels at a constant speed in a vacuum that we denote with the symbol c:[11]

$$v_{\text{light}} = c = 3.000 \times 10^8 \frac{\text{m}}{\text{s}} \tag{6.18}$$

The speed of light in other media depends on a medium-specific constant called the **index of refraction**, n:

$$v = \frac{c}{n} \tag{6.19}$$

Table 6.1 lists the index of refraction for some common media. Notice that the speed of light in air is, to 4 significant figures, still c. If light passes into a different medium, for instance water, then the *speed of the wave* drops and therefore the *wavelength* also drops: $v = \lambda f$ and the frequency is fixed by the source of the wave (Fig. 6.11).[12] We shall have

---

[10]The symbol μ certainly does *not* refer to a coefficient of friction; the same symbol is here being used in a different context.

[11]You may be wondering what, exactly, oscillates in a light wave. The answer is "an electromagnetic wave" – a phrase that very likely has no meaning for you! Never fear, a large portion of Part IV of this book addresses this topic.

[12]Similar statements hold for sound waves and for waves in ropes: the speed of a sound wave increases if it passes into metal, and so the wavelength increases. Similarly, if two ropes of differing densities are tied together, the mass density changes, so the speed changes, so the wavelength changes.

much more to say about light as we proceed, but for now I will point out that *visible light* ranges in wavelength from about 400 nm (deepest violet) to about 700 nm (deepest red).

To summarize, in this section we have discussed the movement of either a *pulse* or a *sine wave* down a taut string. In the case of a sine wave, we have employed the mathematics of SHM to describe the spatial deformation of the string, and we arrived at Equation (6.13) as the general equation describing the position of any point on the string at any time. Meanwhile, Equations (6.14) and (6.16)–(6.18) describe the speed with which any part of a wave (or pulse) propagates down the string.

TABLE 6.1
Indices of refraction.

Medium	n	
Vacuum	1.00	(exactly)
Air	1.00	
Water	1.33	
Corn oil	1.47	
Glass	1.50	
Diamond	2.42	

Example 6.5 Jerking a Rope ⋆

A sturdy rope with a mass density of 10.0 g/cm has been strung down a 120-m-deep mining tunnel. A loaded bucket with a total mass of 15.0 kg has been attached to the bottom of the rope. A worker jerks the bottom of the rope to signal a worker at the top of the tunnel that the bucket is ready to be lifted. How long does it take the pulse to travel to the top of the tunnel?

According to Equation (6.16), the speed of the pulse is[a]

$$v = \left(\frac{F_T}{\mu} \right)^{\frac{1}{2}}$$

$$v = \left(\frac{m_{\text{bucket}} g}{\mu} \right)^{\frac{1}{2}}$$

$$v = \left(\frac{(15.0 \text{ kg}) \left(9.8 \frac{\text{m}}{\text{s}^2} \right)}{1.00 \frac{\text{kg}}{\text{m}}} \right)^{\frac{1}{2}}$$

$$v = 12.1 \frac{\text{m}}{\text{s}}$$

The pulse will travel at this speed until it reaches the top of the rope, a distance of 120 m. Thus we can determine the time it takes for the pulse to reach the top of the mine shaft:

$$v = \frac{\Delta x}{\Delta t}$$

$$\Delta t = \frac{\Delta x}{v}$$

$$\Delta t = \frac{120 \text{ m}}{12.1 \frac{\text{m}}{\text{s}}}$$

$$\Delta t = 9.9 \text{ s}$$

---

[a]In the second line we are noting that the tension in the rope is given by the weight of the bucket, which is correct if we neglect (as usual) the mass of the rope. Here, however, because we have the length and mass density of the rope, we *could* include the weight of the rope itself: the tension at the *bottom* is the weight of the bucket and the tension at the *top* is the weight of the bucket + the weight of the rope. In calculating the average velocity of the pulse you would use the average tension.

Example 6.6 A Traveling Wave   ★★

Figure 6.12 shows two snapshots of a traveling wave, where the same "crest" is marked in both images. Determine the wave equation (Equation (6.13)) for this wave.

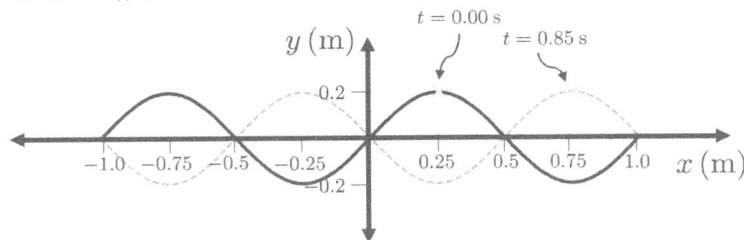

FIGURE 6.12

A portion of a traveling wave is shown at two instants (different shading is used just to differentiate the two images). The same crest is marked in both images.

To fill in Equation (6.13), we need to identify the values of A, k, ω, and ϕ. The amplitude A is straightforward to identify from the figure: 0.2 m. For k, recall $k = \frac{2\pi}{\lambda}$ (Equation (6.11)) and note from the figure that $\lambda = 1.0$ m (for instance, the wave drawn in black completes one cycle between $x = 0$ m and $x = 1.0$ m). Thus

$$k = \frac{2\pi}{\lambda} = \frac{2\pi}{1.0 \text{ m}} = 6.3 \text{ m}^{-1}$$

Determining ω is perhaps trickier because we don't have a position vs. time plot. We are, however, given some temporal information in that we have two different images of the wave at two different instants. The marked crest travels a distance $\Delta x = 0.75$ m $- 0.25$ m $= 0.50$ m in a time $\Delta t = 0.85$ s, so we can determine the wave speed v and then the period (Equation (6.14)):

$$v = \frac{\lambda}{T}$$

$$\frac{\Delta x}{\Delta t} = \frac{\lambda}{T}$$

$$T = \frac{\lambda \Delta t}{\Delta x}$$

$$T = \frac{(1.0 \text{ m})(0.85 \text{ s})}{0.50 \text{ m}}$$

$$T = 1.7 \text{ s}$$

Armed with the period, it is straightforward to determine the magnitude of ω (it has a negative sign when inserted into the wave equation because the wave is traveling to the right):

$$\omega = \frac{2\pi}{T} = \frac{2\pi \text{ rad}}{1.7 \text{ s}} = 3.7 \frac{\text{rad}}{\text{s}}$$

Finally, we need to determine ϕ. Considering the instant when $t = 0.0$ s, we see the amplitude at $x = 0.0$ m is itself 0 m. Thus

$$y(0 \text{ m}, 0 \text{ s}) = A \sin(k(0 \text{ m}) + \omega(0\text{s}) + \phi)$$

$$0 \text{ m} = A \sin(\phi)$$

$$\sin^{-1} 0 = \phi$$

If you plug the above into a calculator you will most likely get 0 as a result. However, if you refer back to Figure 6.2, you'll see that $\sin 0 = \sin \pi = \sin 2\pi = 0$.[a] Thus, much like $\sqrt{4} = \pm 2$, there are multiple solutions and we need to be careful about which one we choose. We see from Figure 6.2 that for $\sin 0$, the value of the sine function *increases* as the angle increases, while for $\sin \pi$ the value of the sine function *decreases* as the angle increases. If we compare this to the figure for this problem, we see that we're dealing with the first case rather than the second case. Thus $\phi = 0$.

Now that we have all of the needed values, we can write down the wave equation:

$$y(x,t) = A \sin(kx \pm \omega t + \phi)$$

$$y(x,t) = (0.2 \text{ m}) \sin\left((6.3 \text{ m}^{-1})\, x - \left(3.70 \frac{\text{rad}}{\text{s}}\right) t\right)$$

---

[a]In the figure we are only showing one complete wavelength; the general formula if we continued the sine wave arbitrarily far in either direction (or traveled an arbitrary number of times around the unit circle, clockwise or counterclockwise) is $\sin n\pi = 0$ for any integer value of n.

Example 6.7 Wave Speeds: Thunder and Lightning   ⋆

During a thunderstorm, you are 15.0 km from a lightning strike and corresponding thunderclap. How long does it take for you to see the lightning? How long does it take for you to hear it?

The time for either wave to reach you is given by

$$t = \frac{|\Delta \vec{x}|}{|\vec{v}|}$$

Here we will use Equations (6.17) and (6.18) to determine the time for each wave to reach you:

$$t_{\text{sound}} = \frac{1.50 \times 10^3 \text{ m}}{343 \text{ m/s}} = 4.37 \text{ s}$$

and

$$t_{\text{light}} = \frac{1.50 \times 10^3 \text{ m}}{3.000 \times 10^8 \text{ m/s}} = 5.00 \times 10^{-6} \text{ s}$$

In other words, the light reaches you almost instantaneously, while several seconds pass for the sound wave to reach you.

6.5   Sound Intensity and Loudness

In Equation (6.17) we defined the speed of sound through air (in typical conditions) to be 343 m/s. This is hardly the only aspect of sound that is relevant, however; how can we describe how *loud* a particular sound wave is?

As a first step in answering this question, let me introduce the **sound intensity**, denoted I, as a measure of *power per area*:

$$I = \frac{P}{A} \tag{6.20}$$

Recall that power itself is a measure of energy per time (1 W = 1 J/s), so sound intensity answers the question "How much energy is being delivered, per second, per unit area?" Practically speaking, the area of interest might be the surface area of your eardrum or of the membrane of a microphone. If we think of a sound wave as traveling as a *sphere* expanding in all directions, then the sound intensity a radial distance r from the source is given by

$$I_{\text{spherical}} = \frac{P}{4\pi r^2} \tag{6.21}$$

If we know the sound intensity due to a spherical wave at one distance r_1, it is straightforward to calculate the sound intensity at some other distance r_2 by taking a ratio of Equation (6.21) at $r = r_1$ and $r = r_2$:

$$\frac{I_1}{I_2} = \frac{r_2^2}{r_1^2} \tag{6.22}$$

Now, while the sound intensity has natural SI units, it unfortunately does not correspond to the human perception of how *loud* the sound is. For instance, if you double the sound intensity, it is *not* the case that people will report that the sound is twice as loud. To quantify the human perception of loudness, then, we instead use the **sound intensity level**, L, measured in decibels (dB). The conversion is

$$L = (10 \text{ dB}) \log_{10}\left(\frac{I}{I_0}\right) \tag{6.23}$$

where

TABLE 6.2

Typical sound intensity levels (L) and intensities (I)

Source	L (dB)	I (W/m^2)
Threshold of hearing	0	10^{-12}
Whispering	20	10^{-10}
Quiet room	40	10^{-8}
Typical conversation	60	10^{-6}
Busy room	80	10^{-4}
Loud traffic	100	10^{-2}
Heavy machinery	120	10^{0}
Near jet taking off	140	10^{2}

$$I_0 = 1.0 \times 10^{-12} \text{ W/m}^2 \tag{6.24}$$

is called the **threshold of hearing**, i.e. the lowest-intensity sound that a human can detect.[13] If you invert Equation (6.23) to solve for the sound intensity, you find (as you can check for yourself):

$$I = I_0 10^{L/(10 \text{ dB})} \tag{6.25}$$

You may not be particularly comfortable visualizing logarithmic relationships such as these; Figure 6.13 shows I as a function of L and L as a function of I.

So what does this have to do with how humans subjectively perceive the loudness of a sound? Well, every 10 dB increase roughly corresponds to *double* the loudness. Values at or near 120 dB are typically described as the "threshold of pain",[14] and hearing protection is typically recommended for sound intensity levels above 80 or so dB. Table 6.2 provides some example sources of sound and their typical decibel rating.

---

[13]Of course, this value varies from person to person; this is just a standard reference value.

[14]I think this would be a good name for a rock band.

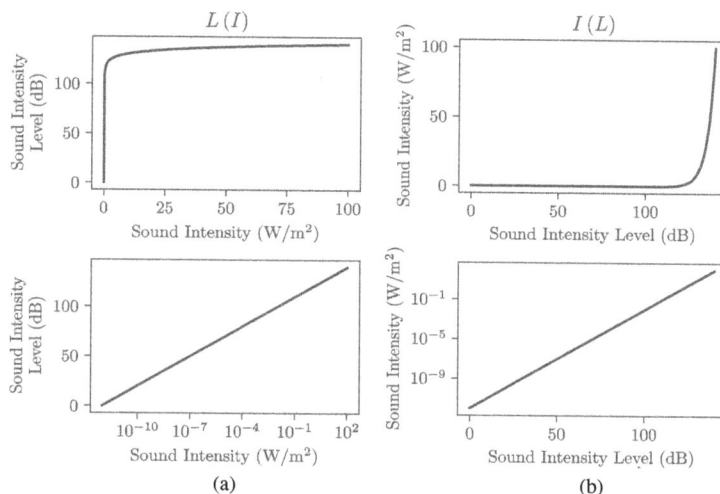

FIGURE 6.13

The relationship between sound intensity and sound intensity level. (a) This depict Equation (6.23) and (b) depict Equation (6.25). The top panels have a *linear* scale while the bottom panels show the same relationships with one axis on a logarithmic scale.

Example 6.8 Speakers: Power and Sound ★

A particular speaker is rated at 650 W. If you assume that the sound energy travels outward in a sphere, then what is the sound intensity level and the sound intensity at a distance of 5.0 m? Would you subjectively describe this as a loud sound?

The sound intensity is given by Equation (6.21):

$$I = \frac{P}{4\pi r^2} = \frac{650 \text{ W}}{4\pi \left(5.0 \text{ m}\right)^2} = 2.1 \text{ W/m}^2$$

We can then convert to Sound Intensity Level with Equation (6.23):

$$L = (10 \text{ dB}) \log_{10} \left(\frac{I}{I_0}\right) = (10 \text{ dB}) \log_{10} \left(\frac{2.1}{1.0 \times 10^{-12}}\right) = 120 \text{ dB}$$

This maps to "Heavy Machinery" according to Table 6.2 and is near typical values for the threshold of pain–so *yes*, it is rather loud!

Example 6.9 Sound from a Jet Engine ★★

A jet engine's sound intensity level is about 140 dB at a distance of 3.0×10^1 m. What is the sound intensity level (in dB) at a distance of 1.0 km? What are the intensities (in W/m^2) at each position? How much softer is the sound to the ear at this distance compared to at the original distance?

We'll convert the given sound intensity level (L_1) to sound intensity (I_1) using Equation (6.25), then determine the second sound intensity (I_2) using Equation (6.22). Once we have I_2, we can determine L_2 using Equation (6.23).

Here we go:

$$I_1 = I_0 10^{L_1/(10 \text{ dB})}$$
$$= \left(1.0 \times 10^{-12}\right) 10^{140 \text{ dB}/(10 \text{ dB})}$$
$$= 1.0 \times 10^2 \text{ W/m}^2$$

Now to find I_2:

$$\frac{I_2}{I_1} = \frac{r_1^2}{r_2^2}$$
$$= I_1 \left(\frac{r_1}{r_2}\right)^2$$
$$= \left(1.0 \times 10^2 \text{W/m}^2\right) \left(\frac{3.0 \times 10^1 \text{ m}}{1.0 \times 10^2 \text{ m}}\right)^2$$
$$= 9.0 \times 10^{-2} \text{ W/m}^2$$

And finally we can convert to L_2:

$$L_2 = (10 \text{ dB}) \log_{10}\left(\frac{I_2}{I_0}\right)$$
$$= (10 \text{ dB}) \log_{10}\left(\frac{9.0 \times 10^{-2} \text{ W/m}^2}{I_0}\right)$$
$$= 110 \text{ dB}$$

We've gone from 140 dB to 110 dB. Recall that every 10 dB corresponds to a factor of 2 in the subjective experience of loudness, so 140 dB to 130 dB would be *half* as loud. 130 dB to 120 dB would be half *again*, and 120 dB to 110 dB would be half for a *third* time. Thus subjectively the sound is $\left(\frac{1}{2}\right)^3 = \frac{1}{8}$ as loud. This assumes, by the way, that there are no other factors (like buildings, trees, or wind) to interfere with the sound wave as it travels this distance.

6.6   The Doppler Effect

If you've ever heard a siren (for example, of a police car, ambulance, or fire truck) approaching you and/or receding away from you, then you know the sound it makes is affected; the frequency you hear is different than if the siren is stationary. Why is this? Well, suppose first that the siren is stationary: it emits waves with some frequency f in all directions. Figure 6.14 shows a top-down view of the situation, with each circle corresponding to a peak of the wave (waves farther from the source were emitted earlier and have therefore been traveling longer), so adjacent circles are separated by one wavelength λ.

Now let's consider motion: suppose the siren is moving to the right, and suppose you're standing off to the right, so the siren is approaching you. If the siren emits a wave peak at

$t = 0$, then when it next emits a wave peak it will have moved to the right so the separation between the peaks (which is λ when the siren is stationary) will be *decreased*. Conversely, if you were standing off to the *left*, then the distance between the wave peaks would *increase* (Fig. 6.15).[15]

Broadly speaking, the same patters emerge if the source is *stationary* and it is the *observer* that is moving (for instance, if you're in a car driving toward a parked ambulance with the siren on). However, the details are different because in the first case the source of the sound is moving relative to the medium (the air) and in the second case the source of the sound is stationary. We will skip the details of the derivation and simply assert the result. Here is the most concise way to express the relationship:

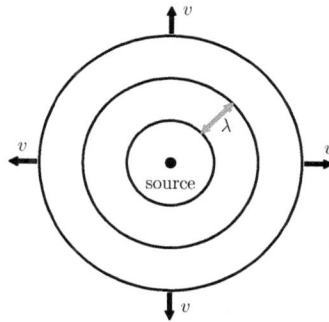

$$f_o = f_s \left(\frac{v \pm v_o}{v \mp v_s} \right) \qquad (6.26)$$

Here f_o is the *observed* frequency (what the observer actually hears), f_s is the *source* frequency (what the observer hears if there is no motion whatsoever), v is the speed of sound (343 m/s), v_s is the speed of the

FIGURE 6.14

A stationary source of a sound wave emits wave crests traveling at the speed of sound v separated by a wavelength λ.

source, and v_o is the speed of the observer. We use the top (bottom) sign in $\pm$ if the observer is approaching (receding from) the source. Similarly, we use the top (bottom) sign in $\mp$ if the source is approaching (receding from) the observer. This gets to be quite a bit to unpack, so we'll list out the four special cases where either the observer or the source are stationary:

$$f_+ = f_s \left(\frac{v}{v - v_s} \right) \qquad \text{source approaching observer} \qquad (6.27)$$

$$f_- = f_s \left(\frac{v}{v + v_s} \right) \qquad \text{source receding from observer} \qquad (6.28)$$

$$f_+ = f_s \left(\frac{v + v_o}{v} \right) \qquad \text{observer approaching source} \qquad (6.29)$$

$$f_- = f_s \left(\frac{v - v_o}{v} \right) \qquad \text{observer receding from source} \qquad (6.30)$$

There is, similarly, a Doppler effect for light. Here, because light does not require a medium we have just two cases: are the observer and source approaching one another or receding from one another? In these cases we refer to their speed of approach (or separation) as v_s and we normally express the shift in terms of the wavelength rather than the frequency:

---

[15]It might help to think of this with balls being thrown rather than sound waves: if I throw a ball at you once a second at a speed of 1.0 m/s, then they'll be separated by 1.0 m, where we can think of this distance as analogous to the wavelength (peak to peak distance) of a wave. But if I start moving toward you at 0.2 m/s and throw the balls such that the frequency and ball velocity through the air are the same, then the separation between the balls will decrease to 0.8 m. Similarly, if I back up instead, the separation between the balls will increase to 1.2 m. In both situations, the frequency with which the balls hit you will change, as well (increasing if I walk toward you and decreasing if I walk away). This isn't completely analogous to actual waves, but the trends described here are the same.

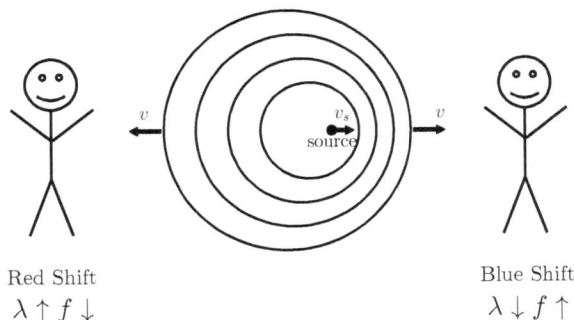

Red Shift
$\lambda \uparrow f \downarrow$

Blue Shift
$\lambda \downarrow f \uparrow$

FIGURE 6.15
A source traveling at a velocity v_s emits wave crests traveling at a speed v e.g. the speed of sound for a sound wave). Due to the motion, an observer that sees the source approaching will observe a decreased wavelength and therefore an increased frequency (to maintain a constant wave speed $v = \lambda f$). For an observer that sees the source receding, the wavelength will appear as increased and therefore the frequency will appear decreased. These situations are respectively called a "blue shift" and a "red shift" because in the case of visible light, a drop in wavelength moves toward the blue (or violet) end of the spectrum and an increase in wavelength moves toward the red end of the spectrum.

$$\lambda_- = \lambda_s \left(\frac{1 - \frac{v_s}{c}}{1 + \frac{v_s}{c}} \right)^{1/2} \qquad \text{source and observer approaching} \qquad (6.31)$$

$$\lambda_+ = \lambda_s \left(\frac{1 + \frac{v_s}{c}}{1 - \frac{v_s}{c}} \right)^{1/2} \qquad \text{source and observer receding} \qquad (6.32)$$

Example 6.10 Doppler Effect with a Diving Bird   ⋆
A particular species of hawk screeches at 8.00×10^2 Hz, but as it dives toward you you perceive the frequency to be 8.80×10^2 Hz. At what speed is the hawk diving? If you heard the screech when the hawk was at an altitude of 3.00×10^2 m, how long do you have to take cover?

This is a situation where the observer (you) is stationary and the source of the sound (the hawk) is moving. We'll pull out the relevant version of the Doppler equation and algebraically solve for the speed of the source, v_s:

$$f_o = f_s \left(\frac{v}{v - v_s} \right)$$

$$v - v_s = v \frac{f_s}{f_o}$$

$$v \left(1 - \frac{f_s}{f_o} \right) = v_s$$

$$v_s = (343 \text{ m/s}) \left(1 - \frac{8.00 \times 10^2 \text{ Hz}}{8.80 \times 10^2 \text{ Hz}} \right)$$

$$v_s = 31.2 \text{ m/s}$$

From here it is straightforward determine the time to impact:

$$\Delta t = \frac{\Delta x}{v}$$

$$= \frac{3.00 \times 10^2 \text{ m}}{31.2 \text{ m/s}}$$

$$= 9.62 \text{ s}$$

This should be enough time for you to get under cover. Phew!

Example 6.11 Doppler Effect with Stars ★

Stars emit light at many different wavelengths both inside and outside the visible spectrum. One wavelength in the visible spectrum has a wavelength of 656.279 nm (red light due to hydrogen in the star traveling through the air of Earth's atmosphere). If the star is *moving* relative to the Earth, the wavelength observed by astronomers is shifted according to the Doppler effect, and the shift allows astronomers to infer the speed of the star. Suppose one such measurement yields a wavelength of 656.400 nm. What is the speed of the star? Is it approaching Earth or moving away from the Earth?

The observed wavelength has *increased*, and we see from Figure 6.15 that this means the star is moving away from the observer. Thus we use the equation for λ_+ (this also follows from the fact that we have an increased wavelength, so we use the equation for λ with the $+$ subscript) and algebraically isolate v_s:

$$\lambda_+ = \lambda_s \left(\frac{1 + \frac{v_s}{c}}{1 - \frac{v_s}{c}} \right)^{1/2}$$

$$\left(\frac{\lambda_+}{\lambda_s} \right)^2 = \frac{1 + \frac{v_s}{c}}{1 - \frac{v_s}{c}}$$

$$\left(\frac{\lambda_+}{\lambda_s} \right)^2 = \frac{c + v_s}{c - v_s}$$

$$(c - v_s) \left(\frac{\lambda_+}{\lambda_s} \right)^2 = c + v_s$$

$$\left(\left(\frac{\lambda_+}{\lambda_s} \right)^2 - 1 \right) c = v_s \left(\left(\frac{\lambda_+}{\lambda_s} \right)^2 + 1 \right)$$

$$\left(\frac{\left(\frac{\lambda_+}{\lambda_s} \right)^2 - 1}{\left(\frac{\lambda_+}{\lambda_s} \right)^2 + 1} \right) c = v_s$$

Gross! Now, though, we can insert our values. $\lambda_+ = 656.400$ nm and $\lambda_s = 656.279$ nm, so the ratio $\lambda_+ / \lambda_- = 1.00018$. Meanwhile, $c = 3.000 \times 10^8$ m/s. Inserting, we find

$$v_s = 2.765 \times 10^4 \text{ m/s}$$

This is over 27 km every second!

By the way, if you make measurements like this for many galaxies, you will find that their speed depends on their distance away from us: more distant galaxies are moving away from us more rapidly than closer galaxies. This is called the **Hubble Law** or the **Hubble-Lemaître Law**, after Edwin Hubble (1931–1953) and Georges Lemaître (1894–1966).

6.7 Standing Waves and Musical Instruments

We now have a sense for how pulses and sinusoidal waves travel when they move unimpeded. This is analogous to understanding how a ball moves when it experiences no net force (i.e. with a constant velocity), but, much like with a ball, things are considerably more interesting when we consider how waves behave when they are not allowed to move on their merry way unimpeded.

Let's consider first the pulse of Figure 6.8. When the pulse reaches the wall, it "bounces back" with the same speed, much like a ball colliding elastically with the wall, but it is also *inverted*, as shown in Figure 6.16. This behavior is perhaps surprising, but it follows directly from Newton's Third Law: as the tip of the wave in Figure 6.16 travels, it is constantly exerting an upward force on the piece of rope to its right. When the wave reaches the fixed wall, the wall (by Newton's Third Law) exerts an equal and opposite *downward* force, which creates an inverted pulse traveling back down the rope.

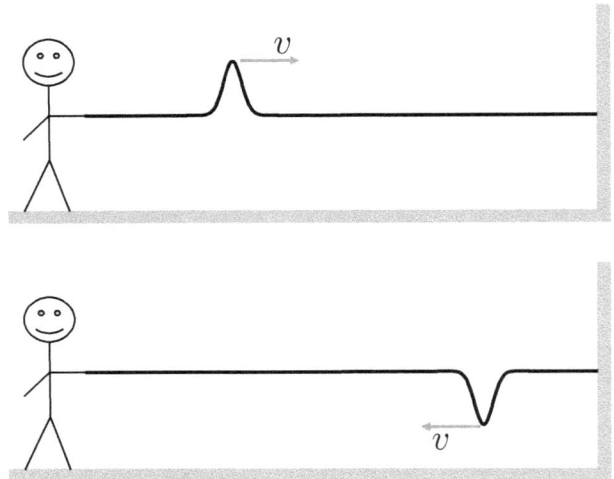

FIGURE 6.16

When a pulse travels down a string and hits a solid barrier (e.g. if the string is tied to a wall), it reflects back with the same speed but reflected about the horizontal axis.

Now suppose we send two pulses, one after another, such that the first pulse reflects from the wall and travels toward the second pulse. As we show in the left column of Figure 6.17, the two pulses will pass through one another; when they overlap the shape of the string correspond to the *sum* of the two pulses. Because the pulses in this cases are mirror images of one another, this means that they will perfectly cancel for one instant, when the middle of the pulses occupy the exact same position (we refer to the process of waves partially/completely cancelling as **destructive interference**).[16] This need not always be the case: as we show in the right column of Figure 6.17, two pulses can lead to a larger pulse when they overlap (we similarly refer to this as **constructive interference**).

---

[16]Destructive interference is an important concept in explaining, for instance, noise-cancellation headphones.

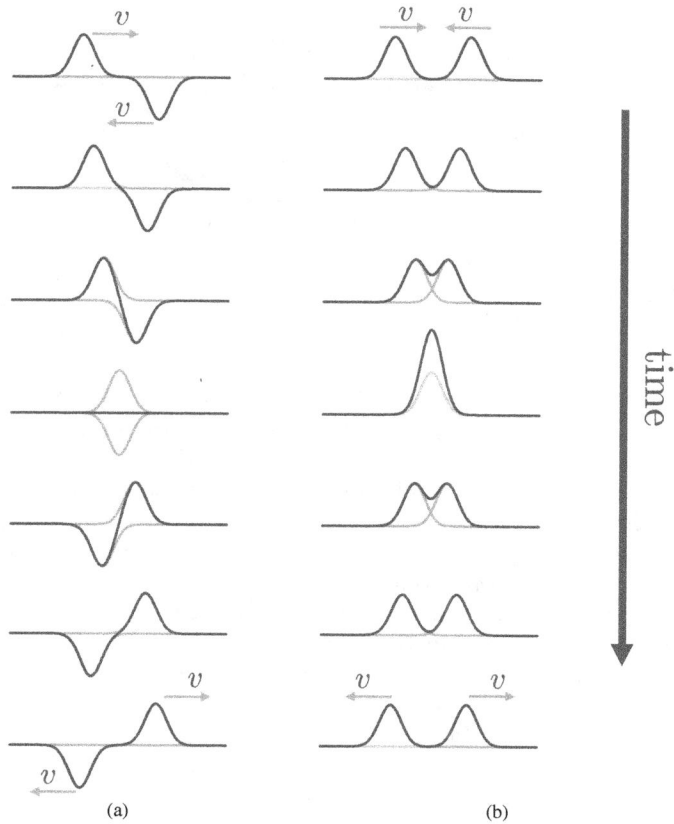

FIGURE 6.17
When two wave pulses "collide" on the same string, each pulse continues at its original velocity and the shape of the string at any instant is the sum of each individual pulse. Here the pulses are shown in gray and their sum (i.e. the actual shape of the string) is shown in black. The axes are omitted for visual clarity; you can think of these as visual snapshots of a string in space, as in Figure 6.16. (a) The pulses are identical except one is negative and the other is positive. When the midpoints perfectly overlap (in the middle panel), they perfectly cancel and the string is completely flat. (b) The pulses are identical; when they perfectly overlap the string has a maximum displacement equal to the sum of the maximum displacement of the two pulses.

Example 6.12 Interference of Wave Pulses ⋆
Figure 6.18 shows two pulses traveling down a taut string at time $t = 0.0$ s. Sketch the string at time $t = 2.0$ s.

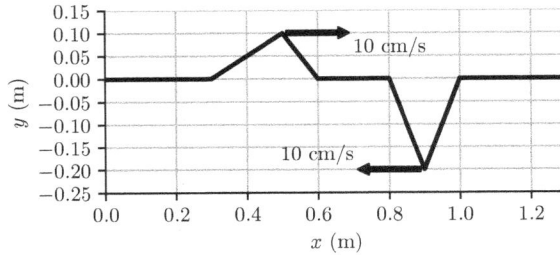

FIGURE 6.18

Both pulses have a speed of $0.10\frac{m}{s}$, so when 2.0 seconds have elapsed, each will have traveled 0.20 m, and the peak of the pulses will overlap at the position marked 0.70 m (Figure 6.19). To determine the shape of the string, we need to add the "desired" displacement due to each pulse for each horizontal position x.

When we have pulses that consist of line segments, as in this example, one efficient way to proceed is to split the horizontal axis into pieces such that within each piece, the slope of both wave pulses is constant. The ends of these segments are marked with circles in Figure 6.19: moving from left to right, each time either pulse experiences a "corner", we add the two pulses and mark a dot at the appropriate vertical position. For instance, at $x = 0.70$ m, the pulses combine to give an amplitude of 0.10 m $+ (-0.20$ m$) = -0.10$ m.

Once we have done this for the entire wave, we can simply "connect the dots" to determine the final shape of the string (between adjacent dots we are adding two line segments, which gives us another line segment).

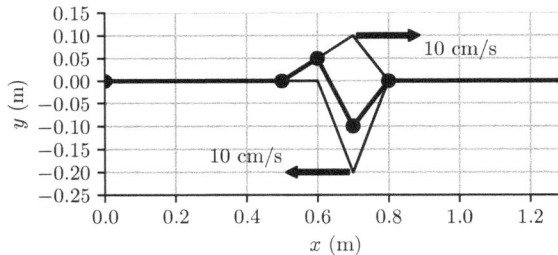

FIGURE 6.19
Here the pulses of Figure 6.18 are shown two seconds later, with thin lines. The resulting shape of the string is shown in black. The amplitude of the resulting wave has been determined at every position where one pulse and/or the other has an edge (shown with black dots); the complete wave is then determined by connecting these points with line segments.

The behavior we have just discussed – how two pulses pass through one another on a taut string – applies in exactly the same way to two sinusoidal waves (Figure 6.20). Visually processing this can be difficult at first, and it may help to think of one crest from one wave interacting with a crest of the other wave, then a trough, then a crest, et cetera. Each of these interactions operates like the pulses in Figure 6.17, including the instants of complete destructive and constructive interference.

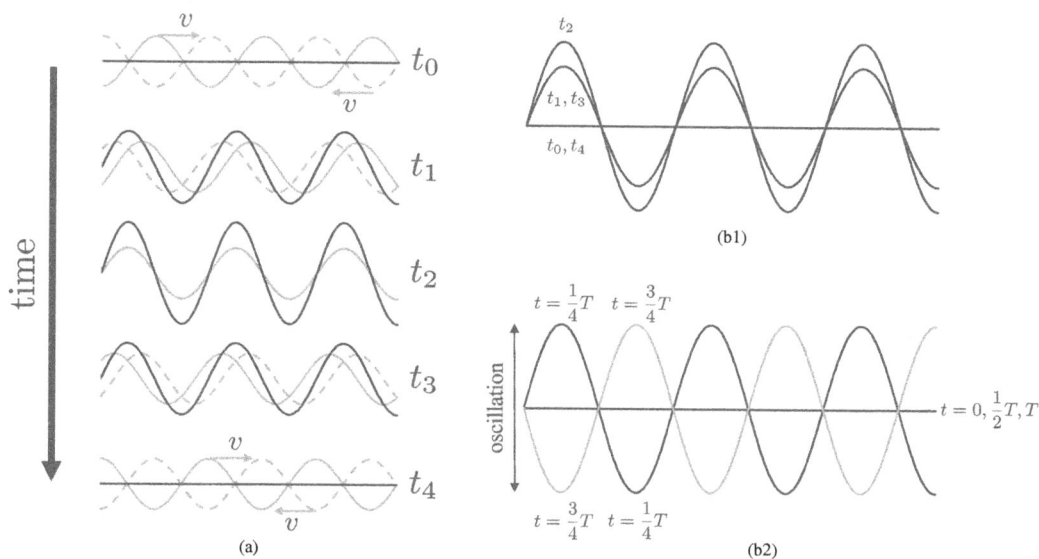

FIGURE 6.20

Two interacting sinusoidal waves interact as a series of overlapping pulses (Figure 6.17 – here we are again omitting the axes for visual clarity). (a) Two sinusoidal traveling waves are shown with solid and dashed gray curves (here it is perhaps best to think of them as a message to the string: *this is what I want you to look like*). The combination of both waves leads the string to take the shape shown in black. Successive instants in time are shown in descending order. (b1) If the string deformations from the left panel are overlaid, we observe that some positions *always* have 0 amplitude (we call these *nodes*). The rest of the string can either be flat (e.g. at $t = t_0$) or takes the form of a sine wave. Notably, the positions of maximum amplitude (*antinodes*) are always the same; there are no traveling wave crests (this is why this is referred to as a *standing wave*). (b2) The other panels consider only a half period; if we look forward in time, the oscillations continue such that the peaks become troughs and vice versa. The instant where the leftmost antinode is at a peak is shown in black, while the instant where it has become a trough is shown in gray (the second peak makes the opposite transition, and so on).

In fact, a remarkable property of these traveling sine waves interacting with one another is that some positions on the string *always* have 0 amplitude (because the traveling waves always cancel at these positions). These positions are referred to as **nodes**. Evenly spaced between adjacent nodes are **antinodes**, which experience oscillations with the largest amplitude of any point on the string. The effect of this is that there are no longer crests obviously moving left or right; all the oscillation is up and down. Thus two traveling sine waves result in a so-called **standing wave**.[17]

This may seem like a whole lot of effort to understand a very particular phenomenon, but it turns out that the idea of a standing wave is incredibly practical and useful. For instance, suppose we have a stringed musical instrument, such as a violin or piano. In such an instrument we have a number of strings of different lengths that are secured on both

---

[17]We're assuming here that the waves travel with the same speed, but in opposite directions. We won't consider the case where they have different speeds, but – as you might expect – things get more complicated if you do!

string (closed/closed)   tube (open/closed)

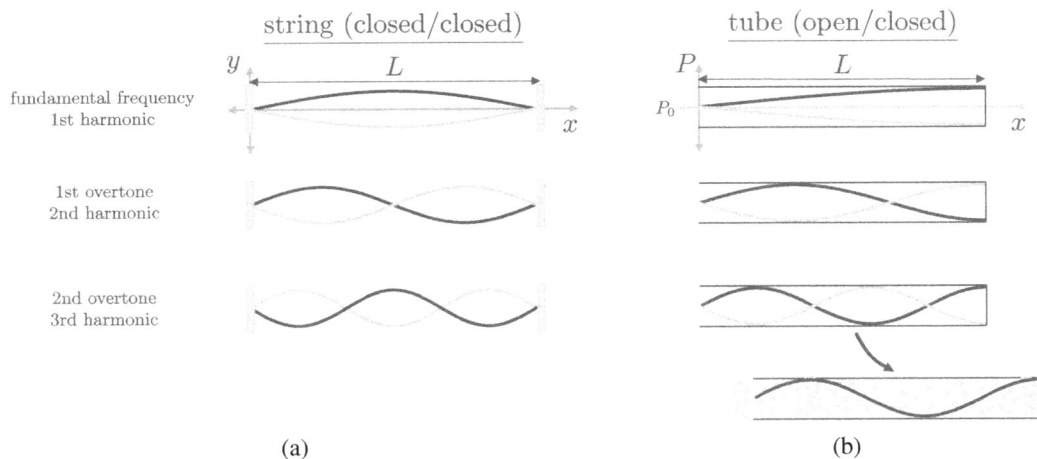

fundamental frequency
1st harmonic

1st overtone
2nd harmonic

2nd overtone
3rd harmonic

(a)   (b)

FIGURE 6.21
Standing waves in a string (a) and a tube that is open on one end and closed on the other
(b). For visual clarity the axes are shown just on the top diagrams. The top diagram in each
case contains the fewest possible nodes and antinodes, and is referred to as the *fundamental
frequency* or the *1ˢᵗ harmonic*. As the number of nodes & antinodes increases, the oscillations
are referring to as the *2ⁿᵈ harmonic*, *3ʳᵈ harmonic* etc., or equivalently the *1ˢᵗ overtone*,
2ⁿᵈ overtone, etc. These are collectively referred to as the **resonant frequencies** of the
string/tube. (a) a string is clamped on both ends, meaning both endpoints are constrained to
be nodes. (b) the open end of the tube is constrained to have atmospheric pressure P_0, so it
is a node. The pressure at the other end, meanwhile, can vary freely and is an antinode. The
bottom diagram depicts air molecules as tiny circles for the 2nd overtone; higher pressure
corresponds to tighter packing.

ends. If a string is pulled and released, a traveling wave will reflect off of the barriers and
the result is a standing wave! Furthermore, because the ends of the string are fixed in place,
they *must* be nodes. As we show in Figure 6.21, this means the string can take the shape of
a half wavelength, a complete wavelength, one and a half wavelengths, and so on. Evidently
for a string of length L pinned at both ends, we have

$$\frac{n}{2}\lambda = L, \ n = 1, 2, 3, ... \tag{6.33}$$

Conversely, many wind instruments are essentially tubes that experience oscillations in
pressure (rather than in the position of a string). If the tube is open at both sides (or closed
at both sides), then the pressure inside also varies according to Equation (6.33). If instead
the tube is closed at one end but open on the other, the open end is a node (fixed at room
pressure) and the closed end is an antinode. This means that a tube of length L contains
$\frac{1}{4}\lambda, \frac{3}{4}\lambda, ...$ (Figure 6.21).[18] We can write this mathematically as

---

[18] If you look at other resources, you may see diagrams for closed/open tubes with the nodes and antinodes
switched. In this case the diagram is talking about how much horizontal displacement air molecules experience,
rather than the air pressure: this displacement behaves in exactly the opposite way as pressure. For instance,
the molecules at a position of maximally varying pressure (i.e. a pressure antinode) don't actually move (it
is a displacement node); rather, the molecules to either side squeeze in or spread out to the sides as the
pressure changes. Mathematically it doesn't matter which quantity we're talking about; Equation (6.34)
works the same in both representations.

$$\frac{2n+1}{4}\lambda = L, \ n = 0, 1, 2, \dots \tag{6.34}$$

When playing, musicians change the effective length of a string (e.g. by pinning it with a finger) or tube (e.g. by placing their finger over a hole in the tube). This, in turn, changes the frequency f of oscillation, which our ears interpret as sound (see Problems 6.97 and 6.98). The next two examples apply these ideas.[19]

Example 6.13 Standing Waves in a Guitar ★★

A particular guitar string is 620.0 mm in length. If the string is allowed to vibrate, it oscillates at its fundamental frequency of 247 Hz.

(a) If the player then pins the string over a fret so as to effectively shorten its length by 67.0 mm, what will be the string's new fundamental frequency?

(b) If the shortened string is forced to oscillate in the 4th overtone, what frequency will it have?

First, recall from Equation (6.16) that the wave velocity on a string depends on its mass density μ and the tension, F_T. Neither of those things change in this problem, so the wave velocity is the same for both the original length and the shortened length. But if you recall that $v = \lambda f$ from Equation (6.14), it follows that

$$v_1 = v_2$$
$$\lambda_1 f_1 = \lambda_2 f_2$$
$$\frac{\lambda_1 f_1}{\lambda_2} = f_2$$

We are given f_1 in the problem, but neither λ_1 nor λ_2. The only other piece of information we have is the length of the string in both cases and the knowledge that the string vibrates in its fundamental frequency. This, as it turns out, is sufficient to determine the wavelengths: when a string is oscillating at its fundamental frequency, its length constitutes $\frac{1}{2}$ of a wavelength. We see this sketched in Figure 6.21 and it follows mathematically from Equation (6.33) if we set $n = 1$, the lowest allowable value:

$$\frac{n}{2}\lambda = L$$
$$\lambda = 2L$$

Initially we have $L_1 = 0.6200$ m, which yields $\lambda_1 = 1.240$ m. When the length of the string is shortened, we have $L_2 = 0.6200$ m $- 0.0670$ m $= 0.5530$ m and so $\lambda_2 = 1.1060$ m.

Thus we can return to our equation for f_2:

$$\frac{\lambda_1 f_1}{\lambda_2} = f_2$$
$$\frac{(1.240 \text{ m})\, 247 \text{ Hz}}{1.1060 \text{ m}} = f_2 = 277 \text{ Hz}$$

---

[19]The combination of traveling waves to yield a standing wave is also very important in quantum mechanics. In Part V of this book we will see that fundamental particles like the electron can behave much like a standing wave!

This is the answer to part (a). For (b), we need to determine the new wavelength. If you compare Figure 6.21 and Equation (6.33), you'll see that the first overtone corresponds to $n = 2$, the second overtone to $n = 3$, and so on. Thus if we are considering the fourth overtone, then $n = 5$, and we can determine the wavelength:

$$\frac{n}{2}\lambda = L$$

$$\lambda = \frac{2L}{n}$$

$$\lambda_3 = \frac{2\,(1.1060\ \text{m})}{5} = 0.44240\ \text{m}$$

Now as we argued above, the quantity λf is always the same, so we can consider this last situation and compare it to either of the previous two:

$$\lambda_1 f_1 = \lambda_3 f_3$$

$$\frac{\lambda_1 f_1}{\lambda_3} = f_3$$

$$\frac{(1.240\ \text{m})\,247\ \text{Hz}}{0.44240\ \text{m}} = f_3 = 692\ \text{Hz}$$

Example 6.14 Resonance with a Tuning Fork   ★★

A tuning fork consists of thin metal prongs that, when struck, vibrate at a specified frequency. If a tuning fork is brought near the open end of a tube and its frequency corresponds to one of the harmonic frequencies of the tube, the tuning fork will induce a standing wave in the tube (this is sometimes referred to as inducing **resonance**). Suppose a particular tube, open at one end and closed at the other, resonates in response to tuning forks of frequencies 786 Hz and 929 Hz, but not at any other frequency between them. What is the tube's fundamental frequency?

First, note that we can solve this example using the result of Problem 6.98; you are asked to do so in Problem 6.99. Here we'll take a different (and more cumbersome) approach. If we recall that $v = \lambda f$ and use this to eliminate λ in Equation (6.34), we find

$$\frac{2n+1}{4}\lambda = L$$

$$\frac{2n+1}{4}\frac{v}{f} = L$$

The quantities v and L (respectively the speed of sound and the length of the tube) are fixed in this problem; we have different values of f that correspond to different values of n. Let's denote these as cases a and b and set the left hand side of the above equation for each case equal to one another (again, because L is the same in both cases):

$$\frac{2n_a+1}{4}\frac{v}{f_a} = \frac{2n_b+1}{4}\frac{v}{f_b}$$

$$\frac{2n_a+1}{f_a} = \frac{2n_b+1}{f_b}$$

$$\frac{2n_a+1}{2n_b+1} = \frac{f_a}{f_b}$$

Now because the two resonant frequencies are adjacent, $n_b = n_a + 1$. If we substitute this into the above equation, we'll have just one equation and one unknown, and we can solve for n_a (the algebra is a bit tedious and worth careful attention, but hopefully the logic that got us to this point seems clear!):

$$\frac{2n_a + 1}{2n_b + 1} = \frac{f_a}{f_b}$$

$$\frac{2n_a + 1}{2(n_a + 1) + 1} = \frac{786 \text{ Hz}}{929 \text{ Hz}}$$

$$\frac{2n_a + 1}{2n_a + 3} = 0.846$$

$$2n_a + 1 = 0.846(2n_a + 3)$$

$$2n_a(1 - 0.846) = 0.846(3) - 1$$

$$n_a = \frac{0.846(3) - 1}{2(1 - 0.846)} = 5$$

We now know that the given frequencies correspond to $n = 5$ and $n = 6$. To determine the fundamental frequency (where $n = 0$), we can return to the equation we worked out above, which applies to any two harmonics (here we set $f_a = f_1$ and $f_b = f_5$):

$$\frac{2n_a + 1}{2n_b + 1} = \frac{f_a}{f_b}$$

$$\frac{2(0) + 1}{2(5) + 1} f_5 = f_1$$

$$\frac{1}{11}(786\text{Hz}) = f_1 = 71.5 \text{ Hz}$$

As you can confirm for yourself, setting $f_b = f_6$ and $n_b = 6$ yields the same answer.

6.8   Problems for Chapter 6

6.2 Two Seemingly Different Versions of the Sine Function

★ **Problem 6.1.** A common trigonometric identity is $\sin^2 \theta + \cos^2 \theta = 1$. Prove this with the help of the unit circle and the "right triangle" definitions of $\sin \theta$ and $\cos \theta$.

★★ **Problem 6.2.** Figure 6.2 shows the mapping of the unit circle to a graph of $\sin \theta$ vs θ. Make a similar figure that shows the relationship between a unit circle and a graph of $\cos \theta$ vs θ. (In Figure 6.2 the graph of $\sin \theta$ is to the *right* of the unit circle. Now you should draw your graph *above* the unit circle, with the θ axis vertical. Why?)

★★ **Problem 6.3.** Sketch $\sin \theta$, $\cos \theta$, and $\cos(\theta + \phi)$ such that $\cos(\theta + \phi) = \sin \theta$. What is ϕ? You might be able to "eyeball" it, but use the trigonometric identities in Appendix B to explain how you arrived at your answer.

★ **Problem 6.4.** The argument of trigonometric functions is an *angle*; if we're working in SI units then the angle is measured in radians. If x and t are distances and times measured in meters and seconds, as usual, then what units must the parameters α, β, and γ have in the following equations to ensure that this is the case?

a. $\sin(\alpha x)$

b. $\cos(\beta t)$

c. $\tan(\gamma)$

★★ **Problem 6.5.** Consider the functions (a) $y(x) = \sin(\alpha x)$ and (b) $y(t) = \cos(\beta t)$. Make a graph of each, starting at $x = 0$ and $t = 0$, respectively, and going far enough along the axes to show two complete cycles. Label the axes in terms of α and β to show how they determine how far along the axes you need to go to complete a full cycle of each function.

★★ **Problem 6.6.** A handy "trick" that can speed up the evaluation of some integrals is the fact that the integral of an *odd* function around a symmetric interval around $x = 0$ (from $x = -a$ to $x = a$, say) is equal to 0. (An odd function obeys $f(x) = -f(-x)$.) Formally, for such a function it follows that

$$\int_{-a}^{a} f(x)\, dx = 0 \tag{6.35}$$

We would describe these functions as *odd* on the interval $(-a, a)$ because the integral on the first half of the interval is equal and opposite to the integral on the second half. One can have functions that obey a similar statement around some *other* interval (from $x = x_0 - a$ to $x = x_0 + a$, say) such that

$$\int_{x_0-a}^{x_0+a} f(x)\, dx = 0 \tag{6.36}$$

These functions are odd on the interval $(x_0 - a, x_0 + a)$.

For each integral below, determine whether or not the function is odd on the specified interval. Support each answer with a sketch of the function. Explain from your sketches of the odd functions *why* the integrals must be 0.

a. $\int_{-10}^{10} x\, dx$

b. $\int_{-\infty}^{\infty} \sin\theta\, d\theta$

c. $\int_{-\infty}^{\infty} \theta \sin\theta\, d\theta$

d. $\int_{-\infty}^{\infty} \cos\theta\, d\theta$

e. $\int_{-\infty}^{\infty} \theta \cos\theta\, d\theta$

f. $\int_{0}^{2\pi} \sin\theta\, d\theta$

g. $\int_{0}^{2\pi} \cos\theta\, d\theta$

h. $\int_{0}^{\pi} \sin\theta\, d\theta$

i. $\int_{0}^{\pi} \cos\theta\, d\theta$

★★ **Problem 6.7.** In addition to *odd* functions (Problem 6.6), one can functions that are *even* about 0, such that $f(-x) = f(x)$ For such a function, an integral with bounds that are symmetric about 0 can be simplified by only evaluating one half of the integral and then doubling it:

$$\int_{-a}^{a} f(x)\, dx = 2\int_{0}^{a} f(x)\, dx \tag{6.37}$$

Similarly, if a function is even around some point other than 0, we have:

$$\int_{x_0-a}^{x_0+a} f(x)\, dx = 2\int_{x_0}^{x_0+a} f(x)\, dx \tag{6.38}$$

Refer back to Problem 6.6. Which integrals are over an even interval? Support your answers with a sketch of each (you may refer to your sketches from Problem 6.6 if you've already worked it) and rewrite the integrals using the properties described here. For this problem you don't need to actually solve the integrals; the point is to identify the symmetry.

6.3 Simple Harmonic Motion

★ **Problem 6.8.** The displacement of an object in simple harmonic motion is given by $x(t) = (5.0\ \text{cm}) \cos(20\pi t)$. Determine the following values for the oscillation of the object:

a. Amplitude

b. Angular frequency

c. Frequency

d. Period

e. Maximum velocity

f. Maximum acceleration

g. The times when the maximum velocity and maximum acceleration occur.

★ **Problem 6.9.** A bungee jumper is hanging from a long bungee cord and oscillating gently in simple harmonic motion. If their period of oscillation is 5.0 s and their mass is 70.0 kg, what is the spring constant of the bungee cord?

★★ **Problem 6.10.** An 85 kg student is hanging from a bungee cord that has a spring constant of 270 N/m. The student is pulled down so the cord is stretched by 5.0 m compared to its equilibrium length, then released. What is the maximum velocity of the student and where does this occur?

★ **Problem 6.11.** A mass spring system oscillates with a frequency of 0.50 Hz. If the spring constant is 150.0 N/m, what is the mass of the object?

★ **Problem 6.12.** A horizontal mass spring system of mass 2.00 kg oscillates with a frequency of 1.00 Hz and amplitude 0.500 m. Write down a possible equation of motion $x(t)$ for this system. If you need additional information to specify any relevant parameters, explain what you'd need to know.

★ **Problem 6.13.** A mass is attached to a spring and undergoing simple harmonic motion on a horizontal, frictionless surface. If the mass is 0.25 kg, the spring constant is 12 N/m, and the amplitude is 15 cm, what is the speed of the mass at 1/3 the maximum displacement from equilibrium?

★ **Problem 6.14.** Two springs have the same spring constant k. The first is attached to an object with mass $4m$ and compressed 10.0 cm from equilibrium. The second is attached to an object of mass m and compressed 5.0 cm from equilibrium. Which mass experiences the larger maximum acceleration?

★★ **Problem 6.15.** A spring is hanging from the ceiling.

a. A 50 kg student grabs onto it and it stretches 12 cm. What is the spring constant?

b. While the student hangs from the spring (in the position described in (a)), another student pulls him down another 5.0 cm and then releases him. What is the period of oscillation?

c. What is the student's maximum speed?

★★ **Problem 6.16.** Consider a particle in simple harmonic motion. For what displacement, as a fraction of the amplitude A, is the energy half kinetic and half potential?

★★ **Problem 6.17.** Consider a 0.15 g disk oscillating in simple harmonic motion (i.e. according to Hooke's Law) at 1.0 MHz. If the maximum force that the disk can withstand is 5.0×10^4 N, then what is the maximum oscillation amplitude that the disk should be subjected to?

★★ **Problem 6.18.** In space, a standard scale can't be used to measure the mass of an astronaut (because astronauts in space are weightless). Instead, they can be attached to a stiff spring and the period of oscillation can be measured. Suppose Figure 6.22 shows the oscillation of an astronaut in space. If the spring constant is 1.14×10^3 N/m, what is the mass of the astronaut? What is the equation $y(t)$ shown in the figure?

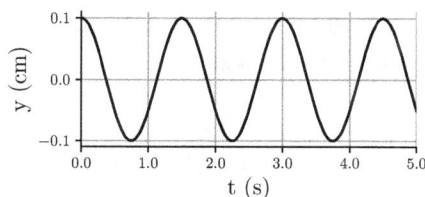

FIGURE 6.22
Problem 6.18

★★ **Problem 6.19.** Suppose an astronaut with a mass of 60.0 kg has been attached to a spring with a spring constant of 1200 N/m.

    a. What is their period of oscillation?

    b. Sketch a position vs. time plot for the astronaut's motion, assuming that at $t = 0$ the astronaut is at their greatest positive displacement from the spring's equilibrium length. Suppose this displacement is 5.0 cm.

    c. What mathematical equation describes the motion you sketched in (b)? Respond using (i) a sine wave and (ii) a cosine wave.

    d. What is the first time after $t = 0$ that the astronaut will have their greatest velocity? What is their x position at this instant?

★★ **Problem 6.20.** Figure 6.23 shows the oscillation of two different mass-spring systems. $m_A = 5.0$ kg and $m_B = 10.0$ kg.

    a. When is the spring potential energy for mass B at a maximum? At a minimum?

    b. What are the periods of oscillation T_A and T_B?

    c. What is the angular frequency ω_A and ω_B?

    d. What are the spring constants k_A and k_B?

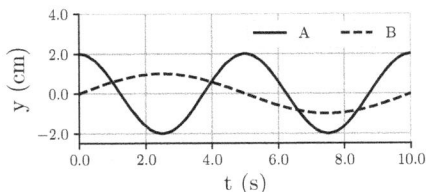

FIGURE 6.23
Problem 6.20

★★ **Problem 6.21.** A 0.200 kg block hangs from a spring with a spring constant 10.0 N/m. At $t = 0$ s the block is 15.0 cm below the equilibrium position moving downward with a speed of 1.00 m/s. What is the amplitude of oscillation?

★★ **Problem 6.22.** Figure 6.24 shows a position vs time plot for a 0.500 kg mass oscillating on a spring. Estimate the spring constant.

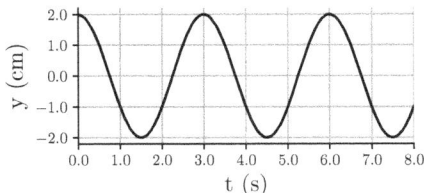

FIGURE 6.24
Problem 6.22

★★ **Problem 6.23.** A spring with an unstretched length of 0.200 m and a spring constant of 5.00×10^2 N/m is attached to the ceiling. A 3.00 kg box is connected to the bottom of the spring.

    a. What is the equilibrium position of the box? Use the ceiling as the origin of your coordinate system.

    The box is then pulled down 2.0 cm and released.

    b. How long does it take for the box to go through one a complete oscillation?

    c. Make a position vs. time graph for the box, using the instant of release as the instant when $t = 0$ s.

d. Write Equation (6.4) for this box (that is, provide numerical values for A, ω, and ϕ).

★★ **Problem 6.24.** Determine the equations for the (a) velocity and (b) acceleration of the box described in Problem 6.23.

★★ **Problem 6.25.** A mass-spring system is in SHM in the horizontal direction. The mass is 0.25 kg, the spring constant is 15 N/m and the amplitude is 0.10 m.

a. What is the total energy in the system?

b. What is the maximum speed of the mass? Where does this occur?

c. What is the speed of the mass at half amplitude?

d. Write an equation for $v(t)$.

★ **Problem 6.26.** A particular wave obeys $y(t) = A\sin(\omega t)$. If you wished to express the same wave with a cosine function, you'd need to introduce a phase shift: $y(t) = A\cos(\omega t + \phi)$. Determine the value of ϕ by considering t at quarter multiples of the period T (i.e. $t = 0$, $t = T/4$, $t = T/2$, $t = 3T/4$, and $t = T$) and determining the value(s) of ϕ that ensures the cosine equation for the wave yields the same displacement as the sine version. Include a graph of both curves.

★ **Problem 6.27.** Starting from Equation (6.4), derive Equations (6.5)–(6.6).

★★ **Problem 6.28.** Consider Equations (6.4)–(6.6) in the context of Figure 6.3.

a. What is ϕ in this case?

b. When in the motion is the *position* a maximum, minimum, or 0 m? Express your answers as multiples of the period, e.g. "0.5T".

c. Repeat (b) for the *velocity*.

d. Repeat (b) for the *acceleration*.

★★ **Problem 6.29.** Consider an object of mass m connected to a spring with a spring constant k. The mass is pulled to the side and released such that it experiences SHM. Show that the total energy of the mass is conserved; i.e. show that it does not vary in time. (The expression for the total energy should include both a *kinetic* term and a *potential* term.)

★★ **Problem 6.30.** A 7.5 kg box sits on a flat, frictionless surface. It is attached on the left side to a spring with spring constant 50.0 N/m. Initially, the box is at rest and the spring is at its equilibrium length. Then, a hammer strikes the right edge if the box such that the box has an instantaneous velocity of 15.0 cm/s to the left.

a. How far will the spring compress from its equilibrium length?

b. How fast will the box be moving when the spring is compressed 1/2 the distance you found in (a)?

c. What acceleration will the box experience when the spring is compressed 1/2 the distance you found in (a)?

d. How long will it take for the box to return to the original position?

★ **Problem 6.31.** A 30.0 kg child is sitting on a playground swing undergoing simple harmonic motion.

a. If the swing is 2.5 m long, what is the period of oscillation?

b. If the child stands up on the swing, will the period of oscillation increase, decrease, or remain the same as the value you found in (a)?

★★ **Problem 6.32.** Suppose the child described in Problem 6.31 is sitting on the swing when they are pulled back 15° by their parent and then shoved forward such that their initial velocity (tangent to the circle formed by the swing's chain) has a magnitude of 0.50 m/s.

a. What will their maximum angular displacement (in degrees) be?

b. What is the child's angular frequency ω?

c. How fast will they be moving as they go through the lowest point of the swing (i.e. when $\theta = 0°$)?

d. What is the child's maximum kinetic energy?

e. What is the child's maximum potential energy?

★★ **Problem 6.33.** Two 25.0 kg children are on playground swings. The swings are of slightly different lengths: $L_A = 2.3$ m and $L_B = 2.6$ m. Their parents pull them back the same angular distance $\theta_i = 12°$ and release them from rest at $t = 0$. (Ignore, as usual, the effect of friction.)

a. What is the period of oscillation of child A?

b. What is the period of oscillation of child B?

c. Write out the equation of motion for each child, $\theta_A(t)$ and $\theta_B(t)$.

d. Graph both equations from (c) on the same graph. (Label which is which!)

e. At what times will the children have the same angular position? Support your answer mathematically but refer also to your graph in (d).

★ **Problem 6.34.** You construct a simple pendulum with a 0.550 kg rubber ball and a light rope with a length of 0.75 m.

a. What will the period of oscillation be for this pendulum?

b. By what factor will the pendulum's period change if the length is doubled? What if it is halved?

c. By what factor will the pendulum's period change if the mass is doubled? What if it is halved?

d. What will the period be if this pendulum is constructed on the Moon?

★ **Problem 6.35.** You'd like to build a simple pendulum with a period of 1.00 s.

a. What length string should you use?

b. What would the period of oscillation be if you used a string with twice the length you found in (a)?

c. What would the period of oscillation be if you used a string with half the length you found in (a)?

★ **Problem 6.36.** Suppose an astronaut takes a 0.56 m long pendulum to Mars and measures the period of oscillation to be 2.45 s. What is the corresponding free-fall acceleration on Mars?

★ **Problem 6.37.** If a simple pendulum has a period of 1.00 s on Earth, what would the period of oscillation be

a. on the Moon?

b. on Mars?

c. on the surface of the Sun?[20]

★★ **Problem 6.38.** Students on another planet measure the period of a simple pendulum with varying lengths and create a graph of the period squared vs. the length using the SI system of units (Fig. 6.25). They perform a linear fit using a computer; the computer provides the equation $y = 3.31x - 0.07$.

a. What are the units of y, 3.31, x, and 0.07?

b. What is the acceleration due to gravity on this planet?

---

[20]I advise against attempting to experimentally validate your answer to this one.

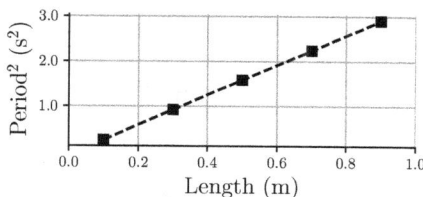

FIGURE 6.25
Problem 6.38

Problem 6.39. A simple pendulum includes a mass m attached to a string of length L; it has a period of oscillation T. If the pendulum is attached to the roof of an elevator and the elevator begins to accelerate up at a rate a, what will the period of oscillation be?

Problem 6.40. In discussing the simple pendulum we made the small angle approximation $\sin \theta \approx \theta$. For what value of θ (measured in degrees, considering integer values only) does the percent difference between $\sin \theta$ and θ exceed

a. 5%?

b. 10%?

Provide your answer by making a table of $\sin \theta$, θ, and the percent difference; identify a few values for θ below and then above each threshold.

Problem 6.41. A common demonstration of simple harmonic motion in physics classrooms involves attaching a bowling ball to a long rope that is tied to the ceiling. The instructor holds the bowling ball in front of their face, releases it, and allows it to swing through a complete oscillation. Because energy is conserved, the ball will come to rest at the same point it was released, just in front of their face. Suppose I am running such a demonstration in a lecture hall where the ceiling is high enough that the pendulum length is 8.0 m.

a. If I want to dramatically scream in feigned panic from the instant I release the ball until the instant it returns to the release point, how long will I need to scream, assuming the small angle approximation is valid?

b. Make a sketch of the situation, assuming that the ball has been pulled to the side such that its initial angular displacement θ_i is sufficiently large that the ball is 2.0 m above its lowest point (in front of my face). From the geometry, determine a numerical value for θ_i.

c. Does your result from (b) seem small enough for the small angle approximation to be valid? Support your answer by referring to the results of problem 6.40.

Problem 6.42. Show that the general solution for simple harmonic motion $x(t) = C_1 \sin(\omega t) + C_2 \cos(\omega t)$ can be rewritten in the form of Equation (6.4). Show the definition of A and ϕ in terms of C_1 and C_2. (For simplicity, set $x_0 = 0$.)

Problem 6.43. An object gently bobbing up and down in a liquid displays simple harmonic motion. It can be shown that for such an object, $\omega^2 = \rho g A/m$, where A is the cross sectional area of the object, m is its mass, ρ is the density of the fluid, and g is the acceleration due to gravity. If you have a cube of brass (density of 8.00×10^3 kg/m^3) which measures 4.0 cm along each edge, what must the density of the fluid be if you want the period of oscillation to be the same as a 0.25 m long simple pendulum?

6.4 Traveling Waves

Problem 6.44. Figure 6.26 shows a snapshot graph at $t = 0.00$ s for a pulse traveling in the $+x$ direction at a speed of 1.00 m/s. (a) When will the pulse finish passing through $x = 0.50$ m? (b) Draw a history graph for $x = 0.50$ m from $t = 0.00$ s until the time you found in (a).

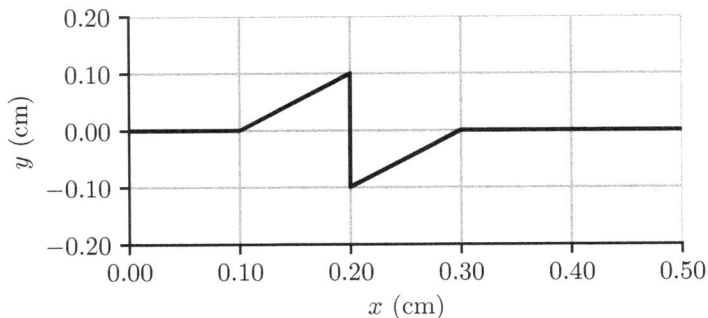

FIGURE 6.26
Problem 6.44

★   **Problem 6.45.** Figure 6.27 shows a snapshot graph at $t = 0.00$ s for a pulse traveling in the $+x$ direction at a speed of 0.50 m/s. (a) When will the pulse finish passing through $x = 0.00$ m? (b) Draw a history graph for $x = 0.00$ m from $t = 0.00$ s until the time you found in (a).

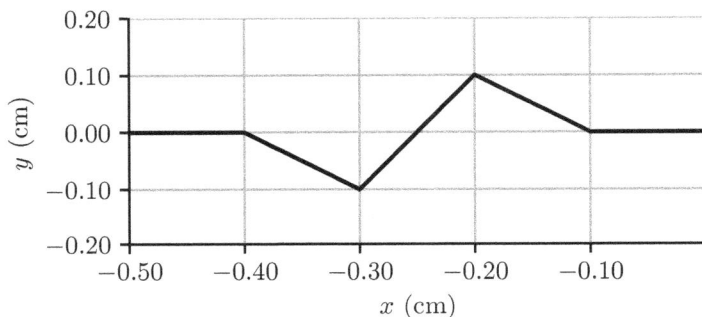

FIGURE 6.27
Problem 6.44

★   **Problem 6.46.** Figure 6.28 shows a history graph for $x = 0.00$ m as a pulse of total width 2.0 cm moving in the $+x$ direction passes through. (a) What is the speed of the pulse? (b) Draw a snapshot graph for $t = 0.00$ s that encompasses the entire pulse.

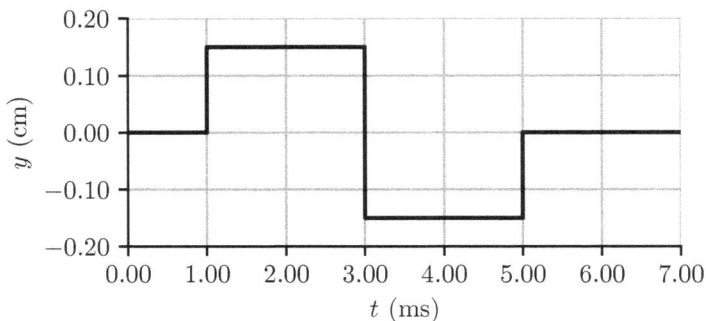

FIGURE 6.28
Problem 6.46

★   **Problem 6.47.** Figure 6.29 shows a history graph for $x = 0.10$ m as a pulse moving at -2.0 m/s passes through. (a) How wide is the pulse? (b) Draw a snapshot graph for $t = 3.00$ ms that encompasses the entire pulse.

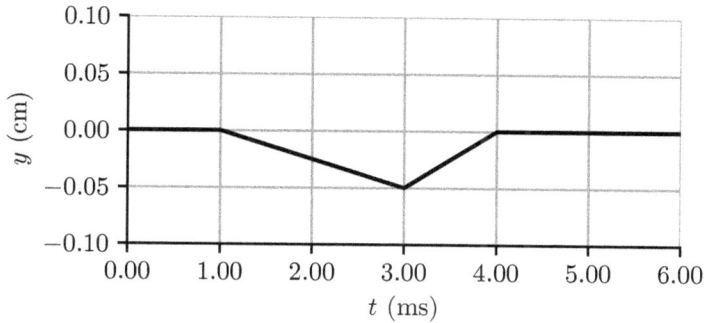

FIGURE 6.29
Problem 6.47

★ **Problem 6.48.** Two ropes of equal length are tied together. The first rope has twice the mass of the second. You grab the free end of the first rope and a friend grabs the free end of the second rope. You pull the ropes taut, then you jerk your side and send a pulse toward your friend. If the pulse travels the length of the first rope in a time Δt_1, then how much time (in terms of Δt_1) will the pulse spend traveling the length of the second rope?

★ **Problem 6.49.** A wave travels down a taut rope at 150 m/s. What is the wave speed if the tension is (a) doubled or (b) halved?

★ **Problem 6.50.** There is talk of one day colonizing Mars, which is an average distance of 2.25×10^8 km from Earth. How long does it take for a 2.00×10^2 MHz radio wave to travel this distance? What is its wavelength?

★ **Problem 6.51.** Green light with a wavelength of $\lambda = 510$ nm passes from air into a 15.0-cm-wide tank of water. (a) What will the wavelength of the light be as it passes through the water? (b) How long will it take the beam to pass through the tank?

★★ **Problem 6.52.** A microphone is placed on one side of a 0.30-m-wide container filled with helium gas at 20.°C. Someone taps the opposite side of the container: two sound pulses are detected, one for the wave that traveled through the helium and one that traveled through the air. The speed of sound in the helium gas is 1.0×10^3 m/s.

a. How much time passed between the two pulses reaching the microphone?

b. The speed of sound in a gas is proportional to the square root of the temperature measured in K, i.e. $v_{\text{sound}} = \sqrt{\alpha T}$ for a gas-specific constant α. With this in mind, repeat (a) for $T = 30.°C$ and $T = 40.°C$.

c. Make a graph of v_{sound}^2 vs. T and include points for each of the three temperatures you considered in (a) and (b). Determine a value of α for helium by using these results. How is the value of α represented in your graph? Why is it useful to plot v_{sound}^2 vs. T instead of v_{sound} vs. $\sqrt{T}$?

★ **Problem 6.53.** You gently strike a tuning fork with the palm of your hand and record the sound with a microphone. Figure 6.30 shows the recording as a function of time.

a. Write out the complete wave equation for this wave.

b. What is the frequency of the tuning fork?

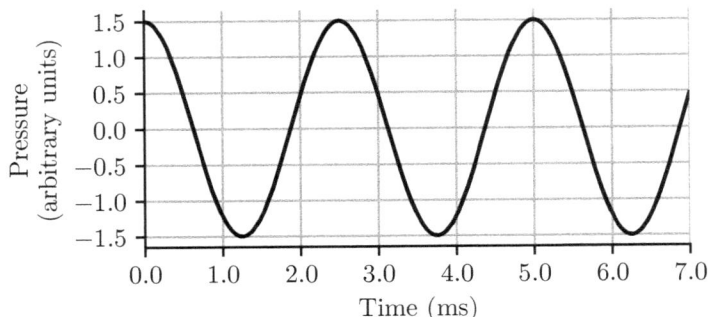

FIGURE 6.30
Problem 6.53

★★ **Problem 6.54.** A microphone records the following signal (measured as an electrical voltage) from a tuning fork:

$$V(t) = (2.0 \text{ mV}) \sin\left(\left(2830\frac{\text{rad}}{\text{s}}\right)t - \pi\right)$$

a. What is the frequency of the tuning fork?

b. What is the wave number?

c. Make a labeled history graph of the wave at the location of the microphone.

★★ **Problem 6.55.** A string is under 50.0 N of tension and has linear mass density of 5.0 g/m. A sinusoidal wave with amplitude 3.0 cm and wavelength of 1.2 m travels along the string.

a. What is the speed with which the wave travels down the string?

b. What is the maximum speed with which a particle on the string moves?

★★ **Problem 6.56.** A traveling wave obeys

$$D(x, t) = (3.0 \text{ cm}) \sin\left(\left(0.25\frac{\text{rad}}{\text{m}}\right)x - \left(700\frac{\text{rad}}{\text{s}}\right)t + \pi\right)$$

Determine the (a) wavelength, (b) frequency, (c) period, (d) angular frequency, (e) amplitude, and (f) wave speed. (g) Make a (neatly labeled) graph of the string at $t = 0$ s. (h) Make a (neatly labeled) graph of the string a half period later.

★ **Problem 6.57.** A 35 g string is under 17 N of tension. A pulse travels the length of the string in 48 ms.

a. How long is the string?

b. If the mass was increased but the tension and length remained the same, would the pulse take more or less time to travel the length of the string?

★ **Problem 6.58.** What is the frequency of an electromagnetic wave traveling through air with a wavelength of 18 cm? Suppose a sound wave in water (where the wave travels at about 1500 m/s) travels with the same frequency; what would its wavelength be?

★ **Problem 6.59.** The musical note C#3 has a frequency of about 140 Hz. What is the corresponding wavelength?

★ **Problem 6.60.** A light string has a mass of 0.100 kg and a length of 2.00 m. What tension is being applied to the string if a pulse travels from one end to the other in 0.020 s?

★ **Problem 6.61.** You see a flash of lightning, then 5.0 seconds later hear a rumble of thunder. How far away from you was the lightning strike?

★ **Problem 6.62.** What is the frequency of visible light with a wavelength of 530 nm?

★ **Problem 6.63.** A blue laser has a wavelength of 480 nm. The beam passes from air into a block of diamond, which has an index of refraction of 2.4. Indicate both the wavelength and frequency of the laser beam (a) before it enters the diamond, (b) in the diamond, and (c) after it leaves the diamond.

★ **Problem 6.64.** Suppose you are calibrating a sonar device used to locate underwater objects. You have a pressure detector just in front of the emitter and have recorded the signal shown in Figure 6.31. What is the wavelength? Assume the speed of sound through the water is 1500 m/s.

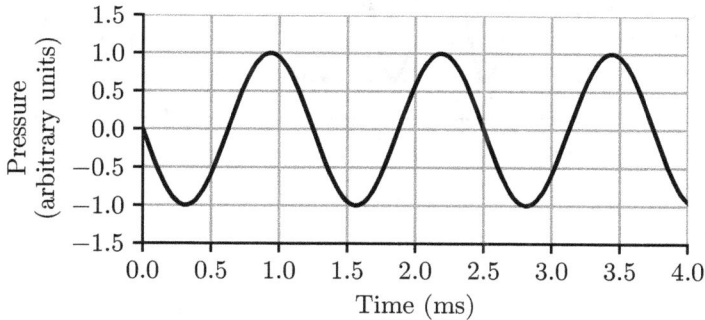

FIGURE 6.31
Problem 6.64

★ **Problem 6.65.** Consider a traveling wave on a taut string that obeys

$$y\,(x,t) = (2.0\ \text{cm})\sin\left((0.50\ \text{rad/m})\,x - (10\ \text{rad/s})\,t + \pi/2\right)$$

a. What is the speed with which the traveling wave moves down the string?

b. Sketch y vs. x when $t = 0$ s.

c. Sketch y vs. x when $t = 2.0$ s.

★★ **Problem 6.66.** Figure 6.32 shows a snapshot of a traveling wave moving to the right at a speed of 100 m/s. Write the complete wave equation $y\,(x,t)$.

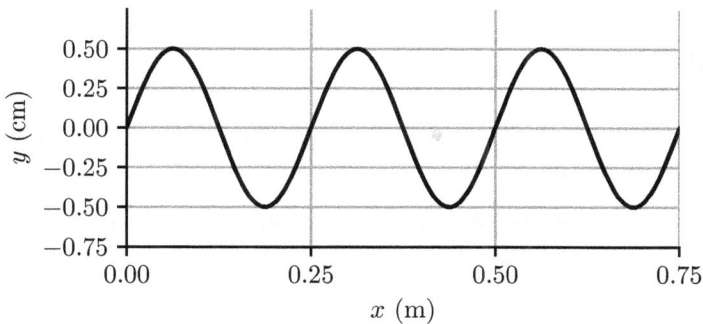

FIGURE 6.32
Problem 6.66

★ **Problem 6.67.**

Figure 6.33 depicts a snapshot of a traveling wave taken at $t = 0$ s. The wave is traveling in the $+x$ direction at 50 m/s.

a. What is the amplitude?

b. What is the wavelength?

c. What is the period?

d. What is the phase? The answer depends on your choice of sine or cosine in expressing the wave; you may choose either but be clear in stating your choice.

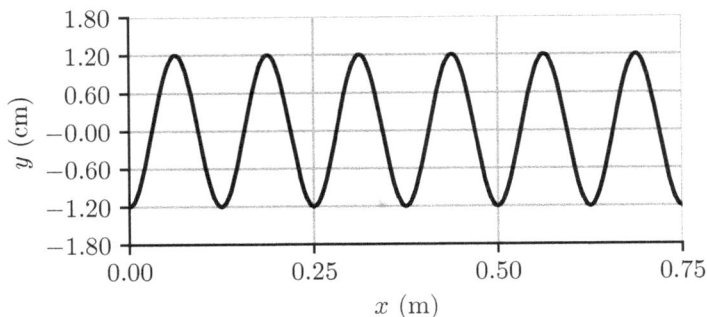

FIGURE 6.33
Problem 6.67

★★   **Problem 6.68.** A traveling wave moves down a string with a speed of 0.50 m/s. Figure 6.34 shows the displacement of one point on the string as a function of time. What is the wavelength of the traveling wave?

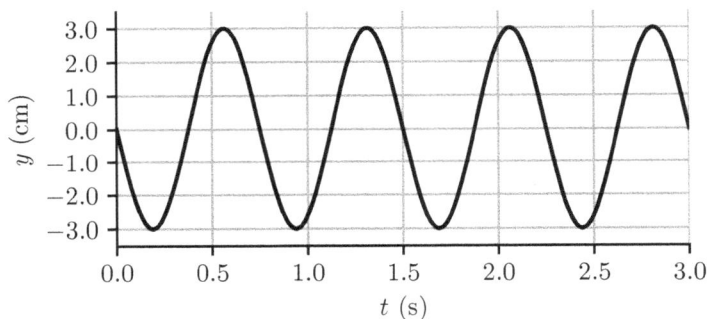

FIGURE 6.34
Problem 6.68

★   **Problem 6.69.** What is the wave number k for a traveling wave with an angular frequency of 20.0 rad/s and a wavelength of 1.3 m?

★★   **Problem 6.70.** A tuning fork is struck and a nearby microphone records the tone. Your data is the voltage (the microphone's way of reporting the sound intensity) over time; you use a computer to fit a sinusoidal wave to the data to yield the following equation:

$$V(t) = (2.50 \text{ mV}) \sin((3770 \text{ rad/s}) t + 3.142 \text{ rad}) + 0.100 \text{ mV}$$

a. What is the frequency of the tuning fork?

b. What is the wavelength of the wave?

c. Sketch and label the wave.

★★   **Problem 6.71.** Power outlets provide electric current that oscillates sinusoidally; the signal (though not the electrons in the wire) travels through electrical wires at around 2.7×10^8 m/s. Figure 6.35 below shows such a current (denoted I) measured at $x = 0$ m.

a. What is the frequency of the current's oscillation?

b. What is the wave equation for the current? Assume the wave is traveling in the positive direction.

c. In addition to the electric current I, analysis of AC circuits often involves the electric potential V. For the · . . . . $I(t)$ and $V(t)$?

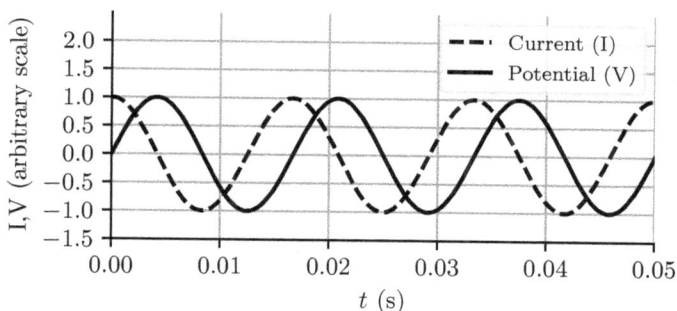

FIGURE 6.35
Problem 6.71

★★ **Problem 6.72.** In discussing Equation (6.13) I stated that the temporal term ωt is negative if the wave is traveling to the right and positive if it is traveling to the left. To see why, suppose $\phi = \pi/2$ for a particular wave.

a. In this case, explain why the term $kx \pm \omega t$ in Equation (6.13) must equal 0 at a maximum.

b. If the wave is traveling to the right, the position x of a maximum should increase as time increases. With this in mind, explain why we should use the negative sign in $kx \pm \omega t$.

c. If the wave is traveling to the left, explain why we should use the positive sign in $kx \pm \omega t$.

d. Your answers for (b) and (c) are general for waves with different values of ϕ. Why?

6.5 Sound Intensity and Loudness

★ **Problem 6.73.** Rank the power delivered by the following three sound waves: wave A delivers 5 J of energy in 5 s, wave B delivers 10 J of energy in 10 s, and wave C delivers 15 J in 5 s.

★ **Problem 6.74.** A typical sound intensity level of a professional football game is 95 dB. At this sound intensity level, what power is being delivered to the surface of a circular ear drum with a radius of 4.0 mm?

★ **Problem 6.75.** A microphone converts mechanical energy (from a speaker's voice) into electrical energy that can be directed to speakers and greatly amplified for an audience. Suppose a person is holding a microphone and speaking at 70. dB. If the microphone's diaphragm (which absorbs the sound) is circular with a diameter of 1.2 cm, then how much power is being absorbed by the microphone?

★ **Problem 6.76.** Your friend takes off their headphones and you hear their music at a conversational volume of 60.0 dB. If you're standing 2.0 m from them when this happens, then what sound intensity level (in dB) is being delivered to their ears with the headphones on? Suppose the distance from the ear drum to the headphones is 2.5 cm when the headphones are being worn.

★ **Problem 6.77.** You are standing 2.0 m from a chainsaw; you use an app on your phone and record the noise at 105 dB. How much sound power, in Joules, is being emitted by the chainsaw every second?

★ **Problem 6.78.** You have been hired to purchase and install audio equipment for an outdoor concert venue. You would like the on-stage speakers to be loud enough for someone standing at the back of the venue to hear the music at 90 dB. If the back of the venue is 80 m from the stage, how much power do the speakers need?

★ **Problem 6.79.** The loudest sound in recorded history is the 1883 volcanic eruption of Krakatoa, an Indonesian island. The explosion was recorded 100 miles (about 161 km) from the source at 172 dB. How far from the explosion would the sound have been recorded at a faint but still clearly audible 50.0 dB? Report your answer in kilometers.

★ **Problem 6.80.** Consider a hallway conversation: assuming that you are speaking with a friend at a normal conversational volume of 60. dB (measured 1.0 m from where you are standing), what sound intensity level (in dB) would a person down the hall hear if they were 15 m away from you? Considering that the sound intensity level of a whisper is about 20. dB, do you think someone at that distance could easily overhear your conversation?

★ **Problem 6.81.** A shotgun being fired can emit sound intensity levels of around 160. dB at a distance of 1.0 m from the muzzle. How far should you stand from the muzzle for the sound intensity level to register at (a) the "threshold of pain", which is 120 dB, and (b) 60 dB, which is the level of a normal conversation?

★ **Problem 6.82.** Someone standing 15 m from the speakers of an outdoor rock concert measures the sound at 135 dB. How far should someone be standing from the speakers to record a whisper-quiet 30. dB? (Your answer will be larger than the actual value because of factors like wind that you are ignoring here.)

★ **Problem 6.83.** Suppose you are standing 1.00 meter from a person screaming at 80.0 dB (a reasonable estimate for the loudest a person can scream). What is the corresponding intensity? What is the power being emitted from a screaming person's mouth? (Estimate the size of their mouth). How far away would you need to be from the person for the sound intensity to fall to a whisper-quiet 20.0 dB?

6.6 The Doppler Effect

★ **Problem 6.84.** At a water park, you're going down a slide when a referee at the bottom of the slide starts blowing a whistle. You'd hear the whistle at a frequency of 3.0 kHz if you were both stationary, but instead you're sliding toward the lifeguard at 15 m/s. What frequency do you hear?

★ **Problem 6.85.** Electrical transformers are found on the electrical power posts that carry electricity to consumers (e.g. businesses and homes) from power stations and substations. The magnetic fields involved in the transformer coils can create a "hum" at a low frequency of around 5.0×10^1 Hz. Suppose you are driving with your windows down toward one such transformer one quiet evening at 22 m/s (about 50 mph).

   a. Do you expect the frequency you hear to be equal to 50 Hz, above 50 Hz, or below 50 Hz? Explain with words and a diagram.

   b. Determine a numerical value for the frequency you will hear.

★★ **Problem 6.86.** You and a friend are performing an experiment where your friend stands on the roof of a building and you stand on the ground below. Your friend activates a small electronic device that emits a pure 4.00×10^2 Hz tone; you hold a microphone below the device as it falls and record the sound. You find that the recording contains frequencies up to 4.20×10^2 Hz. How fast was the emitter traveling when you recorded this frequency?

★ **Problem 6.87.** You analyze an audio recording of a train and determine that the train's whistle had a frequency of 5.30×10^2 Hz as it approached the device recording the audio. When the train is stationary its whistle is supposed to have a frequency of 5.00×10^2 Hz. How fast was the train moving? Is this above or below the speed limit of 49 mph (about 22 m/s)?

★★ **Problem 6.88.** You are standing on a hill when I suddenly zoom past you on my bicycle. An instant later you hear a blood-curdling scream and realize I am being chased by a student on a unicycle. You quickly pull out your phone and measure the frequency of the student's voice to be about 107 Hz as he moves away from you down the hill at a speed of 10.0 m/s. What frequency did you hear when the unicyclist was still approaching you?

★ **Problem 6.89.** White light is consists of an even mixture of all visible wavelengths, which has an average value of around 540 nm. In a movie, white stars are visible through the cockpit of a stationary space ship. The ship then speeds up dramatically, and the stars turn blue. To be specific, at one instant you pause the movie, pull out your electromagnetic spectrum chart, and estimate that the wavelength of the light is about 480 nm. (a) Your friend objects that the stars should appear red, not blue. Are they correct? Why or why not? (b) How fast do you think the ship was traveling at this instant?

Problem 6.90. The star in the Milky Way Galaxy with the highest speed relative to Earth is traveling at 2.30×10^6 m/s. Suppose we would expect to detect electromagnetic radiation from this star with a wavelength of 1.50×10^{-8} m, if the star was stationary. If the star is moving away from us, what wavelength would we expect to detect instead?

Problem 6.91. Consider a distant star that is stationary relative to Earth except for the fact that it is spinning about an axis: one point on the surface of the star has a tangential velocity $\vec{v}$ directed *toward* Earth and the point on the opposite side of the star has a tangential velocity $\vec{v}$ directed *away* from Earth. If scientists would expect to detect light of a particular frequency f_0 if the star were perfectly stationary, what frequency (or frequencies) will they detect instead because of the star's rotation? Suppose $v = 0.24c$ and express your answer as a multiple of f_0.

Problem 6.92. Suppose the light given off by a star includes a wavelength of 501 nm (a blue-green color given off by helium). If the star collides with another star at $0.05c$, what wavelength will this light appear to have from the perspective of the other star?

Problem 6.93. A traffic light emits red light with a wavelength of about 650 nm and green light with a wavelength of about 530 nm. How fast do you have to move relative to a red light for it to appear green? Express your answer as a multiple of c. Do you have to move *toward* or *away from* the light to observe this effect?[21]

Problem 6.94. Scientists claim that the Milky Way galaxy will collide with the Andromeda galaxy (which is about 2.5 million light years away) in about 4 billion years. Skeptical, and more than a little worried, you decide to check their predictions. The velocities of distant galaxies are measured by observing the wavelength of light that they emit. If a distant galaxy is moving relative to the Earth, then this light has a shifted wavelength due to the Doppler effect. One benchmark wavelength is 656.279 nm, which is one wavelength given off by hydrogen. If the scientists are correct that the collision will occur in 4 billion years, what shifted wavelength do you expect to detect? (For the purposes of this problem, assume the distance between the galaxies and the time until collision are *exact*.)

6.7 Standing Waves and Musical Instruments

Problem 6.95. Figure 6.36 shows two pulses approaching each other on a string at $t = 0.0$ s. Sketch the string at (a) $t = 1.0$ s, (b) $t = 1.5$ s, and (c) $t = 2.0$ s.

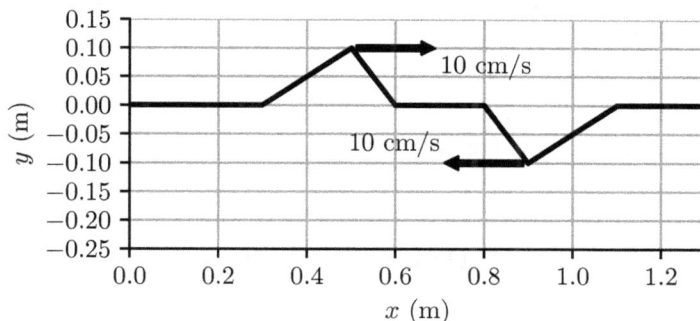

FIGURE 6.36
Problem 6.95

---

[21]Thanks to my student Texas Doehring for suggesting this problem.

Problem 6.96. Figure 6.37 shows two pulses approaching each other on a string at $t = 0.0$ s. Sketch the string at (a) $t = 2.0$ s, (b) $t = 3.0$ s, and (c) $t = 4.0$ s.

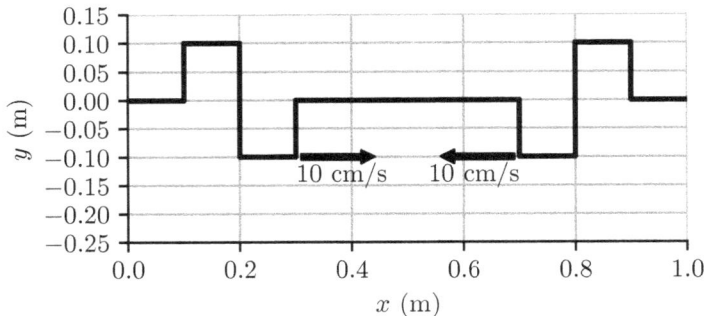

FIGURE 6.37
Problem 6.96

Problem 6.97. Consider a string of length L that is pinned on both sides. Show that standing waves have frequencies given by

$$f_n = n\frac{v}{2L}, n = 1, 2, 3, \ldots \tag{6.39}$$

where v is the wave speed. (This equation also applies to a tube that is open on both sides *or* closed on both sides.)

Problem 6.98. Consider a tube of length L that is closed on one side and open to the air on the other side. Show that standing waves have frequencies given by

$$f_n = (2n + 1)\frac{v}{4L}, n = f_n = (2n + 1)\frac{v}{4L}, n = 0, 1, 2, \ldots \tag{6.40}$$

where v is the wave speed.

Problem 6.99. Solve Example 6.14 using Equation (6.40).

Problem 6.100. Standing waves on a 2.5m-long string that is fixed on both ends are observed at successive frequencies of 36 Hz and 54 Hz. What is the fundamental frequency? What is the wave speed?

Problem 6.101. A piccolo is a musical instrument that resembles a small flute; it is about 32.0 cm long.

a. If we consider the piccolo to be a tube that is open to the air at both sides, then what is its fundamental frequency?

b. Sketch the fundamental standing wave inside the tube.

Problem 6.102. Consider a hollow pipe that is 15 cm long and open to the air at both ends. If you hold a tuning fork with an appropriate frequency near the pipe, you can induce a standing wave in the pipe.

a. What are the two lowest frequencies that will induce such a standing wave?

b. Sketch the lowest-frequency wave (air pressure vs position).

Problem 6.103.

A particular violin string has a mass density of 0.380 g/m and a length of 0.330 m.

a. If the fundamental frequency is supposed to be 660 Hz, then how much tension should be applied to the string?

b. Sketch the standing wave. Label all nodes and antinodes.

★ **Problem 6.104.** A particular guitar string is 0.90 m long and has a mass of 3.8 g. If the string is supposed to have a fundamental frequency of 110 Hz, how much tension should be applied to it?

★★ **Problem 6.105.** A very short piano string has a length of 54.0 mm and should play a note with a frequency of 4186 Hz. The mass density of the string is 5.00×10^{-3} kg/m.

a. How much tension must be applied to the string to ensure it plays the appropriate frequency when it is in its lowest vibrational mode?

b. Sketch the vibrating string: label the nodes, antinodes, and wavelength.

c. With what speed would a pulse travel down this string?

★★ **Problem 6.106.** A student is building a bass guitar. One string has a mass of 2.1 g and length 89.5 cm; the student fixes both ends in place and applies a tension of 1.00×10^2 N to it.

a. What frequency will it have in its fundamental mode of oscillation?

b. Sketch the vibrating string: label the nodes, antinodes, and wavelength.

c. If the student wanted to increase the frequency, should the tension be increased or decreased? Support your answer with appropriate mathematics.

★ **Problem 6.107.** A string is connected to a mechanical oscillator on one side and tied to a post on the other side. The oscillator frequency is adjusted to form a standing wave on the string, as shown below. If the string has a linear mass density of 8.5 g/m and is under 15 N of tension, how long does it take for a point on the string to undergo one complete oscillation?

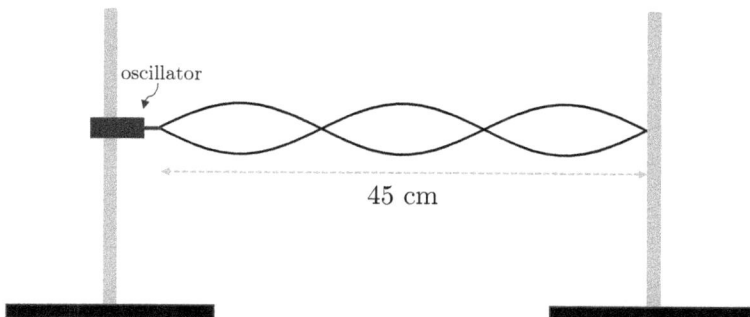

FIGURE 6.38
Problem 6.107

★ **Problem 6.108.**

A particular guitar string has a length of 65 cm and a mass of 1.92 g.

a. How much tension should it be under if its fundamental frequency of vibration is to be 147 Hz?

b. Sketch the standing wave. Label all nodes and antinodes.

★ **Problem 6.109.** A string with a length of 30.0 cm and a mass of 50.0 g is connected to a 600.0 g mass via a pulley on one side and a mechanical oscillator on the other (Fig. 6.39).

a. What is the lowest oscillator frequency that will generate a standing wave?

b. With what speed will a traveling wave move down the string?

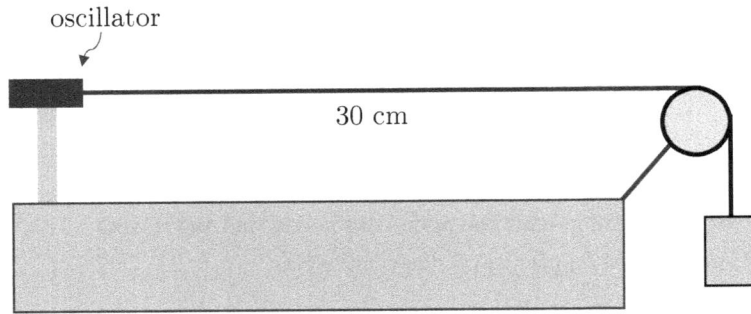

FIGURE 6.39
Problem 6.109

★ **Problem 6.110.** Figure 6.40 shows a standing wave on a string that is being driven at 250 Hz.

   a. What is the wavelength of the standing wave?

   b. What is the speed of the traveling waves that form the standing wave?

★ **Problem 6.111.** Figure 6.41 shows a standing wave on a string. The string has a mass of 6.0 g and the tension in the string is 5.0 N.

   a. What is the wavelength of the standing wave?

   b. What frequency is driving the oscillation?

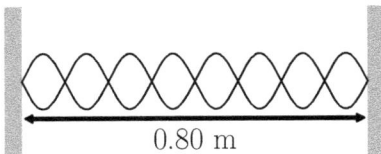

FIGURE 6.40
Problem 6.110

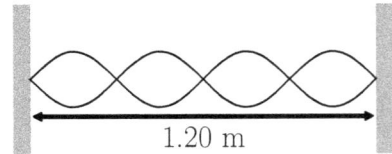

FIGURE 6.41
Problem 6.111

★ **Problem 6.112.** Figure 6.42 shows a standing wave in a tube that is closed to the air on one side and open to the air on the other.[22]

   a. Does the wave depict oscillations in the *pressure* in the tube or the *position* of the air molecules in the tube? How do you know?

   b. What is the wavelength?

   c. What is the frequency of oscillation?

---

[22]These diagrams can be misleading when representing longitudinal waves. The gray "U" depicts the physical location of the tube. For the wave, the horizontal axis similarly represents the position inside the tube. The *vertical axis* of the wave, however, indicates variations in position or pressure. For *pressure*, a node represents steady air pressure (usually *atmospheric* pressure) and an antinode indicates maximal oscillation in air pressure. For *position*, a node represents "molecules at this position don't vibrate due to the wave" whereas an antinode indicates maximal oscillation along the axis of the wave (here, *horizontal*, i.e. left and right oscillations from the molecule's usual position).

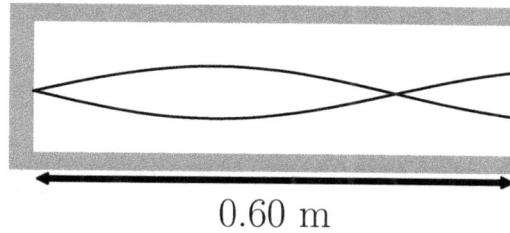

0.60 m

FIGURE 6.42
Problem 6.112

★ **Problem 6.113.** A simple model of a human voice is to consider the "vocal tract" as a pipe that is open to the air on one side (the mouth) and close on the other (the back of the larynx). Suppose the vocal tract for a particular person is 170 mm long. What are the lowest *three* resonant frequencies of this person's voice? Assume the speed of sound in the vocal tract is about 350 m/s (due to the higher temperature).

★★ **Problem 6.114.** A 1.0-m-tall tube is filled with water and held vertically. A 550 Hz tuning fork is held over the top of the tube, which is open to the air, as water is slowly drained through a hole at the bottom of the tube. At what water heights will a standing wave be induced in the tube?

★★ **Problem 6.115.** Vibrations in mechanical equipment can cause damage and make loud noises. Suppose an industrial fan is rotating such that a fan blade on the roof of a factory is passing over the top of an exhaust pipe at a frequency of 2.00×10^2 Hz. Suppose the pipes are straight and open at both sides. Design considerations indicate that you can choose a pipe length from 0.200 m to 1.00 m. What length(s) in this range, if any, should be avoided to prevent a standing wave in the pipe?

Problem 6.116. A long organ pipe might be 2.4 m long (about 8 ft)! If the pipe is open to the air at both sides, what will its lowest two frequencies be? Repeat for the situation where one side is covered. Choose any two of these standing waves and sketch them. Label the nodes and antinodes. (Because the fundamental frequency is fixed for a particular pipe, a complete organ pipes has a wide variety of pipes to sound different musical notes.)

★ **Problem 6.117.** If you put one end of a plastic drinking straw in your mouth, bite down on the tip, and blow just right, you can induce resonance.

   a. What are the lowest three frequencies you'll hear in a 15.0 cm-long straw?

   b. Make a sketch of the standing wave for the largest frequency you reported in (a). Label the nodes and antinodes and explain what is oscillating in your sketch.

★★ **Problem 6.118.** A string has a length of 1.20 m and a mass density of 0.0050 kg/m. One end of the string is tied to an oscillator, which moves the end of the string up and down at a specified frequency f. The other end of the string is tied to a 0.500 kg mass and hung over a pulley such that 1.10 m of the string is stretched out horizontally between the oscillator and the pulley. What are the lowest three frequencies that will drive a standing wave in the string? Make a sketch of each standing wave.

★★ **Problem 6.119.** A 1.50 m-long string is connected to a 0.200 kg hanging mass (via a pulley) on one end and a mechanical oscillator on the other. The oscillator's frequency is slowly increased from 0 Hz to 150 Hz, and a standing wave is observed at three separate frequencies in this range. One of these frequencies is 120 Hz. What are the other two?

★ **Problem 6.120.**

A particular guitar string is 1.10 m long and vibrates with a fundamental frequency of 330 Hz.

   a. What is the wavelength of the standing wave?

b. If you want to construct a wind instrument from a simple tube that is closed at one side and open at the other, what length should it be to have the same fundamental frequency as the guitar string?

★ **Problem 6.121.** Consider a pipe that is open on both sides. For what length L will the fundamental frequency of the pipe be "Concert A", 440 Hz?

★ **Problem 6.122.** Consider a pipe that is open on one side and closed on the other. For what length L will the fundamental frequency of the pipe be "Concert A", 440 Hz?

★ **Problem 6.123.** A simple model of the human ear canal is to consider it to be a tube that is closed at one side (the eardrum) and open at the other. A typical ear canal is about 2.5 cm long. Determine the lowest three sound frequencies that would induce resonance in the ear canal.

★ **Problem 6.124.** If you drive down a highway with one window open, you can induce resonance inside the car. As a simple model, suppose the car is a tube (with the open window on one end and a closed window on the other). If the tube has a length of 2.0 m, then what are the lowest two frequencies that would generate a standing wave inside the tube?

★ **Problem 6.125.** It is possible to induce resonance in a thin tube by blowing across one end. Suppose you do this to a tube that is 12 cm long and open to the air on both ends. Determine the fundamental frequency.

★★ **Problem 6.126.** A high-speed photograph is taken (at time t = 0 s) of a standing wave on a string with a mass density of 2.5 g/m under 30. N of tension. From Figure 6.43, determine a numerical value for (a) the wave number, (b) phase constant and (c) frequency. Assume that the wave is mathematically represented with a sine (not cosine) wave.

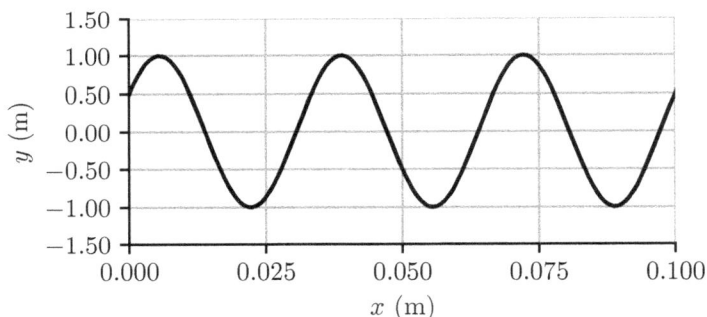

FIGURE 6.43
Problem 6.126

★★ **Problem 6.127.** An interesting demonstration of standing waves involves burning gas: a long, thin metal tube is connected to a propane tank, one end is sealed, and the other end is attached to a speaker cone. Small evenly spaced holes are drilled into the tube and the escaping gas can be lit to form a row of flames. If the speaker cone is driven at an appropriate frequency, the flames form a standing wave. Suppose in one such demonstration the frequency is $f = 1.000$ kHz and the distance between adjacent nodes is 0.15 m.

a. The standing wave is formed by the interference of two traveling waves traveling in opposite directions along the tube, but at the same speed. Determine a numerical value for this speed in m/s.

b. Suppose you increase the speaker cone frequency until the next resonant frequency is found at 1.050 kHz. What is the fundamental frequency of the system?

Additional Problems for Chapter 6

★★ **Problem 6.128.** In a particular microwave oven, standing waves are formed by electromagnetic waves with the following properties:

- The waves have a frequency of 2.45×10^9 Hz

- Heating is most effective at the antinodes of the standing wave.

- The conducting walls of the microwave function like the clamps that hold the ends of a vibrating string (e.g. a piano or guitar string) in place.

Consider an experiment where you take out the "rotating turntable" present in most microwaves and microwave a plate covered in mini marshmallows for about 30 seconds. You'll find that not all the marshmallows melt and that those that do melt are evenly spaced. Figure 6.44 shows a few melted marshmallows near one side of the microwave.

a. Sketch a standing wave over the marshmallows shown below.

b. Determine a numerical value for the distance between the melted marshmallows.

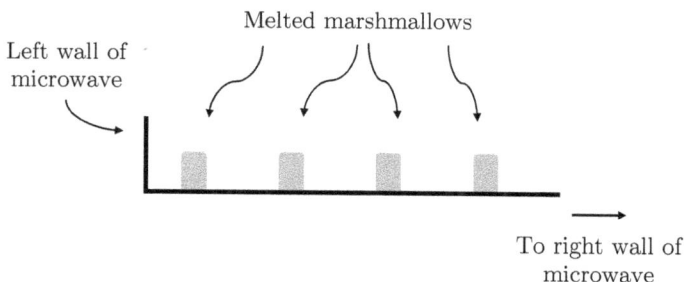

FIGURE 6.44
Problem 6.128.

★★ **Problem 6.129.** When rockets are being launched into space, a portion often detaches partway through the flight: it carries fuel, and when the fuel is spent the so-called "first stage" breaks away to reduce the total mass of the remaining portion of the rocket. Historically these have simply fallen into the ocean, but more recently they have gone through a controlled descent so they can be reused. Suppose the on-board booster engines fire several times in rapid succession as they orient the first stage for its descent: the period of the sound wave is 0.20 s, and the rockets fire when the rocket is 7.0×10^5 m above the earth and traveling toward the ground at 150 m/s.

a. How much time will pass from when you see the pulses to when you hear them?

b. What frequency will the sound waves have to your ear? (Assume that the speed of sound is the usual 343 m/s, though the speed would actually be lower at that altitude.)

Part III

Optics

Our focus turns now to the study of light. In Chapter 7 we will develop the *ray* model of light, which accurately describes many properties of light that we observe every day, such as the formation of an image on the surface of a pond. We will see, also, how an understanding of this model allows us to design useful optical devices, including contact lenses and microscopes.

In Chapter 8 we will explore the limits of the ray model and see that in some cases – namely, when light interacts with one or more narrow slits in an otherwise opaque material – light behaves like a *wave*. This raises the question, "What is oscillating in a light wave?" Fully answering this question will lead us into our study of electric and magnetic fields, which is the topic of Part IV.

7

Ray Optics

Learning Objectives

After reading this chapter, you should be able to:

- Describe the ray model of light.

- Use the ray model of light to analyze the formation of shadows.

- Use the law of reflection to analyze images formed by plane mirrors.

- Use Snell's law to describe how a beam of light travels through transparent materials.

- Use Snell's law to explain the phenomenon of total internal reflection.

- Use ray diagrams and the thin lens equation to describe the images formed by spherical mirrors and lenses (including for instance their magnification).

- Describe how glasses and contact lenses correct nearsightedness and farsightedness.

- Analyze the images formed by systems of two lenses, specifically in the case of the refracting telescope and optical microscope.

7.1   The Ray Model of Light

Consider the situation shown in Figure 7.1: an object near a lamp blocks some light and casts a shadow. We can explain this by thinking of light as a *ray*: a beam that moves in a straight line. In this view, the bulb in the lamp generates rays heading out in all directions, and the task of calculating the extent of a shadow is a relatively straightforward exercise in geometry (see Example 7.1 for a similar situation).

As we shall see, this **ray model of light** is a very useful way of analyzing the behavior of light in many "every day" situations and devices. Keep in mind, however, that this is a description of what light *does*, not what light *is*. In Chapter 8 we shall examine some situations where the ray model of light does *not* work, and we will begin to put the pieces together to find a satisfying description of the fundamental nature of light.

For now, then, let's return to what light *does* according to the ray model. The rays are not affected by one another, but they *do* interact with the rest of their environment: things like tables, eyeglasses, water droplets, and mirrors. How these rays interact with the world before they reach our eye – and what happens *in* the eye – is the topic of this chapter.

DOI: 10.1201/9781003571568-10

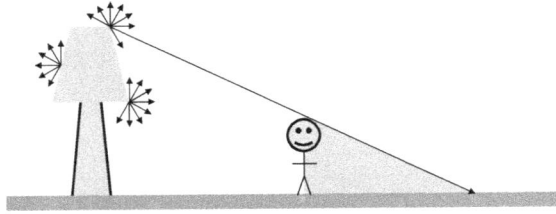

FIGURE 7.1
A lamp casting a shadow over a nearby action figure. Each point on the lamp sends light rays in all directions; some rays from three representative points are shown. The ray that defines the figure's shadow is extended to the table.

To begin, consider that in a bright room, rays of light spread over, and bounce off of, solid objects in such a way that we can think of every point on the object as a *source* of reflected rays, spreading in all directions. This characterizes how we see: the rays that enter our eyes *diverge from an object* and give us a sense of its position in three-dimensional space (Fig. 7.2).

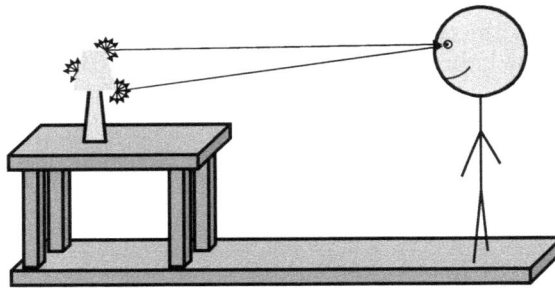

FIGURE 7.2
Rays spread in all directions from an object; the rays that enter our eyes determine what we see.

In our study of mirrors, lenses, and optical devices like microscopes, we will see that these rays can be manipulated to appear as if they diverged from some *other* position. The result is that our eyes record an object where there really *isn't*: an image, like your own reflection in a mirror.

Example 7.1 Shadow in a Pool   ⋆
Consider the empty swimming pool shown in the figure. Sun is low in the sky, $30.°$ above the horizon, and the bottom of the pool is partially cast in shadow, as shown. How far from the left side of the pool is the edge of the shadow?

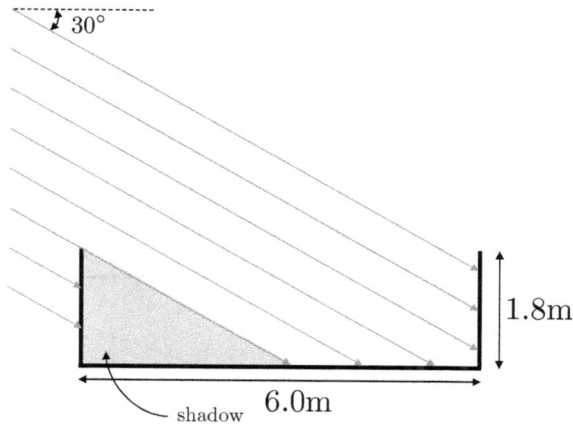

FIGURE 7.3

First note that the rays from the sun are *parallel*. This is because the sun is so far from the earth that all of the rays that reach it are traveling in essentially the same direction. If this is unclear, imagine the person and the lamp of Figure 7.2 being pulled apart: the angle between the rays shown entering the eye *decreases*. In the extreme case that the person and the lamp are infinitely far away, the difference goes to zero, meaning the incoming rays are parallel.

Back to the pool: the lower angle in the shaded triangle must be 30°. If we call the length of the pool that is covered in shadow x, we have

$$\tan(\theta) = \frac{1.8 \text{ m}}{x}$$

$$x = \frac{1.8 \text{ m}}{\tan(30.°)} = 3.1 \text{ m}$$

Example 7.2 The Pinhole Camera ⋆

A simple **pinhole camera** consists of a photographic film placed in the back of a box. The side of the box opposite the film has a small hole that is uncovered to record a photo. Draw some diagrams of this situation, along with representative light rays, to demonstrate that the object will appear (a) *upside down* on the film and (b) *larger* if the object is closer to the hole.

The geometry is shown in Figure 7.4. To travel through the hole, a ray leaving from the top of the object must be angled *down*, which means that it will hit the bottom of the film. Similarly, a ray leaving the bottom of the object will be angled *up* to pass through the hole and so will hit the top of the film. The angles become steeper as the object moves closer to the hole, resulting in a larger image.

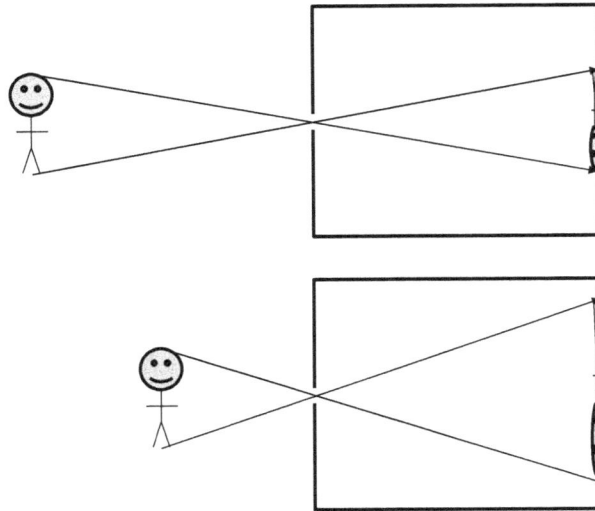

FIGURE 7.4

7.2   Reflection

What happens when a light ray strikes an opaque object like a coffee mug, your shirt, or a mirror?[1] Broadly speaking, there are two phenomena at play:

- The ray can be *absorbed*: the ray vanishes.

- The ray can be *reflected*: the ray bounces off of the object.

In practice, many solid objects do both of these things: the ray is partially absorbed such that the reflected ray is a different *color*: we think of ambient rays from the sun or lightbulbs as a combination of *all* colors (this combination is called *white light*).[2] If a particular shirt absorbs all of the colors except, say, green, then only green rays can move from the shirt to our eye. If the shirt instead absorbed every color, it would appear to our eye as completely black.

We will not consider color/absorption in any more detail for now; our chief concern in this section is the process of reflection. A reflected ray obeys the **law of reflection**: the incident angle θ_i is equal to the reflected angle θ_r (Fig. 7.5):

$$\theta_i = \theta_r \tag{7.1}$$

---

[1]We'll deal with transparent objects in Section 7.6.

[2]In formal language that we'll explore in Chapter 8, we say that the reflected ray also has less *intensity*. Loosely speaking, this just means that a ray becomes more faint when it is reflected.

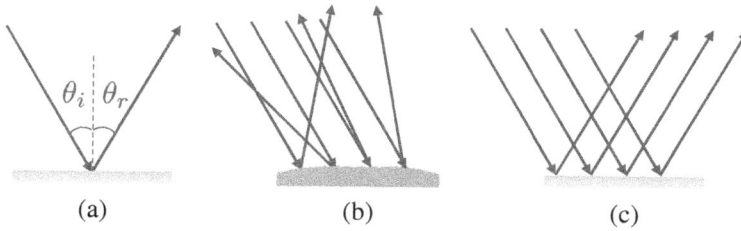

(a) (b) (c)

FIGURE 7.5

(a) A ray reflecting off of a surface obeys the law of reflection: $\theta_i = \theta_r$. (b) Many surfaces are bumpy, so even though each ray individually obeys the law of reflection, incoming parallel rays will scatter in many directions. (c) For a smooth surface such as a mirror, parallel incoming rays are reflected parallel.

Note that these angles are defined relative to a line drawn perpendicular to the surface, which is often referred to as the *normal*.[3]

The surface of most objects are quite bumpy if you zoom in far enough, so light rays from some source tend to be scattered in many directions.[4] If a reflective surface is very smooth, however, then the rays will reflect uniformly (Fig. 7.5). We call such an object a **mirror**.

How does our eye interpret light that has reflected off of a mirror? Well, the eye interprets every light ray *as it is when it enters the eye*; it is not as if each ray brings along a note describing its entire history. So the trick to determining how we interpret light reflected from a mirror is to trace the rays back *as if they have always traveled in a straight line.* Figure 7.6 demonstrates this. We see that the rays appear to have originated from a point behind the mirror that is

1. even with the object along the dimension parallel to the surface of the mirror: $x_o = x_i$.

2. located a distance *behind* the mirror that is equal to the object's perpendicular distance from the mirror: $y_o = y_i$.

Your eye interprets these rays, which *apparently* originate behind the mirror, in the same way that it interprets the rays that *really* originate from the object: it appears as if there is a copy of the object behind the mirror. Of course there isn't *really* an object there; you are seeing a reflected **image**. This is why we used the subscript i above; the subscript o refers to the object.

The two properties listed above are true for any reflection from a flat mirror (or *plane mirror*) to your eye. These rules, along with the law of reflection (which is where these rules come from) can be used to analyze the image(s) formed by one or more plane mirrors.

---

[3]*Normal* is used here in the sense of "perpendicular", the same as when we use the word in a *normal force*.

[4]This is why we can think of an object in a well-lit room as a source of light rays spreading in all directions, as we claimed above. The object doesn't *generate* the rays, of course, but it does *reflect* them in all directions.

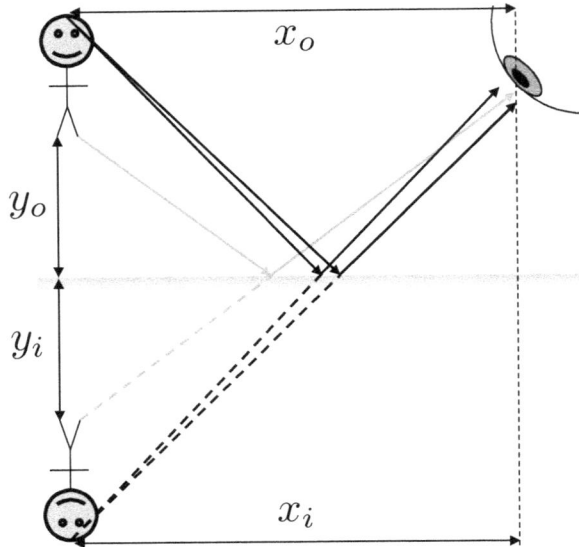

FIGURE 7.6

Image formation due to a plane mirror. Three representative rays are shown leaving the object, reflecting from the mirror, and entering an observer's eye. Dashed lines show the *apparent history* of the rays as they enter the eye. The two black rays diverge from the same point on the object and appear to diverge from the same point on the image; the third ray shows the location of the bottom of the object in the image.

Example 7.3 A Plain Plane Mirror   ★★

A person of height h stands a distance x from a plane mirror hanging vertically on a wall. The mirror extends a distance $y < h$ below the person's eye level. How much of the person's body does he see in the mirror? Determine the answer in (a) the general case (with an algebraic expression) and (b) when $h = 1.80$ m and $y = x = 0.90$ m.

A ray that enters the person's eye after reflecting from the mirror can be traced back to its point of origin according to the law of reflection. If we consider positions that are directly below the person's eye, then we have *similar triangles*: the point of reflection is halfway between the ray's point of origin and the person's eyes.

Now, the limiting case is when the halfway point is on the bottom edge of the mirror: any lower and the ray misses the mirror and does not reflect back into the person's eye. It follows that the person can see a length $L = 2y$ of their body, measured down from eye level.

In part (b), $L = 2(0.90$ m$) = 1.80$ m $= h$, meaning we are at the limiting case where a "half-length" mirror allows the person to just barely see their entire body (actually, the mirror is a bit longer than necessary because our eyes aren't at the very top of our skull!). Interestingly, the horizontal distance x from the mirror does not factor in to our results.

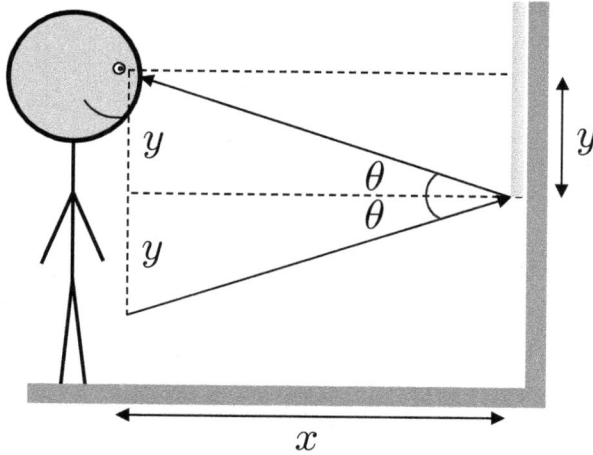

FIGURE 7.7

Example 7.4 Image Formation in a Right Angle Mirror ★★
An observer is looking for images of a small object near two plane mirrors that form a 90° angle, as shown. How many images will the observer see? Support your answer with a ray diagram.

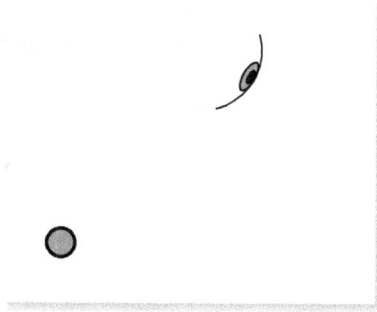

FIGURE 7.8

If we consider each mirror in isolation, then we're back to the situation sketched in Figure 7.6. This accounts for *two* images (one for each mirror). In Figure 7.9 the rays corresponding to these images are shown in black (I am only drawing one ray for each image to avoid cluttering the diagram).

Now, the mirrors are not actually isolated, and as a result it is possible for a ray to leave the object, reflect off of one mirror, then off of the second mirror, and *then* enter the observer's eye. These rays are shown in gray.

Note that the object and three images sit on the corners of a rectangle; the vertex of the mirrors sits at the center (the dashed lines that extend from the vertex of the mirrors are included to guide your eye on this point). This is true for any "right angle mirror"; you can use this fact to find the location of the third image without tracing its rays. Alternatively, you can use the fact that you know where the image will be to assist in drawing the rays.

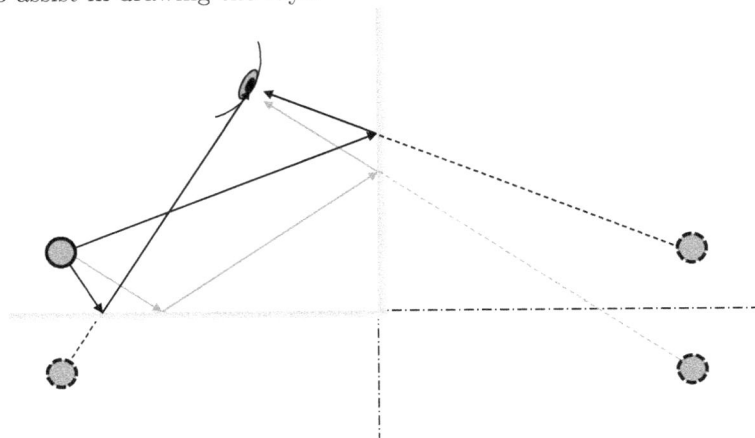

FIGURE 7.9

You may be wondering what happens if the angle between the mirrors is not 90°. Well, if the angle *increases*, then eventually (when the angle is 180°) we're back to a single mirror. Evidently the number of images must *decrease* to 2 and then 1.

What about if the angle *decreases*? The limiting case here is two parallel mirrors: you're looking at your reflection in a mirror, but there is a second mirror on the wall behind you (in this case we obviously have to detach the mirrors so they are no longer joined at a corner). If you've ever done this, then you no doubt recognize that you see many reflections. Evidently decreasing the angle from 90° *increases* the number of images. In fact, in the ideal case where we ignore the fact that a ray loses some energy when it reflects from a mirror, two parallel mirrors generate an *infinite* number of images (see Problem 7.8).

7.3   Magnification

It should come as no surprise when I say that images that you see in a mirror can be of a different size than the actual object. I would like to pause now and briefly explain how we quantify this so-called **magnification**. This may not seem terribly important now, when we've only talked about images being formed by flat mirrors, but we shall see that magnification is very important when we are considering the effect of lenses and curved mirrors – especially if we would like to magnify a distant or tiny object.

There are several useful ways to quantify magnification; they all involve taking a ratio between (1) some quantity associated with an image and (2) some reference quantity associated with the object. Let us first consider **lateral magnification**:

$$m = \frac{h_i}{h_o} \tag{7.2}$$

That is, we simply take the ratio of the image height to the object height. If $|m| < 1$, the image is smaller than the object; if $|m| > 1$, the image is larger than the object. We shall see that in some cases an image can appear upside down (we say that it is *inverted*), where $h_i < 0 \implies m < 0$. For an upright image, $m > 0$.

While the lateral magnification m can be useful, it can also be misleading because a tall image can still appear small to our eye if it is sufficiently far away. Thus, it is sometimes more useful to consider the **angular magnification**, which considers how the image "fills our view" (Fig. 7.10) compared to the original object.

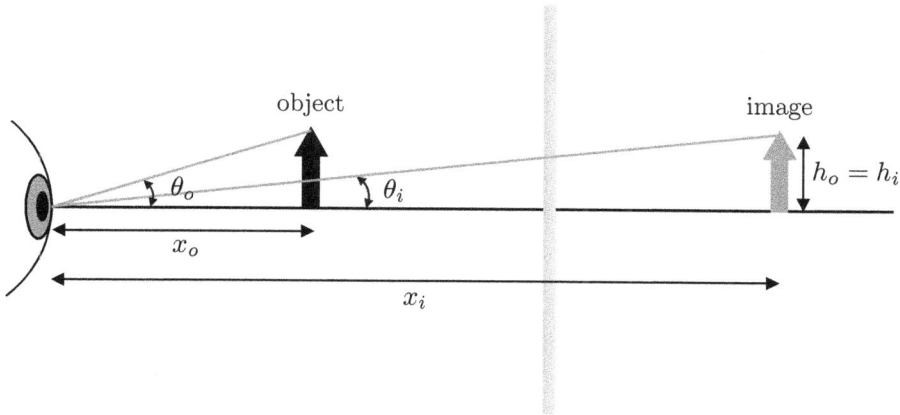

FIGURE 7.10
An object between you and a plane mirror, along with its image. The angular size of the image is smaller than the object ($\theta_o < \theta_i$), even though the object and image have the same height (so $m = 1$).

The angular magnification has two common definitions and they are usually both referred to by the same name. I think this is terribly confusing and so I shall give them different names. First, the **relative angular magnification** compares the angular size of the image to the angular size of the object *at its current position*:

$$\mu_r = \frac{\theta_i}{\theta_o} \tag{7.3}$$

Second, the **absolute angular magnification** compares the angular size of the image to the angular size of the object *when it is as large as it can be without the use of a lens or mirror*, which we denote θ_{NP}:

$$\mu_a = \frac{\theta_i}{\theta_{NP}} \tag{7.4}$$

θ_{NP} is determined by placing the object at the eye's **near point**, which is the closest distance an object can be to the eye without losing focus (we shall have more to say about the near point when we discuss correcting vision in Section 7.9). The near point changes as we age, but the standard value for a healthy adult is 25 cm. Thus $x_o = 25$ cm in Figure 7.10 and we see

$$\tan \theta_{NP} = \frac{h_o}{25 \text{ cm}} \tag{7.5}$$

If the object is sufficiently small, we can apply the small angle approximation $\tan \theta \approx \sin \theta \approx \theta$ so

$$\theta_{NP} \approx \frac{h_o}{25 \text{ cm}} \tag{7.6}$$

Analyzing the equations for angular magnification follows the equation for lateral magnification. If, for instance, $\mu_r < 0$, then $\theta_i < 0$ and we have an inverted image.

Example 7.5 Mirror Magnification ★★

You stand a distance L in front of a mirror attached to a wall. You hold out your thumb, of height h_o, a distance x_o in front of your face. (a) Determine an expression for all three magnifications (m, μ_r, and μ_a) for your thumb's image in the mirror, assuming the small angle approximation is appropriate. (b) Give numerical answers in the case where $L = 1.5$ m, $h = 3.4$ cm, and $x_o = 35$ cm.

The geometry is the same as in Figure 7.10. As we noted above, $h_o = h_i$ and so $m = 1$. To consider angular magnification, we need to express the angular sizes:

$$\theta_o \approx \tan(\theta_o) = \frac{h_o}{x_o}$$

and

$$\theta_i \approx \tan(\theta_i) = \frac{h_i}{x_i}$$

x_i is not a given parameter; as usual we prefer to express our answers in terms of given quantities. Note that the object is a distance $L - x_o$ in front of the mirror. The image is an equal distance *behind* the mirror, so the total distance from the eye to the image is

$$x_i = 2(L - x_o) + x_o = 2L - x_o$$

By applying the small-angle definition for θ_{NP} (Equation 7.6), we can compute the absolute angular magnification:

$$\mu_a \approx \frac{\theta_i}{\theta_{NP}} = \frac{h_i \, (25 \text{ cm})}{x_i h_o} = \frac{(25 \text{ cm})}{(2L - x_o)}$$

In the last step we have noted that $h_i = h_o$ and so $h_i/h_o = 1$. The relative angular magnification is computed similarly:

$$\mu_r \approx \frac{\theta_i}{\theta_o} = \frac{h_i x_o}{x_i h_o} = \frac{x_o}{x_i} = \frac{x_o}{2L - x_o}$$

As above, in the penultimate step we have noted that $h_i/h_o = 1$.

If we plug in the values given in part (b), we find

$$m = 1$$

$$\mu_a = \frac{(0.25 \text{ m})}{2(1.5 \text{ m}) - 0.35 \text{ m}} = 0.094$$

$$\mu_r = \frac{x_o}{2L - x_o} = \frac{0.35 \text{ m}}{2(1.5 \text{ m}) - 0.35 \text{ m}} = 0.13$$

We see that both angular magnifications are less than one (i.e. the image appears smaller than the object). This should not be surprising: the image height is the same as the object height, but the image is much further away than the object. In the next section we shall see how a curved mirror can, in some cases, lead to enlarged images.

7.4 Curved Mirrors: Ray Tracing

We've now dealt with reflection from a *flat* mirror, but what about a *curved* mirror? If you've ever looked at your reflection in a cosmetic mirror or even in a polished spoon, you know the situation is rather different! In this book we will only concern ourselves with the situation where the mirror has *spherical curvature*: the mirror forms part of a sphere (or in a 2D representation, part of a circle). Figure 7.11 shows an object located near such a mirror that is either *concave* or *convex* (these are sometimes called *converging* and *diverging* mirrors, respectively). Our objective here is to determine how rays leaving some object near the mirror will behave: will an image form from the perspective of a nearby observer? Why or why not? If it does, what does it look like?

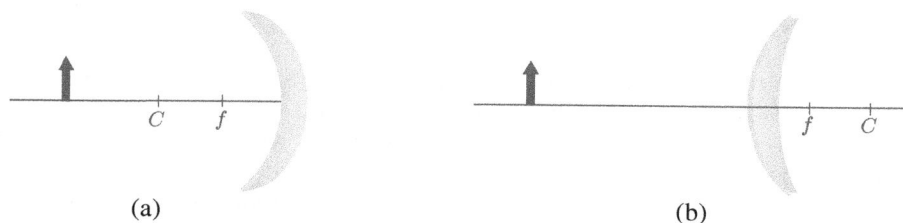

(a)   (b)

FIGURE 7.11

A concave (a) and convex (b) spherical mirror. Each mirror is defined by its center of curvature C and focal point $f = C/2$.

We saw a similar diagram in Example 7.5; let me pause now to describe some important features:

1. The imaginary line running horizontally through the middle of the mirror is referred to as the **optical axis**.

2. The object is always drawn as an upward arrow with its base along the optical axis.[5]

3. The mirror is characterized by its **center of curvature** C and **focal point** f, both of which are located on the optical axis. If you imagine extending the mirror to actually be an *entire* sphere, point C would be at its exact center. Accordingly, the distance to C from the mirror is called the **radius of curvature**. The focal point is exactly halfway between the center of curvature and the middle of the mirror.

How will rays that leave the object interact with the mirror? Well, any ray that strikes a spherical mirror still obeys the law of reflection, but the normal always points to the center of curvature. A consequence of this property is that *rays parallel to the optical axis are always reflected along a line connected to the focal point* (Fig. 7.12).[6]

We can reverse this statement and say that *a ray that approaches the mirror through the focal point will reflect parallel to the optical axis*. This is because the law of reflection doesn't

---

[5]This corresponds to how far the object extends perpendicular from the optical axis. This may seem restrictive, but any object can be represented this way. See Problem 7.18.

[6]The fact that spherical mirrors obey this property is precisely why we are studying them: things get more complicated with other curvatures.

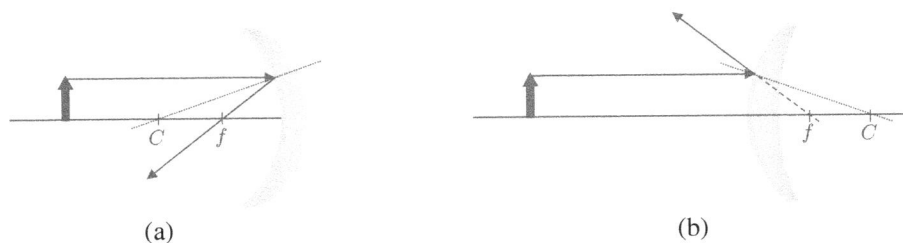

(a)   (b)

FIGURE 7.12
A ray that impacts a spherical mirror parallel to the optical axis is reflected along a line connected to the focal point. For a concave mirror (a), the reflected ray actually moves through the focal point; for a convex mirror (b), the reflected ray *appears to have come from* the focal point.

distinguish the reflected angle from the incident angle: $\theta_i = \theta_r$ is the same as $\theta_r = \theta_i$, so any incoming/outgoing path can just as well be run in reverse.

We shall see that these rays can be used to determine where an image will form, but there will be some distortion if the object is large compared to the curvature of the lens (Problem 7.19). To avoid this complication, we will always assume that the object is small relative to the mirror's curvature, which means that the key rays use a *vertical* reflecting surface, even if the mirror is drawn with sharp curvature to indicate what kind of mirror it is (Fig. 7.13).

We now have two convenient rules for drawing representative rays involving the object and the mirror. It is standard to also use a third rule, which simply involves drawing a ray directed to the center of the mirror: the normal in this case is the optical axis (the tangent surface is vertical) so it is straightforward to draw the reflected angle. For reference, I summarize all three rules below.

Drawing Key Rays for Mirrors with Spherical Curvature

1. Draw a ray from the tip of the object to the mirror, parallel to the optical axis. It will reflect from the mirror along a line connected to the focal point.

2. Draw a ray from the tip of the object to the mirror along a line connected to the focal point. The ray will reflect from the mirror parallel to the optical axis.

3. Draw a ray from the tip of the object to the middle of the mirror. The ray reflects about the optical axis.

The ray diagram is different for a concave mirror depending on whether the object is inside or outside the focal point. Thus, there are a total of three characteristic ray diagrams for spherical mirrors; they are shown in Figure 7.13. In each case the reflected rays form an image: they either cross at a common point (in which case the image is said to be **real**) or *appear to have come from* a common point (in which case the image is said to be **virtual**).

An interesting result of this analysis is that an object that is located beyond the focal point for a concave mirror will be inverted, while an object that is closer to the mirror than the focal point will be upright. You can see this for yourself if you slowly change the distance between an object and a concave mirror: around the focal point, it flips orientation!

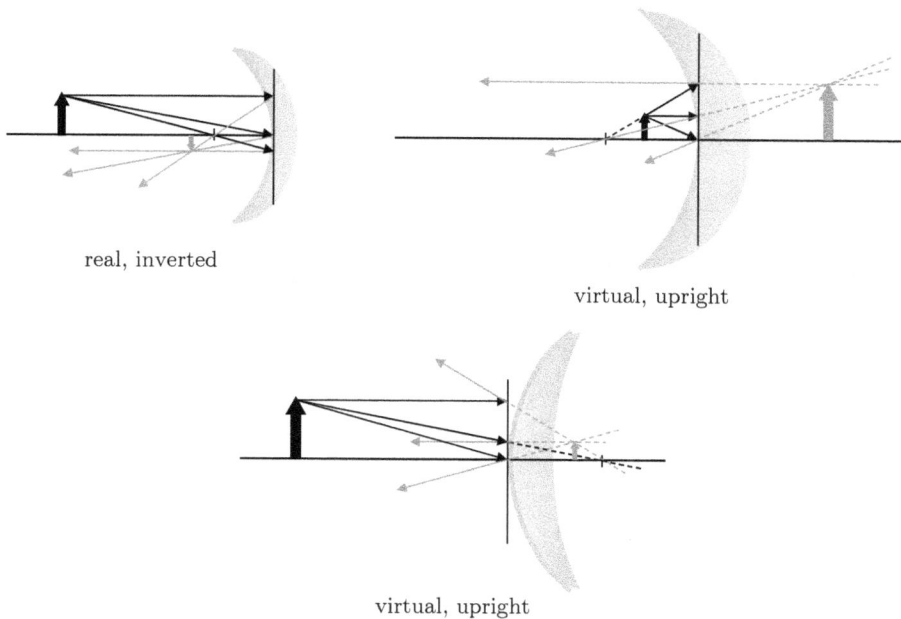

real, inverted

virtual, upright

virtual, upright

FIGURE 7.13
Complete ray diagrams for spherical mirrors. The focal length f is marked with a vertical hash on the optical axis; the radius of curvature C is omitted. Objects are shown in black and their images are shown in gray. Each of the key rays are shown leaving the object in black; they reflect from the mirror in gray. Dashed lines are used to show ray orientation.

Drawing and analyzing ray diagrams can take some getting used to, but it always boils down to correctly applying the rules we listed above; just like the rules we worked out for the plane mirror, they come from the law of reflection.

Example 7.6 Images from Diverging Mirrors ⋆

A convex mirror with a radius of curvature $C = 0.50$ m is positioned in an upper corner of a convenience store. A patron standing 1.0 m from the mirror is holding a 17 cm-tall object. Sketch the situation with a ray diagram, and estimate the location and height of the object's image.

We expect our diagram will look something like the diagram for a convex mirror shown in Figure 7.13, but we need to be careful about using consistent length scales if we are to make estimates from the diagram itself. (In your own work, use a ruler and/or graph paper!)

The diagram is shown in Figure 7.6; to keep it uncluttered I show only two of the three key rays. If you note that $f = C/2 = 25$ cm and measure the figure, then you'll see $x_i \approx 20$ cm and $h_i \approx 3.5$ cm (a box near the bottom of the diagram is provided as reference, to help with scaling).

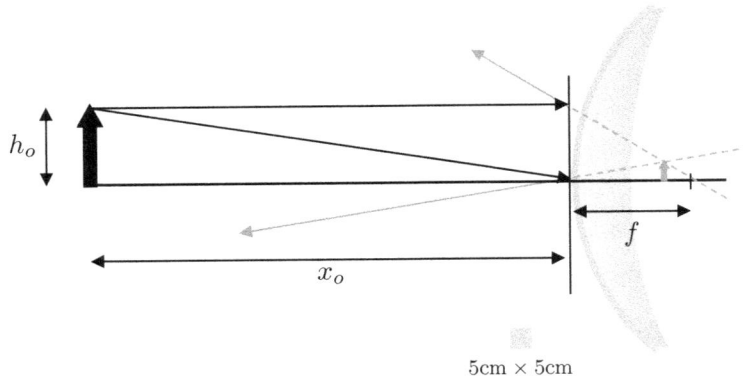

FIGURE 7.14

Example 7.7 Images from Converging Mirrors   ★★
A 0.80-cm-tall insect specimen is held 4.0 cm in front of a concave mirror with
$f = 12.0$ cm. (a) What is the lateral magnification of the image? Is it upright,
inverted, real, or virtual? (b) Determine an equation that relates the focal length f
to the object distance x_o and image distance x_i.

The problem describes an object *inside* the focal length of a concave mirror, so
according to Figure 7.13, the object will be virtual and upright. To determine the
magnification, we need to analyze the geometry in greater detail. Figure 7.15 shows
the ray diagram with some distances and angles labeled; some of the rays have been
removed to reduce clutter.

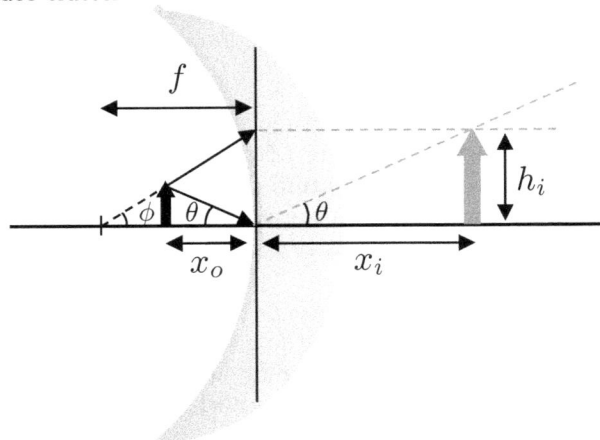

FIGURE 7.15

There are two angles marked θ, and from the triangles they form we see

$$\tan(\theta) = \frac{h_o}{x_o} = \frac{h_i}{x_i} \implies \frac{h_i}{h_o} = \frac{x_i}{x_o} \tag{7.7}$$

There is one angle marked ϕ, but it forms the lower left corner of *two* triangles. We find

$$\tan(\phi) = \frac{h_o}{f - x_o} = \frac{h_i}{f} \implies \frac{h_i}{h_o} = \frac{f}{f - x_o} \tag{7.8}$$

The last term in each of these equations *is* the lateral magnification: $m = h_i/h_o$. Thus we have

$$m = \frac{f}{f - x_o} = \frac{12 \text{ cm}}{12 \text{ cm} - 4 \text{ cm}} = 1.5$$

This completes (a). To move toward the equation requested in (b), we equate the rightmost terms in Equations (7.7) and (7.8) in order to eliminate h_i and h_o. We could stop after the first line below, but instead we'll carry out some algebra to tidy things up:

$$\frac{x_i}{x_o} = \frac{f}{f - x_o}$$

$$\frac{f - x_o}{f} = \frac{x_o}{x_i}$$

$$1 - \frac{x_o}{f} = \frac{x_o}{x_i}$$

$$\frac{1}{x_o} - \frac{1}{f} = \frac{1}{x_i}$$

In this equation, all three of the algebraic quantities are positive; they represent lengths of triangle edges. As we shall see in the next section, however, it is customary to make x_i *negative* if the image exists to the right of the mirror, as it does here. If we make this change, our final equation above can be written

$$\frac{1}{f} = \frac{1}{x_o} + \frac{1}{x_i} \tag{7.9}$$

7.5 Curved Mirrors: The Thin Lens Equation

Much as we developed the kinematic equations to enable us to analyze one-dimensional motion rigorously (and without the need to always draw motion graphs), we would like to develop a mathematical framework to relate quantities involving the object (x_o and h_o), mirror (f), and image (x_i and h_i). As we saw in Example 7.7, we can obtain the relevant relationships from the geometry of our ray diagrams.

In that example, though, we introduced a seemingly arbitrary negative sign in front of the image distance x_i. Why? Well, if you always call all of the quantities positive when analyzing the geometry of the two other situations involving spherical mirrors (Figure 7.13), you get *different equations* (Problem 7.22).

While there is nothing *wrong* with this, it is standard practice to introduce some sign conventions that allow Equation (7.9) to apply to each situation:

1. Vertical quantities (h_o and h_i) are *positive* above the optical axis and *negative* below the optical axis.

2. The object distance x_o is always positive.

3. The focal length f is positive for a converging mirror (where the focal point is in front of the mirror) and negative for a diverging mirror (where the focal point is behind the mirror).

4. The image distance x_i is *positive* in front of the mirror (where the image is *real*) and *negative* behind the mirror (where the image is *virtual*).

With these conventions in mind, Equation (7.9) applies to all spherical mirrors:

$$\frac{1}{f} = \frac{1}{x_o} + \frac{1}{x_i} \tag{7.10}$$

Equation (7.10) is the **thin lens equation**. If you are thinking that this is a horrible misnomer because we're talking about *mirrors* here, rather than *lenses*, then you're right. We shall see, however, that the same equation can be applied to lenses, and I am afraid that this is the common name for this equation.

With these conventions, we can update Equation (7.7) for the lateral magnification in the context of spherical mirrors:

$$m = \frac{h_i}{h_o} = -\frac{x_i}{x_o} \tag{7.11}$$

These results are very useful when mathematically analyzing the behavior of spherical mirrors.

Example 7.8 Thin Lens Equation with Mirrors   ⋆
Refer back to Example 7.6 and mathematically determine m, x_i, and h_i.

We cannot jump straight to Equation (7.11) because only object properties (h_o and x_o) are known. Thus we will turn first to Equation (7.10), which allows us to determine x_i:

$$\frac{1}{f} = \frac{1}{x_o} + \frac{1}{x_i}$$

$$x_i = \left(\frac{1}{f} - \frac{1}{x_o}\right)^{-1}$$

$$x_i = \left(-\frac{1}{25 \text{ cm}} - \frac{1}{1.0 \times 10^2 \text{ cm}}\right)^{-1}$$

$$x_i = -20. \text{ cm}$$

Note that when inserting a value for f we have made it a *negative* quantity, because the focal point lies to the right of the mirror. As we expect, we obtain a negative value for the image distance.

From here we return to Equation (7.11):

$$m = -\frac{x_i}{x_o} = -\frac{(-20. \text{ cm})}{1.0 \times 10^2 \text{ cm}} = 0.20$$

It follows from the same equation that

$$h_i = mh_o = (0.2)(17 \text{ cm}) = 3.4 \text{ cm}$$

These results match well with our estimates from Example 7.6.

7.6   Refraction

It is remarkable (I am remarking about it now, in fact) that almost everything we've done so far in this chapter can be traced back to the law of reflection (Equation (7.1)). However, there is more to the story: in addition to being absorbed or reflected, light rays can also *pass through* some materials, such as water or glass. In passing from one medium to another, light rays change direction in a process called **refraction**.

The equation that describes refraction is called **Snell's law:**[7]

$$n_1 \sin \theta_1 = n_2 \sin \theta_2 \tag{7.12}$$

The law is shown in Figure 7.16. The angles θ_1 and θ_2 are defined relative to the normal; I have drawn it so that θ_1 describes the *incident* angle and θ_2 describes the *refracted* angle.[8] We shall sometimes refer to these as θ_i and θ_r, but the more generic notation is convenient for situations where a single ray refracts more than once.

The unitless value n is the **index of refraction** for the material. We saw in Chapter 6 that the index of refraction has something to say about the speed of light in the medium ($v = c/n$, Equation (6.19) on page 314). We will see *why* this occurs in Chapter 8; for our present purposes it is a number that characterizes refraction according to Snell's law. Table 6.1 (page 315) lists common indices of refraction.

A key observation is that a beam is always *closer to the normal* in the medium with the *higher* index of refraction, regardless of the direction of travel (mathematically, $n_1 > n_2 \implies \theta_1 < \theta_2$). Thus we say that a ray *bends toward the normal* if it moves into a medium with a higher index of refraction; a ray *bends away from the normal* if it moves into a medium with a lower index of refraction.

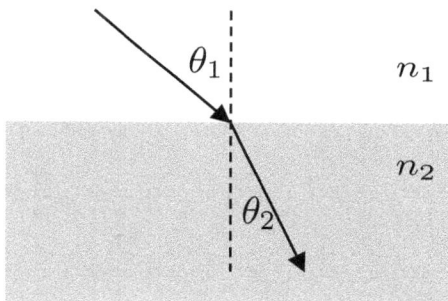

FIGURE 7.16

A ray refracts as it moves from a medium with index of refraction n_1 to a medium with index of refraction n_2.

Examples 7.9 and 7.10 demonstrate Snell's law in some typical situations involving *monochromatic* light, meaning rays of a single color (the simplest example of this is a laser beam). We focus on monochromatic light because the index of refraction for a material actually depends on the color of the light: if you send a narrow beam of white light through, say, a block of glass, it will spread into a rainbow (Problem 7.29). As you may expect, this spreading phenomenon plays a role in the formation of a rainbow that you can sometime see in the sky after it rains: white light from the sun enters water droplets in the atmosphere, spreads into its various colors, and then reflects into your eye (Problem 7.66).

---

[7]After Willebrord Snell (1580–1626).

[8]Just like the law of reflection, Snell's law is symmetric and we could just as well run the ray in the reverse direction.

Example 7.9 Refraction: Laser Through a Slab   ⋆
A laser beam strikes a slab of glass, as shown in Figure 7.17. Describe the subsequent
path of the beam.

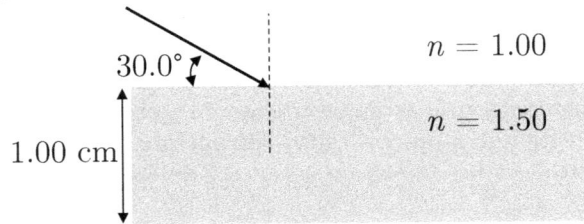

$n = 1.00$

30.0°

$n = 1.50$

1.00 cm

FIGURE 7.17

The ray will bend *toward* the normal as it enters the glass; when it leaves it will
bend *away* from the normal. The path is sketched in Figure 7.9; to determine the
angles and the horizontal distance x, we must apply Snell's Law. Consider first the
refraction that occurs when the beam enters the glass:

$$n_1 \sin \theta_1 = n_2 \sin \theta_2$$

$$\frac{n_1}{n_2} \sin \theta_1 = \sin \theta_2$$

$$\sin^{-1}\left(\frac{n_1}{n_2} \sin \theta_1\right) = \theta_2$$

$$\sin^{-1}\left(\frac{1.00}{1.50} \sin\left(60.0°\right)\right) = \theta_2$$

$$\theta_2 = 35.3°$$

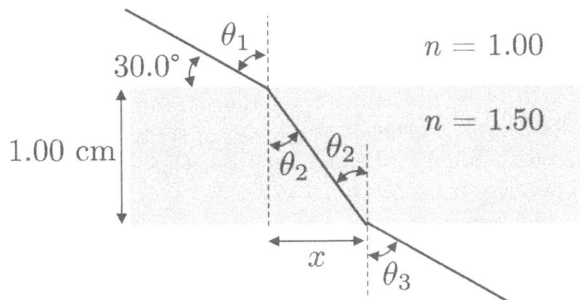

θ_1

30.0°

$n = 1.00$

$n = 1.50$

1.00 cm

θ_2 θ_2

x

θ_3

FIGURE 7.18

With θ_2 identified, we can determine x with a bit of trig:

$$\tan \theta_2 = \frac{x}{1.00 \text{ cm}}$$

$$x = (1.00 \text{ cm}) \tan\left(35.3°\right)$$

$$x = 0.707 \text{ cm}$$

Finally, we would like to know the orientation of the ray once it leaves the glass. We turn again to Snell's Law:

$$n_2 \sin \theta_2 = n_3 \sin \theta_3$$

$$\frac{n_2}{n_3} \sin \theta_2 = \sin \theta_3$$

$$\sin^{-1}\left(\frac{n_2}{n_3} \sin \theta_2\right) = \theta_3$$

$$\sin^{-1}\left(\frac{1.50}{1.00} \sin (35.3°)\right) = \theta_2$$

$$\theta_2 = 60.0°$$

Thus the ray leaves the glass with the same orientation that it entered, but the presence of the glass has shifted its horizontal position.

Example 7.10 Images in a Tank ★★

You look straight into an aquarium and see a fish that appears to your eye to be a distance x_i from the side of the tank. How far from the side of the tank is the fish, really? Neglect the effect of the glass aquarium wall.

Remember that rays leave every point of an object heading in all directions; rays that leave the same point on an object but enter our eyes with slightly different angles help us to ascertain the position of the object. This is possible because our pupils have a width: a ray can enter at the top edge of the pupil and a second ray can enter at the bottom edge of the pupil. And, of course, some rays leaving an object will enter your *left* eye and others will enter your *right* eye, which greatly improves your sense of depth.

Here, the refraction that occurs as the rays move from the water to the air means that the rays *appear* to have come from a position *closer* to the boundary (Fig. 7.19).

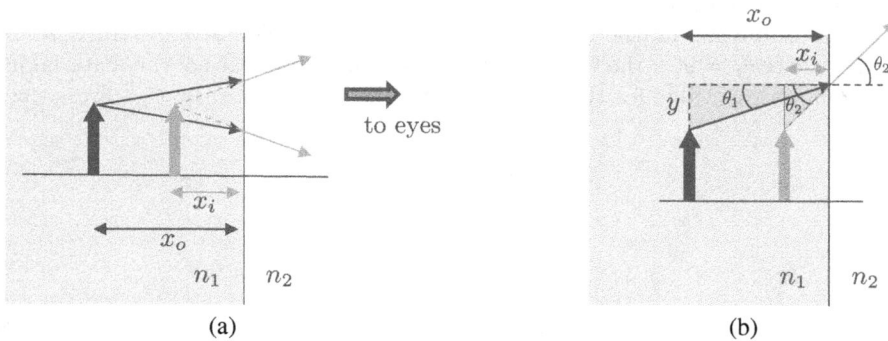

FIGURE 7.19
(a) an object in a medium with a higher index of refraction has an image closer to the boundary than the object. (b) A detailed view with the geometry labeled.

From the figure we see

$$\tan\theta_1 = \frac{y}{x_o} \text{ and } \tan\theta_2 = \frac{y}{x_i}$$

If we take a ratio of these equations, we find

$$\frac{\tan\theta_1}{\tan\theta_2} = \frac{x_i}{x_o}$$

Now, while the figure is drawn so we can see what is going on, in reality these angles tend to be quite small: we are assuming that the rays enter our eyes, and the width of our pupils (and even the distance between our eyes) is small compared to how far from the tank we're likely to stand.

Under the small angle approximation, Snell's Law reduces to $n_2/n_1 = \theta_1/\theta_2$. Applying this to our ratio above, we have

$$\frac{\tan\theta_1}{\tan\theta_2} = \frac{x_i}{x_o}$$
$$\frac{\theta_1}{\theta_2} \approx \frac{x_i}{x_o}$$
$$\frac{n_2}{n_1} \approx \frac{x_i}{x_o}$$

Finally, we express the image distance in terms of the indices of refraction and the object distance:

$$x_i = \frac{n_2}{n_1} x_o \tag{7.13}$$

Before moving on, let's consider an important limiting case of Snell's law. We just pointed out that $n_i > n_r \implies \theta_r > \theta_i$. Is there a limit to how big θ_r can be? Well, if, for instance, $\theta_r > 90°$, then we wouldn't have refraction at all; we would have **total internal reflection** (Fig. 7.20), which I shall sometimes abbreviate TIR.

We can describe this mathematically by varying the incident angle until we find the limiting case where $\theta_r = 90°$: the refracted ray runs along the surface of the material interface. We call this important value for the incident angle the **critical angle**, θ_c.[9] From Snell's law we have

$$n_i \sin\theta_c = n_r \sin(90°)$$
$$\theta_c = \sin^{-1}\left(\frac{n_r}{n_i}\right) \tag{7.14}$$

Total internal reflection is of fundamental importance, for instance, in fiber optics (Problem 7.26).

---

[9]This suggests that TIR behaves like a light switch as θ_i moves above or below θ_c. We will adopt this view in this book to simplify our analysis. You might be interested to know that this isn't the whole story, though: for values of θ_i near θ_c, the ray will split such that it *partially* reflects and *partially* refracts.

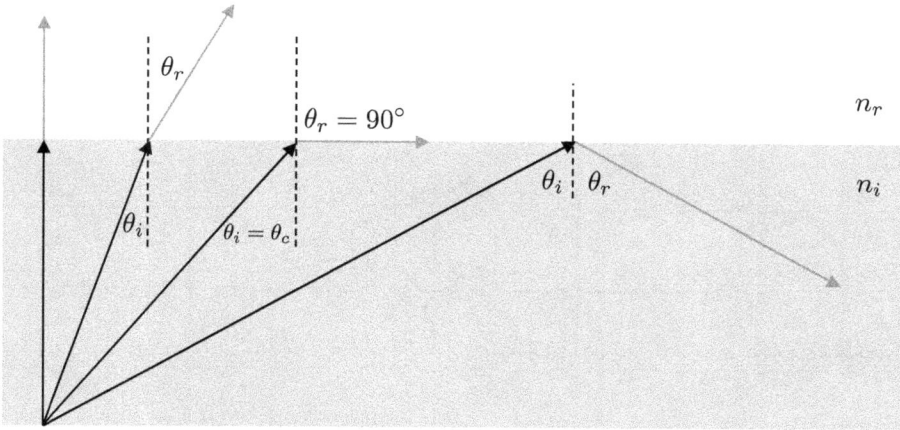

FIGURE 7.20
Rays in a medium with a higher index of refraction $(n_i > n_r)$ can be "trapped" if the angle of incidence exceeds the critical angle θ_c: instead of refracting out of the medium, the ray reflects according to the law of reflection.

Example 7.11 Total Internal Reflection ★★
A laser is directed toward the center of the flat edge of a hemispherical block of glass, as shown in Figure 7.21. What is the smallest possible value of ϕ that results in total internal reflection?

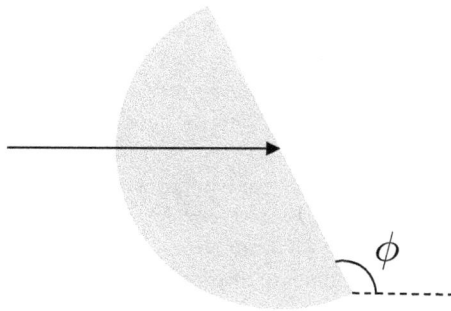

FIGURE 7.21

The fact that the ray impacts the middle of the hemisphere means there is no refraction when it enters the glass: whatever the value of ϕ, the ray is oriented along a radius and so $\theta_i = \theta_r = 0°$. Describing the refraction as the beam *leaves* the glass requires a bit more analysis. The geometry is shown in Figure 7.22. You can analyze the geometry to convince yourself that the second angle marked ϕ is the same as the first, but an intuitive approach is to imagine rotating the hemisphere so the dashed normal line becomes horizontal: clearly both angles marked ϕ become $90°$.

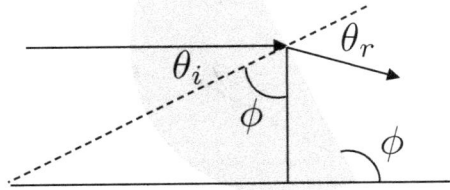

FIGURE 7.22

Now it is clear that $\theta_i + \phi = 90°$. We can use this with Equation (7.14) to determine the value of ϕ when $\theta_i = \theta_c$:

$$\phi = 90° - \sin^{-1}\left(\frac{1.0}{1.5}\right)$$

$$\phi = 48°$$

7.7   Lenses: Ray Tracing

Just as the law of reflection can be used to describe image formation due to mirrors, Snell's law can be used to describe image formation due to a **lens**, which we can define as some transparent material that has been shaped to focus or disperse light rays. Just as in the case of mirrors, we shall consider two kinds of lenses with spherical curvature (Fig. 7.23). Note that, because light can symmetrically pass left-to-right or right-to-left through these lenses, there are necessarily *two* focal points.

Another key difference between lenses and mirrors comes from the fact that a lens must be located *between* the object and the observer (since light passes through the lens via refraction) while for a mirror the object and observer are located on the *same side*. Aside from these important differences, however, our analysis of lenses (via Snell's law) is very

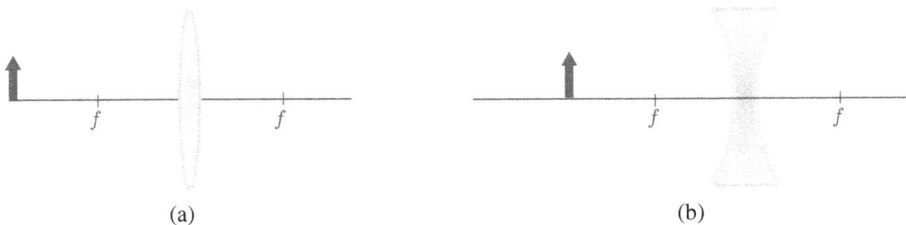

FIGURE 7.23
A converging (a) and diverging (b) thin lens. Each lens is defined by a focal point f on either side.

similar to our analysis of mirrors (via the law of reflection) and the rules we laid out on page 366.

For a lens, a ray that leaves the object parallel to the optical axis will strike the lens and then refract along a line connecting the ray to one of the focal points: a diverging lens uses the focal point on the same side as the object so the rays literally diverge away from the optical axis. Similarly, a converging lens uses the focal point on the far side of the lens so the rays converge to a point (Fig. 7.24). And, just as in the case of mirrors, we can "run the rays in reverse" and see that the rays can *end up* running parallel to the optical axis if they *start* by being directed toward a focal point.

The third rule in the case of mirrors came about from applying the law of reflection around the middle of the mirror, where the normal was aligned with the optical axis. The third rule here is similar, though we're dealing with refraction and (assuming the lens is *thin* enough to treat it like there is a single refraction) we find that a ray can pass through the middle of the lens without being deflected. These rules are summarized below.

Drawing Key Rays for Thin Lenses

1. Draw a ray from the tip of the object to the lens, parallel to the optical axis. It will refract through the lens along a line connected to the *near* focal point for a diverging lens or the *far* focal point for a converging lens.

2. Draw a ray from the tip of the object to the lens along a line that connects the tip of the object to the *far* focal point for a diverging lens or the *near* focal point for a converging lens. The ray will refract through the lens parallel to the optical axis. (Note that because the parallel ray is on the opposite side compared to rule 1, the focal point used for the non-parallel part of the beam is also reversed.)

3. Draw a ray from the tip of the object through the center of the lens.

Depending on the type of lens and the position of the object relative to the lens (i.e. if $x_o > f$ or $x_o < f$), the image can be upright or inverted, and it can be real or virtual. We sketch all of the situations in Figure 7.25.

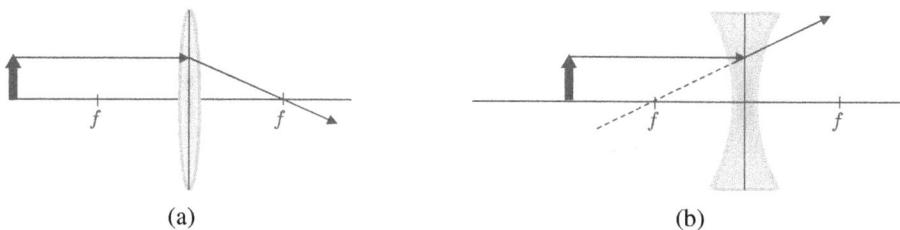

(a)   (b)

FIGURE 7.24
(a) A ray that impacts a converging lens refracts through the *far* focal point. (b) A ray that impacts a diverging lens refracts along a line connected to the *near* focal point.

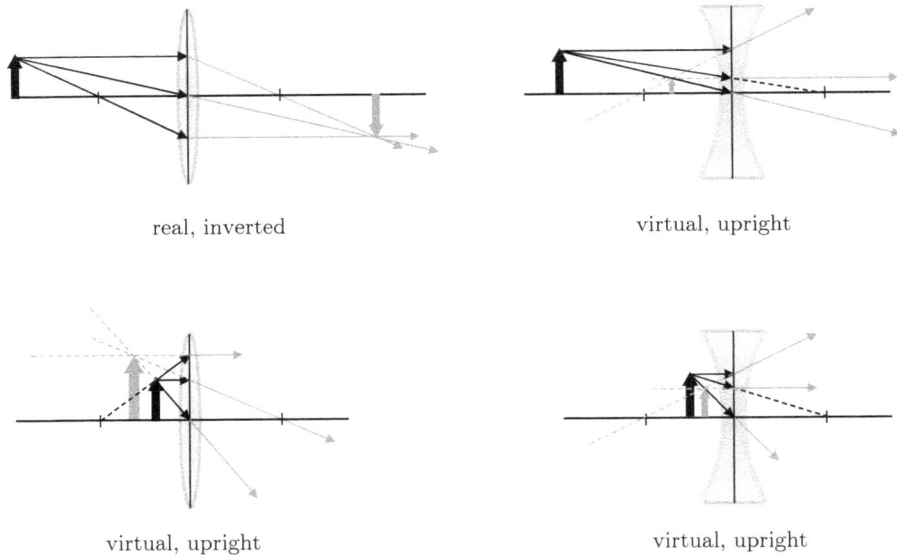

real, inverted   virtual, upright

virtual, upright   virtual, upright

FIGURE 7.25

Complete ray diagrams for thin lenses. The focal length f is marked with a vertical hash. Objects are shown in black and their images are shown in gray. Each of the key rays are shown leaving the object in black; they refract in gray. Dashed lines are used to show ray orientation.

Example 7.12 Images from Converging Lenses   ★★

A converging lens can be used as a magnifying glass. Where should the object to be magnified be located relative to the focal point of the lens to achieve the largest possible absolute angular magnification, μ_a? What is the largest possible value of μ_a? Could a diverging lens be used for the same effect? Why or why not?

From Figure 7.25 we see that for a converging lens the object must be inside the focal point: if it is outside, the image is *inverted*. The ray diagram for the object located inside the focal point indicates that the image is magnified, which is what we want – but where should the object be to maximize the magnification: close to the focal point, close to the lens, or somewhere in between?

Consider the ray that refracts through the far focal point (key ray #1): its orientation does not change as the object move left or right, because it is determined by the height of the object and the position of the focal point, both of which aren't affected by the horizontal position of the object. The top of the image must touch the corresponding dotted line that shows the apparent history of the refracted ray. The line extends up and left, so the image will become taller if it is closer to the focal point. By inspecting the other two key rays, you can see that this occurs if the object moves closer to the focal point.

We conclude, therefore, that the largest magnification occurs when the object is *just inside* the focal point. The ray diagram is shown in Figure 7.12.

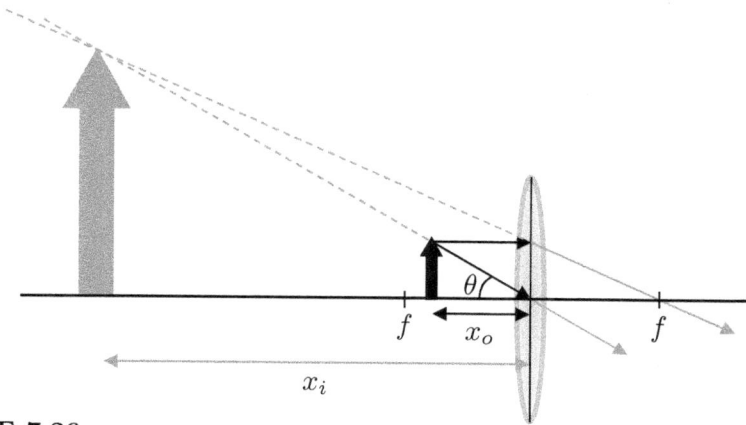

FIGURE 7.26
A magnifier.

We see that the object and the image are both defined by the same angle, i.e.
$\theta_i = \theta_o = \theta$, where

$$\tan \theta = \frac{h_o}{x_o} = \frac{h_i}{x_i}$$

If we apply the small angle approximation $\tan \theta \approx \theta$ and apply Equations (7.4) and
(7.6), we see

$$\mu_a = \frac{\theta}{\theta_{NP}}$$

$$\mu_a = \frac{h_o}{x_o} \frac{25 \text{ cm}}{h_o}$$

$$\mu_a = \frac{25 \text{ cm}}{x_o}$$

We have already argued that the magnification increases if x_o is just inside the focal
point, i.e. $x_o \approx f$. Visually, as the object slides to the left, the angled line that
defines θ *drops*. This pushes its intersection with the top dashed line *left* and *up*,
resulting in a taller image. We conclude that the maximum possible absolute angular
magnification of a simple magnifying glass is given by

$$\mu_a = \frac{25 \text{ cm}}{f} \tag{7.15}$$

This means, for instance, that $\mu_a > 1$ if $f < 25$ cm; if the focal length is larger than
25 cm, then you could get a better view of the object *without* the lens: just hold
the object at your near point! We will return to this point in Section 7.10, when we
discuss optical devices such as the microscope.

Finally, note that a diverging lens always creates an image that is smaller than
the object, and so it cannot be used as an effective magnifier.

Example 7.13 Mathematics of Converging Lenses   ⋆⋆
An object of height h_o is held in front of a converging lens such that $x_o > f$. Analyze
the ray diagram to mathematically relate x_o, x_i, and f.

The ray diagram is shown in Figure 7.25, but we show it again here with more detailed labeling. We see

$$\tan \theta = \frac{h_o}{x_o} = \frac{h_i}{x_i} \implies \frac{h_i}{h_o} = \frac{x_i}{x_o}$$

And

$$\tan \phi = \frac{h_i}{x_i - f} = \frac{h_o}{f} \implies \frac{h_i}{h_o} = \frac{x_i - f}{f}$$

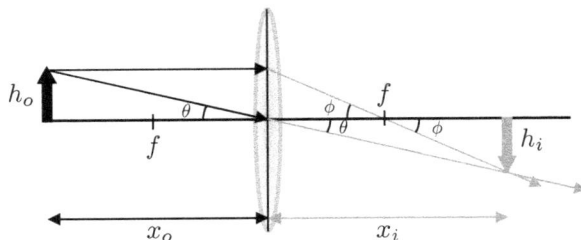

FIGURE 7.27

If we equate the rightmost terms from these equations and perform some simplification, we find

$$\frac{x_i}{x_o} = \frac{x_i - f}{f}$$

$$\frac{x_i}{x_o} = \frac{x_i}{f} - 1$$

$$\frac{1}{x_o} = \frac{1}{f} - \frac{1}{x_i}$$

$$\frac{1}{f} = \frac{1}{x_o} + \frac{1}{x_i}$$

This, of course, is the thin lens equation!

7.8   Lenses: The Thin Lens Equation

In Section 7.5 we derived the thin lens equation (Equation (7.10))

$$\frac{1}{f} = \frac{1}{x_o} + \frac{1}{x_i}$$

by analyzing the geometry of the ray diagrams. As we just saw in Example 7.13, the exact same equation can be derived in the context of thin lenses. However, the geometrical details of the situations we are analyzing differ (Problem 7.36), and so for the thin lens equation to apply, we must once again be careful with our sign conventions:

1. Vertical quantities (h_o and h_i) are *positive* above the optical axis and *negative* below the optical axis. (This is the same as the rules for mirrors.)

2. The object distance x_o is always positive. (This is the same as the rule for mirrors.)

3. The focal length f is positive for a converging lens and negative for a diverging lens. (This is the same as the rule for mirrors, though because mirrors only have one focal point, we could also remember the sign based on it being in front of or behind the mirror.)

4. The image distance x_i is positive to the *right* of the lens (where the image is *real*) and negative to the left (where the image is *virtual*). (This is the same as for mirrors in the sense that $x_i > 0$ for real objects and $x_i < 0$ for virtual objects, but real objects for mirrors are on the *left*, whereas for lenses they are on the *right*.)

It can be hard to remember these rules; Table 7.1 below summarizes the conventions for both mirrors and lenses. Again, these rules come directly from the geometry; we use them so Equation (7.10) can be used for both lenses and mirrors, regardless of context.

Finally, as you can check for yourself (Problem (7.37)), these conventions allow us to use the same equation for lateral magnification that we found when analyzing spherical mirrors (Equation (7.11)):

TABLE 7.1
Sign conventions for the thin lens equation.

Quantity	Symbol	Positive	Negative
Focal Length	f	Converging	Diverging
Object Distance	x_o	Always	Never
Image Distance	x_i	Real Images	Virtual Images

$$m = \frac{h_i}{h_o} = -\frac{x_i}{x_o}$$

Example 7.14 Looking Through a Lens ★★
You and your friend are standing 2.5 m apart. Between you and 0.50 m from your friend there is a table holding a diverging lens with $f = 0.70$ m at eye level. Suppose that both you and your friend's noses are 2.5 cm tall. Describe (a) the image you see of your friend's nose through the lens and (b) the image your friend sees of your nose through the lens (that is, in both cases provide x_i and h_i).

The situation is shown in Figure 7.28; I have set you to the right and your friend to the left. This means that in part (a) we have the usual setup where the object being viewed is to the left of the lens; in part (b) things will be reversed.

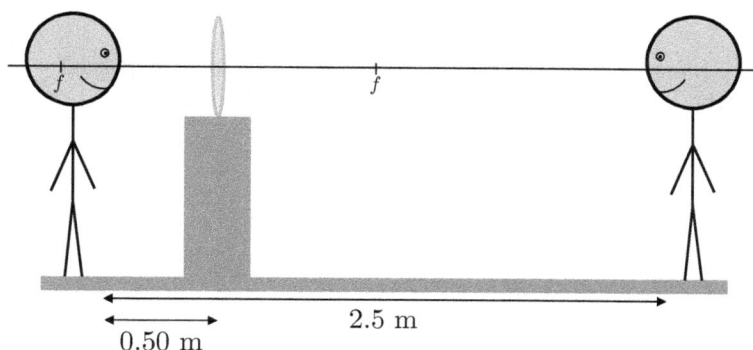

0.50 m   2.5 m

FIGURE 7.28

For (a) we note that your friend (or at least the front of your friend's face) is inside the focal point, so from Figure 7.25 we expect the image to be virtual and upright. Mathematically, we have

$$\frac{1}{f} = \frac{1}{x_o} + \frac{1}{x_i}$$

$$\frac{1}{x_i} = \frac{1}{f} - \frac{1}{x_o}$$

$$x_i = \left(\frac{1}{0.70\text{m}} - \frac{1}{0.50\text{m}}\right)^{-1}$$

$$x_i = -1.75 \text{ m} \approx -1.8 \text{ m}$$

We find a negative image distance for a virtual image, as expected (We round to two significant figures but note the answer to three significant figures for the following calculations). The image height can be found from Equation (7.11):

$$\frac{h_i}{h_o} = -\frac{x_i}{x_o}$$

$$h_i = -\frac{h_o x_i}{x_o}$$

$$h_i = -\frac{(0.025 \text{ m})(-1.75 \text{ m})}{0.50 \text{ m}}$$

$$h_i = 0.0875 \text{ m} \approx 0.088 \text{ m}$$

Thus we see $h_i > 0$ and $|h_i| > h_o$ (i.e. an upright and magnified image), as we expect from the ray diagram.

For (b) we take the same approach, but you are outside the focal point of the lens, so your friend should see a real and inverted image. The thin lens equation gives us

$$\frac{1}{x_i} = \frac{1}{f} - \frac{1}{x_o}$$

$$x_i = \left(\frac{1}{0.70 \text{ m}} - \frac{1}{2.0 \text{ m}}\right)^{-1}$$

$$x_i = 0.93 \text{ m}$$

The positive image distance corresponds to a real image, as expected.

Finally, the height of your image is obtained from Equation (7.11):

$$h_i = -\frac{h_o x_i}{x_o}$$

$$h_i = -\frac{(0.025 \text{ m})(0.93 \text{ m})}{2.0 \text{ m}}$$

$$h_i = -0.011 \text{ m}$$

Thus we see $h_i < 0$ and $|h_i| < h_o$ (i.e. an inverted and shrunken image).

7.9   Correcting Vision

One of the most common applications of ray optics may be literally in front of your eyes right this moment: eyeglasses or contact lenses, which correct vision. How do they work? Well, your eye acts as a converging lens with an *adjustable* focal length; when you focus your vision on a particular object, the focal length shifts such that the image is projected onto the retina. The possible values that f can take for an eye determines what can be seen without distortion (Fig. 7.29).[10]

The closest an object can be to our eye without becoming blurry is referred to as the **near point** (NP); the farthest value is referred to as the **far point** (FP) (Fig. 7.30). The standard values for a healthy adult eye are $NP = 25$ cm and $FP = \infty$.

If $FP < \infty$ for a person, they cannot focus on distant objects (where $x_o > FP$) and are said to be **nearsighted**. If $NP > 25$ cm, they cannot focus on near objects (where $x_o < NP$) and are said to be **farsighted**. Eyeglasses and contact lenses are used to correct these issues. We'll analyze contact lenses, first in the context of nearsightedness.

Our goal is to use a lens so an object located *at infinity* has an *image* located at the person's actual far point (so the eye interprets the object as if it were *at their far point*, where it can be sharply seen). In other words, we "bring in" the image of the object to where the eye can see it. If you refer to Figure 7.25, you'll see that we need a *diverging* lens (for a converging lens the image is either inverted or farther away from the lens than the object). While I'd need an awful lot of paper to draw an object infinitely far away, the geometry in Figure 7.31 demonstrates the idea.

To determine the focal length that places the image at the far point, note that we have $x_o = \infty$ and $x_i = FP$. The thin lens equation becomes

$$\frac{1}{f} = \frac{1}{x_o} + \frac{1}{x_i}$$
$$\frac{1}{f} = \frac{1}{\infty} + \frac{1}{FP}$$
$$\frac{1}{f} = \frac{1}{FP}$$
$$f = FP \tag{7.16}$$

Evidently, to focus on an object that is arbitrarily far away, the focal point should be the same as the person's far point![11]

It is customary to refer to the quantity $1/f$ as the **refractive power** of the lens, P. Thus for a nearsighted person we have

$$P_{\text{near}} = \frac{1}{f} = -\frac{1}{FP} \tag{7.17}$$

In this equation I have made it explicit that the far point is a negative quantity because we have a virtual image. The units of refractive power are m^{-1}, which in this context are called diopters (D).

---

[10]There are other factors that influence vision, of course, such as the actual *shape* of the eye. In fact, the distance between the retina and the lens of the eye can change as you focus on objects that are varying distances away. That said, we will summarize the role of the eye in seeing objects clearly by the range of values that f can take.

[11]Note that by convention both f and x_i are negative in this equation – those negative signs cancel because $1/x_o \to 0$.

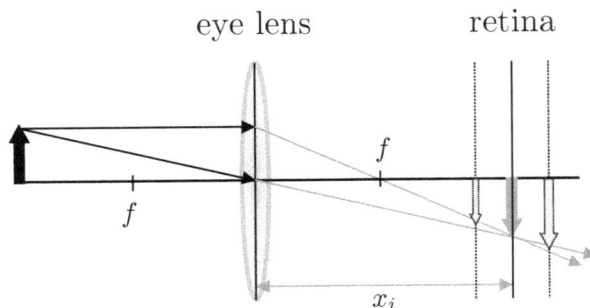

FIGURE 7.29
Our eyes project an inverted image onto the retina, on the back of our eyes (here drawn with a vertical black line). Because a given object and focal length creates an image at a specific distance x_i behind the lens, an incorrect focal length leads to the image forming in front of or behind the retina (the dotted vertical lines). The result is a blurry image on the retina.

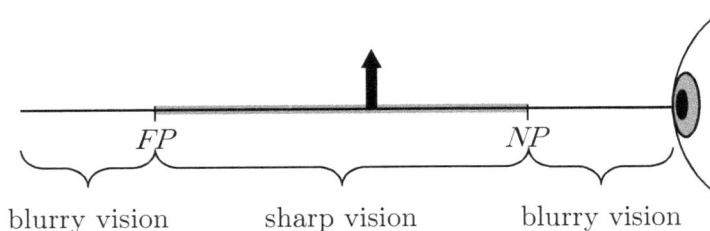

FIGURE 7.30
Our eyes can adjust their focal length such that objects between the near point (NP) and far point (FP) can be viewed clearly.

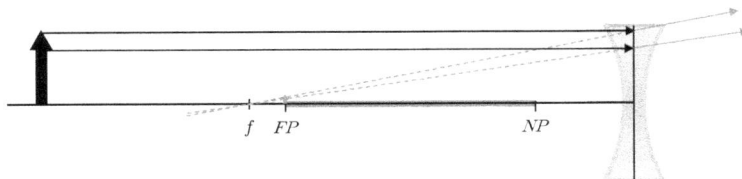

FIGURE 7.31
A distant object's image is formed at the far point (the diagram is clearly not to scale!). Recall that rays that reach our eye from a very distant object are parallel, so to locate the image, two such rays are drawn.

For a farsighted person, the objective is to use a lens so an object located at the *healthy near point* of 25 cm has an *image* located at the person's *actual near point* (where it can be sharply seen). In other words, we "send out" the image of the object to where the eye can see it. Referring back to Figure 7.25, it is clear we need a *converging* lens with the focal point larger than the object distance. Figure 7.32 shows the geometry in this context.

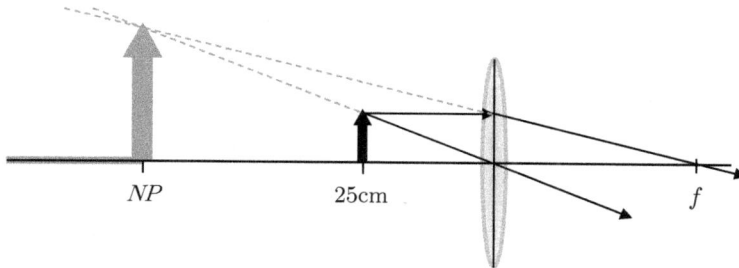

FIGURE 7.32
An object that we would like to see at the healthy near point of 25 cm has an image formed at the person's actual near point.

Numerically, we have $x_o = 25$ cm and $x_i = NP$, so the thin lens equation becomes

$$\frac{1}{f} = \frac{1}{x_o} + \frac{1}{x_i}$$

$$\frac{1}{f} = \frac{1}{0.25 \text{ m}} - \frac{1}{NP} \tag{7.18}$$

In the last line I have introduced a negative sign in front of the near point: we tend to think of the near point as an inherently positive quantity, but we have a virtual image, so by convention the image distance is negative. Making the negative sign explicit allows us to refer to NP as a positive quantity.

In terms of the refractive power, we have for a farsighted person

$$P_{\text{far}} = \frac{1}{f} = \frac{1}{0.25 \text{ m}} - \frac{1}{NP} \tag{7.19}$$

Example 7.15 Correcting Nearsightedness ⋆

A nearsighted person is found to have a far point of 5.0 m. What is the refractive power of a contact lens that will correct their vision?

Directly from Equation (7.17) we have

$$P_{\text{near}} = -\frac{1}{NP} = -\frac{1}{5.0 \text{ m}} = -0.20 \text{ D}$$

Example 7.16 Refractive Power ⋆

Your farsighted friend tells you that their contact lenses have a refractive power of 3.8 D. What is their near point?

If we note that $P = 1/f$, then from Equation (7.18), we see

$$\frac{1}{f} = \frac{1}{0.25 \text{ m}} - \frac{1}{NP}$$

$$NP = \left(\frac{1}{0.25 \text{ m}} - 3.8\frac{1}{\text{m}}\right)^{-1}$$

$$NP = 3.0 \text{ m}$$

7.10 Optical Devices

In this final section of the chapter, we will discuss **telescopes**, which magnify very distant objects, and **microscopes**, which magnify very small (but close) objects. There are many ways to construct these devices; we will focus on the relatively simple case where they use *two converging lenses*. In both devices the lens that is closer to the object is referred to as the **objective lens** and the lens closer to the eye is called the **eyepiece**.

This is the first time we've explicitly analyzed the effect of multiple lenses, so let me emphasize the key point: when the rays enter the second lens (the eyepiece), they appear to be coming from the objective lens image as if it were an actual object (this is, after all, what we mean by the term *image*). Thus, we will proceed by treating the *image* formed by the objective lens as the *object* from the viewpoint of the eyepiece.

With this in mind, let's consider a telescope. Our goal is to magnify a distant object. In Example 7.12 we saw how to magnify an object that is *close* to us. At first glance this may not seem useful in our present context, but we just saw, in the context of correcting nearsightedness, how a lens focuses rays from a distant object at the focal point.[12] Our strategy, therefore is this: the objective lens brings in rays from the object such that the image (which, remember, is the *object* as far as the eyepiece is concerned) can be magnified by the eyepiece.

Now, if the objective lens creates an image of a distant object *on* its focal point, and a magnifier works best when the object is *just inside* its focal point, then for our telescope the focal points of the two lenses should essentially overlap. We sketch the ray diagram for a *non-infinite* object distance in Figure 7.33. Figure 7.34 shows the limiting behavior when the object *is* infinitely far away.

Now, Figure 7.34 doesn't show the final image: where is it, and how large is it? Let's consider first its position. The image formed by the objective lens is on the focal point, so for the eyepiece $x_o \approx f_{\text{eye}}$ and the thin lens equation reduces to

$$\frac{1}{f_{\text{eye}}} = \frac{1}{x_o} + \frac{1}{x_i}$$

$$\frac{1}{f_{\text{eye}}} \approx \frac{1}{f_{\text{eye}}} + \frac{1}{x_i}$$

$$0 = \frac{1}{x_i}$$

The only way this can be a valid equation is in the limit $x_i \to \infty$: the image is at the limit of (healthy) vision, just the same as the actual object.

---

[12]We used a diverging lens in that case, but the same is true for a converging lens.

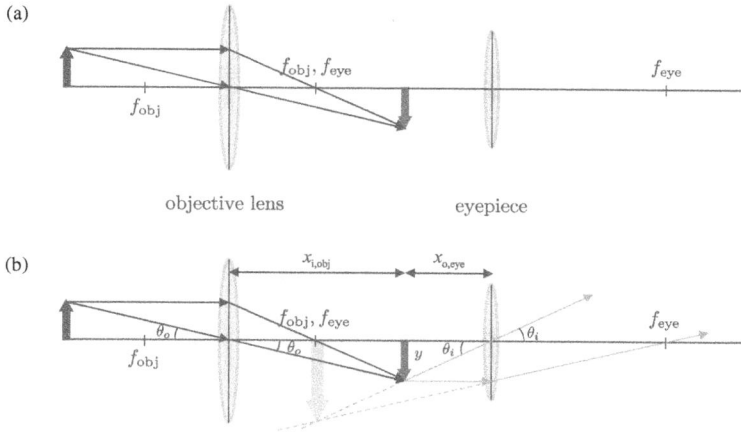

(a)

objective lens   eyepiece

(b)

FIGURE 7.33
Two converging lenses separated such that their focal points overlap in the region between the lenses. (a) This shows the image formation due to the objective lens. (b) This shows the subsequent image formation by the eyepiece. θ_o identifies the angular size of the original (black) object, while θ_i identifies the angular size of the final (pale gray) image. (The other ray that defines the position of the final image will form the same angle in the limiting case that the image is infinitely far away.)

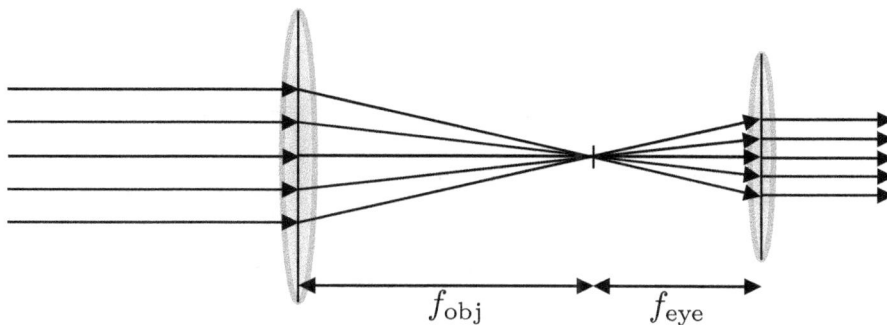

FIGURE 7.34
The ray diagram for a refracting telescope: the objective lens brings parallel incoming rays together at the focal point, and the eyepiece brings them back out to form a magnified image.

What about the magnification? Here we'll use the *relative* angular magnification: how large is the object (a crater on the moon, say) when viewed through the telescope compared to if you looked at it with the naked eye?[13] Referring back to the bottom of Figure 7.33, we see

$$\tan \theta_o = \frac{y}{x_{o,\text{obj}}} \tag{7.20}$$

and

$$\tan \theta_i = \frac{y}{x_{o,\text{eye}}} \tag{7.21}$$

Now if we go to the case of an infinitely distant object, then the image formed by the objective lens shifts to the common focal point: $x_{o,\text{obj}} \to f_{\text{obj}}$ and $x_{o,\text{eye}} \to f_{\text{eye}}$. If we apply the small angle approximation $\tan \theta \approx \theta$, then the equations above become

$$\tan \theta_o \approx \theta_o = \frac{y}{f_{\text{obj}}} \tag{7.22}$$

and

$$\tan \theta_i \approx \theta_i = \frac{y}{f_{\text{eye}}} \tag{7.23}$$

Then the relative angular magnification is given by

$$\mu_r = \frac{\theta_i}{\theta_o}$$

$$\mu_r \approx \frac{y/f_{\text{eye}}}{y/f_{\text{obj}}}$$

$$\mu_r = \frac{f_{\text{obj}}}{f_{\text{eye}}}$$

Or, in other words,

$$\mu_{\text{telescope}} = \frac{f_{\text{obj}}}{f_{\text{eye}}} \tag{7.24}$$

This quantifies the practical effect of the lenses used in this telescope: to maximize the relative angular magnification, increase f_{obj} and/or decrease f_{eye}. Finally, I should mention that this kind of telescope is called a **refracting telescope**. Most "backyard telescopes" follow this design.[14]

Example 7.17 Magnification in a Telescope ⋆
A refracting telescope has an objective lens with $f_{\text{obj}} = 1.00$ m and several eyepieces that can be swapped out to achieve different magnifications. If $f_{\text{eye}} = 4.0$ cm, 6.0 cm, or 10.0 cm, what are the possible values of $\mu_{\text{telescope}}$?

---

[13] Note that the absolute angular magnification is nonsensical in this context: placing Jupiter 25 cm from your eye would obviously be more than sufficient to completely fill your field of view!

[14] Refracting telescopes usually also have a small, flat mirror behind the eyepiece, which serves to redirect the light 90° into your eye. This means that you don't have to squat down to look directly down the length of the tube as it points up into the night sky.

From Equation (7.24), it follows that the possible values are

$$\mu_1 = \frac{1.00 \text{ m}}{0.040 \text{ m}} = 25$$

$$\mu_2 = \frac{1.00 \text{ m}}{0.060 \text{ cm}} = 17$$

$$\mu_3 = \frac{1.00 \text{ m}}{0.100 \text{ cm}} = 10.0$$

Example 7.18 The Moon Through a Telescope ★

The full moon has an angular size of about 0.5° when viewed with the naked eye. Your refracting telescope has $f_o = 1.00$ m and $f_e = 6.0$ cm. What will the angular size of the moon be when you observe it through the telescope?

From Equation (7.24) and the definition of relative angular magnification, we have

$$\mu = \frac{f_{\text{obj}}}{f_{\text{eye}}} = \frac{\theta_i}{\theta_o}$$

It follows that

$$\theta_i = \frac{f_{\text{obj}}}{f_{\text{eye}}} \theta_o$$

$$\theta_i = \frac{1.00 \text{ m}}{0.060 \text{ m}} (0.5°)$$

$$\theta_i = 8°$$

We'll now turn to the **optical microscope**. Here our object is *not* infinitely far away, so *both lenses* can be used to magnify the object. The standard approach is to set the object just *outside* f_{obj} such that the corresponding image forms just *inside* f_{eye}. Figure 7.35 shows the ray diagram where, as usual, the proportions are exaggerated so it is easier to see the geometry. Note that, just as with the telescope, the fact that the image formed by the objective lens is (essentially) on top of the focal point, the final image distance is infinitely far away.

Once again we can quantify the magnification by analyzing the geometry. If we note that $x_{i,\text{obj}} \approx L$ (the focal length of the eyepiece is very small compared to L, the distance between the lenses) and $x_{o,\text{obj}} \approx f_{\text{obj}}$ (the object is just outside the objective lens focal point), then

$$\tan \theta_o \approx \frac{y}{L} \approx \frac{h_o}{f_{\text{obj}}} \tag{7.25}$$

From this result it follows that the lateral magnification due to the objective lens has magnitude

$$|m| = \frac{y}{h_o} = \frac{L}{f_{\text{obj}}}$$

Similarly, we see from the geometry that

$$\tan \theta_i = \frac{y}{f_{\text{eye}}}$$

It is standard here to report the *absolute angular magnification* of the eyepiece under the small angle approximation:

$$\mu_a = \frac{\theta_i}{\theta_{NP}}$$

$$\mu_a = \frac{y/f_{\text{eye}}}{y/25 \text{ cm}}$$

$$\mu_a = \frac{25 \text{ cm}}{y_{\text{eye}}}$$

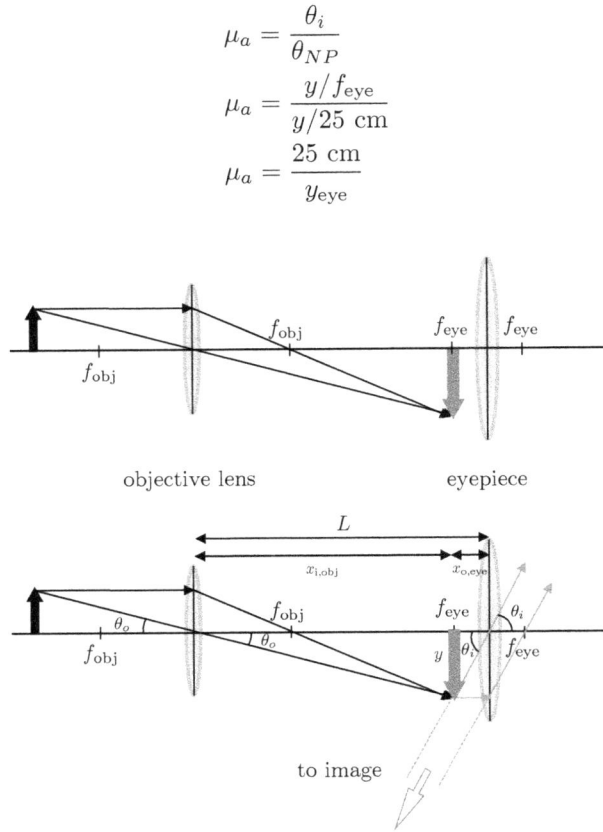

FIGURE 7.35
The ray diagram for a microscope.

Now, the *total* magnification is the net effect (i.e. product) of the two:[15]

$$\mu_{\text{microscope}} = \frac{L(25 \text{ cm})}{f_{\text{obj}} f_{\text{eye}}} \tag{7.26}$$

In words, we are considering the lateral magnification due to the objective lens and then the absolute angular magnification *of that image* due to the eyepiece. This is the conventional equation cited when describing the magnification of a microscope, but it may strike you as somewhat arbitrary: why not, for instance, the product of both absolute angular magnifications?

Perhaps the best answer is that *it doesn't matter*: use any two definitions of magnification that you like, and you will find that you always end up with the product of the focal lengths in the denominator and some constants in the numerator. Thus the effect of changing the focal lengths of a microscope's lenses will be the same regardless of how, exactly, you define magnification.

---

[15] For example, if we triple the apparent size of the object with the objective lens, then double *that* with the eyepiece, we end up with a $2 \times 3 = 6$-fold increase.

Example 7.19 Hair Through a Microscope ⋆

The width of human hair varies, but one approximate value is 0.10 mm. What is the angular size of a human hair held at the healthy near point? Compare this to the size when held under a microscope with $L = 160$ mm, $f_{\text{obj}} = 5.0$ mm and $f_{\text{eye}} = 8.0$ mm.

At the near point (i.e. 0.25 m from your eye) we have

$$\tan\theta_{NP} \approx \theta_{NP} = \frac{1.0 \times 10^{-4} \text{ m}}{0.25 \text{ m}} = 4.1 \times 10^{-7} \text{rad}$$

Using this with Equation (7.26), we find

$$\mu_{\text{microscope}} = \frac{L\,(25 \text{ cm})}{f_{\text{obj}} f_{\text{eye}}} = \frac{\theta_i}{\theta_{NP}}$$

$$\theta_{NP} \frac{L\,(25 \text{ cm})}{f_{\text{obj}} f_{\text{eye}}} = \theta_i$$

$$\left(4.1 \times 10^{-7} \text{rad}\right) \frac{(0.160 \text{ m})\,(0.25 \text{ m})}{(5.0 \times 10^{-3} \text{ m})\,(8.0 \times 10^{-3} \text{ m})} = \theta_i$$

$$4.1 \times 10^{-4} \text{ rad} = \theta_i$$

This is a 1000× increase!

Example 7.20 Structure of a Microscope ⋆⋆⋆

Consider a microscope with an objective lens focal length f_{obj}.

(a) If the objective lens image is to be a distance L from the objective lens, then how far should the object be from the lens? (This describes the distance necessary to observe a sharp image through the microscope, assuming the image needs to be a distance L from the objective lens for the eyepiece to generate a sharp image.)

(b) What is the numerical value if $L = 160$ mm (a typical value for a biological microscope) and $f_{\text{obj}} = 4.0$ mm?

(c) Show that if the object is *exactly* on the focal point, the image is formed infinitely far away.

If you refer back to Figure 7.35, you'll see that the question is really asking us to determine where the two rays leaving the objective lens intersect. Let's define a pair of equations, $y_1(x)$ and $y_2(x)$, that describe the rays, then determine where they intersect.

We'll set the origin of the coordinate system at the center of the objective lens. If we call the object's distance beyond the focal point d, then the top and bottom rays respectively obey

$$y_1(x) = -\frac{h_o}{f_{\text{obj}}} x + h_o$$

$$y_2(x) = -\frac{h_o}{d + f_{\text{obj}}} x$$

Setting these equal at $x = L$, we see

$$-\frac{h_o}{f_{\mathrm{obj}}}L + h_o = -\frac{h_o}{d + f_{\mathrm{obj}}}L$$

Solving this equation for d is a bit tedious. Ready? Here we go:

$$-\frac{L}{f_{\mathrm{obj}}} + 1 = -\frac{L}{d + f_{\mathrm{obj}}}$$

$$\frac{-L + f_{\mathrm{obj}}}{f_{\mathrm{obj}}} = -\frac{L}{d + f_{\mathrm{obj}}}$$

$$\frac{f_{\mathrm{obj}}}{-L + f_{\mathrm{obj}}} = -\frac{d + f_{\mathrm{obj}}}{L}$$

$$\frac{-L f_{\mathrm{obj}}}{-L + f_{\mathrm{obj}}} - f_{\mathrm{obj}} = d$$

$$\frac{-L f_{\mathrm{obj}} - f_{\mathrm{obj}}\left(-L + f_{\mathrm{obj}}\right)}{-L + f_{\mathrm{obj}}} = d$$

$$\frac{\left(f_{\mathrm{obj}}\right)^2}{L - f_{\mathrm{obj}}} = d \tag{7.27}$$

Phew! As you can check for yourself, plugging in our values from (b) yields $d = 0.10$ mm. For (c), note that the only way for the left side of Equation (7.27) to be 0 is if $f_{\mathrm{obj}} = 0$ or $L = \infty$. As the premise of the problem is that $f_{\mathrm{obj}} > 0$, the only possible conclusion is that the image would indeed be infinitely far away.

7.11   Problems for Chapter 7

7.1 The Ray Model of Light

★   **Problem 7.1.** The Washington Monument in Washington, D.C. is about 169 m tall. On a sunny day, you are standing in the shadow of the tip of the Monument, which is 30.0 m from the base. How far above the horizon is the Sun, measured in degrees?

★   **Problem 7.2.** A student wishes to photograph the Washington Monument (which is 169 m tall) with a simple pinhole camera. If the camera has a length of 24 cm and the image is to be 5.6 cm high, how far from the monument should she stand?

★   **Problem 7.3.** A tiny light illuminates a narrow hole that is 1.3 m from the light, which forms a circular patch of light on the wall 0.92 m from the hole. If the patch is 8.0 cm wide, then how large is the hole?

★   **Problem 7.4.** Consider the pinhole camera of Example 7.2. Suppose the person being photographed is 1.8 m tall and they are standing 4.0 m in front of the box. If the box is 15 cm deep, how large will the person's image be on the film?

7.2 Reflection

★   **Problem 7.5.** You are standing in a room that is 4.0 m high with a mirrored ceiling. You are facing a wall that is 12.0 m in front of you. You'd like to shine a laser pointer at the ceiling such that it reflects onto the midpoint of the wall. If you hold the laser pointer 1.5 m above the ground, at what angle (measured up from the horizontal) should you hold the laser?

Problem 7.6. Suppose you are standing 1.5 m in front of a mirror that runs from the ceiling to the floor. Sketch the situation and use the law of reflection to draw the rays that locate your image. How far behind the mirror is it? How far are your (actual) eyes from the image of your toes?

Problem 7.7. Suppose you are standing 1.5 m in front of a mirror that runs from the ceiling to the floor. A second identical mirror runs parallel to the first, 1.5 m behind you. Sketch the situation and use the law of reflection to draw the rays that locate *two* images you see of yourself behind the mirror in front of you. How far is each image behind the mirror?

Problem 7.8. Suppose you and a friend are standing in a room where two parallel walls are mirrored (Fig. 7.36). With clear ray diagrams, determine the locations of *two* images you see of your friend behind the top mirror.

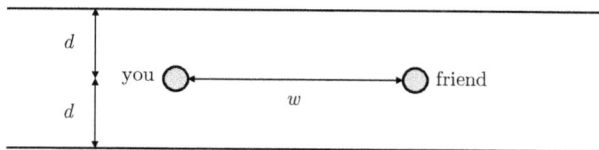

FIGURE 7.36
Problem 7.8

Problem 7.9. Many guns can be fitted with a variety of optical devices (sights, scopes, etc.) to assist with aiming. One relatively inexpensive device is the red dot sight, sketched in Figure 7.37: a narrow beam of red light is emitted from the LED and is reflected off a glass surface into your eye. The glass is transparent (so you can see through it) but is specially treated to be reflective on its inner surface, so for the purpose of the LED light it behaves like a mirror. The net effect is that you see a small red dot in the middle of the sight. If the reflected beam is to travel horizontally into your eye, the glass must be tilted by some angle θ relative to an upright orientation. Determine a numerical value for θ. Explain.

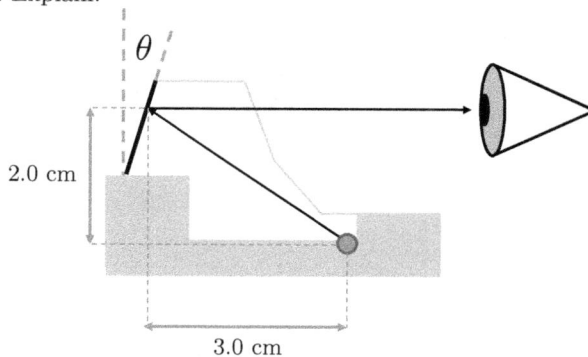

FIGURE 7.37
Problem 7.9

Problem 7.10. Figure 7.38 shows a top-down view of a person driving a car and a light ray traveling from an object directly behind the car to the center of the rear-view mirror. The center of the rear-view mirror is located 0.18 m in front and 0.30 m to the right of the driver's eye.

a. Determine a value for ϕ if the light ray is to be reflected into the driver's eye.

b. Draw the reflected ray on the diagram and determine a numerical value for the angle of reflection, θ_r, assuming ϕ has the numerical value you identified in (a).

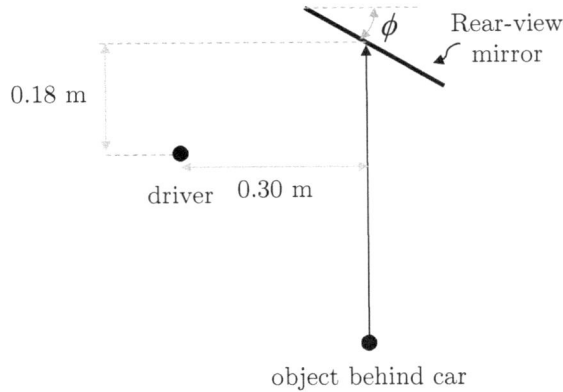

FIGURE 7.38
Problem 7.10

★★   **Problem 7.11.** Movies and video games where the hero explores an ancient tomb or cave often
feature a series of mirrors than can redirect light shining in from the entrance; the hero's goal might
be to illuminate the chamber or activate some mechanism by shining light on it. Figure 7.39 shows
an incoming beam of light, the hero standing by a mirror that can be rotated in place, and a target
below the hero's platform. Determine a numerical value for θ that will reflect the light beam onto
the center of the target.

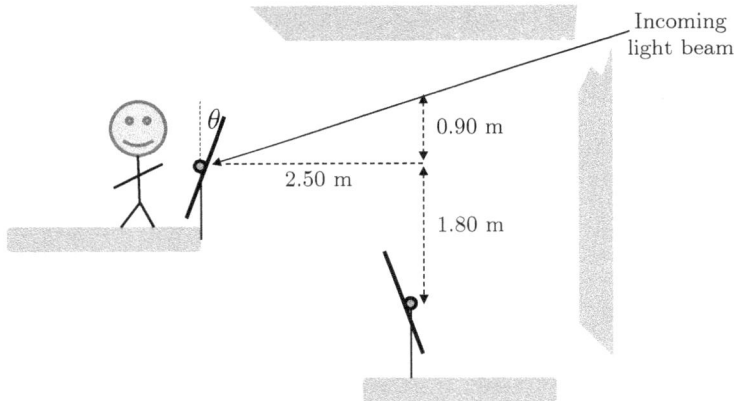

FIGURE 7.39
Problem 7.11

★★★   **Problem 7.12.** You can use a handheld mirror and wall mirror to look at the back of your own
head. Suppose that your favorite teacher is doing this one day when they realize that someone has
stuck a note to the back of their head that has "DORKUS" written on it, as sketched on the left
portion of Figure 7.40.

   a. Carefully sketch a ray that leaves the letter "D", interacts with the mirrors, and ultimately
      enters the person's eyes. You don't need any math, but your diagram should make it clear how
      you applied the law of reflection.

   b. Repeat (a) for the letter "S".

   c. Carefully draw the image of the text "DORKUS" as the person observes it in the mirror.

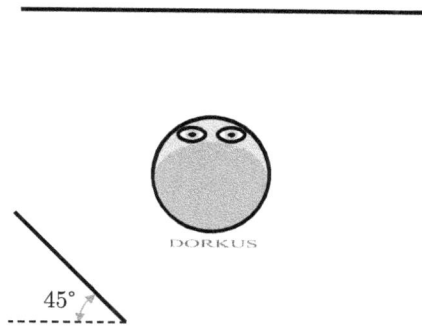

FIGURE 7.40
Problem 7.12

★★ **Problem 7.13.** You are playing laser tag and find yourself in the situation sketched below: your opponent is crouched behind cover but both the ceiling and the wall behind her is reflective. At what angle should you fire your gun so the laser beam finds its way to the target on her back?

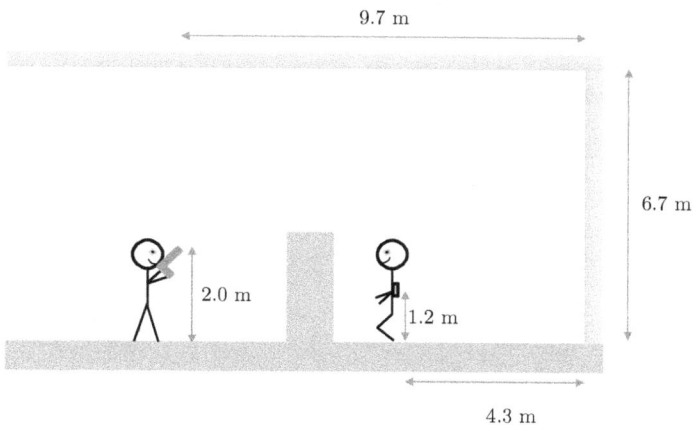

FIGURE 7.41
Problem 7.13

★ **Problem 7.14.** You are standing in a long hallway; one wall is mirrored and a friend of yours is standing to your left and closer to the mirrors, as shown in Figure 7.42. Sketch the figure and include relevant rays to determine where you will see the image of your friend.

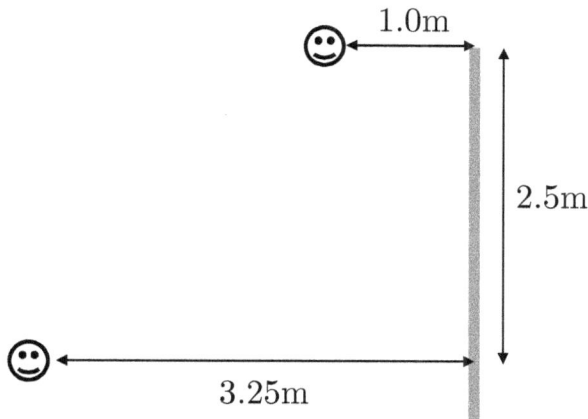

FIGURE 7.42
Problem 7.14

7.3 Magnification

⋆ **Problem 7.15.** You are viewing a bacteria of the genus *Bacillus* through a microscope. Unfortunately, the microscope is old and the labeling on the lenses has worn off. The tube length is 0.20 m and you know that the typical length of the bacteria is about 1.0×10^{-5} m. You estimate that, when looking at the bacteria through the microscope, it has an angular size of about 3.0°. What is the absolute angular magnification of the microscope?

⋆ **Problem 7.16.** Suppose a character in a movie uses a rudimentary telescope to observe a distant person. By watching scenes through the telescope and over the character's shoulder, you estimate the following:

- The telescope is about 20. cm long.
- In the "over the shoulder" clip the angular size of one of the riders is about 1.0° (about the width of your smallest fingernail held at arm's length).
- In the "telescope" clip the rider appears much closer and has an angular size of about 25° (about the angle between your pinky and thumb if they're spread wide).

What is the magnitude of the angular magnification achieved by the telescope? (Does absolute or relative angular magnification seem more appropriate here, or does it not matter?)

7.4 Curved Mirrors: Ray Tracing

⋆ **Problem 7.17.** A small toy with a height of 1.0 cm is placed 8.0 cm in front of a converging mirror with a focal length of 4.0 cm. Neatly sketch a ray diagram to find the image. Is it upright or inverted? Real or virtual?

⋆⋆ **Problem 7.18.** Consider a concave mirror with a focal length of 12 cm. An object is 2.0 cm tall and is held centered on the optical axis so it extends above the optical axis (as usual) 1.0 cm and *below* the optical axis by 1.0 cm. Sketch the key rays from the top of the object *and* from the bottom of the object. In the text I claimed that we can think of an object as just extending above the optical axis without losing any essential information. Does your sketch here support this claim? Why or why not?

⋆⋆⋆ **Problem 7.19.** An object is in front of a converging mirror. The ray begins parallel to the optical axis and reflects around the normal line, according to the law of reflection. The location where the ray crosses the optical axis is a distance f from the middle of the mirror. Show that $f = R/2$ only if θ is a small angle (such that we can approximate $\sin\theta \approx \theta$). How must the height of the object compare to the radius of curvature of the lens if the small angle approximation holds? (*Hint:* It is either $h \gg R$ or $h \ll R$, but which is it–and why?)

7.5 Curved Mirrors: The Thin Lens Equation

⋆ **Problem 7.20.** You stand 3.0 m in front of a convex mirror with a focal length of $f = -15$ cm. Mathematically determine the position and magnification of the image. Is the image upright or inverted?

⋆ **Problem 7.21.** A human skin pore is about 0.40 mm wide. A concave mirror has $f = 40$ cm. You hold the mirror 25 cm from your face. How wide will the image of one of your pores be?

⋆⋆ **Problem 7.22.** Figure 7.13 shows the three characteristic situations for an object held in front of a spherical mirror. Draw each situation and carefully label the geometry. (What is f, x_o, and x_i? What angles are the same? etc.) In each case, geometrically determine an equation that relates the focal length f, object distance x_o, and image distance x_i. Confirm that your equations match the thin lens equation (Equation (7.10)) given the sign conventions discussed just before the equation is defined on page 370.

⋆ **Problem 7.23.** Refer to the situation described in Problem 7.17. Mathematically determine the position and height of the image.

7.6 Refraction

★ **Problem 7.24.** You shine a laser beam into a block of an unknown material, as sketched in Figure 7.43. What is the index of refraction of the material, calculated to two significant figures?

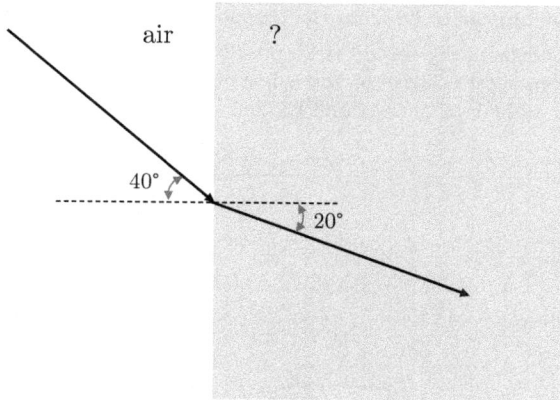

FIGURE 7.43
Problem 7.24

★★ **Problem 7.25.** One sunny day you are lounging by a pool. Right before you go into the pool, the sun is 60.° above the horizon. You then get in the water (with an index of refraction of 1.33). The pool is 2.3 m deep and 6.3 m wide.

a. What angle will the sun *appear* to make with the horizon when you are underwater?

b. What portion (if any) of the bottom of the pool will be in shadow? Give your answer as a percent (0% = none; 100% = all).

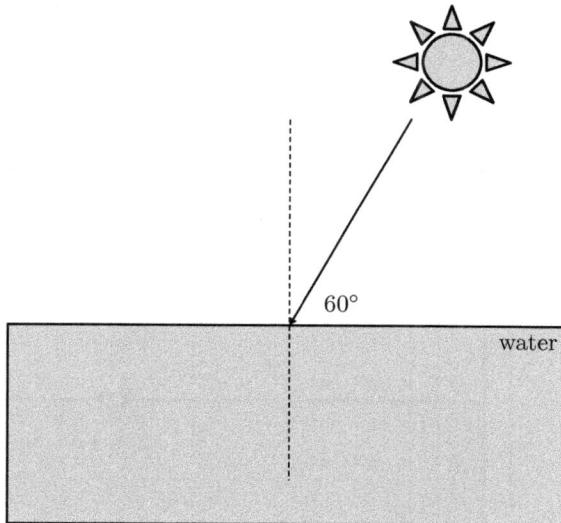

FIGURE 7.44
Problem 7.25

★★ **Problem 7.26.** Consider a long cylinder of glass. If a laser is directed at the center of one of the flat circular ends of the cylinder at an angle of 0°, the beam will refract at an angle of 0°. For an incident angle $\theta > 0°$, the beam will refract and eventually impact the edge of the cylinder. To prevent the beam from refracting back out into the air, suppose the outside of the cylinder is wrapped in another material with an index of refraction of 1.4. How large can θ be to ensure that the beam doesn't refract out of the glass? (This is the basic idea of a fiber optic cable.)

Problem 7.27. A drone is hovering over the surface of an ocean (with an index of refraction of 1.3) near a submerged submarine, as sketched below. Respond to the following questions with clear mathematics and by labeling relevant geometry in a copy of the sketch. You may use the small angle approximation.

a. Where will the drone appear to be from the perspective of the submarine?

b. If the drone moved horizontally, would it be possible for either the drone or the submarine to effectively vanish from sight relative to the other object? If the answer is yes, would the drone need to move left or right in the diagram? Explain.

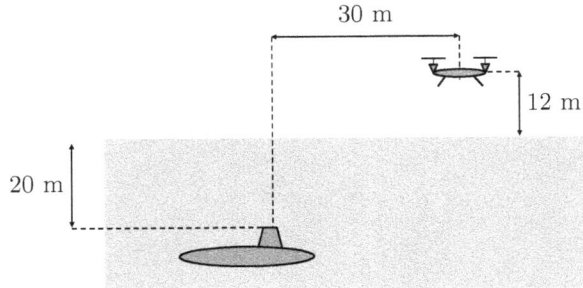

FIGURE 7.45
Problem 7.27

Problem 7.28. A glass of water is half-filled with water and half-filled with oil; you point a laser beam toward the center of the glass, as shown in Figure 7.46. The location where the beam strikes the oil-water interface is marked.

a. What is the index of refraction of the olive oil?

b. The laser will refract into the water; where will it strike the bottom/side of the glass?

c. Is it possible for a laser beam to undergo total internal reflection if it is in oil and traveling toward water? If yes, what is the critical angle? If no, why not?

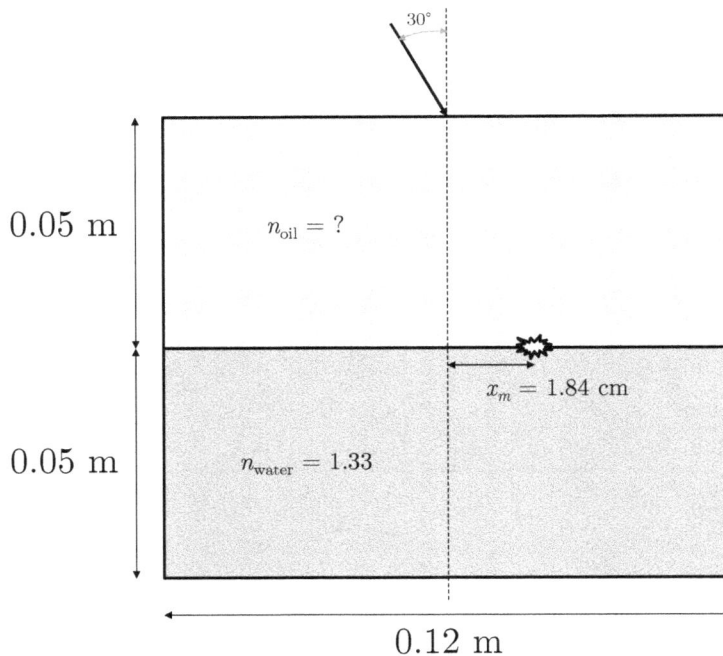

FIGURE 7.46
Problem 7.28

Problem 7.29. While we often think of the index of refraction as a constant value (depending on the type of material), it depends somewhat on the wavelength of the light interacting with the material. For a certain type of glass, suppose the index of refraction for visible light varies from 1.51 for deep red to 1.53 for deep violet. Suppose a narrow beam of white light strikes a flat rectangular slab of this glass at an angle $\theta_i = 30.°$.

   a. What range of refracted angles will be observed?

   b. Sketch the incident beam and the refracted beam. Indicate where you would observe red light and where you would observe violet light.

   c. If the slab is 3.0 cm thick, how wide will the light pattern be when it reaches the opposite side of the slab?

7.4 Lenses: Ray Tracing

Problem 7.30. A lovesick middle schooler is looking at the yearbook photo of his crush through a magnifying glass with a focal length of 8.0 cm. The photo is 4.0 cm tall and the student holds the magnifying glass 6.0 cm above the page.

   a. Sketch the situation with a **neat** ray diagram. Include both the object and the image.

   b. Is the image upright or inverted? Is the image real or virtual?

Problem 7.31. A magnifying glass is held 4.0 cm from a flower. The image has a lateral magnification of 3.0. (a) What is the focal length of the lens? (b) Sketch the situation with a ray diagram.

Problem 7.32. A figurine is 2.5 cm tall and sits 6.0 cm in front of a lens with a focal length $f = -3.0$ cm. The object is 8.0 cm from the lens. Make a neat ray diagram to estimate the location of the image. Is it upright or inverted? Real or virtual?

Problem 7.33. You hold a diverging lens with a focal length $f = -15.0$ cm in front of a 3.0 cm-tall toy figurine. Make a neat ray diagram to estimate the location of the image. Is it upright or inverted? Real or virtual?

Problem 7.34. You set up a two-lens system as shown in Figure 7.47. The object is 1.0 cm tall and 4.2 cm from the left lens, which has a focal length of -2.0 cm. The right lens has a focal length of 0.88 cm and is located 5.4 cm from the first lens.

   a. Where will the image formed by the first lens be located? Show the location with a neat ray diagram.

   b. Where will the final image be located, relative to the second lens? Show the location with a neat ray diagram.

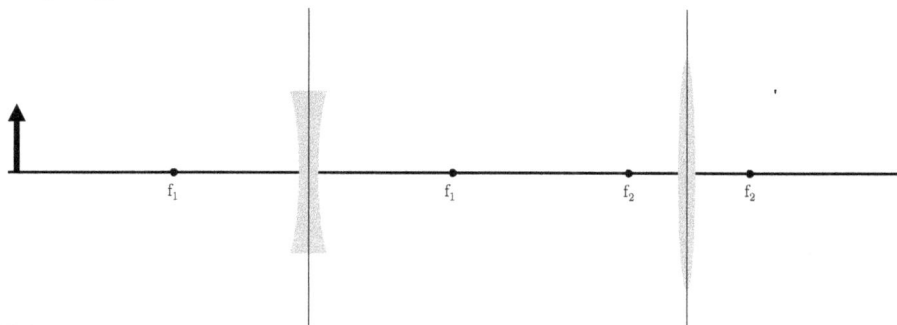

FIGURE 7.47
Problem 7.34

Problem 7.35. You set up a two-lens system as shown in Figure 7.48. The object is 1.0 cm tall and 5.5 cm from the left lens, which has a focal length of 2.0 cm. The right lens has a focal length of 0.88 cm and is located 5.30 cm from the first lens.

a. Where will the image formed by the first lens be located? Show the location with a neat ray diagram.

b. Where will the final image be located, relative to the second lens? Show the location with a neat ray diagram.

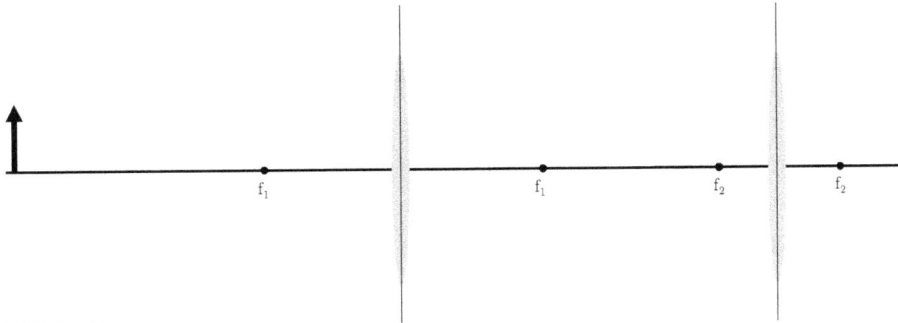

FIGURE 7.48
Problem 7.35

7.8 Lenses: The Thin Lens Equation

★★ **Problem 7.36.** Figure 7.25 shows the four characteristic situations for an object held in front of a thin lens. Draw each situation and carefully label the geometry. (What is f, x_o, and x_i? What angles are the same? etc.) In each case, geometrically determine an equation that relates the focal length f, object distance x_o, and image distance x_i. Confirm that your equations match the thin lens equation (Equation (7.10)) given the sign conventions discussed in the table on page 381.

★ **Problem 7.37.** Using the table on page 381 and the ray diagrams of Figure 7.25 to show that Equation (7.11) correctly indicates the orientation (upright or inverted) of images formed by thin lenses.

★★ **Problem 7.38.** Mathematically determine the position, orientation, and height of the image formed by the lens described in Problem 7.30.

★★ **Problem 7.39.** Mathematically determine the position, orientation, and height of the image formed by the lens system described in Problem 7.34.

★★ **Problem 7.40.** Mathematically determine the position, orientation, and height of the image formed by the lens system described in Problem 7.35.

7.9 Correcting Vision

★ **Problem 7.41.** A person who is farsighted has a near point of 2.3 m. What should the lens power be for contact lens that correct his vision? Give your answer in diopters.

★ **Problem 7.42.** Your friend tells you they have an eyeglass prescription of 3.0 D. Are they nearsighted, farsighted, or both?

★ **Problem 7.43.** Your friend has a near point of 5.0 cm and a far point of 100.0 cm. Are they nearsighted, farsighted, or both?

★ **Problem 7.44.** Your friend has a near point of 30.0 cm and a far point of ∞. Are they nearsighted, farsighted, or both?

★★ **Problem 7.45.** Your friend has an eyeglass prescription of 3.0 D.

a. Are they nearsighted or farsighted?

b. What are your friend's near point and far point? (If either/both are not being treated by the eyeglasses, assume they have the usual healthy value.)

c. Sketch and label a neat ray diagram that demonstrates how the glasses correct their vision. Clearly label the object distance, image distance, and the focal point of the lens.

★★ **Problem 7.46.** Your friend has an eyeglass prescription of −5.0 D.

a. Are they nearsighted or farsighted?

b. Sketch and label a neat ray diagram that demonstrates how the glasses correct their vision.

c. What are your friend's near point and far point? (If either/both are not being treated by the eyeglasses, assume they have the usual healthy value.)

★★ **Problem 7.47.** An ophthalmologist performs an eye exam and finds that their patient's vision becomes blurry when they try to see anything closer than 3.5 m away. What power eyeglasses should the doctor prescribe? Give your answer in Diopters and justify your answer with a diagram and/or a sentence or two.

★★ **Problem 7.48.** Someone who is both nearsighted and farsighted can be prescribed bifocals, which allow the patient to view distant objects when looking through the top of the glasses and close objects when looking through the bottom of the glasses. Suppose a particular bifocal prescription is for glasses with refractive powers +2.5 D and −0.20 D.

a. What is the patient's near point? Support your mathematics with a clear ray diagram.

b. What is the patient's far point? Support your mathematics with a clear ray diagram.

★★ **Problem 7.49.** For each of the following patients, (i) identify if they are nearsighted or farsighted, (ii) draw a clear ray diagram of an appropriate corrective lens, and (iii) determine the appropriate prescription, in diopters, to correct their vision.

a. Patient A cannot clearly see objects more than 2.5 m away.

b. Patient B cannot clearly see objects closer than 3.0 m away.

7.10 Optical Devices

★★ **Problem 7.50.** A microscope is designed so that the image formed by the objective is located very close to the focal point for the eyepiece. Why? A complete answer should include a diagram and a sentence or two; no math is needed.

★ **Problem 7.51.** A microscope has a tube length of 180 mm and has a 40× objective and a 30× eyepiece. How far is the sample from the objective lens if it is in focus?

★★ **Problem 7.52.** A microscope has a 0.20 m tube length. What focal length objective will give a total magnification of 500× when used with an eyepiece with a 4.0 cm focal length?

★ **Problem 7.53.** You have lenses of focal lengths 2.0 cm, 4.0 cm, 8.0 cm, and 16.0 cm. If you could use any two of these to build a telescope, which ones should you select as the eyepiece and as the objective to provide an angular magnification with the maximum possible magnitude? What magnification would you achieve?

★★ **Problem 7.54.** Mars is about 6800 km in diameter. Suppose it is 1.1×10^8 km from the Earth when you view it through a telescope. Its angular size when viewed through the telescope is 0.50°, about the same as the Moon viewed with the naked eye. If the eyepiece focal length is 25 mm, how long is the telescope (measured from one lens to the other)?

★ **Problem 7.55.** What relative angular magnification can be achieved in a telescope with an objective lens focal length of 0.500 m and an eyepiece focal length of 0.010 m? What would the value be if the lenses were switched?

★ **Problem 7.56.** A bee is about 1.2 cm long. What angular size does it have if it is sitting on a leaf 14.0 m away and you

a. look at it with the unaided eye?

b. look at it through a telescope with an objective lens focal length of 0.750 m and an eyepiece focal length of 0.020 m?

★★ **Problem 7.57.** Some astronomy enthusiasts take impressive photos of space (this is often called *astrophotography*). Suppose a photo of Mars rising above the horizon of the Moon is taken with a two-lens zoom consisting of an objective lens with a focal length of 15.0 cm and an eyepiece with a focal length of 8.0 mm. The lenses are spaced such that the image formed by the first lens is 7.0 mm in front of the second lens. The diameter of Mars is 6.8×10^6 m and suppose it is about 8.3×10^{10} m away from Earth when the photo is taken.

a. Make a neat ray diagram showing how rays from Mars interact with both lenses to form a final image. (You can't do it to a proper scale since Mars is so far away; just make a correct diagram for some object that is "far away".)

b. Use your ray diagram: is the final image upright or inverted?

Mathematically determine the answers to the following:

c. What is the distance between the objective lens and the image it creates?

d. How tall is the final image of Mars? (This is the size of Mars on the camera's CCD chip, which is analogous to the retina of the eye.)

★ **Problem 7.58.** A microscope has a tube length of 180 mm and has a 40× objective and a 20× eyepiece. The microscope is focused for viewing with a relaxed eye. How far is the sample from the objective lens?

Additional Problems for Chapter 7

★★ **Problem 7.59.** *Optical tweezers* can move tiny objects by subjecting them to a focused laser beam. Reflection and refraction plays a central role in explaining this behavior. Figure 7.49 shows a spherical bead made of a material with an index of refraction $n = 1.5$. A beam of light is directed through a lens and onto the bead. In the figure, two light rays are shown impacting the edge of the bead.

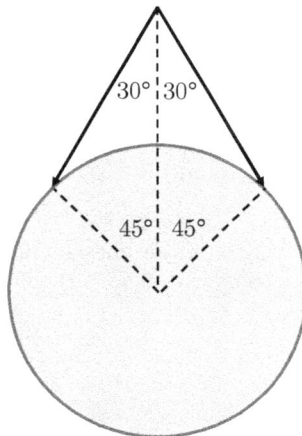

FIGURE 7.49
Problem 7.59

a. Suppose each ray *reflects* off of the surface. Sketch the paths of the reflected rays. Determine a numerical value for the angles of reflection and label them on the diagram.

b. Suppose each ray *refracts* into the bead. Sketch the paths of the rays after they refract into the bead. Determine a numerical value for the angle of refraction and label them on the diagram.

c. Continue your sketches from (b) and show the path of the ray if it refracts again as it *leaves* the bead. (No numerical values needed for this part, but make the sketch neat.)

d. Light rays carry momentum. If the *magnitude* of their momentum doesn't change (from before they interact with the bead to when they move away from the bead) and if the bead was originally at rest, then in what direction is the bead moving due to the reflection described in (a)? What about for the refraction described in (b) and (c) (from before the rays enter the bead to when they leave)?

Problem 7.60. A ruler is partially submerged in a tank filled with a transparent liquid. As a result, the submerged portion of the ruler has an angular size that has been increased by 30%. The left side of Figure 7.60 shows a "head on view" of what you see when you look at the tank. The right side of the figure shows a side view; the dot represents a single point on the submerged portion of the ruler (the schematic below is a reminder of how the angular size of an object changes if the image is at a different position). You are $L = 0.80$ m in front of the tank and the ruler has been submerged $s = 3.5$ m behind the wall. The index of refraction of the liquid (call it n_1) is unknown.

a. Make a neat and labeled ray diagram that demonstrates where the image will be formed.

b. Determine an expression for the image's distance behind the front wall of the tank.

c. Determine a numerical value for the index of refraction of the liquid.

If necessary, you may make use of the small angle approximation $\sin \theta \approx \tan \theta \approx \theta$.

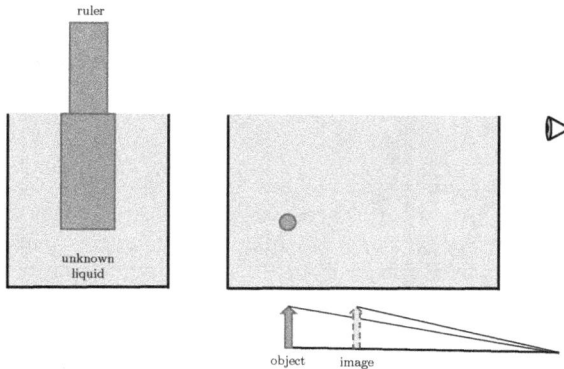

FIGURE 7.50
Problem 7.60

Problem 7.61. You lean over the edge of a pool and look down (Fig. 7.51). How deep does the pool appear to be? Assume the index of refraction of the water is 1.3. Use both a diagram and mathematics to support your answer. You may make use of the small angle approximation $\tan \theta \approx \sin \theta \approx \theta$.

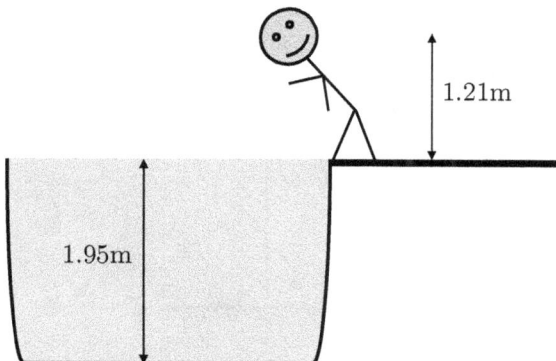

FIGURE 7.51
Problem 7.61

★★★ **Problem 7.62.** Consider a hemisphere of plastic ($n = 1.2$), with radius R, resting with its flat side down on a table. A ray leaving a point on the edge travels through the glass and refracts out of the glass tangent to the boundary, as shown. A person's eye is to be held a distance h above the top of the hemisphere such that the refracted ray enters the eye.

  a. Determine an algebraic expression for h in terms of R. Support your answer with a clear diagram.

  b. If the hemisphere is placed atop something interesting, like a map, then the effect is that the portion under the hemisphere is magnified. Explain why this is so with the help of a diagram (no formulas or calculations needed).

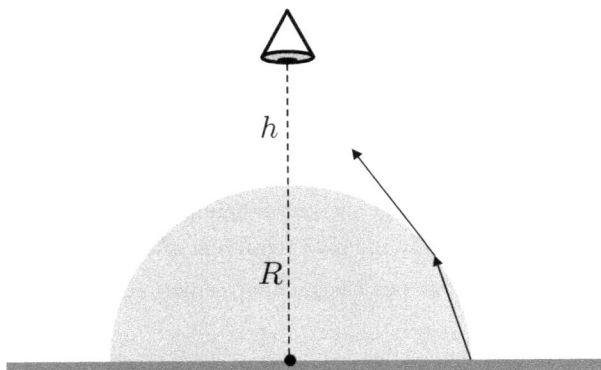

FIGURE 7.52
Problem 7.62

★★ **Problem 7.63.** You have a 5.0-cm-tall bobblehead of a favorite TV, movie, or comic character (if you're solving this problem to be graded by someone else, be sure to indicate the character. Maybe include a sketch.) and place it 0.15 m to the left of a converging lens with a focal length of 0.10 m. A second, diverging lens with a focal length of -0.10 m is placed 0.45 m to the right of the first lens.

  a. Make a neat diagram of the object and two lenses.

  b. Draw principle rays to estimate the location of the image formed by the first lens. Is it real or virtual? Upright or inverted?

  c. Draw principle rays to estimate the location of the image formed by the second lens. Is it real or virtual? Upright or inverted?

  d. Mathematically determine the position, height, and magnification m of the final image.

★★★ **Problem 7.64.** A hungry cat looks at a fish through a flat-sided aquarium (Fig. 7.53). How far from the right side of the aquarium does the fish appear to be according to the cat? Support your answer mathematically and with a clear diagram of relevant rays. Recall that the index of refraction for air is 1.0 and the index of refraction for water is 1.33. You may apply the small angle approximation.

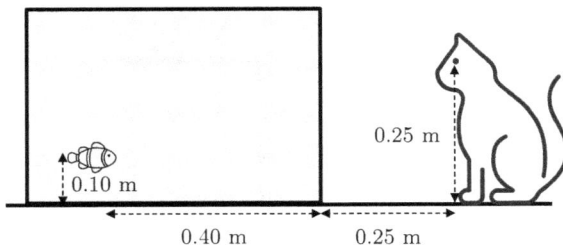

FIGURE 7.53
Problem 7.64

★★ **Problem 7.65.** Your friend holds up a magnifying glass to their eye. You estimate the angular magnification is $\mu_a \approx 3.0$.

   a. What is the focal length of the magnifier?

   b. Suppose you had two of these magnifiers and you'd like to use them to construct a rudimentary microscope. If the object is to be positioned 1.0 cm beyond the objective lens's focal point, how far should the eyepiece be from the objective lens? Support your mathematical work with a clear ray diagram.

★★ **Problem 7.66.** One happy spring morning you go for a walk. It recently rained, and as you look up into the sky you see a rainbow. This happens when light from the sun enters a drop of water suspended in the air such that it is eventually redirected into your eye: the index of refraction for violet light in water is slightly larger than for red light; every other (visible) color falls somewhere in between.

   Treat the incoming ray from the sun as a red ray and a violet ray stacked on top of one another; white light is a mixture of all colors. Using Figure 7.54 as a template, sketch the resulting path of both rays until they leave the droplet and head toward your eye (the rays must refract–with different angles of refraction–as they enter your eye, but what happens next?). When you see a rainbow, is it "red on top, violet on the bottom", or the reverse? Use your diagram to justify your answer.

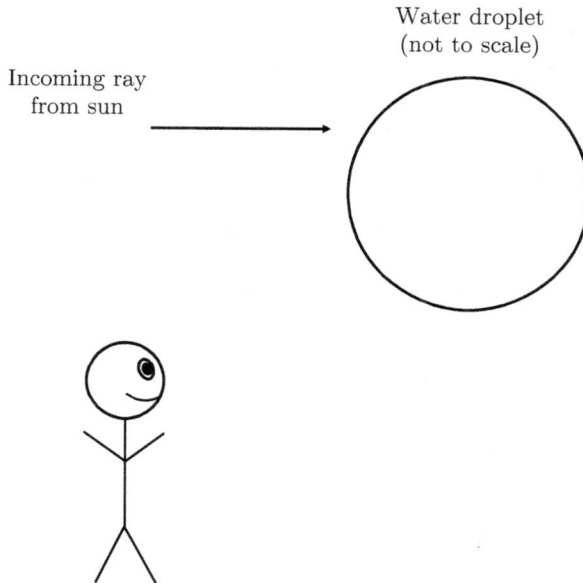

FIGURE 7.54
Problem 7.66

★★ **Problem 7.67.** A laser beam is traveling horizontally through a piece of glass, as shown in Figure 7.55. Assume the index of refraction for the glass is 1.5 and suppose $\phi = 60.°$.

   a. Sketch the path of the ray as it leaves the glass.

   b. Determine the angle of refraction.

   c. For what value of ϕ would the entire beam experience total internal reflection?

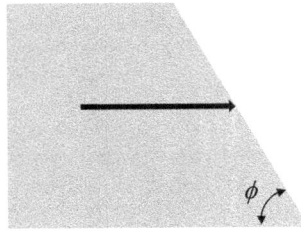

FIGURE 7.55
Problem 7.67

★ **Problem 7.68.** A magnifier has a magnification of 5.0. How far from the lens should an object be held so that its image is seen at the near-point distance of 25 cm?

★ **Problem 7.69.** An object is 0.20 m in front of a converging lens that has a focal length of 0.10 m.

  a. Make a neat ray diagram and use ray tracing to estimate the location of the image. Include all three principle rays.

  b. Is the image upright or inverted? Is it real or virtual?

  c. Mathematically determine the image position and magnification.

★ **Problem 7.70.** An object is 0.30 m in front of a convex mirror with a focal length of −0.20 m.

  a. Make a neat ray diagram and use ray tracing to estimate the location of the image. Include all three principle rays.

  b. Is the image upright or inverted? Is it real or virtual?

  c. Mathematically determine the image position and magnification.

★★ **Problem 7.71.** An object is 17 cm in front of a convex mirror. The mirror creates an image that is 70% as tall as the object.

  a. Mathematically determine the focal length and image distance.

  b. Use ray tracing to estimate the location of the image.

  c. Is the image upright or inverted?

  d. Is the image real or virtual?

★★ **Problem 7.72.** Oh no–you have a splinter in your finger! Armed with a trusty pair of tweezers, you'd like to extract the splinter. To make it easier to grab, you use a (converging) cosmetic mirror with a focal length of 25 cm and hold your finger 5.0 cm in front of it. If the tip of the splinter sticking out from your finger is 0.50 cm, how large will it appear in the mirror? Will it be upright or inverted?

★★ **Problem 7.73.** You look at a small object that is 4.0 cm behind a converging lens with a 5.0 cm focal length. 6.0 cm behind the object is a converging mirror with a 0.50 m focal length.

  a. How many images of the object do you see?

  b. Determine the location of each image you found in (a). Set $x = 0$ cm on the object and call *toward you* the positive x direction.

  c. Determine the orientation (upright or inverted) of each image you found in (a).

★ **Problem 7.74.** A 1.3 cm tall object is 25 cm in front of a mirror with a -45 cm focal length.

   a. Sketch the situation, including the object, image, and principal rays.

   b. Where is the image?

   c. What is the height of the image?

★★ **Problem 7.75.** A 2.0 cm tall object is at $x = 0.0$ cm. A lens with $f = 10.0$ cm is at $x = 30.0$ cm and a lens with $f = 5.0$ cm is at $x = 60.0$ cm.

   a. Where is the final image?

   b. How tall is the final image? Is it upright or inverted?

   c. If the lenses are removed and replaced with a new lens at $x = 10.0$ cm. What focal length would produce an image at the same location that you found in (a)?

   d. How tall is the image formed in (c)? Is it upright or inverted?

★★ **Problem 7.76.** A 0.75 cm tall object is 15 cm to the left of a lens with $f_1 = 7.5$ cm. 50 cm to the right of the first lens is a second lens with $f_2 = 25$ cm. (a) How far from the first lens is the final image? (b) How tall is it?

★★ **Problem 7.77.** Two converging lenses with focal lengths of 0.40 m and 0.20 m are 1.00 m apart. A 2.0-cm-tall object is between the lenses and 15 cm in front of the lens with a focal length of 0.40 m.

   a. Sketch the situation and make a clear ray diagram to estimate the position, size, and orientation of the image(s) formed when viewing the object through both lenses.

   b. Calculate the position, size, and orientation and compare to your estimates in (a).

8

Wave Optics

Learning Objectives

After reading this chapter, you should be able to:

- Describe the principle of interference in the context of two dimensional waves.

- Analyze the wavelike behavior of light in the double split, single slit, and diffraction grating experiments.

- Describe the principle of reflection and refraction using the wave model of light.

- Describe light as an electromagnetic wave.

- Characterize the electromagnetic spectrum in terms of the energy, wavelength, and frequency of light.

8.1   Interference from Path Length Differences

In Chapter 6 we saw that traveling waves can *interfere*; in particular, we saw that identical waves traveling in opposite directions can form standing waves, which can lead, for instance, to a piano string vibrating at a frequency corresponding to a certain musical note. We will begin this chapter by considering a related situation. Suppose that you have two sources of traveling waves, like audio speakers, that are connected such that they emit identical waves. If the speakers are right next to one and a detector (e.g. a microphone) can pick up the combined signal from both sources, then the total signal will obey:

$$D_{\text{tot}}(x,t) = D_1(x,t) + D_2(x,t)$$
$$= A\sin(kx - \omega t) + A\sin(kx - \omega t)$$
$$= 2A\sin(kx - \omega t)$$

The result is a signal with twice the amplitude of either input.[1] But what if one speaker is slightly in front or behind the other speaker (Fig. 8.1)? We need to adjust the effective origin for one of the speakers to compensate:

$$D_{\text{tot}}(x,t) = A\sin(k(x + \Delta x) - \omega t) + A\sin(kx - \omega t)$$
$$= A\sin(kx - \omega t + k\Delta x) + A\sin(kx - \omega t)$$
$$= A\sin(kx - \omega t + \phi) + A\sin(kx - \omega t)$$

In the last line I've set the constant $k\Delta x$ equal to a *phase shift* ϕ to represent the wave in familiar terms. Now if Δx happens to be exactly one wavelength,[2] then the wave from the

---

[1]We saw perfectly overlapping and identical sine waves combining to yield a larger amplitude in the context of traveling waves moving in opposite directions; see t_2 on the left side of Figure 6.20.

[2]Equivalently, if $\phi = 2\pi$; recall $k = 2\pi/\lambda$.

DOI: 10.1201/9781003571568-11

Constructive Interference   Destructive Interference

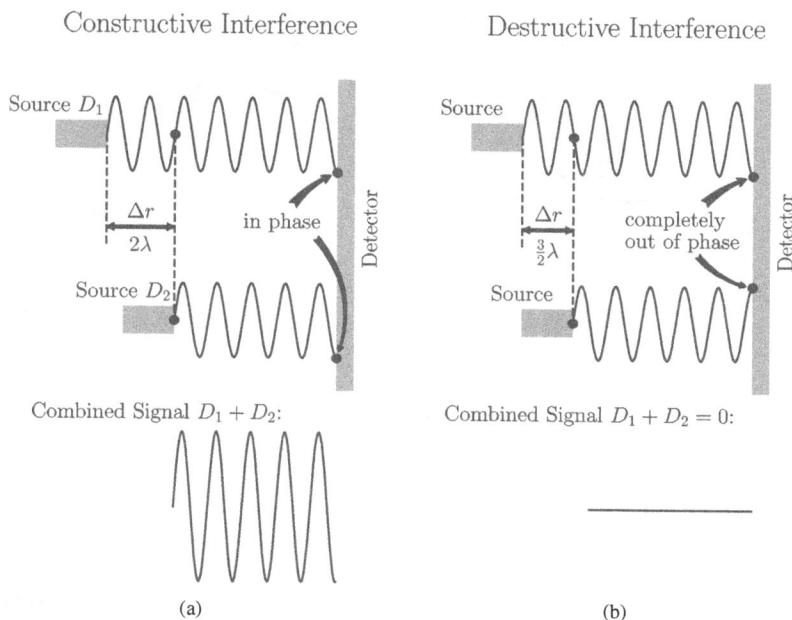

(a)

(b)

FIGURE 8.1

(a) Two synchronized sources of identical waves are offset by a distance $\Delta r = 2\lambda$. When the wave from the more distant source reaches the location of the closer source, the waves are precisely in phase. From there until the waves reach the detector, they continue to travel in phase and yield a high-amplitude signal on the detector, regardless of how far it is from the sources. (b) The synchronized sources are separated by a distance $\Delta r = \frac{3}{2}\lambda$. The signals now reach the detector perfectly out of phase (at the instant shown, one is a maximum and the other is a minimum) and the detector detects no net signal.

more distant source undergoes exactly one oscillation as it "catches up" to the closer source; from then on the waves are synchronized as if they were right next to each other from the beginning. But this is *also* true if the sources are separated by exactly two wavelengths, three wavelengths, or indeed *any integer multiple of the wavelength*! In general, then, the waves *constructively interfere* if

$$\Delta r = m\lambda, m = 0, 1, 2, \dots \tag{8.1}$$

In contrast, if the sources are separated by a half-integer multiple of the wavelength, the waves *destructively interfere* (Fig. 8.1):

$$\Delta r = \left(m + \frac{1}{2}\right)\lambda, m = 0, 1, 2, \dots \tag{8.2}$$

I am using r rather than x here to extend these statements to the more general two-dimensional case; we shall see that the results follow directly.

To move to two dimensions, suppose that you sinusoidally disturb the surface of a pool with the tip of a pen (or a speaker cone). You'll see that the ensuing ripples can be characterized by a wavelength and positions of maximum and minimum displacement. Because the ripples on the pool are spreading along a flat surface (instead of along a line), we call them a **two-dimensional wave**.

$$D_{\text{tot}}(r, t) = D_1(r, t) + D_2(r, t)$$

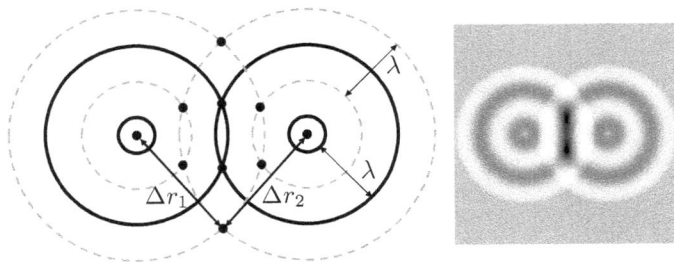

FIGURE 8.2

Two two-dimensional waves spreading radially from nearby positions. (Left) The black dots at the center of each set of circles represent the source of each wave. Black circles mark a wave crest, and gray dashed circles mark a wave trough. The wavelength is labeled, and positions of maximum constructive interference (peak-peak and trough-trough) are marked with black dots. The path length from one such dot to each source, Δr_1 and Δr_2, are labeled by way of example (in this case $\Delta r = 0$ and Equation (8.1) indicates the interference is, indeed, constructive). Points where a black line and a dashed line intersect would, conversely, be locations of destructive interference. (Right) A computer-generated image of the same situation; the equilibrium surface height is gray and displacements due to the waves are shown in increasingly darker (for positive displacements) and lighter (for negative displacements) tones.

FIGURE 8.3

A perspective view of the simulated waves shown in Figure 8.2.

Suppose, for instance, that we use *two* pens and look at a snapshot of the pool a short time after we start disturbing the water. As Figures 8.2 and 8.3 show, at some positions we will observe maximum constructive interference (positions where each of the constituent waves are at a maximum, or, alternatively, positions where each is at a minimum) or perfectly destructive interference (where the displacements cancel). As we argued above, these locations are given by Equations (8.1) and (8.2), respectively.

I've used the intuitive example of ripples in water, but this kind of interference is typical of *any* 2D wave. A second example is sound: even though it naturally spreads in three dimensions, we can measure interference on a plane by moving a small microphone near a pair of speakers.[3] A third example is light.

---

[3]While not immediately important for our present circumstances, I should mention that the amplitude of a wave spreading in two or three dimensions will decrease as it spreads from the point of origin: the wave carries a certain amount of energy, and as it expands, the energy is spread over a larger perimeter (for a 2D wave) or area (for a 3D wave). See Problems 8.38–8.39.

This is important, so I'll repeat the point: light can display interference patterns that are best explained by supposing that *light is a wave*. In Sections 8.2–8.4 I will describe three experiments that demonstrate this. Then, in Section 8.5, we will reconcile this *wave-like* behavior with the *ray* model we introduced in Chapter 7. Finally, in Section 8.6 we will address a fundamental question: if light is a wave, then what, exactly, is oscillating?

8.2   Interference from Two Points: The Double Slit

Our first experiment is very similar to the situation shown in Figure 8.2: we would like to generate waves of wavelength λ that spread radially from *two points* that are separated by some distance d. This can be done by creating a **plane wave** (think of the long waves along an ocean shore) and then sending it through a pair of narrow slits. If the gap width a is smaller than the wavelength λ, then the plane wave will spread in a circular fashion after passing through the gap (Fig. 8.4). We will explore this behavior in more detail in Section 8.4; for now, we will take it as an experimental fact. Waves that spread radially from a point like this are sometimes called **wavelets**, and the process of spreading from a narrow gap is called **diffraction**.

How do you generate plane waves in the first place? Well, in a water tank you can use a flat piston to sinusoidally oscillate the water. Plane sound waves can be generated in the same way (think of a large speaker cone), though for sound it is often easier to just use small speakers to generate wavelets directly instead of pairing a plane wave with narrow slits. Plane light waves can be generated by a laser.

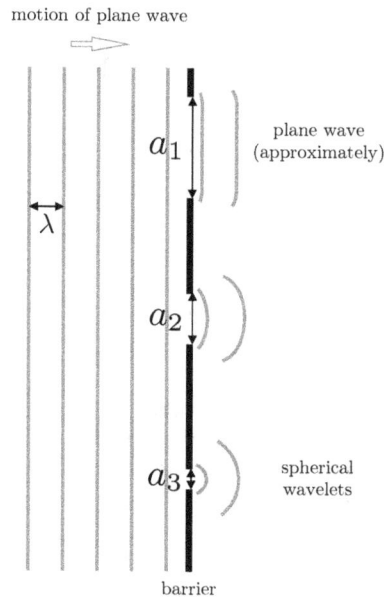

FIGURE 8.4

A plane wave approaches a barrier from the left; the crests are marked in gray. It interacts with a barrier that has several gaps of varying widths, $a_1 > a_2 > a_3$. For a large gap $a_1 > \lambda$, the transmitted wave is a plane wave except for some distortions near the edges. For a small gap $a_3 < \lambda$, the transmitted wave spreads radially from the center of the gap.

Now, if you send plane waves through two narrow slits separated by a distance d, then you're back to (half of) the situation sketched in Figure 8.2. For water waves we can observe the ensuing interference patterns directly, but to investigate what happens with light waves, we introduce a *viewing screen* opposite the barrier.[4] We sketch this approach, which is called

---

[4] For sound you could use a microphone that can move back and forth along the screen to record the sound intensity at each position.

the **double slit experiment** or sometimes **Young's Double Slit Experiment**,[5] in Figure 8.5. Before we analyze the geometry in detail, let me also draw your attention to Figure 8.6, which gives two examples of the complete interference pattern between the slits and the screen. In the case of light, the amplitude of the wave when it interacts with the screen corresponds to its brightness.

Let us return our attention now to Figure 8.5. We would like to determine the amplitude of the overall wave at an arbitrary position y along the viewing screen. Importantly, because the two wavelets are formed from the same incoming plane wave, we know that they are *in phase*: if the top wave is at a maximum some distance r from the *top* slit, then the bottom wave will also be at a maximum a distance r from the *bottom* slit.[6]

Indeed, if you refer back to Figure 8.2, you'll see that there are four positions of constructive interference marked along a vertical line between the center of the two waves. Keep in mind, though, that Figure 8.2 shows just one instant in time. If you imagine the waves continuing to spread radially (and new waves being generated by each source), then the crest intersection points along that central line will shift, eventually reaching *every* point along the line. We conclude that *any* position that is

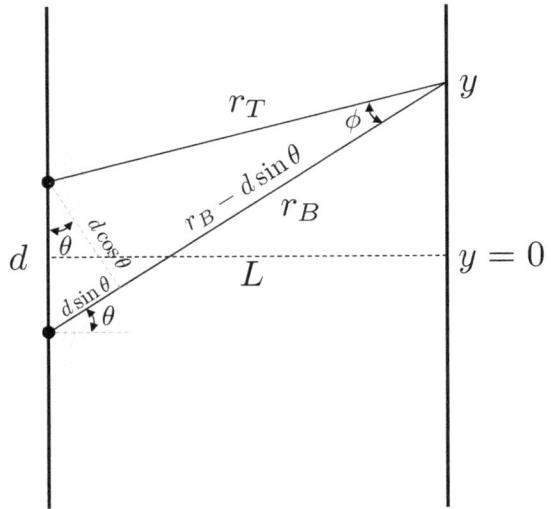

FIGURE 8.5

The double slit experiment. A plane wave has struck a barrier with two small slits separated by a distance d. These slits act as sources for radially spreading wavelets; a few crests are drawn in pale gray. A distance L from the barrier is a screen. The geometry describes the distance from each slit to a position y on the screen.

equidistant from the two slits will be a position of constructive interference.[7]

In the context of the double slit experiment, the only position on the viewing screen that is equidistant from both slits is directly opposite the midpoint between the slits (marked $y = 0$ in Figure 8.5). Where else will we see constructive interference? Well, refer again to Figure 8.2. There are four positions of constructive interference marked that *aren't* on the vertical line between the two sources. This occurs, for instance, where the *first* trough from one source intersections the *second* trough from the other source. If we make the figure larger by adding a third crest from each source (Fig. 8.7), we see that they each intersect with both of the existing crests from the other source.

Now, if we're talking about, say, the *third* crest from the left source intersecting with the *second* crest from the right source, then it follows that the *difference* in the distances (from the source of each wave to the point of interference) is simply λ. If we match the third

---

[5]After Thomas Young (1773–1829)

[6]If we're dealing with two small speakers in the case of sound waves, they'll be in phase as long as they're both connected to the same source.

[7]This is analogous to the antinodes in a 1D standing wave: at some instant the displacement from equilibrium might be 0, but *averaged over time* they have the largest displacement of any position on the wave.

$$\lambda = 5\text{cm}$$

viewing screen

$$L = 50\text{cm}$$

$$d = 20\text{cm}$$

$$\lambda = 1\text{cm}$$

viewing screen

$$d = 10\text{cm}$$

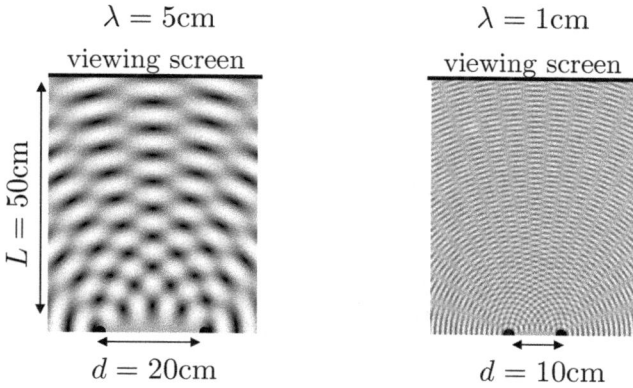

FIGURE 8.6
Two examples of interference from the double slit experiment.

crest with the first crest, then the difference is 2λ. In general, then, maximum constructive interference occurs where

$$|r_T - r_B| = m\lambda, m = 0, 1, 2, \ldots \qquad (8.3)$$

where r_T and r_B are the distances from the top and bottom slits to the point in question, respectively. In words, this means that if the difference in the path lengths for the two waves is an *integer number of wavelengths*, they will be in phase, constructively interfere, and oscillate with maximum amplitude. Similarly, we observe *destructive* interference when

$$|r_B - r_T| = \left(m + \frac{1}{2}\right)\lambda, m = 0, 1, 2, \ldots$$
$$(8.4)$$

With all of this in mind, let's return to the geometry of Figure 8.5. We see

$$\sin\phi = \frac{d\cos\theta}{r_T}$$
$$r_T = d\frac{\cos\theta}{\sin\phi}$$

and

$$\tan\phi = \frac{d\cos\theta}{r_B - d\sin\phi}$$
$$d\cos\theta = \tan\phi\left(r_B - d\sin\phi\right)$$
$$d\frac{\cos\theta}{\tan\phi} + d\sin\theta = r_B$$

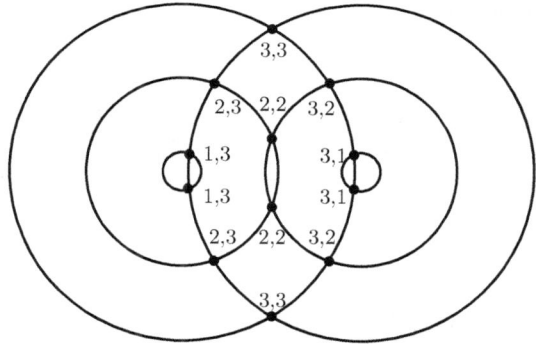

FIGURE 8.7
Interactions between two spherically spreading waves. This follows Figure 8.2, though here troughs are omitted for clarity and a third crest is shown from each source. Positions of constructive interference are marked with a black dot and labeled by which pair of crests are overlapping (e.g. (3, 1) indicates the third crest from the left source is intersecting the first crest from the right source).

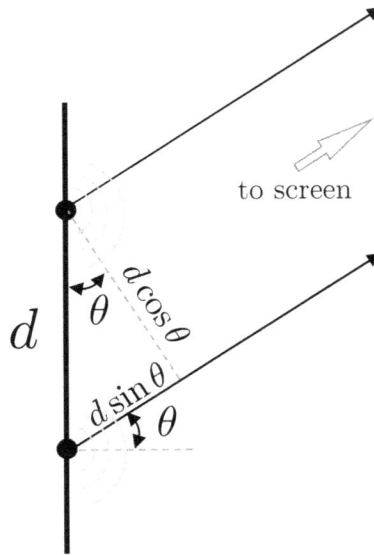

FIGURE 8.8
In the double slit experiment, if the viewing screen is sufficiently far away, then lines leaving each slit and heading toward the same point on the screen are essentially parallel.

It follows that the difference in paths is given by

$$r_B - r_T = d\left(\frac{\cos\theta}{\tan\phi} + \sin\theta - \frac{\cos\theta}{\sin\phi}\right) \qquad (8.5)$$

Now, in practice $L \gg d$: in the case of a laser, the screen might be a few meters away, while the slits are so close as to be imperceptible. This means ϕ is a small angle, so $\sin\phi \approx \tan\phi$ and the first and last terms on the right side of the above equation cancel. We find:

$$r_B - r_T \approx d\sin\theta \qquad (8.6)$$

We see this result more directly if we note that ϕ being a small angle means that r_T and r_B are essentially parallel (Fig. 8.8).

Referring back to Equations (8.3) and (8.4), our conclusion is that constructive interference is observed on a distant screen when

$$m\lambda = d\sin\theta_m, m = 0, \pm 1, \pm 2, \pm 3, \ldots \qquad (8.7)$$

and destructive interference is observed when

$$\left(m + \frac{1}{2}\right)\lambda = d\sin\theta_m, m = 0, \pm 1, \pm 2, \pm 3, \ldots \qquad (8.8)$$

In these equations, the use of "$\pm$" when referring to the possible values of m simply means that we can move up *or* down from the $y = 0$ position. We note furthermore that for a distant screen, any position y is related to L and θ according to $y = L\tan\theta$ and so for positions of constructive interference,

$$y_m = L \tan \theta_m \tag{8.9}$$

In a typical double slit experiment, $d \gg \lambda$ (and $L \gg y$) so it is valid to make the small angle approximation. Thus, for instance, Equation (8.7) is often approximated $\theta_m \approx m\lambda/d$.

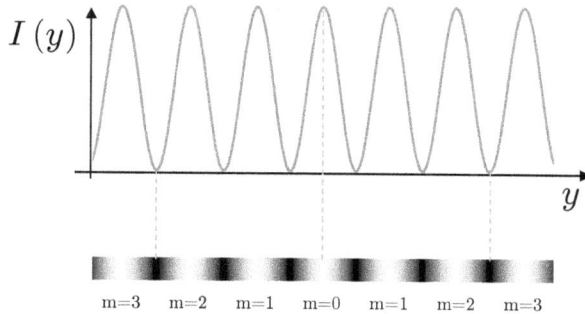

FIGURE 8.9
The interference pattern from a double slit experiment shown with the corresponding Intensity graph (see Problem 8.7). Dashed vertical lines show the alignment between (1) maxima in intensity and bright spots on the screen and (2) minima in intensity and dark spots on the screen.

A double slit interference pattern is shown and the corresponding **intensity** function (the brightness) are shown in Figure 8.9. In an actual experiment the fringes tend to be packed closely together, as the following example demonstrates.[8] Before moving on, though, let me emphasize again that if you shine a laser beam through two sufficiently narrow slits like this, you'll see an interference pattern that agrees with these predictions; this provides compelling evidence that light is a wave.

Example 8.1 The Double Slit ★★
A green laser beam typically has a wavelength of 532 nm. Suppose such a beam is directed through two slits separated by a distance $d = 0.150$ mm. If a screen is $L = 1.20$ m from the slits, then
(a) How far from the central ($m = 0$) maximum are the first three adjacent maxima?

(b) How many bright lines are visible on the screen, assuming it is arbitrarily long?

(c) How many bright lines are visible on the screen if it is 2.50 m wide?

For (a) we can use Equation (8.7) to determine θ_1, θ_2, and θ_3, then Equation (8.9) to determine the coordinates y_1, y_2, and y_3. Beginning with Equation (8.7), we

---

[8]If you'd like to think mathematically about how the brightness varies between the bright and dark spots, see Problem 8.7.

have

$$m\lambda = d\sin\theta_m$$

$$\theta_m = \sin^{-1}\left(\frac{m\lambda}{d}\right)$$

$$\theta_m = \sin^{-1}\left(\frac{m\left(532\times 10^{-9}\text{ m}\right)}{1.50\times 10^{-4}\text{ m}}\right)$$

From this result we find $\theta_1 = 3.55\times 10^{-3}$ rad, $\theta_2 = 7.10\times 10^{-3}$ rad, and $\theta_3 = 1.06\times 10^{-2}$ rad. These are all very small; each is less than 1°! (This demonstrates the fact that with light you can often get away with using the small angle approximation when considering positions near the central maximum.)

We can determine where they are on the screen with Equation (8.9):

$$y = L\tan\theta$$

Plugging in our values for θ_1, θ_2, and θ_3, we find $y_1 = 4.26\times 10^{-3}$ m, $y_2 = 8.51\times 10^{-3}$ m, and $y_3 = 1.28\times 10^{-2}$ m.

Moving on to (b), we note from Equation (8.9) that as we look farther and farther away from the midpoint of the screen, the angle θ increases. What do we mean when we say the screen is "arbitrarily" long? Well, the natural limit is $y = \infty$, where $\theta = 90°$. This cannot be done in the real world, of course, but if we have a very broad screen, this can be a reasonable approximation – and in any event it gives us an upper limit on what we can observe in practice.

Now, if $\theta = 90°$, Equation (8.7) reduces:

$$m_{\max}\lambda = d\sin\left(90°\right)$$

$$m_{\max}\lambda = d$$

$$m_{\max} = \frac{d}{\lambda}$$

$$m_{\max} = \frac{1.50\times 10^{-4}\text{ m}}{632\times 10^{-9}\text{ m}}$$

$$m_{\max} = 237$$

If you check the math yourself, you'll see the result is actually a bit more than 237; I've rounded down since m can only take integer values. Now, every value of $m > 1$ is repeated on either side of the central ($m = 0$) bright fringe, so the *total* number of bright fringes is given by $2m_{\max} + 1 = 475$.

Now, finally, for (c) we have a different limiting value of θ:

$$\theta_{\max} = \tan^{-1}\left(\frac{1.25\text{ m}}{1.20\text{ m}}\right) = 46.2°$$

Here I've made use of the fact that if the entire screen is 2.50 m wide, then it *extends* 1.25m to either side of its midpoint. If you repeat our analysis from part (b) with this new limiting value of θ, you'll find $m_{\max} = 171$. It follows by the same logic that there are **343** bright fringes in total.

8.3   Interference from Many Points: The Diffraction Grating

A **diffraction grating** consists of many narrow slits, evenly spaced and packed tightly together. Such a grating is usually described in terms of its number of "lines per mm"; the distance between successive slits is simply the inverse of this quantity. For example, a grating with 200 "lines per mm" is equivalent to $200 \times 10^3 = 2.00 \times 10^5$ lines per meter; the slit separation is then $d = 1/ \left(2.00 \times 10^5 \right)$ m $= 5.00 \times 10^{-6}$ m.

Our experimental setup is the same as for the double slit: We will send a plane wave through a diffraction grating and observe the interference pattern on a viewing screen some distance L away. We will concern ourselves only with identifying the positions of maximum constructive interference (i.e. the bright spots). It turns out that this is surprisingly simple: simply treat every pair of adjacent slits like the double slit experiment (Fig. 8.10)! The fact that different pairs of slits are spread out side-to-side will lead to some blurring, but the effect, as you will see for yourself if you perform this experiment, is minute.[9]

We conclude, therefore, that Equations (8.7)–(8.9) apply just as well to the double slit experiment, where we shouldn't simplify further by assuming $\sin \theta_m \approx \theta_m$: θ_m is no longer a small angle because it is no longer the case that $d \gg \lambda$. This means (as you ought to convince yourself by analyzing Equation (8.7)) that the angular separation between bright fringes is *larger* when using a diffraction grating.

I should be careful to note, however, that while the locations of the minima and maxima are the same in a double slit and in a diffraction grating, the full intensity function, which I am showing without derivation, is quite different (compare Figure 8.9 to Figure 8.11).

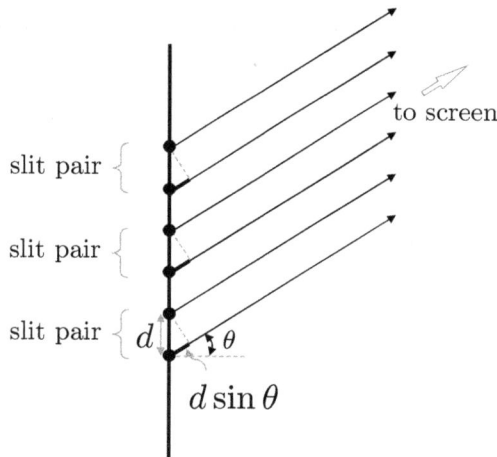

FIGURE 8.10
The slits in a diffraction grating; the geometry is greatly exaggerated. Adjacent slits are paired and analyzed in the framework of the *double* slit experiment; the locations of the minima and maxima on the viewing screen are the same.

---

[9]You may be wondering what happens if there is an *odd* number of slits, so they can't all be paired off. The answer is that the effect is too small to notice.

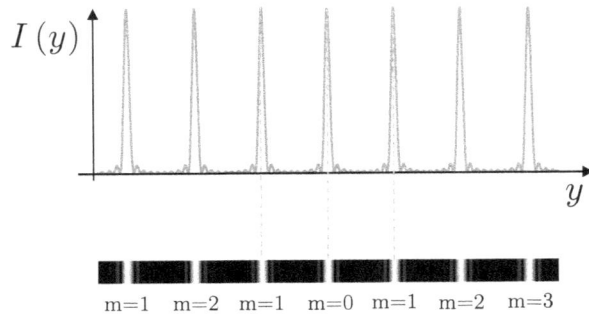

FIGURE 8.11

A sample intensity function and a visualization of a diffraction pattern for a diffraction grating. The vertical gray lines show the alignment between the narrow bright spots and the maxima of the intensity function.

Example 8.2 The Diffraction Grating ⋆

A red laser has a wavelength of 650. nm. It is directed through a diffraction grating rated at 150. lines per mm. The viewing screen is 0.20 m from the grating. How many bright spots are visible within 75 cm of the center of the screen?

We'll begin by determining the separation between adjacent slits. Because we're inverting a quantity that is itself in units of "per mm", we can bring "mm" directly to the numerator (dividing by a fraction is equivalent to multiplying by the reciprocal):

$$d = \frac{1}{150.}\,\text{mm}$$

$$d = \frac{1}{150.} \times 10^{-3}\,\text{m}$$

$$d = 6.67 \times 10^{-6}\,\text{m}$$

The maximum angle that we can accommodate with the experimental setup described in the problem is

$$\theta_{\max} = \tan^{-1}\left(\frac{0.75\,\text{m}}{0.20\,\text{m}}\right) = 1.3\,\text{rad}$$

We will use this value in Equation (8.7) to determine the value of m that corresponds to this maximal angle:

$$m_{\max}\lambda = d\sin\theta_{\max}$$

$$m_{\max} = \frac{d\sin\theta_{\max}}{\lambda}$$

$$m_{\max} = \frac{\left(6.67 \times 10^{-6}\,\text{m}\right)\sin\left(1.3\,\text{rad}\right)}{650 \times 10^{-9}\,\text{m}}$$

$$m_{\max} = 9.9$$

But, because m can only take integer values, the maximum value is instead $m = 9$. Because the situation is symmetric about the midpoint, we once again have a total of $2m_{\max} + 1 = 19$ bright lines in total. This is far more manageable than the values we obtained in Example 8.1!

8.4 Interference from Infinite Points: The Single Slit

Thus far we have considered a slit to be a *point source*, meaning it is arbitrarily narrow. We will now consider what happens when we take the more realistic view that a slit has a narrow but finite width a. The behavior we observe in the lab is explained by **Huygen's Principle**[10], which states that every point in the slit acts as point source of the spherical wavelets we've been considering. In other words, we analyze the situation as if there were an *infinite* number of the point sources packed into a physical slit of width a (Fig. 8.12).

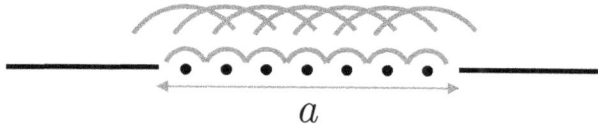

FIGURE 8.12
According to Huygen's Principle, any point on a wave can be understood by supposing that every point is the source of radially expanding wavelets. Here a plane wave travels up through a gap of width a. A few representative wavelets are centered on the black dots; the cumulative effect is a plane wave traveling up and through the gap.

Now, a consequence of packing our point sources infinitely close together is that adjacent points effectively overlap – the $d \sin \theta$ term sketched in Figure 8.10 approaches 0 as d shrinks to 0. The main question we've asked so far in analyzing these experiments, "where on the viewing screen do adjacent points constructively interfere?", isn't particularly useful in this context, because if the points essentially *overlap*, they'll *always* constructively interfere!

To proceed, then, we need to take a broader view – what conditions lead to destructive or constructive interference when assessing the entire slit? Well, as we move along the viewing screen away from the central $y = 0$ position, the distance traveled between the *top* point on the slit and the *bottom* point on the slit increases according to $a \sin \theta$ (Fig. 8.13). At certain values of θ, this difference will be equal to λ, then later $2\lambda, 3\lambda$, etc. It turns out that these values correspond to the positions of destructive interference.

To see why, consider the angle θ where $a \sin \theta = \lambda$, so the wavelet coming from the top of the slit is *in phase* with the wavelet coming from the bottom (the bottom wavelet travels an additional distance λ compared to the wavelet from the top when they meet on the viewing screen). In other words, the two wavelets differ by exactly one cycle and have a *phase difference* of 2π. It follows that the phase difference between the top of the slit and *other points* within the slit varies between 0 and 2π. For instance, if a traveling wave leaving the top of the slit impacts the screen with a phase of 0, then a wave leaving the middle of the slit will impact the same location with a phase of π.[11]

So the sum of a wave leaving the top of the slit and some other wave leaving another point in the gap varies like a sine curve when expressed as a function of the separation between the wave sources. (Fig. 8.14). However, what we're really concerned with is the *cumulative effect* of all of the wavelets from every position within the slit. You can see that for any pair of points that are equidistant from the middle, the displacements perfectly

---

[10]After Christiaan Huygens (1629–1695)
[11]More specifically, if the top wave $D = D_0 \sin (kr + \omega t + \phi)$ evaluates to $D = D_0 \sin (2\pi n)$ for some integer n, then a wave leaving from the middle of the slit evaluates to $D_2 = D_0 \sin (2\pi n + \pi/2)$.

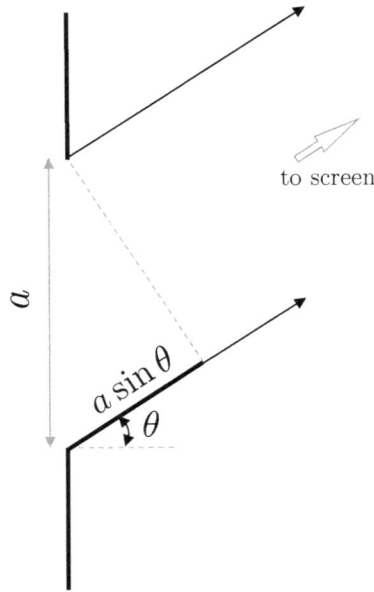

FIGURE 8.13
A single slit of width a; the total distance traveled from the top of the slit differs from the bottom of the slit according to $a \sin \theta$.

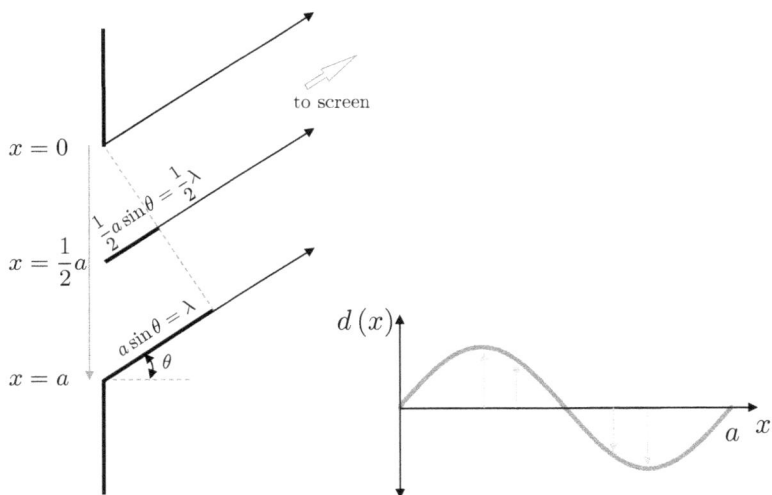

FIGURE 8.14
A single slit where $a \sin \theta = \lambda$. The graph on the bottom shows the relative phase of waves based on their position within the slit. When evaluating all of the wavelets (i.e. the integral of the curve) we see there is as much positive displacement as negative displacement (as indicated by the two sets of arrows, which indicate where two pairs of positions add to yield 0 overall displacement). Thus the wavelets destructively interfere and a dark spot will be visible on the screen.

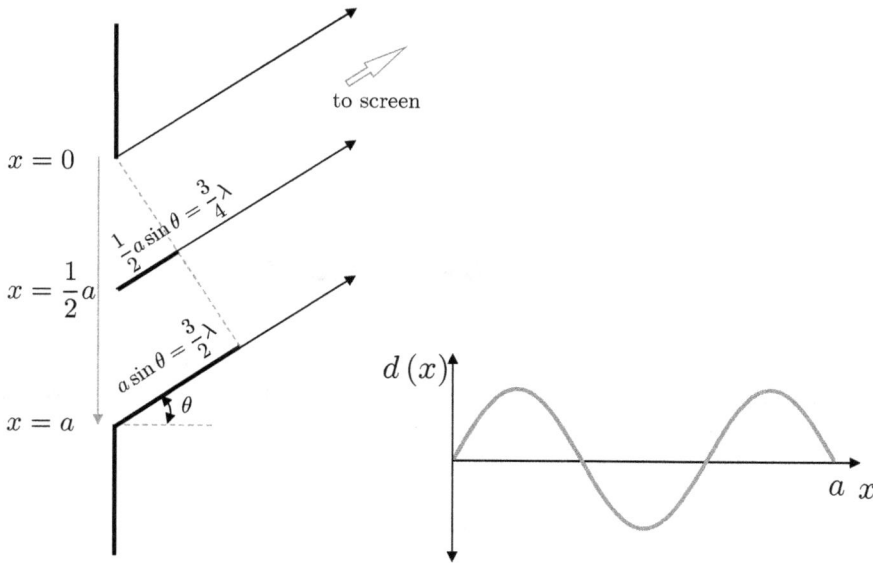

FIGURE 8.15
A single slit where $a \sin \theta = \frac{3}{2}\lambda$. The graph on the bottom shows the relative phase of waves based on their position within the slit. The differences in the wavelets can be evaluated in terms of "two thirds positive, one third negative", yielding a nonzero integral and a bright spot on the screen.

cancel – leading to overall destructive interference! The same phenomenon occurs when the total distance is any integer multiple of the wavelength (see Problems 8.33 and 8.34).[12]

We conclude, then, that in a single slit we observe destructive interference when

$$m\lambda = a\sin\theta_m, m = \pm 1, \pm 2, ... \tag{8.10}$$

Where do we observe *constructive* interference? If your intuition is "at angles midway between positions of destructive interference", you're right. It is a little clunky to write this mathematically because there is a position of constructive interference at $\theta = 0$ (the middle of the viewing screen) and then at the half-integer multiples of the wavelength after the first minima:

$$\begin{cases} \theta = 0 \\ \left(m + \frac{1}{2}\right)\lambda = a\sin\theta_m, m = \pm 1, \pm 2, ... \end{cases} \tag{8.11}$$

This is sketched in Figure 8.15; the overall pattern is shown in Figure 8.16. After exploring single slit diffraction in Example 8.3, we consider the cumulative effect of a double slit experiment where we also consider the finite width of each slit in Example 8.4.

---

[12]Mathematically, we say that these are *odd functions* measured about their midpoint; their integral is 0.

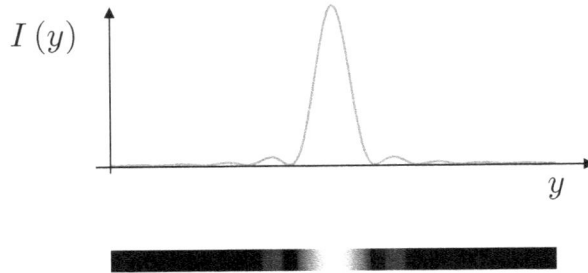

FIGURE 8.16

A sample intensity function and a visualization of a diffraction pattern for a single slit. The vertical gray lines show the first few minima, though the fact that the central peak is so much higher (i.e. brighter) than the others means that the other minima are washed out in the visualization.

Example 8.3 The Single Slit   ⋆

You measure the width of the central bright spot in a single slit experiment to be 6.0 cm wide. If the viewing screen is 35 cm from the slit and you're using a standard red laser ($\lambda = 650.$ nm), what is the width of the slit?

From Figure 8.16, we know that the central bright spot runs from $m = -1$ to $m = 1$ according to Equation (8.10):

$$m\lambda = a \sin \theta_m$$

Let's first relate this to the position on the viewing screen, which is given by Equation (8.9):

$$y = L \tan \theta$$

If the *total* width of the bright spot is 6.0 cm, then the distance from the middle of the screen to one edge is just half of that, 3.0 cm. The corresponding angle, where $m = 1$, (or equivalently where $m = -1$, if we go to the other side), is given by

$$\theta_1 = \tan^{-1}\left(\frac{y}{L}\right)$$

$$\theta_1 = \tan^{-1}\left(\frac{3.0 \text{ cm}}{35 \text{ cm}}\right)$$

$$\theta_1 = 0.0855 \text{ rad}$$

Returning to Equation (8.10), we find

$$m\lambda = a \sin \theta_m$$

$$\frac{\lambda}{\sin \theta_1} = a$$

$$\frac{650. \times 10^{-9} \text{ m}}{\sin(0.0855 \text{ rad})} = a$$

$$a = 7.61 \times 10^{-6} \text{ m}$$

Example 8.4 Intensity of a Double Slit ★★

a. The intensity of the double slit (Figure 8.9) is given by $I = I_0 \cos^2\left(\frac{\pi d \sin\theta}{\lambda}\right)$ where I_0 characterizes the overall amplitude. Check that the local minima of this function agree with Equation (8.8).

b. The intensity of the single slit (Figure 8.16) is given by $I = I_0 \left(\frac{\sin\alpha}{\alpha}\right)^2$, where $\alpha = \frac{\pi a \sin\theta}{\lambda}$. Check that the local minima of this function agree with Equation (8.10).

c. If we consider the finite width of each slit in a double slit experiment, the *overall* intensity is given by the product of the individual intensities given above (though the I_0 term is *not* squared). Sketch this intensity pattern and describe its main features.

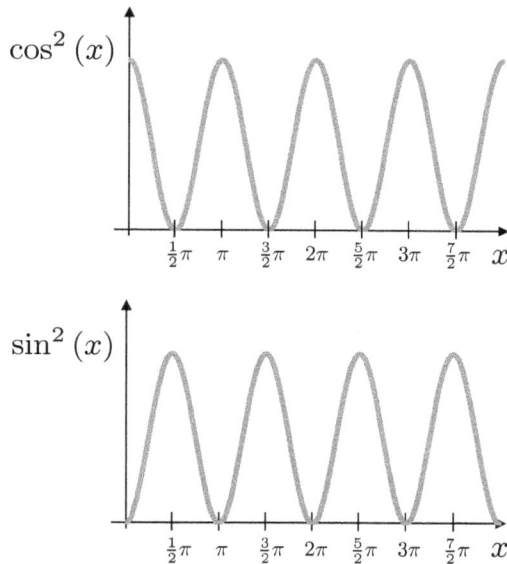

FIGURE 8.17

For both parts (a) and (b) we *could* take the usual approach from differential calculus: look for positions where the first derivative of the function is 0 and the second derivative is positive. However, I probably don't need to convince you that this gets rather ugly (try it for yourself if you like!). Instead, let's combine our equations. For (a), we are predicting that when

$$\left(m + \frac{1}{2}\right)\lambda = d\sin\theta_m, m = 0, \pm 1, \pm 2, \pm 3, \dots$$

the function

$$I = I_0 \cos^2\left(\frac{\pi d \sin\theta}{\lambda}\right)$$

is at a local minimum. If we solve the first equation for $d\sin\theta/\lambda$ and substitute it in to the second, then we pinpoint the predicted local minima at the positions

$$I = I_0 \cos^2\left(\pi\left(m + \frac{1}{2}\right)\right), m = 0, \pm 1, \pm 2, \pm 3, \dots$$

A quick sketch should convince you that $\cos^2(x)$ does indeed have a local minimum at odd half-integer multiples of π (Fig. 8.17).

We can treat (b) in the same way; we predict that when

$$m\lambda = a \sin \theta_m, m = \pm 1, \pm 2, \ldots$$

the function

$$I = I_0 \left(\frac{\sin \alpha}{\alpha} \right)^2, \alpha = \frac{\pi a \sin \theta}{\lambda}$$

is at a local minimum. Solving the first equation for $a \sin \theta_m$ and substituting into the second, we see the predicted minima are at

$$I = I_0 \left(\frac{\sin (m\pi)}{m\pi} \right)^2$$

Once again, sketching $\sin^2 x$ should convince you that the minima are located at multiples of π (Fig. 8.17).

Finally, for (c) we are asked to sketch the following equation:

$$I = I_0 \cos^2 \left(\frac{\pi d \sin \theta}{\lambda} \right) \left(\frac{\sin \alpha}{\alpha} \right)^2$$

You could rely on a computer to do this for you, but it is worth your time to develop some intuition. We know what each term individually looks like (Figures 8.9 and 8.16); graphing their product involves multiplying the two intensity values at every point. For instance, if either value is 0, the product will be 0; if both values are a local maximum, their product will also be a local maximum.

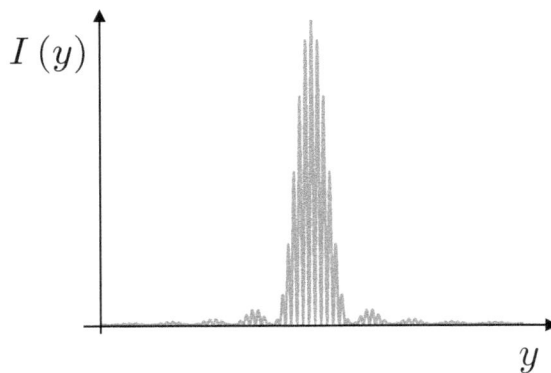

FIGURE 8.18

Both functions have a maximum at their midpoint, and the single slit interference pattern has a wide peak that drops off to 0. While the double slit interference pattern oscillates steadily, then, the amplitude of the product will drop off according to the behavior of the single slit interference pattern. The result, which you can double check yourself with your favorite graphing software, is shown in Figure 8.18; the single slit intensity pattern acts as an "envelope" over the double slit intensity pattern.

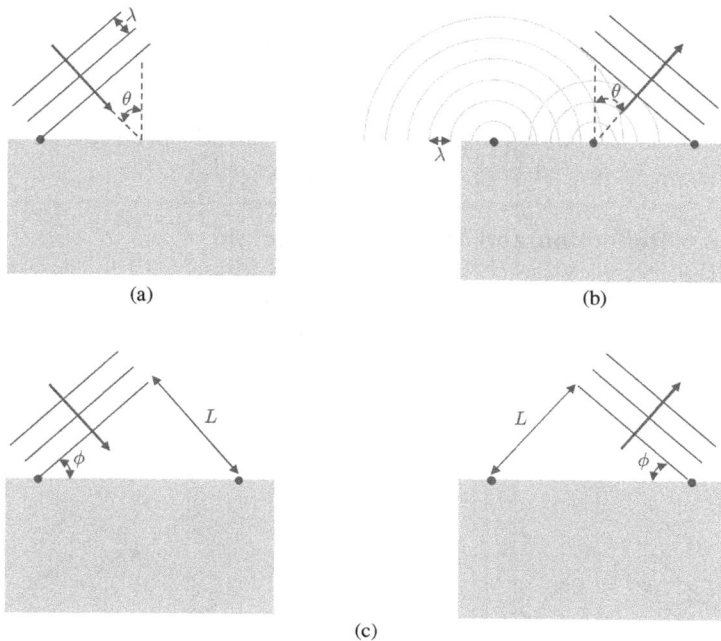

FIGURE 8.19
A wave reflects from a flat surface; a ray indicates the direction of wave propagation. (a) The instant when one edge of the wave has just impacted the surface. According to Huygen's Principle, a wavelet is created at this point and begins to radially spread waves of the same wavelength. (b) Later, when the other edge of the incident wave strikes the surface, the already-formed wavelets along the length of the surface have spread waves such that the wave front (lines tangent to the waves from each wavelet; only two are drawn) travels away from the surface according to the law of reflection: the incident angle is the same as the reflected angle. (c) The same situation. Between the two panels the right edge of the wave travels a distance L, and therefore so too does the wavelet that spreads from the left edge, as it forms part of the reflected wave. The fact that the two angles marked ϕ are identical is an equivalent way of stating the law of reflection (see Example 8.5).

8.5 Reflection and Refraction in the Wave Model

If you've read Chapter 7, you know that many practical properties of light can be described by treating light as a narrow beam that travels in a straight line. We've seen in this chapter, however, that when light interacts with a narrow gap, it behaves like a wave. These are rather different ways of looking at things; can we reconcile these models?

The answer, happily, is yes: a light ray is oriented in the same direction as the direction of a wave propagation. As long as the light does not interact with a narrow gap (recall Figure 8.4 on page 411), the ray model is an accurate simplification of the wave model.

But if the ray model is a *simplification* of the wave model, then we should be able to explain the main mechanisms discussed in the context of the ray model – reflection and refraction – in the framework of the *wave* model. Indeed, we can explain both of these phenomena by applying Huygen's Principle. We'll consider reflection first.

Suppose we have a wave approaching a flat surface of some reflective material (Fig. 8.19). Because the wave is moving at an angle relative to the normal, one edge of the wave strikes the glass first; when this happens, a wavelet at this point begins to spread waves radially. As the original wave continues to propagate, more wavelets along the surface begin spreading their waves, from the leading edge of the wave over to the far edge. Importantly, there is a time delay between the creation of the first and the last wavelet, so when the last one begins to send out its waves, the waves from the first have already propagated. The cumulative effect is an outgoing wave that obeys the law of reflection![13]

Example 8.5 Reflection and Huygen's Principle   ⋆

Show that the geometry sketched in the bottom of Figure 8.19 is equivalent to the law of reflection, i.e. that the angle formed by the incoming ray and the normal is equal to the angle between the outgoing ray and the normal.

The geometry is shown in Figure 8.20. You may notice that the situation is very similar to what we discussed in the context of the normal force on an incline plane (Example 3.9 on page 119); the angle marked ϕ is actually equivalent to the angle of reflection: $\phi = \theta_i = \theta_r$.

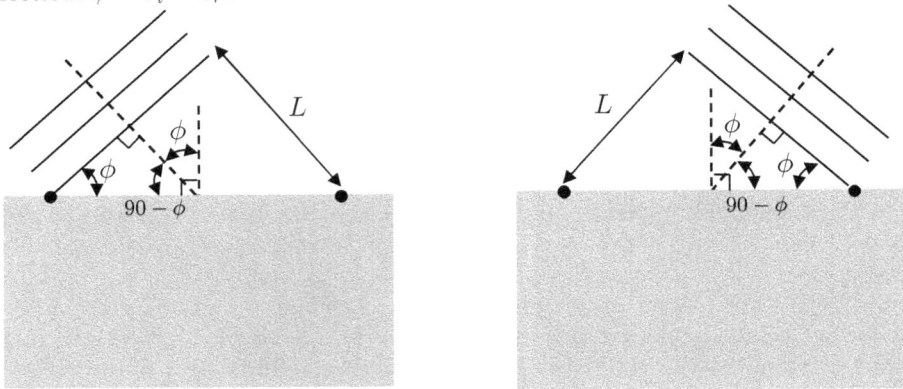

FIGURE 8.20

Now, there is no reason why the radially spreading waves should be restricted to "up and away" from the surface; waves are also sent *in to* the material. These waves are affected by the matter that makes up the material, and in some cases (for opaque materials) they're absorbed almost immediately. In other cases, however, the waves are able to propagate through the material, albeit at a *lower speed* than when they were moving through air.

This lower speed affects the orientation of the wave front that spreads from the wavelets along the surface, as sketched in Figure 8.21. Remember that the speed of light in a vacuum is given by

$$c = 3.00 \times 10^8 \frac{\text{m}}{\text{s}} \tag{8.12}$$

and that the speed of light in air is almost identical to the speed of light in a vacuum. We quantify the speed of light v in some other material with the material-dependent **index of refraction** n such that

---

[13]A more fundamental explanation is that the incident wave causes atoms on the surface of the glass to oscillate; it is these oscillations that form the "wavelets" of Huygen's Principle.

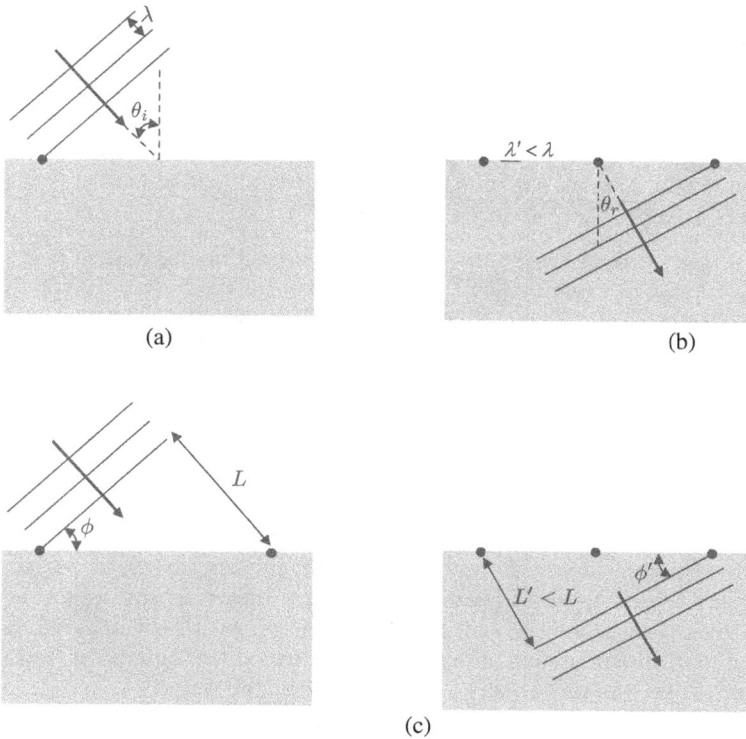

FIGURE 8.21
A wave refracts through a flat surface; a ray indicates the direction of wave propagation. (a) The instant when one edge of the wave has just impacted the surface. According to Huygen's Principle, a wavelet is created at this point. Importantly, the waves from the wavelet travel *more slowly* through the material than through air, resulting in a decreased wavelength. (b) Later, when the other edge of the incident wave strikes the surface, the already-formed wavelets along the length of the surface have spread waves such that the wave front (lines tangent to the waves from each wavelet; only two are drawn) travels away from the surface; because the waves move slower in the material, $\theta_r < \theta_i$. (c) The same situation. Between the two panels the right edge of the wave travels a distance L through air; in the same time the wave spreading into the material from the leftmost wavelet travels slower and therefore a smaller distance L'. The two angles ϕ and ϕ' differ according to Snell's Law.

$$v = \frac{c}{n} \tag{8.13}$$

Thus, if part of the wave travels a distance L through the air in some time interval, then the wave will travel a smaller distance $L' = \frac{L}{n}$ through the material in the same time interval. This accounts for the difference between the angle of refraction and the angle of incidence and explains Snell's Law in the context of the wave model (Fig. 8.21):

$$n_1 \sin \theta_1 = n_2 \sin \theta_2 \tag{8.14}$$

Of course, we came across the index of refraction in Chapter 7. We see, now, in the context of the wave model, that the "bending of light" described by the index of refraction is more fundamentally understood to be a description of the material-dependent speed of light.

8.6   Electromagnetic Waves

Before you started reading this book you already knew that light *exists*; at this point you hopefully understand the main rules that describe what light *does*. In this final section of our discussion of optics, I'd like to dig a bit deeper into what light *is*. We've seen that it behaves like a wave, and every other type of wave we've considered involves the distortion of some physical medium: for a wave to move down a string (i.e. for energy to travel along the string), the string must be *distorted*; the distortion of the medium and the transport of energy are co-implicated. So too with other types of waves we've discussed: sound involves the distortion of air molecules, and a water wave involves a disturbance in the surface of the water.

It is natural, then, to wonder what medium is being distorted by light. We know that light can travel through many different materials, including air, water, glass, and even the vacuum of space. Moreover, it appears that light can move in any direction (so long as it isn't blocked by some opaque object). For a long time scientists theorized the existence of some hard-to-detect medium that permeates space, called the *ether*, and supposed that this was the medium that was tied to the propagation of light waves.

Significant effort was expended trying to detect the ether, but no one was ever able to do it because it *doesn't exist*. We now understand that light waves don't transport energy via the distortion of some medium; rather, they are *self-propagating*. More specifically, light is composed of both an **electric field** and a **magnetic field**. These are new terms, which we will study in depth in Part IV; for now you can think of them simply as a pair of vector quantities that can vary from position-to-position and from instant-to-instant.

It turns out that light consists of these two fields ($\vec{E}$ for the electric field and $\vec{B}$ for the magnetic field; this is the conventional notation), each of which oscillate sinusoidally in time and space in a specific way. If we call the direction of motion the $\hat{x}$ direction, then one way to describe the fields is

$$\vec{E}(x,t) = E_0 \sin\left(\frac{2\pi}{\lambda}x - 2\pi f t\right)\hat{y} \tag{8.15}$$

$$\vec{B}(x,t) = B_0 \sin\left(\frac{2\pi}{\lambda}x - 2\pi f t\right)\hat{z} \tag{8.16}$$

Set up this way, the changing electric field *induces* the changing magnetic field, and the changing magnetic field *induces* the changing electric field (Fig. 8.22). In this manner light is able to carry itself from a computer screen to your eye just as it travels across the vastness of interstellar space.

Now, in a vacuum, *all* light travels at the same speed, $c = 3.00 \times 10^8$ m/s. We also know that the speed of a traveling wave is equal to the product of its wavelength and frequency, so for light in a vacuum we have

$$c = \lambda f \tag{8.17}$$

Thus, if you know the wavelength of light it is straightforward to calculate its frequency ($f = c/\lambda$) and if you know its frequency it is straightforward to calculate its wavelength ($\lambda = c/f$). While their product is fixed (in a given medium defined by the index of refraction),

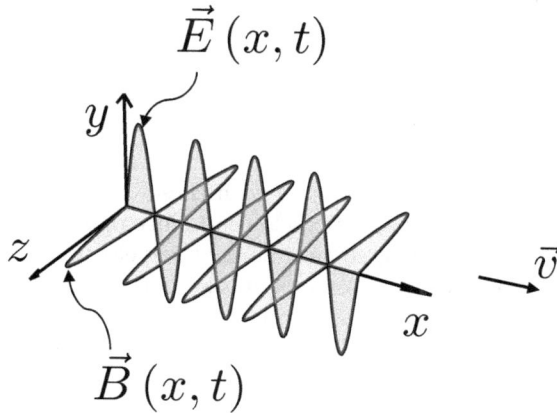

FIGURE 8.22
An electromagnetic wave consists of an oscillating electric field and an oscillating magnetic field. Here, the wave travels along the x axis, the electric field oscillates in the $y - x$ plane, and the magnetic field oscillates in the $z - x$ plane.

the wavelength and frequency can take any positive value. Thus light exists on the so-called **electromagnetic spectrum** (Fig. 8.23).

Our eyes are adapted to detect specific range of wavelengths (or frequencies, if you prefer), but radio waves, X-rays, and microwaves are in fact different instances of the same phenomenon. The most important difference between otherwise identical[14] electromagnetic

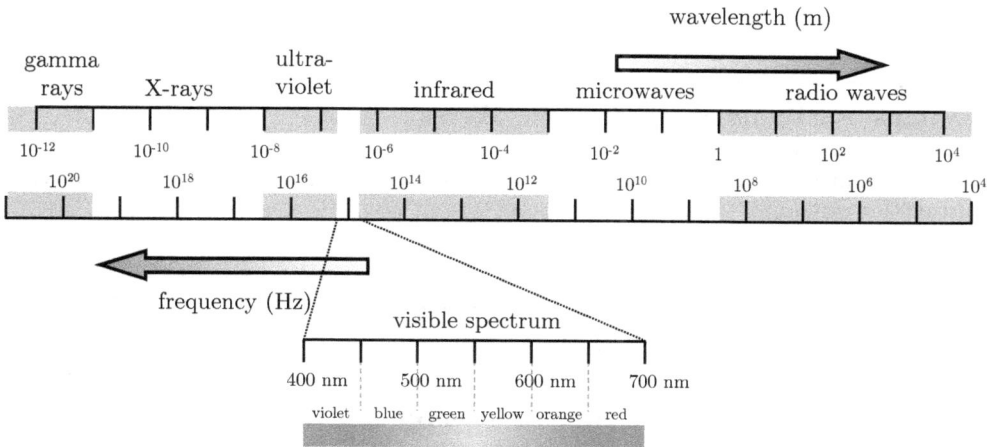

FIGURE 8.23
The electromagnetic spectrum, shown in terms of both the wavelength and frequency. The visible spectrum is enlarged and labeled according to color.

---

[14]The brightness, or intensity, of the waves could also differ in principle – we're here supposing that they are the same for the two waves.

waves from different regions of the electromagnetic spectrum is their energy, which is proportional to their frequency and therefore inversely proportional to their wavelength:

$$\frac{E_1}{E_2} = \frac{f_1}{f_2} = \frac{\lambda_1}{\lambda_2} \tag{8.18}$$

The following examples give some practice with these new ideas. We will return to the peculiar way in which electric and magnetic fields co-induce one another to form light at the end of Chapter 12, once we have spent some time studying electric fields and magnetic fields in isolation. We will also return to the topic of energy in electromagnetic waves, in Part V, when we will explore how light – even though it can behave like a wave, as we've discussed here – can also behave like a stream of discrete energy packets called photons.

Example 8.6 Oscillations in Electromagnetic Waves   ⋆

The most accurate clocks in the world measure electron oscillations in caesium-133 atoms: one second is defined as the time needed for about 9.19×10^9 of these oscillations. Consider an electromagnetic wave with the same oscillation frequency. What type of wave is it (visible, infrared...)? Interpret in a sentence or two what the oscillation frequency means in the context of the electromagnetic wave.

The atomic oscillation frequency is just under 10^{10} Hz. Referring to Figure 8.23, we see that this falls in the microwave region. What this means is that if you were able to measure the electric field strength and/or the magnetic field strength at a particular position in space with arbitrary precision, you would see that the fields go through about 10^{10} complete cycles every second!

Example 8.7 Radio and Ultraviolet Waves   ⋆

A typical FM radio wave is around 100 MHz. Estimate how the energy of such a wave compares to the highest-energy light given off by the Sun, which falls in the ultraviolet range.

The UV range approximately corresponds to the $f = 10^{16}$ Hz region of the electromagnetic spectrum. Referring to Equation (8.18), we see

$$\begin{aligned}
\frac{E_{\text{UV}}}{E_{\text{FM}}} &= \frac{f_{\text{UV}}}{f_{\text{FM}}} \\
&\approx \frac{10^{16} \text{ Hz}}{100 \times 10^6 \text{ Hz}} \\
&= 10^8
\end{aligned}$$

Thus, UV light carries about 100 million times as much energy as a radio wave of equal intensity.

8.7   Problems for Chapter 8

8.1 Interference from Path Length Differences

⋆ **Problem 8.1.** Two speakers are lined up facing a microphone, but one is slightly closer to the microphone than the other (as in Figure 8.1). Suppose the speakers emit identical waves with $\lambda = 0.10$ m. Initially, one speaker is 1.00 m from the microphone and the other speaker is 0.90 m from the microphone.

   a. Will the speaker record constructive interference, total destructive interference, or something in between?

   b. Repeat (a) but suppose the microphone is moved 5.0 cm closer to the speakers.

   c. Repeat (b) but suppose the microphone is moved *another* 5.0 cm closer to the speakers.

⋆ **Problem 8.2.** Repeat Problem 8.1 but suppose that one speaker is initially 1.00 m from the microphone (as before) and the other speaker is 0.95 m from the microphone.

⋆⋆ **Problem 8.3.** Two identical speakers are side-by-side, as in Figure 8.24, and are emitting synchronized sound waves. For each of the points (a)–(c), determine if the wave will be constructive, completely destructive, or somewhere in between. (The figure is drawn twice just to reduce visual clutter.)

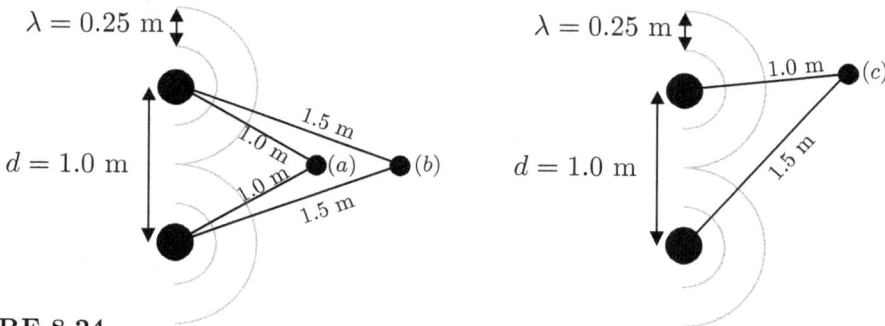

FIGURE 8.24
Problem 8.3

⋆⋆ **Problem 8.4.** Consider two synchronized speakers that are set near one another but staggered by a small distance $\Delta x = 0.020$ cm, as in Figure 8.1. Suppose the frequency of the speakers is 170 Hz.

   a. What is the wavelength of the waves?

   b. What is the period of oscillation of the waves?

   c. Determine numerical values for k and ω for these waves.

   d. As we discussed, the combined signal of these waves as detected by a microphone is given by $D_{\text{tot}} = A \sin(kx - \omega t + \phi) + A \sin(kx - \omega t)$. What is ϕ in this case? Support your answer with a sketch of the situation. (Your origin should be on top of one of the speakers, but is the other speaker in front of the origin or behind it? Both choices are valid but your choice will influence the answer, so be clear in your logic.)

   e. Make a graph of the combined signal as a function of x for $t = 0$ s. Extend your x axis far enough to show at least two complete oscillations.

   f. Repeat (e) for $t = T/2$, where T is the period you found in (b).

8.2 Interference from Two Points: The Double Slit

★ **Problem 8.5.** Consider a double slit experiment. Explain how the distance between fringes would change if you...

   a. ...increased the slit spacing.

   b. ...decreased the wavelength of the light.

   c. ...submerged the experiment in oil.

   d. ...increased the distance to the viewing screen.

★ **Problem 8.6.** In a double slit experiment, light of wavelength $\lambda = 589$ nm creates an interference pattern with fringe spacing equal to 4.0 mm on a screen 180 cm behind the slits. What is the slit spacing?

★★ **Problem 8.7.** The functional form of the intensity due to the double slit interference pattern (Figure 8.9) can be rather complex to derive, but the procedure is simplified considerably if we simply assert that the intensity of a single wave depends on the square of the wave and that the wave (as we would expect) oscillates sinusoidally. For the two interfering waves of the double slit experiment, it follows that the total intensity at some point on the screen obeys

$$I = I_0 \cos^2\left(\alpha\theta\right) \tag{8.19}$$

where I_0 is the maximum amplitude and α is a term that accounts for the even spacing of maxima (recall a cosine curve is just a phase-shifted sine curve) along the viewing screen. Show that in the small angle regime $\alpha = \pi d/\lambda$ where d is the slit spacing and λ is the wavelength. *Hint:* Consider the conditions where $\cos^2\left(\alpha\theta\right)$ is a maximum and compare to Equation (8.7).

★ **Problem 8.8.** A laser is directed through a double slit experiment. Then, the entire experiment is submerged in water. How would the original diffraction pattern change? Support your response with words, equations, and a sketch.

★★ **Problem 8.9.** A 635 nm laser illuminates a double slit experiment, as sketched below. The bright spots on the viewing screen are marked.

   a. Determine a numerical value for the distance between the two slits.

   b. Consider the topmost bright spot marked on the screen. How much farther does the wave leaving the bottom slit travel to reach this location compared to the wave leaving the top slit?

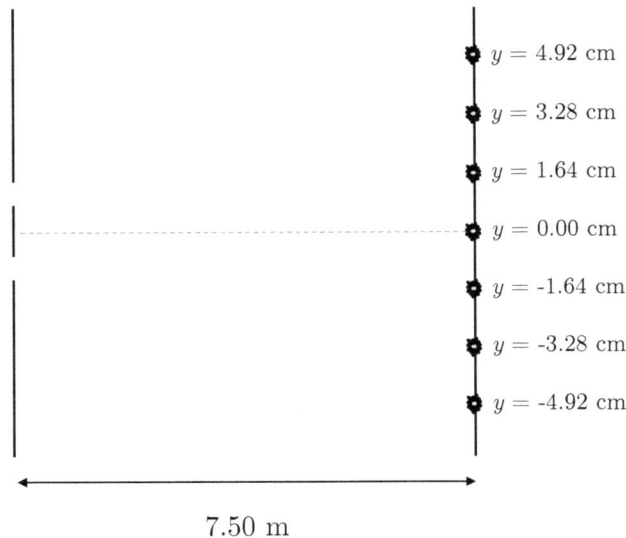

FIGURE 8.25
Problem 8.9

★★ **Problem 8.10.**

To amplify the signal generated by an antenna, a technician proposes constructing a second antenna 8.00 m next to the first. Both antennae will emit identical radio waves with a wavelength of 3.00 m. Figure 8.26 shows a circle with a 1.00 km radius around the two antenna.

a. Determine the positions on the perimeter of the circle where you would expect to find the strongest possible signal. Identify the angular position for each and be clear where you're measuring the angle from (e.g. up from the +x axis).

b. How many spots are there, total?

c. Describe (no math needed) where you would expect to find "dead zones" with a particularly weak signal.

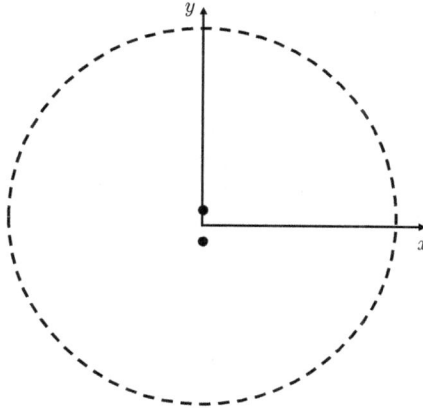

FIGURE 8.26
Problem 8.10

★★ **Problem 8.11.**

Two speakers are a fixed distance d apart and emit identical sound waves (Fig. 8.27). The small dot between them marks the origin of a coordinate system $(x, y) = (0, 0)$. Consider some other point (x, y) above the two speakers $(y > 0)$ where the sound waves interfere.

a. With a sketch and a few sentences, explain what it means for the sound waves to destructively interfere at the point (x, y).

b. For a given value of x, what value(s) of y define a point of destructive interference? Your answer will be an algebraic expression including x, y, and some or all of d, an integer m, the velocity of sound v and the frequency of the sound wave f. You need a valid equation but you do **not** need to simplify or solve the equation for y.

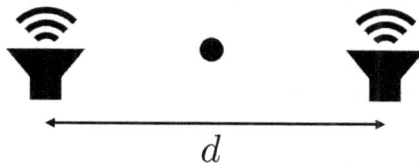

FIGURE 8.27
Problem 8.11

★★ **Problem 8.12.** The double slit experiment is usually performed with a laser beam, but it can be observed with other waves, as well. Consider two parallel rails that are $L = 0.30$ m apart. On one rail, two speakers are $d = 4.0$ cm apart and generate identical sinusoidal sound waves at a frequency *f.* A microphone can move side-to-side on the other rail. The microphone records a high-intensity signal at positions $y = 0.0, \pm 2.0$, and ± 4.0 cm where $y = 0$ is directly across from the midpoint between the speakers. What is the frequency of the sound waves?

8.3 Interference from Many Points: The Diffraction Grating

⋆ **Problem 8.13.** A diffraction grating rated at 6.00×10^2 lines/mm is illuminated by light with a wavelength of 647 nm. If the viewing screen is a distance of 2.00 m from the grating and the $y = 0$ position is directly across from the grating (as usual), then where will the $m = 2$ bright spots be located?

⋆ **Problem 8.14.** A 5.0-cm-wide diffraction grating has 2000 slits. It is illuminated by light of wavelength 550 nm. What are the angles, in degrees, of the first two diffraction orders?

⋆ **Problem 8.15.** A laser is directed through a diffraction grating. What happens to the distance between the bright fringes if the entire apparatus is submerged in water?

⋆ **Problem 8.16.** A laser with $\lambda = 633$ nm is directed at a diffraction grating. The screen is 2.2 m behind the grating and you observe that the distance between the $m = 2$ fringes is 65 cm. What is the separation between adjacent slits in the grating? How many slits exist in each millimeter of the grating? (This is often reported as the number of "lines per millimeter".)

⋆ **Problem 8.17.** A diffraction grating rated at 500 lines/mm is used with a 650 nm laser to generate a diffraction pattern on a wall 1.5 m from the grating. How many bright spots will you see on the wall? What are their angular positions relative to "directly across from the laser"?

⋆⋆ **Problem 8.18.** A diffraction grating is illuminated by light of wavelength 660 nm and also by light of some other wavelength λ.

   a. The fourth-order maximum of the unknown wavelength overlaps the third-order maximum of the 660 nm wave. What is λ?

   b. Consider the bright spots discussed in (a). If the distance from this point to one slit is r_1 and the distance to the other slit is r_2, then what is the difference $|r_1 - r_2|$ for the 660 nm laser? (In other words, how much farther did light from one of the slits travel compared to the other?)

   c. Repeat (b) for the second laser.

Support your answers to (b) and (c) with a sketch of the experiment.

⋆ **Problem 8.19.** A diffraction grating with 800 lines/mm is illuminated by a 530 nm laser. On a wall 2.3 m away, a diffraction pattern is observed. How far apart are the two bright spots closest to the central bright spot?

⋆ **Problem 8.20.** Light of wavelength 520 nm is directed at a diffraction grating. The second order maximum is at an angle of 43°. What is the distance between adjacent slits in this grating? How many slits exist in each millimeter of the grating?

⋆⋆ **Problem 8.21.** A blue laser with wavelength 445 nm is directed through a diffraction grating that is 5.0 cm wide and contains 2500 slits. A very long viewing screen is located 2.2 m from the grating.

   a. Describe the positions of the *four* positions of constructive interference closest to the position on the screen directly across from the grating. Include a sketch.

   b. You replace the laser with a 650 nm red laser. Will the values you found in (a) increase, decrease, or remain the same? You don't need to provide numerical answers but briefly explain your logic.

⋆⋆ **Problem 8.22.** A diffraction grating is rated at 300 lines/mm and is illuminated by light from a biological sample that consists of a mixture of 600. nm light and 300. nm light. A screen is held 1.5 m from the grating and has a total width of 1.2 m. Where on the screen do you expect to detect points of constructive interference from each wavelength that are (a) overlapping and (b) separate? Include a sketch of the viewing screen and mark the locations of constructive interference for each wavelength.

★ **Problem 8.23.** A diffraction grating rated at 700 lines/mm is illuminated by monochromatic light with a wavelength of 488 nm (a green laser beam). If the $m = 0$ bright spot is at 0° (as usual), at what angle (in degrees) is the $m = 2$ bright spot?

★★ **Problem 8.24.** Molecular spectroscopy involves the study of radiation that is absorbed or emitted by molecules. For example, a diatomic molecule like H_2 (hydrogen gas) can give off electromagnetic radiation as it vibrates (much like two masses connected by a spring can stretch/compress through simple harmonic motion). The frequency of the vibrational radiation given off by H_2 is 1.3×10^{14} Hz. In an experiment, the electromagnetic radiation given off by a container of hydrogen gas is directed through a diffraction grating with adjacent slits separated by 3.33×10^{-6} m. The detection screen is 0.75 m from the grating.

    a. What is the "lines per mm" rating of this grating?

    b. Sketch the diffraction grating and screen.

    c. Where on the screen would you expect to detect the radiation described above? Determine a numerical value and label the position(s) on your diagram.

★★ **Problem 8.25.** "Diffraction Grating Glasses", are (as the name suggests) glasses where the lenses are diffraction gratings. While an interesting fashion statement, their primary purpose is educational: if you put on a pair and look at a white light (like ceiling lights in a well-lit room), you'll see a rainbow. Suppose the gratings are rated at 500 lines/mm and the distance from the grating to your retina (the "viewing screen") is 2.0 cm.

    a. Sketch the geometry of the situation. (Ignore the effects of the lens of your eye.)

    b. How far will the principle maximum for red light (wavelength of about 700 nm) be from the principle maximum for purple light (wavelength of about 400 nm)?

★★ **Problem 8.26.** The light given off by a chemical sample is directed through a diffraction grating to generate a complicated interference pattern. Your colleague notices a peak that looks suspiciously like it might actually be two peaks, and suggests that the sensitivity of your device can be improved (i.e. points of constructive interference will spread apart and thereby become more easily distinguishable) if you modify it so the light travels through oil rather than air after moving through the diffraction grating. Is your colleague correct? Clearly explain, using words, mathematics, and perhaps a diagram.

★★ **Problem 8.27.** In molecular biology, samples can be given a *fluorescent tag* so certain molecules give off light of a known wavelength: light from the sample can be analyzed to determine the abundance of the tagged molecules. Suppose you are analyzing such a sample by allowing its light to pass through a diffraction grating with 500 lines/mm. The viewing screen is located 0.40 m from the grating; what you observe is shown in Fig. 8.28. What is the wavelength of the light giving this signal?

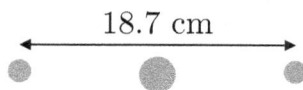

18.7 cm

FIGURE 8.28
Problem 8.27

★★ **Problem 8.28.** White light is a mixture of all wavelengths. Suppose a narrow beam of white light is directed through a diffraction grating rated at 500 lines/mm to make a pretty rainbow pattern on a viewing screen located 2.5 m from the grating. How far from the first order maximum for red light (7.00×10^2 nm) will the first order maximum for violet light (4.00×10^2 nm) be?

8.4 Interference from Infinite Points: The Single Slit

★ **Problem 8.29.** You cut a 0.10 mm-wide slit in a sheet of aluminum foil and direct a green laser pointer (that is advertised to have a wavelength of 543 nm) through it. On a wall 1.8 m away, you observe a broad bright spot. How wide is it, in meters?

⋆ **Problem 8.30.** Your cell phone broadcasts at 8.00×10^2 MHz, but you are blocked by radio-wave-absorbing buildings separated by 15 m. What is the angular width of the central maximum of your phone's signal, in degrees, after it passes through the gap?

⋆⋆ **Problem 8.31.** Water waves with a wavelength of 3.0 m are parallel to the shoreline when they slam into a rocky barrier (also parallel to shore) with a 0.50 m wide gap. The barrier is 15 m from the shore.

    a. In a few sentences, explain the principle of interference in the context of this situation: what happens to the water that passes through the gap? Use a diagram along with your written explanation and refer to appropriate mathematical equations, but no calculations are necessary.

    b. Observers on the beach will see a broad wave impact the shore directly opposite the gap. How wide will it be?

⋆⋆ **Problem 8.32.** A water park has a "wave pool" sketched in Figure 8.29: plane waves with a wavelength of 1.3 m are sent through an opening of width 4.0 m. The pool is a square with edge length 20.0 m. Identify all the locations along the back wall where the water will be calm (i.e. where you can stand and not be buffeted by waves).

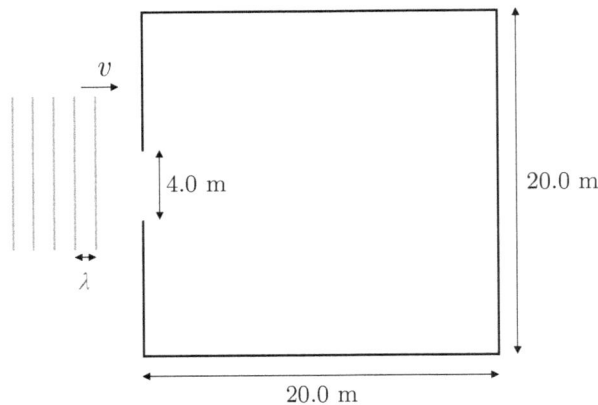

FIGURE 8.29
Problem 8.32

⋆⋆ **Problem 8.33.** In Figure 8.14 we depict Equation (8.10) for $m = 1$. Make a corresponding sketch for the case $m = 2$. Comment on the symmetry in the inset graph that shows the phase difference $d(x)$. What about this graph tells you that the waves destructively interfere on the viewing screen? How does this result change for $m > 2$?

⋆⋆ **Problem 8.34.** Consider the sketch you set up in Problem 8.33. Set up and evaluate the integral $\int_0^a d(x)\, dx$. Comment on the result.

Additional Problems for Chapter 8

⋆⋆ **Problem 8.35.** Lasers are commonly used in medical applications (e.g. to correct vision or for minimally invasive spine surgery). One medical-grade laser operates at a wavelength of 10,600 nm. Suppose such a laser is directed through a narrow slit; the patient's body is 0.20 m from the slit. If the central maximum of the beam is to be 3.0 mm wide, how wide should the slit be?

Problem 8.36. A set of two-way communication devices uses a frequency of 465 MHz. In an urban setting, interference from buildings can prevent these devices from working. Suppose one person is broadcasting from the location indicated. Identify how many "dead spots" (locations where the signal would not be received) will exist on the road in the map in Figure 8.30, neglecting the width of the buildings.

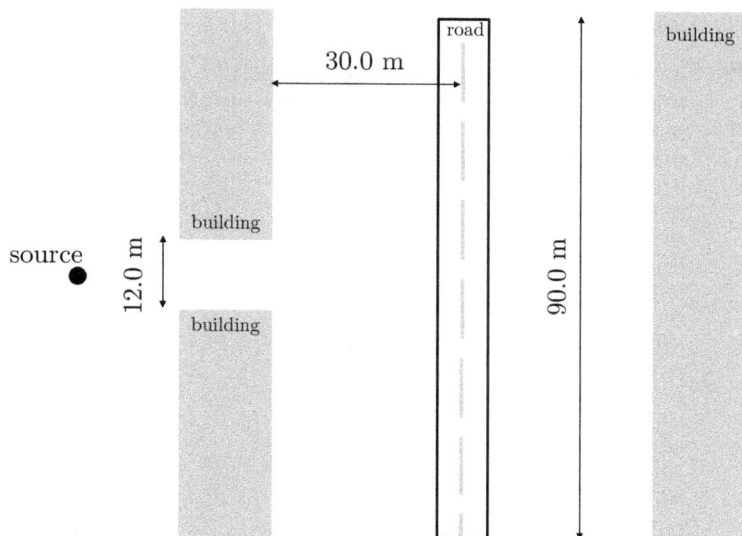

FIGURE 8.30
Problem 8.36

Problem 8.37. In an amusement park pool, water waves with a 1.0 m wavelength crash into a barrier that has two small openings, as shown in the "looking down from above" sketch shown in Figure 8.31. Identify three locations along the right edge of the pool where the waves will be tallest. Give your answers in terms of the coordinate y. In a sentence or so (and maybe a sketch), briefly explain why these locations have tall waves.

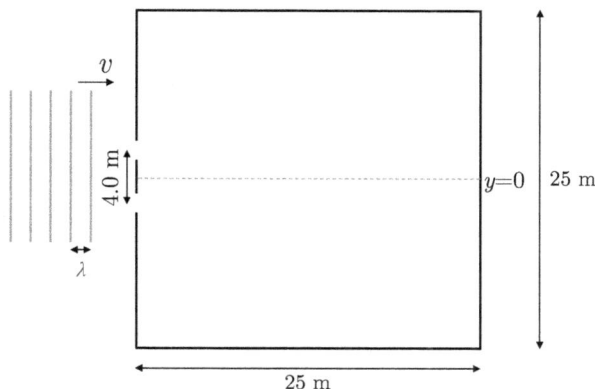

FIGURE 8.31
Problem 8.37

Problem 8.38. A piston is sinusoidally inserted into the surface of a pond, creating waves that spread radially away. The piston delivers power P to the pond. You put your finger, which has a width x, into the water a distance r from the piston. Determine an algebraic expression for the energy your finger absorbs every second due to the spreading waves. Comment on how your answer depends on r. Is this obvious, at least in retrospect?

★★ **Problem 8.39.** A 20.0 W light bulb emits energy evenly in all directions. If you have a square sensor that measures 1.00 cm to a side, then how much energy will it absorb if you hold it (a) 15.0 cm, (b) 30.0 cm, or (c) an arbitrary distance r from the bulb? Comment on how your answer depends on r. Is this obvious, at least in retrospect?

★★ **Problem 8.40.** Shining a laser through 1 or more narrow slits provides evidence that light has wave-like properties. With a diagram corresponding to any one experiment and a sentence or two, explain why. (This problem essentially requests that you summarize, briefly and in your own words, the arguments from the experiments discussed in this chapter.)

Part IV

Electricity and Magnetism

We trace the motion of the heavenly bodies back to the force of gravity: the particles that make up matter have *mass*, and mass advertises its existence to other mass in the universe according to Newton's Law of Gravitation. But gravity is not the only long-range force. Some fundamental particles (specifically, the proton and electron) also have *electric charge*, and the interactions between charged particles are fundamentally important to our understanding of the atom (and thereby chemistry and biology) and a host of topics ranging from lightning strikes to the electrical circuits that pervade the modern world. Our goals now are to explore what charge *does* and learn how we can *use* charge to our benefit.

We shall see that there are *two* forces that relate to charge. The simpler of the two is the *electric force*, which plays a somewhat analogous role to Newton's Law of Gravitation for mass. But we shall also see that, in addition to the electric force, *a moving charge exerts a force on other moving charges*. This *magnetic force* explains how a refrigerator magnet works – but it is also critically important for the conversion of mechanical energy (like a rotating turbine) into electrical power.

To analyze how these forces reach across space – and to explore how they are bound up with one another – we will introduce the important concept of a *vector field*. As perhaps the most striking application of this framework, we will find that when we talk about light, we are referring to very specific interactions between electric fields and magnetic fields.

9

Electric Charge: Fundamentals

Matter is composed of atoms, and atoms are made up of protons, neutrons, and electrons. Thus when we measure the mass of an object – the mug of tea sitting next to my computer right now, for instance – we are measuring the total mass of the each of the protons, neutrons, and electrons that make up the object. But these fundamental particles are endowed not only with a mass: the proton and electron also carry an *electric charge*: the proton carries a *positive* charge denoted e, while the electron carries an equal *negative* charge $-e$. Just as there is a *gravitational* force between particles because of their mass, there is an *electric* force because of their charge.

We shall see that the situation with charge is very rich compared to mass. In this chapter we will carefully build new terminology to describe the forces and energy associated with interacting charges. We will draw close parallels with gravity to build your intuition, and in this chapter we will mostly focus our attention on **point charges**: one or more protons or electrons that are, for practical purposes, located at a single position in space. Examples of this include an isolated proton, an isolated electron, and atomic nuclei (where protons and neutrons are clustered together). On a larger scale, metal balls that do not have roughly equivalent numbers of protons and neutrons are said to be *charged* and (as we shall prove in Chapter 10) can also be treated as point charges.

Once we have analyzed point charges and established the language used to describe their interactions, we will be in an excellent position to scale up to interesting phenomenon involving charge and more complicated *charge distributions* in Chapter 10.

Learning Objectives

After reading this chapter, you should be able to:

- Use Coulomb's Law to analyze the electric forces between point charges.
- Analyze the force acting on a point charge in terms of an electric field..
- Characterize the electric field created by one or more point charges..
- Analyze the energy associated with a configuration of point charges..
- Characterize the potential field created by one or more point charges..
- Graphically and mathematically relate the concepts of electric potential, electric potential energy, electric potential fields, and electric fields, and compare/contrast to their gravitational analogs.

9.1 An Analogy: Gravitational Fields

Recall that the gravitational force between two objects is given by Newton's Law of Gravity (Equation (3.4)):

$$\vec{F}_G = G\frac{Mm}{r^2} \text{ (attractive)} \tag{9.1}$$

where M and m are the masses of the objects and r is the distance between them; G, is a fundamental constant of nature called the gravitational constant ($G = 6.67 \times 10^{-11} \text{m}^3\text{kg}^{-1}\text{s}^{-2}$). The "(attractive)" label simply means that the force felt by M points toward m and vice versa. A more mathematically precise way to say this is that the gravitational force of M on m is given by

$$\vec{F}_{\text{G, M on m}} = G\frac{Mm}{r^2}\left(-\hat{r}\right) \tag{9.2}$$

where $\hat{r}$ points radially away from M.

It is natural to wonder *how* the gravitational force is transmitted if the objects aren't touching. After all, every other force we've studied so far involves some sort of *contact*: you push on a box, or pull on a rope, or the ground rubs the side of a book as it slides, etc. In the case of gravity, an object must evidently do *something* to the space around it, such that a second object some distance away experiences a gravitational force as a result.

Now, *do something* is not very precise language; we say instead than an object *creates a gravitational field* in the space around it. An object experiences a gravitational force, then, if it is immersed in a gravitational field created by some other object.

Mathematically, the gravitational field created by an object of mass M is

$$\vec{g}_M = \frac{GM}{r^2}\left(-\hat{r}\right) \tag{9.3}$$

If this seems familiar, good: on and near the surface of the Earth, we know this value works out to be $\vec{g} = 9.8$ m/s^2 directed down to the surface of the Earth (Section 3.4), regardless of where you are.[1] Far out in space, though, it is no longer the case that r in Equation (9.3) is approximately equal to the radius of the earth. Instead, the magnitude of the $\vec{g}_M$ decays with the square of the distance from the object, just as with the gravitational force (Fig. 9.1).

We can visualize this with a set of vectors all throughout space, each indicating the magnitude and orientation of $\vec{g}_M$ at that location. To reduce visual clutter, we sometimes draw the field not with a set of vectors, but with continuous curves that point in the direction of the field at any position. In this representation the *magnitude* of the field roughly corresponds to how tightly packed the field lines are (essentially, you connect adjacent vectors, tail to tip, to make a continuous curve; see the bottom right of Figure 9.1). We refer to these continuous curves as **gravitational field lines**.

If you know the field created by mass M, you obtain the force on mass m by multiplying the two terms:

$$\vec{F}_G = m\vec{g}_M \tag{9.4}$$

If you combine Equation (9.4) with Equation (9.3) you simply get back to Newton's Law of Gravity:

$$\vec{F}_G = m\vec{g}_M = m\left(G\frac{M}{r^2}\left(-\hat{r}\right)\right) = G\frac{Mm}{r^2}\left(-\hat{r}\right) \tag{9.5}$$

Now, if an object is experiencing *multiple* gravitational forces (for instance, if we care about the combined force of the Sun and Jupiter on Earth), then the *net* gravitational field is determined in the same way as the net force:

$$\vec{g}_{\text{NET}} = \vec{g}_1 + \vec{g}_2 + \dots \tag{9.6}$$

---

[1]Strictly speaking $\vec{g}$ *does* vary, even near the surface of the Earth, but the variations are minor and as usual we'll consider it to be a constant.

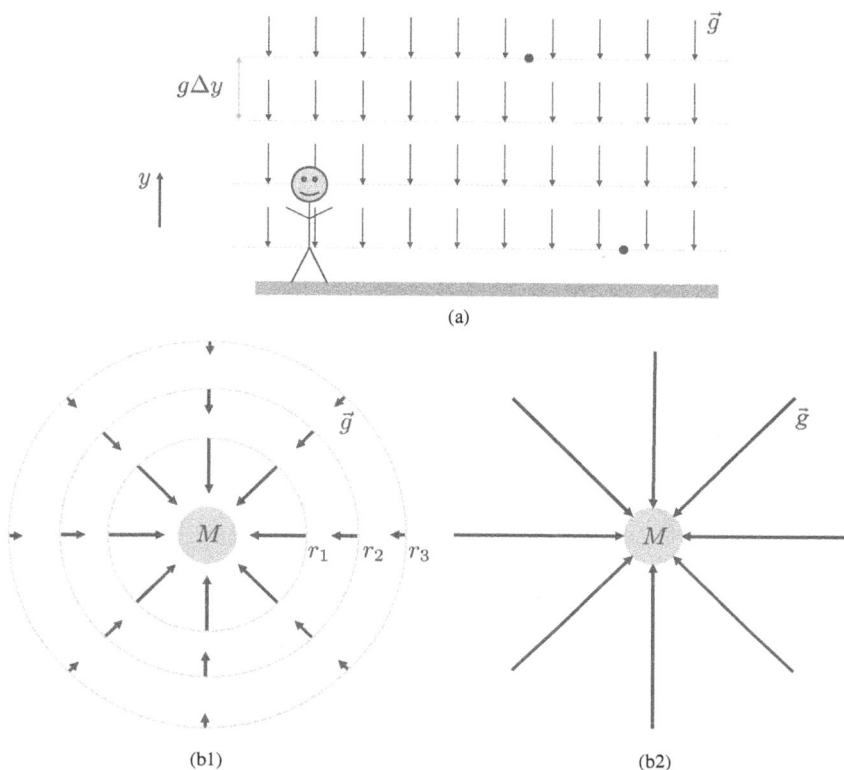

FIGURE 9.1

(a) Near the surface of the Earth, objects feel a gravitational force $\vec{F} = m\vec{g}$ toward the center of the Earth. Thus we say that there is a constant *gravitational field* $\vec{g}$ everywhere near the surface of the Earth, which we here represent with a few vectors of equal length and uniform direction. The stick figure is smiling because it is happy to be in a physics diagram. (Bottom) Far from the surface of a planetary body of mass M, the gravitational field is no longer constant. (b1) At every position, the gravitational field vector points in toward the mass, and the magnitude of the vector decreases with the square of the radial distance (Equation (9.3)). (b2) The field can also be shown with continuous curves (in the case of the single mass shown here, the curves are just straight lines) such that the field at any position is tangent to the curve (here, radially in). Here again we can only draw a few representative curves or else the entire figure would be black. The fact that the curves are more tightly packed near the mass indicates that the field has greater magnitude in that region.

Another way of thinking about the gravitational field is that it makes it straightforward to answer the question, "What gravitational force *would* an object experience if it was located somewhere in this region?" We shall see in the following section that charged particles create *electric* fields in an analogous way to the *gravitational* fields we are considering here.[2]

---

[2]You may be wondering why we bother defining a field at all: why not just stick with the force? For gravity, there isn't much reason to favor fields over forces (at least for our purposes in this book), which is why we didn't introduce the idea of a field when we originally discussed the gravitational force. But electric fields and forces are more complicated, and we will see that working with fields in this context makes our lives *easier* rather than needlessly more complex.

Example 9.1 The Gravitational Field of the Earth ⋆

What is the gravitational field created by the Earth? What about for the Moon? Why is it that even though these fields are different, the magnitude of the force of the *Earth acting on the Moon* is the same as the magnitude of the force of the *Moon acting on the Earth*?

From Equation (9.3) we have for the Earth

$$\vec{g}_{\text{Earth}} = \frac{GM_{\text{Earth}}}{r^2}(-\hat{r})$$

$$\vec{g}_{\text{Earth}} = \frac{\left(6.67 \times 10^{-11} \text{ m}^3\text{kg}^{-1}\text{s}^{-2}\right)\left(5.97 \times 10^{24} \text{ kg}\right)}{r^2}(-\hat{r})$$

$$\vec{g}_{\text{Earth}} = \frac{3.98 \times 10^{14} \text{ m}^3\text{s}^{-2}}{r^2}(-\hat{r})$$

If you do the same calculation for the Moon, with a mass of 7.35×10^{22} kg, you'll find

$$\vec{g}_{\text{Moon}} = \frac{4.90 \times 10^{12} \text{ m}^3\text{s}^{-2}}{r^2}(-\hat{r})$$

These equations tell you the field for any r; if you want to know the value for the field at a specific distance from the center of the mass, you just need to insert the value of r into the equation and simplify.

Now, the magnitude of the force on the Earth due to the Moon is given by $g_{\text{Moon}}M_{\text{Earth}}$ and the magnitude of the force of the Moon due to the Earth is given by $g_{\text{Earth}}M_{\text{Moon}}$. You can verify for yourself that these give the same number, but hopefully that isn't shocking in light of the discussion surrounding Equation (9.5).

Example 9.2 The Gravitational Field and Force Due to Two Bodies ⋆⋆

Suppose three stars are arranged as shown in Figure 9.2; A has mass m and the other two stars have mass M. What net gravitational field is mass A experiencing? What is the corresponding net force?

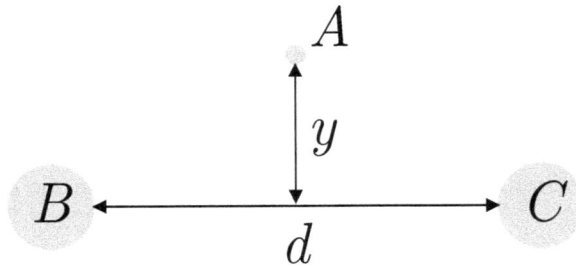

FIGURE 9.2

First, note that the gravitational fields due to masses B and C at mass A partially cancel: if we break each into its horizontal and vertical components, you'll see that the horizontal forces have the same magnitude but point in opposite directions (the components are drawn in gray in Fig. 9.3). Thus, there is no net field (and therefore no net force) in the horizontal direction.

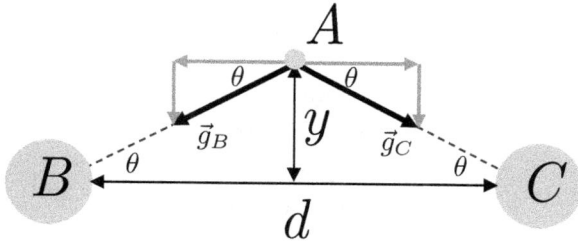

FIGURE 9.3

What about in the vertical dimension? Here the forces have equal magnitudes but the point in the same direction, so the net field is just double the field (in the vertical dimension) due to one of the masses. With this in mind, let's work out the vertical component of the field due to mass B:

$$g_{B,y} = \frac{GM}{r^2}\sin\theta$$

$$g_{B,y} = \frac{GM}{r^2}\frac{y}{r}$$

$$g_{B,y} = \frac{GMy}{\left(\left(\frac{d}{2}\right)^2 + y^2\right)^{\frac{3}{2}}}$$

In the first line I included $\sin\theta$ to pick out just the vertical component; in the last line I've applied the Pythagorean Theorem to express r in terms of the variables labeled in the problem statement diagram. Doubling this field to find the net field gives

$$g_{NET,y} = 2g_{B,y} = 2\frac{GMy}{\left(\left(\frac{d}{2}\right)^2 + y^2\right)^{\frac{3}{2}}}$$

This is the magnitude of the net field; the direction is *down*.

From this result we simply multiply by the mass of A to determine the force:

$$\vec{F}_{NET} = M_A\vec{g}_{NET}$$

$$\vec{F}_{NET} = 2\frac{GMmy}{\left(\left(\frac{d}{2}\right)^2 + y^2\right)^{\frac{3}{2}}}(-\hat{y})$$

In this equation I've set up the y axis to be positive pointing *up*, which is why the net force is negative.

It is always a good idea to *check* that our results make sense in special cases where we know what the answer *should* be. In this case, we should expect *no force* if $y = 0$. Why? Well, in this case A is directly between the other two planets: the masses are the same and the distances are the same, so the fields (and forces) are equal and opposite. You can see that our expression in this case becomes

$$\vec{F}_{\text{NET}} = \frac{GM^2 (0)}{\left(\left(\frac{d}{2} \right)^2 + (0)^2 \right)^{\frac{3}{2}}} (-\hat{y}) = 0 \text{ N}$$

So far, so good. What about if we're *really* far away? In this case B and C look like one single mass of magnitude $2M$. If $y \gg d$, then

$$\left(\frac{d}{2} \right)^2 + (y)^2 \approx y^2$$

and the net field is

$$\vec{F}_{\text{NET}} \approx 2 \frac{GM^2 y}{(y^2)^{\frac{3}{2}}} (-\hat{y}) = 2 \frac{GM^2}{y^2} (-\hat{y})$$

exactly as we expect from Newton's Law of Gravitation for masses m and $2M$ separated by a distance y.

Example 9.3 Visualizing the Gravitational Field: Binary Star Systems ★★
Consider two stars of equal mass separated by some distance (this describes a binary star system, though in general the masses need not be equal). Make a rough sketch of the gravitational field in a style similar to Fig. 9.1.

At any location in space, the gravitational field will be determined by the vector sum of the gravitational fields due to each of the two masses:

$$\vec{g}_{\text{NET}} = \vec{g}_1 + \vec{g}_2$$

If you read the preceding example then you already know that *strictly from symmetry*, the horizontal components of the field cancel for any position along the line that cuts perpendicular through the midpoint between the stars. We also know that the magnitude of the field at the exact midpoint should be 0 N/kg.

What about along the line that runs through the two stars? Well, if we're *between* the stars then the forces will oppose one another, but the force due to the closer star will be larger (because r is smaller for the closer star, which means a larger field according to Equation (9.3)) and so the net force will point toward the closer star. In contrast, if both stars are to one side of the position we're considering (e.g. if we're considering the region to the left of both stars), then the forces align and the net field points toward the masses (and is larger at that position than it would be with just one star present.)

At positions *not* on one of these two lines, there isn't any nice symmetry. Since we're just after a rough sketch, we can get by with a few general observations:

1. Because the two constituent fields always point to the two stars, the *net* field will always point somewhere *between* the two stars.

2. Because a field is stronger the closer we are to the star creating the field, the net field will point closer to the nearest star.

3. In general, the farther we are from the stars, the weaker the net field will be.

Figure 9.4 shows a few of the net field vectors (in black) along with the constituent vectors due to each star. Figure 9.5 shows a more complete set of net field vectors, and Figure 9.6 shows the field lines.

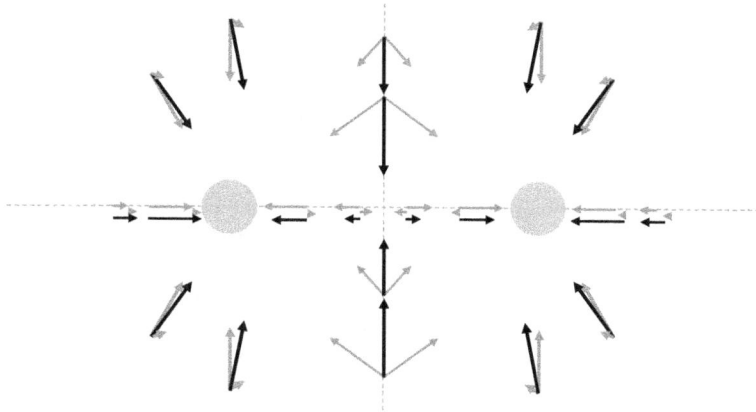

FIGURE 9.4

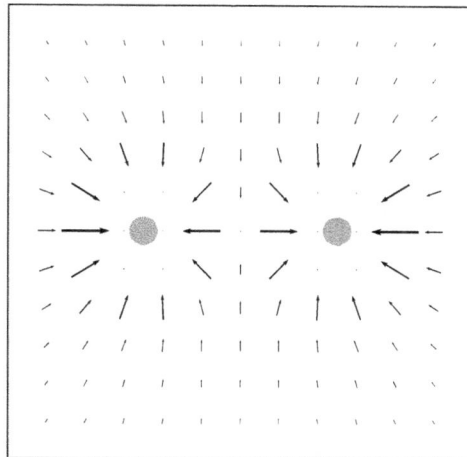

FIGURE 9.5

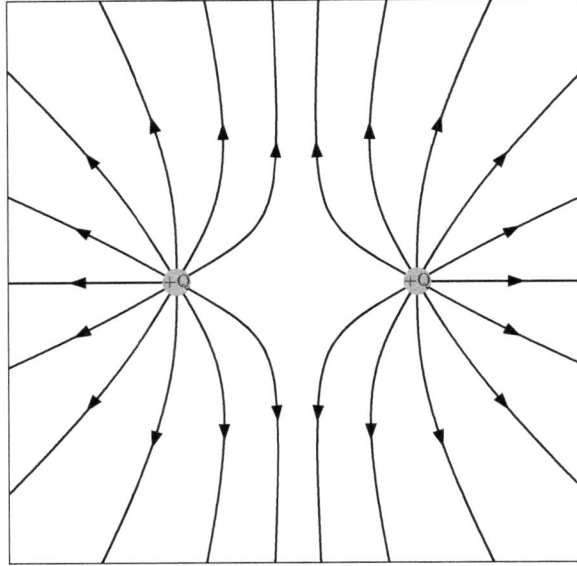

FIGURE 9.6

9.2   The Electric Force and the Electric Field

As I pointed out at the beginning of this chapter, the fundamental particles are defined not only by their masses (we generally denote a chunk of mass as M or m) but also by their charges (we'll use the conventional notation and denote a chunk of charge – many protons clustered together in an atomic nucleus, for instance – with a Q or q). I began this chapter by discussing gravity because the basic interaction force between two charges looks surprisingly similar to the gravitational force. It is referred to as **Coulomb's Law**[3] or sometimes simply as the **electric force**:

$$\vec{F}_{\text{E, Q on q}} = K\frac{Qq}{r^2}\hat{r} \tag{9.7}$$

K is called the **electrostatic constant**:[4]

$$K = \frac{1}{4\pi\varepsilon_0} = 8.99 \times 10^9 \frac{\text{Nm}^2}{\text{C}^2} \tag{9.8}$$

Here "C" stands for a **Coulomb**, the SI unit of charge and ε is a constant of nature called the **permittivity of free space**.[5] Compare this to Equation (9.4): if you swap out G for K and the masses (M and m) for the charges (Q and q), then the two equations are *the same* except for a negative sign.

---

[3]After Charles–Augustin de Coulomb (1736–1806).
[4]Or sometimes the **Coulomb constant**.
[5]The numerical value is $\varepsilon_0 = 8.85 \times 10^{12}$ $\text{C}^2\text{N}^{-1}\text{m}^{-2}$.

What does the lack of a negative sign mean for Coulomb's Law? Well, Newton's Law of Gravitation says that the gravitational force is *attractive*, so if Coulomb's Law carries the opposite sign, you might conclude that the electric force is *repulsive*. This logic is sound but the conclusion is incomplete, because while mass is *always positive*, charge is *positive **or** negative*.

With this in mind, If you refer back to Equation (9.7), you'll see that the Qq term is *positive* (like the Mm term for gravity) if both charges are positive *or* if they're both negative. If this is the case, then Coulomb's Law says that the force is *repulsive*, as we reasoned above. However, if one charge is positive and the other is negative, then the Qq term introduces a negative sign into Coulomb's Law, and the force becomes *attractive*. (This follows straightforwardly from the mathematics, but it can be an elusive point. This is very important; make sure you follow the argument!)

We summarize all of this by saying that

Like charges repel; opposite charges attract.

This is shown graphically in Figure 9.7.

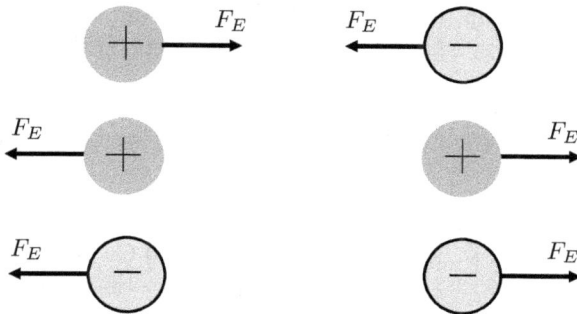

FIGURE 9.7
According to Coulomb's Law, two charges with opposite sign are attracted to one another; two charges with the same sign are repelled.

A second major difference between the electric force and the gravitational force is one of *scale*. As we show in Example 9.4, the electric force is staggeringly larger than the gravitational force: for two protons the forces differ by a factor of about 10^{36}. For this reason, in situations where we are considering electric forces it is safe to ignore gravitational forces.[6]

Now, just as the *gravitational field* $\vec{g}$ answers the question, "What *gravitational force* would an object experience if it was located somewhere in this region?", the **electric field** $\vec{E}$ answers the question, "What *electric force* would a charge experience if it was located somewhere in this region?" As a direct analogy to the gravitational case ($\vec{F}_g = m\vec{g}$), we express this mathematically like so:

---

[6]You might be wondering why the gravitational force *ever* matters, if the electric force is so much more powerful. The answer is that the electric force is incredibly important at small scales – inside atoms and molecules – but these structures tend to be overall neutral, meaning that the total charge is 0. Thus at larger scales only the *mass* of the objects generally contributes to the forces acting upon them. Of course, there are some exceptions where the electric force is important at the scale of everyday life; the effects of static electricity and the design of electric circuits are obvious examples.

$$\vec{F}_E = q\vec{E} \tag{9.9}$$

Where F_E is the *net* electric field acting on charge q. By comparing this equation to Coulomb's Law (Equation (9.7)), we see that the electric field of a charge Q is given by

$$\vec{E}_Q = K\frac{Q}{r^2}\hat{r} \tag{9.10}$$

This means that the electric created by a positive charge points *away* from it, while the electric field created by a negative charge points *toward* it (Fig. 9.8). It follows, then, that

Positive charges feel a force in the direction of the net electric field; negative charges feel a force *against* the net electric field.

If all of this seems daunting, keep in mind that what we've essentially done is introduce a new force (that acts between *electric charges*) and a way of thinking about how a charge "primes" its surroundings to apply that force to other nearby charges (the electric field). All of the tools we've built up so far in this book concerning forces, including for instance kinematics, torque, and energy (via the work-energy theorem) still apply, as the next sequence of examples demonstrates.

Before moving on to the examples, though, I should provide you with the numerical values for the masses and charges of the fundamental particles (these values are also provided in Appendix B):

$$
\begin{array}{llll}
m_{\mathrm{p}} & = & 1.67 \times 10^{-27}~\text{kg} & \quad q_{\mathrm{p}} & = & 1.60 \times 10^{-19}~\text{C} \\
m_{\mathrm{n}} & = & 1.67 \times 10^{-27}~\text{kg} & \quad q_{\mathrm{n}} & = & 0~\text{C} \\
m_{\mathrm{e}} & = & 9.11 \times 10^{-31}~\text{kg} & \quad q_{\mathrm{e}} & = & -1.60 \times 10^{-19}~\text{C}
\end{array}
$$

Note that the charge on the proton and the charge on the electron are equal and opposite. We refer to the magnitude of their charge as the **fundamental unit of charge**, e:

$$e = 1.60 \times 10^{-19}~\text{C} \tag{9.11}$$

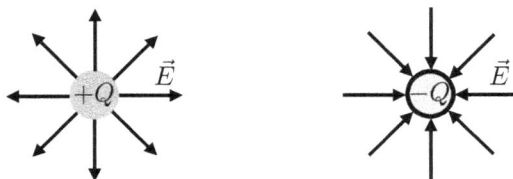

FIGURE 9.8
The electric field created by a positive charge points radially *out*; the electric field of a negative charge points radially *in*. This is shown mathematically in Equation (9.10).

Example 9.4 Comparing the Gravitational and Electric Forces ⋆

Confirm that the electric force between two protons is larger than the gravitational force by a factor of about 10^{36}, as I claimed in the main text.

Let's express the ratio and simplify algebraically before plugging in any numbers:

$$\frac{F_E}{F_G} = \frac{\frac{Ke^2}{r^2}}{\frac{Gm_p^2}{r^2}}$$

$$= \frac{Ke^2}{Gm_p^2}$$

$$= \frac{\left(8.99 \times 10^9 \, \frac{\text{Nm}^2}{\text{C}^2}\right)\left(1.60 \times 10^{-19} \text{ C}\right)^2}{\left(6.67 \times 10^{-11} \, \frac{\text{m}^3}{\text{kgs}^{-2}}\right)\left(1.67 \times 10^{-27} \text{ kg}\right)^2}$$

$$= 1.24 \times 10^{36}$$

To give you a sense of how large this number is, consider that the diameter of the observable Universe divided by the diameter of the hydrogen atom is roughly 10^{34}!

Example 9.5 Electric Force and Field with Three Charges ⋆⋆

Consider the three charges shown in Figure 9.9. (a) Determine the electric force experienced by the top charge. What electric field does this charge experience? (b) Sketch the electric field created by just the two lower charges. (c–d) Repeat parts (a) and (b) in the case where the bottom right charge is $-Q$.

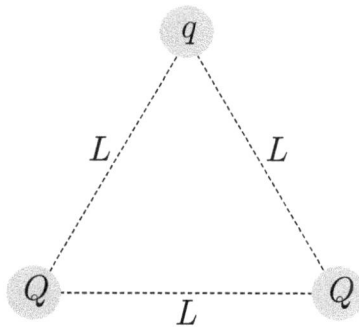

FIGURE 9.9

First note that this is rather similar to Example 9.2, where here we have charges instead of masses. For (a) all of the charges are positive, which means that the interaction forces are all *repulsive*. If we draw the two forces acting on the top charge, we see that their horizontal components cancel and the net force is *up* (Fig. 9.10).

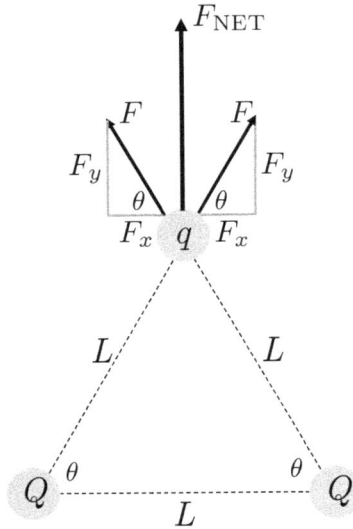

FIGURE 9.10

To quantify this we will use our usual coordinate system with the positive y axis directed up the page. The net force is thus

$$\vec{F}_{\text{NET}} = 2F_y \hat{y}$$
$$\vec{F}_{\text{NET}} = 2F \sin \theta \hat{y}$$
$$\vec{F}_{\text{NET}} = 2 \left(K \frac{Qq}{L^2} \right) \sin \theta \hat{y}$$
$$\vec{F}_{\text{NET}} = 2 \left(K \frac{Qq}{L^2} \right) \sin (60°) \hat{y}$$
$$\vec{F}_{\text{NET}} = \sqrt{3} K \frac{Qq}{L^2} \hat{y}$$

In the last step I've made use of the fact that $\sin 60° = \sqrt{3}/2$. The fact that $\theta = 60°$ follows from the fact that we're dealing with an equilateral triangle, where each interior angle is $60°$.

It follows from Equation (9.9) that the field at this location is given by

$$\vec{E} = \frac{\vec{F}_{\text{NET}}}{q} = \sqrt{3} K \frac{Q}{L^2} \hat{y}$$

For (b), we can apply the same approach as in Example 9.3. Indeed, in this case the vector field should look the same, except every vector points in the *opposite direction* because the electric field points radially away from positive charge while the gravitational field points radially in (Figs. 9.11–9.12).

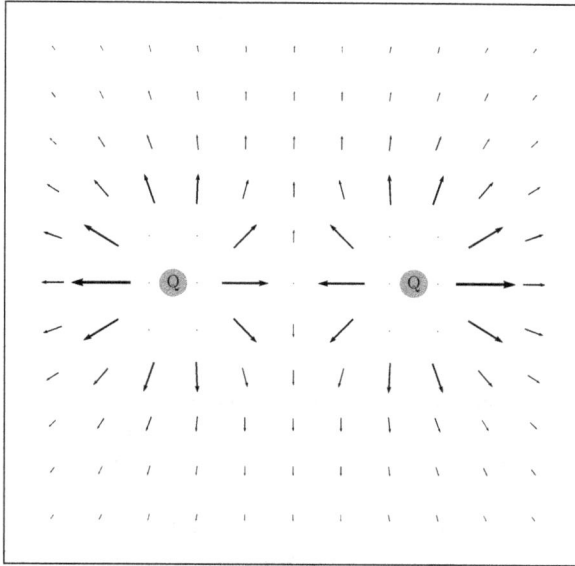

FIGURE 9.11

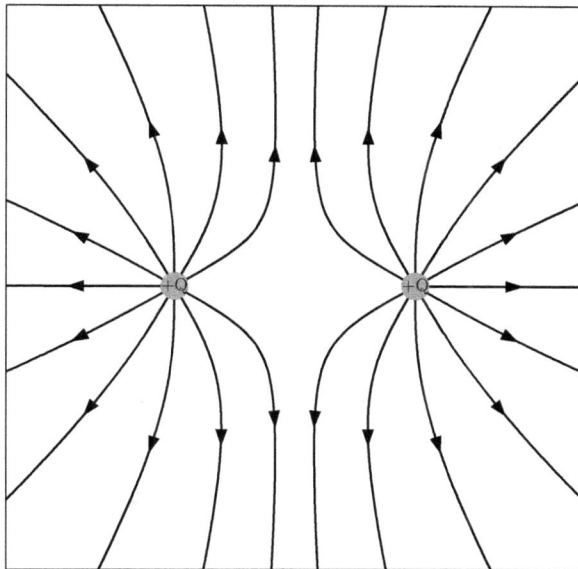

FIGURE 9.12

Turning now to (c), we find that one of the constituent forces points in the opposite direction such that the *vertical* components of the forces cancel and the net force points to the *right* (Fig. 9.13).

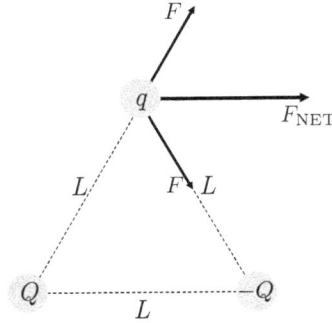

FIGURE 9.13

While we *could* go through the detailed mathematics as in (a), it is faster to recognize that everything is the *same* except we need to pick out the horizontal components of the constituent forces instead of the vertical components (since it is the two F_x components that add in this case). Thus we want to pick out the $\cos\theta$ components of the two electric forces $\vec{F}$, rather than the $\sin\theta$ components:

$$\vec{F}_{\text{NET}} = 2\left(K\frac{Qq}{L^2}\right)\cos\left(60°\right)\hat{x}$$

$$\vec{F}_{\text{NET}} = K\frac{Qq}{L^2}\hat{x}$$

Where we've noted that $\cos 60° = 1/2$. It follows that the electric field in this case is just $\vec{E} = \vec{F}_{\text{NET}}/q = K\frac{Q}{L^2}\hat{x}$.

Finally, for (d) we can go through a similar approach to Example 9.3. For instance, *between* the two stars the two constituent fields align (away from the positive charge and toward the negative charge), while in the upper right region of the diagram we would expect the attraction toward the (closer) negative charge to have a larger effect than the repulsion from the (farther) positive charge. The complete diagram is shown in Figures 9.14–9.15.

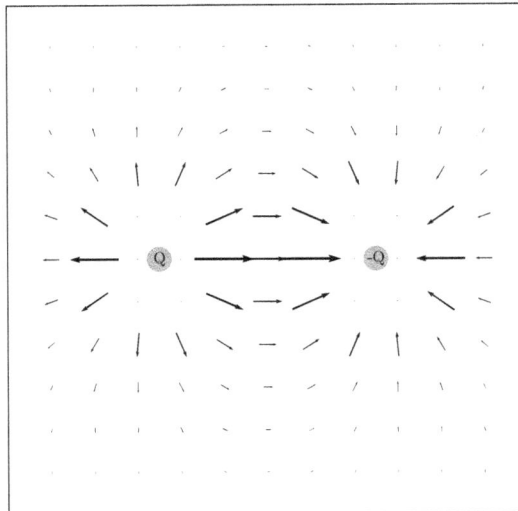

FIGURE 9.14

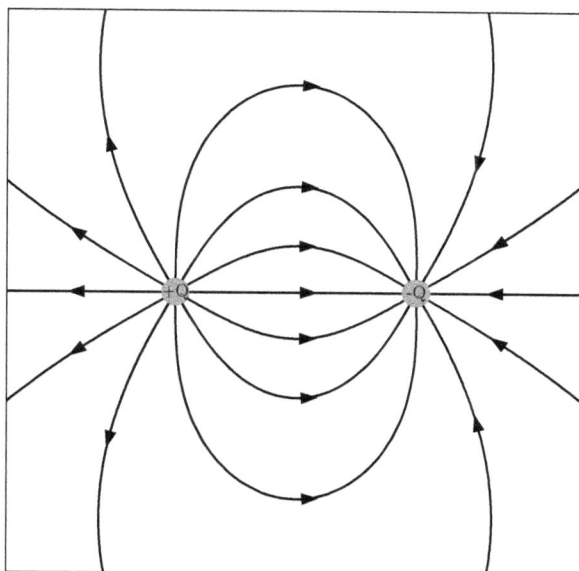

FIGURE 9.15

Before moving on, I should mention that the second charge configuration we considered, where we have two charges of equal magnitude and opposite sign separated by some fixed distance, is referred to as an **electric dipole**. The electric dipole is an excellent model for molecules, such as water, that have a net positive charge on one side and a net negative charge on the other (see Example 9.9).

Example 9.6 Finding Electric Equilibrium on a Line   ⋆⋆
Consider two positive charges of magnitudes Q and $3Q$ separated by a distance L. Where on the line that runs through the two charges could a third charge be placed such that it experiences no net force?

First, note that the answer can *only* be somewhere *between* the two charges: they're both *positive*, so the field due to each is *away*. If we arrange the two charges on a horizontal line and we consider a point to the left of both of them, then both field components point left. Similarly, if we're to the right of both of them then the net field points right. Only between the two charges do the field components point in opposite directions. If this is unclear, refer back to Example 9.5.

Let's focus our attention, then, on the region between the charges. Because $F_E = qE$, we can interpret this question as asking "where between the charges is the electric field 0 N/C?" Consider a position a distance x from charge Q, which means that it is a distance $L - x$ from the charge $3Q$ (Fig. 9.16). If the net electric field at this position is 0 N/C, then we have

$$\vec{E}_{\text{NET}} = \vec{E}_Q + \vec{E}_{3Q}$$

$$0 = K\frac{Q}{x^2} - K\frac{3Q}{(L-x)^2}$$

$$\frac{3}{(L-x)^2} = \frac{1}{x^2}$$

$$3x^2 = (L-x)^2$$

$$\sqrt{3}x = L - x$$

$$x = \frac{L}{1 + \sqrt{3}} \approx 0.37L$$

(In the second to last line I took the *positive* square root on both sides of the equation. As you can check for yourself, taking a negative root leads to a nonphysical negative solution for x.) Thus we're *closer* to the charge with *smaller* magnitude, which must be the case if the field due to the weaker charge is to cancel the field due to the larger charge.

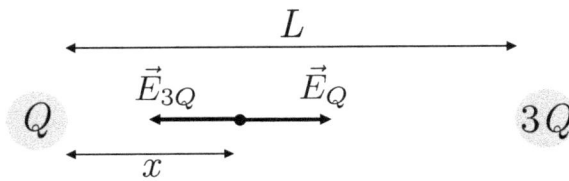

FIGURE 9.16

Example 9.7 Pith Ball Separation   ★★

While we've mostly been thinking of small collections of charge, like in atomic nuclei, we shall see later that it is possible to move *many* electrons onto objects or, conversely, to remove many electrons from objects, giving them an overall charge. With this in mind, consider two small balls of mass m that each carry an unknown charge Q (small balls that carry charge well are called **pith balls**). They are connected to two light strings of length L, which are attached to a common support above the balls. The electric repulsion between the balls pushes them away until the strings form an angle θ, as shown in Figure 9.17. (a) Determine an expression for Q in terms of m, L, θ, and g. (b) Determine a numerical value for Q if $m = 10.0$ g, $L = 20.0$ cm, and $\theta = 10.0°$. (c) How many electrons need to be moved onto (or taken away from) each ball to account for the charge you found in (b)?

   The balls are in equilibrium, so there must be 0 net force acting on each. As we identify in Figure 9.17, each ball experiences a gravitational force, a tension force along the string, and an electric force due to the other ball. We are asked to *not* invoke the tension force in our equation for θ, and so we will use Newton's 2nd Law (expressed in the horizontal direction and again in the vertical direction) to eliminate it.

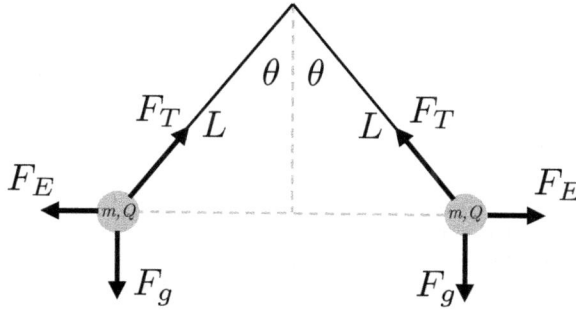

FIGURE 9.17

Starting with the horizontal forces acting on the left ball, we have

$$\vec{F}_{\text{NET,x}} = F_T \sin\theta - QE$$
$$0 = F_T \sin\theta - QE$$
$$F_T = \frac{QE}{\sin\theta}$$

And in the vertical direction we have

$$\vec{F}_{\text{NET,y}} = F_T \cos\theta - mg$$
$$0 = F_T \cos\theta - mg$$
$$F_T = \frac{mg}{\cos\theta}$$

Setting these equations equal to one another yields

$$\frac{QE}{\sin\theta} = \frac{mg}{\cos\theta}$$
$$\frac{QE}{mg} = \frac{\sin\theta}{\cos\theta}$$

We don't want to leave E in our expression, so let's spell it out explicitly. Note from the force diagram that the distance between the two balls is $d = 2L\sin\theta$, so the electric field experienced by either ball has magnitude

$$E = K\frac{Q}{d^2} = K\frac{Q}{4L^2\sin^2\theta}$$

Our above result then becomes

$$\frac{KQ^2}{4L^2 mg \sin^2\theta} = \frac{\sin\theta}{\cos\theta}$$
$$Q = \left(\frac{4mgL^2\sin^3\theta}{K\cos\theta}\right)^{1/2}$$
$$Q = \left(\frac{4mgL^2\sin^2\theta\tan\theta}{K}\right)^{1/2}$$

For (b) we insert the given values:

$$Q = \left(\frac{4\,(0.0100\text{ kg})\left(9.8\text{ m/s}^2\right)(0.200\text{ m})^2 \sin^2(10.0°)\tan(10.0°)}{8.99 \times 10^9 \frac{\text{Nm}^2}{\text{C}^2}} \right)^{1/2}$$

$$Q = 9.63 \times 10^{-8}\text{ C} = 96.3\text{ nC}$$

In the last step I have converted to nanoCoulombs, i.e. 10^{-9} C $= 1$ nC.

Finally, for (c) we are asked to determine *how many* electrons we need to generate this amount of charge. Now we don't know if there the balls have *extra* electrons or *missing* electrons (i.e. we don't know if Q is negative or positive, respectively; the balls will repel as long as they're *both* positive or negative). With this in mind, let's work with absolute values: the magnitude of the charge on an electron is $e = 1.60 \times 10^{-19}$ C, so if we have N electrons (extra or missing compared to the usual "overall neutral" case) then the magnitude of the total charge is just $Q = Ne$. To determine N in this case we just plug in the values:

$$Q = Ne$$

$$N = \frac{Q}{e}$$

$$N = \frac{9.63 \times 10^{-8}\text{ C}}{1.60 \times 10^{-19}\text{ C}}$$

$$N = 6.02 \times 10^{11}$$

It is worth taking a moment to appreciate how large this number is: over 600 billion! The phenomenon we're describing here isn't all that impressive, really: a couple of tiny balls carry enough charge to separate a few centimeters. Clearly the charge on a *single* electron is miniscule, as far as your daily experience is concerned!

Example 9.8 Charge Entering an Electric Field   ★★
An electron is traveling in the $+x$ direction with a speed $v = 5.0 \times 10^5 \frac{\text{m}}{\text{s}}$ when it enters a region of space with a constant electric field $\vec{E} = -1.00 \times 10^2 \frac{\text{N}}{\text{C}}\hat{x}$. The electric field covers a region that is $d = 2.5$ m wide. Will the electron make it through the electric field? If it doesn't, where will it stop? If it does, how fast will it be going when it exits the field? Answer the question using (a) kinematics and (b) energy analysis.

The situation is sketched in Figure 9.18. Recall that electrons feel a force directed *opposite* the direction of the net electric field. Thus, because the electric field is directed in the $-x$ direction, the electron feels a force in the $+x$ direction. It will therefore increase in speed as it moves through the electric field and it will certainly make it to the far side.

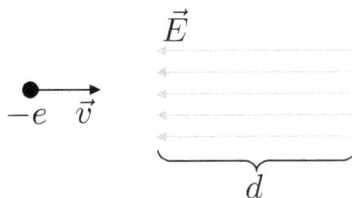

FIGURE 9.18

To determine its final speed using kinematics, we first note its acceleration from force analysis (recall that the electron's charge is denoted $-e$):

$$\vec{F}_{\text{NET}} = m\vec{a}$$
$$-e\vec{E} = m\vec{a}$$
$$\vec{a} = \frac{-e\vec{E}}{m}$$
$$\vec{a} = \frac{\left(-1.60 \times 10^{-19}\text{C}\right)\left(-1.00 \times 10^2 \frac{\text{N}}{\text{C}}\hat{x}\right)}{9.11 \times 10^{-31} \text{ kg}}$$
$$\vec{a} = 1.76 \times 10^{13}\frac{\text{m}}{\text{s}^2}\hat{x}$$

(Note that the positive direction of the acceleration falls out of our mathematical analysis.) Then we turn to the kinematic equations. If we recall that we neither know nor care to know how much time it takes for the electron to pass through the electric field, then it becomes clear that we can solve for the final velocity using the kinematic equation that omits time:

$$v_f^2 = v_i^2 + 2a\Delta x$$
$$v_f = \left(v_i^2 + 2ad\right)^{1/2}$$
$$v_f = \left(\left(5.0 \times 10^5\frac{\text{m}}{\text{s}}\right)^2 + 2\left(1.76 \times 10^{13}\frac{\text{m}}{\text{s}^2}\right)(2.5 \text{ m})\right)^{1/2}$$
$$v_f = 9.4 \times 10^6\frac{\text{m}}{\text{s}}$$

For (b), we refer to the work-energy theorem and recall how to define work in terms of a constant force applied over some displacement (Equations (4.6) and (4.8)):

$$W = \vec{F} \cdot \vec{d} = \Delta K$$

If we rewrite the electric force in terms of the electron charge and the electric field E and note that F and d point in the same direction (so $\vec{F} \cdot \vec{d} = Fd$), we find

$$Fd = \frac{1}{2}mv_f^2 - \frac{1}{2}mv_i^2$$
$$eEd = \frac{1}{2}mv_f^2 - \frac{1}{2}mv_i^2$$
$$v_f = \left(\frac{2eEd}{m} + v_i^2\right)^{1/2}$$
$$v_f = \left(2ad + v_i^2\right)^{1/2}$$

I'll stop here and simply note that this equation is identical to what we worked out in the context of our kinematics analysis; clearly we'll get the same numerical result.

Example 9.9 Force and Torque on an Electric Dipole ★★★

Consider an electric dipole consisting of two charges of magnitudes $-Q$ and Q connected by a light rigid rod of length L. The dipole is placed in a uniform electric field as shown in Figure 9.19. (a) What is the net force on the dipole? (b) What is the net torque on the dipole? (c) Demonstrate that if the dipole is rotated 90° counter-clockwise, then there is no net force or net torque acting on the dipole. (d) How much rotational kinetic energy will the dipole have when it has rotated 90°, assuming it is released from rest?

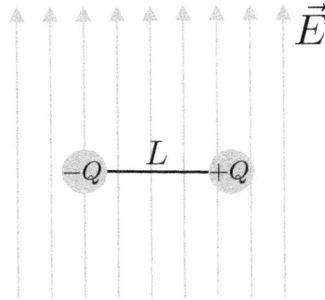

FIGURE 9.19

(a) The positive charge will feel a force $\vec{F} = Q\vec{E}$, which is to say a force with magnitude QE directed up the page. Meanwhile the negative charge will feel a force with equal magnitude but *down* the page (because negative charges feel a force in the direction opposite the electric field). The net force on the entire dipole is the sum of these two forces, i.e. $\vec{F}_{\text{NET}} = \vec{F}_Q + \vec{F}_{-Q} = 0$.

(b) While the *forces* cancel, the *torques* add: they both tend to rotate the dipole in a counter-clockwise fashion (Fig. 9.23). We will use the center of the dipole as the reference point, so the net torque is given by

$$\tau_{\text{NET}} = \sum \vec{\tau}$$
$$\tau_{\text{NET}} = (F_Q)\frac{L}{2} + (F_{-Q})\frac{L}{2}$$
$$\tau_{\text{NET}} = 2QE\frac{L}{2} = QEL$$

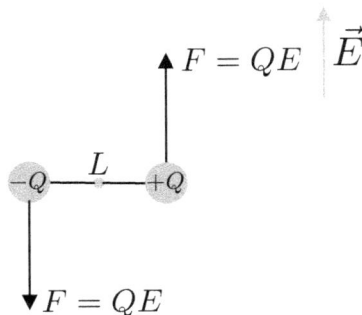

FIGURE 9.20
The forces acting on dipole in a uniform electric field directed up the page (only one vector is drawn for $\vec{E}$ just to prevent the diagram from being too cluttered).

(c) Consider first the fact that as soon as the dipole begins to rotate, the magnitude of the torque begins to decrease: the forces will always be directed *up* the page for the $+Q$ charge and *down* the page for the $-Q$ charge, but only the component of the force perpendicular to the lever arm contributes to the torque (Fig. 9.21). As an extreme case, when the dipole has rotated 90°, the forces are entirely parallel to the lever arm; it follows that there is no net torque on the dipole.

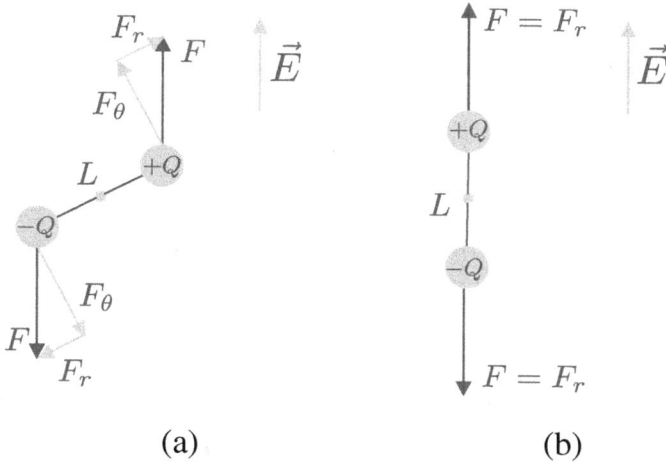

(a)   (b)

FIGURE 9.21
(a) As the dipole begins to rotate, the torque decreases because only a component of the force on each charge is perpendicular to the lever arm (i.e. in the $\hat{\theta}$ direction). (b) When the dipole is aligned with the field, there is no force component in the $\hat{\theta}$ direction and so there is no torque due to the electric forces.

(d) We're asked here to think about the kinetic energy picked up by the dipole as it rotates. This should call to mind the work-energy theorem, which we also considered in the preceding example. The most general equation for the work done by a force is given by Equation (4.7):

$$W = \int \vec{F} \cdot d\vec{r}$$

In the previous example we saw that the electric field and the displacement vector were both *constant*, so the integral simplified to $W = \int \vec{F} \cdot d\vec{r} = \vec{F} \cdot \vec{d}$ where $\vec{d}$ was the displacement of the charge.

Here, the electric field is still constant but the *displacement* vector is not because each charge takes a *curved* path. The dot product means that we want the displacement *parallel* (or antiparallel) to the force for each little step inside the integral. Because the electric field is constant and vertical, this means we only need to identify the *net vertical displacement*. (If the particle's path was curved *and* we had an electric field that was changing as the particle moved, we'd have to evaluate the dot product on every infinitesimal step of the integral.)

If you refer back to Figure 9.9 you'll see that because the dipole has rotated $90°$, each charge has moved a vertical distance $L/2$. Thus for each charge we have

$$W = \int \vec{F} \cdot d\vec{r} = QE\left(\frac{L}{2}\right)$$

and so the total work is just

$$W_{\text{TOT}} = 2W = QEL$$

According to the work-energy theorem, this total work equals the change in kinetic energy:

$$W = \Delta K$$
$$QEL = K_f - K_i$$
$$QEL = K_f$$

In the last step I noted that the initial kinetic energy is 0 J since the dipole starts from rest.

If this all seems very abstract to you, consider the fact that what we've analyzed here is very similar to attaching a string to a post on one end and to a ball on the other, then pulling the ball horizontal, releasing it, and analyzing its kinetic energy when it swings through the vertical position (I'm describing Problem 4.45). The answer is straightforward from energy conservation:

$$\Delta U = -\Delta K \implies mgh = K_f$$

The result here is directly analogous, where we have swapped the mass for the charge, the gravitational field g for the electric field E, and the height h for the dipole separation L (really each charge swings through half of L, but because there are two charges the result is doubled).

In writing down the solution for the ball and string analogy, we've invoked conservation of energy and, in particular, the conversion of gravitational potential energy to kinetic energy. This raises an interesting question: is there an analogous *electric* potential energy that simplifies the problem we've just analyzed? The answer is *yes*, as we shall see in the following section.

9.3   Electric Potential Energy and Electric Potential

If a charged particle of mass m and charge q is immersed in an electric field, it will experience a force $\vec{F} = q\vec{E}$. If the particle moves (for instance, if there are no other forces then its acceleration is just $\vec{a} = q\vec{E}/m$) we can calculate the work done by the electric force (Equation (4.7)):

$$W = \int \vec{F} \cdot d\vec{r} = q \int \vec{E} \cdot d\vec{r} \tag{9.12}$$

Before proceeding, I would like to pause to once again draw a close analogy to gravity: I am going to restate some familiar results in a slightly different way, in order to introduce some ideas that will be very useful when we return to work in the context of electric fields.

The Analogy to Gravitation

If we release a ball of mass m near the surface of the Earth, then it will experience a force $\vec{F} = m\vec{g}$ and at any height y it has gravitational potential energy $U_g = mgy$. Now, just as the gravitational field tells us the gravitational *force* an object would feel at any location (i.e. $\vec{g} = \vec{F}_g/m$, or analogously for an electric field, $\vec{E} = \vec{F}_E/q$), we introduce the **gravitational potential** to tell us the *energy* an object would have at any location:[7]

$$V_g = gy = \frac{U_g}{m} \tag{9.13}$$

Put another way, if we know the gravitational potential in some region, then it is straightforward to determine the energy of a mass m:

$$U_g = mV_g \tag{9.14}$$

From Equation (9.13) it is also possible to express how the gravitational potential changes with the position (call up from the surface of the earth $\hat{y}$):

$$\vec{g}_{\text{avg}} = -\frac{\Delta V_g}{\Delta y}\hat{y} \implies \vec{g} = -\frac{dV_g}{dy}\hat{y} \tag{9.15}$$

The negative sign here accounts for the fact that the gravitational field points *down*, but the gravitational potential increases as we move *up*. I am providing here both the *average* gravitational field (corresponding to a finite step Δy and the corresponding changing in gravitational potential) and the *exact* gravitational field in the limit of an infinitesimally small displacement dy.

I would like to emphasize a few points here:

- While $\vec{g}$ is a *vector field*, the gravitational potential V_g is a **scalar field**: it defines a scalar value everywhere in space. This comes directly from the fact that the gravitational field is used to determine the *force* (a vector) an object experiences, while the gravitational potential is used to determine what *energy* (a scalar) it has.[8]

- The gravitational potential energy only changes if the displacement of the mass has a *component* that is either up or down (i.e. parallel or antiparallel to the gravitational field): any component of the displacement perpendicular to the field has no effect on the gravitational potential energy. Thus if we mark off positions that are perpendicular to the gravitational field $\vec{g}$ (i.e. if we mark off horizontal lines), we know that moving from any position on one line to any position on another line results in the *same* change in gravitational potential energy. We therefore call these lines **equipotentials** (Fig. 9.22).

- As noted above, the gravitational field points down (in the $-\hat{y}$ direction), while the gravitational field increases in magnitude as you move up (i.e. in the $+\hat{y}$ direction). Thus we say that the field points in the direction of decreasing potential.

- As a direct result of the preceding point, we also say that a mass feels a force in the direction of decreasing potential.

Hopefully all of this analysis of a constant *gravitational* field strikes you as familiar (and perhaps even redundant). Let's map this analysis over to the case of a constant *electric* field.

---

[7]Notice I didn't say "gravitational potential *energy*". I know the similarity in the terms are confusing but I am afraid this is standard notation.

[8]Later we'll consider the combination of multiple potentials, so now is as good a time as any to point out that because we're talking about scalar fields, we combine them with a scalar sum: $V_{\text{TOT}} = V_1 + V_2 + ...$ This is in contrast to the *vector* sum we're concerned with for vector fields.

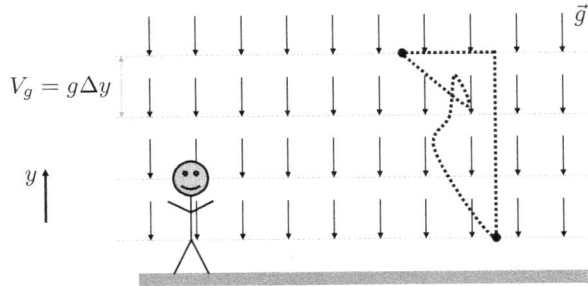

FIGURE 9.22

A gravitational field near the surface of the earth. The horizontal dashed lines represent *equipotentials*, i.e. locations of constant gravitational potential $V_g = gy$ and therefore constant gravitational potential energy for a mass m, $U_g = mV_g = mgy$. The thick dotted lines show two possible ways to move a mass between two positions. These paths make no difference to the overall change in gravitational potential (or gravitational potential energy); only the initial and final y coordinate matters.

Electric Potential in a Constant Electric Field

It isn't hard to find an example of a (relatively) constant electric field in nature: near the surface of the Earth there is a downward electric field (its magnitude is about 100 N/C), just as there is a downward gravitational field! Thus, if we think of some positive charge q near the surface of the Earth, it will experience a downward force $\vec{F}_E = q\vec{E}$ and as it moves the electric field will do work, resulting in a decrease in **electric potential energy** U. By directly comparing to Equations (9.14)–(9.15), we find

$$U = qV \tag{9.16}$$

$$\Delta U = q\Delta V \tag{9.17}$$

$$\vec{E}_{\text{avg}} = -\frac{\Delta V}{\Delta r}\hat{r} \implies \vec{E} = -\frac{dV}{dr}\hat{r} \tag{9.18}$$

In these equations, V is the **electric potential** and I am using the coordinate r to represent changes in position to account for the fact that, generally speaking, electric fields can be oriented arbitrarily in space.[9] The units of electric potential are "energy per charge", or J/C. 1 J/C is equivalent to 1 volt (1 V), which is the SI unit of electric potential.[10] We see from Equation (9.18) that an equivalent way of expressing the units of an electric field is in "volts per meter", or V/m.

The points I emphasized in the list above carry over directly, though we must keep track of the fact that charge can be positive or negative while mass is only positive. In brief:

- $\vec{E}$ is a vector field, while V is a scalar field.

- The equipotential surfaces are perpendicular to the electric field lines.

- The electric field points in the direction of decreasing electric potential.

---

[9]I could include E subscripts here (e.g. $U_E = qV_E$) but I am leaving these off to keep the equations a bit simpler. We'll be focusing on the electric case from now and the meaning should be clear from context.
[10]After Alessandro Volta (1745–1827).

- A positive charge feels a force in the direction of *decreasing* electric potential, while a negative charge feels a force in the direction of *increasing* electric potential.[11]

As we shall see, these results are *always* true, even when the electric field is far more complicated. Before considering more complicated fields, though, let's apply what we've done so far.

Example 9.10 Potential Difference of a Charge Moving in a Field ★
An electron is located near the surface of the Earth, where there is an electric field of 1.00×10^2 V/m directed down. If it starts from rest what electric potential difference will it have moved through after it has traveled 3.0 m? How much kinetic energy will it have?

We'll start from Equation (9.18) and, for the moment, worry only about magnitudes by considering the absolute value of both sides of the equation:

$$\left| \vec{E} \right| = \left| \frac{\Delta V}{\Delta r} \right|$$

$$|\Delta V| = \left| \vec{E} \Delta r \right|$$

$$|\Delta V| = \left| \left(1.00 \times 10^2 \frac{V}{m} \right) (3.0 \text{ m}) \right|$$

$$|\Delta V| = 3.0 \times 10^2 \text{ V}$$

The direction of motion is *up*, since the negatively charged electron feels a force opposite the downward electric field. This also corresponds to the direction of increasing electric potential.

Knowing the change in electric potential allows us to determine the change in electric potential energy (Equation (9.17)):

$$\Delta U = q \Delta V$$

$$\Delta U = (-e) \left(3.0 \times 10^2 \text{ V} \right)$$

If you insert $e = 1.60 \times 10^{-19}$ C into this equation, you'll find

$$U = -4.8 \times 10^{-17} \text{ J}$$

However, when we're talking about the energy associated with small charges (and it doesn't get any smaller than a single electron!), we sometimes use the **electron volt** (eV) as a unit of energy. One electron volt is simply the amount of energy an electron picks up when it moves through a potential difference of 1 volt. Thus we have

$$\Delta U = (-e) \left(3.0 \times 10^2 \text{ V} \right) = -3.0 \times 10^2 \text{ eV}$$

Hopefully you see the appeal of these units: you simply "absorb" the factor of e into the potential to convert to potential energy. More formally, the conversion between electron volts and Joules is

$$1 \text{ eV} = 1.60 \times 10^{-19} \text{ J}$$

---

[11]This follows because $U = q \Delta V$ and the electric force points in the direction of decreasing U. For a positive q this means the energy decreases as V decreases, but for a negative q the energy decreases as V increases.

However we express this change in potential energy, the fact that the electron experiences a *decrease* in potential energy means that it experiences an equivalent *increase* in kinetic energy. If we note that $K_i = 0$ eV $= 0$ J because it starts from rest, then

$$\Delta K = K_f = -\Delta U = 3.0 \times 10^2 \text{ eV} = 4.8 \times 10^{-17} \text{ J}$$

Example 9.11 Energy of a Rotating Dipole   ⋆
Consider part (d) of Example 9.9 again: A dipole is initially perpendicular to a constant electric field and then is allowed to rotate $90°$. How much kinetic energy does it gain during the rotation? Answer the question now that we are armed with the concept of electric potential.

If the dipole is of length L, then when the dipole has rotated $90°$, each charge has moved through a potential difference given by

$$|\Delta V| = |E\Delta r| = \frac{EL}{2}$$

The motion for each charge is in its preferred direction (i.e. in the direction of the electric force, so as to decrease its potential energy). Thus, the change in potential energy for each charge is

$$U = -\frac{QEL}{2}$$

and so the total change in potential energy for the dipole is just twice this,

$$U_{\text{TOT}} = -QEL$$

It follows by conservation of energy that this loss of potential energy corresponds to an equal increase in kinetic energy,

$$\Delta K = QEL$$

The geometry for this procedure is shown in Figure 9.23. Note that the "0 point" for electric potential in this problem is arbitrary: all we care about is how potential (energy) *changes*. Thus, here I've set the position of the dipole's midpoint to sit on the $V = 0$ equipotential.

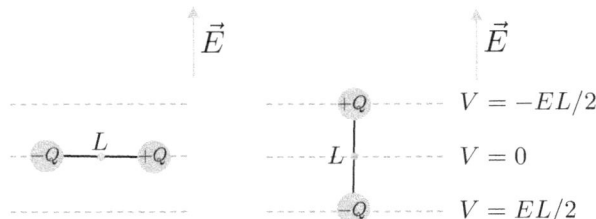

FIGURE 9.23

While we're here considering a rotation of $90°$, where we've found that $U_{\text{TOT}} = -QEL$, we would like a general expression for the potential energy of a dipole in an external electric field for *any* orientation. To express this mathematically we often represent the dipole with a vector called the **dipole moment**, $\vec{p}$:

$$\vec{p} = Q\vec{L} \tag{9.19}$$

where Q is the magnitude of each charge and $\vec{L}$ runs from the negative charge to the positive charge (Fig. 9.24). (The vector $\vec{L}$ is sometimes denoted $\vec{s}$, as well.) With this notation the electric potential energy of a dipole $\vec{p}$ in an external electric field $\vec{E}$ is given by

$$U = -\vec{p} \cdot \vec{E} = -pE\cos\theta \tag{9.20}$$

You are invited to explore the validity of this equation in Problem 9.47. As a quick check, however, note that if $\theta = 0$, then Equation (9.20) states that $U = -pE = -QEL$, as we found above.

FIGURE 9.24
A dipole may be represented by the dipole moment vector $\vec{p}$. The angle formed between $\vec{p}$ and the external field $\vec{E}$ determines the potential energy of the configuration according to Equation (9.20).

If the idea of a dipole moment seems a bit odd to you, consider Figure 9.25, which shows a water molecule consisting of one oxygen atom bonded to two hydrogen atoms (H_2O): the response of the molecule to an external electric field can be neatly summarized by representing the molecule with a single dipole moment with the appropriate magnitude.

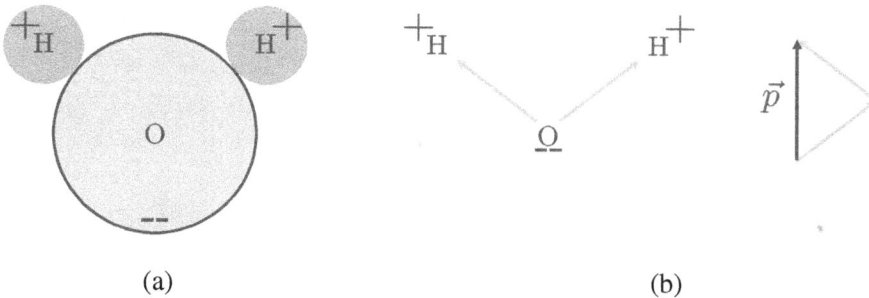

(a)   (b)

FIGURE 9.25
(a) A water molecule represented by circles for each of the three atoms. A positive charge (+) sits near each hydrogen atom and a negative charge (−−) sits on the opposite side of the oxygen atom such that the total charge on the molecule is 0 C. (Middle) The atom can be represented by a pair of equal magnitude dipole moments (gray arrows) from each negative charge to each positive charge. (b) The vector sum of the constituent dipole moments yields the *net* dipole moment $\vec{p}$, which runs along the line of symmetry of the molecule.

Let me make one final comment. If you refer back to Example 9.9, you'll see that we worked through the mathematics necessary to directly relate work done by the electric force to the energy. By working through the general relationships between these quantities (and introducing the concept of electric potential), we are equipped to apply the results directly without going through the derivation time and again in each problem.

Electric Potential in Non-Constant Electric Fields

Thus far we've considered electric potential energy by comparing electric fields with the gravitational field near the surface of the Earth. While this has hopefully helped you build some intuition, the fact is that we deal with electric fields that are *far* more complex than "constant magnitude, constant direction". We need to think through how to handle a more general electric field. While the situation can become quite complicated in the most general case, in this book we'll mostly worry about fields that are constant in *time* and vary in only one *spatial* dimension.

Suppose, then, that that a charge q moves through an electric field that fills some region in space and can vary in the $\hat{r}$ direction, which might refer to a standard Cartesian coordinate (e.g. *up* or *right*) or perhaps to the radial dimension in polar coordinates (e.g. *radially away* from the nucleus of some atom). The work done by the electric field is

$$W = \int \vec{F} \cdot d\vec{r} = q \int \vec{E} \cdot d\vec{r} = \Delta K = -\Delta U \tag{9.21}$$

I factored the q out of the integral because it is a constant, and at the end I have reminded you that work changes kinetic energy (this is the work-energy theorem, where strictly speaking I am referring here to the *total* work) and that because the electric force is conservative and energy is conserved, the electric potential energy necessarily decreases by the same amount. If we pick out just the potential energy term and the integral from this series of equations, we have

$$\Delta U = -q \int \vec{E} \cdot d\vec{r} \tag{9.22}$$

This integral (and the negative sign) defines the change in **electric potential**, ΔV:

$$\Delta V = -\int \vec{E} \cdot d\vec{r} \tag{9.23}$$

Inserting Equation (9.23) into Equation (9.22) yields

$$\Delta U = q\Delta V \tag{9.24}$$

or just

$$U = qV \tag{9.25}$$

as we found above (Equation (9.16)). Note also that if we take the derivative of both sides of Equation (9.23), we find

$$\vec{E} = -\frac{dV}{dr}\hat{r} \tag{9.26}$$

Thus we see that the key points we emphasized above in the context of a *constant* electric field still hold here in the context of a *varying* electric field. In particular, Equation (9.26) says that the electric field is equal to the derivative of the electric potential and points in the local "downhill" direction (Fig. 9.26), which supports the mapping we made for the previous result (Equation (9.18)).

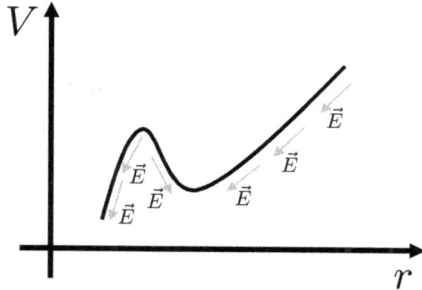

FIGURE 9.26
According to Equation (9.26), the electric field points in the direction of decreasing electric potential. Here we show some arbitrary potential and a series of electric field vectors for varying values of the spatial coordinate r.

The most obvious example of a non-constant electric field is that of a point charge, where $\vec{E} = \left(KQq/r^2 \right) \hat{r}$. In Example 9.12 we'll determine the corresponding electric potential for a point charge, which is of fundamental importance. The following examples make use of this relationship and consider some other important uses of electric potential and electric potential energy.

Example 9.12 Sketching Equipotentials $\qquad\qquad \star\star \int$
Determine the electric potential for (a) a positive point charge q and (b) a negative point charge $-q$. Graph the potentials as functions of r, and make a sketch of the equipotential field lines surrounding the charge in each case.

We'll start from Equation (9.23) and insert the equation for the electric field of a point charge. To be specific, we'll perform the integration from some initial position r_i to some final position r_f:

$$\Delta V = -\int_{r_i}^{r_f} \vec{E} \cdot d\vec{r} = -\int_{r_i}^{r_f} \frac{Kq}{r^2} \hat{r} \cdot d\vec{r}$$

$$= -Kq \int_{r_i}^{r_f} \frac{dr}{r^2}$$

$$= \left. \frac{Kq}{r} \right|_{r_i}^{r_f}$$

$$= Kq \left(\frac{1}{r_f} - \frac{1}{r_i} \right)$$

It follows that if we want the electric potential at a single position (i.e. V) instead of the change in potential between two positions (i.e. ΔV), then we have the **electric potential of a point charge**:

$$V_q = K\frac{q}{r} \qquad\qquad (9.27)$$

If we combine this with $U = qV$, we see that the energy associated with two interacting point charges (q and Q, say) is given by

$$U = K\frac{qQ}{r} \qquad (9.28)$$

These results hold regardless of the sign on the charges, so we've answered both (a) and (b) here. The potential in each case is sketched in Figure 9.27. Note in particular that $V \to 0$ when $r \to \infty$, which hopefully matches your intuition: if a second charge Q is infinitely far away from charge q, then we would *expect* to have no electric potential energy associated with the interaction. Mathematically, $U = QV_q = Q\left(0 \text{ V}\right) = 0 \text{ J}$.

On the other hand, Equation (9.27) says that if $r \to 0$, then $V \to \infty$, which means that it takes *infinite* energy to force two charges (of the same sign) to sit directly on top of one another! Now, in reality even the smallest charges – electrons and protons – have some *size*, so at some point these particles begin to spatially overlap and we need to invoke quantum mechanics to fully understand what is going on.

For our purposes, the point here is that when two charges of the same sign are very close together, there is a lot of potential energy stored in the interaction. The following example explores this.

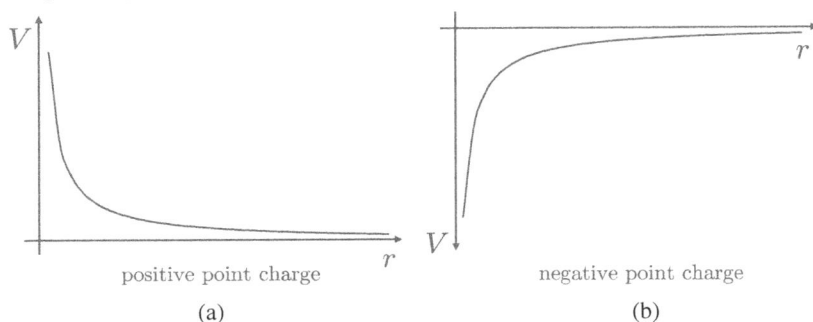

positive point charge

(a)

negative point charge

(b)

FIGURE 9.27
The electric potential of a (a) positive and (b) negative point charge.

Our result (Equation (9.27)) shows that the potential is constant for any value of r, which is to say for any radial distance from the charge. The equipotential lines are just circles (or, in three dimensional space, *spheres*), as shown below.

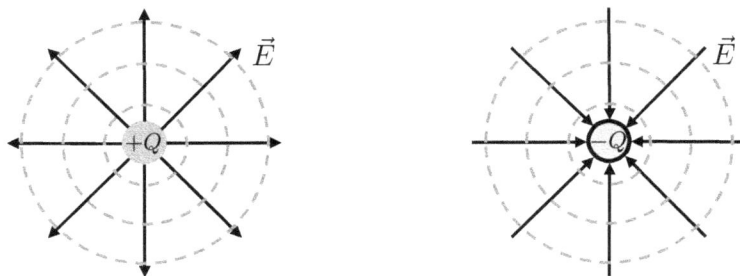

FIGURE 9.28
Equipotential surfaces (dashed lines) and electric field streamlines (solid arrows) around a (left) positive (b) negative point charge.

Example 9.13 Potential Energy in Hydrogen ⋆

A simple model of an atom consists of a positively charged nucleus and electrons that orbit around the nucleus, much like a planet orbits around a star. While the complete quantum-mechanical picture is much more complicated, there is just one piece of additional information that is relevant for us now: the electrons can only exist in discrete states, and each state has a characteristic radius (really the electron lives in a sort of *probability cloud* and the radius we're describing is a most probable radius for the state) and a corresponding electric potential energy for the atom in the state. For hydrogen (one electron orbiting around one proton), the smallest possible radius is called the **Bohr radius**, $a_B = 5.29 \times 10^{-11}$ m. (When the electron orbits at this radius, it is in the so-called *ground state*). How much electric potential energy is stored in the atom in this situation?

The situation is sketched in Figure 9.29. From Equation (9.28) we have

$$U = K\frac{qQ}{r}$$

$$U = -K\frac{e^2}{a_B}$$

$$U = -\left(8.99 \times 10^9 \frac{\text{Nm}^2}{\text{C}^2}\right)\frac{\left(1.60 \times 10^{-19}\ \text{C}\right)^2}{5.29 \times 10^{-11}\ \text{m}}$$

$$U = -4.35 \times 10^{-18}\,\text{J}$$

Or, if we prefer to express the energy in electron volts instead (remember 1 eV $= 1.60 \times 10^{-19}$ J), we find $U = -27.2$ eV.

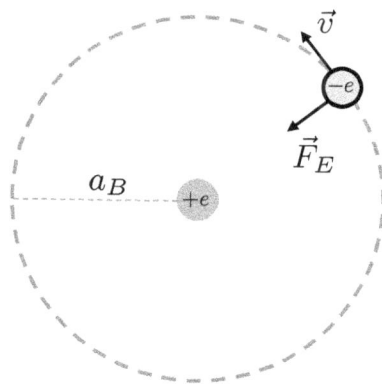

FIGURE 9.29

You might think that we must therefore supply energy equal to 27.2 eV to tear an electron away from the hydrogen nucleus (i.e. to *ionize* it), but this approach doesn't account for the *kinetic* energy of the electron or some important details that come from a more complete quantum mechanical analysis. We'll return to this example in Part V, but for now the main point is that we have correctly analyzed the electric potential energy associated with the hydrogen atom in its ground state.

Example 9.14 Repulsion in Radioactive Decay   ★★

Heavy elements are unstable: their nuclei tend to break apart into smaller and more stable atoms. One mechanism that describes this is **alpha decay**, where two neutrons and two protons break off from the parent nucleus as a single unit (this is just a helium nucleus, though in this context it is often called an **alpha particle**). The two positively charged particles (generically Q for the larger nucleus and $q = 2e$ for the alpha particle) then separate because of their Coulombic repulsion. Consider the alpha decay of uranium 238 (which consists of 92 protons and 146 neutrons) into an alpha particle and a thorium 234 nucleus. If the alpha particle starts a distance $R = 7.4 \times 10^{-15}$ m from the center of the thorium 234 nucleus (approximately its radius) and the thorium 234 nucleus remains essentially motionless, then how fast will the alpha particle be moving when it is very far away?

Before we dive in to the particulars, let me address a question that may have occurred to you in reading the problem: why do we assume the thorium nucleus remains motionless? The answer is *because it has much more mass than the alpha particle* (234 nucleons vs. 4 nucleons). Recall from Equations (4.26)–(4.27) and Problem 4.107 that when two particles elastically interact and one is much more massive than the other, the massive particle essentially acts like a motionless wall.

With this in mind, let's describe the situation from the context of energy conservation. The electric potential energy that arises from the proximity of these particles is converted into the kinetic energy of the alpha particle:

$$U_i + K_i = U_f + K_f$$

$$qV_Q + 0 \text{ J} = 0 \text{ J} + \frac{1}{2}m_\alpha v^2$$

$$K\frac{qQ}{R} = \frac{1}{2}m_\alpha v^2$$

$$\left(2K\frac{qQ}{m_\alpha R}\right)^{1/2} = v$$

In the second line I have described the interaction energy as qV_Q, i.e. as the alpha particle interacting with the electric potential of the thorium nucleus. I could have written this instead as the interaction between the thorium nucleus and the alpha particle (QV_q), but this is just a matter of bookkeeping: plugging in the expression for V gives the same result in the following line regardless of which route we take.

All that remains is plugging in numerical values:

$$v = \left(2\frac{\left(8.99 \times 10^9 \frac{\text{Nm}^2}{\text{C}^2}\right)(2e)(90e)}{(4\,(1.67 \times 10^{-27}\text{ kg}))(7.4 \times 10^{-15}\text{ m})}\right)^{1/2}$$

$$v = 430 \text{ m/s}$$

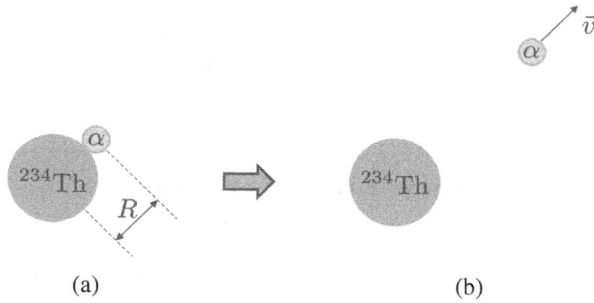

(a)   (b)

FIGURE 9.30
(a) The alpha particle when it first breaks off from the uranium 238 nucleus (which becomes a thorium 234 nucleus as a result). (b) The alpha particle loses electric potential energy and gains kinetic energy as it is repelled from the thorium 234 nucleus.

In the last line I've plugged in $e = 1.60 \times 10^{-19}$ C to obtain the final answer (I left it out above to help keep track of where the values come from). Note also that the alpha particle contains two neutrons and two protons, but because their masses are the same (to three significant figures, at least) we can obtain the mass of the alpha particle as $4m_p$.

Example 9.15 Field Lines and Equipotentials of a Dipole ⋆
Consider an electric dipole: make a sketch of its electric field lines and its equipotential surfaces.

We've already sketched the field lines of a dipole (Example 9.5); what we're adding here is the equipotential surfaces. This is actually rather straightforward if you have the field lines: the equipotential surface is always perpendicular to the field lines, so all we need to do is choose a point out in space and carefully start drawing an equipotential surface perpendicular to the local field lines. Extend the equipotential to either side, curving as necessary to keep it perpendicular to the field, and stop once you've made a complete circuit or you've moved out of the region of interest.

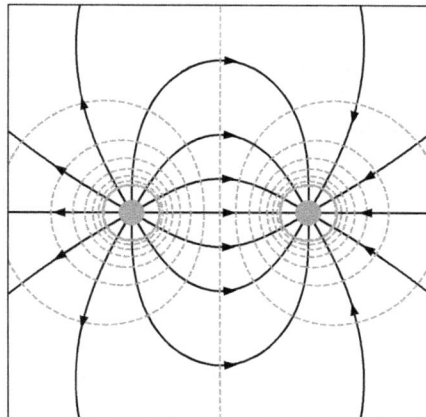

FIGURE 9.31

In Figure 9.31 I have drawn the equipotentials as dashed lines superimposed over the field lines. I've additionally spaced them so the distance between adjacent equipotentials corresponds to the same ΔV. This way you can see that the field is quite large close to the charges, and drops rapidly as you move farther away – as you would expect given that V falls off with the inverse of the radial distance r.

Example 9.16 Moving Charge Near Other Charges ★★

Consider four positive charges Q placed at the corners of a square with edge length L. Determine how much energy would be required to bring a fifth positive charge q from very far away to (a) the center of the square and (b) the middle of one of the edges.

The two situations for (a) and (b) are shown in Figure 9.32.

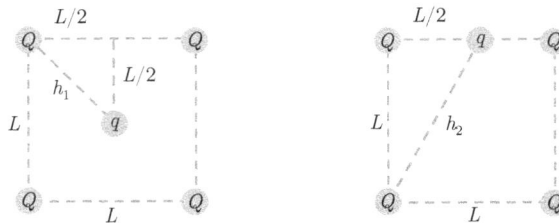

FIGURE 9.32

Wherever the fifth charge ends up, there will be some potential energy involved with its interaction with *each* of the other charges: in the case where all of the charges are positive, then bringing a new charge in involves overcoming the resistive Coulomb force that exists between the new charge and *every other* charge that is already present. The *total* energy will just be the sum of each interaction term:

$$U_{\text{TOT}} = U_1 + U_2 + U_3 + U_4$$

Or, because $U = qV$ for a charge q in a potential field V,

$$U_{\text{TOT}} = qV_1 + qV_2 + qV_3 + qV_4$$
$$U_{\text{TOT}} = qV_{\text{TOT}}$$

In other words, the overall potential field that charge q experiences is just the sum of the fields due to each of the pre-existing charges. Note that this is a *scalar* sum since potential is a scalar; this makes potential fields a bit easier to work with than electric fields, which are *vector* fields.

Each of the constituent potentials (V_1 to V_4) is due to one of the pre-existing charges on the corner of the square, and we know from Equation (9.27) that the potential due to a point charge Q is given by

$$V = K\frac{Q}{r}$$

It follows that the total potential anywhere in space due to the four point charges Q is

$$V_{\text{TOT}} = KQ\left(\frac{1}{r_1} + \frac{1}{r_2} + \frac{1}{r_3} + \frac{1}{r_4}\right)$$

where each r coordinate is the distance from a source charge Q to the location where we're evaluating the overall potential (that is, the final location of the new charge q).

This is a generic result that applies equally to both (a) and (b), and to proceed we need to determine the values for the r coordinates in each situation. In (a) the r values are all the same; the distance is labeled h_1 in Figure 9.16. We see from the geometry that this is the hypotenuse of a right triangle with equal edge lengths $L/2$. From the Pythagorean theorem we have

$$h_1^2 = \left(\frac{L}{2}\right)^2 + \left(\frac{L}{2}\right)^2$$

$$h_1 = \left(2\left(\frac{L}{2}\right)^2\right)^{1/2}$$

$$h_1 = (2)^{1/2}\frac{L}{2}$$

$$h_1 = \frac{L}{2^{1/2}}$$

Inserting this for r_1, r_2, r_3, and r_4 in our equation for V_{TOT}, we have

$$V_{\text{TOT}} = \frac{4\left(2^{1/2}\right)KQ}{L}$$

From which it follows that

$$U_{\text{TOT}} = qV_{\text{TOT}} = \frac{4\left(2^{1/2}\right)KQq}{L}$$

For (b) we must determine V_{TOT} on the middle of an edge. We see that this position is a distance $L/2$ from the two adjacent charges and a distance labeled h_2 from the other two charges. Once again we can use the Pythagorean theorem do determine an expression for h_2 in terms of the edge length L:

$$h_2^2 = \left(\frac{L}{2}\right)^2 + L^2$$

$$h_2 = \left(\frac{5}{4}L^2\right)^{1/2}$$

$$h_2 = \frac{5^{1/2}}{2}L$$

We can then determine V_{TOT}:

$$V_{\text{TOT}} = KQ\left(\frac{1}{r_1} + \frac{1}{r_2} + \frac{1}{r_3} + \frac{1}{r_4}\right)$$

$$V_{\text{TOT}} = KQ\left(\frac{2}{L} + \frac{2}{L} + \frac{2}{5^{1/2}L} + \frac{2}{5^{1/2}L}\right)$$

$$V_{\text{TOT}} = KQ\left(\frac{4}{L} + \frac{4}{5^{1/2}L}\right)$$

$$V_{\text{TOT}} = \frac{4KQ}{L}\left(1 + \frac{1}{5^{1/2}}\right)$$

Once again, knowing the total potential tells us how much energy is required to bring in a charge q from far away to this location:

$$U_{\text{TOT}} = qV_{\text{TOT}} = \frac{4KQq}{L}\left(1 + \frac{1}{5^{1/2}}\right)$$

While this might strike you as a rather esoteric problem, the idea of bringing charged particles together is very important in particle physics experiments where charged particles are deliberately slammed together: you need to know how much (initial kinetic) energy to give to your particles so they come sufficiently close together! In the simplest case we are concerned with just *two* charges.

Example 9.17 Building a Charge Square   ★★
Refer back to the four charges described in Example 9.16: they have equal magnitude Q and sit on the corners of a square of edge length L. How much energy was required to arrange the charges in this way?

First, note that one application of this question is to consider how much electric potential energy is required to *assemble a nucleus*. The geometry in a nucleus is not quite so simple as this, of course, and to properly analyze the situation one must invoke quantum mechanics, but the electric potential energy associated with bringing charges together is certainly an important component of a more thorough analysis.

To proceed, keep in mind the underlying principle that two nearby charges have some interaction energy ($U = qV = KQq/r$). Thus, here we need to identify all of the *unique pairs* of charges, because there is some energy associated with each of those interactions. As we sketch in Figure 9.33, there are a total of 6 such pairs in this situation.

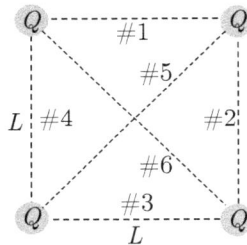

FIGURE 9.33
Each dashed line connects a pair of charges. The lines are labeled (in arbitrary order); there are 6 total.

Now, each of these six interaction terms involves two point charges Q, so the total potential energy will be given by

$$U_{\text{TOT}} = KQ^2\left(\frac{1}{r_1} + \frac{1}{r_2} + \frac{1}{r_3} + \frac{1}{r_4} + \frac{1}{r_5} + \frac{1}{r_6}\right)$$

What are the r values? Well, four of them will be L: this corresponds to interactions 1–4 in Figure 9.33. The other two correspond to the diagonal interactions, which have length $(2)^{1/2}L$ from the Pythagorean theorem, as you can check for yourself.

Plugging in, we have

$$U_{\text{TOT}} = KQ^2 \left(\frac{4}{L} + \frac{2}{(2)^{1/2} L} \right)$$

$$U_{\text{TOT}} = \frac{KQ^2}{L} \left(4 + (2)^{1/2} \right)$$

This is our desired result and we'll stop here. While we will mostly concern ourselves with situations where we can count interaction pairs directly from a diagram, you may be curious about how to do this in general. What we're talking about here is *combinations*, which we encountered in a rather different context back in Equation (5.34) on page 283. Here we've directly enumerated "M choose 2" for $M = 4$ charges.

9.4 Problems for Chapter 9

9.1 An Analogy: Gravitational Fields

★ **Problem 9.1.** Consider the Earth and Moon on a one-dimensional coordinate system where the center of the Earth is at the origin and the Moon is at $x = x_M$, the radius of the Moon's orbit around the Earth. For what value(s) of x is the net gravitational field 0 N/kg? Express your answer(s) (a) in meters and (b) as a fraction of the distance from the surface of the Earth to the surface of the Moon.

★★ **Problem 9.2.** Calculate the magnitude of the gravitational field on the surface of the Earth due to (a) the Earth, (b) the Moon, and (c) the Sun.

★★ **Problem 9.3.** Set up a one-dimensional coordinate system with the Sun at the origin and Earth at x_E, the radius of the Earth's orbit around the Sun. Consider the Moon at (a) $x = x_E + x_M$ and (b) $x = x_E - x_M$ where x_M is the radius of the Moon's orbit around the Earth. In each case, determine the net gravitational field on Earth. (Use your answers from Problem 9.2.)

★★ **Problem 9.4.** Three objects of mass M are on the corners of an equilateral triangle of edge length L. Sketch the gravitational field lines inside and outside the triangle. What force will each mass experience?

★★ **Problem 9.5.** Four objects of mass M are on the corners of a square of edge length L. Sketch the gravitational field lines inside and outside the square.

9.2 The Electric Force and the Electric Field

★ **Problem 9.6.** Two charges, of magnitude q and $3q$, are a distance r apart. (a) Write down an expression for the force between them, F. (b) Is the force attractive or repulsive? (c) If the charges are then moved apart until their separation is $4r$, what will the new force be? Express your answer as a multiple of F.

★ **Problem 9.7.** Three charges are arranged in a line: $Q_A = -1.0$ nC is at $x = 0.0$ cm, $Q_B = 4.5$ nC is at $x = 2.0$ cm, and $Q_C = 9.0$ nC is at $x = 3.0$ cm. Sketch the situation, draw a force diagram for each charge, and determine the net force experienced by each charge.

★ **Problem 9.8.** Two equal and opposite charges of magnitude Q are separated, as shown in Figure 9.34. At each position marked with a small dot, draw a vector indicating the force that a third charge of magnitude $+q$ would feel if it were placed at that location. (Don't worry too much about the length of the vector – but get the direction!)

FIGURE 9.34
Problem 9.8

★★   **Problem 9.9.** Two objects of equal mass m each carry 7.5 μC of charge. Initially, they are 0.50 m apart and the magnitude of their acceleration is 1.0 m/s^2.

   a. Will they accelerate *toward* each other or *away* from each other? Explain.

   b. What is m?

   c. Suppose the charge of one of the objects is *quadrupled*. What would the acceleration of each object be in this case?

★   **Problem 9.10.** Honeybees can accumulate electric charge as they fly through the air. Suppose a 1.00×10^2 mg bee carries a 35 pC (i.e. 35×10^{-12} C) charge. The Earth's electric field is about 1.00×10^2 N/C (pointed down) near the surface of the Earth.

   a. What is the electric force (magnitude and direction) experienced by the bee?

   b. What is this force as a fraction of the bee's weight?

   c. Does the bee feel heavier or lighter than when it is uncharged?

★   **Problem 9.11.** The nucleus of a lead atom has 82 protons. Suppose an alpha particle (consisting of two protons and two neutrons) is 5.0×10^{-10} m from the center of a lead nucleus. What is the alpha particle's acceleration at this instant?

★★   **Problem 9.12.** An *alpha particle* is a Helium nucleus: two protons and two neutrons bound together. Suppose that at some instant an electron is a distance r from an alpha particle with a speed v perpendicular to the line that connects the particles.

   a. Sketch the electric field created by the alpha particle.

   b. Determine an expression for the force felt by the electron.

   c. Sketch a few possible paths that the electron might take, depending on the relative values of the parameters (e.g. if v is essentially 0 m/s). Explain.

★   **Problem 9.13.** A metal paperclip has a mass of about 0.50 g. Suppose two paperclips are given equal charges q and separated by 1.0 mm, one above the other. (a) What value of q will allow the upper paperclip to hover in place without being touched? (b) How many electrons are necessary to obtain this charge?

★   **Problem 9.14.** A massless, 6.5 cm long spring is fixed to a wall on one end and is attached to a 3.0 μC charge on the other end. A -4.5μC charge is slowly brought toward the spring along its axis. When the charges are 5.0 cm away from one another, the spring's length is 8.0 cm. What is the spring constant of the spring?

★   **Problem 9.15.** Consider the charge configuration shown in Figure 9.35. Determine the electric field at the point marked with a small dot, which is in the center of the rectangle formed by the four charges.

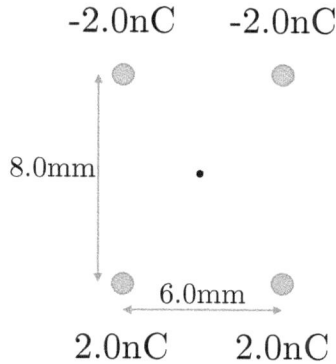

FIGURE 9.35
Problem 9.15

★★ **Problem 9.16.** Consider three point charges of equal magnitude q positioned on the vertices of an equilateral triangle. Sketch this configuration and draw representative electric field vectors in and around the triangle. If the triangle has edge length L, what magnitude force will each charge experience?

★ **Problem 9.17.** A $+15$ nC charge is placed at the origin.

a. Sketch a few representative electric field vectors around the charge.

b. What is the electric field at $(x, y) = (5.0 \text{ cm}, 0.0 \text{ cm})$, in component form?

c. What is the electric field at $(x, y) = (-5.0 \text{ cm}, 5.0 \text{ cm})$, in component form?

d. What is the electric field at $(x, y) = (5.0 \text{ cm}, 5.0 \text{ cm})$, in component form?

★ **Problem 9.18.** A charge with positive magnitude $4q$ is placed at the origin and a charge with positive magnitude q is placed at a position $x_0 > 0$.

a. Is there a position in the region $x < 0$ where the electric field has 0 magnitude? If so, where is it? If not, why not?

b. Repeat part (a) for the region $0 < x < x_0$.

c. Repeat part (a) for the region $x_0 < x$.

★ **Problem 9.19.** What force is felt by charge $+q$ in Figure 9.36? Express your answer in (a) "magnitude and angle" form and (b) "component" form.

★★ **Problem 9.20.** Two 15.0 g beads each carry a charge of 150 nC. They are attached to two 0.75 m-long strings and suspended from a common point on a support rod. The charges repel such that each string forms an angle θ with the vertical (Fig. 9.37). What is θ?

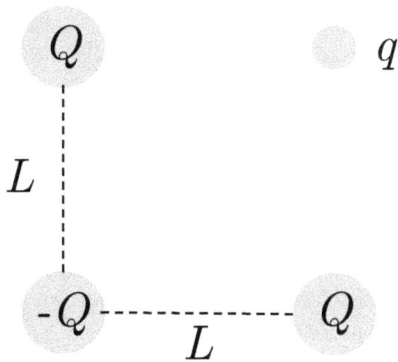

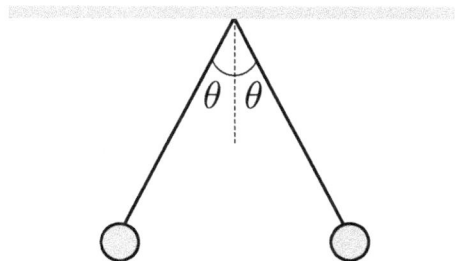

FIGURE 9.36
Problem 9.19

FIGURE 9.37
Problem 9.20

★★ **Problem 9.21.** Three charges are located on the corners of a right triangle, as shown in Figure 9.38. The force on the top charge has magnitude $F = 1.5$ N. What is the numerical value of charge q?

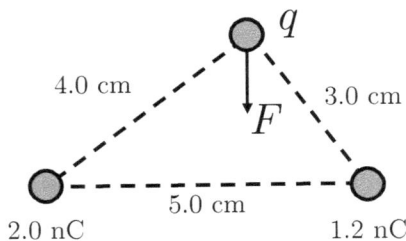

FIGURE 9.38
Problem 9.21

★★ **Problem 9.22.** A proton moving in the $+x$ direction with a speed 1.2×10^6 m/s enters a uniform electric field $\vec{E} = 1.3$ N/C$\hat{x}$. If the electric field has a total width of 3.0 m, will the proton emerge on the other side of the field? If yes, what speed will it have? If not, where will it stop?

★★ **Problem 9.23.** A ball with charge Q and mass m is hanging vertically from a thin insulating thread when a horizontal electric field with magnitude E is turned on. Sketch the new equilibrium position of the ball. If its original angular position was $\theta = 0°$, what is the new equilibrium angular position?

★★ **Problem 9.24.** A point charge with magnitude $+q$ is fixed at the origin of a coordinate system. A point charge with magnitude $-5q$ is fixed a distance r from the origin along the positive x axis.

  a. Determine the position(s) along the x axis between the charges where the electric field is 0 N/C. If there are no such positions, briefly explain why.

  b. Repeat (a) for the position(s) where $x < 0$.

  c. Repeat (a) for the position(s) where $x > r$.

★★ **Problem 9.25.** Two positive charges of magnitudes $4Q$ and $3Q$ are separated by a distance d, as shown in Figure 9.39. Where, if anywhere, on the line through the charges is the electric field 0 N/C? Include a sketch of E vs. x and explain your reasoning with words as well as mathematically. Be sure to consider each of the three regions $x < 0$, $0 < x < d$, and $x > d$.

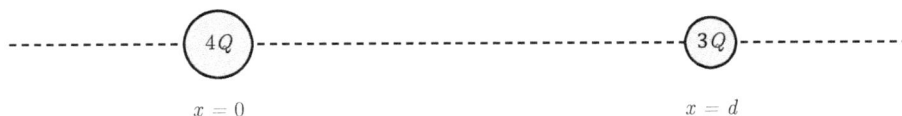

FIGURE 9.39
Problem 9.25

★★ **Problem 9.26.** A $m = 15.0$ g ball carries charge of $q = 2.0 \times 10^2$ nC. It is suspended from the ceiling by an insulating wire. A second ball that carries a charge $Q = 120$ nC is attached to an insulating rod and is slowly brought toward the hanging ball, as shown in Figure 9.40.

  a. Draw a force diagram for the hanging ball.

  b. What electric field is the hanging ball experiencing? Provide a numerical value and a direction.

  c. What is θ?

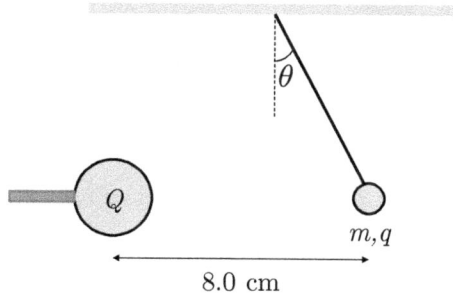

FIGURE 9.40
Problem 9.26

★★ **Problem 9.27.** Three 8.0 nC charges are arranged as shown in Figure 9.41.

a. What is the electric field at the location indicated?

b. If a 1.5 nC charge was placed at the location indicated, what force would it experience?

c. If β-boy delivers a total of 2.0 C, how many electrons does γ-girl absorb?

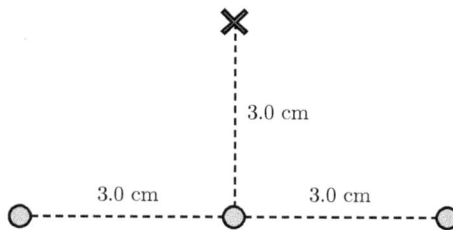

FIGURE 9.41
Problem 9.27

★★ **Problem 9.28.** Suppose a hydrogen atom consists of a stationary proton and an electron moving at some speed v in a circular orbit of radius r. If $r = 5.29 \times 10^{-11}$ m, what is v? Give a numerical answer in m/s.

★★ **Problem 9.29.** A proton enters a parallel plate capacitor as shown in Figure 9.42. Suppose $v = 5.0 \times 10^5$ m/s. The electric field between the plates of a capacitor is constant and points from the positive plate toward the negative plate.[12]

a. If the proton is to exit the right side of the capacitor with an upward velocity such that its *total* velocity forms an angle of 10.0° with the horizontal, then what is the strength of the electric field between the capacitor plates?

b. What vertical distance will the proton cover as it travels through the capacitor? Is it possible for it to exit the capacitor as described in (a), or will it collide with one of the capacitor plates first?

★★ **Problem 9.30.** A metal sphere of radius R and positive charge Q is held directly below a small bead of mass m and charge q. The bead is released from rest a height h above the top of the sphere, as shown in Figure 9.43.

---

[12]The field is more complicated near the edges of the capacitor, but here we'll ignore those for the sake of simplicity. Thus we are considering a so-called *ideal capacitor*.

a. Draw a force diagram for the bead at the instant it is released.

b. Use your answer from (a) to describe the conditions necessary if the bead is to begin falling toward the sphere when it is released.

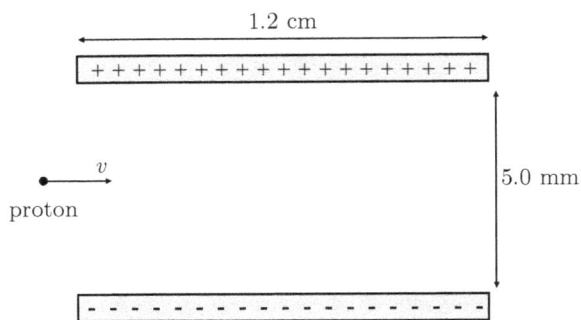

FIGURE 9.42
Problem 9.29

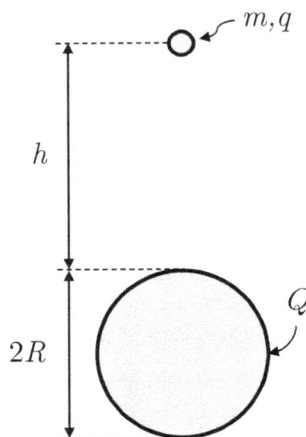

FIGURE 9.43
Problem 9.30

★★   **Problem 9.31.** Air filtration systems (and N95 masks) make use of charged fibers to attract particles out of the air. Consider a long straight fiber with a uniform charge density $\lambda > 0$.[13] In Figure 9.44 a tiny object–for example, a speck of dust or airborne virus–is drawn as a dipole with charges $\pm q$ and length d; it has total mass m and is initially at rest with its midpoint a distance z from the fiber.

a. Why is it reasonable to consider the object to be a dipole? Explain in a sentence or two.

b. The figure does not show which side of the dipole carries the positive charge and which carries the negative charge. Sketch the situation with the appropriate labels and explain.

c. Determine an expression for the *magnitude* of the object's acceleration at the instant shown.

d. Consider your answer to (c) in the situation where $d \ll z$. What does this mean, physically?

e. Determine the *direction* of the object's acceleration at the instant shown. Explain.

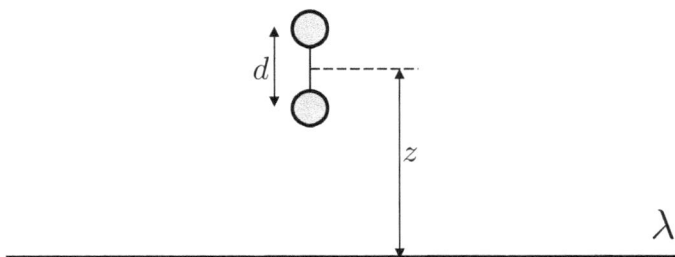

FIGURE 9.44
Problem 9.31

---

[13]We will find in Equation (10.30) that the magnitude of an electric field generated by such a fiber obeys $E = 2K\lambda/r$ where r is the perpendicular distance from the wire. λ is the amount of charged packed onto a given length of wire, $\lambda = Q/L$.

★★ **Problem 9.32.** A dipole consists of two point charges $\pm q$ separated by a distance s. The dipole is fixed in place while a point charge Q is positioned to sit on the plane that bisects the dipole, a distance $r \gg s$ from its center. The dipole is then released. What is the net torque experienced by the dipole?

★★ **Problem 9.33.** The dipole moment $(\vec{p} = q\vec{L})$ of a water molecule has magnitude 6.1×10^{-30} Cm. If a water molecule is placed in a uniform electric field with magnitude 150 N/C, what is (a) the maximum and (b) the minimum torque it can experience? Explain the orientation of the dipole in the field in each case. (c) The moment of inertia of a water molecule is about $I = 2 \times 10^{-47}$ kg m^2. Do you expect that a water molecule will align with the Earth's electric field?

★★ **Problem 9.34.** A long line of charge carries a positive charge density λ (see the footnote to Problem 9.31) and is held vertically near a small ball of mass m and charge $-q$. The ball is suspended from a rod by a thin string of length L, as shown in Figure 9.45. The distance between the end of the rod and the line of charge is d and the angle formed between the string and the vertical is θ.

   a. If d is increased, do you expect θ to increase or decrease? Explain.

   b. Draw all relevant forces acting on the ball and write out two Newton's 2nd Law equations (one for the vertical dimension and one for the horizontal dimension) for the ball.

   c. Determine an equation involving θ, d, and other given parameters and fundamental constants, as needed.

   d. Is the equation you found in (c) consistent with your prediction in (a)? Explain.

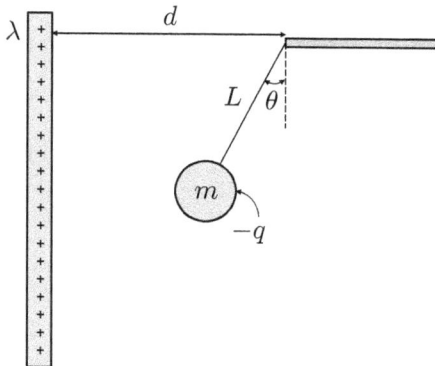

FIGURE 9.45
Problem 9.34

★★ **Problem 9.35.** A large metal ball has a radius R and an unknown charge Q. To measure the charge, you attach a small metal bead with mass m and charge $-q$ to a string of length L. The bead swings in toward the ball, forming an angle θ with the vertical, as shown in Figure 9.46; the bead is a horizontal distance r from the center of the ball. (All of these parameters are known except for Q.)

   a Draw a force diagram for all of the forces acting on the bead.

   b Determine an expression for the electric field experienced by the bead. Your answer should be in terms of some or all of m, g, q, L, and θ.

   c If the metal ball is treated as a point charge, then what charge does it carry? (Use your result from (b).)

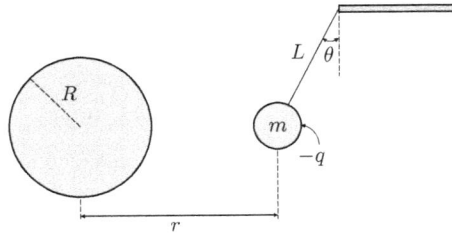

FIGURE 9.46
Problem 9.35

Problem 9.36. A plane of charge has surface area A and charge Q.[14] It is held very close to a small piece of paper. We consider the paper molecules to be tiny dipoles that, at least at first, have random orientations. Draw a force diagram for both dipoles sketched in Figure 9.47 and determine an equation for the net force and net torque on each.

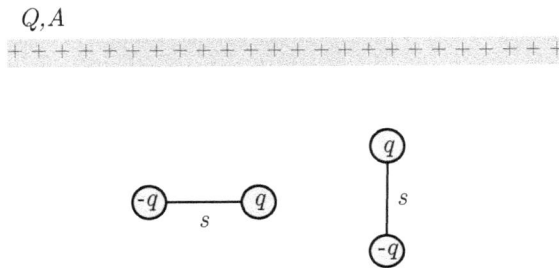

FIGURE 9.47
Problem 9.36

Problem 9.37. An electric dipole is of foundational importance for understanding how an antenna works. Here, we will consider the electric field created by a dipole with charges $\pm q$ separated by a distance s. In Figure 9.48 the three "x" positions are at angular positions $0°$, $45°$, and $90°$ measured counterclockwise from the $+x$ axis.

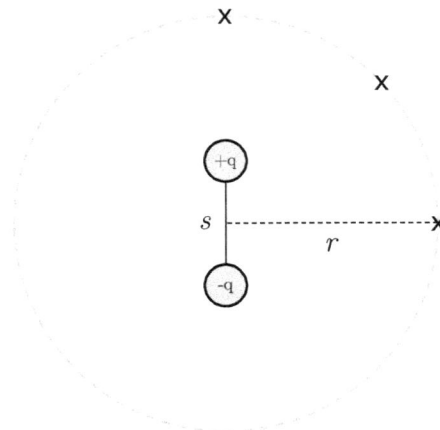

FIGURE 9.48
Problem 9.37

---

[14]We will see in Equation (10.25) that the electric field of such a plane obeys $E = Q/(2A\varepsilon_0)$.

a. Sketch the electric field due to each charge at each of the locations marked with an "x". (Just the relative length and orientation matters.)

b. Use your sketches from (a) to also sketch the *net* electric field at each of the marked locations. Which of the three positions has the largest magnitude? Explain your reasoning in words.

c. Give an algebraic expression for the magnitude of the electric field at the rightmost position.

d. Give an algebraic expression for the magnitude of the electric field at the top position.

e. Give an algebraic expression for the magnitude of the electric field at the middle position.

Problem 9.38. Three identical dipoles are immersed in an electric field, as shown in Figure 9.49. Sketch the figure and draw force vectors on the charges. Indicate if each dipole will rotate clockwise, counterclockwise, or will experience no torque. Rank the magnitudes of the net force acting on the dipoles and the magnitudes of the net torque acting on the dipoles. Briefly justify your answers.

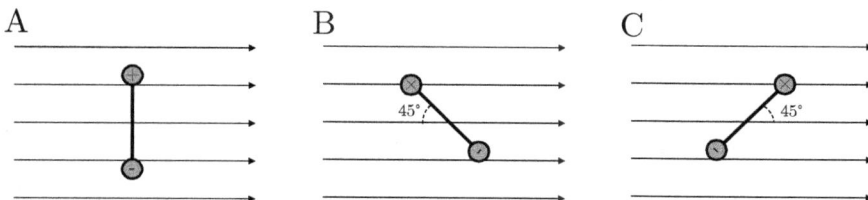

FIGURE 9.49
Problem 9.38

9.3 Electric Potential Energy and Electric Potential

Problem 9.39. Two particles with mass M and charge Q are arranged at the edges of an equilateral triangle of edge length L along with a third charge of mass m and charge q, as shown in Figure 9.50.

a. Determine an algebraic expression for the force experienced by charge q.

b. If $m \ll M$ and the charges are free to move, how fast will charge q be moving when it is far from the other two charges?

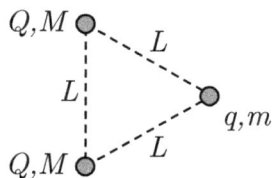

FIGURE 9.50
Problem 9.39

Problem 9.40. How much energy do you need to take two charges of equal magnitude that are initially very far away and place them directly on top of one another? Explain.

Problem 9.41. The nefarious superhero villain β-boy is battling the heroic γ-girl.12 β-boy is attempting to shoot a stream of electrons (i.e. a bolt of lightning) at γ-girl. For lightning to arc through the air, an electric field with a magnitude of about 3.0 million N/C is needed. Suppose β-boy has generated such a field and γ-girl is 2.5 m away from his fingers.13

a. Should the electric field be pointing toward γ-girl or away from her if an electron is to travel from β-boy's fingers to γ-girl? Explain.

b. What is the change in electric potential energy of an electron that moves from β-boy's fingers to γ-girl?

★★   **Problem 9.42.** How much energy do you need to assemble an equilateral triangle of edge length L with equal charges $+q$ on each vertex?

★   **Problem 9.43.** Consider the region of space shown in Figure 9.51, where there is a constant electric field of magnitude E directed up the page. Points 1 and 2 are separated by a distance a and points 1 and 3 are separated by a distance b. Determine expressions for ΔV_{12}, ΔV_{13}, and ΔV_{23}.

★★   **Problem 9.44.** Consider the region of space shown in Figure 9.52, where the dashed curves are equipotentials. Suppose a proton is moving along the following paths, assuming a "straight line" path unless otherwise noted. For each, determine the proton's change in electric potential energy. Report your answer both in Joules and in **electron volts**, where 1 eV is the energy required to move a charge of magnitude e through 1 volt.

   a. $a \rightarrow b$

   b. $a \rightarrow b \rightarrow c$, where the second part of the path is along the equipotential.

   c. $d \rightarrow e$, along the equipotential.

   d. $d \rightarrow c \rightarrow e$

   e. $e \rightarrow f$

   f. The complete path from a through every point up to and including f, then back to a. Suppose $b \rightarrow c$, $d \rightarrow e$, and $f \rightarrow a$ are along the equipotentials.

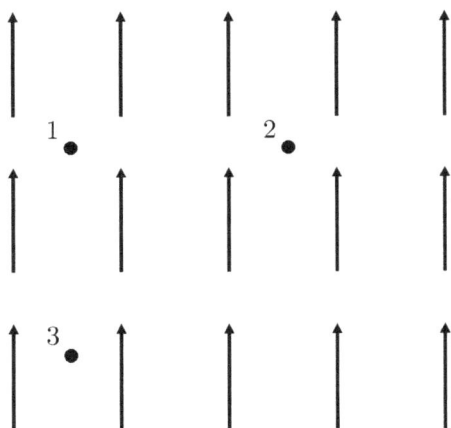

FIGURE 9.51
Problem 9.43

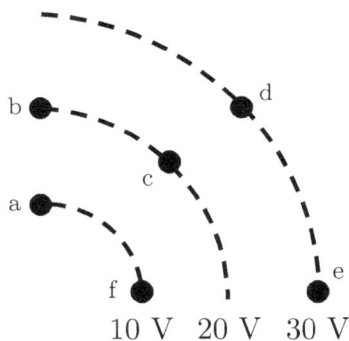

FIGURE 9.52
Problem 9.44

★★   **Problem 9.45.** Use the work energy theorem and the definition of electric force to verify Equation (9.20).

★★   **Problem 9.46.** Two positive point charges of magnitude $+q$ are located on the y axis at $\pm\frac{1}{2}s$. Determine an expression for the electric potential anywhere on the x axis, and sketch a graph of V vs. x.

★★   **Problem 9.47.** Consider a charge $q = 4.2$ nC. What is the electric potential at each of the distances $r_A = 2.0$ cm and $r_B = 4.0$ cm from the charge? What is the electric potential difference between the points, $V_B - V_A$? Explain the sign of your answer.

★   **Problem 9.48.** Consider a constant electric field $\vec{E} = 5.0\hat{x}$ N/C. What is the change in electric potential between $x = -2.0$ cm and $x = 3.0$ cm?

★   **Problem 9.49.** A proton's speed as it passes a point in space where the electric potential is 15 V is 5.0×10^4 m/s. If it later passes a point in space where the electric potential is -25 V, what speed will it have?

★ **Problem 9.50.** A charge $2Q$ is at $x = 0$ cm and a charge $-Q$ is at $x = 3.0$ cm. Where on the x axis is the electric potential 0 V?

★ **Problem 9.51.** A charge $q_A = 2.0$ nC is at $(x, y) = (0, 0)$ and a charge $q_B = -3.0$ nC is at $(x, y) = (4.0, 0)$ cm. What is the electric potential at $(x, y) = (4.0, 3.0)$ cm? Include a sketch!

★★ **Problem 9.52.** In a super-exciting superhero battle, a hero grabs two metal balls and holds them 1.0 cm apart. Using his superhero powers, he delivers 1.0 C of charge to each 1.0 kg ball. He releases them and they fly apart. How fast will they be going when they're very far from one another? (Neglect gravity.)

★★ **Problem 9.53.** In proton beam therapy, protons are shot into cancerous tumors to break apart the tumor's DNA and kill the cells. Suppose it is desired to deliver 0.10 J to a tumor, and protons are accelerated through a 1.0×10^5 kV potential difference. (a) What is the total charge of the protons that must be fired at the tumor? (b) How many protons is this?

★★ **Problem 9.54.** Consider the three charges in Figure 9.53. What is the electric potential at the fourth corner of the rectangle?

★★ **Problem 9.55.** A proton is shot directly at a gold nucleus, which has a charge of $Q = 79e$. If the proton has an initial speed of $v = 5.0 \times 10^4$ m/s, how close will it get to the nucleus?

★★ **Problem 9.56.** You measure the electric potential at three locations: $V(x = 0.0$ cm$) = -20.$ V, $V(x = 4.0$ cm$) = 30.$ V, and $V(x = 10.$ cm$) = 0$ V. Sketch $V(x)$ for 0 cm $< x <$ 10 cm and determine a value for the electric potential at $x = 2.0$ cm and $x = 8.0$ cm. Assume the electric potential varies linearly in the regions between your measurements.

★★ **Problem 9.57.** Repeat Problem 9.24 but this time consider the electric *potential* rather than the electric *field*.

★★ **Problem 9.58.** Repeat Problem 9.25 but this time consider the electric *potential* rather than the electric *field*.

★★ **Problem 9.59.** Three protons are located on the corners of an equilateral triangle of edge length 2.0 nm.

  a. How much energy was required to assemble the charges in this position?

  b. What net force is acting on any one proton? Sketch the triangle and indicate the direction as well as the magnitude.

  c. If the protons are free to move, how fast will they be moving a long time after they are released?

★★ **Problem 9.60.** What is the energy associated with the charge configuration shown in Figure 9.54? The charges are on the vertices of an equilateral triangle.

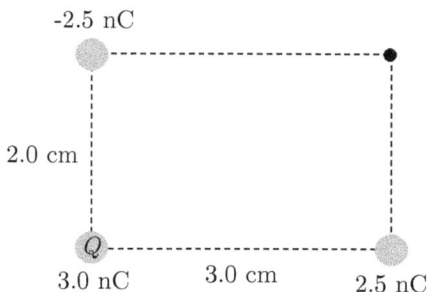

FIGURE 9.53
Problem 9.54

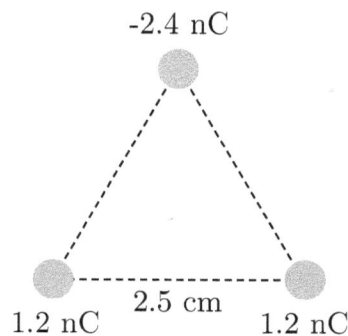

FIGURE 9.54
Problem 9.60

★★ **Problem 9.61.** Four 2.0 g spheres are each charged to 15 nC and placed on the corners of a 0.50 cm square. They are released and allowed to move away from one another. What is the speed of each of the spheres when they have moved very far from one another?

★★ **Problem 9.62.** Two point charges $q_1 = -1.5$ μF and $q_2 = 1.5$ μF are placed on the x axis at $x_1 = -2.0$ mm and $x_2 = 2.0$ mm, as shown in Figure 9.55.

   a. What is the electric potential at points A and B?

   b. How much energy is required to move a proton from point A to point B?

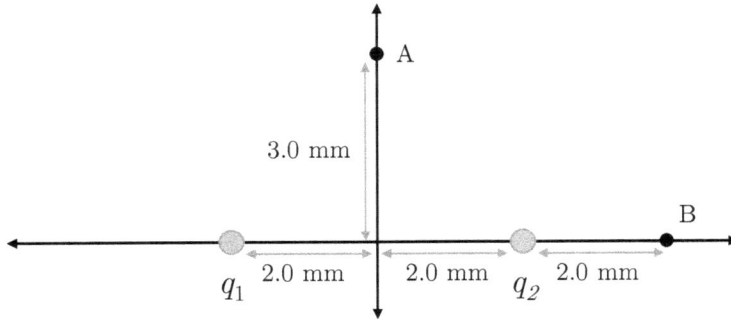

FIGURE 9.55
Problem 9.62

★★ **Problem 9.63.** Two circular pieces of metal are pinned to a board and attached to a power supply that maintains a potential difference between them, as shown in Figure 9.56. Sketch the electric field lines and equipotential surfaces in the region around and between the plates.

★★ **Problem 9.64.** A water molecule has a permanent dipole moment due to the structure of its chemical bonds. One way to model this behavior is to represent the oxygen atom with an effective charge $-2q$ and to represent each hydrogen atom with an effective charge $+q$, as shown in Figure 9.57. The angle between the two O-H bonds is θ and the distance between the oxygen atom and either hydrogen atom is L. Determine an expression for the electric potential energy of this configuration in terms of the parameters described above and fundamental constants, as needed.

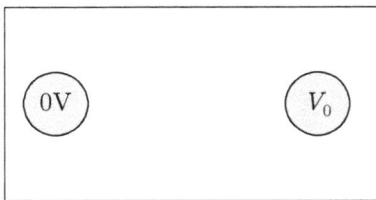

FIGURE 9.56
Problem 9.63

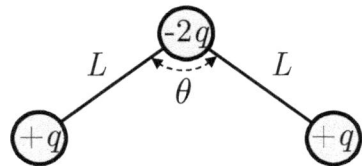

FIGURE 9.57
Problem 9.64

★★ dx **Problem 9.65.** What value of θ minimizes the energy of the water molecule according to the model described in Problem 9.64? Compare to the accepted value of 104° for $L = 95.8$ pm (and $q = e$). How well does this simple model do?[15]

----

[15]A better model would account for the position of the negative charge around the perimeter of the oxygen nucleus, which would move as the hydrogen atoms rotated.

★ **Problem 9.66.** A circular ring with a diameter of 5.0 cm lies flat on a horizontal table. A 25 g bead is held 8.0 cm directly above the middle of the ring and dropped. (It undergoes one-dimensional motion as it falls.) The bead is charged to −15 nC and the ring has a charge of 150 nC uniformly distributed around its perimeter. What is the speed of the bead as it impacts the table?[16]

★ **Problem 9.67.** Consider a lab apparatus where two metal plates are separated by a distance d and maintained at a specific potential difference ΔV. An electron is placed near one of the plates and given a speed v directed toward the opposite plate.

a. If the goal is to slow down the electron as it moves, which of the plates should be at the higher potential: the one near the electron or the one opposite the electron?

b. Determine an expression for the initial speed such that the electron stops just before impacting the plate.

★ **Problem 9.68.** Consider the charge configuration shown in Figure 9.35. Determine the electric potential at the point marked with a small dot, which is in the center of the rectangle formed by the four charges.

★★ **Problem 9.69.** Consider the situation described in Problem 9.30. Determine an algebraic expression for m in terms of other quantities if the bead is to just barely impact the top of the sphere.

★★ **Problem 9.70.** Suppose you want to fire a proton at a gold atom with enough energy that the proton contacts the edge of the nucleus, which contains 79 protons and has a radius of about 7.5 fm (7.5×10^{-15} m).

a. How much kinetic energy does the proton need?

b. What is its corresponding speed, in m/s?

c. It isn't possible for a proton (or anything with mass) to travel faster than the speed of light, $c \approx 3 \times 10^8$ m/s. Compare your answer to (b) to this limit. Does it seem possible to force a proton to impact a gold nucleus?[17]

★★ **Problem 9.71.** A copper nucleus (with charge $+29e$) is fixed in place; an electron is initially stationary a distance $d = 1.8 \times 10^{-9}$ m away. Suppose the nucleus is sufficiently massive that we can reasonably approximate it as remaining motionless in the ensuing motion.

a. How fast will the electron be traveling when it collides with the nucleus?

b. What electric field (magnitude and direction – *you* sketch the situation and define the coordinate system) would prevent the electron from moving from its initial position?

★★ **Problem 9.72.** In the Bohr model of the Hydrogen atom, an electron moves in a circular orbit around a proton, much like the Earth orbits around the Sun.

a. What is the electric potential energy of an electron a distance r from the proton?

b. What electric field does the electron experience a distance r from the proton?

c. The Bohr model predicts that the radius of the electron's orbit can only take certain values. The lowest two values are $r_1 = a_B$ and $r_2 = 4a_B$, where $a_B = 0.0529$ nm is the **Bohr radius**. What is the difference between the electric potential energies of an electron at each of these radii? Give a numerical answer in Joules and in electron volts (eV; the conversion is provided in Appendix B).[18]

---

[16]This is the basic idea of a coil gun.

[17]For speeds approaching the speed of light one needs to make use of *relativistic* dynamics, which will likely be a topic you'll study if you continue your education in physics beyond the scope of this book. For now, take the results of this problem with a grain of salt!

[18]While this problem leaves out some important details (your answer will differ from the actual answer by a factor of two because we are ignoring how kinetic energy is affected), an electron really does make such a transition by absorbing or emitting a photon with the same magnitude energy (so the total energy is conserved). This explains the spectrum of hydrogen. Historically, the Bohr model was an important early step in the development of quantum mechanics.

** **Problem 9.73.** Fun with atomic energy:

   a. A helium nucleus contains two protons and two neutrons. Suppose each is a point particle and the separation between any pair of particles is about 1.4×10^{-15} m. How much electric potential energy is stored in the nucleus of helium?

   b. Suppose a helium nucleus is to be formed by giving each of the nucleons the same speed v and smashing them together. Determine a numerical value for v. (This is a very crude model for nuclear fusion.)

   c. Consider an electron orbiting around the nucleus like a planet around a star. If the radius of the electron's orbit is $r_1 = 2.65 \times 10^{-11}$ m, what is the electric force acting on the electron? Sketch the nucleus and orbiting electron: label the force vectors, electric field vectors, and equipotentials.

*** **Problem 9.74.** Suppose an **alpha particle** (two protons and two neutrons stuck together) with kinetic energy K is moving parallel to a nucleus with Z protons, as shown in Figure 9.58.

   a. Sketch the figure and include a vector describing the force felt by the alpha particle due to the nucleus (just the orientation of the vector matters). Briefly explain how you arrived at your answer.

   b. Make a rough sketch of one possible path of the alpha particle. Briefly explain.

   c. Now suppose so the alpha particle is traveling directly toward the nucleus (i.e. $b = 0$ in the figure). If the alpha particle is supposed to stop at some point, should its initially velocity be directed toward the nucleus or away from the nucleus? Where will the alpha particle be when it stops? Express your answer in terms of the initial separation R, the kinetic energy K, and any other parameters needed.

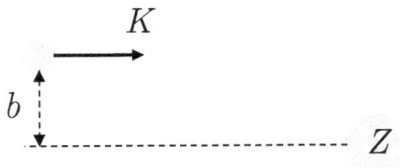

FIGURE 9.58
Problem 9.74

Additional Problems for Chapter 9

** **Problem 9.75.** Suppose a charge $+Q$ is fixed in place and a second charge $-q$ is a distance R from it. The second charge is then oscillated sinusoidally so its radial distance from the first charge is given by $r(t) = R(1 + 0.01 \sin(\omega t))$ for some angular frequency ω. How much net electric work is done on the charge after (a) 1 or (b) 10 complete cycles?

*** **Problem 9.76.** Consider a dipole consisting of two charges $\pm q$, each with mass m, fixed a distance L apart. Initially the dipole is held perpendicular to an electric field of strength E.

   a. What is the dipole's angular velocity when it has rotated to be parallel to the electric field?

   b. Show that if the axis of the dipole is aligned with E, rotated a small angle θ, and then released from rest, then it will undergo simple harmonic motion.

** **Problem 9.77.** Figure 9.59 shows the electric field along a one-dimensional region of space.

   a. What is the change in electric potential of a particle that starts from rest at $x = 0.0$ m and ends at $x = 5.0$ m?

   b. Describe the force(s) (magnitude and direction) it experiences along the way. What sign (positive or negative) does the charge have?

c. If an electron experiences this electrical field and it is at rest when it reaches $x = 5.0$ m, what was its velocity at $x = 0.0$ m?

★★ **Problem 9.78.** Some atoms can be broken apart in a process called nuclear fission; one standard example involves firing a neutron into the nucleus of uranium, causing it to split into (1) a krypton atom containing 36 protons and 56 neutrons and (2) a barium atom containing 56 protons and 85 neutrons. Suppose that just after the uranium atom splits apart, the center of the krypton atom is 1.2×10^{-14} m from the center of the barium atom. Both atoms are at rest at this instant and the krypton atom is later measured to have a kinetic energy of 2.34×10^{-11} J. If you measured the kinetic energy of the barium atom at the same time as the krypton atom, what would you expect to find?

★★ **Problem 9.79.** Figure 9.57 sketches the geometry of a water molecule. If $L = 96$ pm and the angle between the hydrogen atoms (with the oxygen atom at the vertex) is 105°, then:

a. What is the net force acting on the oxygen nucleus due to the nuclei of the hydrogen atoms? Specify a magnitude and direction. (Ignore the molecular electrons, which balance nucleus-nucleus interactions to ensure the molecule is in equilibrium.)

b. How much energy is needed to assemble all three nuclei in this configuration? (Again, ignore the role of the electrons.)

★★ **Problem 9.80.** Consider the electric potential vs. position graph shown in Figure 9.60.

a. Draw the corresponding electric field vs. position graph.

b. What force would a proton feel at $x = 0.5$ cm and $x = 2.5$ cm?

c. Repeat (b) for an electron.

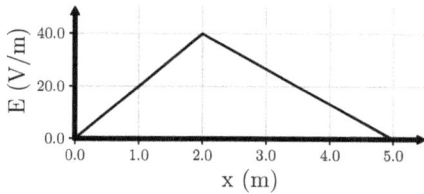

FIGURE 9.59
Problem 9.77

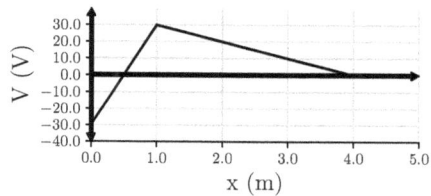

FIGURE 9.60
Problem 9.80

10

Electric Charge: Cumulative Effects

In Chapter 9 we established the basic laws that govern interactions between charged particles, and we considered the behavior of small collections of point charges in isolation and/or in the presence of an external electric field. This is hardly the complete picture, however: how can we apply what we've learned to understand what happens when we're dealing with more than just a *few* interacting charges? If you've ever received an electrical shock when you grabbed a doorknob, for instance, then the electrical phenomenon at play involves something like 10^{14} electrons jumping from the handle, across the air, and into your hand! Moreover, how do we *create* a constant electric field – or any other, for that matter?

The general question that we shall address in this chapter is "How do macroscopic objects create, and interact with, electric fields?" We will *qualitatively* explore this question in Section 10.1.

Of course, we can't be totally satisfied with only a qualitative approach: if we would like to effectively manipulate electric interactions to do something useful, we must also be *quantitative*. In Sections 10.2–10.3 we will consider how enormous quantities of charge create electric fields – a topic that will be very relevant to our study of electric circuits in Chapter 11.

Learning Objectives

After reading this chapter, you should be able to:

- Describe the behavior of conductors and insulators when interacting with one another and with externally applied electric fields.

- Characterize the electric field and electric potential due to charge distributions (e.g. a ring, disk, or line of charge) by setting up and solving appropriate integrals.

- Calculate the electric field due to a charge distribution from the electric potential and vice versa.

- Determine the electric flux passing through some surface.

- Calculate the electric field of a charge distribution by using Gauss's Law.

10.1   Insulators and Conductors

In this section we will explore the following ideas:

1. Many objects will transfer electrons when they are rubbed together, meaning one of the objects acquires a net positive charge and the other a net negative charge. This process is referred to as the **triboelectric effect** and can be used to perform many simple experiments with electric charge.

2. We can classify materials as **insulators** (like glass or rubber) and **conductors** (metals). In an insulator, the electrons tend to be tightly bound to their nuclei, which makes them respond rather weakly to external electric fields. In contrast, in a conductor there are many electrons that can move throughout the material in response to an external electric field, which allows for much more dramatic effects.

Once we've explained these points, we'll consider several interesting experiments involving electric charge.

The Triboelectric Effect

Most materials are overall *neutral*, meaning there are an equal numbers of electrons and protons spread throughout. In other words, it has a net charge of 0 C, and as a result it creates no electric field.[1] If you would like to make an object a source of an electric field, evidently you need to *transfer* some charged particles onto, or away from, it. You may already be familiar with some ways to do this:

- Rub your hair with a balloon.

- Rub a glass rod with a silk cloth.

- Rub a rubber rod with some fur.

In each case you are rubbing two initially neutral objects together, and as a result there is a transfer of charge: one will be overall *positive* and the other will be overall *negative*. As mentioned above, this phenomenon is called the **triboelectric effect**. Because the charge isn't *moving* on the objects, we sometimes refer to it as **static charge**. In the examples listed above the balloon and the rubber rod become negatively charged and the glass rod becomes positively charged.[2]

Why does this occur, and what determines which pairs of materials will display this effect? The situation is somewhat analogous to a ball free to roll along hilly terrain: it tends to find a low point, i.e. a point that has the lowest (gravitational) potential energy in the region. In the same way, when rubbing objects together, in some cases the electrons on one object are able to go to lower (electrochemical) potential energy by moving to the other object.[3] We won't worry about this process is any more detail in this book.

Insulators

As we discussed above, an insulator is characterized by the fact that its electrons are bound to their parent nuclei. This does not mean, however, that *nothing* happens if an insulator is exposed to an electric field. There are two major phenomena at play.

First, some molecules (like water) can be modeled as dipoles: there is a permanent separation between the positive and the negative charge. We sometimes say that water has a **permanent dipole moment**. Usually the dipoles tend to align randomly, meaning there is no net electric field generated by the water (Fig. 10.1). However, molecular dipoles, can, at least in principle, rotate in response to an external electric field, as we discussed in

---

[1]In practice the number of electrons and protons might not be *exactly* equal; we're here referring to the case where any difference is negligible.

[2]It is also worth noting that the charge separation doesn't last, in large part because water molecules in the air slowly neutralize the charges. This is why physics instructors dread demonstrating static charge on a humid day.

[3]Note that the protons don't move; the atomic nuclei, where the protons reside, are in general tightly bound to the object and won't jump to another object in the same way as an electron.

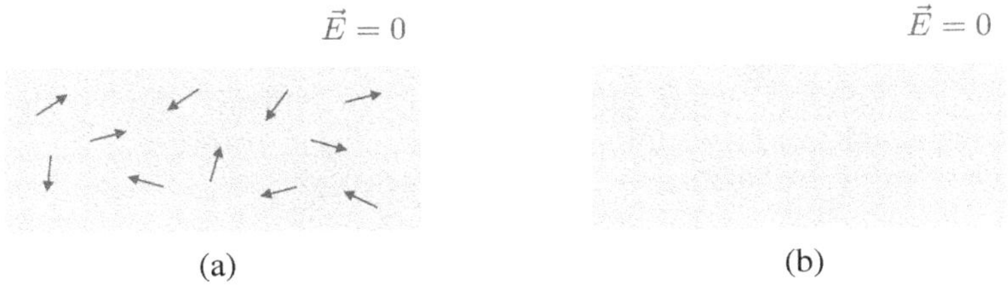

$$\vec{E} = 0 \qquad\qquad\qquad \vec{E} = 0$$

(a)   (b)

FIGURE 10.1
A rectangular piece of an insulator composed of molecules with a permanent dipole moment. (a) In the absence of an external electric field, the dipoles tend to be randomly oriented. (b) The electric fields of the dipoles tend to cancel on average, leaving no net electric field inside the insulator.

Example 9.9 and sketch in Figure 10.2. I say *in principle* because the extent of the response is material-dependent. Liquid water molecules, for instance, are much more responsive than water molecules in *ice*!

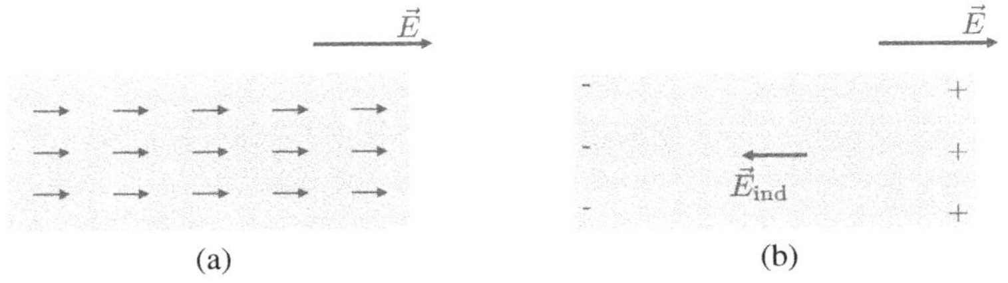

(a)   (b)

FIGURE 10.2
(a) In the presence of an external electric field, the dipoles in an insulator tend to align with the field. (b) Because the dipoles line up end-to-end, the positive side of a dipole tends to be located near the negative side of a nearby dipole, which causes the interior of the insulator to be overall neutral. The alignment of the dipoles leads to an unbalanced accumulation of charge only on the edges of the insulator, which in turn generates a relatively weak electric field $\vec{E}_{\text{ind}}$ opposed to the external field $\vec{E}$.

The dipole alignment happens throughout the material; suppose that the external field is directed to the right, so the dipoles tend to rotate horizontally, with their positive charge on the right. The center of the dipoles tend to be fixed in place, and so the *right* side of some dipoles tend to line up with the *left* side of adjacent dipoles (and vice versa). The overall electric field of these adjacent charges tends to cancel, on average. However, there are two locations where this cancellation *doesn't* occur: on the far left side of the insulator and the far right side, where there are no more adjacent dipoles to cancel out the effect.

The net effect of all of this is some excess positive charge on the far right side of the insulator, and some excess negative charge on the far left side. Now, when we previously discussed dipoles it was in the context of a *uniform* electric field, where the net force was 0 N (even if the net torque was *not* 0 Nm). If you imagine an experiment where we bring a charged rod near some liquid water, though, the field *won't* be constant: it will be stronger close to the rod and weaker further away.[4] Thus the surface of the water closer to the rod feels a stronger *attractive* force than the other end, which feels a *repulsive* force. The net

---

[4]We'll explore this in mathematical detail in the next section, but for now this hopefully strikes you as intuitive: you don't feel any effect from a charged rod on the other side of the planet, after all!

effect is a small attractive force. This is strikingly demonstrated if you bring a charged rod near a faucet that is turned on low, so a narrow stream of water falls into the basin: the path of the stream can be visibly deflected.

I said above there where *two* main mechanisms used to describe an insulator's response to an external electric field. The second is this: even if the molecules inside some object do not have a permanent dipole moment, an electric field can create an **induced dipole moment**, as we sketch in Figure 10.3.

$$\vec{E} = 0 \qquad\qquad\qquad\qquad \xrightarrow{\quad\quad\quad} \vec{E}$$

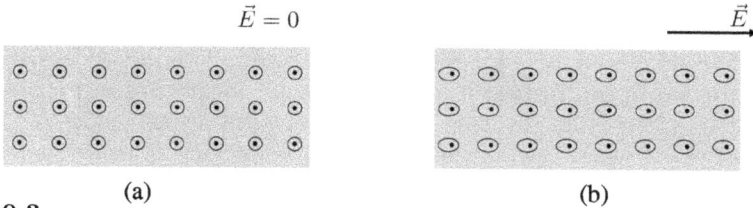

(a) (b)

FIGURE 10.3
Here molecules in an insulator are represented by a positively charged nucleus (black dots) surrounded by electrons (black curves) that orbit the nuclei in a planetary model. (a) In the absence of an external field, the center of the electrons' orbit overlaps with the nuclei, giving no overall charge separation. The situation leads to no net electric field, as in the case of Figure 10.1 for molecules with a permanent dipole moment. (b) In the presence of an external electric field, the electrons' orbits are perturbed, which shifts the centers of their orbits away from the nuclei. The electrons therefore spend more time on one side of the nuclei than the other, leading to an *induced* dipole moment. A net electric field is therefore induced in the insulator in much the same way as is sketched in Figure 10.2 for the case of molecules with a permanent dipole moment.

A good example of this is if you "charge up" a glass rod and a plastic rod (which carry positive and negative charge as a result, respectively) and then bring them near two piles of small pieces of paper. You will observe that *both* piles of paper are attracted to the rods. In fact, if they are small enough (and if the rods carry enough charge), the paper will lift up and cling to the rods! This same principle is used in some air filtration systems: particles in the air are attracted to charged fibers and are thereby removed from the air.

Finally, I should mention that the extent to which insulators respond to external electric fields turns out to be very useful in the context of electric circuits, and specifically when designing strong capacitors. We will take up a more quantitative analysis of insulators, then, in Chapter 11.

Conductors

When a conductor is exposed to an external electric field $\vec{E}$, electrons in the conductor feel a force $\vec{F} = -e\vec{E}$, just like in an insulator. What is different is that the some of the electrons are weakly bound to their atoms and can readily "detach" from their nuclei and move freely inside the conductor.[5] The result is sketched in Figure 10.4: negative charges pile up on one side, leaving a net positive charge on the other side. This process is referred to as **induced polarization**.

---

[5]I say *some* because typically only the outermost or *valence* electrons are free to move about; the nuclei maintain a tight hold on the electrons that "live" close to the nuclei. That said, for the electric field strengths we're concerned with in this book, no conductor will ever "run out" of valence electrons that can move about in the conductor in response to an external electric field.

$$\vec{E} = 0$$
$$\vec{E}$$

(a)   (b)

FIGURE 10.4

(a) In the absence of an external field, the electrons in a conductor are evenly distributed, giving no local accumulation of charge. (b) In the presence of an external field $\vec{E}$, electrons in the conductor move, leading to an accumulation of negative charge on one end of the conductor (due to an *excess* of electrons) and an equivalent accumulation of positive charge (due to the *absence* of electrons) on the other side. Unlike in an insulator, in a conductor this process continues until the induced electric field perfectly balances the external field, leaving no *net* field inside the conductor.

Now, as soon as the induced polarization starts to occur, the charge separation creates a *second* electric field, which we'll call the *induced* field: $\vec{E}_{\text{ind}}$. Notice that the orientation of this field is *opposite* the external field. As a result, the *net* field inside the conductor ($\vec{E}_{\text{NET}} = \vec{E} + \vec{E}_{\text{ind}}$) *decreases*. As more and more electrons move to the edge of the conductor, the magnitude of E_{ind} increases until eventually it perfectly balances the external field. When this happens, the electrons inside the conductor experience no net electric field and therefore no net force: the polarization stops. The entire process is essentially instantaneous.

The fact that the net field inside the conductor goes to 0 N/C because of polarization is an important result that holds in general. To see why, suppose that at any time (and for any reason) there *is* a net field in the conductor. This means the free electrons feel a force and, as we've seen, they will move in such a way as to weaken the field; this process will continue until there *isn't* a field. We summarize this behavior like so:

The net electric field in a conductor in electrostatic equilibrium is 0 N/C.

We specify that the conductor is in *electrostatic equilibrium* because we're assuming the process we just described has already finished, and no electrons are moving anymore.

What can we say about the electric *potential* inside the conductor? Recall that it is related to the electric field by Equation (9.26):

$$\vec{E} = -\frac{dV}{dr}\hat{r}$$

This equation indicates that if we graph the potential V as a function of r, then the electric field has a magnitude equal to the *slope* of the curve (and points "downhill" as sketched in Figure 9.26). But this means that if $E = 0$ N/C, then the slope of the V vs. r curve is *also* 0 N/C and the electric potential must therefore be some *constant* value.[6] Thus we say in general

The electric potential in a conductor in electrostatic equilibrium is constant.

These two statements are very useful in analyzing the behavior of conductors. Before applying these principles in some examples, I would like to provide a third statement:

---

[6]We see this also from equation (9.23): $\Delta V = -\int \vec{E} \cdot d\vec{r}$. If the field is 0 N/C, then we have $\Delta V = \int 0 \, dr = c$ for a constant c.

Any excess charge in a conductor in electrostatic equilibrium exists on its *surface*.

Note that this is different from the induced polarization discussed above, where the conductor was overall neutral even though electrons were moving about to give a charge separation. Here we're talking about the case where the conductor is *not* overall neutral.

The idea that the excess charge *all* goes to the surface may surprise you: the excess charge is mutually repulsive, and it seems reasonable to think that it might, for instance, *evenly distribute* itself through the conductor such every charge is as far as possible from every other charge. This simply isn't the case; the charge goes to the surface. The most straightforward proof of this statement invokes *Gauss's Law*, which we'll get to in section 10.3. Until then I hope you'll take my word for it.

The properties we have discussed are very useful in describing experiments involving static electricity, as the next sequence of examples demonstrates.

Example 10.1 Polarization in Conducting Spheres ★

Two identical metal spheres are initially uncharged and are touching. A positively charged rod is brought near the spheres, as shown in Figure 10.5. Determine if each sphere is positively charged, negatively charged, or neutral for each of the following situations: (a) The spheres are separated and then the rod is removed. (b) The rod is removed and then the spheres are separated.

FIGURE 10.5

When the conducting spheres are touching, they effectively behave as a single conductor because electrons are free to move from one sphere to the other. Thus we have induced polarization, as we sketch in Figure 10.6. In (a), this polarization is maintained while the spheres are separated because the rod is still present: it is the field created by the rod that causes the polarization in the first place. We conclude that the sphere closer to the rod will carry a net negative charge and the other sphere will carry an equal positive charge. (The charges must be equal in magnitude because the system was overall neutral to begin with.) This is a new way to create a charged object in addition to the triboelectric effect we have already discussed. This new process is referred to as **charging by induction**.

FIGURE 10.6

In situation (b) we remove the rod while the spheres are still touching. In order for the net field in the conductor to be 0 N/C, then, the field due to the induced polarization must *also* vanish. This occurs when the electrons redistribute themselves evenly through the spheres such that they are both overall neutral, just as they were before the rod was brought in in the first place. (In other words, without the nearby charged rod, the electrons only feel a force of attraction toward the excess positive charge *in the spheres*, which causes them to move back to where they were originally, thus canceling the net charge distribution.)

Example 10.2 The Electroscope   ⋆
An **electroscope** consists of a conducting tip that is attached to two thin conducting plates hinged at one end, as sketched in Figure 10.7. (You can make one yourself by getting three strips of aluminum foil: stack them on top of one another, pull the middle one almost all the way out, then staple them together where they overlap. Attach a string to the non-stapled side of the middle strip and dangle the entire contraption by the string.) A negatively charged rod is brought near the tip but does not touch it, as shown on the left side of the figure. Describe what happens by indicating the presence of positive and/or negative charge on the electroscope.

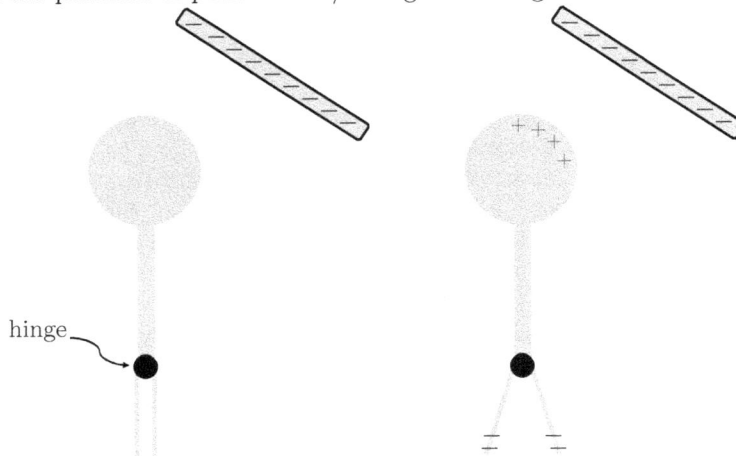

FIGURE 10.7

Just as we saw in Example 10.1, the presence of some external charge induces polarization in our contraption. In this case the rod is negatively charged, so we find polarization in the electroscope such that positive charge is clustered near the top and negative charge is clustered on the plates at the bottom. The negative will accumulate on *both* plates, but of course negative charge is mutually repulsive. Because the plates are hinged, this repulsion results in the plates swinging away from one another! This is sketched on the right side of Figure 10.7.

Example 10.3 Dragging a Can with the Electric Force   ★★
A charged rubber rod is held above and to the side of an empty aluminum can that rests, on its side, on a table. As a result, the can *rolls* along the table toward the rod. The process is repeated with a charged glass rod and the aluminum can still rolls toward the rod. Explain this behavior.

We'll consider the can to be a hollow cylinder with thin walls, and to make our lives a bit easier we'll imagine that the ends are chopped off so the center is hollow. Note that, as usual, there is a gravitational force and a normal force acting on the can.

What happens when we bring in a charged rod? To be specific, let's first consider the negatively charged rubber rod. When it is brought near the can, the electrons in the aluminum are repelled from the negative charge on the rod. As a result, they begin to cluster as far from the rod as possible, leaving a net positive charge on the part of the can closest to the rod. This is another example of *induced polarization* that we discussed above. And, because the excess positive charge on the can is closer to the rod than the excess negative charge, the *attractive* force is larger than the *repulsive* force (Fig. 10.8).

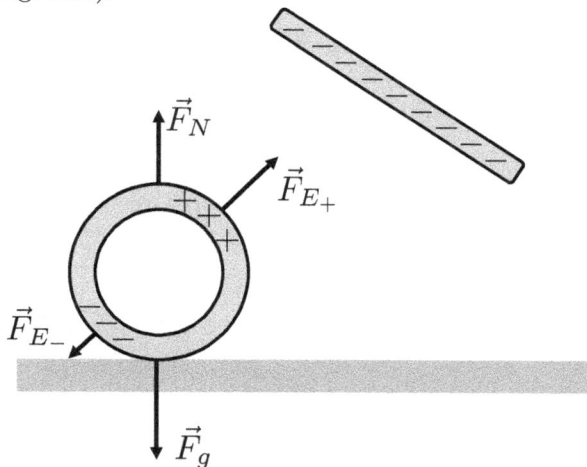

FIGURE 10.8

Now, assuming that you don't end up lifting the can off of the table entirely, there will be no net force in the vertical direction (the normal force adjusts itself to balance all other forces in the vertical direction). What about in the horizontal direction? Well, a *component* of each electric force exists in the horizontal direction, but (as we've seen) the attractive force toward the rod is larger than the repulsive force away from the rod.

We conclude, then, that there is a net horizontal force acting on the can in the direction toward the rod that can cause it to roll. Importantly, if we bring in a *positively* charged rod, the polarization reverses: the edge of the can closer to the rod carries excess *negative* charge while the far side is left with an excess *positive charge*. The force diagram, however, is identical because the sign of the charge on the rod has also reversed. As a result, our conclusions are identical.

Example 10.4 Charges on Conducting Spheres   ⋆

Two identical metal spheres carry charge Q and $-Q/2$. They briefly touch and are then separated. How much charge is on each sphere now?

Before the spheres touch, they each carry their charge distributed evenly along their surfaces. When they are placed into contact, they behave as a single larger conductor with an irregular shape. The key principles are still that there is no field and a constant potential inside the conductor; the charge still resides on the surface.

For complicated shapes the way that the charge spreads along the surface to achieve a constant potential can be tricky to work out. In the situation where we have two identical objects being brought together, though, we know from the symmetry of the problem that *half of the total charge* will be on each object. In this case the total charge is $Q + (-Q/2) = Q/2$. Thus, each sphere has half of this, or $Q/4$. They will retain this charge when they are pulled apart.

This process of self-averaging charges can be repeated for multiple charged objects; see for instance Problem 10.2.

Example 10.5 Electric Ping Pong   ⋆⋆

Two neutral circular metal plates face one another and have been separated a short distance relative to their diameter. Between the plates hangs a small neutral metal ball, which is suspended by a thin insulating wire. A negatively charged rubber rod is then gently touched to the left plate. As a result, the metal ball violently swings into the left plate, then back and forth between the two plates, until slowing back to a stop. Explain *why* this happens with a sequence of diagrams indicating the presence of charge as the ball swings back and forth. (The effect, which is sometimes called *electric ping pong*, is rather dramatic. When I demonstrate it to students I typically scream in feigned shock to heighten the effect and also because I'm more than a little bit odd.)

We sketch the situation in three steps in Figure 10.9. The left panel shows the initial situation: the excess charge on the left plate polarizes the right plate and the hanging sphere. Because the field is stronger closer to the plate (schematically represented with the two field vectors near the top of the region), the overall force on the sphere is attractive.

In the center panel, the sphere has made contact with the plate, and as a result the plate-sphere system temporarily acts as a single conductor: some of the excess negative charge flows from the plate and onto the sphere to ensure the potential in the sphere is the same as in the plate. When this occurs, the force on the sphere becomes repulsive.

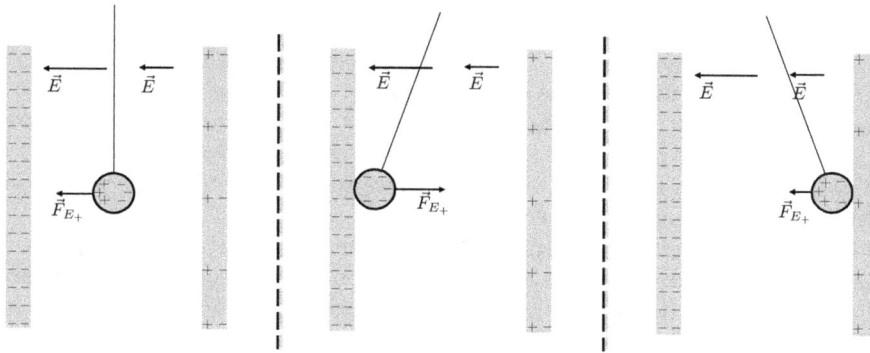

FIGURE 10.9
The ball at three different positions in its motion. The dashed vertical lines separate the three panels.

Finally, on the right panel the sphere has swung to the other side and is touching the right plate. When this occurs its excess charge disperses onto the plate by the same reasoning as above. Because the sphere is now very nearly neutral again, it polarizes in response to the still-strong field from the left plate and the system swings back to the left, as we sketched in the left panel.

This process repeats itself, though in each cycle the charge on the left plate decreases and the overall charge on the right plate increases. This results in an overall decrease in the strength of the electric field, so the sphere, which is behaving like a sort of *charge shuttle*, slows as the charge on the plates begins to equalize.

10.2 Fields from Charge Distributions: Integration

While I certainly hope you found the preceding section interesting and informative, I think you'll agree that it wasn't *complete*, in the sense that we didn't form many conclusions in the precise language of mathematics. Consider bringing a charged rod near an uncharged metal sphere: we know, for example, that (1) the sphere becomes polarized, (2) all of the excess charge is on the surface, and (3) the net electric field inside the sphere is 0 N/C. But *how much* charge is driven to the surface of the sphere, and how is it *distributed* along the surface?

These are complicated questions; as a first step we would need to determine what the electric field is *everywhere in space* due to the charged rod. Then, we'd need to consider what that field looked like specifically *inside the sphere*. Finally, we'd need to analyze how the electrons inside the sphere distribute themselves along the surface of the sphere so the ensuing field perfectly *cancels* the rod's field at every point in the sphere (or, equivalently, to ensure that the electric potential inside the sphere is a constant value).

To answer these questions fully we would need to invoke mathematics beyond the scope of this book.[7] We can, however, tackle a number of simpler problems that are informative,

---

[7]For instance, if the electric potential is a three-dimensional function, which we might write in Cartesian coordinates as $V(x, y, z)$, then the equation $\vec{E} = -\frac{dV}{dr}\hat{r}$ is clearly no longer valid, since it assumes that the potential is a function of just one spatial coordinate. It turns out that the general form of this

useful, and will serve to prepare you for a more advanced treatment if you go on to study electromagnetism at a higher level.

Let's consider, then, how we can determine the electric field and the electric potential at *one point in space* when we have lots of charge spread out somewhere nearby, such as along the surface of a rod (Fig. 10.10). The basic idea here is really quite simple: determine the field (or potential) at the point in question separately for every nearby charge, and then *add them all up*! (Keeping in mind that for the field we're talking about a *vector sum* and for the potential we're talking about a *scalar sum*.)

Actually *doing* this can be challenging. We'll be dealing with integral calculus, but even if you haven't been trained in it, I still encourage you to read through what follows – you can still appreciate how to *set up* the integrals, which is where the interesting physics is involved!

To proceed, recall that we need to determine the cumulative field and potential of a nearby charge distribution. We're talking about a *sum* here, but because each charge is tiny (we're talking about protons and/or electrons, remember) and we're considering situations where we have *lots* of them (typically $\gg 10^{10}$), we can treat these sums as *integrals*. (Remember when I said we'd be doing calculus, one paragraph up? Well, here we go.) Thus for the electric field we have

$$\vec{E}_{\text{NET}} = K \sum \frac{Q_i}{r_i^2}\hat{r}_i \rightarrow \vec{E}_{\text{NET}} = K \int \frac{dQ}{r^2}\hat{r} \tag{10.1}$$

The $\hat{r}$ in this integral indicates that we're adding up a series of vectors, and they'll each have their own orientation since each piece of charge is in a different location (e.g. in Fig. 10.10 the value of θ is different for every charge dQ). To deal with this mathematically we will split this integral up into the components according to whatever coordinate system is convenient for a particular problem. If the field due to some charge dQ is given by $d\vec{E}$, then (for example) in a 2D Cartesian coordinate system we have:

$$d\vec{E} = K\frac{dQ}{r^2}\hat{r} = dE_x\hat{x} + dE_y\hat{y} \tag{10.2}$$

The details of how we determine the mathematical form of dE_x and dE_y is context-dependent, but in the example of Figure 10.10 we just need to pick out a $\cos\theta$ and $\sin\theta$, respectively:

$$dE_x = dE\cos\theta = K\frac{dQ}{r^2}\cos\theta \tag{10.3}$$

$$dE_y = dE\sin\theta = K\frac{dQ}{r^2}\sin\theta \tag{10.4}$$

Once we've split the field into components, the integral for the total electric field can be expressed in terms of each component:

$$\vec{E}_{\text{NET}} = \int d\vec{E} = \int dE_x\hat{x} + \int dE_y\hat{y} \tag{10.5}$$

The process of breaking the field into its components is context-dependent, and so I will refer to the integral in Equation (10.1) when discussing the electric field due to a charge distribution in general.

The electric *potential*, meanwhile, is simpler than the field because there are no vector components to work out:

---

relationship invokes a sort of three-dimensional derivative called the **gradient**, which is written ∇. Thus we have $\vec{E} = -\nabla V$. Calculating the gradient of a three-dimensional scalar field can become mathematically challenging quite quickly.

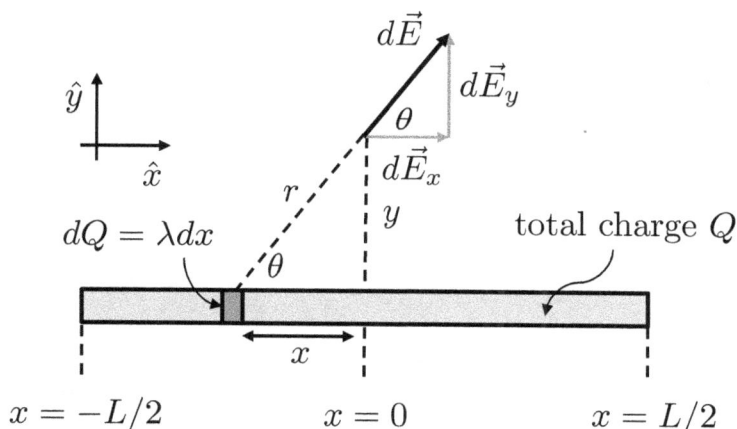

FIGURE 10.10
A rod of length L has charge Q evenly distributed along its length. Each infinitesimally small piece of charge dQ creates an infinitesimally small electric field $d\vec{E}$ out in space. Here we consider the field at a single point a perpendicular distance y from the middle of the rod. The overall field at this position is the vector sum (really an *integral*) of all of the field due to each of the dQ charges. Similarly, the electric potential V at this position is the sum (again, really *integral*) of the potential due to each charge dQ.

$$V_{\text{NET}} = K \sum \frac{Q_i}{r_i} \rightarrow V_{\text{NET}} = K \int \frac{dQ}{r} \qquad (10.6)$$

Importantly, in all these equations r refers to the distance from each little chunk of charge to the point where we want to know the overall field and potential, and this quantity *changes* for different chunks of charge because they are spread out in space. In other words, r is a function of the charge chunk dQ, and we need to make the dependence explicit if we have any hope of actually carrying out the integrals.

To address this, we will express the integrals not in terms of the charges dQ, but in terms of *where they are*. In a 1D problem (like a long straight rod), we might consider the charge dQ located in a narrow region dx, as in Figure 10.10. The value of r for any particular length chunk dx is then clear from the geometry.

How do we move from an integral in terms of dQ to an integral in terms of dx? Well, if the *total* charge on the rod is Q and it is spread evenly over the rod's length L, then we can introduce the **linear charge density** λ:

$$\lambda = \frac{Q}{L} = \frac{dQ}{dx} \qquad (10.7)$$

The final term above follows from the fact that if the charge is evenly spread along the surface, then the *overall* charge-to-length ratio Q/L is the same as the ratio on an infinitely small piece of the rod, dQ/dx.

This set of equations allows us to express dQ in terms of dx:

$$dQ = \lambda dx \qquad (10.8)$$

Inserting this into the integral expressions for $\vec{E}$ and V (Equations (10.1) and (10.6)) gives

$$\vec{E}_{\text{line}} = K\lambda \int_{-L/2}^{L/2} \frac{dx}{r^2}\hat{r} \qquad (10.9)$$

and

$$V_{\text{line}} = K\lambda \int_{-L/2}^{L/2} \frac{dx}{r} \qquad (10.10)$$

Instead of integrating over the *charge*, we're now integrating over the *position* along our rod.[8] The last step is to express r in terms of the position coordinate dx; this is something that we'll do on a case-by-case basis depending on the geometry of the problem.

We'll demonstrate how to do this with some examples, but first let me make an additional comment. Equations (10.9) and (10.10) assume a *one-dimensional* distribution of charge, which is why we invoked the *linear* charge density, λ. But we can adapt what we've done to two or three dimensions very quickly.

If we have a charged *surface*, like a thin square of metal, then the charge Q is distributed over an *area a* and we have a **surface charge density** σ:

$$\sigma = \frac{Q}{a} = \frac{dQ}{da} \qquad (10.11)$$

Here da refers to a differential piece of area; in Cartesian coordinates this would be given by $da = dx \times dy$ (Fig. 10.11).

Similarly, if we have a charged *volume*, like an insulating sphere that carries some charge throughout, then the charge Q is distributed over the *volume v* and we have a **volume charge density** ρ:[9]

$$\rho = \frac{Q}{v} = \frac{dQ}{dv} \qquad (10.12)$$

Here dv refers to a differential chunk of volume; in Cartesian coordinates we have $dv = dx \times dy \times dz$ (Fig. 10.11).

With these definitions in mind, the 1D integral equations for the field and potential (Equations (10.9) and (10.10)) carry over directly to the 2D and 3D cases:

$$\vec{E}_{\text{surface}} = K\sigma \int_0^a \frac{da'}{r^2}\hat{r}, \qquad \vec{E}_{\text{volume}} = K\rho \int_0^v \frac{dv'}{r^2}\hat{r} \qquad (10.13)$$

and

$$V_{\text{surface}} = K\sigma \int_0^a \frac{da'}{r}, \qquad V_{\text{volume}} = K\rho \int_0^v \frac{dv'}{r} \qquad (10.14)$$

---

[8]I am writing the bounds here as $-L/2$ to $L/2$; this says "start adding at $x = -L/2$ and stop at $x = L/2$". This is in alignment with the coordinate system set up in Figure 10.10. The general principle is that you need to integrate over all of the charge; if your origin was on one end of the rod then the bounds would be from 0 to L.

[9]I am using lowercase a and v for area and volume, respectively, because an *uppercase* V refers to electric potential.

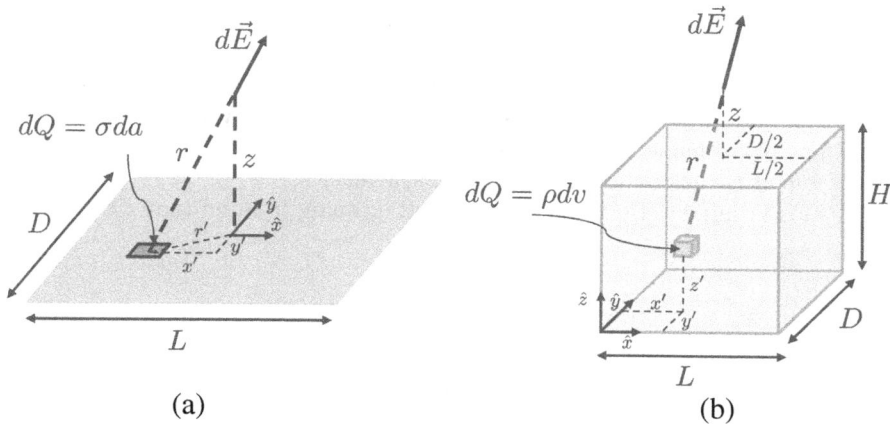

FIGURE 10.11
(a) An example of a surface charge distribution. Here we are concerned with the field a distance z above the middle of a rectangular sheet that carries a charge density $\sigma = Q/A$, where $A = LD$ is the area of the sheet. (b) An example of a volume charge distribution. Here we have a cube of volume $v = LDH$ and volume charge density $\rho = Q/v$. In both cases the location of an arbitrary chunk of charge dQ is drawn, along with the corresponding field $d\vec{E}$. In general we wish to determine the *total* field and/or potential.

Here I am using primes (da' and dv') to distinguish the variable of integration from the upper limit of the integral (e.g. we're adding up little chunks of area da' until we have covered the entire area a.)

This is abstract material and it is especially important to work through some examples and problems to cement your understanding. In Example 10.6 we will "finish the job" in the case of the rod that we've been considering as we set up the 1D analysis. In Examples 10.7–10.8 we will consider some other interesting charge distributions. Then, in Examples 10.9–10.10, we will *apply* these results to quantitatively analyze some interesting situations. Example 10.10 in particular introduces the **parallel plate capacitor**, which we will analyze in further detail in Chapter 11.

Example 10.6 Electric Field of a Line of Charge $\qquad \star\star \int dx$
Consider a thin rod of length L carrying a uniform charge Q (Fig. 10.10). (a) Determine, through integration, the electric field and electric potential a distance y perpendicular to the midpoint of the rod. (b) Use a computer (or graphing calculator) to make graphs of $\vec{E}$ and V as functions of y.

We have already sketched the geometry in Figure 10.10. The electric field component $d\vec{E}$ in the figure has two components, so in general we would need to calculate two integrals; one for the x component of the overall field and one for the y component. However, in this case we can call out from the geometry that $E_x = 0$ N/C.

Why? Well, for any charge dQ a distance x to the *left* of the midpoint of the rod, there will be a charge dQ the same distance to the *right* of the midpoint. The field components $d\vec{E}$ will have the same magnitude and their y components will *add* (they both point up) while their x components will *cancel* (they point in opposite directions).

The overall field will therefore point in the y direction and so we can evaluate just one integral. In fact, we can evaluate just *half* of the integral for the y component (due to the right side of the rod, say) and then double the result, since the left side and the right side will make identical contributions to the overall field. (This is the same argument we made when analyzing the electric dipole.) We don't *have* to do it this way, but it turns out to simplify the mathematics.

With this in mind, let's set up our integral, keeping in mind that $dE_y = dE \sin \theta$:

$$\vec{E} = \int dE_y \hat{y}$$

$$\vec{E} = \int K \frac{dQ}{r^2} \sin \theta \hat{y}$$

$$\vec{E} = 2K\lambda \int_0^{L/2} \frac{\sin \theta dx}{r^2} \hat{y}$$

Here we've substituted $dQ = \lambda dx$, inserted the bounds of integration (we've called the midpoint of the rod $x = 0$, so we start at $x = 0$ and end at $x = L/2$), and inserted a factor of 2 in front to account for the fact that our integral determines just half of the overall field.

To proceed, we need to write $\sin \theta$ and r in terms of x: both of these quantities change as we consider different charges spread along the rod, so the functional dependence on our spatial coordinate x must be made explicit. From the geometry of our figure we have

$$\sin \theta = \frac{y}{r}$$

and (from the Pythagorean theorem)

$$r^2 = y^2 + x^2 \implies r = \left(y^2 + x^2\right)^{1/2}$$

Inserting both results into our integral gives

$$\vec{E} = 2Ky\lambda \int_0^{L/2} \frac{dx}{r^3} \hat{y}$$

$$\vec{E} = 2Ky\lambda \int_0^{L/2} \frac{dx}{(y^2 + x^2)^{3/2}} \hat{y} \tag{10.15}$$

The problem is now completely set up: everything in the integral is either a constant (y) or the spatial coordinate we're integrating over (x). We now have an integral that could be passed off to someone who hasn't studied physics but who *has* studied integral calculus. If you're in the opposite situation and aren't familiar with how to solve complicated integrals like this, then this is a reasonable place to stop. (You might take the same approach when solving problems; the ones at the end of this chapter generally ask you to *set up* an integral like this in one part and to *solve* it in another part.)

If you *have* studied integral calculus, I'll just note that this integral can be solved with the substitution $x = y \tan u$. Working through the details yields

$$\vec{E}_{\text{rod}} = \frac{KL\lambda}{y \left(y^2 + (L/2)^2\right)^{1/2}} \hat{y} \tag{10.16}$$

Finally, if we note that $\lambda L = Q$, we can express our result in terms of the overall charge on the rod:

$$\vec{E}_{\text{rod}} = \frac{KQ}{y\left(y^2 + (L/2)^2\right)^{1/2}}\hat{y} \qquad (10.17)$$

where $\hat{y}$ refers to the direction perpendicular to the rod. This equation is plotted in Figure 10.12, where to keep things generic I've plotted E relative to KQ on the vertical axis and y as a fractional value of L on the horizontal axis.

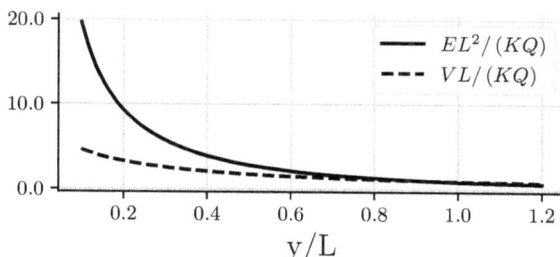

FIGURE 10.12
The electric field and electric potential of a positively charged rod of length L and charge Q, for any position located a perpendicular distance y from the rod's center. Both quantities have been scaled so the axes are in dimensionless units (see Problem 10.30).

As always, it is a good idea to check our results in limiting cases, where we know what the result *should* be. Here, for example, if $y \gg L$, we would expect the rod to look like a point charge because the length of the rod is insignificant compared to how far we are away from *all* of the charge.

In this case it is reasonable to approximate

$$\left(y^2 + (L/2)^2\right)^{1/2} \approx y$$

because $(L/2)^2 \ll y^2$. Thus Equation (10.17) becomes

$$E_{\text{rod}} \approx K\frac{Q}{y^2}$$

as we would expect!

Phew. Let's move on to calculating V. The potential is, of course, a scalar, so there are no components to worry about. The overall integral will still be symmetric, though, since every charge dQ has a partner charge that is equidistant from the midpoint of the rod and therefore has the same contribution to the overall potential. So, like before, we'll calculate just half of the overall field and then double the result:

$$V = K \int \frac{dQ}{r}$$

$$V = 2K\lambda \int_0^{L/2} \frac{dx}{r}$$

The substitutions $dQ = \lambda dx$, the insertion of the bounds of integration, and the introduction of a factor of 2 all mirror what we did when calculating $\vec{E}$. Inserting for r gives

$$V = 2K\lambda \int_0^{L/2} \frac{dx}{(y^2 + x^2)^{1/2}} \tag{10.18}$$

Once again we've *set up* the integral. Solving it (again with trig substitution) yields

$$V_{\text{rod}} = 2K\lambda \ln\left(\left(\frac{L}{2y}\right) + \left(\left(\frac{L}{2y}\right)^2 + 1\right)^{1/2}\right) \tag{10.19}$$

or, in terms of Q and L instead of λ:

$$V_{\text{rod}} = \frac{2KQ}{L} \ln\left(\left(\frac{L}{2y}\right) + \left(\left(\frac{L}{2y}\right)^2 + 1\right)^{1/2}\right) \tag{10.20}$$

This ugly looking equation is plotted in Figure 10.12. If we consider the case where $y \gg L$, we find that $V \approx KQ/y$, as expected. Showing this gets a bit messy; I walk you through the process in Problem 10.31.

A related problem is the case where the distance y is very small compared to L (i.e. $y \ll L$). I leave this one to you, but one approach is to take our solutions here (Equations (10.17) and (10.20)) and simplify them in the case $y \ll L$ (Problem 10.32).

A second approach is to recognize that if $y \ll L$, we're thinking about a position so close to the rod that it might as well extend off infinitely far in either direction. If this seems absurd to you, refer back to Figure 10.10 on page 503. The angle θ is *smallest* at the end of the rod, and the longer the rod is the smaller it will be. If we have a "long" rod, what we mean is that the smallest value of θ is so close to 0 that we might as well just make it *exactly* 0. This, in turn, corresponds to the case where x extends from $-\infty$ to ∞. Thus we should find the same result by simplifying our general results in the case of $y \ll L$ (Problem 10.32) *or* by re-evaluating the integrals we set up here with these different limits. Problem 10.33 asks you to carry out this latter approach. We will also tackle this problem a different way, using Gauss's Law, in Example 10.12.

Example 10.7 Electric Field of a Ring of Charge $\qquad \star\star \int dx$

Consider a thin ring of radius R and uniform charge Q. (a) Set up and solve an integral to determine the electric potential anywhere along the line that runs through the center of the ring and perpendicular to the plane of the ring. (b) Use your result from (a) to determine an expression for the electric field of the ring along the same line. (c) Use a computer to plot both $\vec{E}$ and V.

We sketch the geometry in Figure 10.13. Because the charge is spread out in a circle, it is appropriate to use *polar* coordinates here rather than *Cartesian* coordinates. You may remember from Example 3.28 on page 164 that the length of a piece of the circumference of a circle of radius R is given by $x = R\phi$, where ϕ is some angle between 0 radians (for no arc length at all) and 2π radians (for the entire circumference). It follows that an infinitely small chunk of the circumference is given by $dx = R d\phi$. Thus the electric potential at a distance z above the center of the ring is given by

$$V = K \int \frac{dQ}{r} = K\lambda \int_0^{2\pi} \frac{R d\phi}{r}$$

Where r is, as usual, the distance from each chunk of charge to the point where we're evaluating the potential. λ is the charge density, which is here given by $\lambda = Q/(2\pi R)$ (i.e. the charge divided by the circumference of the ring).

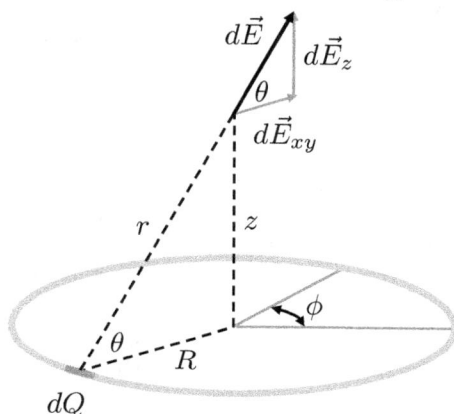

FIGURE 10.13
Perspective view of a ring with radius R and uniform charge Q. The electric field $d\vec{E}$ due to a piece of charge dQ a distance z above the middle of the ring is shown, along with its components.

From the geometry we see r is *constant* with respect to ϕ (it depends on R, a constant, and z, another constant). This means we can factor it out of the integral, leaving us with

$$V = \frac{K\lambda R}{r} \int_0^{2\pi} d\phi$$

Evaluating just the integral yields 2π. Inserting this and substituting $r = \left(R^2 + z^2\right)^{1/2}$, we obtain

$$V = \frac{2\pi K \lambda R}{\left(R^2 + z^2\right)^{1/2}}$$

Once again we can simplify by noting $2\pi R\lambda = Q$ because λ is the charge divided by the linear distance along which the charge is spread (here, the circumference of the ring). Thus our final result is:

$$V_{\text{ring}} = \frac{KQ}{\left(R^2 + z^2\right)^{1/2}} \tag{10.21}$$

Now, we *could* set up a similar integral for $\vec{E}$: once again symmetry plays a role and only the component of $\vec{E}$ along the z axis survives. You are asked to do this in Problem 10.34, but here we are asked to compute $\vec{E}$ directly from V. This turns out to be somewhat quicker than computing the integral, and it is worth keeping in mind that, generally speaking, you have more than one way to obtain expressions for $\vec{E}$ and V.

To proceed we use

$$\vec{E}_{\text{ring}} = -\frac{dV_{\text{ring}}}{dz}\hat{z}$$

$$\vec{E}_{\text{ring}} = -\frac{d}{dz}\left(\frac{KQ}{(R^2 + z^2)^{1/2}}\right)\hat{z}$$

If we apply the chain rule to actually take the derivative of this expression, we find

$$\vec{E}_{\text{ring}} = \frac{KQz}{(R^2 + z^2)^{3/2}}\hat{z} \qquad (10.22)$$

Both of these equations are plotted in Figure 10.14. Notice that the relationship $\vec{E} = -(dV/dz)\hat{z}$ is cleanly evident insofar as the location where V is at a maximum (where its derivative is 0) corresponds to no electric field. This should also make intuitive sense: the center of the ring is the position *closest* to the charge, so the *scalar* quantity V should be the largest there.

At the same time, the *vector* quantity $\vec{E} = 0$ for $z = 0$: every charge dQ has a partner on the opposite side of the ring such that their fields point in opposite directions and therefore cancel. In contrast, for any value of z other than 0, there will be some components of the fields that point in the $\pm z$ direction and therefore *add* when we integrate over the entire ring. This is why $\vec{E}$ isn't 0 *everywhere*.

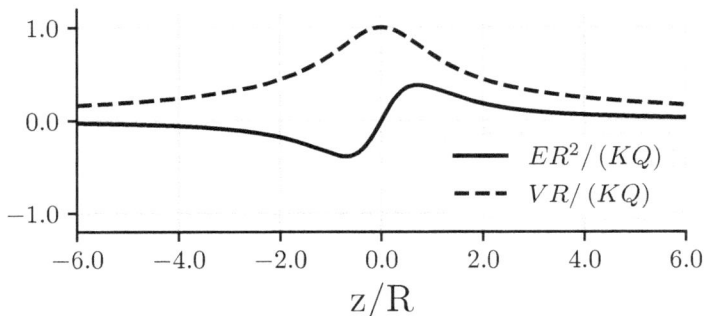

FIGURE 10.14
The electric field and electric potential of a ring of radius R and uniform charge Q, for any distance z from the center of the ring and perpendicular to its plane. Both quantities have been scaled so the axes are in dimensionless units.

Confirming that these equations reduce to the "point charge" results for $z \gg R$ is left as an exercise (Problem 10.36).

Example 10.8 Potential and Field Above a Charged Plate   ★★★ $\int dx$
Consider a point a distance z above a circular disk of radius R and uniform surface charge density σ. (a) Determine an expression for the electric potential V. (b)

Determine the corresponding expression for $\vec{E}$. (c) Simplify your result for (b) in the case where $z \ll R$. If a particle with charge q moves a small distance Δz in this field, how much work will the field do on the particle?

I show the geometry in Figure 10.15. We'll proceed similar to the previous example by calculating V via integration and then determining $\vec{E}$ from V. From Equation (10.14) we have

$$V = K\sigma \int_0^a \frac{da'}{r}$$

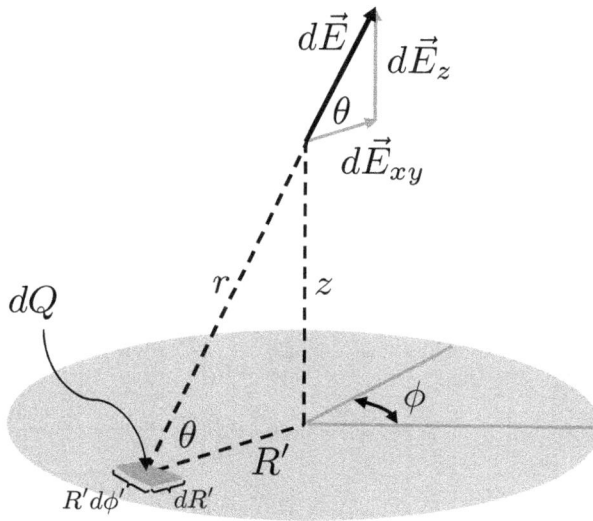

FIGURE 10.15
A circular plate carries a uniform surface charge $\sigma = Q/A$ for a charge Q and area A. The electric field above the center will point in the $\hat{z}$ direction with a value determined by integrating the field components over the entire area.

I noted before that in Cartesian coordinates, $da = dxdy$. What about in polar coordinates? We have seen the "Cartesian to polar" conversion $dx \to Rd\phi$, which accounts for a differential arc length (i.e. a variation in the coordinate ϕ). To get an *area* we need to give the arc some thickness, i.e. we need to multiply the differential arc length by a differential change in the radius. We are also going to switch from calling the radius R to R', because now when we integrate the radius is *changing* from 0 (the center of the disk) to R (the edge of the disk), and we should call "the radius we're considering right now, as we integrate" something besides R. Thus

$$da' = R'dR'd\phi'$$

Because we're integrating over a two-dimensional surface, we can express the integral in two pieces: one over the coordinate ϕ' and one over the coordinate R'. You can visualize this like so: the first integral "walks over" a ring of a given radius, and the second integral "walks over" a series of these rings to cover the entire disk. Mathematically we write this as

$$V = K\sigma \int_0^R \int_0^{2\pi} \frac{R'}{r} d\phi dR'$$

$$V = K\sigma \int_0^R \left(\int_0^{2\pi} \frac{R'}{(R'^2 + z^2)^{1/2}} d\phi \right) dR'$$

In the second line I've made the substitution (from the geometry) that $r = \left(R'^2 + z^2\right)^{1/2}$ and put in parentheses to make the notation here clearer: this is an example of a **double integral**, which may be something you haven't seen before even if you have studied integral calculus.

Once again we might stop here, but if you've studied integral calculus, evaluating this integral isn't as awful as it may sound. We solve these by treating them as two regular (i.e. *single*) integrals. When we evaluate an integral over ϕ' we treat R' as a constant, and vice versa.

The inner integral, over ϕ, gives a factor of 2π and we are left with

$$V = 2\pi K\sigma \int_0^R \frac{R'}{(R'^2 + z^2)^{1/2}} dR'$$

The remaining integral over R' can be solved with the substitution $u = R'^2 + z^2$. Working things through, we find

$$V_{\text{disk}} = 2\pi K\sigma \left(\left(z^2 + R^2\right)^{1/2} - z \right) \tag{10.23}$$

On to (b): What is the electric field, $\vec{E}$, for this disk? Well,

$$\vec{E} = -\frac{dV}{dz}\hat{z}$$

$$\vec{E} = -\frac{d}{dz}\left(2\pi K\sigma \left(\left(z^2 + R^2\right)^{1/2} - z \right) \right)\hat{z}$$

Taking the derivative gives

$$\vec{E}_{\text{disk}} = -2\pi K\sigma \left(\frac{z}{(z^2 + R^2)^{1/2}} - 1 \right)\hat{z} \tag{10.24}$$

I leave graphing these as an exercise (Problem 10.37).

For (c), we evaluate this result in the case where $z \ll R$. The first term in parentheses goes to 0 because the denominator is huge compared to the numerator, so we're left with

$$\vec{E}_{\text{plane}} = 2\pi K\sigma\hat{z} = \frac{Q}{2\varepsilon_0 A}\hat{z} \tag{10.25}$$

(In the last step I used $K = 1/4\pi\varepsilon_0$ and $\sigma = Q/A$.) I'm calling this the electric field of a *plane* because when we're very close to the surface of the disk, the edge geometry doesn't matter (we'd get the same result in this case if we started by analyzing a square plate, for instance).

This is a *constant* electric field, so the force acting on a charge q is constant and the work $W = \int \vec{F} \cdot d\vec{z}$ is just

$$W = F\Delta z = qE\Delta z = q\Delta V$$

as we discussed in Chapter 9. In fact, if you wondered how we can *create* a constant electric field – something we invoked quite a bit in the last chapter – you now have your answer. Of course, this analysis only works for positions close to the surface of our charged plate. (If you're wondering how close is "close", see Problem 10.40!) We'll see in Example 10.10 how to do even better, but in Example 10.9 we'll first apply what we've learned here to a practical situation.

Example 10.9 Charged Particle Near a Charged Air Filter ★
A dust particle with a mass of 1.5×10^{-6} kg and charge $q = 2.0$ nC is 1.0 cm from a square air filter with a surface area of $A = 1.6$ m^2 and total charge $Q = -5.0 \times 10^{-5}$ C. (a) What is the electric force acting on the dust particle? (b) If you ignore the effects of air resistance and assume the particle starts from rest, what speed does the particle have when it impacts the surface of the filter? How much work has the filter done?

We did the hard work of finding the electric field of the filter in the previous example (Equation (10.25)). Here we simply need to plug in some values. Let's first determine the electric field:

$$\vec{E} = q\vec{E}_{\text{plane}} = 2\pi K\sigma\hat{z} = 2\pi K\frac{Q}{A}\hat{z}$$

Plugging in our numerical values (and dropping the "plane" subscript), we find

$$\vec{E} = 2\pi \left(8.99 \times 10^9 \text{ Nm}^2\text{C}^{-2}\right)\left(\frac{-5.0 \times 10^{-5} \text{ C}}{1.6 \text{ m}^2}\right)\hat{z}$$

$$\vec{E} = \left(-1.8 \times 10^6 \text{ N/C}\right)\hat{z}$$

The negative sign here indicates that the electric field points *toward* the plate, as we'd expect from its negative charge. Now to calculate the force on the dust particle:

$$\vec{F} = q\vec{E}$$

$$\vec{F} = \left(2.0 \times 10^{-9} \text{ C}\right)\left(-1.8 \times 10^6 \text{ N/C}\right)\hat{z}$$

$$\vec{F} = -3.5 \times 10^{-3} \text{ N}\hat{z}$$

(As usual I kept additional digits in my value for E when evaluating the following calculation for F, to avoid rounding error.) Again, the negative sign indicates that the positively charged particle is attracted toward the negatively charged plate. Because the force is constant and in the direction of the displacement, the work done is just $W = F\Delta z$, for $\Delta z = 0.010$ m. The work done is then

$$W = 3.5 \times 10^{-5} \text{ J}$$

This work corresponds to the change in kinetic energy of the particle. If it starts from rest and ends with a velocity v, then

$$W = \Delta K \implies W = \frac{1}{2}mv^2 \implies v = \left(\frac{2W}{m}\right)^{1/2}$$

Again inserting our values, we find $v = 6.9$ m/s.

Example 10.10 Field and Potential of a Parallel Plate Capacitor   ★★
Consider two large plates of equal area A that are parallel to one another and separated by a distance d that is much smaller than the edge length of the plates. The plates carry equal and opposite charges Q and $-Q$. (a) What is the electric field at an arbitrary distance x from the positively charged plate somewhere far from the plate's edge? (b) What is the potential difference ΔV between the two plates?

   This arrangement, which is referred to as a **parallel plate capacitor**, is sketched in Figure 10.16. We already know the electric field due to *one* of the plates, from Equation (10.25):

$$\vec{E}_{\text{plane}} = 2\pi K \sigma \hat{z}$$

Here we have *two* plates, and because one is negatively charged and the other is positively charged, between the plates the fields *add*: "away from the positive plate" is in the same direction as "toward the negative plate", so we have

$$\vec{E}_{\text{pp cap}} = \vec{E}_Q + \vec{E}_{-Q}$$
$$\vec{E}_{\text{pp cap}} = 4\pi K \sigma \hat{z} = \frac{Q}{\varepsilon_0 A}\hat{z} \qquad (10.26)$$

where $\sigma = Q/A$. Meanwhile, as we found in Example 10.8, a constant field means an electric potential that changes *linearly* with position:

$$\Delta V = E\Delta z$$

Here, $\Delta z = d$ between the two plates, so

$$\Delta V = Ed = 4\pi K \sigma d = \frac{Qd}{\varepsilon_0 A} \qquad (10.27)$$

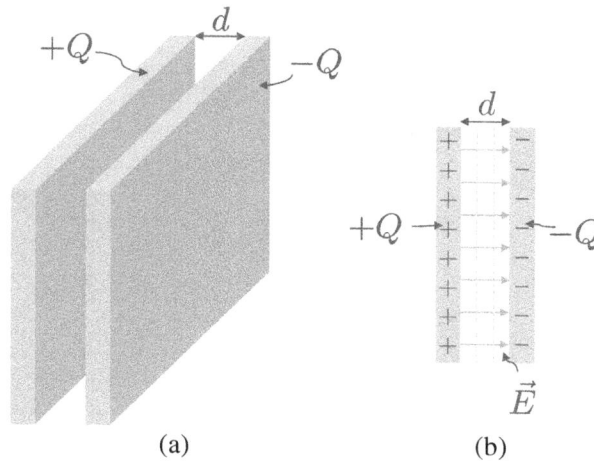

(a)   (b)

FIGURE 10.16

(a) The parallel-plate capacitor consists of two plates separated by a distance d that is small compared to the size of the plates. (The plates are here rectangular, but our analysis applies to any geometry so long as the separation is indeed small and we consider regions far from the edges of the plates.) (b) Electric field vectors (gray arrows) and equipotential surfaces (dashed lines).

I should emphasize that even though our analysis has considered a position at the center of the plate, it holds just as well for any position between the plates that isn't terribly close to an edge: we found that the condition $z \ll R$ means that the edge geometry does not affect our analysis at the *center*, but this same argument holds throughout the interior of the capacitor.

The parallel plate capacitor is a particularly important example because it is commonly used in electrical circuits. We shall see in Chapter 11 that a battery can readily be used to "charge up" two metal plates such that they carry equal and opposite charges, as we have considered here. A charged capacitor, in turn, can be used (for example) in a radio receiver and in a car's windshield wiper.

10.3   Fields from Charge Distributions: Flux and Gauss's Law

You now know how to calculate the electric field and electric potential of a charge distribution *directly*, which is to say by adding up (integrating) the contribution due to each tiny bit of charge. I started the last section by saying that the *idea* of doing this is fairly straightforward, but as we've seen, the reality of actually *setting up* and *doing* the calculations for a particular problem can be very complicated. In this section we will discuss an alternative approach to calculating electric fields which is simpler to carry out, at least in some situations – though it also tends to be harder to understand conceptually.

To begin, let's return to the simple case of a positive point charge q. We know from Coulomb's Law that the electric field at a radial distance r is given by

$$\vec{E} = K\frac{q}{r^2}\hat{r}$$

but let's assume for the moment that we *don't* know this equation – just that the charge creates *some* outward radial electric field that we would like to characterize without using Coulomb's Law.

To do this, let's consider a spherical surface of radius R centered on the point charge. The field lines pass through this surface (Fig. 10.17), and we can quantify *how much field passes through the surface* with the **electric flux**, Φ_E:

$$\Phi_E = \int \vec{E} \cdot d\vec{a} \qquad (10.28)$$

The vector quantity $d\vec{a}$ represents a tiny chunk of area on our surface; it is perpendicular to our spherical surface and points *away* from the enclosed charge. The dot product in this integral, therefore, means that we only want the *component* of the electric field that passes *perpendicularly through* the surface. The fact that we're integrating this quantity means that we're evaluating the orientation between the vectors $\vec{E}$ and $d\vec{a}$ with a dot product *everywhere on the surface*, and then adding up the results. If we can find a situation where the electric field is *constant and* the orientation between the field and $d\vec{a}$ is constant, then our equation for flux simplifies nicely:

$$\Phi_E = \int \vec{E} \cdot d\vec{a} = EA \cos(\theta)$$

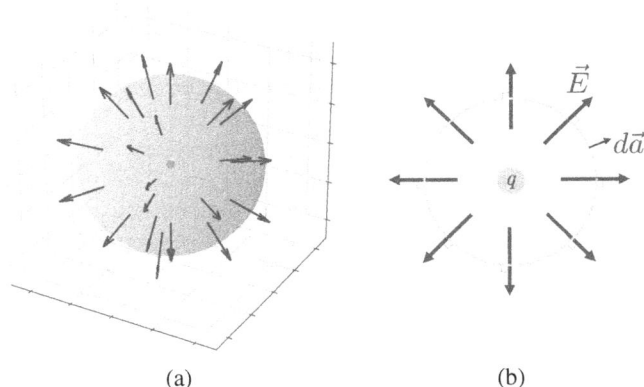

(a)   (b)

FIGURE 10.17
A point charge and an imaginary spherical surface surrounding it. (a) A three-dimensional view. As usual the field permeates all of space and we are drawing only a few field vectors to represent it. (b) A cross-sectional view. The field penetrates the sphere at right angles. The differential area vector $d\vec{a}$ points radially away from the surface of the sphere; just one such vector is drawn to reduce clutter.

where A is the total area of the surface. Now, in general we can imagine all sorts of weird surfaces surrounding our charge (e.g. a cubical box with one side bent in), and for most of them calculating the flux won't simplify like this and will instead be very complicated (see e.g. Fig. 10.18). But if we consider a concentric sphere, as suggested above, then *all* of the field vectors are perpendicular to the *spherical* surface we're considering because the charge is *spherically symmetric*: you can imagine spinning the point charge around in 3D space but

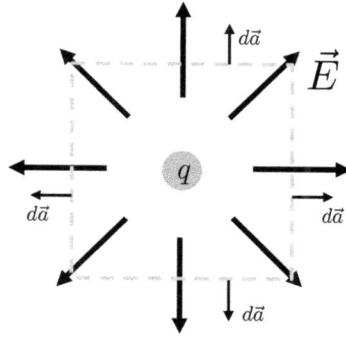

FIGURE 10.18
While we are free to consider the flux passing through any surface enclosing some charge, some surfaces are not convenient to work with mathematically. Here we show a cross-section of a box surrounding a point charge. The orientation of the electric field with the differential area vector $d\vec{a}$ is not constant, and mathematically evaluating the total flux becomes complicated as a result.

nothing about the charge distribution changes if you do.[10] This means there is no way the field can be anything *but* a function of how far you are from the charge, which is to say the radial distance r.

Mathematically, this means that at every position on the sphere $\vec{E} \cdot d\vec{a} = Eda \cos(0°) = Eda$. (If the charge at the center of the sphere was negative, then the field would point *in* and the dot product would be negative.) The equation for the flux therefore simplifies:

$$\Phi_E = \int Eda$$

Furthermore, the electric field is *constant* at a fixed radius (again, by symmetry), so we can pull it out of the integral:

$$\Phi_E = E \int da$$

We're left with an integral over the surface of our sphere, which gives the standard result $A = 4\pi R^2$ for a sphere of radius R. Plugging this in, we find

$$\Phi_E = E\left(4\pi R^2\right)$$

Let's pause for a moment to consider this in the context of the equation for the electric field of a point charge, which we already know from Coulomb's Law. With this in mind, the equation for the flux involves most of the terms we expect; the only term we're missing is the charge q. In fact, it must be the case that

$$\Phi_E = E\left(4\pi R^2\right) = q/\varepsilon_0$$

If this is unclear, consider just the right two terms and divide each by $4\pi R^2$; you obtain $E = \frac{q}{4\pi\varepsilon_0 R^2} = K\frac{q}{R^2}$, as expected.

---

[10]In contrast, if you have a long rod of charge and you swivel it around so it is vertical instead of horizontal, the field will *definitely* look different!

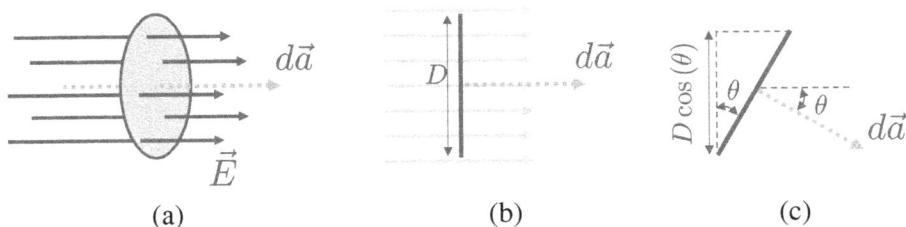

(a)   (b)   (c)

FIGURE 10.19
Electric flux can be defined through any surface. Here we show a circular hoop. (a) a perspective view. (b) An edge view demonstrates that here the electric field is perpendicular to the face of the hoop, which maximizes the flux because $\Phi_E = \int \vec{E} \cdot d\vec{a} = EA\cos(0°) = EA = E\pi D^2/4$, where D is the diameter of the hoop. (c) If the hoop is tilted relative to the field, $\Phi_E = EA\cos(\theta) = E\pi D^2 \cos(\theta)/4$, which can be visualized as the *projection* of the hoop's area that *is* perpendicular to the field. (Here we don't show the electric field vectors to reduce clutter, but because of the tilt, *fewer* field lines pass through the hoop–hence the reduction in flux.)

We've now established a relationship between this new quantity, the *electric flux*, and our old friends charge and electric field. While we've set this up in the context of a *point charge*, it turns out that these relationships apply to *any* charge distribution. The result is called **Gauss's Law** and it is formally written like so:

$$\Phi_E = \oint \vec{E} \cdot d\vec{a} = \frac{Q_{\text{enc}}}{\varepsilon_0} \tag{10.29}$$

Let me unpack this equation with two comments:

- The symbol $\oint$ means that we're integrating over a *closed* surface. In the example of a point charge we considered a sphere that completely enclosed the charge: if we considered, for instance, a half sphere then half of the electric field lines wouldn't pass through the surface and we certainly would not obtain the correct result. (That being said, you can still calculate the flux through other surfaces; it just can't be used in Gauss's Law. See for example Figure 10.19 and Problem 10.43.)

- The Q_{enc} term refers to the *enclosed charge*. Any charge *outside* of the surface will contribute just as much *positive* flux as *negative* flux (Fig. 10.20). We see from this immediately that if $E = 0$ everywhere along our Gaussian surface, then there is no net flux and therefore no enclosed charge. A prominent example where we would find no field is inside a conductor, so we see right away that Gauss's Law supports the argument we made near the beginning of this chapter: because the electric field in a conductor in electrostatic equilibrium is 0 N/C, Gauss's Law tells us that any excess charge resides on the surface.

I do not pretend to have *derived* Gauss's Law in the general case, though I hope it strikes you as plausible enough given how we've set it up.[11] The power of Gauss's Law is that if the

---

[11]If you believe that what we've done for a point charge is valid, then you get most of the rest of the way by the superposition principle: in a charge distribution the net flux is just the sum of the flux due to all of the charges. That said, a *complete* derivation requires invoking properties of vector fields that are beyond the scope of this book.

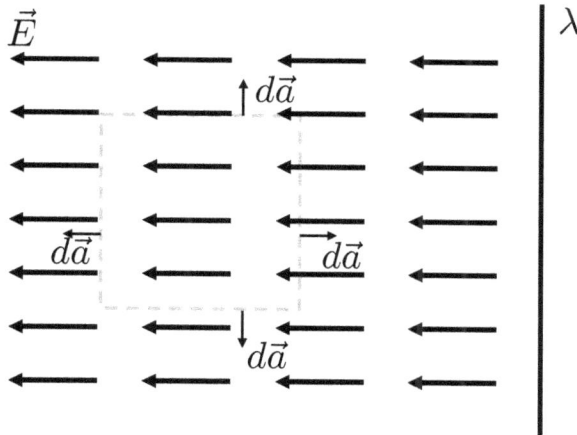

FIGURE 10.20
The electric field near a line of charge with linear charge density λ. A Gaussian box nearby has 0 net flux due to the charge, which is entirely outside the surface: $\vec{E} \cdot d\vec{a} = -E\,da$ on the *right* vertical surface (because $d\vec{a}$ and $\vec{E}$ are antiparallel), $\vec{E} \cdot d\vec{a} = E\,da$ on the left surface (the vectors are parallel), and $\vec{E} \cdot d\vec{a} = 0$ on the top and bottom horizontal surfaces (the vectors are at right angles). When adding up these terms over all four surfaces, the overall result is 0 Nm^2/C.

integral over $\vec{E} \cdot d\vec{a}$ can be simplified, then you can determine the electric field much more easily than with the nasty integrals we invoked in the previous section.

How do we simplify the integral in Gauss's Law in other situations? Well, $\vec{E}$ needs to have a consistent orientation relative to our surface so the dot product is *the same* everywhere. We also want E to be *constant* along the so-called **Gaussian surface** that we use to evaluate the field; this allows us to factor it out of the integral entirely. There are three situations where this symmetry exists (Fig. 10.21):

- We have seen that a point charge is spherically symmetric, but (at the risk of stating the obvious) so too is a *sphere*. Example 10.11 demonstrates such a situation, where our *Gaussian surface* is a sphere.

- A long line of charge is symmetric if you rotate it around its long axis, and the field falls off as a function only of the radial distance from the line. We therefore say that it is has *cylindrical symmetry*. We therefore choose our Gaussian surface to be a cylinder (Example 10.12).

- As we've seen, a large plane of charge has a field that points perpendicular to its surface and must fall as a function only of the distance above the plane. Here we use a *Gaussian box* (Example 10.13).

You can conceive of other situations where there *isn't* a convenient symmetry argument (e.g. a line of charge that is tied into a complicated knot.) In these cases Gauss's Law is still *true*, but evaluating the electric flux becomes very complicated. We won't worry about applying Gauss's Law to such problems in this book.

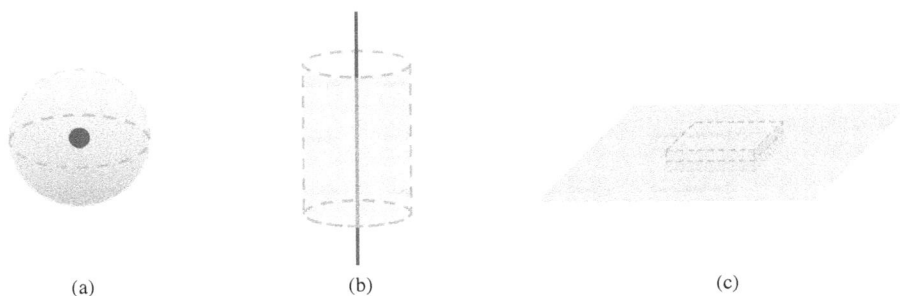

(a)   (b)   (c)

FIGURE 10.21
Examples of charge distributions and the corresponding Gaussian surfaces that simplify the mathematical analysis. (a) The charge is shown in black and has *spherical* symmetry. For instance, it might be on the outside of a spherical shell, or perhaps evenly distributed throughout the sphere. The Gaussian surface, in gray, is also a sphere. Example 10.11 examines this in detail. (b) The charge is distributed along a straight line and has *cylindrical* symmetry. The Gaussian surface is a cylinder centered on the wire. This allows us to characterize the field "per unit length", since we are here assuming it is very long. Example 10.12 examines this in detail. (c) The charge is distributed on a large flat plane and has *planar* symmetry. The Gaussian surface has surfaces above and below the plane that are parallel to the plane. Example 10.13 examines this in detail; we characterize the electric field per unit area.

Example 10.11 Gauss's Law with a Spherical Shell $\star\star\int$
Consider a spherical shell of radius R that carries charge Q uniformly distributed on its surface. Use Gauss's Law to determine the electric field at an arbitrary distance r from the center of the sphere.

A cross-section of the sphere is shown in Figure 10.22. There are two distinct regions: inside the sphere ($r < R$) and outside ($r > R$). Because the shell is spherically symmetric, it is most convenient to use a spherical Gaussian surface, centered at the middle of the charged shell.

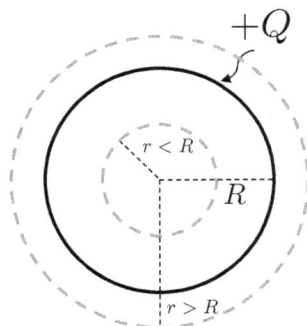

FIGURE 10.22

Now, in the region $r < R$ there is no enclosed charge, Gauss's law tells us right away that there is no field, just like inside a charged conductor. Gauss's Law states that

$$\oint \vec{E} \cdot d\vec{a} = \frac{Q_{\text{enc}}}{\varepsilon_0}$$

But if $Q_{\text{enc}} = 0$ then it follows that $E = 0$, as well (by symmetry $\vec{E} \cdot d\vec{a} = E\,da$, so $E = 0$ for the integral to be 0). This is a surprising result, perhaps: each bit of charge is indeed contributing *some* field at every point inside the shell, but evidently the *net* effect of every bit of field perfectly cancels, *everywhere* inside.

What about *outside* the sphere? Here $Q_{\text{enc}} = Q$, the entire charge on the sphere. We have set up the problem such that the electric field is constant along the sphere and is aligned with the differential area vector $d\vec{a}$, so the integral simplifies just as with a point charge:

$$\oint \vec{E} \cdot d\vec{a} = \frac{Q}{\varepsilon_0}$$

$$E \oint da = \frac{Q}{\varepsilon_0}$$

$$E\left(4\pi r^2\right) = \frac{Q}{\varepsilon_0}$$

$$E = \frac{Q}{4\pi\varepsilon_0 r^2} = K\frac{Q}{r^2}$$

"Wait a minute!" you might say – this is the exact same field that we'd get if all of the charge was concentrated at a *point* at the middle of a sphere. Yes indeed–this remarkable result is very challenging to show through direct integration (I invite you to try and set it up if you don't believe me). The overall field is graphed in Figure 10.23.

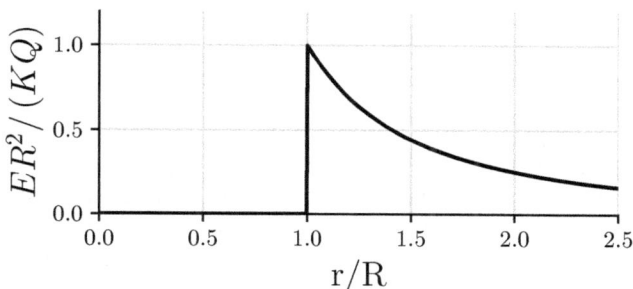

FIGURE 10.23
The electric field of a spherical shell of radius R that carries charge Q on its surface. The field is 0 inside the sphere and outside the field obeys $E = KQ/r^2$

Example 10.12 Gauss's Law with a Long Wire $\star\star \int$
Consider a long straight wire that carries a uniform linear charge density λ. Use Gauss's Law to determine the electric field at an arbitrary perpendicular distance r from the wire.

Note first that we analyzed a similar situation through direct integration in Example 10.6. There, though, we were considering a rod of finite length. Here we have a wire that is *long*, i.e. we're thinking about a position so close to the wire that it might as well extend off infinitely far in either direction (refer back to the end of Example 10.6 if this is unclear). We shall see here that Gauss's Law allows us to solve the problem much more easily.

To begin, we need to recognize that whatever the electric field due to this wire is, it must point radially away from the wire. This follows from the symmetry of the system: if we rotate the rod around the axis running along its middle and/or flip the entire thing around it by 180° (so for instance the top becomes the bottom, or the left side becomes the right side), *nothing has changed* about how the charge is situated: the wire looks just as it did before the rotation. Thus, the electric field must also be the same – but that is only possible if the field is radial, as we have said.

Thus the convenient Gaussian surface to use here is a cylinder with an arbitrary radius r and length L, located such that the wire runs through its middle (Fig. 10.24). There are two pieces to the Gaussian surface: the "end caps", which consist of two circles with area $A_1 = \pi r^2$, and the tube that connects them, with area $A_2 = 2\pi r L$ (just the circumference of the end cap multiplied by the length of the cylinder).

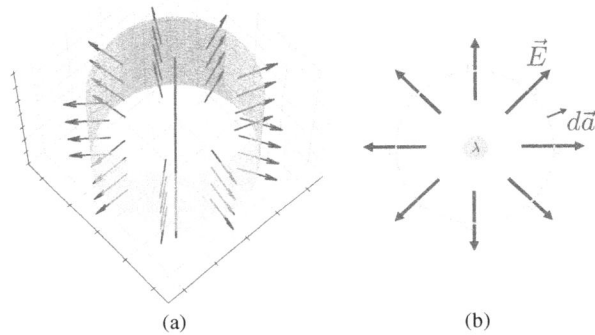

(a)   (b)

FIGURE 10.24
A line of charge and an enclosing Gaussian surface. (a) A perspective view. The line of charge is shown as a vertical line, and the outside edge of the Gaussian cylinder is drawn in gray with some penetrating electric field vectors. (b) a side view looking down the wire, with a slice of the Gaussian surface shown as a circle.

When we consider $\oint \vec{E} \cdot d\vec{a}$ for Gauss's Law, we must consider each component of the Gaussian surface separately. For the two end caps, $\vec{E}$ is perpendicular to $d\vec{a}$ (the first is perpendicular to the wire and the second points along the axis of the wire), so the dot product yields a 0 and, therefore, this portion of the Gaussian surface makes no contribution to the integral.

With this in mind, let's apply Gauss's Law to the tube portion of the surface, where $\vec{E}$ and $d\vec{a}$ are *parallel* (they both point radially away from the wire):

$$\oint \vec{E} \cdot d\vec{a} = \frac{Q_{\text{enc}}}{\varepsilon_0}$$

$$E \oint da = \frac{Q_{\text{enc}}}{\varepsilon_0}$$

$$E\left(2\pi r L\right) = \frac{Q_{\text{enc}}}{\varepsilon_0}$$

How much charge is the Gaussian tube, of length L, enclosing? Well, $\lambda = Q_{enc}/L$, so if we solve the equation above for E and make this substitution, we find

$$E_{line} = \frac{\lambda}{2\pi\varepsilon_0 r} = \frac{2K\lambda}{r} \tag{10.30}$$

I am referring to this as the electric field of a *line* of charge to distinguish the "very long" situation we're considering here from the "finite length" situation of a *rod*, which we have considered separately.

Example 10.13 Gauss's Law with a Plane ⋆⋆∫

Consider a large flat surface that carries a uniform surface charge density σ. Use Gauss's Law to determine the electric field at distance z above the surface.

I'll move through this one a bit faster, in an effort to demonstrate how quickly you can solve a problem with Gauss's Law once you understand the idea. A cross-sectional view of the sheet is shown in Figure 10.25. By the same logic as the preceding example, only the two parts of the Gaussian surface that are parallel to the plane (in the figure this is the top and the bottom portion of the box; each has area A, say) will accept any electric flux, so Gauss's Law gives

$$\oint \vec{E} \cdot d\vec{a} = \frac{Q_{enc}}{\varepsilon_0}$$

$$E(2A) = \frac{Q_{enc}}{\varepsilon_0}$$

If we note that $\sigma = Q_{enc}/A$, then we find

$$E_{plane} = \frac{\sigma}{2\varepsilon_0} \tag{10.31}$$

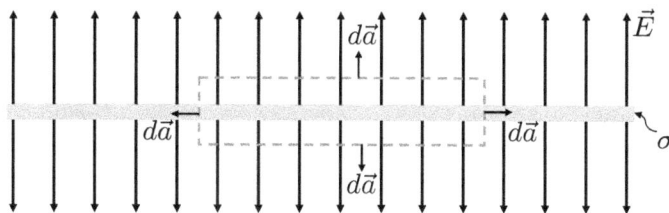

FIGURE 10.25

A cross-sectional view of a thin sheet carrying charge density σ. The Gaussian box (with gray dashes) is centered on the plane.

10.4   Problems for Chapter 10

10.1 Insulators and Conductors

★   **Problem 10.1.** A plastic ball carries a charge of -25 nC before it briefly touches a metal plate. After, it carries a charge -10 nC.

   a. What kind of particles (electrons or protons) were transferred, and in which direction (ball to plate or plate to ball)? Explain.

   b. How many charged particles were transferred?

★   **Problem 10.2.** Three identical metal spheres are hung by insulated wires from the ceiling. Initially, ball A (on the left) carries a charge $+Q$, ball B is neutral, and ball C (on the right) carries a charge $-Q/2$.

   a. Ball B is given a gentle shove to the left so it bumps into ball A. What charge does each of the balls have right after ball B swings away from ball A?

   b. Ball B then swings into ball C. What charge does each of the balls have right after ball B swings away from ball C?

   c. Ball B swings back into ball A one more time. What charge does each ball have right after ball B swings away from ball A?

★   **Problem 10.3.** An electroscope is touched with a positively charged glass rod. Use a series of diagrams and text to explain what happens to the leaves.

★   **Problem 10.4.** A metal plate is given a large negative charge, held vertically, and brought from the side until it is near (but not touching) a metal ball that is hanging from the ceiling by a thin string. (a) Sketch the situation and mark the locations of any excess positive and negative charge. (b) Will the ball move? Why or why not?

★   **Problem 10.5.** A conductive ball is dropped onto a metal plate that is sitting flat on the ground. Both the ball and the plate are neutral and the ball bounces back to a height h_0 above the ground. Suppose this experiment was repeated where the ball remains neutral but the plate now carries a positive charge. The ball bounces to a height h. Is h larger than, smaller than, or the same as h_0? No mathematics is necessary, but explain with a diagram and a sentence or two.

★★ **Problem 10.6.** A large metal plate is given a very large positive charge, Q. A small metal ball carries no charge and is to be dropped onto the floor next to the plate, as shown in Figure 10.26. Your friend confidently declares, "The charge on the plate doesn't matter; the ball will fall straight down".

   a. You gently explain to your friend that they are incorrect. Will the ball be deflected toward or away from the plate as it falls? Explain with a diagram and a sentence or two.

   b. If your answer to (a) is that the ball deflects toward the plate, suppose the deflection is so strong that the ball hits the plate before it hits the ground. What happens next? If, on the other hand, your answer is that it deflects *away*, consider the ball's kinetic energy as it strikes the ground. Will it be less than, equal to, or larger than the kinetic energy if the experiment was repeated without the plate? Explain with a diagram and a sentence or two.

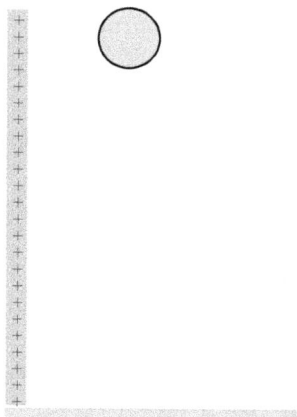

FIGURE 10.26
Problem 10.6

★ **Problem 10.7.** A spherical piece of metal has radius R and charge $+Q$. What is the potential difference between the exact center and a point inside the sphere at a distance $R/2$ from the center?

★★ **Problem 10.8.** A Newton's Cradle is a desktop toy where five identical metal balls are suspended by light strings; they rest side-by-side in a horizontal row. If the leftmost ball is pulled to the side and released, it will impact the row and the rightmost ball will swing away; the process then repeats in reverse. Suppose the leftmost ball is pulled to the side and given a positive charge Q (Fig. 10.27). a. What charge will the rightmost ball have as it swings away from the other balls? Explain any reasonable approximations you make in arriving at your answer. b. Consider the maximum height the right ball reaches in its swing. Will it be higher, lower, or the same as its height would be if the leftmost ball was not charged? Explain.

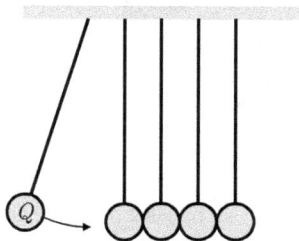

FIGURE 10.27
Problem 10.8

10.2 Fields from Charge Distributions: Integration

★★ **Problem 10.9.** Consider a quarter ring of radius R that carries a charge Q uniformly spread over its length.

a. Sketch the situation and *set up, but do not solve* an integral to determine the electric field at the center of the ring. Your equation should be solvable by someone unaware of the physical context of the problem: any dependence on the variable of integration should be explicit, and the bounds should be clearly marked. Explain all parameters, simplifications, etc. Indicate on your sketch the *direction* of the electric field and briefly explain.

b. Repeat (a) for the electric potential rather than the electric field.

∫ **Problem 10.10.** Solve the integrals (a) and (b) you set up in Problem 10.9.

★★ **Problem 10.11.** Repeat Problem 10.9 for a half ring.

★★ ∫ **Problem 10.12.** Solve the integrals you set up in Problem 10.11.

★★ **Problem 10.13.** Repeat Problem 10.9 for a complete ring.

★★ **Problem 10.14.** Solve the integrals you set up in Problem 10.13 by analyzing the symmetry of the problem (no calculus allowed!).

★★ ∫ **Problem 10.15.** Check your answer to Problem 10.14 by evaluating your integrals from Problem 10.13.

★★ **Problem 10.16.** A thin conducting rod of length L carries charge Q.

  a. Set up (but don't solve!) an integral for the component of the electric field that is perpendicular to the rod, a distance y perpendicular to one end of the rod.

  b. Repeat (a), but for the component of the electric field that is parallel to the rod.

  c. Now set up an integral for the electric potential at the same location considered in (a) and (b).

★★ ∫ **Problem 10.17.** Solve the integrals (a)–(c) you set up in Problem 10.16. Check that your results simplify as you'd expect when $y \gg L$.

★ **Problem 10.18.** An air filter is in the shape of a 0.30 m × 0.50 m rectangle. It carries a 5.0×10^{-6} C charge evenly distributed across its surface. What is the electric field just above the surface of the filter?

★★ **Problem 10.19.** A capacitor is formed by placing two square metal plates close to each other and charging them with equal and opposite charges of ±1.5 nC. If the plates have an edge length of 1.0 cm, what is the magnitude of the electric field inside the capacitor? What would the field be if the edge length of the plates was doubled?

★★ **Problem 10.20.** A thin flat sheet has an area A and carries a total charge $+Q$ evenly distributed along its surface.

  a. Derive an equation for the electric field a small distance y above the surface of the sheet. Include a diagram.

  b. A particle with charge $-q$ and mass m starts a distance y above the sheet and is given an initial velocity $\vec{v}$ directly away from the sheet. How far from the plate is it when $\vec{v} = 0$?

★★ **Problem 10.21.** Consider the shocking superhero situation described in Problem 10.30.

  a. This time, suppose β-boy holds up his palm 1.0 cm from γ-girl to shock her. What net charge must exist on β-boy's hand for an electrical discharge to occur? Consider his palm to be a flat rectangular surface with an area of 75 cm².

  b. Suppose instead the charge is concentrated at a small point with an area of 0.50 cm² at the tip of β-boy's finger. How much charge is needed to shock the hero, still 1.0 cm away, in this situation?

★★ **Problem 10.22.** Science museums for children often have interactive demonstrations. As a demonstration of electric fields, consider a metal ball suspended by a thin rope between the plates of a large capacitor, as sketched in Figure 10.28. The rope is long enough that it is possible for the ball to swing into contact with both plates. Initially the ball carries a small positive charge, and the capacitor is not charged. Museum attendees have three buttons to push that do the following:

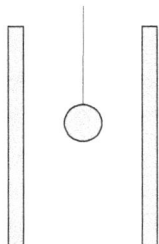

FIGURE 10.28
Problem 10.22

I. Make the left capacitor plate positively charged and the right plate negatively charged.

II. Make the right capacitor plate positively charged and the left plate negatively charged.

III. Make the charge on the plates sinusoidally oscillate between I and II.

In addition, a dial can be turned to increase/decrease the strength of the electric field between the capacitor plates.

Consider the following:

a. Describe what happens when button (I) is pushed when the dial is set to a low (but nonzero) capacitor voltage. Include a diagram that indicates all forces acting on the ball and the location(s) of electric charge on the ball and/or plates (use "+" and "-" symbols).

b. Repeat (a) in the case where the voltage is relatively high.

c. Describe what happens when button (III) is pushed with a relatively low voltage.

No mathematical equations are needed in your responses, but be clear in your explanations and diagrams!

⋆⋆ **Problem 10.23.** An "electron gun" refers to a device that launches electrons in a narrow beam. Suppose such a gun is built by setting up a parallel plate capacitor with a narrow hole in one plate. Electrons are released from a hot cathode near the negatively charged plate.

a. Assuming the electrons start from rest, what potential difference is needed across the plates of the capacitor to ensure the electrons exit the capacitor with a speed of 3.0×10^5 m/s?

b. If the capacitor plates are separated by 2.0 mm, what is the electric field inside the capacitor?

⋆ **Problem 10.24.** X-rays are commonly used in medical imaging (e.g. so dentists can evaluate the health of your teeth). Figure 10.29 sketches the basic idea of how to generate X-rays: free electrons are initially at rest near a negatively charged plate, speed up as they move through a large potential difference, then rapidly slow down as they slam into a block of metal (this energetic collision is what creates the X-rays; the block is angled to control their direction). Suppose the potential difference is 9.0×10^5 V and is generated with a capacitor consisting of square plates of edge length 0.60 m and plate separation 0.50 mm.

a. What is the excess charge on each capacitor plate?

b. What is the electric field (magnitude and direction) between the capacitor plates?

c. Is it possible to design a new capacitor with the same potential difference but a weaker electric field? If *yes*, how? If *no*, why not? Explain your answer (both words and equations needed).

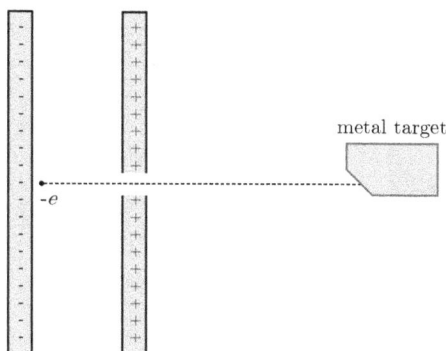

FIGURE 10.29
Problem 10.24

★　**Problem 10.25.** Three large metal plates are stacked together with a very small gap between the top and middle plates and also between the middle and bottom plates. The top and middle plates carry a charge density of 2.0×10^{-6} C/m^2; the bottom plate carries a charge density of -2.0×10^{-6} C/m^2. What is the electric field at the midpoint between (a) the top two plates and (b) the bottom two plates? Specify a magnitude and a direction for each. (These positions are marked by a dot on the right side of Figure 10.30.)

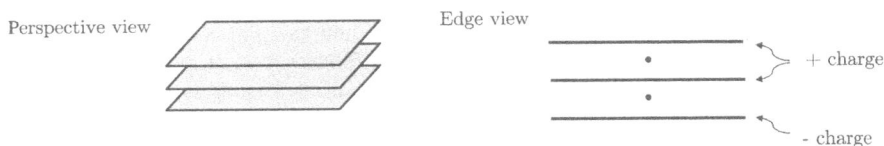

FIGURE 10.30
Problem 10.25

★★　**Problem 10.26.** Two long parallel wires have charge densities of $\pm\lambda$.

a. Set up an equation for the electric field at a point M equidistant from each wire. Your equation should involve at least one integral. You do not need to solve it, but, as usual, simplify it to the point that it would be easy for a veteran of integral calculus to "finish the job".

b. You should be able to state the result of the equation set up in (a) without solving the integrals (use an appropriate result worked out in the main text.) What is the electric field at M?

★★★　**Problem 10.27.** Two long wires, each with length L and charge densities λ_1 and λ_2, meet at one end to form a right angle (Fig. 10.31).

a. Sketch the electric field and equipotential surfaces in the region between the lines in the case where $\lambda_1 = \lambda_2$. How does the sketch change if $\lambda_1 > \lambda_2$ or $\lambda_1 < \lambda_2$?

b. Determine (but do not solve) an integral for the horizontal component of the electric field at P.

c. Determine (but do not solve) an integral for the vertical component of the electric field at P.

d. Show that if $a = b$ and $\lambda_1 = \lambda_2$, the integrals you found in (b) and (c) are identical (except for the fact that one wire is rotated by 90°). Based on your diagram in (a), which way does the field point in this case?

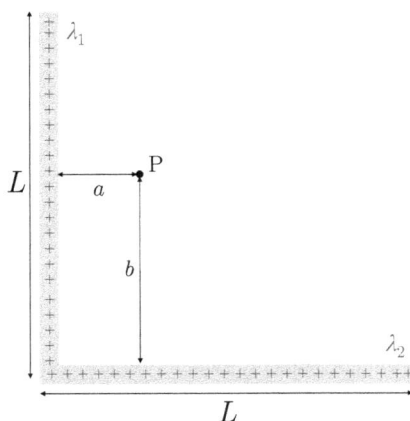

FIGURE 10.31
Problem 10.27

★★★　**Problem 10.28.** Solve the integrals you set up in parts (b) and (c) of Problem 10.27.

** **Problem 10.29.** Consider a thin conducting ring of radius R and charge Q.

    a. Set up (but don't solve) an integral for the electric potential at the middle of the ring. What is the answer? (You don't need calculus to figure this one out.) If your friend says, "I know the potential is 0 V because of symmetry!" how would you gently explain why they are wrong?

    b. What is the electric field at the center of the ring? How do you know?

** **Problem 10.30.** The axis labels in Figure 10.12 may strike you as confusing. To get a handle on this:

    a. Start with Equation (10.17). Divide both sides by KQ, factor L^2 out of the denominator on the right hand side, and then multiply both sides by L^2.

    b. Introduce a term $x = y/L$. Rewrite the right hand side of the equation you ended with in (a) in terms of x.

    c. Use a computer to graph $EL^2/(KQ)$ vs. x. Compare to Figure 10.14.

    d. Repeat the process described in (a)–(c) for Equation (10.20). (The algebraic steps won't be exactly the same!)

** **Problem 10.31.** Equation (10.20) for the electric potential of a finite rod is messy. If we're far from the rod (i.e. if $y \gg L$), the potential *should* reduce to the potential of a point charge, $V = KQ/y$. Show that this is the case by referring to Appendix B and using an appropriate series expansion (keeping only the leading order term) to show that $V_{\text{rod}} \approx KQ/y$.[12]

** **Problem 10.32.** Consider the thin rod of Example 10.6 in the situation where $y \ll L$.

    a. Show that, if we set $V = 0$ V at the appropriate location, Equation (10.20) for the electric potential reduces to

$$V_{\text{line}} = -2K\lambda \ln(r) \qquad (10.32)$$

    b. Show that Equation (10.17) for the electric field reduces to Equation (10.30).

** **Problem 10.33.** Consider the thin rod of Example 10.6, but in the case where the rod is an infinitely long *line* of charge.

    a. Set up and evaluate an integral for the electric *field* a perpendicular distance y from the line of charge.

    b. Set up and evaluate an integral for the electric *potential* a perpendicular distance y from the line of charge.

Example 10.6 does much of the set up for you; the challenge here is to modify the limits of integration and then actually *evaluate* the integrals.

** **Problem 10.34.** Consider the charged ring described in Example 10.7. Set up an integral for the electric field a distance z above the center of the ring and perpendicular to the plane of the ring, as in Figure 10.13.

** **Problem 10.35.** Solve the integral you set up in Problem 10.34.

** **Problem 10.36.** Consider the charged ring described in Example 10.7.

    a. Show that Equation (10.21) reduces to the equation for the potential of a point charge when $z \gg R$.

    b. Show that Equation (10.22) reduces to the equation for the electric field of a point charge when $z \gg R$.

** **Problem 10.37.** Consider the charged plate of Example 10.8.

    a. Use a computer to graph Equation (10.24). Scale the equation so each axis can be expressed in unitless coordinates.

---

[12]Keeping only the term that decays *most slowly* in the limit we're considering is called *keeping the leading order behavior*.

b. Use a computer to graph Equation (10.23). Scale the equation so each axis can be expressed in unitless coordinates.

★★★   **Problem 10.38.** Show that Equation (10.25) can be obtained by considering a charged *square plate* of edge length L instead of a disk of radius R.

★   **Problem 10.39.** Make a graph of the electric potential between the plates of a parallel plate capacitor, starting from $x = 0$ on one plate, where $V = 0$ V, and ending on the other plate at $x = d$, where $V = V_0$.

★★   **Problem 10.40.** Consider the ratio of Equation (10.24) divided by Equation (10.25).

a Graph the ratio as a function of z/R.

b For what value of z/R is the ratio equal to 0.99? What about 0.9?

★   **Problem 10.41.** Two flat, parallel plates are separated by 2.0 cm and are at potentials 25 V and 45 V. Estimate the electric field strength between the plates.

★   **Problem 10.42.** Two flat circular plates each have a 2.5 cm diameter and are separated by 3.2 mm. The electric field between the plates is 6.5×10^4 V/m. What is (a) the potential difference across the capacitor and (b) the charge on each plate?

10.3 Fields from Charge Distributions: Flux and Gauss's Law

★   **Problem 10.43.** A circular loop of radius r is in a region of space with a constant electric field with magnitude E, as shown in Figure 10.32. What is the electric flux through the loop? Provide an algebraic answer.

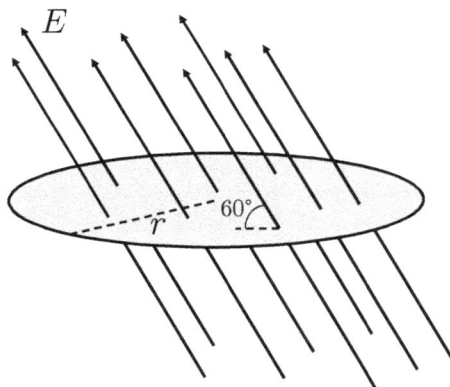

FIGURE 10.32
Problem 10.43

★   **Problem 10.44.** Figure 10.33 shows cross-sectional views of three identical cubes. In each figure, the electric field is in the plane of the page.

a Is the net charge enclosed in each cube positive, negative, or is there no enclosed charge at all? Explain.

b For any of the cubes that *do* contain some charge, suppose the charge has a magnitude of 15.0 nC. If the area of any one face of the cube is 15 cm^2, then what is the magnitude of the electric field?

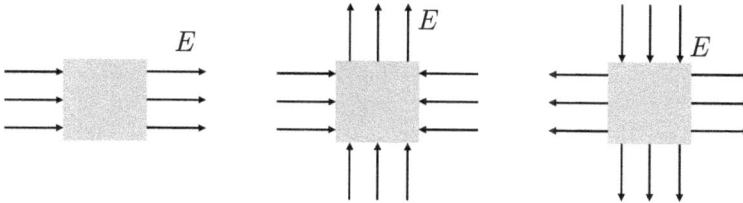

FIGURE 10.33
Problem 10.44

★★ **Problem 10.45.** An uncharged metal ball contains a hollow cavity in which there is a 35.0 nC point charge.

    a. Sketch the ball. What charge is on the inner surface of the ball? What charge is on the outer surface? What charge is spread in between? Explain.

    b. −25.0 nC is then transferred to the ball. Repeat (a).

★ **Problem 10.46.** A plastic straw has been uniformly charged so it carries a net surface charge density σ. If you think of the straw as a cylindrical tube of radius r and length L, then what is the electric field at its exact center?

★★ **Problem 10.47.** A long, thin wire with linear charge density λ runs down the center of a thin, hollow metal cylinder of radius R. The cylinder has a linear charge density of 2λ. What is the electric field for (a) $r < R$ and (b) $r > R$?

★ **Problem 10.48.** The electric field just above a thin metal plate is measured to be 3500 N/C. What is the surface charge density (i.e. charge/area)?

★★ **Problem 10.49.** The electric field just above a thick metal plate is measured to be 3500 N/C. What is the surface charge density (i.e. charge/area) on the surface? Compare to Problem 10.48.

★★ **Problem 10.50.** The Earth maintains an electric field of about 100 N/C directed down toward the surface of the Earth. What excess charge exists on the surface of the Earth?

★★ **Problem 10.51.** Consider a large rectangular sheet that carries a surface charge density σ. Use Gauss's Law to show that the electric field just above the sheet is given by $E = \frac{\sigma}{2\varepsilon_0}$. (I am asking you here to re-derive Equation (10.25).)

★ **Problem 10.52.** A **Van de Graaf generator** is a device used for placing large amounts of electric charge on a hollow metal sphere. Suppose one such sphere has a 15 cm radius and the electric potential on the surface is 2.0×10^5 V.

    a. How much excess charge is on the sphere?

    b. Use Gauss's Law to determine an equation for the magnitude of the electric field some radial distance r from the center of the sphere. You should end up with an algebraic equation. Include a diagram and briefly explain parameters and key mathematical steps.

    c. Use your result from (b) to determine a numerical value for the field at a point 10 cm from the middle of the sphere and again for a point 20 cm from the middle of the sphere.

★★ **Problem 10.53.** We can model the nucleus of an atom as a sphere with evenly distributed charge. The nucleus of a Fermium atom contains 100 protons and has a radius of $R = 7.6 \times 10^{-15}$ m.

    a. Use Gauss's Law to determine an equation for the magnitude of the electric field for some point inside the nucleus, i.e. for $r < R$ where r is the distance from the center of the nucleus.

    b. At what radius r is the electric field the largest?

    c. Based on your answer to (b), if a Fermion nucleus is to split apart (i.e. undergo fission), do you expect the fracture to occur deep within the nucleus or near its surface?

★★ **Problem 10.54.** Consider a sphere of radius R that carries a charge Q uniformly distributed through its volume. Uses Gauss's Law to determine an expression for the electric field at a radial distance (a) $r < R$ and (b) $r > R$ from its center. Briefly explain each step. (c) Repeat (a) and (b) for the case where the ball is a conductor.

★ **Problem 10.55.** Consider the sphere described in Problem 10.54. If the entire sphere is enclosed in a cubical box of edge length $2R$, then what electric flux passes through the box?

★★★ **Problem 10.56.** Consider a simple model of an atom that consists of a positive point charge (the nucleus) of charge $+Ze$ surrounded by a uniform sphere of charge $-Ze$ and radius R. Show that the magnitude of the electric field inside the atom (that is, for $r < R$) is given by

$$E = \frac{Ze}{4\pi\varepsilon_0}\left(\frac{1}{r^2} - \frac{r}{R^3}\right)$$

Which way does it point? What happens when $r > R$?

★★ **Problem 10.57.** While a parallel plate capacitor involves two flat plates, an ideal *cylindrical capacitor* involves two long coaxial cylindrical tubes of different radii, $R_1 < R_2$. Suppose the inner tube carries charge per unit length $+\lambda$ and the outer tube carries charge per unit length $-\lambda$. Use Gauss's Law to determine an algebraic expression for the electric field at the following radii:

a. $r < R_1$ (inside the inner tube)

b. $R_1 < r < R_2$ (between the tubes)

c. $R_2 < r$ (outside the outer tube)

Explain key mathematical steps with a phrase or sentence.

★★ **Problem 10.58.** A **coaxial cable** consists of a metallic "shield" wrapped around an inner core; the two conductors are separated by an insulator and the whole thing is wrapped in plastic. Figure 10.34 shows a schematic view and labels the radius of the core, insulation, and metallic shield as r_1, r_2, and r_3, respectively. Suppose the inner core carries a charge per length of λ and the metallic shield carries a charge per length of $-\lambda$. Use Gauss's Law to determine the electric field at the following radii:

a. $r < r_1$ (i.e. inside the core)

b. $r_1 < r < r_2$ (i.e. in the insulator)

c. $r_2 < r < r_3$ (i.e. in the shield)

d. $r_3 < r$ (i.e. outside the cable)

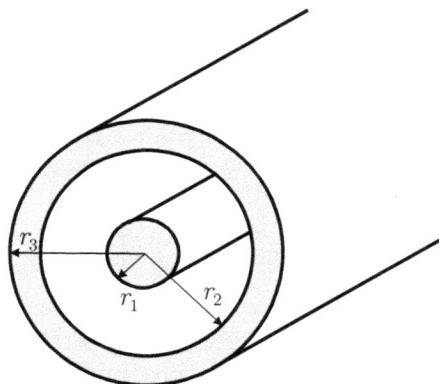

FIGURE 10.34
Problem 10.58

★★ **Problem 10.59.** A charged rod can be held beneath a light charged ball (or balloon) such that the ball levitates above the rod. Suppose the rod has length L and charge Q and the ball has charge q and mass m.[13] How far above the middle of the rod will the ball be if it is in equilibrium? (Use the result from Gauss's law for the field of a long wire, which assumes the ball hovers very close to the rod. If you're feeling bold, relax this assumption and use the equation for the field of a finite rod, and show that the result reduces to what you'd expect if equilibrium is both near and far from the rod.)

★★ **Problem 10.60.** Two 2.80-cm diameter disks with charge ± 30.0 nC have their centers aligned and are 1.50 mm apart.

a. What is the magnitude of the electric field between the disks?

b. A proton is shot from the negative disk toward the positive disk. What launch speed must it have to just barely reach the positive disk?

★ **Problem 10.61.** A 2.0 m × 1.5 m carpet has a uniform charge of -15 μC. If a dust particle has a mass of 12 μg and is suspended above the carpet at some position far from the edges, then what charge is on the dust particle?

★★ **Problem 10.62.** Three planes of charge are parallel with small gap between adjacent planes. The top plane carries a charge density σ, the middle plane carries a charge $-\sigma$, and the bottom plane carries a charge density σ. There are four distinct regions (above all the planes, above just the middle two, above just the bottom plane, and below all of the planes). Use Gauss's Law to determine the electric field in each of these regions.

---

[13]In reality the ball tends to move perpendicular to the axis of the rod, so the demonstrator needs to constantly move the rod to keep it beneath the ball and prevent it from falling.

11

Moving Charge: Electric Circuits

Learning Objectives

After reading this chapter, you should be able to:

- Simplify resistors to determine their *equivalent resistance*. Similarly, simplify capacitors to determine their *equivalent capacitance*.

- Use Ohm's Law and Kirchhoff's Rules to analyze the current, potential difference, and power supplied to resistors connected to one or more batteries.

- Determine the charge and potential difference of capacitors in capacitor circuits.

- Characterize the time-varying behavior of resistor-capacitor (RC) circuits.

- Characterize the resistance of a piece of metal (e.g. a wire) in terms of its resistivity and physical dimensions.

- Describe the *drift velocity* and *electron flow rate* in a wire.

11.1 Introduction: Current, Resistance, and Power, Oh My!

We learned in Section 10.1 that when two charged conductors are placed into contact, the electrons very quickly distribute themselves so as to ensure that the electric field everywhere inside the conductors is 0 N/C. We refer to the phenomenon of *moving charge* as **electric current**. In this chapter we will examine current: how do we create it, and how do we harness it to do something useful? I will begin, in this section, by introducing some important ideas.

As a specific example of electric current, consider a charged capacitor whose plates (which carry charge $\pm Q$) have just been connected by a thin metal wire (Fig. 11.1). The wire has joined the two plates so the entire system is effectively one large conductor. The *net* charge on it is $Q + (-Q) = 0$ C, which correctly suggests that the electrons will flow from the negatively charged plate to the positively charged plate until both plates carry a net charge of 0 C. Once this has occurred we say that the capacitor has *discharged*.

The current, I, that flows through the wire during this process is mathematically defined in terms of the amount of charge that flows through some cross-section in a time interval Δt:

$$I = \frac{\Delta Q}{\Delta t} \tag{11.1}$$

DOI: 10.1201/9781003571568-15

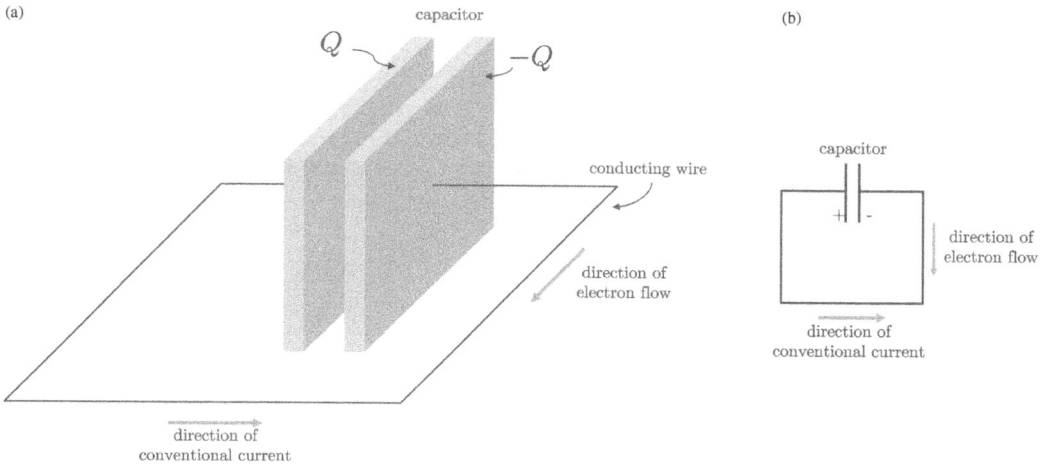

FIGURE 11.1

A charged capacitor's plates are connected by a conducting wire. The excess electrons on the negatively charged plate will quickly move onto the wire, and electrons in the wire will simultaneously move onto the positively charged plate, so in short order both plates will become neutral. The traditional or *conventional* direction of current flow supposes that it is instead the positively charged protons that move, yielding the same neutralizing effect. (a) A perspective view. (b) A **circuit diagram** version of the same situation. The two parallel lines of equal length represent a capacitor and the line segments represent the wire.

The SI unit of current is the **amp**, which is abbreviated A (note that 1 A = 1 C/s).[1] In the language of calculus (i.e. in the limit of Δt being infinitely short), current is simply the time derivative of charge passing through the cross-section:

$$I = \frac{dQ}{dt} \tag{11.2}$$

The current is the same everywhere in the wire, so it doesn't matter *where* the cross-section is.[2]

While we know that it is *electrons* that move as the capacitor equilibrates, it is historical convention to define the direction of current *backwards*, as if it is the positively charged protons that move (the fact that it is, in fact, the negatively charged particles that move was unknown when the convention was established). This is confusing at first but utterly harmless in terms of understanding and predicting the behavior of circuits, and it is such a widely used convention that I would be doing you a disservice by failing to adopt it myself. Thus, as we sketch in Figure 11.1, the electrons flow from the negative plate to the positive plate but we refer to the direction of current flow as from the positive plate to the negative plate.

The capacitor, which began with some potential difference ΔV_i, ends with a potential difference $\Delta V_f = 0$ V. For typical capacitors, this discharge happens very quickly, perhaps on the order of a millionth of a second.[3] We shall see that this behavior can be extremely

---

[1]After André–Marie Ampère (1775–1836)

[2]We'll investigate this in more detail later, but for now a convenient mental image is a long line at the entrance to a concert or sporting event: the people play the role of electrons, and everyone marches along at a constant speed as they move down the line.

[3]The current, in similar fashion, starts out very high and then falls off as the capacitor discharges. We will analyze the details of charging and discharging capacitors in more detail later in this chapter.

useful, but in many cases – if we'd like to keep a lightbulb brightly lit, say – we would prefer to have a *constant* source of electric potential. The most obvious device that does this is, of course, a *battery*. A battery provides an electric potential in a circuit that is sometimes called the *electromotive force*, or *emf*, which is represented with ε. (For instance, a AA battery has an emf of 1.5 V.) Typical batteries use a chemical reaction to maintain a constant potential difference across its ends, which we refer to as the positive and negative *terminals*. (Real batteries eventually drain, of course, but we won't worry about that in this book.)

A simple example of an electric circuit involves connecting a battery to a lightbulb with a pair of wires (Fig. 11.2). The battery acts like a "charge pump" that steadily pushes current from the positive terminal, through the bulb, and to the negative terminal.[4]

Now, it is intuitive that a stronger battery (i.e. one rated at a higher potential difference, say 9 V instead of 1.5 V) will drive a larger current and that this will, in turn, make the bulb glow more brightly. This is correct, but the *bulb* plays a role, too. It is characterized by its electrical **resistance**, R, which has the SI unit of the Ohm, Ω (this symbol is the capital Greek letter Omega).[5] In fact, if we refer generically to a "resistor" with some resistance R, then it could correspond to any number of useful devices, including the heating element of a toaster or air dryer, the fan inside a computer, or a digital chip called an *integrated circuit* that is being powered by our battery. Even the wires themselves have some resistance, but it turns out that the resistance of a wire is very small compared to other circuit elements, so we typically assume a wire's resistance is 0 Ω.

The potential $\Delta V = \varepsilon$ across the bulb (which is provided by the battery), the current I, and the resistance R of the bulb are related according to **Ohm's Law:**[6]

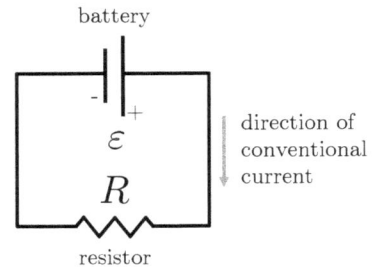

FIGURE 11.2

A circuit diagram of a battery and a resistor, which can represent a wide variety of circuit elements that consume electrical power (e.g. a lightbulb). The battery, which provides an electric potential ε to the resistor, is represented by two parallel lines of unequal length; the longer is the positive terminal and the shorter is the negative terminal. Conventional current flows from the positive terminal to the negative terminal. The resistor, of strength R, is represented by the "squiggle" on the bottom section of the circuit.

$$\Delta V = IR \qquad (11.3)$$

Thus, for instance, a bulb with a *larger* resistance results in a *smaller* current (assuming equivalent batteries).

How does all of this correspond to the bulb's *brightness*? Well, it depends on the rate at which *energy* is being delivered to the bulb. We refer to this as the electric **power** delivered to the resistor, P, and analyzing the units of the relevant quantities (ΔV, I, and R) correctly suggests that
$$P = I\Delta V \qquad (11.4)$$

---

[4]Again, this is according to the (wrong) convention where we assume it is positive charge that moves through the wires. While I have you down here in a footnote, let me also point out that the term "charge pump" might call to mind the flow of *water*. This is a very useful analogy for the purposes of visualizing circuits; you can also consider electric *current* as analogous to water being pushed through pipes by the pump/battery.

[5]After Georg Ohm (1789–1854)

[6]I should point out that Ohm's Law isn't really fundamental, and there are some so-called *non-Ohmic* materials that don't obey it. That said, in this book we'll assume it is always valid.

Why? Well, insert Equation (11.1) for I and you have

$$P = \frac{\Delta Q \Delta V}{\Delta t}$$

The numerator is a measure of energy: some amount of charge ΔQ moves through a potential difference ΔV (refer back to Equation (9.16) on page 464 if you like). Thus we have the desired units of "energy per time", which has the SI unit of a watt (W), as we originally introduced back on page 216.

If you use Ohm's Law with Equation (11.4), you can write the power in terms of any two of the three variables ΔV, I, and R:

$$P = I\Delta V = I^2 R = \frac{(\Delta V)^2}{R} \tag{11.5}$$

Let's briefly recap:

- Electric current refers to the flow of charge. Devices that *produce* current include a charged capacitor and a battery.

- A resistor refers to some device, such as a lightbulb or a computer fan, that *consumes* electrical energy. Its power consumption can be expressed in terms of Equation (11.5).

- Ohm's Law (Equation (11.3)) relates the essential quantities of electric potential (ΔV), current (I), and resistance (R).

These are the fundamental ideas; in the rest of this chapter we will explore them in more depth. For instance, what happens when you have more than one battery and/or resistor? With more circuit elements there are more ways to connect them together; for a given situation, how does the current flow and how much power is delivered to each resistor? Similarly, how can we predict how long it takes to charge or discharge a capacitor, and why might we want to use a capacitor in a circuit in the first place? Before we move on, though, let's apply what we've covered so far in an example.

Example 11.1 Ohm's Law and Power in a Fan ⋆

Desktop computers use fans to blow warm air out of the chassis and to draw cool outside air in. The operation of the fan (e.g. how fast it spins) is controlled by the computer itself, which monitors the internal temperature of the computer. But suppose you have removed such a fan and attached it to a 12 V battery. You use a device called an **ammeter** that has a digital display to report the current through the wire; it reads 250 mA. (Ammeters are schematically represented with an "A" in a circle as in Fig. 11.3. A **voltmeter** is represented with a "V".) (a) What is the fan's resistance, in ohms? (b) How much power, in watts, is the fan consuming?

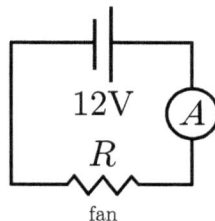

FIGURE 11.3

Because we know both the electric potential provided to the battery and the current running through it, we can determine its resistance from Ohm's Law:

$$\Delta V = IR$$
$$R = \frac{\Delta V}{I}$$
$$R = \frac{12 \text{ V}}{0.250 \text{ A}} = 48 \ \Omega$$

For (b), we can use any of the expressions in Equation (11.5) to determine the power, since we now have all three variables. If we use current and potential, then we find

$$P = I\Delta V$$
$$P = (0.250 \text{ A}) (12 \text{ V}) = 3.0 \text{ W}$$

I invite you to confirm that you obtain the same result with the other relations, i.e. with $P = I^2 R$ and $P = (\Delta V)^2 / R$.

11.2 Resistor Circuits

In this section we will discuss how to analyze *any number* of batteries and resistors that form a circuit. One such example, consisting of one battery and three resistors, is shown on the left of Figure 11.4. Suppose that we know the emf of the battery and the value of each of the resistors, but we *don't* know the current in each section of the wire. There are two main principles involved in analyzing these so-called *resistor circuits*.

Kirchhoff's Rules

First, observe that there are *junctions* in the wire: if we're dragging our finger along the wire, then at a junction we have to choose which path to follow. (In contrast, in Figure 11.2 you just have one path to follow, across the battery and the resistor.) In general, each section of wire will have its own current – which is why they each have their own label in the figure – and since no electrons are created or destroyed in the circuit, it follows that the total current going *toward* a junction must be the same as the total current going *away* from the junction. This is known as **Kirchhoff's junction rule**,[7] which is mathematically expressed as

$$\sum I_{\text{in}} = \sum I_{\text{out}} \tag{11.6}$$

For instance, in Figure 11.4 the junction rule applied to the top junction states

$$I_1 = I_2 + I_3$$

---

[7]After Gustav Kirchhoff (1824–1187). If you think of electrical current as analogous to current in a river, then a junction is just a confluence of multiple rivers and/or a fork: the total amount of water coming in equals the total amount going out, even if the exact amount of water in each branch isn't immediately obvious.

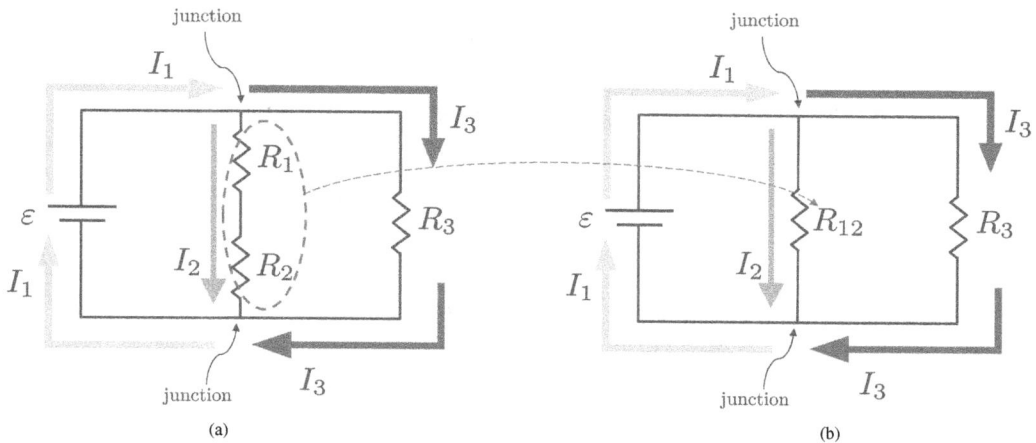

FIGURE 11.4

(a) One way of connecting a battery with emf ε and three resistors with resistances R_1, R_2, and R_3. There are three distinct pieces of wire and the current through each has its own label; the direction of the current is indicated by color-coded gray arrows. The locations where the wires join are referred to as *junctions*. (b) As a first step in simplifying the circuit, the resistors R_1 and R_2 can be represented by an *equivalent resistor* denoted R_{12}.

and applied to the bottom junction we have the equivalent result

$$I_2 + I_3 = I_1$$

Second, notice that there are multiple *loops* in this circuit: there are a total of *three* ways to put your finger down, drag it along a wire, and end up where you started without retracing any wire.[8] We've already seen that an electron moving along a circuit experiences a change in electric potential, but if it makes a complete loop and finds itself back where it started, it must have the *same* potential as when it started (because the potential at some position is determined *only* by the position, and not the past history of any particular charge at that position). In other words, the *net change* in potential along a complete loop must be 0 V. This is analogous to a person walking along a hill: their gravitational potential energy changes as they move up or down relative to the base of the hill, but if they end up where they started they have a net change of 0 J.

Applied to circuits, this principle is known as **Kirchhoff's loop rule**, which is mathematically stated as

$$\sum_{\text{loop}} \Delta V = 0 \text{ V} \qquad (11.7)$$

To apply this rule properly, you need to be careful about the *direction* you move around the loop – generically, *clockwise* or *counter-clockwise*. It doesn't matter which you choose, but your choice affects the way you write down the equation so you have to be thoughtful about how you analyze your loop.

For instance, suppose we walk along the outside perimeter of the circuit of Figure 11.4, starting at the bottom left and moving clockwise, in the direction of the current. The

---

[8]If this is unclear, the three loops are: the outside loop (skip R_1 and R_2), the left inner loop (skip R_3), and the right inner loop (skip ε).

potential *increases* as we move from the negative terminal of the battery to the positive terminal, so for this step $\Delta V = \varepsilon$. The potential then *decreases* as energy is consumed by R_3, so according to Ohm's Law, $\Delta V = -I_3 R_3$ (notice the introduction of the negative sign to make it explicit that ΔV is negative). This brings us back to where we started, so over the entire loop we have

$$\sum_{\text{loop}} \Delta V = \varepsilon - I_3 R_3 = 0 \text{ V} \implies I_3 = \varepsilon / R_3$$

If we had instead walked *counter-clockwise* along this same perimeter, then we'd go *against* the current over R_3 and in the direction of *decreasing* potential across the battery. The sign on both terms reverses as a result and we'd find

$$\sum_{\text{loop}} \Delta V = I_3 R_3 - \varepsilon = 0 \text{ V} \implies I_3 = \varepsilon / R_3$$

We get the same result for I_3, of course, but if we made an error on the sign of one our terms we would find $I_3 = -\varepsilon / R_3$. This negative sign erroneously suggests that the direction of the current marked in Figure 11.4 is backward.

Analyzed correctly, we now have an expression for I_3 in terms of known quantities (the emf and resistances). Example 11.2 finishes the analysis by determining I_1 and I_2.

Example 11.2 Kirchhoff's Laws   ⋆
Determine expressions for I_1 and I_2 in the circuit shown on the left panel of Figure 11.4.

Let's apply the loop rule to the inner left loop, which covers the battery and resistors R_1 and R_2. If we start, as before, at the bottom left of the circuit and move clockwise, we find

$$\sum_{\text{loop}} \Delta V = 0$$
$$\varepsilon - I_2 R_1 - I_2 R_2 = 0$$
$$I_2 = \frac{\varepsilon}{R_1 + R_2}$$

We now have expressions for I_2 and, from the main text, I_3. Thus we can solve for I_1 with the junction rule, which yields

$$I_1 = I_2 + I_3$$
$$I_1 = \frac{\varepsilon}{R_1 + R_2} + \frac{\varepsilon}{R_3}$$

If we perform a bit of algebra to tidy this up, we find

$$I_1 = \varepsilon \frac{(R_1 + R_2 + R_3)}{R_3 (R_1 + R_2)}$$

Reducing Resistors in Series

Kirchhoff's rules, along with Ohm's Law, suffice to fully characterize any circuit composed of batteries and resistors. This can get quite messy for larger circuits, though; Kirchhoff's

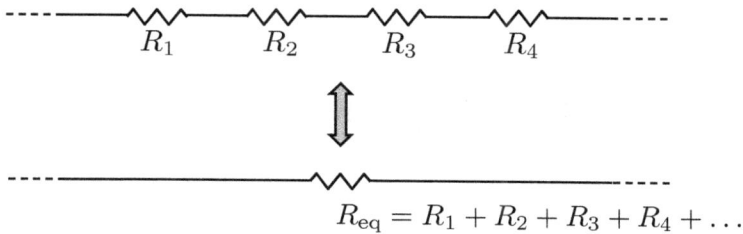

$$R_{\text{eq}} = R_1 + R_2 + R_3 + R_4 + \ldots$$

FIGURE 11.5

A collection of resistors are said to be in *series* if they are connected in an end-to-end chain with no junctions between any pair. The dashed lines indicate that additional resistors may be included and that the chain may be connected to a larger circuit. (Note that while the chain here is laid out in a straight line, there can be *bends* in the wire. If you pick up the four resistors shown here and twist the wire so R_1 and R_2 are aligned vertically, nothing has changed about how the resistors are *electrically connected.*) The chain behaves identically to a single resistor with a resistance equal to the sum of all of the constituent resistances.

rules yield a *system of equations* that can be difficult to handle algebraically. It turns out that there is an additional technique that can simplify the analysis. It is called **circuit reduction**.

The general idea is to re-draw the circuit diagram where some resistors are combined into a so-called **equivalent resistor**: as far as the rest of the circuit is concerned, nothing about the currents or changes in electrical potential change as a result of the reduction. We can iteratively do this, simplifying the circuit diagram at each step, ideally until we have just one battery and one equivalent resistor. This allows us to determine, from Ohm's law, the current passing through the wires connected to the battery, which then makes it easier to analyze the original circuit.

An example will (hopefully) make this clear; we'll stick with the three-resistor circuit shown in Figure 11.4. When we analyzed the left loop with the loop rule, we found

$$\varepsilon - I_2 R_1 - I_2 R_2 = 0$$

Some algebra yields

$$\varepsilon = I_2 \left(R_1 + R_2 \right)$$

This looks suspiciously like Ohm's law for a single battery with emf ε and resistor with resistance R, for which have $\varepsilon = IR$. The only difference is that the resistance R is now the *sum* of the two resistors R_1 and R_2.

This is a general result: if we have any number of resistors connected end-to-end (i.e. with no junctions or other circuit elements between them), then they have the same current and the total potential difference across the entire chain is $\Delta V = I \left(R_1 + R_2 + \ldots \right)$. Thus the *equivalent resistance* is given by the sum of each individual resistance (Fig. 11.5):

$$R_{\text{eq,series}} = \sum R_i = R_1 + R_2 + R_3 + \ldots \tag{11.8}$$

This means that the circuit on the left panel of Figure 11.4 can be equivalently represented by the right panel of Figure 11.4, where resistors R_1 and R_2 are represented by a single resistor $R_{12} = R_1 + R_2$.

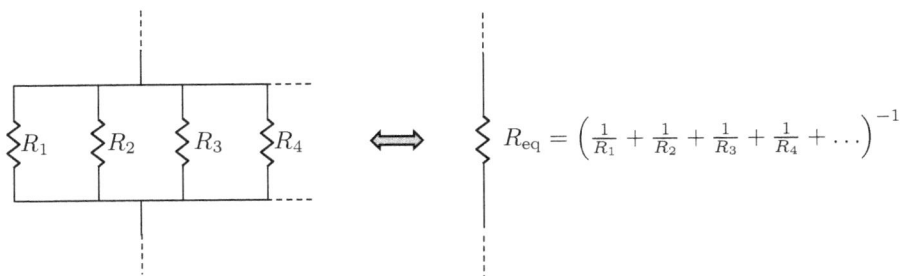

FIGURE 11.6
A collection of resistors are said to be in *parallel* if they are connected by a common junction on both sides. The dashed lines indicate that the wires may be connected to a larger circuit or that additional resistors may be connected in parallel with the ones shown. The collection behaves identically to a single resistor with a resistance equal to the "inverse of the sum of the inverses".

Reducing Resistors in Parallel

Continuing with the same example, notice that we can't combine R_{12} and R_3 in a similar fashion because they're *not* connected in series. Instead, they exist on *parallel* tracks, because the circuit splits into two paths on one side of the resistors (at the top junction, say), with each resistor on one side, before re-joining on the other side (at the bottom junction).

We'd like to determine a general rule for combining resistors that are connected in parallel like this. Conveniently, we have already done so, in Example 11.2. The key points here are (1) the electric potential across each resistor is *the same* (because of the loop rule) and (2) the currents passing through the resistors obey the junction rule. We found

$$I_1 = \frac{\varepsilon}{R_{12}} + \frac{\varepsilon}{R_3}$$

Algebra once again puts this in a form that resembles Ohm's Law:

$$\varepsilon = I_1 \left(\frac{1}{R_{12}} + \frac{1}{R_3} \right)^{-1}$$

This is Ohm's Law if we write $\varepsilon = I_1 R_{123}$ where $R_{123} = \left(\frac{1}{R_{12}} + \frac{1}{R_3} \right)^{-1}$. This is, once again, a general result that can be applied to any number of resistors connected in parallel (Fig. 11.6):

$$R_{\text{eq,parallel}} = \left(\sum \frac{1}{R_i} \right)^{-1} = \left(\frac{1}{R_1} + \frac{1}{R_2} + \frac{1}{R_3} + \dots \right)^{-1} \tag{11.9}$$

Returning to our example of Figure 11.4, we can apply this rule to the resistors R_{12} and R_3 to find

$$R_{123} = \left(\frac{1}{R_{12}} + \frac{1}{R_3} \right)^{-1}$$

$$R_{123} = \left(\frac{1}{R_1 + R_2} + \frac{1}{R_3} \right)^{-1}$$

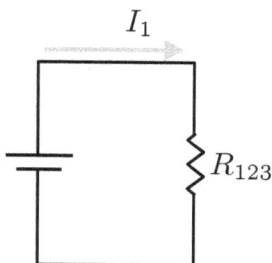

FIGURE 11.7
A further reduction of the circuit shown in Figure 11.4, where the two parallel resistors R_{12} and R_3 have been combined into a single resistor R_{123}. There is now only one current, I_1, that travels around the loop; this current is equivalent to the current leaving the battery in the original (non-reduced) circuit.

What this means is that we can draw a reduced circuit diagram where we have just resistor R_{123} connected to the battery (Fig. 11.7). The current passing through the battery (which is the *same* as in the original circuit – this is the whole point and the reason why we've been using the term *equivalent*) is given by

$$I_1 = \frac{\varepsilon}{R_{123}} = \varepsilon \left(\frac{1}{R_1 + R_2} + \frac{1}{R_3} \right)$$

This is precisely the same result we found *without* circuit reduction in Example 11.2. The next example will give some additional practice with this, and demonstrate some cases where circuit reduction is a handy way to analyze circuits.

Example 11.3 Finding Equivalent Resistance ★★
Separately consider each circuit shown in Figure 11.8. In both cases, determine the power dissipated by each resistor, assuming $R = 5.0\ \Omega$ for each resistor and $\varepsilon = 9.0$ V for both batteries.

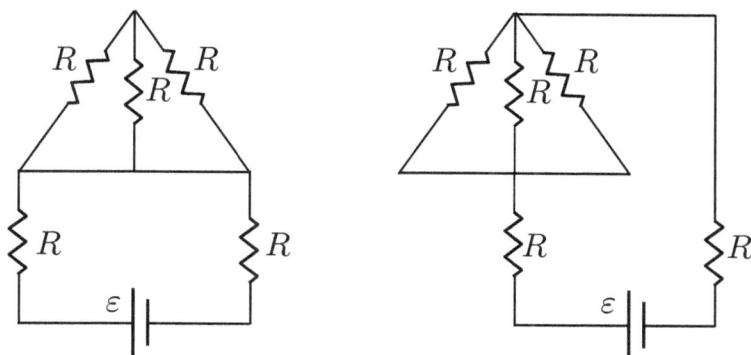

FIGURE 11.8
Two circuits consisting of a battery with emf ε and five resistors with equal resistances R.

Let's first consider the circuit on the left. This circuit can be confusing because some of the resistors are drawn at angles (instead of horizontally or vertically), making it difficult to determine which sets of resistors are in series or parallel with one another. An important insight here is that only the connections made by wires matter; if those stay the same then it doesn't matter how much wire is used to do the connecting or if a resistor is drawn vertically, horizontally, or at an angle. So a *completely* equivalent way of showing circuit (a) involves making all of the resistors vertical, as in the left panel of Figure 11.9.

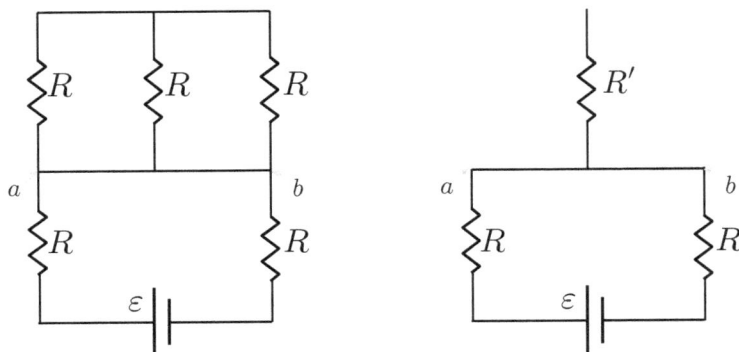

FIGURE 11.9

In this configuration it is more obvious that the top three resistors are all in parallel with one another: they are each directly connected at their top (the topmost horizontal wire) and their bottom (the other horizontal wire). Now, we *could* reduce these three resistors:

$$R' = \left(\frac{1}{5\ \Omega} + \frac{1}{5\ \Omega} + \frac{1}{5\ \Omega} \right)^{-1}$$

$$R' = \left(\frac{3}{5\ \Omega} \right)^{-1}$$

$$R' = \frac{5}{3}\Omega$$

However, while this is good practice it turns out it isn't necessary in this problem. Why? Well, if we redraw the circuit with R' instead of the original three resistors (the right panel of Figure 11.3), we see that it is still connected to the other two resistors on its bottom end but there is *no connection* anymore at the top: the resistor is just dangling off of the rest, and there is therefore no way for current to flow through it.

What do we make of this? Well, the three resistors in question have been *shorted out*: the horizontal path $a \rightarrow b$ experiences *no resistance* whatsoever, while the longer routes (for example, up from a and then through the center or right resistor) involves passing through some resistors. Now, we know from the loop rule that two parallel paths must experience the same drop in potential, so if one path has a *lower* resistance it must have a *higher* current (because the product $IR = \Delta V$ must be the same on each branch).

In the limiting case that one path has negligible *resistance* (and a wire has a resistance that is so tiny that we often think of it as being $0\ \Omega$), then the *other* path must have negligible *current* to compensate.

We conclude, then, that the three central resistors are ignored entirely and dissipate no power. The other two resistors are in series for a net resistance $2R = 10\ \Omega$ and experience a current $I = 9.0\text{V}/10\Omega = 0.90$ A. They each dissipate power

$$P = I^2 R = (0.90\ \text{A})^2 (5.0\ \Omega) = 4.1\ \text{W}$$

What about the circuit on the right side of Figure 11.8? Well, here the set of three resistors, which we calculated to have an equivalent resistance of $R' = \frac{5}{3}\Omega$, are *in series* with the other two resistors (Fig. 11.10), so

$$R_{\text{eq}} = \left(5 + \frac{5}{3} + 5\right)\Omega = \frac{35}{3}\Omega$$

Then it follows from $I = \varepsilon/R_{\text{eq}}$ that $I = 0.77$ A.

This current runs through the resistors directly connected to the battery, and in between it splits into three equal branches while going through the central three resistors. Why *equal*? Well, the potential drop must be the same across each of them (again, from the loop rule), so if the resistances are the same so too must be the currents. Thus each of these interior resistors experiences a current of $I_{\text{interior}} = I/3 = 0.26$ A.

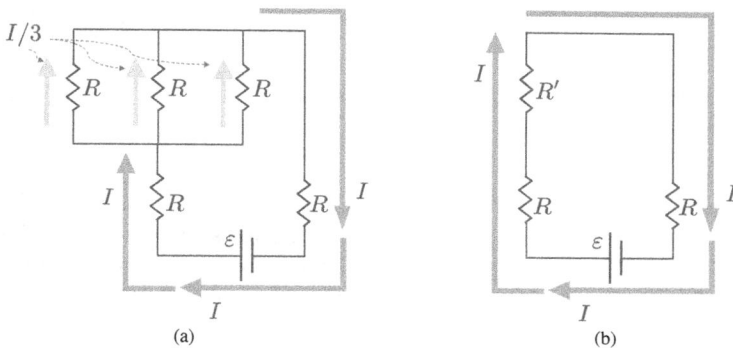

(a)   (b)

FIGURE 11.10
(a) Circuit (b) can be redrawn in the same way as Circuit (a), with each resistor aligned vertically. Here we label the current running through each section of wire with adjacent arrows. (b) The three parallel resistors can be represented with an equivalent resistor R', which is then connected in series with each of the remaining resistors. The reduced circuit has a single loop and Ohm's Law can therefore be used straightforwardly to calculate the current I.

In summary, for (b) the two resistors immediately adjacent to the battery each dissipate power

$$P = I^2 R = (0.77\ \text{A})^2 (5.0\ \Omega) = 3.0\ \text{W}$$

And the three other resistors each dissipate power

$$P = I^2 R = (0.26\ \text{A})^2 (5.0\ \Omega) = 0.33\ \text{W}$$

11.3 Capacitor Circuits

We've now seen how to analyze circuits that include one or more batteries and one or more resistors. Let us now turn our attention to connecting one or more batteries to one or more *capacitors*. To begin, recall Equation (10.27) (from page 514), which describes the potential difference across the plates of a parallel-plate capacitor:

$$\Delta V = 4\pi K \sigma d$$

In this equation, d is the separation between the places and $\sigma = Q/A$, the *charge density*, or simply the magnitude of the charge on one of the plates divided by its surface area A. If we recall that the electrostatic constant K is related to the permittivity of free space ε_0 by $K = 1/4\pi\varepsilon_0$, then we can write the above equation as

$$\Delta V = \left(\frac{d}{\varepsilon_0 A}\right) Q$$

$$\frac{A\varepsilon_0}{d} = \frac{Q}{\Delta V}$$

The left hand side of the equation above is *constant* for a given parallel-plate capacitor; it depends only on the plate area and the plate separation. We refer to this quantity as the **capacitance**, C:

$$C = \frac{A\varepsilon_0}{d} = \frac{Q}{\Delta V} \tag{11.10}$$

Just as a resistor is characterized by its resistance R, a capacitor is characterized by its capacitance C. The unit of capacitance is the **farad**, symbolized F.[9] The capacitance can be increased by filling the space between the plates with a suitable insulator; see Problem 11.33. A typical capacitance for a laboratory capacitor is measured in microFarads ($1\,\mu\text{F} = 1 \times 10^{-6}\,\text{F}$).

The energy in a charged capacitor can be expressed in terms of its capacitance, charge, and potential difference(see Problem 11.27):

$$U = \frac{1}{2}C\left(\Delta V\right)^2 = \frac{Q^2}{2C} = \frac{1}{2}Q\Delta V \tag{11.11}$$

What happens when we connect multiple capacitors and batteries together to make an electrical circuit? Well, when we analyze *resistor* circuits, we make use of three principles:

- Kirchhoff's junction rule ($\sum I_{\text{in}} = \sum I_{\text{out}}$)

- Kirchhoff's loop rule ($\sum \Delta V = 0$)

- Circuit reduction (grouping resistors in series/parallel to determine an equivalent resistance)

There are similar principles at play when we analyze capacitor circuits:

---

[9]After Michael Faraday (1791–1867)

- **The total charge on electrically isolated plates cannot change.** In other words, just as *current* cannot be created or destroyed just because it is flowing through a junction of wires, *charge* cannot be created or destroyed just because it is redistributing itself on isolated capacitor plates.

  As a specific example of this, we find that capacitors in series have the same *charge* (Fig. 11.11), just as resistors in series have the same *current* (So if one capacitor in series has, say, ± 12 nC on its plates, every other capacitor in the series also has ± 12 nC on their plates).

  For capacitors in parallel, we find that (just as the current in a resistor circuit splits at a junction based on the resistance along each track), the charge will split among capacitors in parallel based on the capacitance along each track. I'll demonstrate how this works in practice in the examples at the end of this section.

- Kirchhoff's loop rule ($\sum \Delta V = 0$) still applies, though for a capacitor we have (from Equation (11.10)) $\Delta V = Q/C$ instead of $\Delta V = IR$ (Ohm's Law) for a resistor.

- Circuit reduction can be applied to capacitors, though the rules are different (we'll see what they are and why just as soon as we're done with this bullet point). The key idea, though, is the same: in a resistor circuit you can determine the *current* pushed out by the battery, while in a capacitor circuit you can determine the *charge* pushed out by the battery.

We'll apply these principles to some capacitor circuits, but first let me explain the details of that last bullet point. A key result from our analysis of resistor circuits was that we can *reduce* resistors that are connected in series ($R_{\text{eq,series}} = R_1 + R_2 + ...$) or in parallel ($R_{\text{eq,parallel}} = \left(\frac{1}{R_1} + \frac{1}{R_2} + ... \right)^{-1}$). The corresponding result for capacitors is that *the rules are reversed* (Figs. 11.11 and 11.12):

$$C_{\text{eq,series}} = \left(\sum \frac{1}{C_i} \right)^{-1} = \left(\frac{1}{C_1} + \frac{1}{C_2} + ... \right)^{-1} \tag{11.12}$$

$$C_{\text{eq,parallel}} = \sum C_i = C_1 + C_2 + ... \tag{11.13}$$

Why? Let's consider a battery with just two capacitors in *parallel* (Fig. 11.13). We can draw a complete loop by moving across the battery (from the negative terminal through the positive terminal, say, for a potential difference of ε) and then *any one* of the capacitors, for a potential difference we'll call $-V_C$ (negative because we'll be moving from the positive plate to the negative plate as we continue our loop, which corresponds to a decrease in potential). The loop rule then says

$$\sum V_i = 0$$
$$\varepsilon - V_C = 0$$
$$\varepsilon = V_C$$

Because this holds for *any* of the parallel capacitors, we summarize by saying they *all* have a potential difference equal to that of the battery, ε. In order to establish this potential difference across each capacitor, the battery has to provide a charge to each, which is given by

$$Q_i = C_i \Delta V_C = C_i \varepsilon$$

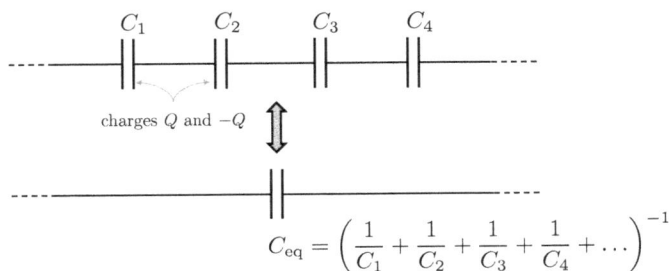

FIGURE 11.11

A collection of capacitors in *series* along with an equivalent capacitor. Compare to Figure 11.5 on page 541, which considers *resistors* in series. Here, neighboring plates from adjacent capacitors are electrically isolated: no wires connect them to the rest of the circuit (plates from C_1 and C_2 are marked as an example). No charge can move onto or away from this "plate pair", so if they *begin* neutral they must *remain* neutral: if electrons move to give a charge $-Q$ on one plate, then the other plate must have a charge $+Q$. If you combine this with the fact that every capacitor (C_1, say) has equal and opposite charge on its plates, then it follows that every capacitor in the series has the same magnitude charge.

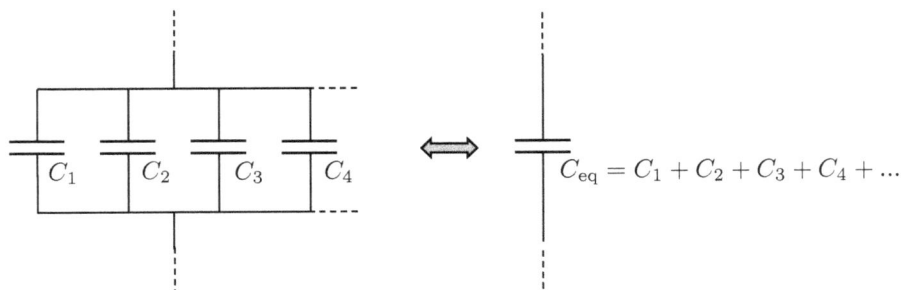

FIGURE 11.12

A collection of capacitors in *parallel* along with an equivalent capacitor. Compare to Figure 11.6 on page 542, which considers *resistors* in parallel.

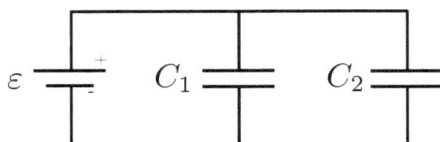

FIGURE 11.13

A battery in parallel with two capacitors.

Thus the *total* charge delivered by the battery is

$$Q_{\text{tot,parallel}} = \varepsilon \left(C_1 + C_2 \right)$$

An *equivalent* capacitor – which plays a role identical to our set of parallel capacitors – looks the same to the battery in the sense that it provides the same emf (ε) and delivers the same amount of charge ($Q_{\text{tot,parallel}}$). Mathematically,

$$Q_{\text{eq}} = \varepsilon C_{\text{eq,parallel}}$$

Comparing the equations above yields

$$C_{\text{eq,parallel}} = C_1 + C_2$$

in agreement with Equation (11.13). The analysis of capacitors in series is left as an exercise (Problem 11.31). The approach is very similar, though the key point is that the *total* potential delivered to the capacitors is ε, rather than *each* of the capacitors receiving that potential difference.

We've just discussed the rules for capacitor circuit reduction. As we discussed in the bullet point list above, we can use this technique along with the principle of charge conservation and Kirchhoff's loop rule to characterize capacitor circuits. The next set of examples demonstrates how.

Example 11.4 Charging and Discharging a Capacitor ★★
Consider the circuit shown in Figure 11.14, where $\varepsilon = 9.0$ V, $C_1 = 1.00 \times 10^2$ μF, $C_2 = 2C_1$ and $C_3 = 3C_1$. The switch is initially in position a long enough for capacitor C_1 to become completely charged. (a) How much charge is on its plates, and what is the potential difference across it? (b) The switch is then moved to position b. What is the charge and potential difference across each of the capacitors once the circuit has re-equilibrated?

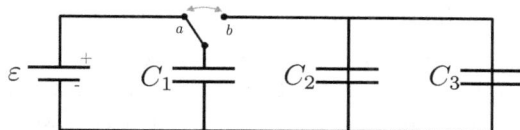

FIGURE 11.14

When the switch is in position a, no charge whatsoever will flow to capacitors C_2 and C_3: just as a resistor requires a closed path for current to flow through it, so too does a capacitor require a closed path for charge to accumulate on its plates.

This means we have a simple circuit of a battery and a single capacitor. The loop rule says, then, that whatever potential is *provided* by the battery is *consumed* by the capacitor:

$$\sum \Delta V = 0$$
$$\varepsilon - \Delta V_{1a} = 0$$
$$\varepsilon = \Delta V_{1a} = 9.0 \text{ V}$$

Then, from Equation (11.10) we have

$$Q_{1a} = C_1 \Delta V_{1a} = \left(1.0 \times 10^{-4} \text{ F}\right)(9.0 \text{ V}) = 9.0 \times 10^{-4} \text{ C}$$

I am using a $1a$ subscript to indicate that we're talking about capacitor 1 during part (a): the charge and potential (but not the capacitance!) will change in part (b).

When the switch is flipped to b, the battery is cut off from the circuit and so it no longer acts as a charge pump. Instead, the charge that we already have on C_1 will distribute itself along the plates of C_2 and C_3. Specifically, the top plate, which initially carries charge $+Q_{1a}$, will spread along the top plate of all three capacitors. At the same time, the bottom plate of C_1, which initially carries charge $-Q_{1a}$, will spread along the bottom plate of all three capacitors.

This is the idea; we can proceed through the mathematics a few different ways. Below I will show you two approaches.

(b) The easier way:

We have three capacitors in parallel, which means we can determine the equivalent capacitance:

$$C_{eq} = C_1 + C_2 + C_3 = 6.0 \times 10^{-4} \text{ F}$$

Then the potential across this equivalent capacitor (which is the same as across each constituent capacitor separately – they're all in parallel, so the loop rule says they must have the same potential) is

$$\Delta V = \frac{Q}{C_{eq}} = \frac{9.0 \times 10^{-4} \text{ C}}{6.0 \times 10^{-4} \text{ F}} = 1.5 \text{ V}$$

Notice I used the *total* charge, which we worked out in part (a): however it distributes itself among the three capacitors, all we care about for this equivalent capacitor is the *total*. From here we can work out the charge on each of the three constituent capacitors with $Q = C\Delta V$. You can verify for yourself that we find $Q_{1b} = 1.5 \times 10^{-4}$ C, $Q_{2b} = 3.0 \times 10^{-4}$ C and $Q_{3b} = 4.5 \times 10^{-4}$ C. If you add all of these up, you find Q_{1a}, which verifies that charge has not been created or destroyed in this process.

(b) the harder way:

We can mathematically express the fact that the charge that is on C_1 in part (a) spreads across all three capacitors in part (b) like so:

$$Q_{1a} = C_1 \Delta V_{1a} = Q_{1b} + Q_{2b} + Q_{3b}$$

Now (as we noted above), the loop rule applies here as well; the potential difference across each of the capacitors in part (b) must be the *same* (you can make a loop with any two of the three capacitors). This gives us the following set of equations:

$$\Delta V_{1b} = \Delta V_{2b} \implies \frac{Q_{1b}}{C_1} = \frac{Q_{2b}}{C_2}$$

$$\Delta V_{1b} = \Delta V_{3b} \implies \frac{Q_{1b}}{C_1} = \frac{Q_{3b}}{C_3}$$

$$\Delta V_{2b} = \Delta V_{3b} \implies \frac{Q_{2b}}{C_2} = \frac{Q_{3b}}{C_3}$$

We're essentially done with the *physics*; we've arrived at a total of four equations and three unknown variables (the three charges – once we know those we can go back and solve for the potentials straightaway using $\Delta V = Q/C$). From here it is just algebra, but it is sufficiently tricky that I'll walk you through one way you can solve these equations.

Let's first use our loop rule equations to determine expressions for Q_{2b} and Q_{3b} in terms of Q_{1b}:

$$Q_{2b} = \frac{C_2}{C_1} Q_{1b}$$

$$Q_{3b} = \frac{C_3}{C_1} Q_{1b}$$

We can then insert these results into our charge conservation equation:

$$C_1 \Delta V_{1a} = Q_{1b} + \left(\frac{C_2}{C_1}Q_{1b}\right) + \left(\frac{C_3}{C_1}Q_{1b}\right)$$

From there we can solve for Q_{1b}:

$$Q_{1b} = \frac{C_1 \Delta V_{1a}}{1 + \frac{C_2}{C_1} + \frac{C_3}{C_1}}$$

$$Q_{1b} = \frac{9 \times 10^{-4} \text{ C}}{1 + 2 + 3}$$

$$Q_{1b} = 1.5 \times 10^{-4} \text{ C}$$

Now it is straightforward to go back and compute Q_{2b} and Q_{3b}:

$$Q_{2b} = \frac{C_2}{C_1}Q_{1b} = 2Q_{1b} = 3.0 \times 10^{-4} \text{ C}$$

$$Q_{3b} = \frac{C_3}{C_1}Q_{1b} = 3Q_{1b} = 4.5 \times 10^{-4} \text{ C}$$

We can check our results: is the total charge on all three plates in (b) the same as what we found on C_1 in (a)? Well:

$$Q_{1b} + Q_{2b} + Q_{3b} = (1.5 + 3.0 + 4.5) \times 10^{-4}\text{C} = 9 \times 10^{-4} \text{ C}$$

Which is indeed what we found for Q_{1a}.

Finally, we can determine the potential across each capacitor. For instance,

$$\Delta V_{3b} = \frac{Q_{3b}}{C_3}$$

$$= \frac{4.5 \times 10^{-4} \text{ C}}{300 \times 10^{-6} \text{ F}}$$

$$= 1.5 \text{ V}$$

You can verify for yourself that the same potential is obtained for the other two capacitors – as we *must*, again according to the loop rule.

Example 11.5 Equivalent Capacitance ★★
Consider the circuit shown in Figure 11.15, where $\varepsilon = 1.50$ V, $C_1 = 1.50 \times 10^2$ μF, $C_2 = 2C_1$, and $C_3 = 3C_1$. (a) What is the equivalent capacitance of this circuit? (b) What is the charge on each capacitor? (c) What is the potential difference across each capacitor?

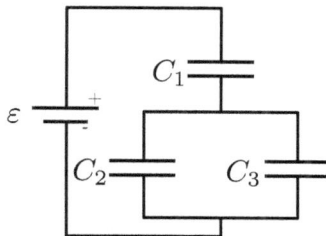

FIGURE 11.15

We can reduce the circuit in two stages: C_2 and C_3 are in parallel and can be combined to yield $C_{23} = C_2 + C_3$. This equivalent capacitor is in series with C_1, and so we can reduce one more time:

$$C_{123} = \left(\frac{1}{C_{23}} + \frac{1}{C_1} \right)^{-1} = \left(\frac{1}{C_2 + C_3} + \frac{1}{C_1} \right)^{-1}$$

This process is shown in Figure 11.16 (reading from top left to bottom right). We could simplify this result algebraically, but instead let's just plug in the given values (with the answer accurate to three significant figures):

$$C_{123} = \left(\frac{1}{300 \ \mu\text{F} + 450 \ \mu\text{F}} + \frac{1}{150 \ \mu\text{F}} \right)^{-1} = 125 \ \mu\text{F}$$

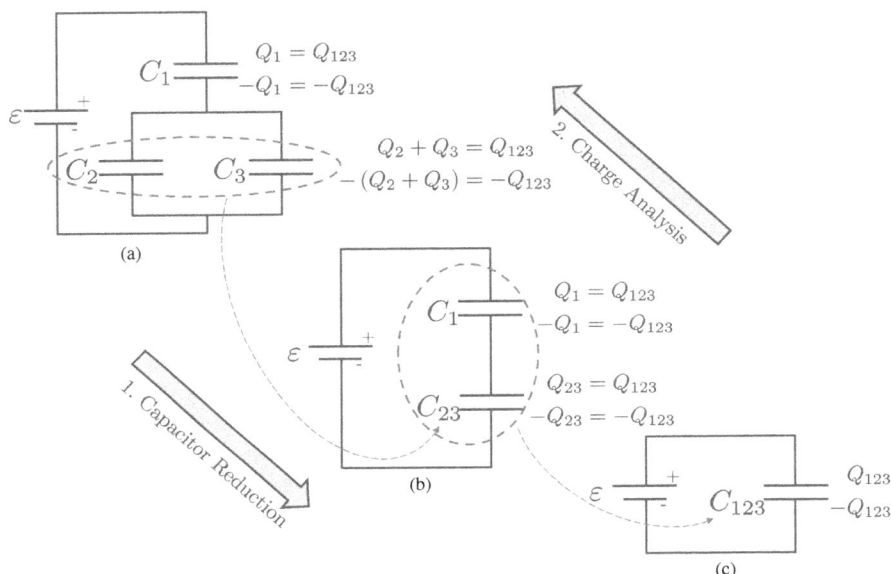

FIGURE 11.16

The original circuit is shown in (a). We first reduce the circuit by grouping C_2 and C_3, which are in parallel, and re-drawing the circuit with the equivalent capacitor C_{23} in the middle diagram (as indicated with the dashed ellipse around C_2 and C_3 and the dashed arrow pointing toward C_{23}). Similarly, in (b) we see that C_1 and C_{23} are in series and can be further reduced to the equivalent capacitor C_{123}, as shown in (c). Once the capacitors are fully reduced, the corresponding plate charges, $\pm Q_{123}$, can be calculated from the loop rule and $C = Q/\Delta V = Q/\varepsilon$. From there we can "work back"' to the original circuit by noting that the top and bottom plates of C_{123} correspond to the top and bottom plates of C_1 and C_{23}, respectively, in the middle diagram. Finally, C_{23} really consists of two capacitors, and so the charge Q_{23} is split between C_2 and C_3 in the original circuit. Along with the loop rule, as discussed in the text, this analysis suffices to determine the charge and potential difference across each of the capacitors in the original circuit.

This takes care of part (a). We now need to analyze the original circuit. To begin, let's work out the potential and charge on the equivalent capacitor C_{123}. Just like in the previous example, if we have a battery connected to just one capacitor, the potential difference across the capacitor is the same as the emf of the battery:

$$\Delta V_{123} = \varepsilon$$

It follows that

$$Q_{123} = \Delta V_{123} C_{123} = (1.50 \text{ V}) \left(125 \times 10^{-6} \text{ F}\right) = 1.88 \times 10^{-4} \text{ C}$$

Knowing the total charge pushed out by the battery is helpful in the same way that knowing the current pushed out by a battery in a resistor circuit is helpful. Here, we know that the top plate of C_1 has a charge Q_{123} and the bottom plates of C_2 and C_3 together have a charge $-Q_{123}$.

Because opposite plates in a capacitor must have opposite charges, we conclude also that the bottom plate of C_1 carries charge $-Q_{123}$ and the top plates of C_2 and C_3 collectively carry charge Q_{123}. (Note also that these three plates are electrically isolated from the rest of the circuit, so they must be overall neutral.)

We know enough to finish our analysis of C_1, since knowing the charge and the capacitance allows us to determine the electric potential:

$$\Delta V_1 = \frac{Q_1}{C_1} = \frac{1.88 \times 10^{-4} \text{ C}}{150 \times 10^{-6} \text{ F}} = 1.25 \text{ V}$$

Now that we know ΔV_1, we can apply the loop rule to the entire circuit. Suppose for instance that we start at the negative terminal of the battery and go clockwise through C_1 and then C_2:

$$\sum \Delta V = 0$$
$$\varepsilon - \Delta V_1 - \Delta V_2 = 0$$
$$\varepsilon - \Delta V_1 = \Delta V_2$$
$$1.50 \text{ V} - 1.25 \text{ V} = \Delta V_2 = 0.25 \text{ V}$$

We could set up an identical loop through C_3 instead of C_2, which means that it must have the same potential difference (equivalently, because these two capacitors are in parallel we could set up a loop just through the two of them; either way their potential differences must be the same):

$$\Delta V_2 = \Delta V_3 = 0.25 \text{ V}$$

Finally, we can determine the charges on their plates:

$$Q_2 = \Delta V_2 C_2 = (0.25 \text{ V}) \left(3.0 \times 10^{-4} \text{ F}\right) = 7.5 \times 10^{-5} \text{ C}$$
$$Q_3 = \Delta V_3 C_3 = (0.25 \text{ V}) \left(4.5 \times 10^{-4} \text{ F}\right) = 1.1 \times 10^{-4} \text{ C}$$

As a final check, we can see if $Q_2 + Q_3 = Q_{123}$ as we claimed above. If you plug in the numbers for yourself you'll see that this is indeed the case!

Finally, I'll note that just as in the previous example, you could have approached this differently. If you have a different strategy in mind, I encourage you to give it a shot!

11.4   Resistor-Capacitor (RC) Circuits

In the previous sections we dealt with circuits consisting of batteries and *either* resistors *or* capacitors. You probably agree that this can get messy, but one convenient property of resistor circuits is that the current in any section of wire is *constant*. In contrast, when we have capacitor circuits we argued that the charging process occurs very quickly, meaning that we don't worry about the details of the current *at all*. Instead, we concerned ourselves with the final distribution of *charge* on the plates of the capacitor(s) in the circuit (and the corresponding potential difference across the plates of each capacitor).

In this section we would like to consider a circuit consisting of all three elements: batteries, capacitors, *and* resistors. Here we shall see that the current is *not* constant.[10]

To see why, consider a battery ε connected in series with a resistor R and initially uncharged capacitor C (Fig. 11.17, when the switch is in position a). Because the capacitor starts with no charge, there is no potential difference across its plates (remember $\Delta V = Q/C$), so the loop rule on this circuit states

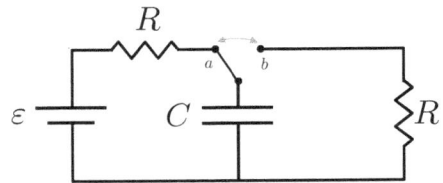

FIGURE 11.17

The "standard" RC circuit: when the switch is in position a, the battery (with emf ε) is connected in series with a capacitor C and resistor R. If the capacitor is initially uncharged, then current flows and charge accumulates on the plates of the capacitor. If the switch is moved to position b when the capacitor carries charge, then charge will leave the plates and pass through the second resistor until the capacitor is completely discharged. The presence of resistance in each situation means that the charge and discharge processes are *not* instantaneous.

$$\varepsilon + \Delta V_R + \Delta V_C = 0$$
$$\varepsilon - IR + 0 \text{ V} = 0$$
$$\varepsilon = IR$$

In other words, we can completely ignore the capacitor (i.e. consider it like a piece of wire with no resistance) *when it has no charge*. However, as soon as current starts to flow, charge will begin pile up on the capacitor plates. Consequently, the above result no longer holds. Instead, we have

$$\varepsilon - \frac{dQ}{dt}R - \frac{Q}{C} = 0 \qquad (11.14)$$

Here I have taken care to express the current running through the resistor in terms of the charge passing through a cross section of the wire (Equation (11.2)).[11] This is a *differential equation* because it involves both a variable (here, Q) and its derivative (dQ/dt).

We would like to determine a function $Q(t)$ that solves this equation, meaning that if we plug the function in for Q and its derivative in for dQ/dt, the left side of the equation simplifies to give 0 (i.e. the right side of the equation).

If you pursue your education in physics and mathematics far enough, you'll eventually receive quite a bit of formal training in solving differential equations. I won't delve too deeply into the realm of *derivation* in what follows, though I invite you to follow along and,

---

[10]Before proceeding I should point out that every capacitor circuit is *really* a resistor-capacitor circuit because (as noted above) the wires that connect the circuit elements have some small resistance. When we say that a capacitor circuit charges "very quickly" we're ignoring this small resistance to argue that the capacitors charge so fast that for practical purposes it might as well be instantaneous.

[11]Notice that Equation (11.14) reduces to $\varepsilon = IR$ when $Q = 0$, as we argued above.

if you've studied calculus, to *check* it! To proceed, let's solve Equation (11.14) for dQ/dt:

$$\frac{dQ}{dt} = \frac{\varepsilon}{R} - \frac{1}{RC}Q \tag{11.15}$$

This equation says that we need a function whose derivative gives the same function multiplied by a negative coefficient $(-1/RC)$ and an additive term (ε/R). The solution to this equation (which, to be clear, applies to a **charging capacitor**, and which you can check for yourself by differentiating and comparing to the above) is

$$Q\left(t\right) = C\varepsilon\left(1 - e^{-t/RC}\right) \tag{11.16}$$

Because the potential difference across a capacitor is given by $\Delta V = Q/C$, the potential difference across the capacitor follows directly:

$$\Delta V\left(t\right) = \varepsilon\left(1 - e^{-t/RC}\right) \tag{11.17}$$

Finally, if you differentiate the equation for $Q\left(t\right)$, you obtain the current:

$$I\left(t\right) = \frac{dQ}{dt} = \frac{\varepsilon}{R}e^{-t/RC} = I_0 e^{-t/RC} \tag{11.18}$$

In the above, I_0 refers to the *initial* current; $I_0 = \varepsilon/R$. Figure 11.19 shows the current and electric potential across the capacitor as a function of time. We see that the current starts very high and decays exponentially as more and more charge piles onto the capacitor. Eventually, the current decays to 0 A; we say that a fully charged capacitor acts like a *break* (or open switch) in the wire because no current flows through it or anything in series with it.

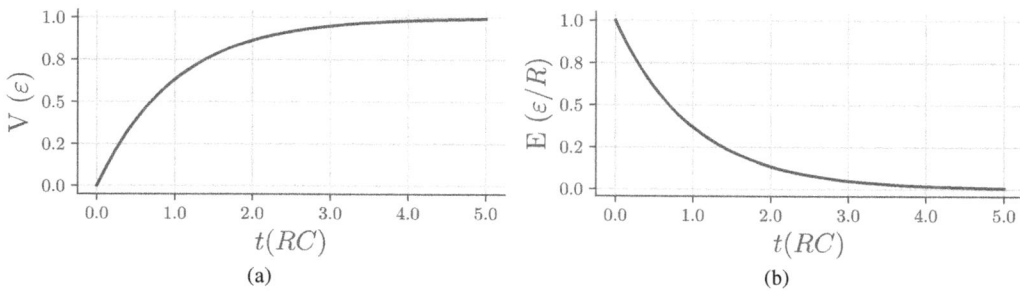

(a)   (b)

FIGURE 11.18
In a charging RC circuit consisting of a single capacitor and resistor (Fig. 11.17), the electric potential across the capacitor (a) obeys Equation (11.17). The vertical axis is shown in units of the battery emf ε and the horizontal axis is shown in units of the time constant RC. Similarly, the current running through the resistor (b) obeys Equation (11.18). The net effect is that the total potential provided to the circuit by the battery changes from the resistor (high $I \implies$ high ΔV_R) to the capacitor (high ΔV_C) as the capacitor charges.

In contrast, the electric potential *starts* at 0 V and increases quickly before leveling off to its final value, where $\Delta V = \varepsilon$: when there is no current, no potential is lost through the resistor and so the loop rule says that *all* the potential provided by the battery is lost through the capacitor, as we saw in the preceding section.

We see for both ΔV and I that it is the quantity RC that characterizes how rapidly these processes occur; it is sometimes called the **time constant** for the circuit. The time constant is often denoted τ:

$$\tau \equiv RC \qquad (11.19)$$

For example, when $t = \tau = RC$ the current will be $I = I_0 e^{-1} \approx 0.37 I_0$, when $t = 2\tau$ the current will be $I = I_0 e^{-2} \approx 0.14 I_0$, and when $t = 3\tau$ the current will be $I = I_0 e^{-3} \approx 0.05 I_0$. The following example provides some practice with these ideas.

Example 11.6 Timing in RC Circuits   ★★

A complicated circuit uses a 1.0×10^2 μF capacitor to control the flashing of a small LED (a *light-emitting diode*): the diode will be turned on when the capacitor has a potential of at least 0.75 V. The capacitor goes through a cycle of charging and then discharging, but let's just consider the portion of the cycle where the capacitor is charging. If it is initially uncharged and then is connected in series to a power supply that provides $\varepsilon = 2.0$ V and a resistor with a resistance of $R = 150$ Ω, how much time will pass before the diode turns on?

We must work with Equation (11.17) and determine the time t that corresponds to a specific ΔV:

$$\Delta V(t) = \varepsilon \left(1 - e^{-t/RC}\right)$$

$$e^{-t/RC} = 1 - \frac{\Delta V(t)}{\varepsilon}$$

$$-\frac{t}{RC} = \ln\left(1 - \frac{\Delta V(t)}{\varepsilon}\right)$$

$$t = -RC \ln\left(1 - \frac{\Delta V(t)}{\varepsilon}\right)$$

In moving to the third line, I took the natural log of both sides of the equation to pull t out of the exponent (remember $\ln(e^x) = x$). All that remains is to plug in our numerical values:

$$t = -(150\ \Omega)\left(1.0 \times 10^{-4}\ \text{F}\right)\ln\left(1 - \frac{0.75\ \text{V}}{2.0\ \text{V}}\right) = 7.1 \times 10^{-3}\ \text{s} = 7.1\ \text{ms}$$

As we move on to consider discharging capacitors, we will become equipped to complete the analysis of this circuit: if the capacitor stops charging at some point and then begins to discharge, eventually the electric potential on the capacitor will drop below the threshold of 0.75 V and shut off. Controlling the potential provided to the capacitor and when it switches from charging to discharging provides fine control over the behavior of the LED. Doing this simultaneously with many LEDs in a grid allows the panel to display, for instance, scrolling text or animations of dancing frogs.

We have now described *charging* a capacitor in an RC circuit. The other scenario that we would like to analyze is the case where we are *discharging* a capacitor (Fig. 11.17 when the switch is in position b). Here the loop rule states

$$\Delta V_C + \Delta V_R = 0$$

$$\frac{Q_C}{C} - \frac{dQ_R}{dt}R = 0$$

I am being careful here to distinguish between the charge on the capacitor at some instant (Q_C) and the charge passing through the resistor (Q_R). Why not call them both Q like before? Well, when we have a *charging* capacitor, the battery provides the charge: it passes through the resistor and accumulates on the capacitor plates, so $dQ_C = dQ_R$. Here, however, the *capacitor* is providing the charge to the resistor, so whatever is gained by the resistor is *lost* by the capacitor: $dQ_C = -dQ_R$. For convenience we'll revert back to the general Q notation, but we need to account for the additional negative sign, which gives

$$\frac{dQ}{dt} = -\frac{1}{RC}Q \tag{11.20}$$

Once again we can solve this differential equation; we find for a **discharging capacitor**

$$Q(t) = C\varepsilon e^{-t/RC} \tag{11.21}$$
$$\Delta V(t) = \varepsilon e^{-t/RC} \tag{11.22}$$
$$I(t) = \frac{\varepsilon}{R}e^{-t/RC} = I_0 e^{-t/RC} \tag{11.23}$$

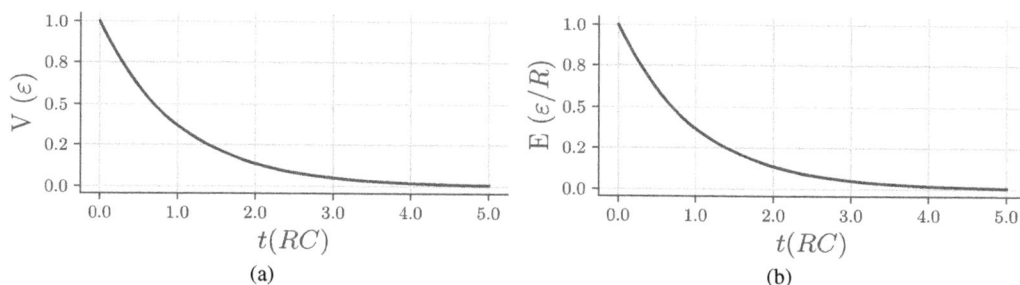

(a)   (b)

FIGURE 11.19
In a discharging RC circuit consisting of a single capacitor and resistor (Fig. 11.17), the electric potential across the capacitor (a) obeys Equation (11.22). and the current running through the resistor (b) obeys Equation (11.23). The charged capacitor initially provides a large current to the resistor, but as the capacitor loses charge the potential difference across the capacitor and the current decay exponentially.

Interestingly, the equation for the current is the *same* as for a charging capacitor. Here, though, the electric potential *also* decays exponentially. These relationships are shown in Figure 11.19 and the following example demonstrates one situation where a capacitor discharge can be particularly useful.

Example 11.7 Power in RC Circuits ★★
Suppose a particular light bulb has a resistance of 8.0 Ω. (a) If it is connected to a 9.0 V battery, what power will be dissipated by the bulb? (b) Now suppose a 4.0×10^2 μF capacitor is charged by a 27 V battery and then allowed to discharge across the lightbulb. What is the *initial* power dissipation through the bulb? How long does it take for the power to drop to the answer you found in (a)? (This is the idea of how a camera's *flash* works.)

The first part of the problem describes a simple resistor circuit; we can use Equation (11.5) to determine the power:

$$P_a = \frac{(\Delta V)^2}{R} = \frac{(9.0 \text{ V})^2}{8.0 \text{ }\Omega} = 10. \text{ W}$$

In (b) we are dealing with a discharging capacitor; using Equation (11.23) when $t = 0$ allows us to determine the initial current passing through the resistor:

$$I(t = 0 \text{ s}) = \frac{\varepsilon}{R} e^{-t/RC}$$
$$= \frac{\varepsilon}{R} e^0$$
$$= \frac{\varepsilon}{R} = \frac{27 \text{ V}}{8.0 \text{ }\Omega} = 3.4 \text{ A}$$

From here we determine the initial power:

$$P_b = I^2 R = (3.4 \text{ A})^2 (8.0 \text{ }\Omega) = 91 \text{ W}$$

As the capacitor discharges, the current and therefore the power will drop. What current corresponds to the power we found in (a)? Well:

$$P_a = I_a^2 R \implies I_a = \left(\frac{P_a}{R}\right)^{1/2} = \left(\frac{10. \text{ W}}{8.0 \text{ }\Omega}\right)^{1/2} = 1.1 \text{ A}$$

Then we can determine the time necessary to reach this current using Equation (11.23):

$$I_a = \frac{\varepsilon}{R} e^{-t/RC}$$
$$\frac{R}{\varepsilon} I_a = e^{-t/RC}$$
$$-RC \ln\left(\frac{R}{\varepsilon} I_a\right) = t$$
$$-(8.0 \text{ }\Omega)(4.0 \times 10^{-4} \text{ F}) \ln\left(\frac{8.0 \text{ }\Omega}{27 \text{ V}} (1.1 \text{ A})\right) = t$$
$$3.5 \times 10^{-3} \text{ s} = 3.5 \text{ ms} = t$$

Thus when we charge the capacitor with a 27V emf, it starts off about 9 times brighter (91 W vs. 10. W) than when it is powered by a steady 9.0 V emf. Furthermore, it takes about 3.5 milliseconds for the power provided by the capacitor to drop down to the "steady 9.0 V" level.

11.5   A Closer Look at Current

When we set up the idea of electrical current in a wire, I gave you the analogy of a line of people steadily marching: the people correspond to electrons, and current is a measure of how many electrons pass a certain point in the line in a certain interval of time. In this section I would like to take a closer look at this analogy. In particular, we would like to determine *how fast* electrons are moving down a wire. You might think this speed is incredibly fast–lights turn on pretty much instantly when we flip a switch, for instance–but *hold on*.

To start, suppose that we have attached a wire of length L to the terminals of a battery. The edges of our wire are at different electric potentials, which we might generically call V_0 and 0 volts at the left ($x = 0$) and right ($x = L$) ends, respectively (Fig. 11.20).

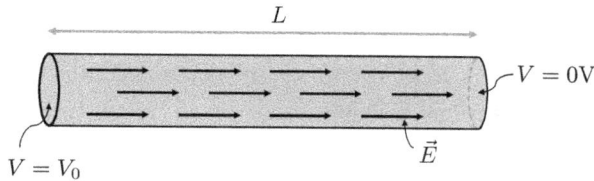

FIGURE 11.20
A straight section of conducting wire has a fixed potential at each end. The result is a constant electric field (indicated with black vectors) through the length of the wire.

The fact that we have a potential difference across the wire suggests that there is an *electric field* in the wire, because

$$\vec{E} = -\frac{dV}{dx}\hat{x} \tag{11.24}$$

This is indeed the case, but it isn't immediately obvious what the functional form of this electric field is because we haven't specified the functional form of the potential: we know that it has *boundary values* of V_0 and 0 volts, but there are many ways to move from $V = V_0$ at the left boundary to $V = 0$ V at the right boundary.

If you move on to study electromagnetism at an advanced level, you will learn to love and hate these so-called boundary-value problems. To get a handle on how to determine the solution to this problem we would need some theoretical results I don't want to get into, so for our purposes I will simply state the outcome: the electric potential decays *linearly* as we move from one end to the other, and this means that the electric field is *constant* in the wire:

$$V = -\frac{V_0}{L}x + V_0 \implies \vec{E} = \frac{V_0}{L}\hat{x} \tag{11.25}$$

Remarkably, this linear decrease in potential (and the corresponding constant field) occurs *even if the wire is bent*, not just if it is a straight line as I've drawn in Figure 11.20.

Okay, so the electrons in our wire experience a constant electric field, much like if the ends of the wire were capacitor plates (though here electrons are free to move through the conducting wire). This means that they experience an electric force, and if that was the complete story they would *accelerate* down the length of the wire. This would mean different velocities at different pieces of wire, which would mean in turn that the current would *not* be the same everywhere in the wire. This conflicts with our argument that current is, in fact, the same at any point in an isolated section of wire. Evidently there must be more to the story.

The piece we're missing is the fact that the electrons aren't moving isolated out in space (or in air), they're moving through *matter*, and the matter includes stationary nuclei. The electrons, then, *do* accelerate because of the external electric field, as we argued above – but they also frequently bang into nuclei.[12] We won't worry about mathematically formalizing

---

[12]In fact, these collisions happen even if the wire isn't attached to a circuit in the first place: there is some thermal energy associated with the fact that the wire is at a temperature above 0 K, and this manifests itself in the form of atomic vibrations and random movement of the conduction electrons (i.e. those that are free to move) through the wire.

the *mean collision time* that passes between conduction electrons and nuclei; instead we'll content ourselves by simply labeling it τ. (We'll get a handle on how large τ is shortly.)

During one of these collisions, the electron will lose some or all of its kinetic energy (the energy becomes *thermal* energy, which is why wires will heat up over time), and its velocity vector can be re-oriented after the collision. You can think of this loosely like a large, stationary ball (the nucleus) being struck by a very small ball (the electron) at some random angle.

Because the velocity vector is re-oriented essentially at random, the net effect is that, on *average*, the electron starts from rest after a collision. This means that the electrons go through an "accelerate - stop - accelerate - stop ... " cycle as they haphazardly make their way down the wire. One cycle happens very rapidly in practice, and so it is useful to "smooth over" the motion and define an average velocity, which we call the **drift velocity**, to characterize the rate at which the electrons move down the wire. Recall the basic kinematic formula $v = v_i + at$: this tells us the velocity of a ball that has been accelerating for some amount of time, t, at a rate a, given that it started with some initial velocity v_i. On *average*, the balls start from rest, so if we call the average time that has passed since the last collision $\bar{t}$, then we have

$$v_d = a\bar{t} \tag{11.26}$$

We would like to define both of the terms on the right side of this equation in terms of quantities that we can easily measure. First, the acceleration in this context is that of an electron with charge e and mass m in an electric field of magnitude E,

$$F = eE = ma \implies a = \frac{eE}{m}$$

What about the time that has passed since the last collision, $\bar{t}$? It turns out, somewhat counterintuitively, that this is the *same* as the mean time *between* collisions, i.e. $\bar{t} = \tau$.[13] Putting it all together yields

$$v_d = \frac{eE\tau}{m} = \frac{e\Delta V\tau}{mL} \tag{11.27}$$

In the last expression I've replaced the electric field with the electric potential and the length, L, of the wire we're concerned with.

Let me push this just a bit further. Once we know v_d, it is natural to want to relate it to the current that we can measure with an ammeter. In other words, given the drift velocity, how much *charge* passes through a cross-section of the wire in some interval of time (Fig. 11.21)? To answer this we need to know not just the electrons' *speed* (v_d), but also *how many* are doing the moving. This is quantified in terms of the **conduction electron density**, a material-specific quantity we'll denote n_e. The units correspond to a *number per cubic meter*, and since a number by itself is unitless, the SI units for the conduction electron density are m^{-3}. This quantity can be calculated fairly straightforwardly if you know other properties for the metals such as its usual *mass* density and its atomic mass; see for instance Problem 11.63. Table 11.1 lists some values for a few metals at room temperature.

---

[13]Showing this rigorously requires the machinery of probability and *statistical mechanics*. We will avoid going too far into the weeds here, but the general idea is that not all of the electrons travel for exactly τ: some travel for more time, and some travel for less time. For the electrons that travel longer, the additional time spent accelerating disproportionately increases their velocity, and the net effect – which can be shown mathematically – is that the mean time *since* the last collision is the same as the mean time *between* collisions: τ. I should also mention that the idea of electrons and atomic nuclei interacting like two balls isn't really correct; the interactions are really quantum-mechanical. That said, the classical approach we're outlining here is sufficient for our purposes.

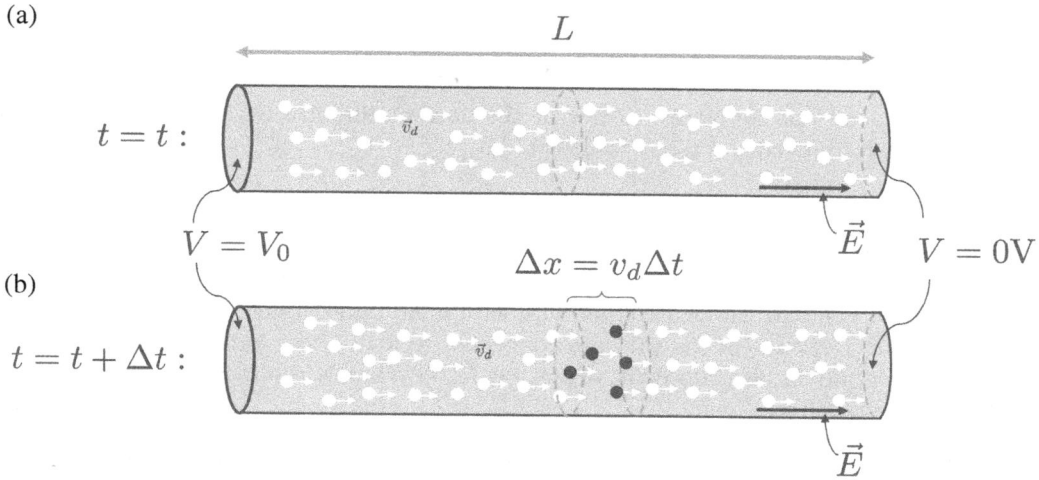

(a)

(b)

FIGURE 11.21

Electrons (circles) traveling down a wire. Here each electron is shown traveling at a uniform drift velocity $\vec{v}_d$, though this is only true on average. (a) We mark an arbitrary cross-section near the middle of the wire (the dashed surface) and, at time t, begin marking the electrons that pass through. (b) At time $t + \Delta t$, all of the electrons in a cylinder of length $\Delta x = v_d \Delta t$ will be marked and will therefore determine the current through this section of the wire during this interval. (The key relationship is summarized in Equation (11.30).)

Now, if n_e is the conduction electron density, then the total number of electrons (call it N_e) in some volume of wire V is given by

$$N_e = n_e V$$

We're considering a cylindrical wire with a cross-section A, and in some interval of time Δt the electrons will move a distance $\Delta x = v_d \Delta t$. We say that during this time the electrons *sweep out* a volume

TABLE 11.1

Conduction electron densities

Material	$n_e \left(\text{m}^{-3} \right)$
Copper	8.5×10^{28}
Iron	8.5×10^{28}
Aluminum	6.0×10^{28}
Gold	5.9×10^{28}

$$V = A\Delta x = Av_d \Delta t$$

Combining the two equations above yields an expression for how many electrons pass through a cross section in a given time interval:

$$N_e = n_e A v_d \Delta t \tag{11.28}$$

Or, if we want to express this in terms of an *electron flow rate*,[14] we can simply divide over the Δt:

$$\frac{N_e}{\Delta t} = n_e A v_d \tag{11.29}$$

We're almost there! We want to connect this ultimately to *current*, which is a measure of *charge per time* ($I = \Delta Q / \Delta t$), rather than the above quantity, which has units of *number of electrons per time*. How are these quantities related? Well, each electron carries a charge e, so the total charge of N_e of them is just $\Delta Q = N_e e$. Thus finally we have

---

[14]This is sometimes called the *electron current*, but I think this can be confusingly similar to the regular *current*.

$$I = \frac{\Delta Q}{\Delta t} = \frac{N_e e}{\Delta t} = n_e e A v_d \tag{11.30}$$

Sometimes we are concerned not with the current directly but rather the **current density**, which we denote J (see Problem 11.60):

$$J = \frac{I}{A} \tag{11.31}$$

Example 11.8 Drift Velocity in a Wire   ★★

Consider a 1.0×10^1 m-long copper wire with a 2.0 mm diameter and a resistance of 0.054 Ω. The ends of the wire are connected to a 12 V car battery. (a) What is the corresponding electron drift velocity v_d? (b) What is the mean time between atomic-electron collisions?

For (a), we can refer to Equation (11.30). To do so, though, we need to know the current, I, which we can find from Ohm's Law:

$$I = \frac{\Delta V}{R} = \frac{12 \text{ V}}{0.054 \text{ } \Omega} = 220 \text{ A}$$

This is a *huge* current because we have a relatively large potential and a very small resistance. Using this in Equation (11.30), we can determine v_d:

$$v_d = \frac{I}{n_e e A}$$

$$v_d = \frac{220 \text{ A}}{\left(8.5 \times 10^{28} \text{ m}^{-3}\right)\left(1.60 \times 10^{-19} \text{ C}\right)\left(\pi \left(2 \times 10^{-3} \text{ m}/2\right)^2\right)}$$

$$v_d = 5.2 \times 10^{-3} \text{ m/s}$$

When plugging in above I have made us of the fact that the area of the cross section, A, is just πr^2 where r is half of the given diameter. I also dropped the negative sign on the charge of an electron since v_d is just capturing the *magnitude* of the velocity – if we wanted to know the direction of flow, then we'd need to know which side of the wire has higher potential; the electrons would be moving toward that side.

It is worth noting that this is quite slow; it works out to be about 1 foot per minute. The reason why flipping a light switch results in a lightbulb turning on almost instantly is because electrons along the entire length of the wire begin moving at the same time. Thus there is current almost immediately *everywhere* in the wire, even if it takes an individual electron a very long time to move very far.

OK, on to (b). Here we refer to Equation (11.27) and solve for τ:

$$\tau = \frac{v_d m L}{e \Delta V}$$

$$\tau = \frac{\left(5.2 \times 10^{-3} \text{ m/s}\right)\left(9.11 \times 10^{-31} \text{ kg}\right)\left(1.0 \times 10^1 \text{ m}\right)}{\left(1.60 \times 10^{-19} \text{ C}\right)\left(12 \text{ V}\right)}$$

$$\tau = 2.5 \times 10^{-14} \text{ s}$$

Again, it is worth reflecting on the scale of this number. If it takes something like 10^{-14} s for each collision, then there are 10^{14}, or 100 *trillion*, collisions total every second! Clearly our decision to "smooth over" these collisions and work with the drift velocity (which is really just an average) is sufficient for most purposes!

In the preceding example I told you the physical dimensions of a copper wire *and* its resistance. It turns out that I was giving you more information than was strictly necessary because you can calculate the resistance R of a wire given the substance it is made from and its physical dimensions. The relationship is (see Problem 11.73)

$$R = \frac{\rho L}{A} \tag{11.32}$$

where L is the wire's length, A is its cross-section area, and ρ is a material-dependent quantity called the **resistivity**.[15] If you consider the units in Equation (11.32), you'll find the SI units for ρ are Ωm. Some typical resistivities are provided in Table 11.2.

TABLE 11.2
Resistivities of some metals

Material	$\rho\,(\Omega\text{m})$
Copper	1.70×10^{-8}
Iron	10.0×10^{-8}
Aluminum	2.82×10^{-8}
Gold	2.44×10^{-8}

Example 11.9 Resistivity and Resistance ⋆
Refer back to Example 11.8 and confirm that the copper wire's dimensions yield a resistance of 0.054 Ω.

This involves a straightforward application of Equation (11.32); from Example 11.8 we know $L = 1.0 \times 10^1$ m and $A = \pi r^2 = \pi \left(1.0 \times 10^{-3}\text{m}\right)^2 = 3.1 \times 10^{-6}$ m². We also know that the wire is copper, so from the table above we have $\rho = 1.70 \times 10^{-8}$ Ωm. Plugging everything in, we have

$$R = \frac{\rho L}{A}$$
$$R = \frac{\left(1.70 \times 10^{-8}\ \Omega\text{m}\right)\left(1.0 \times 10^1\ \text{m}\right)}{3.1 \times 10^{-6}\ \text{m}^2}$$
$$R = 0.054\ \Omega$$

as expected!

Example 11.10 Building a Resistor ⋆
Circuit boards – which are used to control the behavior of many machines – can include many electrical components and connections. Suppose you are on a submarine when an important machine malfunctions. You determine that the issue is that a

---

[15] In other books and resources you may hear about a material's **conductivity**, σ, which is just the inverse of the resistivity.

resistor on a circuit board needs to be replaced. The resistor that broke was in the form of a cylinder, with a length of 0.75 cm long and a diameter of 0.50 cm. You need a replacement with the same resistance, but you don't have any exact duplicates handy. However, you *do* have several long sticks of the same *material* as the broken resistor. If the sticks have a square cross-section with a length of 1.0 cm to a side, then how long of a piece should you cut to obtain a suitable replacement?

We want the two resistors to be equivalent, so

$$R_{\text{new}} = R_{\text{old}}$$
$$\frac{\rho L_{\text{new}}}{A_{\text{new}}} = \frac{\rho L_{\text{old}}}{A_{\text{old}}}$$
$$L_{\text{new}} = \frac{L_{\text{old}} A_{\text{new}}}{A_{\text{old}}}$$

In the above I have made note of the fact that the material is the same for both resistors, so ρ algebraically cancels. (This is convenient because we don't have a numerical value for it!) All that remains is to plug in our numerical values for the known dimensions of the resistors:

$$L_{\text{new}} = \frac{\left(0.75 \times 10^{-2} \text{ m}\right)\left(0.010 \text{ m} \times 0.010 \text{ m}\right)}{\pi \left(0.25 \times 10^{-2} \text{ m}\right)^2} = 0.038 \text{ m} = 3.8 \text{ cm}$$

11.6   Problems for Chapter 11

11.1 Introduction: Current, Resistance, and Power, Oh My!

★ **Problem 11.1.** Lightning involves a rapid delivery of electrons from the bottom of a cloud to the surface of the Earth. Suppose a certain lightning bolt delivers 15 C in 0.50 ms (these are typical values). What is the average current of the bolt?

★ **Problem 11.2.** In your own words, what do each of the following circuit elements *do* in a circuit? (a) batteries, (b) wires, (c) resistors, and (d) capacitors. (Your answers should invoke terms such as current, energy, power, etc.)

11.2 Resistor Circuits

★ **Problem 11.3.** Three resistors and one battery are wired as shown in Figure 11.22.

  a. To three significant figures, what is the equivalent resistance of the circuit?

  b. Identify the electric potential delivered to each resistor.

  c. If the resistors are all fans, which do you expect to spin the fastest?

★ **Problem 11.4.** What current (to two significant figures) will flow through the ammeter in the circuit shown in Figure 11.23?

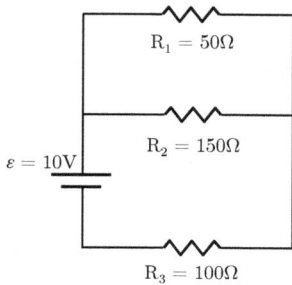

FIGURE 11.22
Problem 11.3

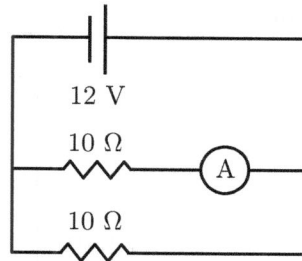

FIGURE 11.23
Problem 11.4

★ **Problem 11.5.** Strings of Christmas tree lights used to be wired such that when one bulb went out, all the bulbs went out. Describe how these bulbs were wired. How might they be wired instead to alleviate this problem?

★ **Problem 11.6.** Suppose you have a 1.0×10^2 V power supply and three resistors: $R_1 = 2.0$ kΩ, $R_2 = 3.0$ kΩ, and $R_3 = 4.0$ kΩ. What is the total power delivered to the circuit if the resistors are connected in (a) series and (b) parallel? (c) Take the ratio of these values, P_a/P_b, to quantify how these values differ from one another.

★★ **Problem 11.7.** You have a 1.7 kV power supply and a bucket of 1.00×10^2 kΩ resistors. Is it possible to construct a circuit using no more than 3 of the resistors such that at least one of the resistors receives a current of 17 mA? If so, draw a circuit diagram and indicated the placement of an ammeter that would record this current. If not, clearly explain why.

★★ **Problem 11.8.** You have a 30.0 V power supply and a bucket of 50. Ω resistors. Is it possible to construct a circuit such that at least one of the resistors receives a current of 0.10 A? If so, draw a circuit diagram and indicated the placement of an ammeter that would record this current. If not, clearly explain why.

★ **Problem 11.9.** Three identical fans, each with resistance R, are connected to a battery with emf ε as shown in Figure 11.24. Determine algebraic expressions (in terms of ε and R) for the current through, and electric potential drop across, the top resistor if...

a. ...the switch is open.

b. ...the switch is closed.

c. Considering your responses to (a) and (b), determine if the fan will spin faster when the switch is open or when the switch is closed.

★ **Problem 11.10.** Consider the circuit shown in Figure 11.25.

a. What is the current through resistor R_2?

b. If a dog chews on the wire feeding in to R_1, breaking it, then what is the current in resistor R_2?

c. Consider the total power that is provided by the battery in situation (a) and (b). Which is larger? How do you know?

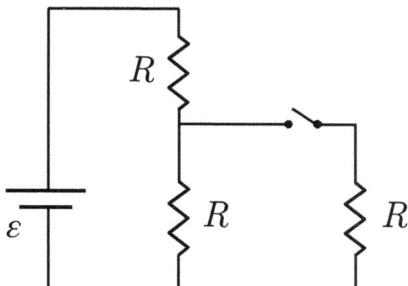

FIGURE 11.24
Problem 11.9

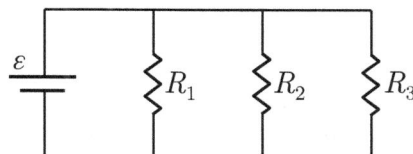

FIGURE 11.25
Problem 11.10

Problem 11.11. How much power is dissipated in resistor R_2 in Figure 11.26 if the switch is (a) open and (b) closed?

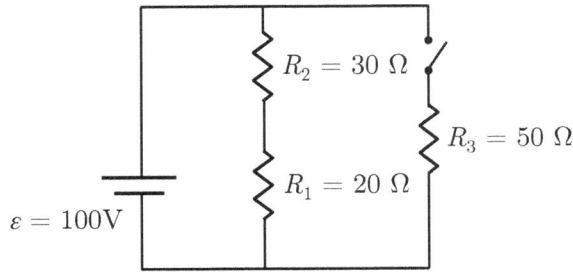

FIGURE 11.26
Problem 11.11

Problem 11.12. Determine the current through each of the three batteries and both resistors in the circuit shown in Figure 11.27. Note the direction of current flow in all wires by sketching and labeling the circuit diagram.

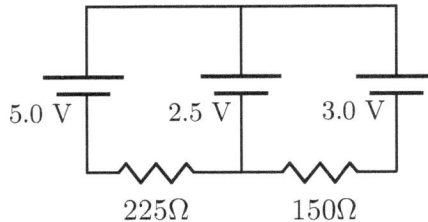

FIGURE 11.27
Problem 11.12

Problem 11.13. Consider the circuit diagram shown in Figure 11.28. Determine the current through, and potential drop across, each resistor (provide all answers with three significant figures). Write your final answers in a table.

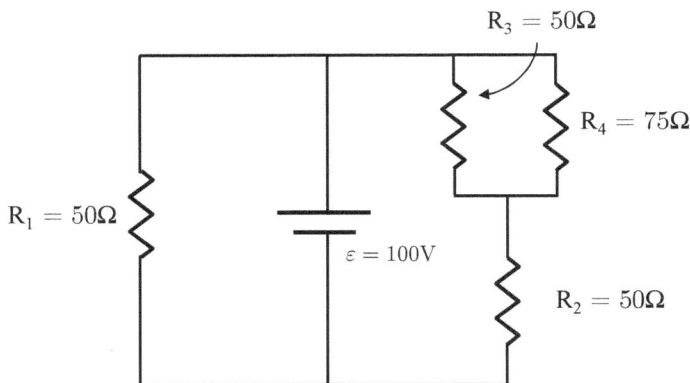

FIGURE 11.28
Problem 11.13

Problem 11.14. Three identical fans, each with resistance R, are connected to a battery with emf ε as shown in Figure 11.29. The fourth resistor is not a fan and has resistance r.

a. Determine an expression for the current running through each section of the wire (sketch the circuit and label appropriately so your responses are clear). Your answers will be algebraic expressions in terms of ε, R, and r.

b. Which fan(s) will spin the fastest if $r = \frac{1}{2}R$? What about if $r = 2R$? Explain.

Problem 11.15. In the circuit shown in Figure 11.30, $\varepsilon = 10.0$ V, $R_1 = R_3 = 20.0\ \Omega$, and $R_2 = R_4 = 40.0\ \Omega$.

    a. What is the equivalent resistance of the circuit?

    b. What is the potential difference across each resistor?

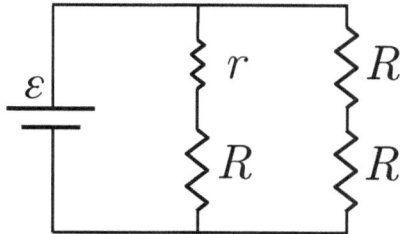

FIGURE 11.29
Problem 11.14

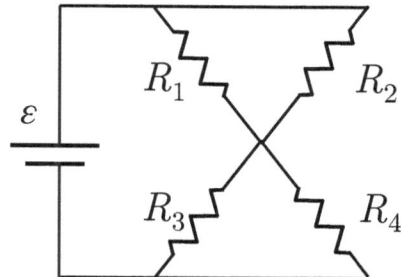

FIGURE 11.30
Problem 11.15

Problem 11.16. During a laboratory exercise, you are asked to construct the circuit sketched in Figure 11.31. You know that you're supposed to use a 9.0 V battery and four resistors that each have a resistance of 5.0 Ω. Unfortunately, the two meters you are supposed to attach to the circuit aren't labeled with an "A" for ammeter or "V" for voltmeter, like usual. Rather than waiting for help from the lab TA or instructor, you think for a moment and realize that there is only one correct choice for each meter.

    a. Sketch the figure and draw an "A" or "V" in each circle to indicate which kind of meter it is. Briefly explain.

    b. What value do you expect to read in the top meter?

    c. What value do you expect to read in the side meter?

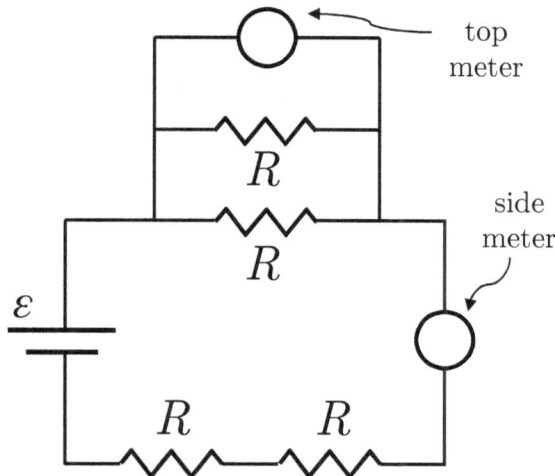

FIGURE 11.31
Problem 11.16

Problem 11.17. What is the potential difference across and current through each of the resistors in the circuit shown in Figure 11.32?

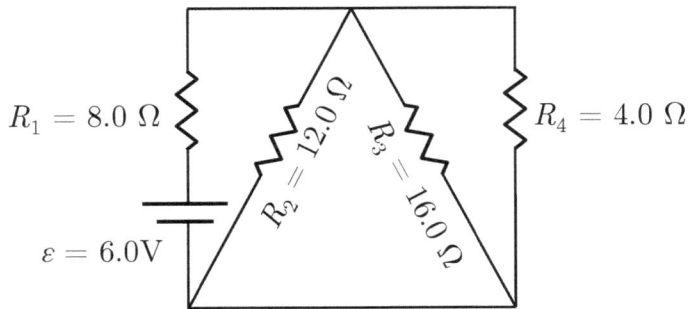

FIGURE 11.32
Problem 11.17

★   **Problem 11.18.** Consider the circuit shown in Figure 11.33. $\varepsilon = 9.0$ V and all the resistors have a resistance of 50.0 Ω. Determine the power dissipated by R_3 when the switch is (a) opened and (b) closed.

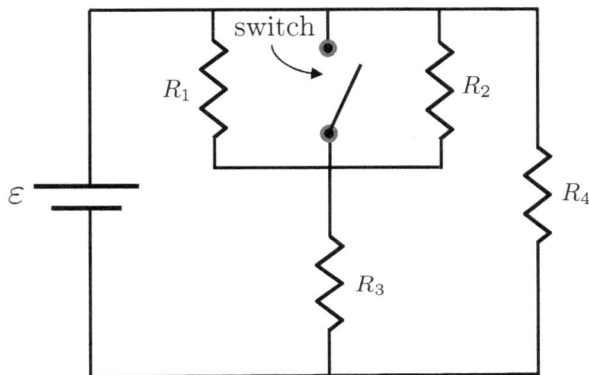

FIGURE 11.33
Problem 11.18

★   **Problem 11.19.** What power is dissipated by each resistor in Figure 11.34 when the switch is (a) open and (b) closed?

★★   **Problem 11.20.** Consider the circuit diagram shown in Figure 11.35. Determine the current through, and potential drop across, each resistor. Provide your answers with three significant figures and write your final answers in a table.

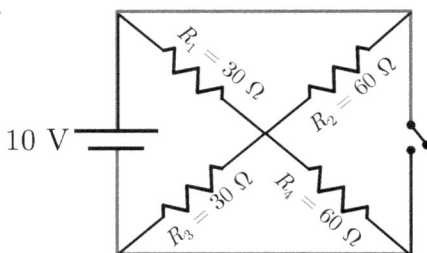

FIGURE 11.34
Problem 11.18

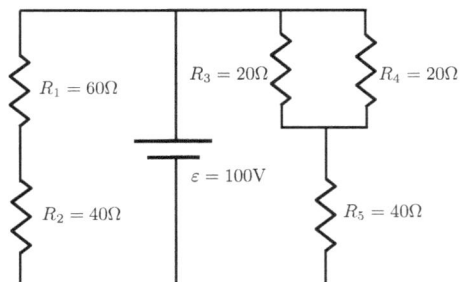

FIGURE 11.35
Problem 11.20

★★ **Problem 11.21.** Consider the circuit shown in Figure 11.36. Assume the provided values are exact and provide answers to three significant figures.

a. What is the equivalent resistance?

b. What current runs through the wire directly above the battery?

c. What power is dissipated by each resistor?

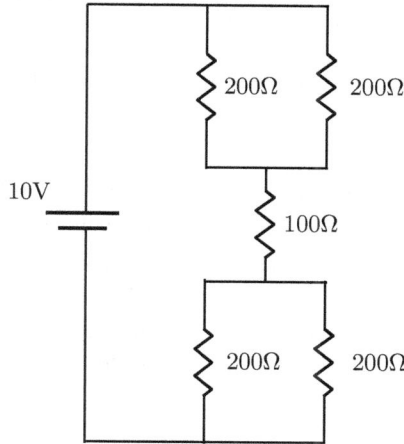

FIGURE 11.36
Problem 11.21

★★ **Problem 11.22.** What current will the ammeter read in the circuit shown in Figure 11.37 if $\varepsilon = 5.0$ V and $R = 10.0$ Ω?

★★ **Problem 11.23.** Consider the circuit shown in Figure 11.38. $\varepsilon = 9.0$ V and all the resistors have a resistance of 1.0×10^2 Ω. Determine (a) the current through, (b), potential difference across, and (c) power dissipated by each resistor.

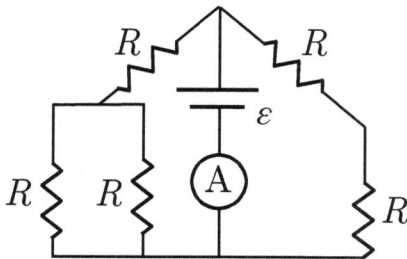

FIGURE 11.37
Problem 11.22

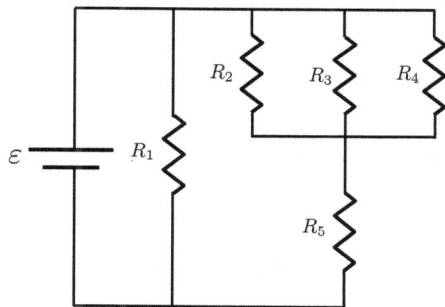

FIGURE 11.38
Problem 11.23

★ **Problem 11.24.** Consider a string of holiday lights as a single resistor with a resistance of 55 Ω. Suppose you have three such strings that you will power with a 120 V power supply.

a. If you connects the three strings of lights in *series*, how much power will be dissipated by the circuit?

b. If you connect the three strings of lights in *parallel*, how much power will be dissipated by the circuit?

c. Draw a circuit diagram showing the configuration in both (a) and (b). Be sure to indicate which is which!

★★★   **Problem 11.25.** Suppose a car battery has died, meaning it can no longer drive sufficient current through the starter motor (due to a drop in its potential difference and an increase in its internal resistance). You have attached a fresh battery with a set of jumper cables so the entire circuit consists of the *good* battery, the *dead* battery, and the *starter motor* all in parallel with one another. Suppose the good battery has a potential of 12 V and an internal resistance (effectively a resistor just adjacent to the battery) of 0.05 Ω, the bad battery has a potential of 6.0 V and an internal resistance of 1.0 Ω, and the starter motor has a resistance of 0.10 Ω. The batteries are oriented so the positive terminals are directly connected and the negative terminals are directly connected.

   a. Sketch the circuit.

   b. Determine how much current the good battery could drive through the starter motor alone.

   c. Determine how much current the bad battery could drive through the starter motor alone.

   d. Determine how much current passes through the starter motor with both batteries connected as described above.

   e. How much current (including direction) passes through the dead battery when everything is connected as described above? Do you think the dead battery can be charged in this way?

11.3 Capacitor Circuits

★   **Problem 11.26.** Three capacitors with capacitances 6.0 μF, 10.0 μF, and 16.0 μF are connected series. What is the equivalent capacitance? What is the equivalent capacitance if they are connected in parallel? Sketch the connections in each case.

★★   **Problem 11.27.** Starting from the definition of capacitance and the energy required to move a charge q through a potential difference ΔV, derive Equation (11.11). Your solution should include at least one graph showing the relationship between C, Q, and ΔV.

★   **Problem 11.28.** To what potential should you charge a 10.0 μF capacitor if you need to store 5.0 J of energy?

★   **Problem 11.29.** Two parallel square metal plates are 2.0 cm to a side. They are 2.0 mm apart and are connected to a 40.0 V power supply. What is (a) the capacitance and (b) the charge on each plate?

★★   **Problem 11.30.** Consider the capacitor described in Problem 11.29. What is the energy stored in the capacitor

   a. originally?

   b. if the capacitor plates are pulled apart until their separation is 4.0 mm *while still connected to the battery*?

   c. if the capacitor plates are pulled apart until their separation is 4.0 mm *after first being disconnected from the battery*?

★★★   **Problem 11.31.** Consider a set of N capacitors connected in series to a battery with emf ε. Show that the equivalent capacitance obeys Equation (11.12):

   a. Consider two adjacent capacitors, C_i and C_{i+1}. One plate from C_i is electrically connected to one plate from C_{i+1}. They must have the same magnitude charge, Q: why? Explain with the help of a diagram.

   b. The total potential drop across the capacitors is $\Delta V_{\text{tot}} = \Delta V_1 + \Delta V_2 + \cdots \Delta V_N$. Rewrite this in terms of the capacitance of each capacitor and the charge on the capacitor (use Equation (11.10)).

c. We'd like to define C_{eq} such that $C_{eq} = Q/\Delta V_{tot}$. Given your work above, show that Equation (11.12) is the appropriate choice.

Problem 11.32. You are building a strobe light with a 9.0 V battery and up to three 21 μF capacitors. What is (a) the minimum and (b) the maximum energy you can store with the capacitors? You are free to use one, two, or all three of the capacitors and you can connect them however you wish. For both (a) and (b), show a circuit diagram that includes the battery, wires, and all of the used capacitors.

Problem 11.33. In discussing capacitance, we assumed the space between the capacitor plates is filled with a vacuum (or air). If the space between the plates is instead filled with an insulator (sometimes called a *dielectric*), the capacitor plates can induce polarization in the insulator, which changes the electric field between the plates. Show that the capacitance can be expressed as

$$C = \kappa C_0 \tag{11.33}$$

where κ is a material-specific *dielectric constant* and C_0 is the usual (unfilled) capacitance.

Problem 11.34. Suppose you have a parallel plate capacitor consisting of two circulars disks with radii 5.0×10^{-2} m and plate separation 1.5×10^{-4} m. The capacitor is filled with a dielectric with a dielectric constant of 300 and the potential difference between the plates is 12 V.

a. How much charge is on each capacitor plate?

b. What is the electric field at the exact center of the capacitor?

c. How much energy is stored in the capacitor?

Problem 11.35. Capacitors can be destroyed if they are exposed to excessively large electric fields. Suppose a 45 μF capacitor embedded in a complex (and very expensive) circuit board has been destroyed in this way. Rather than replacing the entire board, you seek to replace just the capacitor. Unfortunately, you only have 30. μF, 60. μF, and 90. μF capacitors on hand (but you do have a lot of each variety, at least). Is it possible to combine some number of these capacitors such that they have the desired equivalent capacitance? If *yes*, explain how they should be connected. If *no*, explain why.

Problem 11.36. The keys on many keyboards are capacitors; when you press down on the key you are pushing down on the top plate, which moves down toward the bottom plate. Suppose the plates of the capacitor are squares separated by 2.0 mm. Before pushing down on the plates, they have a potential difference of 5.0×10^{-2} V. If pushing the key changes the distance between the plates by 1.0 mm and the charge on the plates does not change, then what is the new potential difference across the capacitor? (This change is what the computer detects in order to recognize the fact that you pushed a key.)

Problem 11.37. You have 3 capacitors each rated at 5.0 μF. Determine all possible equivalent capacitances that you can obtain by using all three capacitors in a circuit. Explain your mathematics with a circuit diagram in each case.

Problem 11.38. Suppose you have an 8.45 pF capacitor. A potential difference of 3.5 V is applied across the capacitor plates. How much charge is each plate? How much energy is stored in the capacitor?

★   **Problem 11.39.** What is the equivalent capacitance of the circuit shown in Figure 11.39?

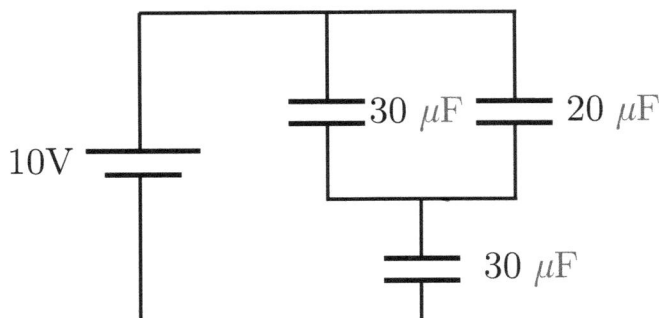

FIGURE 11.39
Problem 11.39

★   **Problem 11.40.** Four capacitors are connected to a battery as shown in Figure 11.40. What is the equivalent capacitance of the circuit?

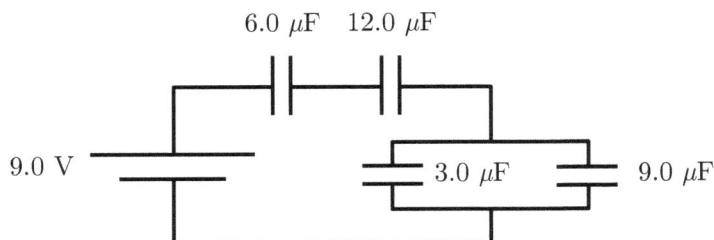

FIGURE 11.40
Problem 11.40

★   **Problem 11.41.** Your electronics lab partner built a circuit, but unfortunately the label on one of the capacitors has worn off and neither you nor your partner recall its capacitance. To make matters worse, the circuit has been permanently affixed to the board, so you can't take the circuit apart. Instead of waiting to see if the instructor can help, you run some tests to determine that the equivalent capacitance of the pair is 0.45 μF. What is the unknown capacitance?

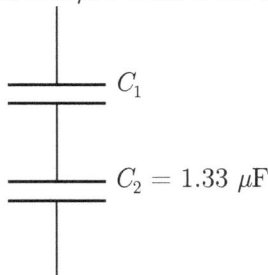

FIGURE 11.41
Problem 11.41

★   **Problem 11.42.** Consider a large parallel plate capacitor consisting of two circular plates with a diameter of 0.20 m separated by 0.0050 m. Suppose the capacitor is connected to a power supply set to a potential difference of 1.00×10^3 V.

   a. How much charge, in Coulombs, exists on each plate of the capacitor once the capacitor is fully charged?

   b. If the power supply is then disconnected and the plates are pushed together until they are separated by 1.0 mm, how much charge will be on each plate What is the potential difference now?

11.4 Resistor-Capacitor (RC) Circuits

★★ **Problem 11.43.** A 3.0 V battery is used to charge a 250 μF capacitor. The battery is then removed and, at t = 0, the capacitor is connected to a 5.00×10^2 Ω resistor so it can discharge. What is the current through the resistor at (a) $t = 0.0$ s (b) $t = 3.0$ ms, and (c) $t = 1.0 \times 10^2$ ms? (d) Make a sketch of current vs. time.

★ **Problem 11.44.** The switch in Figure 11.42 shown below has been in the left position for a long time. At $t = 0$, the switch is flipped to the right position. How much power (to two significant figures) is being dissipated by the resistor at $t = 150 \times 10^{-3}$ s?

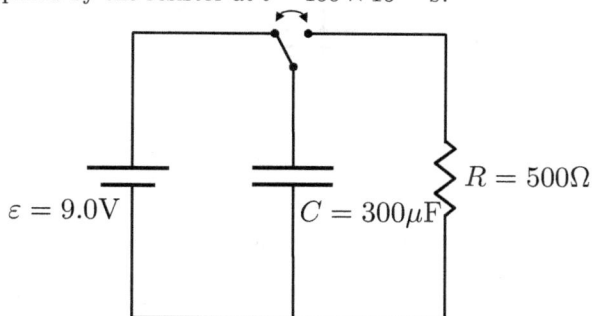

FIGURE 11.42
Problem 11.44

★ **Problem 11.45.** Consider the circuit diagram shown in Figure 11.43. Initially, the switch has been in the left position for a long time. At $t = 0$ s the switch is flipped to the right position. Suppose $\varepsilon = 15$ V, $C = 3.0 \times 10^{-4}$ F, $R_1 = 150.$ Ω, and $R_2 = 50.$ Ω.

   a. How much charge is on the capacitor plates just before the switch is flipped?

   b. What current will flow out of the capacitor immediately after the switch is flipped?

   c. How much power will each resistor be dissipating 2.0 ms after the switch is flipped?

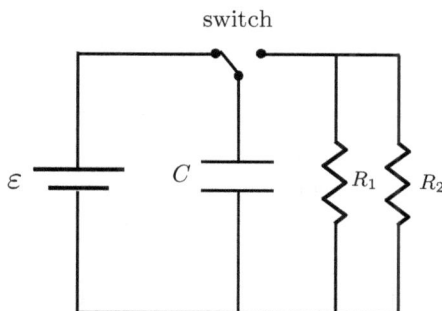

FIGURE 11.43
Problem 11.45

★★ **Problem 11.46.** An electrical circuit is designed so a 9.0 V battery can charge a 4.00×10^2 μF capacitor through a 5.00×10^2 Ω resistor when a switch is in position "A". When the switch is toggled to position "B", the capacitor discharges through a 2.0 kΩ resistor. Suppose that at $t = 0$ s the capacitor is fully charged and the switch has just flipped to position "B".

   a. How long will it take for the current leaving the capacitor plates to drop to 10% of its initial value? What is the voltage across the capacitor plates when this occurs?

   b. Once the current drops to 10% of its initial value (the time you found in (a)), the switch is flipped to "A". How long does it take for the capacitor to charge back to 99% of its original value?

   c. How long does it take for a complete charge/discharge cycle?

★ **Problem 11.47.** The switch in Figure 11.44 has been in the left position for a long time. At $t = 0$ it is flipped to the right position.

a. What is the initial potential difference across the resistor?

b. At what time is the potential difference across the resistor 5% of its initial value?

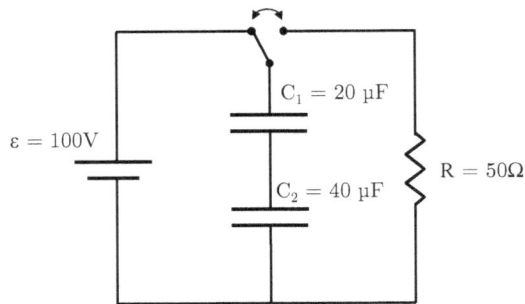

FIGURE 11.44
Problem 11.47

★★	**Problem 11.48.** A camera contains a 120 μF capacitor that retains a 150 V potential difference when an unwary student removes it by touching one hand to each plate. Suppose the resistance of his body (measured between his hands) is 1.8 kΩ. For how long will the current through his body exceed 0.050 A (the "danger level")?

★★	**Problem 11.49.** A parallel plate capacitor is formed with two square sheets with edge length 1.5 cm separated by a distance of 1.0 mm. At $t = 0$ the capacitor is attached to a 750 Ω resistor and a 15 V battery.

a. What maximum charge can the capacitor carry?

b. At what time will the capacitor have 99% of the charge you identified in (a)?

c. What is the magnitude of the electric field between the capacitor plates once it is fully charged?

★	**Problem 11.50.** A resistor, capacitor, and battery are connected as shown in Figure 11.45. Initially, the switch is held in the position shown for a long time. Then, at $t = 0$ s the switch is thrown to the other position. Assume all values are exact and provide your numerical answers to three significant figures.

a. What current will the ammeter record right after the switch is thrown?

b. What current will the ammeter record at $t = 0.20$ s?

c. Suppose the switch is toggled back and forth like this every 0.20 s. Sketch the ammeter current as a function of time for to $t = 0$ s to $t = 1.0$ s. Explain.

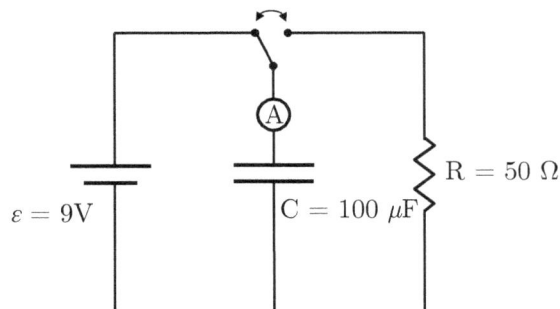

FIGURE 11.45
Problem 11.50

★	**Problem 11.51.** Resistors and capacitors can be used to generate periodic signals that can be used in everything from pacemakers to flashing traffic lights. The circuit diagram in Figure 11.51 demonstrates the idea: The Neon lamp acts like an open switch until the voltage across the capacitor is 80 V. Then, it acts like a piece of low-resistance wire that allows the capacitor to discharge, causing an essentially instantaneous flash in the lamp. (a) How much time passes between bulb flashes? Assume the values in the diagram are exact and provide your answer to three significant figures. (b) Sketch $V(t)$ in the capacitor through a few charge/discharge cycles.

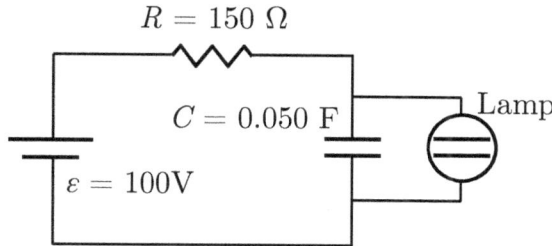

$$R = 150 \; \Omega$$

$$C = 0.050 \; F$$

Lamp

$$\varepsilon = 100V$$

FIGURE 11.46
Problem 11.51

★★ **Problem 11.52.** Many electronic devices (DVD players, TVs, etc.) have warning labels that essentially say "don't open this and mess around with the interior electronics because you could break the machine and/or die". A friend of yours contemplates a broken TV and reasons, "It is unplugged from the wall outlet, so how could it hurt me?" Unfortunately, while fiddling with the insides of the TV, the fingers of his hands touch the plates of a charged capacitor such that it discharges through his body. Suppose he's dealing with a 150 μF capacitor that was charged with a 25 V power supply. The resistance of his body (measured from fingertip to fingertip) is $1.00 \times 10^3 \; \Omega$.

a. What is the highest current that passes through his body?

b. If you answer to (a) is above 5 mA (a typical upper limit for a "safe" current), then how long does it take for the current to fall to 5 mA? If your answer to (a) is below 5 mA, then how long does it take for the current to fall to 10% of its initial value?

★★ **Problem 11.53.** Electric signals pervade the human body. One example is the propagation of an electric potential through a nerve cell (e.g. the nerves in your fingers telling your brain that you are touching something hot). The signal moves by "hopping" across a series of myelin sheaths that coat the axons of your nerve cells.[16] Each myelin sheath behaves like a charging RC circuit where $R = 25 \times 10^6 \; \Omega$ and $C = 4.0 \times 10^{-12} \; F$; the "battery" in this context is called the action potential and has a value of 0.100 V. A myelin sheath activates its neighbor (so the neighbor begins charging) once it reaches 90% of the action potential.

a. How long does it take for one myelin sheath to activate its neighbor?

b. If it takes 3.5 ms for a signal to cross a nerve cell, how many myelin sheaths does the cell have?

★★ **Problem 11.54.** A heart defibrillator is used to resuscitate someone whose heart has stopped by discharging a capacitor through the trunk of their body. Suppose the 32 μF capacitor is initially charged to a potential difference of 10.0 kV and the patient's body has a resistance of 1.0 kΩ.

a. How long does it take for the capacitor to discharge to 5% of its initial charge?

b. Make a sketch of the electric potential delivered to the patient as a function of time.

11.5 A Closer Look at Current

★ **Problem 11.55.** If you have purchased electrical cables for a stereo, headphones, or computer speakers, then you may be aware that some high-end cables have "jacks" that are made out of gold instead of a more common metal such as copper. Suppose you have a friend who is a big fan of stereos. He confidently tells you that the gold plating is preferred because "gold is a better conductor than copper". Is he correct? How do you know?

---

[16]Nerve disorders such as multiple sclerosis (MS) involve the degradation of myelin.

★   **Problem 11.56.** 1.2×10^{20} electrons flow through a cross section of a copper wire that is 2.5 mm in diameter in 3.5 s. What is the electron drift speed?

★   **Problem 11.57.** Suppose a current is flowing through a thin gold film that is 3.0 μm thick and 80.0 μm wide. If the current density is 8.0×10^5 A/m^2, then how much charge flows through the film every minute? How many electrons is this?

★   **Problem 11.58.** A 5.0 cm long aluminum wire is connected to a AA battery, which has a 1.5 V emf. (a) What is the electric field in the wire? (b) What is the diameter of the wire if the current is 2.5 A? (c) How would the current change if the wire's diameter increased?

★   **Problem 11.59.** A 1.5 m long copper wire is 0.25 mm in diameter. (a) What is its resistance? (b) If it is attached to the terminals of a 0.50 V battery, what current will be in the wire? (c) What power will be dissipated by the wire?

★   **Problem 11.60.** A particular material will melt if the current density exceeds 6.00×10^2 A/cm^2. What diameter should the material have if you *want* it to melt if the current exceeds 1.2 A? (This is the basic idea of a *fuse*).

★★ **Problem 11.61.** If a motor draws 150 A through a copper wire that is 6.0 mm in diameter and 1.5 m long for a total of 0.75 s, then (a) how much charge passes through a cross section of the wire? (b) How far does an electron move, on average, while the motor is on?

★   **Problem 11.62.** The electron drift velocity in a wire tends to be quite small – often near 0.1 mm/s. However, if you connect the plates of a capacitor with a conducting wire, it will discharge very quickly, usually in just a few nanoseconds. How can both of these statements be true?

★★ **Problem 11.63.** The density of gold is 19.3 g/cm^3 and its atomic mass is 197 amu (the conversion from atomic mass units to kg is provided in Appendix B). Gold has 1 valence electron per atom.[17]

   a. How many gold atoms exist in 1 cubic meter of gold?

   b. How many valence electrons exist in 1 cubic meter of gold?

★★ **Problem 11.64.** The density of aluminum is 2.71 g/cm^3 and its atomic mass is 27.0 amu (the conversion from atomic mass units to kg is provided in Appendix B). Aluminum has 3 valence electrons per atom.

   a. How many aluminum atoms exist in 1 cubic meter of aluminum?

   b. How many valence electrons exist in 1 cubic meter of aluminum?

★★ **Problem 11.65.** The resistance of the human body from head to feet varies from around 5.0×10^2 kΩ (when dry) to 1.0 kΩ (when wet). Humans can tolerate a current of about 5.0 mA (much above that and the current can induce fatal muscular contractions).

   a. What is the maximum potential difference that can be applied to a person who is dry?

   b. What is the maximum potential difference that can be applied to a person who is wet?

   c. Approximate the resistivity of a dry person. How does it compare to metals, which typically have $\rho \approx 10^{-8}$ Ωm?

★   **Problem 11.66.** A cylindrical metal rod of length L and diameter D is connected to a battery (one end of the rod to the positive terminal and the other to the negative terminal). A current I is measured through the rod. What is the new current (in terms of I) if

   a. the length is doubled?

   b. the diameter is halved?

   c. the length is doubled *and* the diameter is halved?

---

   [17]While not immediately relevant, it is worth noting that the **electron configuration** of neutral gold's outer shell is written 6 s^1, meaning 1 electron is in the 6th *s* orbital. The meaning behind this notation becomes quite important when studying atomic and molecular interactions.

★ **Problem 11.67.** 16 gauge copper wire has a diameter of 1.29 mm. If such a wire carries a 5.50 mA current, how many electrons pass by any given cross-section of the wire each second?

★★ **Problem 11.68.** A gold wire is 6.40 m long and has a 0.840 mm diameter. What is the potential difference across the ends of the wire if the current in the wire is 1.15 A?

★★ **Problem 11.69.** A ductile metal wire has resistance R. You pull on the wire so as to increase its length from L to $1.1L$. Assuming that the density and resistivity of the metal does not change, what is the new resistance? (*Hint:* how does the cross-sectional area change?)

★★ **Problem 11.70.** Suppose you attach a 60.0 mm long × 0.50 mm diameter gold wire to a 9.0 V battery. Determine

   a. the resistance of the wire.

   b. the current, electron flow rate, and current density (the current divided by the cross-sectional area).

   c. the electron drift velocity.

   d. the mean time between electron-nucleus collisions.

★ **Problem 11.71.** You measure the current through a wire to be some value I. If the wire later becomes thicker, how will the following quantities change? Briefly justify your answers.

   a. Current

   b. Current Density (i.e. the current divided by the cross-sectional area of the wire, I/A)

   c. Electric field

   d. Drift velocity

   e. Electron flow rate

   f. Resistance (per unit length of wire)

   g. Conductivity (per unit length of wire)

★★ **Problem 11.72.** Repeat Problem 11.71, but this time provide numerical values for the case where:

   • The wire is made of copper.

   • In the initial portion of the wire the current is 0.20 A and the diameter is 3.0 mm.

   • When the wire thickens, the diameter is 6.0 mm.

★★ **Problem 11.73.** Show that the resistivity of a piece of wire is given by Equation (11.32):

   a. Start with Equation (11.30) for current in a wire. Eliminate the drift velocity in this equation in favor of the electric field by way of Equation (11.27).

   b. Show that your result in (a) can be written $I = EA/\rho$ for an appropriate definition of ρ in terms of other constants. Which of the terms are fundamental constants? Which are constants specific to a given material (copper, aluminum, rubber, etc.)?

   c. Eliminate E in your result from (b) by noting $E = \Delta V/L$ for a constant electric field in a wire of length L subjected to a potential difference ΔV.

   d. Show that your result to (c) can be written as $\Delta V = IR$ if $R = \rho L/A$.

★★ **Problem 11.74.** Suppose that the terminals of a 1.5 V battery are connected to an iron nail. Consider the nail to be a cylinder that has a length of 2.5 cm and a radius of 3.0 mm.

   a. What is the resistance of the nail?

   b. What current will flow through the nail?

   c. At what speed will electrons move through the nail, on average?

★ **Problem 11.75.** An aluminum paperclip is unfolded such that its ends can be attached to the terminals of a 9.0 V battery. Suppose the paperclip has a diameter of 0.75 mm and a length of 15.0 cm.

   a. What electrical power is dissipated by the paperclip?[18]

   b. What is the magnitude of the electric field in the paperclip?

★   **Problem 11.76.** A 9.0 V battery is connected to a 12 cm copper wire with a diameter of 2.0 mm.

   a. What is the resistance of the wire?

   b. How long does it take for an electron to travel the length of the wire?

Additional Problems for Chapter 11

★★   **Problem 11.77.** An electrical circuit is sometimes made with solder, a malleable metal that can be melted and then "drawn" onto a circuit board to establish electrical connections. A line of solder can be approximated as a half-cylinder of radius r and length L, as shown in Figure 11.47. Suppose the resistivity is 1.7×10^{-7} Ωm. On a particular board, $r = 3.0$ mm, $L = 2.0$ cm, and a fixed current is directed through the solder. the electric potential across the solder is 4.0×10^{-5} V. What would the electric potential across the solder be if the solder was twice as thick (that is, if $r = 6.0$ mm) but the current was unchanged?

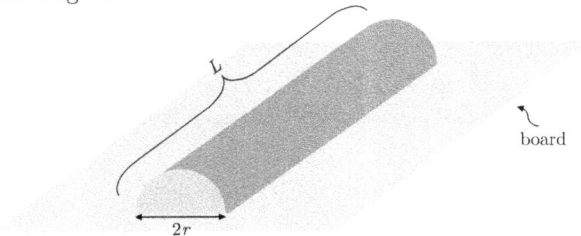

FIGURE 11.47
Problem 11.77

★★   **Problem 11.78.** When he was in 4th grade, one of my sons built the card shown below: a watch battery is glued to some strips of copper foil that are connected to a LED light. When the card is folded (along the vertical line in the middle of the card) the battery touches the rightmost strip of foil and the LED turns on. Suppose the copper has a total length of 0.20 m, a width of 0.010 m, and a thickness of 1.6×10^{-5} m. The LED has a resistance of 13.00 Ω and the battery has a voltage of 1.5 V.

   a. What is the total resistance of the copper foil?

   b. What current will flow through the circuit? Consider the resistance of the strips in your calculation.

   c. On average, how long will it take for a single electron to complete the journey through the entire circuit?

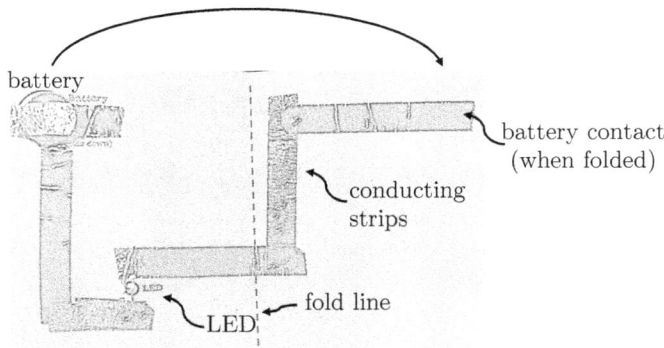

FIGURE 11.48
Problem 11.78

---

[18]This energy is dissipated primarily in the form of heat. Don't try this at home!

★★ **Problem 11.79.** Students in an electronics class decide to construct a rudimentary windshield wiper by designing a circuit where switch toggles between (1) charging three identical capacitors (each with capacitance C) through a resistor (resistance R) and power supply (emf ε) and (2) discharging the capacitors through a motor (also resistance R) that moves the wiper blades. A digital sensor (with essentially no resistance) automatically toggles the switch when the current through the sensor falls below 5% of the maximum current it is expected to experience.

a. The diagram below does not show how the capacitors are attached to the circuit. Fill in the gaps to indicate how the capacitors should be connected to make the charging process as fast as possible. Explain.

b. What is the maximum current that the motor will experience?

c. Make a sketch of the current through the sensor for a complete "charge, then discharge" cycle. How long does this take?

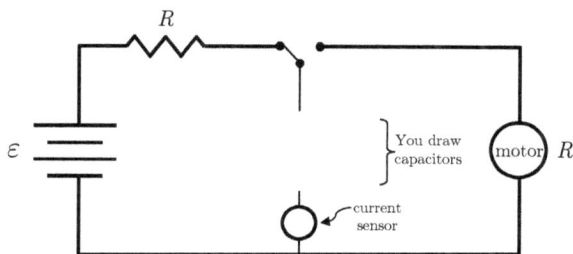

FIGURE 11.49
Problem 11.79

★★ **Problem 11.80.** Fun with current:

a. When a nerve cell "fires", about 9.0 pC of charge moves across the cell membrane in 0.5 ms. What is the average current in the cell during this process?

b. If the electron drift speed in a 1.6-mm-diameter copper wire is 0.20 mm/s, then how many electrons pass through a cross section of the wire each day?

c. Lightning bolts can carry up to 50 kA of current and might deliver the charge in about 100 microseconds. How much charge is delivered by such a lightning bolt?

★ **Problem 11.81.** Ohm's Law is sometimes written in terms of the conductance, $G = \sigma A/L$.

a. Write down Ohm's Law in terms of ΔV, I, and G.

b. What is G for a 8.5-cm-long, 0.25-mm-diameter gold wire?

c. A 1.5 A current flows through the wire described in (b). What is the potential difference between the ends of the wire?

★★ **Problem 11.82.** The electrical properties of materials are temperature dependent; this is an important consideration when designing precision circuits and/or considering superconductivity, which states $\rho \approx 0$ at $T = 0$ K for most metals. The temperature dependence of the resistivity of metals is given by

$$\rho = \rho_0 \left(1 + \alpha \left(T - T_0\right)\right)$$

where ρ_0 is the resistivity at $T_0 = 20°C$ and α is a metal-dependent constant.

a. Graph $\rho(T)$ from $0°C$ to $2T_0$. Clearly label the diagram with slopes, intercepts, etc. as appropriate.

b. For copper, $\alpha = 3.9 \times 10^{-3°}C^{-1}$. If the resistance of a particular copper wire at $20°C$ is 0.50 Ω, then at what temperature is the resistance 0.60 Ω? At what temperature is the resistance 0.40 Ω?

★★   **Problem 11.83.** Digital circuits rely on timers to control their behavior. One common circuit element is called the 555 timer. The timer outputs a time-varying voltage such that it repeatedly switches between a high voltage and a low voltage; it is in the high voltage state for a time T_H and in a low voltage state for a time T_L such that the overall period is $T = T_H + T_L$ (and the frequency is $f = 1/T$ as usual). The values of T_H and T_L are controlled by two resistors and one capacitor that are connected to the timer:

$$T_H = (R_1 + R_2) \, C \ln 2$$
$$T_L = R_2 C \ln 2$$

Suppose you want the overall frequency of oscillation to be 15 MHz and you want 55% of each cycle to be in the high voltage state. If the only capacitor you have available has a capacitance of 3.00×10^2 pF, then what values do you need for R_1 and R_2?

12

Moving Charge: Magnetism

Learning Objectives

After reading this chapter, you should be able to:

- Calculate the cross product of two vectors.

- Use the Lorentz Force to describe the forces acting on charged particles and current carrying wires in magnetic field.

- Apply the Biot Savart Law to determine the magnetic field created by charges and current carry wires.

- Apply Ampere's Law to relate magnetic fields and currents.

- Describe the sources of magnetism.

- Calculate the magnetic flux through a loop of wire.

- Apply Faraday's Law to determine the magnitude of an EMF induced by a change in magnetic flux.

- Describe how a motional EMF can be created by moving a conductor through a magnetic field.

- Analyze the relationship between magnetic flux and induced current (Lenz's Law).

12.1   Introduction

We've now seen that charged particles interact with on one another via the Coulomb force; we can characterize the *force landscape* and corresponding *energy landscape* in some region of interest by respectively analyzing the *electric field* and *electric potential* due to the nearby charge. This is important for understanding atomic and molecular interactions, and (as we saw in the last chapter) it is also useful in the context of electric circuits; electrical current is driven by an electric field (which is created by, for example, a battery with a specified potential difference across its terminals).

There is, however, a *second* force associated with electric charge; it is called the **magnetic force**. Just as we can describe an electric force in terms of an electric field, we can describe the magnetic force in terms of a **magnetic field**, measured in Tesla (T).[1] We can summarize the "big ideas" with two sentences:

---

[1] After Nikola Tesla (1856–1943). To give a sense of scale, the Earth's magnetic field is around 10^{-4} T and a strong magnetic field found inside a magnetic resonance imaging (MRI) machine might be around 1.5 T. I should also note that there is a "magnetic potential" that is analogous to electric potential, but the math gets to be pretty complicated and we won't worry about it in this book.

DOI: 10.1201/9781003571568-16

A moving charge creates a magnetic field.

and

A moving charge in an external magnetic field experiences a magnetic force.

Thus, while two nearby charged particles *always* exert an *electric* force on one another, it is *also* the case that *if they're both moving*, they exert a *magnetic* force on one another as well.

After some preliminary mathematics that we will cover in Section 12.2, in Section 12.3 we will explore the magnetic force for *one* moving charge and then *many* charges moving in a current-carrying wire. Then, in Section 12.4 we will see how we a magnetic field is *generated* in the first place. We will also consider the manifestation of magnetism with which you're probably the most familiar: bar magnets.

We'll explore the relationship between electric fields and magnetic fields more closely in Section 12.5. We will see how a *change* in one kind of field can create, or *induce*, the other kind of field. This gets to be pretty abstract material, but it is very important for three reasons that we will investigate:

1. *Light is an electromagnetic wave*, which refers to a special situation where these waves co-induce each other.

2. We will see that this same principle explains how using mechanical energy to rotate a loop of wire that is immersed in a magnetic field can generate electrical energy. This is the fundamental principle used to generate electricity in modern power plants; it is difficult to overstate its importance.

3. We will see that we can run the "mechanical energy to electrical energy" process described above *in reverse*: we can convert stored electrical energy (in a battery, say) back into mechanical energy. I am here describing, for instance, a *motor*, the importance of which is also hard to overstate.

12.2   The Cross Product

I'd like to pause for a moment and note that we've made it to page 582 of this book without worrying much about vectors interacting in three dimensions.[2] Now, however, we *need* to stop thinking two-dimensionally.

To do so we will work with the Cartesian coordinate system; our usual unit vectors $\hat{x}$ and $\hat{y}$ are joined now by $\hat{z}$. Just as $\hat{y}$ and $\hat{x}$ must be perpendicular, our new third coordinate $\hat{z}$ must be perpendicular to *both* $\hat{x}$ and $\hat{y}$. What does this mean? Well, if you think of $\hat{x}$ and $\hat{y}$ as pointing along the edges of a piece of paper, then $\hat{z}$ must rise perpendicular to the plane formed by the paper (Fig. 12.1).

"Wait a minute", you might object. "There are two vectors that are perpendicular to a plane!" Indeed: if $\hat{x}$ and $\hat{y}$ form a plane that is horizontal (flat on a desk, say), then our new perpendicular vector $\hat{z}$ could point *straight up* or *straight down*. We resolve this choice with the **right hand rule for vectors**. (I'll sometimes abbreviate this as "the RHR".) I'll describe this rule in two equivalent ways:

---

[2]We *did* worry about three dimensional space in the context of Gauss's Law, but in those circumstances our vectors ($\vec{E}$ and $d\vec{a}$) were always either parallel or perpendicular, so the vector analysis was not as complicated as it could have been.

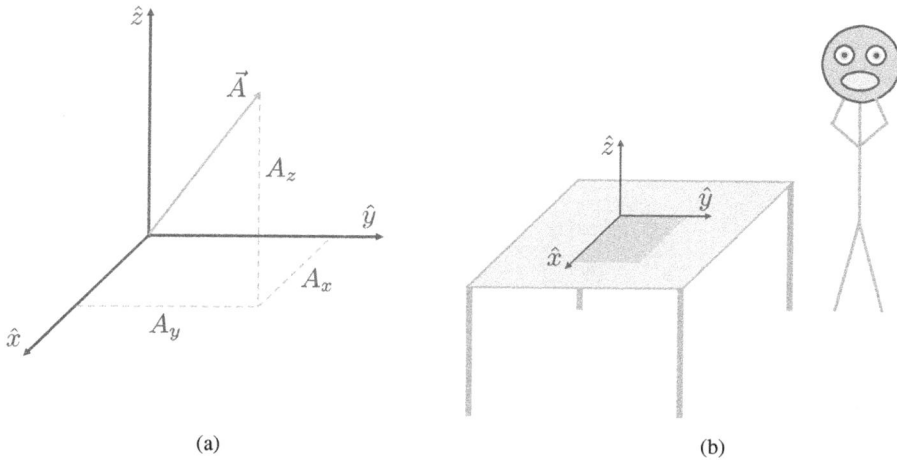

(a)   (b)

FIGURE 12.1

(a) A three-dimensional (3D) vector on a Cartesian coordinate system. The vector $\vec{A}$ has a length in each of the three dimensions labeled $\hat{x}$, $\hat{y}$ and $\hat{z}$ and can be expressed mathematically as $\vec{A} = A_x\hat{x} + A_y\hat{y} + A_z\hat{z}$. (b) An example of a 3D Cartesian coordinate system superimposed on a desk. The $\hat{x}$ and $\hat{y}$ vectors define a plane on the surface of a desk (assuming it is a well made desk, the plane is horizontal). Meanwhile, $\hat{z}$ is perpendicular to the plane (i.e. *vertical*, again assuming your pencil isn't likely to roll off the edge). The observer is shocked and delighted.

- Open your right hand and align it with $\hat{x}$, with your fingertips at the tip of the vector and your wrist at its base. Make sure your hand is rotated so the $\hat{y}$ vector is rising straight out the surface of your palm. Now extend your thumb so it forms an "L" with your index finger. The direction of your thumb is the direction of $\hat{z}$ (Fig. 12.2).

- Imagine standing on the origin with your right arm extended along $\hat{x}$ and your left arm extended over $\hat{y}$. The unit vector $\hat{z}$ runs up your torso and points out of your head (Fig. 12.3).

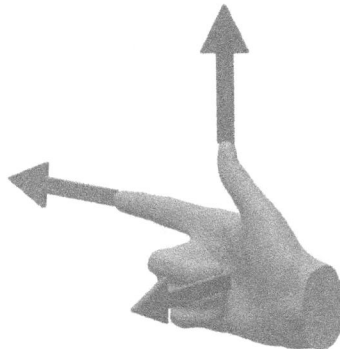

FIGURE 12.2

The arrangement of the unit vectors in a 3D Cartesian coordinate system obey the right hand rule for the cross product $\hat{x} \times \hat{y} = \hat{z}$. This is demonstrated by the hand: if the pointer finger is aligned with the first vector ($\hat{x}$) and a line coming straight out of the palm is aligned with the second vector ($\hat{y}$), then the resulting vector ($\hat{z}$) points up along the extended thumb.

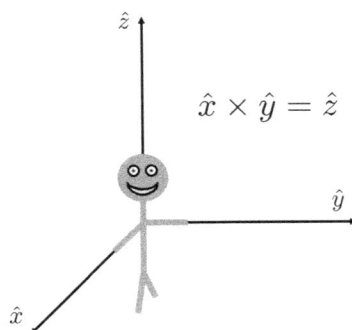

FIGURE 12.3
The arrangement of the unit vectors in a 3D Cartesian coordinate system obey the right hand rule for the cross product $\hat{x} \times \hat{y} = \hat{z}$. This is demonstrated by the stick figure: if the right arm is aligned with the first vector ($\hat{x}$) and the left arm is aligned with the second vector ($\hat{y}$), then the resulting vector ($\hat{z}$) points up along the figure's back. (As an example, suppose the vectors $\hat{x}$ and $\hat{y}$ switched places: the poor stick figure would have to stand on its head to align its arms correctly, which would result in $\hat{z}$ pointing *down* rather than *up*!)

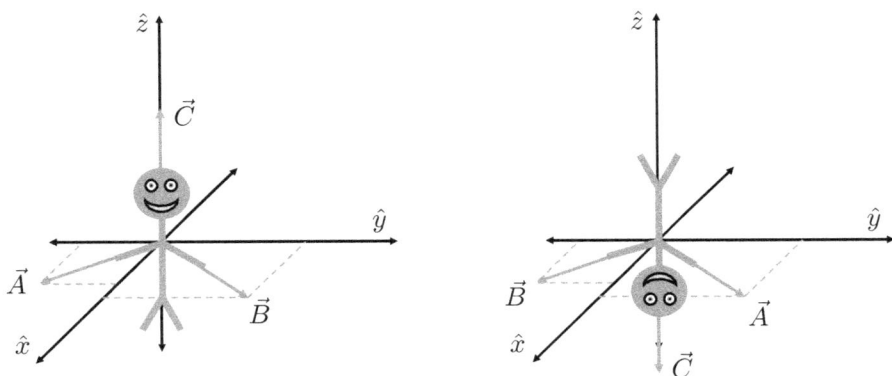

FIGURE 12.4
Two examples of a cross product $\vec{A} \times \vec{B} = \vec{C}$. Here we use the stick figure to orient the relationship between the vectors; Figure 12.5 shows some more complicated examples without it.

Now, this can seem like quite a bit to worry about just to set up a coordinate system, but this procedure is actually very useful because *other* sets of three vectors can be described in the same way.[3] Consider applying the RHR to two arbitrary vectors we think of as forming a "V" when they are arranged to have overlapping tails: you can find the direction for a third vector that is perpendicular to the plane formed by the first two by applying the RHR (Figs. 12.4-12.5). If we call the first two vectors $\vec{A}$ and $\vec{B}$ and the third vector $\vec{C}$, we mathematically denote this procedure with a **cross product**:

$$\vec{C} = \vec{A} \times \vec{B} \tag{12.1}$$

---

[3]We'll see in the next section that one such set involves the velocity of a charged particle, the local magnetic field, and the resulting magnetic force acting on the particle.

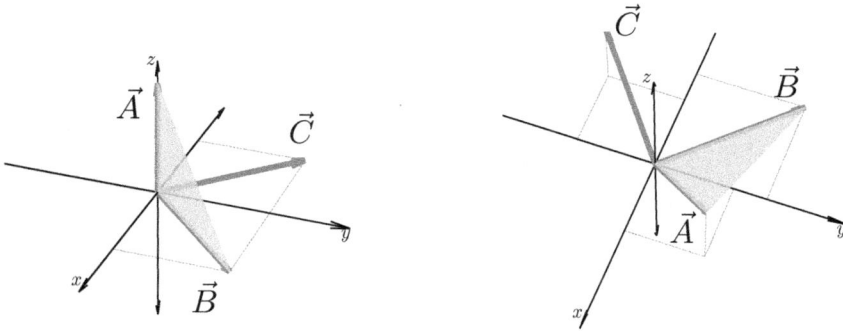

FIGURE 12.5
Two more examples of a cross product $\vec{A} \times \vec{B} = \vec{C}$. (The vectors look a bit different only because I am using plotting software better suited to complicated 3D graphics.) The transparent triangle connects vectors $\vec{A}$ and $\vec{B}$ to show the plane they form; $\vec{C}$ is perpendicular to it. The thin lines show the x, y, and z components of each vector. In the right panel we have a particularly complicated situation where *none* of the three vectors are aligned with an axis. We will not deal with these cases *mathematically*, though it is a good idea to be able to *visualize* them.

We read the right hand side of this equation as "A cross B". It tells us to use the right hand rule to determine the direction of the resulting vector $\vec{C}$.[4]

In addition to the *direction* of $\vec{C}$, the cross product also tells us how to determine the *magnitude* of $\vec{C}$: multiply the *perpendicular components* of $\vec{A}$ and $\vec{B}$ together.[5] As I show in Figure 12.6, one way to denote this mathematically is to identify the angle in the "V" orientation as θ and then recognize that the "$\sin\theta$ component" of one of the vectors is perpendicular to the other. Mathematically,

$$\vec{C} = \vec{A} \times \vec{B} \implies C = AB\sin\theta \tag{12.2}$$

This means that the cross product is *largest* when $\vec{A}$ and $\vec{B}$ are perpendicular, i.e. when $\theta = 90°$ and $\sin 90° = 1$. In contrast, if $\vec{A}$ and $\vec{B}$ are parallel (i.e. if $\theta = 0°$), then $\sin 0° = 0$ and the cross product yields 0 (Fig. 12.6).[6]

Working with three vectors related according to a cross product will be very important when we get to the *physics* in the next section; let's pause now for an example.

Example 12.1 The Cross Product   ⋆⋆
Consider a cross product $\vec{A} \times \vec{B} = \vec{C}$. In each situation, identify the missing magnitude and/or direction for the incomplete vector; sketch all three vectors. (a) $\vec{A}$ has magnitude 0.50 and is aligned with the negative x axis; $\vec{B}$ has magnitude 1.0 and

---

[4]Incidentally, we shall always use a so-called *right handed coordinate system* where $\hat{x} \times \hat{y} = \hat{z}$.

[5]This is in contrast to the *dot product*, which we introduced in the context of the *work* done by applying a force to some object and considering its effect as the object experiences some displacement: we multiplied the *parallel* components of the force and the displacement to determine the work done. Note also that the dot product takes two vectors and returns a *scalar*, while the cross product takes two vectors and returns a *vector*.

[6]This is all in contrast to the dot product, which depends on $\cos\theta$ and is therefore maximized with the vectors are parallel ($\theta = 0°$) rather than perpendicular. Incidentally, it is worth noting that the magnitude and direction of a cross product can also be worked out using Equation (4.38).

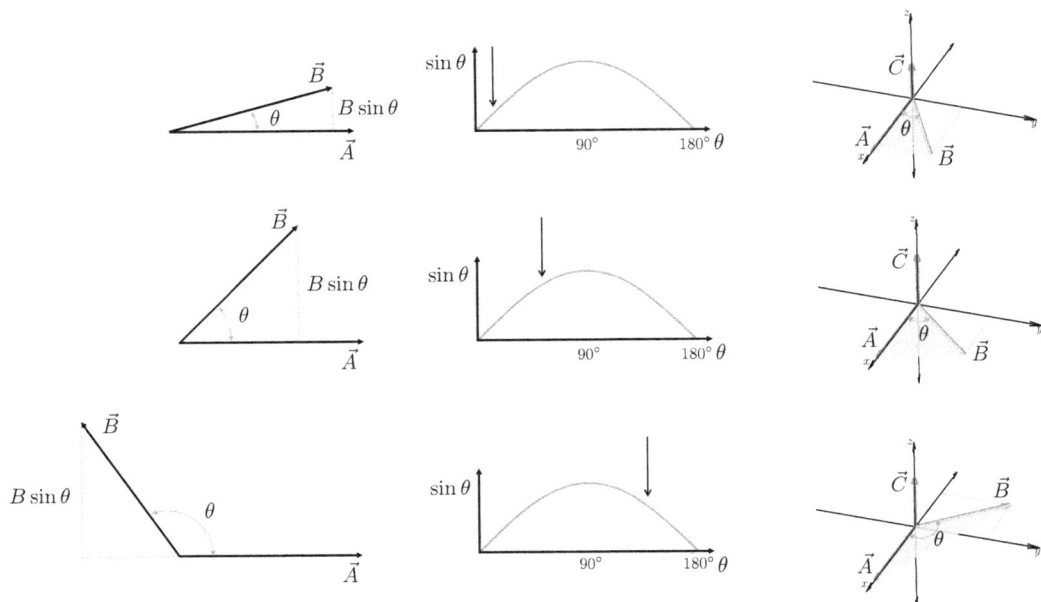

FIGURE 12.6

When computing the magnitude of a cross product $\vec{A} \times \vec{B}$, multiply the components of the vectors $\vec{A}$ and $\vec{B}$ that are *perpendicular*. This can be defined in terms of the angle θ formed between them, as the left column shows for three different angles θ. The middle column indicates the corresponding value of $\sin\theta$; the cross product is largest when $\theta = 90°$ and it equals 0 when $\theta = 0°$ and $\theta = 180°$. (To go back to the stick figure analogy, at $\theta = 0°$ the arms are closed, like it is saying "stay away!" As the angle increases to $90°$, the arms open and the magnitude of the cross product increases; the figure is happy to see you and asking for a hug. Beyond $90°$ the magnitude of the cross product begins to decrease; evidently if the figure's arms are *too* wide, the invitation for a hug is sarcastic.) The right column shows the same two vectors on a 3D Cartesian coordinate system, with $\vec{A}$ aligned with the x axis and $\vec{B}$ on the xy plane. The height of $\vec{C}$ changes based on θ, though note that the second two examples actually give the *same* magnitude because they are equidistant from the midpoint at $90°$.

is at a 45° angle between the positive x and positive z axes. (b) $\vec{A}$ is straight *down* the page with a magnitude of 2.0 and $\vec{C}$ is *to the left* with a magnitude of 1.5. $\vec{B}$ is perpendicular to $\vec{A}$.

(a) The vectors $\vec{A}$ and $\vec{B}$ lie on the xz plane, so $\vec{C}$ will point along either the positive or the negative y axis to be perpendicular to both $\vec{A}$ and $\vec{B}$. To determine *which* of these choices is the correct orientation, we make a sketch and apply the RHR. With the "stick figure arms" approach we have the right arm along the $-x$ axis and the left arm between the $+x$ and $+z$ axes, which means the head points along the $+y$ axis (Fig. 12.8).

For the *magnitude*, note that the angle between $\vec{A}$ and $\vec{B}$ is $90° + 45° = 135°$, so

$$C = AB\sin\theta = (0.50)(1.0)\sin(135°) = 0.35$$

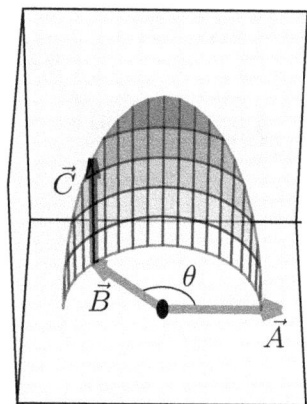

FIGURE 12.7
Another view of the cross product $\vec{A} \times \vec{B} = \vec{C}$. The shaded surface shows the magnitude of $\vec{C}$ ($= AB \sin \theta$) for values of θ ranging from 0° to 180°. The largest value occurs for $\theta = 90°$, and the cross product has no magnitude at all when $\theta = 0°$ or $\theta = 180°$.

To conclude, then, we have $\vec{C} = 0.35\hat{y}$.

(b) Here we're avoiding explicit references to a coordinate system and are instead orienting the vectors relative to the page, which is perhaps a more intuitive choice. We know the angle between $\vec{A}$ and $\vec{B}$ is 90°, so we can work out the magnitude of $\vec{B}$ directly:

$$C = AB \sin \theta \implies B = \frac{C}{A \sin \theta} = \frac{1.5}{2.0 \sin(90°)} = 0.75$$

To determine the direction of $\vec{B}$, note that the resultant vector $\vec{C}$ (which is perpendicular to the plane formed by the other two vectors) points to the left, so the other two vectors must lie on a plane that is oriented "up/down the page" and "in/out of the page". Now $\vec{A}$ points *down* and it is perpendicular to $\vec{B}$, so $\vec{B}$ must lie either *in to the page* or *out of the page*. Using the RHR with our imaginary stick figure, we have the right arm ($\vec{A}$) down and the head ($\vec{C}$) to the left, which means the left arm ($\vec{B}$) must point *out* of the page.

In Figure 12.9 I show the vectors on a 2D plane (in these cases it is customary to use a × symbol in a circle to indicate a vector pointing in to the page and a · symbol in a circle to indicate a vector pointing out of the page) and also with a perspective view of the 3D Cartesian system, where the yz plane is the page ($-y$ points left and z points up) and x is out of the page.

Thus finally we have $\vec{B} = 0.75\hat{x}$.

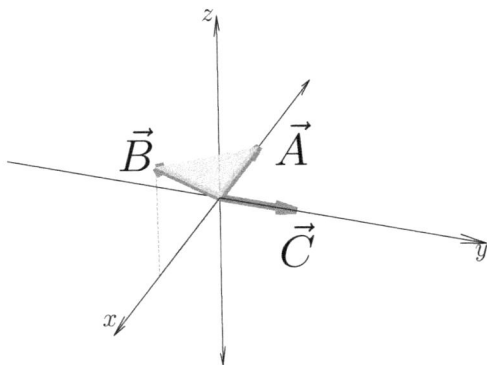

FIGURE 12.8

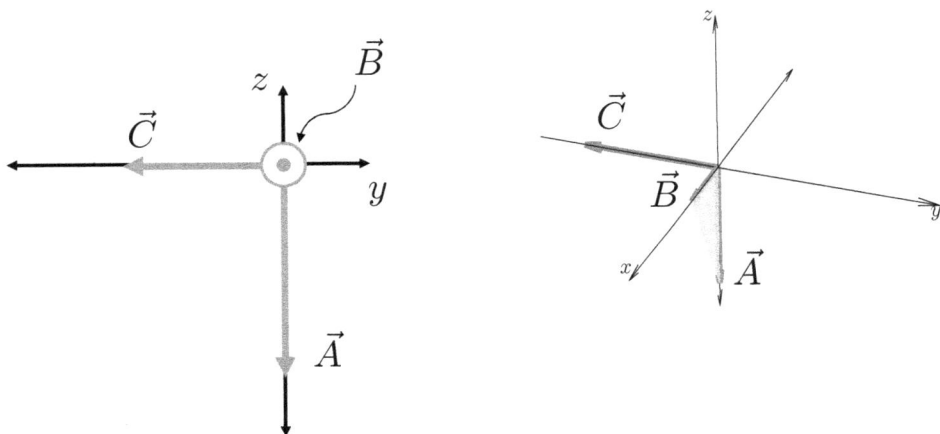

FIGURE 12.9

12.3   The Magnetic Force

In this section I would like to consider the *second* bolded statement I made in the Introduction: a particle with electric charge q feels a force as it moves with velocity $\vec{v}$ through a magnetic field $\vec{B}$ (we don't use an M or m when referring to magnetic fields to avoid confusion with *mass*). As you probably expect after going through the last section, expressing the relationship mathematically requires a cross product:

$$\vec{F}_B = q\left(\vec{v} \times \vec{B}\right) \tag{12.3}$$

In words, we take the cross product of the particle's velocity $\vec{v}$ and the magnetic field at the particle's position $\vec{B}$. The resulting vector (what we generically called $\vec{C}$ in the last section)

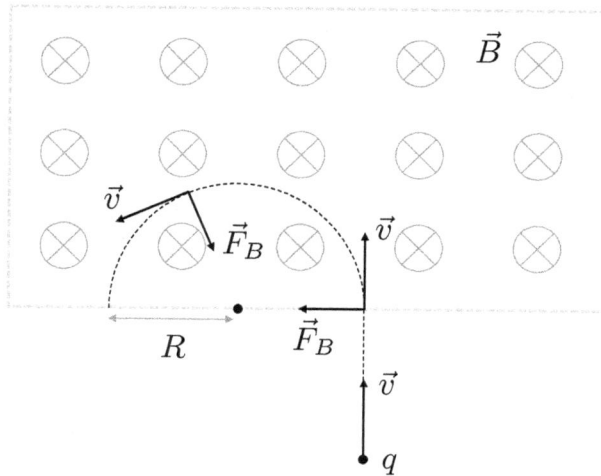

FIGURE 12.10

A positively charged particle enters a rectangular region where a magnetic field is directed in to the page, as indicated by the "× in a circle" symbols. Applying the right hand rule identifies the initial direction of the magnetic force as to the left; because the magnetic force is always perpendicular to the velocity vector, the particle moves in a circular path with some radius R. One later set of $\vec{v}$ and $\vec{F}_B$ vectors are shown by way of example.

is multiplied by the charge q to determine the magnetic force $\vec{F}_B$. Multiplying by q has two effects: first, it ensures $\vec{F}_B$ has the right magnitude; second, if q is *negative* then $\vec{F}_B$ points in the *opposite* direction of $\vec{v} \times \vec{B}$.[7]

This equation tells us that the magnetic force is always perpendicular to both the particle's velocity and the magnetic field at the particle's location. Importantly, the fact that the force and velocity are perpendicular means that a charged particle in a magnetic field can undergo *circular motion*. To see why, consider the situation of Figure 12.10: a magnetic field inside a rectangular region of space is directed in to the page, and a particle with positive charge q located below the region travels upward with a speed v.

Immediately as it enters the magnetic field, Equation (12.3) tells us the charge experiences a magnetic force directed to the *left*. Because $\vec{v}$ and $\vec{B}$ are perpendicular, their cross product simplifies:

$$\vec{F}_B = q\left(\vec{v} \times \vec{B}\right)$$
$$F_B = qvB\sin\left(90°\right)$$
$$F_B = qvB \tag{12.4}$$

Thus, the upward velocity vector shifts slightly to the left – but as it does, the magnetic force *also* shifts to remain perpendicular to the velocity. The net effect is a *circular path* of some radius R that the proton travels at a *constant speed*, just as we discussed when introducing circular motion.

---

[7]This is because multiplying a vector by -1 reverses its direction, so if $q < 0$ then multiplying by q reverses the direction of the vector *and* scales the magnitude by a multiplicative factor $|q|$. For example, if a particle with charge $+q$ experiences a magnetic force to the left, then a particle with charge $-q$ in the exact same situation (i.e. same $\vec{v}$ and $\vec{B}$) would experience an equal-magnitude force to the right.

All of the tools we developed there apply just as well here, but we can go a bit further because Equation (12.4) gives us additional information about the force causing the motion. For instance, if we recall the centripetal acceleration is given by $a_c = mv^2/R$ (Equation (2.18) on page 84), then we can apply Newton's Second Law to determine a useful expression for its radius of motion:

$$\vec{F}_{NET} = m\vec{a}$$
$$F_B = ma_c$$
$$qvB = m\frac{v^2}{R}$$
$$R = \frac{mv}{qB} \tag{12.5}$$

The circular motion of a charged particle in a magnetic field is sometimes called **cyclotron motion**, and so the radius R defined in Equation (12.5) is sometimes called the cyclotron radius. In the next two examples we consider useful applications of cyclotron motion: mass spectrometers and particle accelerators.

Example 12.2 The Mass Spectrometer   ★★

Here is the basic idea behind a **mass spectrometer**: a particle with an unknown mass m and/or charge q starts from rest and accelerates through a potential difference ΔV, giving it a speed v. From there it enters a perpendicular magnetic field $\vec{B}$, as sketched in Figure 12.10. A detector outside the field collects the particles when they exit the field and records the radius R of the particle's motion. Determine an expression for the mass-to-charge ratio m/q for the particle in terms of ΔV, B, and R.

We can determine the particle's speed in terms of the potential difference it accelerates through by invoking conservation of energy: the particle moves through the potential difference and picks up kinetic energy in an analogous way to a ball picking up kinetic energy as is falls toward the ground. Mathematically,

$$U_i = K_f \tag{12.6}$$

$$q\Delta V = \frac{1}{2}mv^2 \tag{12.7}$$

$$v^2 = \frac{2q\Delta V}{m} \tag{12.8}$$

I've left this in terms of v^2 instead of solving for v to make the following algebra a bit easier. We can insert this result in to Equation (12.5) and then solve for m/q, but (again, to make the math a bit easier) we'll first square both sides of the equation:

$$R^2 = \left(\frac{m}{qB}\right)^2 v^2$$

Now let's insert the result for v^2 we found above and then algebraically solve for m/q:

$$R^2 = \left(\frac{m}{qB}\right)^2 \frac{2q\Delta V}{m}$$

$$R^2 = \left(\frac{m}{q}\right) \frac{2\Delta V}{B^2}$$

$$\frac{m}{q} = \frac{(RB)^2}{2\Delta V} \tag{12.9}$$

Equation (12.9) is our desired result. Problems 12.18–12.19 provide some specific scenarios where it is quite useful in characterizing particles, for instance molecules from a biological sample.

Example 12.3 Particle Accelerators ★★

Particle accelerators are used to bring charged particles up to very high speeds – in some cases very close to the speed of light! One example where this is useful is when performing experiments in the so-called field of high-energy particle physics. A second example is proton-beam therapy for treating deep-seated tumors (this was previously discussed in Problem 9.53). Figure 12.11 shows one way to achieve these high speeds: low-speed charged particles enter a magnetic field $\vec{B}$ near the center of the device and begin undergoing cyclotron motion. In addition, a narrow strip of width d contains an electric field $\vec{E}$ that accelerates the particle every time it passes by. The result is a *growing spiral* (really a series of half-circles of increasing radius); once the particles have the desired speed $\vec{v}$, they exit the fields and proceed to travel toward the target in a straight line. (a) What exit radius R should be used if protons are to be accelerated to 8.0×10^7 m/s in a 1.5 T magnetic field and a 5.0×10^4 V/m electric field? Assume the width of the accelerating strip is 1.0 cm. (b) How many revolutions will each proton go through before exiting the device?

For (a) we can identify the needed radius directly with Equation (12.5):

$$R = \frac{mv}{qB}$$

$$R = \frac{(1.67 \times 10^{-27} \text{ kg}) (8.0 \times 10^7 \text{ m/s})}{(1.60 \times 10^{-19} \text{ C}) (1.5 \text{ T})}$$

$$R = 0.57 \text{ m}$$

OK, so how many revolutions will a proton make before reaching the desired speed and corresponding radius? Well, it accelerates a bit every time it passes through the strip containing an electric field, and the *total* distance it needs to travel while in the electric field, which we'll call D, can be determined from energy conservation:

$$q\Delta V = \frac{1}{2}mv^2$$

$$q\left(ED\right) = \frac{1}{2}mv^2$$

$$D = \frac{mv^2}{2qE}$$

$$D = \frac{\left(1.67 \times 10^{-27} \text{ kg}\right)\left(8.0 \times 10^7 \text{ m/s}\right)^2}{2\left(1.60 \times 10^{-19} \text{ C}\right)\left(5.0 \times 10^4 \text{ V/m}\right)}$$

$$D = 6.7 \times 10^2 \text{ m}$$

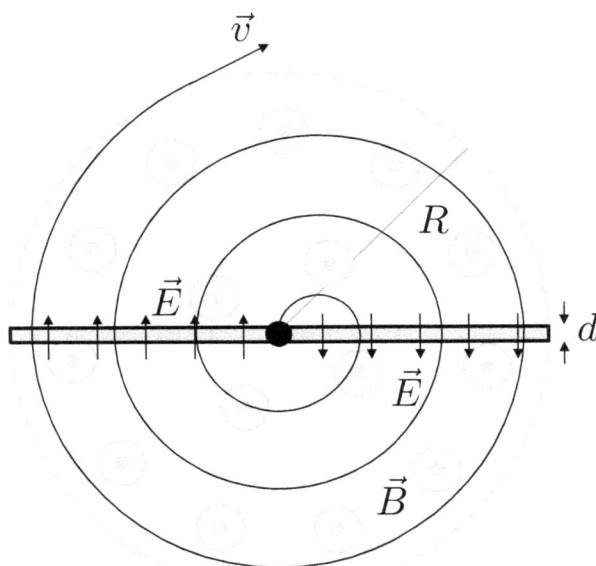

FIGURE 12.11

This is the *total* distance it needs to travel in the electric field. Now note that every revolution involves passing through the strip twice (on Fig. 12.11, this is once as it passes on the right and once on the left), for a total distance of $2d = 2.0$ cm in the electric field per revolution. Thus the number of revolutions N it will go through is determined by

$$N\left(2d\right) = D$$

$$N = \frac{D}{2d}$$

$$N = \frac{6.7 \times 10^2 \text{m}}{0.020 \text{ m}} = 3.3 \times 10^4$$

(Be careful with rounding intermediate calculations too far if you check the numerical value here.)

Before moving on to discuss how magnetic *fields* are *created*, let me make one more comment on magnetic *force*. So far we've thought about a single charged particle moving in some magnetic field, but an obvious extension is to consider a current-carrying wire: after all, it contains a whole *stream* of moving charged particles! In particular, suppose some length L of wire is immersed in a magnetic field while carrying a current I, as sketched in Figure 12.12. What force is acting on the wire?

First, note that the *direction* of the force is obtained in the usual way: we think of current as the direction of motion for positively charged particles, so we have $q\left(\vec{v} \times \vec{B}\right) = \vec{F}_B \implies$ *"right cross in = up"*, i.e. the magnetic force is up the page.

As for the magnitude, consider the force for just one charge in the wire:

$$\vec{F}_B = q\left(\vec{v} \times \vec{B}\right)$$

To extend this to the wire, consider an infinitesimally small piece of the wire with some length dx; it contains a correspondingly small charge dq and experiences a force $d\vec{F}_B$. Then

$$d\vec{F}_B = dq\left(\frac{d\vec{x}}{dt} \times \vec{B}\right)$$

$$d\vec{F}_B = \frac{dq}{dt}\left(d\vec{x} \times \vec{B}\right)$$

In the second step I pulled the dt out of the parentheses. Now dq/dt is just the current, which is constant in the wire. When we integrate this equation over the length of the wire, the $d\vec{x}$ becomes $\vec{L}$ (the vector's direction comes from $\vec{v}$ for a single charge, so it points in the direction of the current) and we find the magnetic force on the current-carrying wire is

$$\vec{F}_B = I\left(\vec{L} \times \vec{B}\right) \tag{12.10}$$

As we will see in Example 12.5, a current generating a force is the essential feature of an **electric motor**.

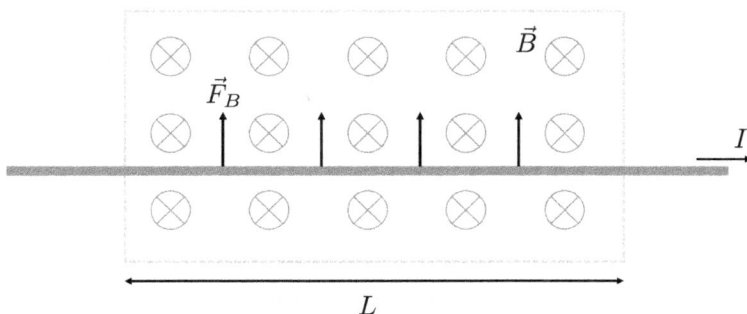

FIGURE 12.12
A long wire carries a current I to the right. A length L is immersed in a magnetic field $\vec{B}$ pointing in to the page. As a consequence, the wire feels a force directed up the page (black vectors).

Example 12.4 Magnetic Force on a Wire   ★★
(a) A straight wire has length L, mass m, and carries current I to the right. Consider
a magnetic field with strength B. Determine (a) the *direction* of $\vec{B}$ necessary for the
wire to experience an *upward* magnetic force. (b) Determine an algebraic expression
for I that allows the wire to experience no net force (i.e. for the magnetic force to
perfectly cancel the gravitational force) assuming $\vec{B}$ and $\vec{L}$ are perpendicular. (c) Use
the result from (b) to calculate a numerical value for I if $B = 0.50$ T, $m = 2.0$ g, and
$L = 30$ cm.

To determine the direction we need to think through the right hand rule with
$\vec{F}_B = I\left(\vec{L} \times \vec{B}\right)$: here $\vec{L}$ is to the right and $\vec{F}_B$ is up, so we have "up = right cross
?" This means our stick figure's head is pointing up and its right arm points to the
right. If the arms start closed ($\theta = 0°$) then the left arm also points to the right; if
the left arm then swings open, then it is swinging in to the page. If $\theta = 90°$, as we
assume in (b), then it points straight in to the page (This is the same arrangement
shown in Figure 12.12).

For (b) the fact that $\vec{L}$ is perpendicular to $\vec{B}$ means the force equation simplifies:

$$\vec{F}_B = I\left(\vec{L} \times \vec{B}\right) \implies F_B = ILB$$

Now the only other force acting on the wire is its weight, so we can straightforwardly
apply Newton's Second Law to determine the equilibrium condition (as usual we will
call up the positive direction):

$$\vec{F}_{\text{NET}} = m\vec{a}$$
$$F_B - mg = ma = 0$$
$$F_B = mg$$
$$ILB = mg$$
$$I = \frac{mg}{LB}$$

Finally for (c) we will plug in the given numerical values:

$$I = \frac{\left(2.0 \times 10^{-3} \text{ kg}\right)\left(9.8\frac{\text{m}}{\text{s}^2}\right)}{(0.30 \text{ m})(0.50 \text{ T})} = 0.13 \text{ A}$$

Example 12.5 Magnetic Force on a Square Loop   ★★★
A wire has been twisted to form a square of edge length w, as shown in Figure 12.13.
Current I enters and then exits along a narrow "handle" at the middle of one side of
the square. (a) The wire is held horizontally in a vertical magnetic field, as shown in
the figure. What is the magnetic force acting on each of the four sides of the square?

If the wire is free to rotate by twisting along the "handle", will it? Repeat (a) for the cases where the loop is rotated clockwise about its "handle" (as viewed from the $+y$ axis) by (b) 90°, (c) 180°, and (d) 270°.

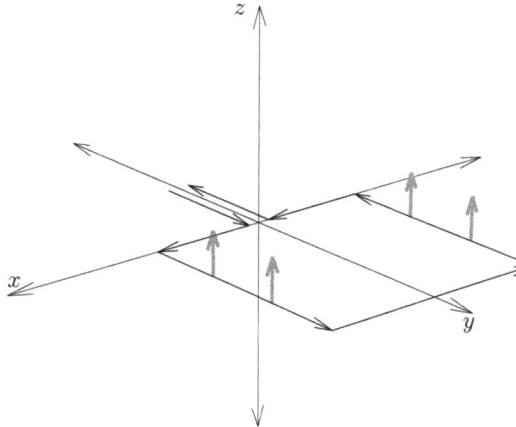

FIGURE 12.13

Determining the direction of the force on a section of the wire requires us to apply the right hand rule with $\vec{F}_B = I\left(\vec{L} \times \vec{B}\right)$. With four sides to a square and four configurations to consider, we need to apply the RHR a total of 16 times! Figure 12.14 shows the results with a pair of identical force vectors along each edge of the square. (The forces are shown in a "dash-dot" style just to make them easier to distinguish from the magnetic field vectors, which always point up along the z axis.)

Each wire either has no magnetic force (if $\vec{L}$ and $\vec{B}$ are parallel) or a force with magnitude $F_B = IwB$ (when $\vec{L}$, with length w, and $\vec{B}$ are perpendicular). Notably, the *net force* on the square is *always 0*, but the *net torque isn't*. In (b), the forces on the top and bottom sections of the wire both exert a counter-clockwise torque, while in (d) they both exert a clockwise torque.

This can seem like a contrived example, but we're actually most of the way to describing an **electric motor**, where current is used to induce motion. While the situation we've described here involves opposing torques, we can eliminate one half of the torque if we arrange for no current to flow for half of the rotation. Alternatively, if the direction of current flow changes as the loop rotates, it is possible for all the torques to point in the same direction.

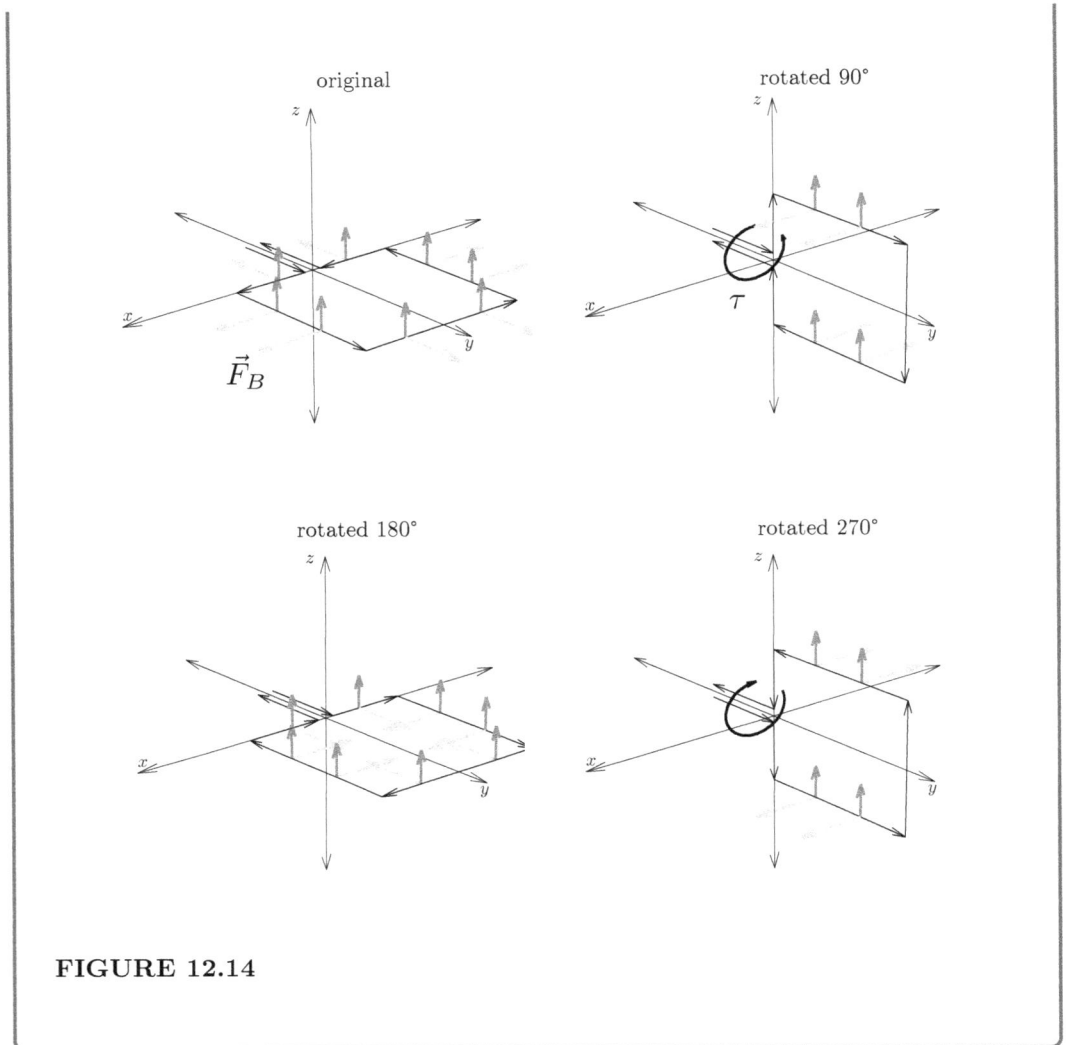

FIGURE 12.14

12.4   The Magnetic Field: The Biot–Savart Law and Ampere's Law

For a charged particle to experience a magnetic *force*, it must be in a magnetic *field* (and have an appropriate velocity). We've now talked quite a bit about the magnetic force, but where does the requisite magnetic field come from?

Well, at the beginning of the chapter I said that a magnetic field is created by a *moving charge*. Let's dig in to the details by considering a charge q moving at some velocity $\vec{v}$. We'll also consider a vector $\vec{r}$ that points from the charge out to the position in space where we want to know the magnetic field $\vec{B}$ (Fig. 12.15). These quantities are related mathematically with (surprise!) a cross product:

$$\vec{B} = \frac{\mu_0}{4\pi} \frac{q\,(\vec{v} \times \hat{r})}{r^2} \tag{12.11}$$

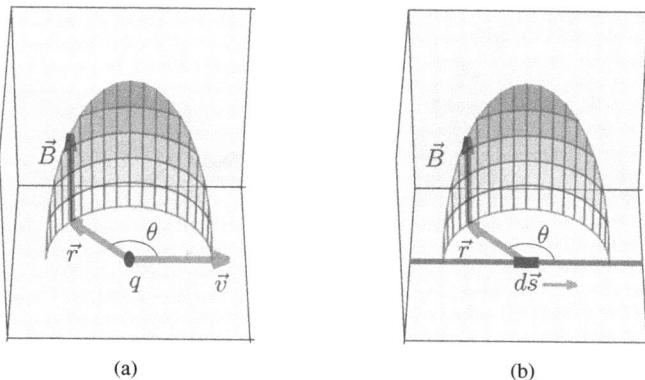

(a)   (b)

FIGURE 12.15
The magnetic field created by one or more moving charges involves a cross product. (a) A single moving charge creates a magnetic field according to Equation (12.11). (b) A wire carrying a current creates a magnetic field according to the Biot–Savart Law (Equation (12.13)). Here we consider the field due to just one infinitesimally small piece of wire of length ds is shown in black; the direction of the vector $d\vec{s}$ is the same as the (conventional) current.

Here μ_0 is the **permeability of free space**, a fundamental constant of nature with value

$$\mu_0 = 1.26 \times 10^{-6} \frac{\text{m kg}}{\text{s}^2 \text{A}^2} \qquad (12.12)$$

Just as the *permittivity* of free space ε_0 comes up when discussing *electric* fields, the *permeability* of free space μ_0 comes up when discussing *magnetic* fields.

The cross product in Equation (12.11) is between the charge's velocity $\vec{v}$ and the *unit vector* $\hat{r}$ that points from the charge to the point where we'd like to know the magnetic field. As you know, $\hat{r}$ has length 1 (as all unit vectors do) and points in the direction of $\vec{r}$. Thus the effect of the cross product is to identify the *component* of $\vec{v}$ that is perpendicular to $\vec{r}$.

The net effect is that *any* magnetic field vector created by this moving charge is tangent to a circle centered on the line formed by the charge's velocity vector (and perpendicular to it). In fact, we commonly just *draw a circle* to show the direction of the magnetic field, as in Figure 12.16 and the bottom panel of Figure 12.17 (though we don't see how *strong* the field is; in the figure we give some sense of relative strength with the *color* of the arc).

This is all well and good, but usually we aren't concerned with the magnetic field created by just one moving charge. Instead we'd like to consider the effect of a *current-carrying wire*. We can think of this as simply adding up the effect of many individual charges, which entails *integrating* Equation (12.11) over the length of the wire. The details are summarized mathematically in the **Biot–Savart Law**:[8]

$$\vec{B} = \frac{\mu_0}{4\pi} \int \frac{I\left(d\vec{s} \times \hat{r}\right)}{r^2} \qquad (12.13)$$

To move from Equation (12.11) to Equation (12.13), note that $q\vec{v} = q d\vec{x}/dt$ has become

---

[8] After Jean–Baptiste Biot (1774–1862) and Félix Savart (1791–1841). I should mention too that the connection between the "single charge" and "current" equations is more complicated than I am suggesting here. You'll go through the specifics if you study electromagnetism at a more advanced level.

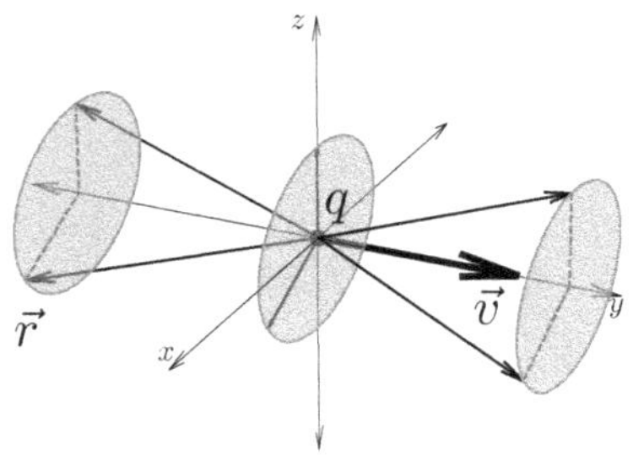

FIGURE 12.16
A charge q with a velocity $\vec{v}$ directed along the positive y axis. We would like to determine the magnetic field at a point at a position $\vec{r}$ that we can consider to be on the perimeter of a circle centered on the y axis.

$(dq/dt)\,d\vec{x} = I\,d\vec{s}$, where we've substituted $d\vec{s}$ (a chunk of wire pointing in any direction) for $d\vec{x}$ (which suggests the wire is entirely along the x axis). The integral runs over the entire length of the current-carrying wire that we're considering. Much as when we determined the electric potential or the electric field from a distribution of charge, we're back into the realm of *integrating* to determine the *cumulative effect* of many charges (in this case, *moving* charges creating a magnetic field).

By applying the Biot–Savart Law to different arrangements of current-carrying wire, we can characterize the magnetic field formed by it. The most common examples are summarized in Figure 12.18. The equations for the corresponding magnetic fields are as follows:

- For a long straight wire carrying current I, the magnetic field a distance R from the wire has magnitude

$$B_{\text{wire}} = \frac{\mu_0 I}{2\pi R} \tag{12.14}$$

- For a circular loop of wire with radius R and carrying current I, the magnetic field at the center has magnitude[9]

$$B_{\text{loop}} = \frac{\mu_0 I}{2R} \tag{12.15}$$

- For a **solenoid**, which has N loops coiled in a stack over length L, the magnetic field throughout the middle of the solenoid (i.e. until you're close to the edges) has magnitude

$$B_{\text{solenoid}} = \frac{\mu_0 N I}{L} \tag{12.16}$$

In each case the *direction* of these magnetic fields can be determined by analyzing the direction of the cross product $d\vec{s} \times \hat{r}$. A shortcut is to use the **right hand rule for fields**, which says to *point the thumb of your right hand in the direction of the current*. Then, *the magnetic field wraps around the thumb in the same direction your fingers prefer to rotate when you bend your wrist forward*. For example, if you point your right thumb directly at your eye[10], your fingers naturally want to curl counter-clockwise. If you try this with the examples in Figure 12.18, you'll see that this is perfectly consistent with the longer "vector right hand rule" $d\vec{s} \times \hat{r}$.[11]

---

[9]A tight coil of N loops simply increases the magnetic field of a single loop by a factor of N.
[10]Don't poke yourself, that would be embarrassing!
[11]For example, for the circular loop carrying counter clockwise current, consider any section of the wire

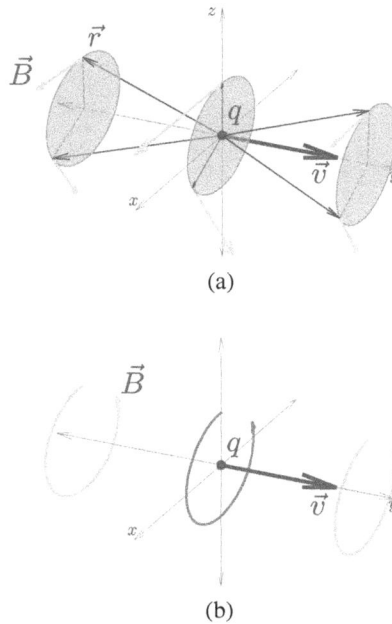

(a)

(b)

FIGURE 12.17
(a) An update of Figure 12.16, now with each of the $\vec{B}$ lines drawn in according to the right hand rule. Each magnetic field vector is tangent to the circle. Note that the magnitude is largest on the center circle, which is centered on the charge (as this corresponds to the special case where $\vec{r}$ and $\vec{v}$ are perpendicular). (b) It is tidier to omit the $\vec{r}$ and $\vec{B}$ vectors and instead show magnetic field lines, where the field at any point is tangent to the circle. The lighter coloring of the front and back arcs indicates the magnitude of the field is weaker than the central arc. While only three arcs are shown here, any number could be drawn; they collectively wrap around a cylinder centered on the charge and velocity $\vec{v}$ (as such this is an example of so-called *cylindrical symmetry*.)

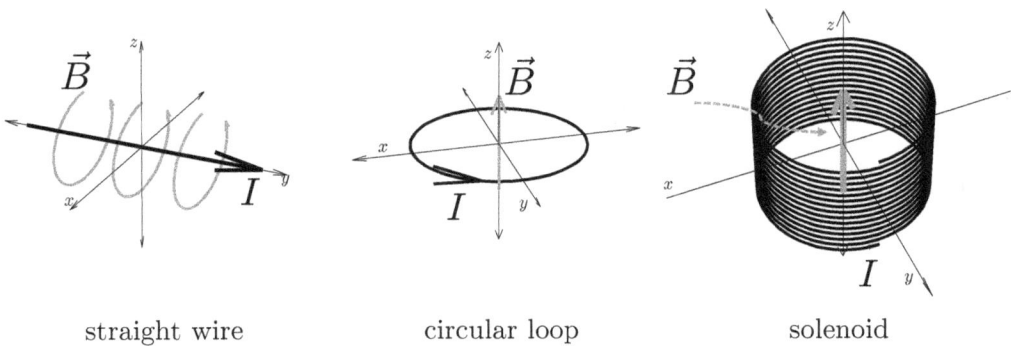

straight wire   circular loop   solenoid

FIGURE 12.18
Three configurations for a current carrying wire: a straight wire, a circular loop, and a solenoid (i.e. a coil of wire in the form of a spring or Slinky).

Figure 12.19 shows a more detailed view of the magnetic fields created by the configurations shown in Figure 12.18 (which shows just one or a few characteristic field lines). In the next two examples we will *derive* Equations (12.14) and (12.15) by applying the Biot–Savart Law.

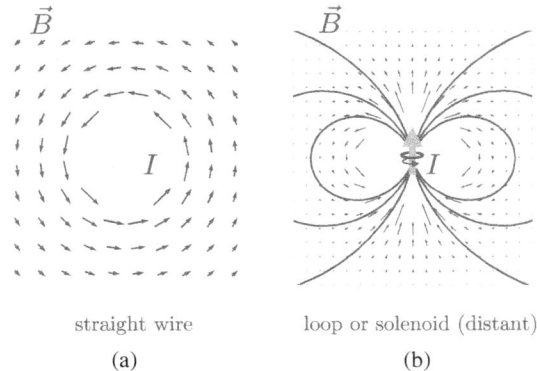

straight wire	loop or solenoid (distant)
(a)	(b)

FIGURE 12.19
A *vector field* representation of the magnetic fields created by (a) a straight wire (viewed with the current coming straight out of the page) and (b) a ring or solenoid viewed from a great distance. In this second example the solid arcs show the magnetic field lines (i.e. a curve whose tangent always points in the direction of the field) and the large gray arrow indicates the direction of the field at the center of the loop/solenoid. You can think of the magnetic field vectors as "spraying" up and out, then falling back down along the sides before being sucked back up into the bottom of the loop/solenoid. In both panels the vectors very close to the current are omitted because they would be very long.

Example 12.6 Magnetic Interactions Between Wires ⋆
Two long parallel wires are separated by a distance r. The wires carry equal currents I in the same direction. (a) Determine the force per unit length (magnitude and direction) on each wire. (b) Give a numerical result if $I = 25.0$ A and $r = 3.0$ cm.

The force on a current-carry wire is given by Equation (12.10). Here, each wire experiences a force due to the magnetic field created by the *other* wire. If we label the wires 1 and 2, we have

$$\vec{F}_{\text{on 1}} = I_1 \left(\vec{L}_1 \times \vec{B}_{\text{from 2}} \right)$$

and

$$\vec{F}_{\text{on 2}} = I_2 \left(\vec{L}_2 \times \vec{B}_{\text{from 1}} \right)$$

The magnetic field created by each wire is given by Equation (12.14). R in that equation refers to how far from the wire we'd like to know the field; here we have

---

and point your thumb in the corresponding current direction. As you rotate your fingers (or imagine doing so if you're contorted awkwardly), your fingers will always point *up* when they're inside the circle – which is where we want to know the direction of the field. If the current were flowing in the opposite direction, then your fingers, and the magnetic field, would point *down* inside the circle. Incidentally, this explains the cross product notation we used for torque (Section 3.13): there we just talked about "clockwise" and "counter-clockwise", but in reality a torque will point, say, out of the page, with the corresponding direction of motion given by the RHR (i.e. counter-clockwise in this example).

labeled that distance r (we want to know the field from one wire at the location of the *other* wire). In Figure 12.20 we suppose the wires are running horizontally and the currents are running to the right. This means the field due to the bottom wire at a position above the bottom wire is directed *out* of the page (from the Biot–Savart law, the direction of $\vec{B}$ is given by $\int d\vec{s} \times \hat{r}$, which here is "right cross up = out of the page".) Or, using the right hand rule for fields, stick your right thumb to the right (the direction of the current) and allow your fingers to wrap in a circular fashion. Above the wire, your fingers are pointing out of the page.

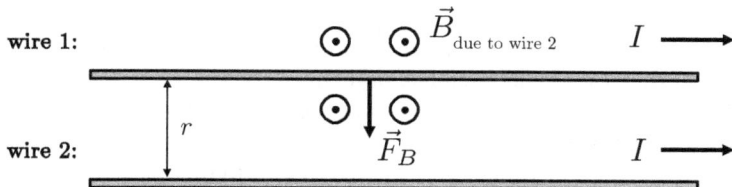

FIGURE 12.20

Similarly, the direction of the magnetic field on the bottom wire due to the top wire points *in* to the page (Fig. 12.21).

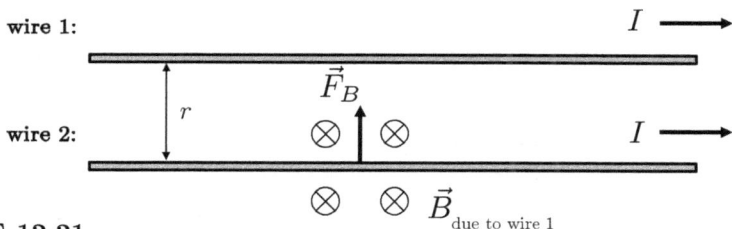

FIGURE 12.21

Because we now know the direction of the currents in the wire and the direction of the magnetic fields at the location of the wires, we can determine by direction of the force on each wire by evaluating the cross product in $\vec{F} = I\left(\vec{L} \times \vec{B}\right)$. For the top wire in Figure 12.20, we have "right cross out = down", while for the bottom wire (Fig. 12.20) we have "right cross in = up". Thus the wires are *attracted* to one another!

OK, what about the magnitude of the force? Well, the wire is perpendicular to the field (the wire carries current to the right, while the field is either in or out of the page), so the $\sin\theta$ terms that fall out of the cross product in our force equations are simply $\sin(90°) = 1$. With this in mind, the magnitude of the force on wire 1 is

$$\vec{F}_{\text{on }1} = I_1\left(\vec{L}_1 \times \vec{B}_{\text{from }2}\right)$$

$$F_{\text{on }1} = IL\frac{\mu_0 I}{2\pi r}$$

$$\frac{F_{\text{on }1}}{L} = \frac{\mu_0 I^2}{2\pi r}$$

As you can check for yourself, the magnitude of the force on wire 2 is the same.

For (b), we just need to insert numerical values into this result:

$$\frac{F}{L} = \frac{\mu_0 I^2}{2\pi r}$$

$$\frac{F}{L} = \frac{\left(1.26 \times 10^{-6} \frac{\text{m kg}}{\text{s}^2 \text{A}^2}\right)(25.0 \text{ A})^2}{2\pi (0.030 \text{ m})}$$

$$\frac{F}{L} = 4.2 \times 10^{-3} \text{ N/m}$$

Thus if you have two wires carrying currents near one another, they can flex toward one another or, if the currents point in opposite direction (Problems 12.27 and 12.28) they can flex away from one another. A common demonstration of this effect involves connecting a car battery to two wires that are strung parallel to one another. Because the resistance of a wire is quite small, the current is quite large and the effect can be rather dramatic.

Example 12.7 Finding the Magnetic Field of a Long Straight Wire   ⋆⋆
A long straight wire carries current I to the right. Use the Biot–Savart Law to determine the magnetic field a radial distance R directly above the wire.

The geometry is shown in Figure 12.22. Note that I have identified the direction of $\vec{B}$ as *out of the page* above the wire, which follows from the RHR. To see this, note $d\vec{s} \times \hat{r} =$ "right cross up = out". Alternatively, with the newer version of the RHR for fields, point your right thumb to the right (the direction of the current); this means your fingers point toward you (out of the page) when they're above your thumb (i.e. above the wire, where we want to know the field). To be mathematically precise we'll call the "out of the page" direction the $\hat{z}$ direction.

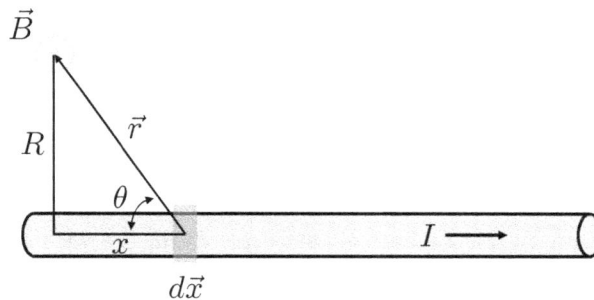

FIGURE 12.22

Every piece of current moving through every piece of the wire makes a contribution to the field at this point, though (as we saw in Figure 12.17) the more distant the charge, the weaker the contribution. Let's work through the Biot–Savart Law (we'll use $d\vec{x}$ instead of $d\vec{s}$ because in this case the wire *does* move in a simple 1D line):

$$\vec{B} = \frac{\mu_0}{4\pi} \int_{-\infty}^{\infty} \frac{I\,(d\vec{x} \times \hat{r})}{r^2}$$

$$\vec{B} = \frac{\mu_0 I}{4\pi} \int_{-\infty}^{\infty} \frac{(\sin\theta\,dx)}{r^2} \hat{z}$$

In this first step I pulled the current I out of the integral (it is constant) and carried out the cross product, which gives us a $\hat{z}$ unit vector (as we argued above) and picks out the piece of $d\vec{x}$ that is perpendicular to $\hat{r}$, the unit vector pointing from the wire to the field point. Note also that the bounds of integration are from $-\infty$ to ∞, which means that we're considering the wire to be arbitrarily long. This is of course ridiculous in practice; what we really mean is that we're considering the strength of the magnetic field very close to the wire, i.e. where $L \gg R$ where L is the length of the wire.

Before we can actually compute an integral, everything inside either needs to be a *constant* or clearly expressed in terms of the variable of integration, which here is x. Thus we will substitute $\sin\theta = R/r$ and then $r = \left(x^2 + R^2\right)^{1/2}$:

$$\vec{B} = \frac{\mu_0 I}{4\pi} \int_{-\infty}^{\infty} \frac{(\sin\theta\,dx)}{r^2} \hat{z}$$

$$\vec{B} = \frac{\mu_0 I}{4\pi} \int_{-\infty}^{\infty} \frac{R}{r^3}\,dx\,\hat{z}$$

$$\vec{B} = \frac{\mu_0 I R}{4\pi} \int_{-\infty}^{\infty} \frac{1}{(x^2 + R^2)^{3/2}}\,dx\,\hat{z}$$

Now we're "done with the physics" and have a problem in pure calculus. I won't go through the details, but if you've studied integral calculus you can confirm that the result is what we claimed in Equation (12.14) (if you want to do it, try trig substitution):

$$B_{\text{wire}} = \frac{\mu_0 I}{2\pi R}$$

Example 12.8 Finding the Magnetic Field of a Circular Loop $\quad \star\star \int dx$

A circular loop of wire of radius R is centered at the origin and lies on the xy plane; it carries current I. Use the Biot–Savart Law to determine the magnetic field at its center.

The geometry is shown in Figure 12.23; we have identified the direction of the field as *up* (perpendicular to the plane of the loop with direction determined, as usual, by the RHR). We'll refer to it as the $\hat{z}$ direction. Here $d\vec{s}$ is always perpendicular to the edge of the loop, which means that it is perpendicular with the vector $\vec{r}$ (which has length R, the radius of the circle) pointing toward the center. Thus $d\vec{s} \times \hat{r} = ds$

and the Biot–Savart Law simplifies:

$$\vec{B} = \frac{\mu_0}{4\pi} \int \frac{I\,(d\vec{s} \times \hat{r})}{r^2}$$

$$\vec{B} = \frac{\mu_0 I}{4\pi} \int \frac{ds}{R^2}\,\hat{z}$$

In the second step I've also changed the denominator inside the integral from r to R, since (as noted above) the distance from every chunk of wire to the field point is just the radius of the circle.

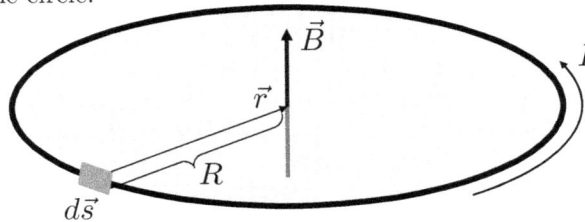

FIGURE 12.23

Because we're walking along the edge of a circle, it is natural to express the differential step length ds in polar coordinates, i.e. $ds = R\,d\theta$. With this substitution the limits of integration are 0 to 2π (i.e. a full 360° or, in SI units, 2π radians). Plugging in, we have

$$\vec{B} = \frac{\mu_0 I}{4\pi R} \int_0^{2\pi} d\theta\,\hat{z}$$

Once again we're "done with the physics", but this time the integral is (if you've studied integral calculus) rather easy. We find

$$\vec{B} = \frac{\mu_0 I}{4\pi R}\,(2\pi)\,\hat{z}$$

$$\vec{B} = \frac{\mu_0 I}{2R}\,\hat{z}$$

in agreement with Equation (12.15).

Before moving on, I should mention that there is another way to determine the magnetic field due to a current-carrying wire. The approach is called **Ampere's Law**:[12]

$$\oint \vec{B} \cdot d\vec{s} = \mu_0 I_{\text{enc}} \tag{12.17}$$

Here the loop on the integral sign means we are performing a **path integral**: we perform the integral over $d\vec{s}$ along some path that wraps around some current and starts back where it ends. At every step along the way we take the dot product of the magnetic field at that location, $\vec{B}$, with our differential step $d\vec{s}$ (Figure 12.24; note that this means we need to

---

[12]After Andre–Marie Ampere (1775–1836). Ampere's Law is somewhat analogous to Gauss's Law for electric fields insofar as both laws offer an alternative method for identifying a field ($\vec{E}$ or $\vec{B}$) than directly integrating the contribution of every charge.

know the *direction* of $\vec{B}$, likely through the RHR). The I_{enc} term on the right side refers to the current *enclosed* by the loop.

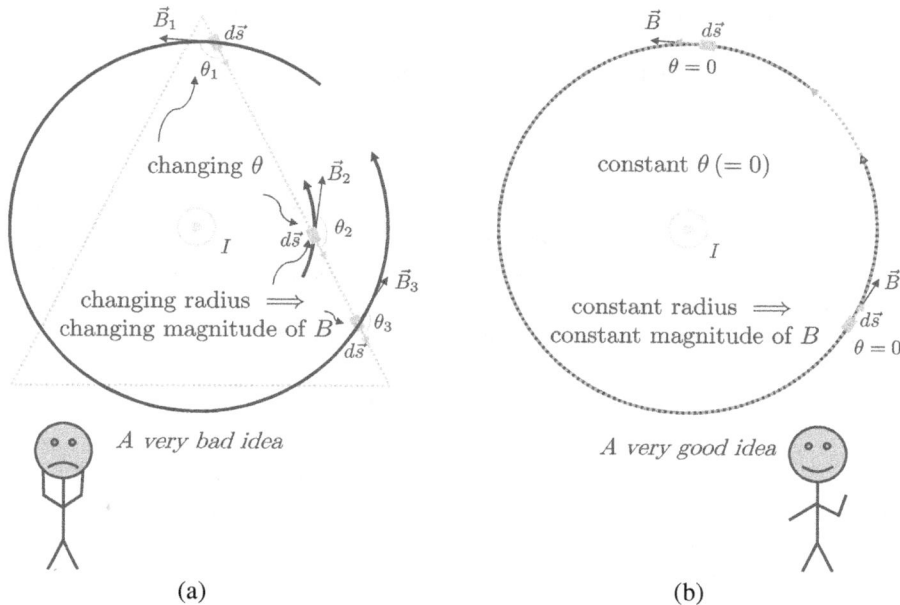

(a) (b)

FIGURE 12.24
A long straight wire carries current I out of the page, yielding a magnetic field $\vec{B}$ that wraps counter-clockwise around the wire. For Ampere's law we consider walking along some path of our choice that ends where it starts. (a) We choose a triangle: suppose we start at the top vertex, then begin by walking down and right. This is a bad idea: the angle between $\vec{B}$ and $d\vec{s}$ changes as we walk along the triangle; furthermore, the magnitude of $\vec{B}$ also changes. Working out the integral $\oint \vec{B} \cdot d\vec{s}$ is therefore quite challenging. (b) We instead choose a circular path that overlaps exactly with the $\vec{B}$ field lines at that radius. $\vec{B}$ has constant magnitude and always points in the same direction as $d\vec{s}$, which simplifies the integral (see Example 12.9).

As Figure 12.24 demonstrates, this can be quite complicated for an arbitrary path in an arbitrary field, because the angle between the vectors and/or the magnitude of $\vec{B}$ is constantly changing. However, in some cases we can be clever and Ampere's Law becomes quite simple to implement:

- If we choose our path so that $\vec{B}$ and $d\vec{s}$ are *perpendicular* for some portion of our path, then the dot product gives 0 for that portion.

- Similarly, if the vectors are *parallel* for some portion, then the dot product simplifies to $\int \vec{B} \cdot d\vec{s} = \int B ds \cos 0 = \int B ds$. If we *also* know that $\vec{B}$ is *constant* along the portion of the path we're considering, then it comes out of the integral: $\int B ds = B \int ds$. This final integral is just the length of the section of the path we're considering.

Our "long straight wire" example meets all of these criteria, as the following example shows. We can also derive Equation (12.16) quickly with Ampere's Law (see Problem 12.34); the process of determining the equation with the Biot–Savart Law is considerably more complex!

Example 12.9 The Field of a Long Wire Revisited $\star\star \int dx$
Derive Equation (12.14) with Ampere's Law.

We use the circular path shown in the right panel of Figure 12.24. $\vec{B}$ points in the same direction as $d\vec{s}$ for any position on the circle, so

$$\oint \vec{B} \cdot d\vec{s} = \oint B ds$$

Now the magnitude of B is constant at any point on the path: the radius is constant and the magnitude of B can't depend on any other coordinate (i.e. the angular position) by symmetry: if we twist the wire along its long axis then nothing actually changes about the situation. (This is similar to arguments we made from symmetry concerning Gauss's Law.) Thus B comes out of the integral:

$$\oint B ds = B \int_0^{2\pi} R d\theta$$

In the last step I made the usual switch from ds, a differential arc length along a circle, to its equivalent expression in polar coordinates. If we evaluate this integral we are left with $(2\pi R) B$.

Now all of this has been the left hand side of Ampere's Law. Plugging in:

$$\oint \vec{B} \cdot d\vec{s} = \mu_0 I_{\text{enc}}$$
$$B(2\pi R) = \mu_0 I_{\text{enc}}$$
$$B = \frac{\mu_0 I}{2\pi R}$$

In the last step I've replaced I_{enc} with I, the current we've enclosed with the loop. This last result is indeed Equation (12.14).

Magnetic Materials

During all of this business about currents creating magnetic fields, you may have been wondering about *magnets* in the everyday sense of the term: the things that stick to refrigerators and that we use in our compasses. I've held off addressing this until now in part because discussing magnetism in the context of currents has provided us with the tools we need to understand these so-called *magnetic materials*.

Refer back to the right side of Figure 12.19 (page 600): a *current loop* creates a magnetic field. With this in mind, consider the following fundamentally important idea: protons and electrons each behave like little current loops due to a quantum mechanical property called the **spin**. In an atom, the revolution of electrons around the nucleus *also* constitutes a sort of current loop, which creates another magnetic field (Fig. 12.25).

In most materials these tiny magnetic fields are randomly oriented and cancel on average. In some materials, however, it is possible for many of the vectors to align, forming a nonzero magnetic field. In a **ferromagnet**, it is possible for this alignment to be essentially permanent, resulting in what we commonly refer to as a bar magnet or permanent magnet. In a **paramagnet**, the effect is only temporary and in response to an external magnetic field: most refrigerators will become magnetized in the present of a permanent magnet and thereby cling to it, but they are not (permanently) magnetic themselves. Other materials (even some metals!) aren't magnetic at *all* and simply won't respond to an external magnetic field.[13]

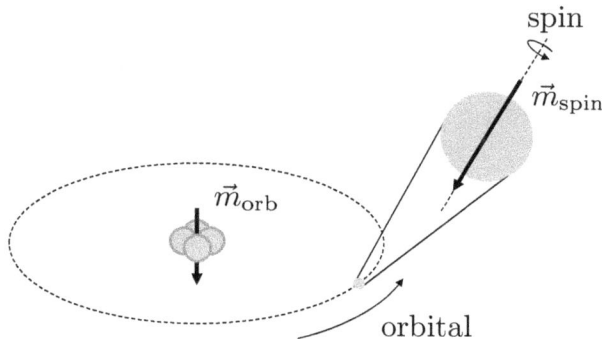

FIGURE 12.25

An atom visualized as an electron orbiting around the nucleus much like how the earth orbits around the sun (this is the so-called *planetary model.*) The electron creates a magnetic field, represented by a pair of vectors $\vec{m}$ called *magnetic dipole moments*, because it behaves like a sort of *current loop* in two ways. First, because of its rotation around the nucleus (cf. the 365 days of the Earth's year); second, because of its rotation, or spin, around its own central axis (cf. the 24 hours of Earth's day; just as the Earth's crust is really moving as it rotates, we consider the charge distributed through the electron as moving in a similar way). If the direction of the two vectors seems to be backward, keep in mind that we are here considering the motion of a *negative* charge–an electron–rather than the usual (conventional but wrong) approach of considering charge as due to the movement of *positive* charge. I should stress, though, that this convenient model is ultimately wrong: an electron is *not* a "tiny sphere" and the full details are quantum mechanical in nature. They simply do not have an analog to our common experiences and intuition.

We sometimes omit drawing the current loop entirely and just draw a vector representing the orientation and strength of the created field; it is called the **magnetic dipole moment**, $\vec{m}$ (Fig. 12.25).[14] We call $\vec{m}$ a *dipole* moment for good reason: there are important parallels with the *electric* dipole we discussed in Chapter 9:

- An electric dipole has a positive side (where the field lines *leave* the dipole) and a negative side (where they *enter*). Similarly, a magnetic dipole has a *north* side, N (where the field lines *leave*), and a *south* side, S (where they *enter*) (Fig. 12.26).

- In both types of dipoles, opposite poles attract and like poles repel. That is, there is a Coulomb attraction/repulsion between the charges in an electric dipole; similarly, a N pole is attracted to a S pole in a magnetic dipole, but two N poles or two S poles will repel (Fig. 12.27).

---

[13]Understanding *why* a particular material has the magnetic properties it does (e.g. aluminum is a paramagnet) is rather complicated and we won't worry about the details in this book. That said, you might be curious to know that temperature plays an important role: a material that is ferromagnetic at room temperature can become paramagnetic at high temperatures!

[14]Early in the chapter I explained that we use $\vec{B}$ for magnetic *fields* to avoid confusion with mass. Unfortunately, we've betrayed that principle here – but I am following the usual convention.

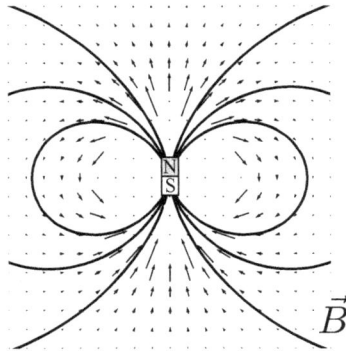

FIGURE 12.26
A *bar magnet* is characterized by a north pole (N) and a south pole (S); it creates a magnetic field much like a current loop or solenoid (Fig. 12.19) with the field lines leaving the N pole and wrapping around to the S pole.

- Both types of dipoles experience a *torque* in an external field (Fig. 12.28). For an electric dipole $\vec{p}$ we have $\vec{\tau} = \vec{p} \times \vec{E}$; for a magnetic dipole we have

$$\vec{\tau} = \vec{m} \times \vec{B} \qquad (12.18)$$

This means, for instance, that a small magnet that is free to rotate due to the Earth's magnetic field will end up with its N pole pointing to the Earth's S pole: *geographic* north is *magnetic* south (Problem 12.35)!

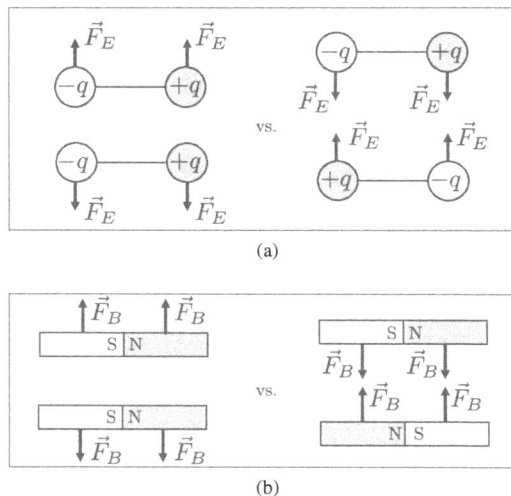

FIGURE 12.27
A comparison of the interaction forces between a pair of (a) electric and (b) magnetic dipoles.

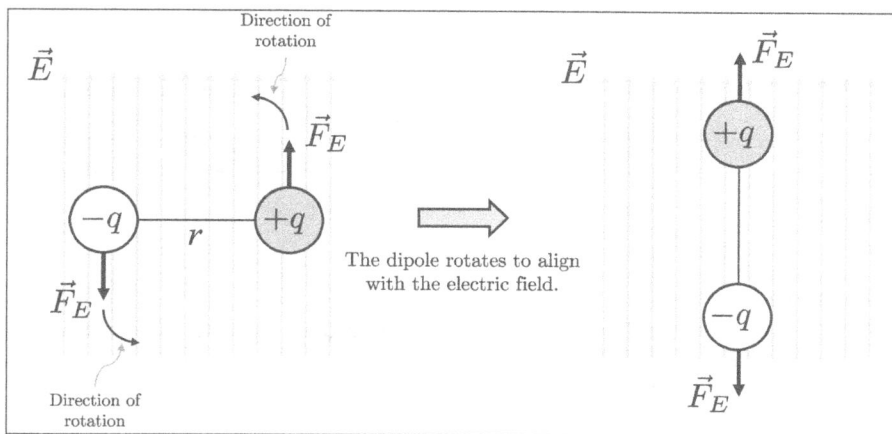

(a)

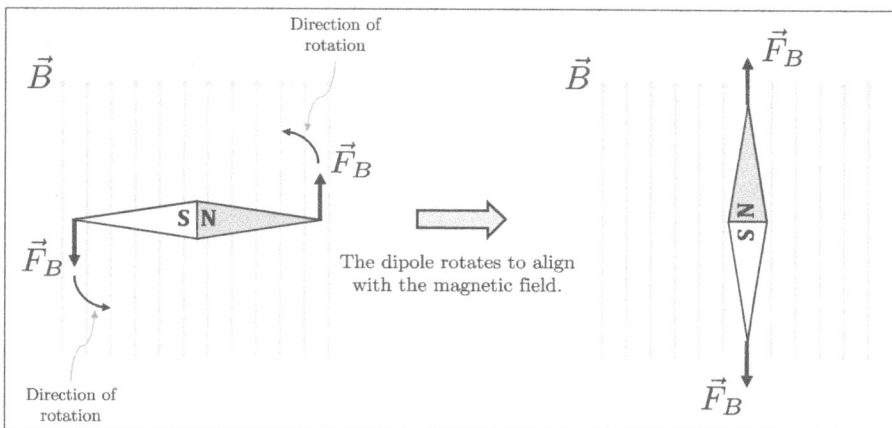

(b)

FIGURE 12.28

A comparison of the forces and torques acting on (a) electric and (b) magnetic dipoles when they are immersed in an external field.

- Just as an electric dipole in an external electric field has a potential energy (Equation (9.20)) $U = -\vec{p} \cdot \vec{E}$, a magnetic dipole in a magnetic field has a potential energy

$$U = -\vec{m} \cdot \vec{B} \qquad (12.19)$$

A critical difference between these two types of dipoles, though, is that *you can't split a magnetic dipole into two isolated poles*. This is because the fundamental quantity in an electric dipole are the constituent charges: cut the dipole in half and you're left with a positive charge and a negative charge. In a magnetic dipole, though, $\vec{m}$ is the fundamental thing, formed by a little current loop (the destruction of which destroys the entire field) or the quantum mechanical spin (which is intrinsic to the particle).

As an example, consider cutting a bar magnet in half (Fig. 12.29). To start, the magnetic dipole moments point right (at least on average), meaning the N pole is at the right of the magnet and the S pole is at the left. If you cut the magnet in half, then in each of the smaller magnets you have *the exact same thing*: You can separate the magnetic dipoles into two

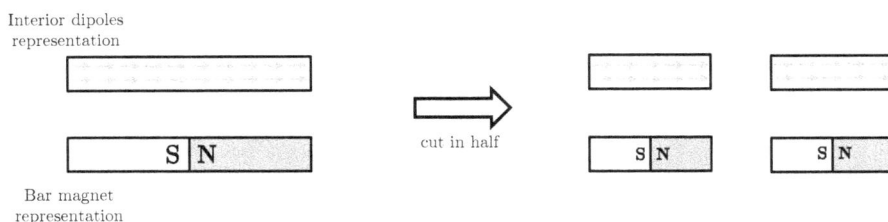

Interior dipoles
representation

cut in half

Bar magnet
representation

FIGURE 12.29

Cutting a bar magnet in half results in two smaller bar magnets: the interior magnetic dipoles (top view) still have a uniform alignment (at least on average) that results in a N pole and a S pole (bottom view).

Example 12.10 Alignment of Magnetic Dipoles   ⋆

Three equal-magnitude magnetic dipoles are arranged in a line, as shown. The two on the sides are fixed in place and the one in the middle is free to rotate. What will its final orientation be?

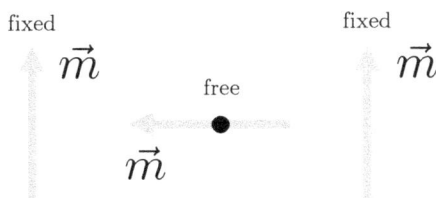

FIGURE 12.30

The magnetic field at the location of the central dipole is *down* due to the contribution of both other dipoles (refer back to Figure 12.19 if this is not clear). Thus the central dipole experiences a torque given by a "left cross down = out" cross product. According to the RHR for fields, this means the dipole rotates counter-clockwise. As it does so, the magnitude of the torque decreases as the vector $\vec{m}$ begins to align with the external magnetic field $\vec{B}$. Indeed, the torque becomes 0 precisely when the dipole points directly down, aligned with the field (Fig. 12.31). (If the system is frictionless, it will actually oscillate forever around the "straight down" position, somewhat like a pendulum – but real systems will have some level of friction, meaning the dipole slows and eventually settles pointing down.)

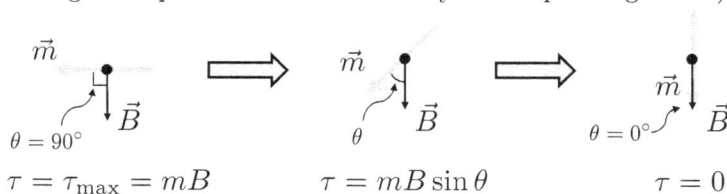

$$\theta = 90°$$
$$\tau = \tau_{\max} = mB$$

$$\theta$$
$$\tau = mB \sin\theta$$

$$\theta = 0°$$
$$\tau = 0$$

FIGURE 12.31

In a ferromagnet, the tendency is for the dipoles to *align* (due to quantum mechanical effects we are neglecting here). Perfect alignment of every magnetic dipole $\vec{m}$ in a ferromagnet does not occur for a variety of reasons, but one reason involves a confounding alignment of neighbors; see Problem 12.37 for a simple example. In both ferromagnets and paramagnets, this alignment can also be driven by a strong *external* magnetic field; this is in fact how unmagnetized iron can become magnetized.

12.5 Magnetic Flux

Faraday's Law

We've now discussed both of the major statements I made back at the beginning of the chapter: magnetic fields are created by moving charges (though for magnetic materials the "movement" is really due to quantum mechanical properties of atomic electrons), and moving charges will experience a force if they're moving through an external magnetic field.

If this was the whole story, then one thing that *wouldn't* be possible is for a magnetic field to interact with stationary charges to *create* a current: we've only seen that magnetic fields can *affect existing currents* by applying a force to the wire ($\vec{F}_B = I\left(\vec{L} \times \vec{B}\right)$). However, if you perform the experiment sketched in Fig. 12.32, where a magnet is quickly dropped toward a conducting loop of metal attached to an ammeter, you'll observe a current – even though there wasn't one to begin with. Evidently we're dealing with something new here![15]

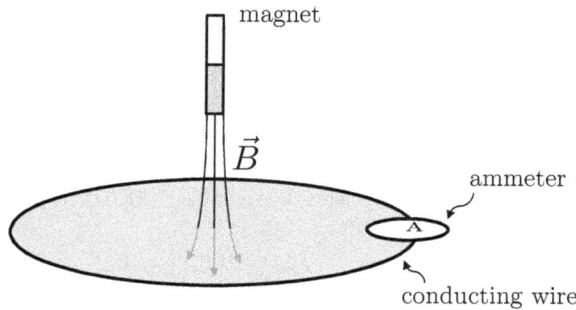

FIGURE 12.32
A bar magnet is held, N pole down, above a conducting loop of wire. A few representative magnetic field lines are drawn passing through the face of the loop. If the magnet is moved suddenly, the ammeter attached to the loop will register a current.

The key principle turns out to be this:

> **If the magnetic flux through a closed conducting loop *changes*, an emf will be induced.**

---

[15]You might say "Isn't this the same as moving the wire up toward a stationary magnet, in which case we *do* have moving charges?" This is a bit of a thorny topic and I invite you to analyze this scenario in Problem 12.49. For now, though, let's stick with the situation I am describing: at least from *our* point of view, the charges in the ring are indeed stationary as the magnet moves.

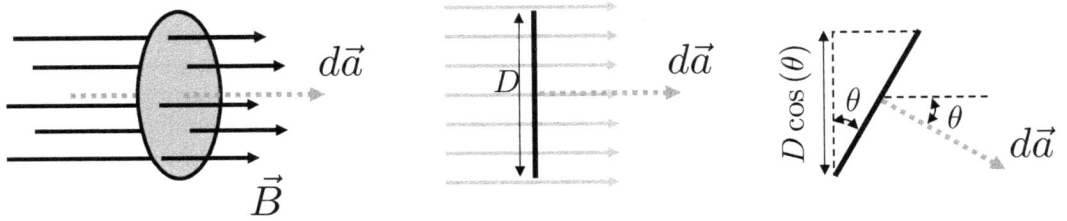

FIGURE 12.33
Magnetic flux characterizes the amount of magnetic field passing perpendicularly through a closed surface, such as that formed by a loop of conducting wire (see Equation (12.22)). The dot product in Equation (12.22) is between the magnetic field $\vec{B}$ and the vector $d\vec{a}$ that is perpendicular to its surface, so the flux is maximized (to BA where A is the area of the surface) when the angle θ shown in the right panel is equal to 0.

This is mathematically formalized by **Faraday's Law:**[16]

$$\left| \frac{d\Phi_B}{dt} \right| = \varepsilon \qquad\qquad (12.20)$$

If you prefer to avoid calculus, Faraday's Law can be written as

$$\left| \frac{\Delta\Phi_B}{\Delta t} \right| = \varepsilon \qquad\qquad (12.21)$$

though as usual the calculus version (Equation (12.20)) is correct *at any instant* and the algebraic version (Equation (12.21)) is only an average across some time interval Δt.

The emf created by the changing flux behaves just like the emf provided by a battery and can drive a current according to Ohm's Law and Kirchhoff's Loop Rule (which together yield $\varepsilon = IR$ for the wire's resistance R). Meanwhile the **magnetic flux** is defined just like electric flux back in Chapter 10 (Fig. 12.33):

$$\Phi_B = \int \vec{B} \cdot d\vec{a} \qquad\qquad (12.22)$$

In words, the magnetic flux (measured in T m^2 or equivalently in webers, Wb[17]) describes the "amount of magnetic field" passing perpendicular through some surface. For our purposes in this chapter, we only care about the flux inside some conducting loop, so the surfaces we consider will always have said conducting loop wrapped around its perimeter.[18] As an example, refer back to Figure 12.32: a bar magnet is dropped, N pole down, through a circular loop of conducting wire. As the bar magnet approaches, the magnetic field passing through the loop gets *stronger* (since the magnet is *closer*), and so the magnetic flux passing

---

[16]After Michael Faraday (1791–1867). You might see this law presented elsewhere with the absolute value replaced with a negative sign. That version of Faraday's Law "folds in" Lenz's Law, which considers the *direction* of current flow. We'll get to Lenz's Law shortly and for now we're only worried about the *magnitude* of the current – I believe taking the topics in stages is much easier when you're first broaching this topic.

[17]After Wilhelm Eduard Weber (1804–1891).

[18]They will also be two dimensional, like a square or a circle. This is in contrast to electric flux and Gauss's Law, where we considered for instance a three-dimensional *sphere*.

through the loop *increases*. According to Faraday's Law, this results in an emf in the loop and, therefore, a current that can (at least in principle) be detected by our ammeter. Example 12.11 considers a related situation in detail.

Example 12.11 Current with Faraday's Law ★★

A constant magnetic field of magnitude B points out of the page in a region of space where a U-shaped piece of wire (with negligible resistance) is connected to a conducting "slider" of resistance R and width w that connects the arms of the "U" to form a closed loop (Fig. 12.34). You grab the slider and pull it to the right at a constant speed v. Determine an expression for the current I that will be induced in the loop (a) *without* and (b) *with* Faraday's Law.

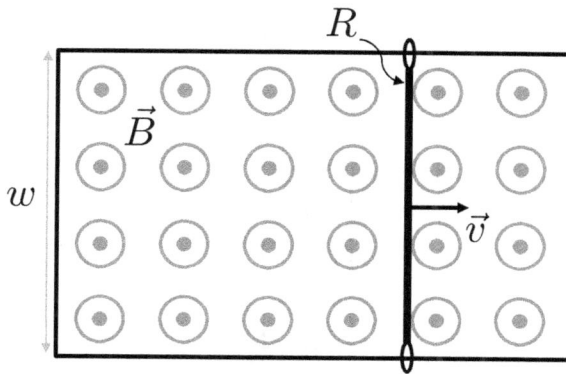

FIGURE 12.34

If we ignore Faraday's Law and we *start* with no current, then it may seem like a hopeless task. But note that the slider, a conductor, is *moving*. This means its electrons carry a velocity of magnitude v in a magnetic field and the magnetic force $\vec{F}_B = q\left(\vec{v} \times \vec{B}\right)$ therefore comes in to play. Here the magnetic field is out of the page and the velocity is to the right, which gives a downward force on a positive charge (we consider the force on a *positive* charge in keeping with the conventional current – even though, as always, it is really the negatively charged *electrons* that do the moving.)

The velocity of the slider is perpendicular to the magnetic field, so the cross product simplifies:

$$\vec{F}_B = q\left(\vec{v} \times \vec{B}\right)$$

$$F_B = qvB$$

This magnetic force causes charges in the conductor to begin moving, but as soon as they do so the slider becomes polarized (Fig. 12.35), creating a potential difference ΔV along the width w of the slider. This, of course, corresponds to an electric field of magnitude $E = \frac{\Delta V}{w}$. Almost immediately the potential difference grows until the magnetic force acting on a charge in the slider is equal and opposite to the electric force, i.e.

$$F_B = F_E$$
$$qvB = qE$$
$$vB = \frac{\Delta V}{w}$$
$$\Delta V = vwB$$

It is common practice to refer to this potential difference ΔV as a **motional emf** ε, because the current in the ring can be thought of as *caused by* this potential difference (just like a battery's chemical emf drives a current through a flashlight), and as we have seen, the potential difference is caused by the *motion* of the slider through the magnetic field.

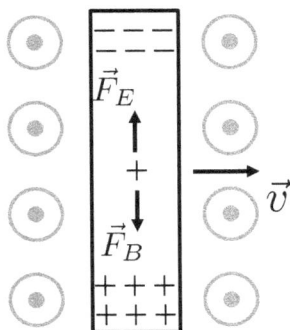

FIGURE 12.35

In any event, if we make the usual leap using Ohm's Law and the Loop Rule to say that $\varepsilon = IR$, then we finally have

$$IR = vwB$$
$$I = \frac{vwB}{R} \qquad (12.23)$$

As we have seen, the current flows clockwise.

Phew! Now in (b) we are asked to find the *same* result by considering the magnetic flux and using Faraday's Law. Consider the situation shown in Fig. 12.34: The area of the loop is $A = wx$ for some width x. The magnetic field passes perpendicularly through the loop with magnitude B, so

$$\Phi_B = \int \vec{B} \cdot d\vec{a} = BA$$

If we plug this in to the left side of Faraday's Law, we have

$$\left| \frac{\Delta \Phi_B}{\Delta t} \right| = \left| \frac{\Delta (BA)}{\Delta t} \right|$$

Now in this case B is a constant, so

$$\Delta (BA) = B (A_f - A_i) = B\Delta A$$

What we've just done is set $\Delta(BA) = B\Delta A$ because B is constant. Well, $A = wx$ and w is constant, so by the exact same logic we have $\Delta(BA) = B\Delta A = Bw\Delta x$. Plugging in,

$$\left|\frac{\Delta\Phi_B}{\Delta t}\right| = \left|\frac{\Delta(BA)}{\Delta t}\right| = Bw\left|\frac{\Delta x}{\Delta t}\right| = vwB$$

In this last step I've noted that $\Delta x/\Delta t$ is simply the speed v at which we're pulling the slider.

According to Faraday's Law, this is all equal to the induced emf, ε, which (from the Loop rule and Ohm's Law) is equal to IR. Thus

$$IR = vwB$$
$$I = \frac{vwB}{R} \tag{12.24}$$

exactly as we noted above.

While the preceding example might make it seem like Faraday's Law is largely redundant with force analysis, keep in mind our original example: a magnet moving through a stationary loop creates a current, and this simply *can't* be explained by considering a magnetic force on charges in the wire because (unlike with our slider example) they *aren't moving*! The problems at the end of the chapter provide some additional examples of interesting applications of Faraday's Law.

Lenz's Law

Faraday's Law tells us about the *magnitude* of the current but not its *direction*. To get at this, let me summarize the chain of logic that we go through with Faraday's Law, but with one additional comment at the end:

1. A magnetic field passes through a conducting loop, generating some magnetic flux.

2. The area of the loop and/or the strength of the magnetic field changes, which in turn changes the flux.

3. According to Faraday's Law, the change in flux (however it comes about) induces an emf in the loop. Assuming the loop's resistance isn't infinite, this also creates an induced current.

4. This induced current creates (as we saw in Section 12.4) a magnetic field. We called this the *induced magnetic field* to distinguish it from the external magnetic field that started the entire process in step 1.

I am stressing the fact that we have an induced magnetic field because the orientation of the induced magnetic field tells us the direction of the induced current: for instance, if the induced magnetic field points *out* of the page, the current is on the plane of the page and traveling counter-clockwise; if the magnetic field points *in*, the current travels clockwise. (The RHR for fields says that in both cases, point the thumb of your right hand in the direction of $\vec{B}_{\text{ind}}$ and your fingers curl in the direction if I_{ind}.)

With this in mind, the orientation of the induced magnetic field is determined by **Lenz's Law:**[19]

---

[19]After Emil Lenz (1804–1865)

An induced magnetic field is oriented so as to oppose the change in flux due to an external magnetic field.

This can be rather tricky to grasp, so let me break it down into two cases:

1. If the flux is increasing (because B_{ext} is increasing and/or the area of the loop is increasing), then B_{ind} points *opposite* B_{ext}.

2. If the flux is decreasing (because B_{ext} is decreasing and/or the area of the loop is decreasing), then B_{ind} points *in the same direction* as B_{ext}.

In either case, the induced field orients itself as if to say to the flux "Don't change, you're perfect just as you are!"[20]

Example 12.12 Direction of Current with Lenz's Law ★
In Example 12.11 we applied force analysis to determine that the current was *clockwise*. Confirm that this is consistent with Lenz's Law.

The starting point here is "what is the flux doing?" In this case it is increasing because the area of the conducting loop is increasing ($\Phi_B = BA$ with B constant and A increasing). Because the external field is pointing *out* of the page, the *induced* field points *in* to the page ("No, no, go back!" sobs the induced field). This, according to the RHR, means the induced current flows *clockwise*, consistent with our previous result.

Example 12.13 Lenz's Law with a Generator ★
In Example 12.5 we considered a simple motor, where current could be used to cause rotational motion. (The setup involved a square loop immersed in a constant magnetic field: running a current I through the loop can cause rotation.) Now consider a **generator**, which reverses the process: rotational motion creates a current. Our setup is the same: a square loop of wire is in a constant upward magnetic field. Here, though, we force the loop to rotate about the axis of its handle (due to wind pushing on the arms of a windmill, for instance). Use Lenz's Law to describe the direction of the induced current as it rotates.

Figure 12.36 shows the outcome of our analysis for eight different orientations of the loop relative to the field: the rightmost panel, which is labeled 0°, shows the plane flat on the xy plane. As you work your way counterclockwise around the panels, the loop rotates 45° per panel in the direction of the rotation. Notice that the 0° and 180° panels correspond to Φ_B being a *maximum* ($\vec{A}$ points along the z axis, so $\vec{B} \cdot d\vec{a} = B\,da$), and that the 90° and 270° panels correspond to $\Phi_B = 0$ (the dot product is 0: no component of *any* magnetic field lines pass perpendicularly through the loop).

---

[20]If you want to start an interesting and/or awkward and/or brief conversation with someone, you might tell them that induced magnetic fields are *very romantic* for this very reason.

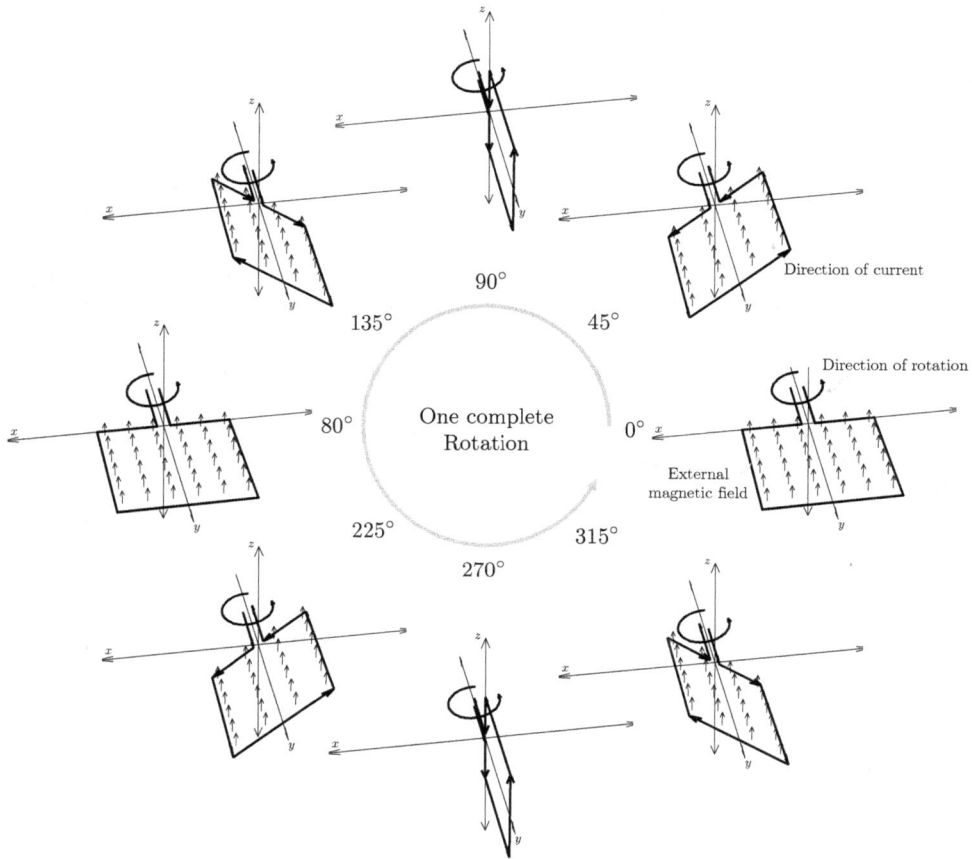

FIGURE 12.36

As the loop goes through its first quarter-rotation, from 0° to 90°, the flux is *decreasing*. The induced magnetic field therefore points in the same direction as the magnetic field (or rather, the piece of $\vec{B}_{\text{ext}}$ that is perpendicular to the face of the loop). Applying the right hand rule yields the current shown on the figure: for instance, the frontmost piece of the wire has a current running to the right. If this is not clear, point the thumb of your right hand to the right, then curl your fingers: when they are farther away from you than your thumb (corresponding to inside the loop), they will point in the direction of $\vec{B}_{\text{ind}}$.

Now consider the second quarter-rotation: the flux goes from 0 back to a maximum. Lenz's Law says the induced magnetic field opposes this increase, meaning the current will point to the left along the front wire. The analysis of the second half of the rotation is analogous and I will leave it to you to think through the details.

If the rotation occurs at a constant frequency, the induced current will look like a sine wave, as you may have intuited from this discussion. Problem 12.52 walks you through some of the details.

12.6   Electromagnetic Waves

Faraday's Law says that a changing magnetic flux can drive a current. However, if the force acting on a charged particle is given by

$$\vec{F} = q\left(\vec{E} + \left(\vec{v} \times \vec{B}\right)\right) \tag{12.25}$$

and the charge is initially at rest (i.e. $\vec{v} = 0$), then the only way that the charge could accelerate to start a current is if there is an *electric* field.[21]

This correctly suggests that **a changing magnetic field creates an electric field**. Now, we have also seen (from the Biot–Savart Law) that **a changing electric field creates a magnetic field**.[22] There is a lovely bit of symmetry here, and in fact much of what we have covered in the last few chapters can be summarized in the famous **Maxwell's Equations**,[23] which express the major laws of electromagnetism (Gauss's Law, Ampere's Law, and Faraday's Law) and succinctly describe the interplay between electric fields, magnetic fields, charges, and their motion (i.e. currents) in three-dimensional space.

Understanding Maxwell's equations requires one to think hard about fields in three dimensions and multivariable calculus at a level far beyond the scope of this book, so I won't write them down (the standard notation alone would take quite a bit of time to explain). They are among the most beautiful equations in physics, however, and you can look forward to studying them if you pursue your education in physics.[24]

I would like to describe one important *prediction* of Maxwell's equations, however. The symmetry described above suggests that it is perhaps possible for a changing magnetic field and a changing electric field to *co-induce* one another–perhaps, at least in principle, forever. Maxwell's equations predict that this is indeed possible through appropriate acceleration of charge (e.g. up and down an antenna); the configuration of the resulting waves is shown in Figure 12.37. In a vacuum, Maxwell's equations predict that these waves propagate, locked together in this configuration, at a speed that is precisely equal to $c = 3.00 \times 10^8$ m/s! Indeed, these so-called **electromagnetic waves** *are* light, as we alluded back in Chapter 8.

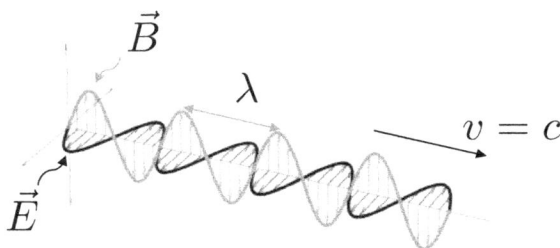

FIGURE 12.37

An electromagnetic wave (light) consists of a sinusoidal traveling electric wave and a sinusoidal traveling magnetic wave. The waves are in phase and oscillate at right angles (e.g. on the xy plane and zy plane). The direction of travel is along the shared axis (e.g. the y axis); in a vacuum the wave travels at the speed of light c.

---

[21] Equation (12.25), which describes the combined electric and magnetic force on a charged particle, is called the **Lorentz force** after Hendrik Lorentz (1853–1928).

[22] We didn't quite phrase it this way, but the electric field created by a charge will change over time if the charge is *moving*, so the magnetic field is coupled in some sense to the changing electric field.

[23] After James Clerk Maxwell (1831–1879)

[24] The **Navier–Stokes equations** that describe fluid flow use similar notation and, in some cases, the *mathematics* is identical to Maxwell's equations even though the *physical context* is wildly different.

The study of electricity is therefore coupled with the study of magnetism, and the joined field of electromagnetism underlies a huge variety of modern topics that are, at first glance, very different from one another–including for instance optics, astronomy, medical imaging, and wireless communications.

12.7 Problems for Chapter 12

12.2 The Cross Product

⋆ **Problem 12.1.** Is it possible for two vectors $\vec{A}$ and $\vec{B}$ to have a cross product of 0? If *no*, explain why. If *yes*, explain and provide at least one example.

⋆ **Problem 12.2.** You're designing an experiment where a vector quantity $\vec{A}$ associated with a particle is directed straight up, toward the roof of the room you're in. You want vector quantity $\vec{C}$ to point to the left as you face the device. How should you arrange vector quantity $\vec{B}$ if the three quantities are related by $\vec{A} \times \vec{B} = \vec{C}$?

⋆ **Problem 12.3.** You are designing an experiment as in Problem 12.2, but now $\vec{B}$ is directed away from you and $\vec{C}$ is directed to your right. In what direction should $\vec{A}$ point if the three quantities are related by $\vec{A} \times \vec{B} = \vec{C}$?

⋆ **Problem 12.4.** Three vectors are related by $\vec{A} \times \vec{B} = \vec{C}$. Below, the direction of two of these three vectors are provided; your task is to determine the direction of the third and to explain your reasoning with a diagram. Your answer will always be one of: *left, right, up, down, in,* or *out*. As needed, indicate a vector pointing in to the page with a cross (×) and a vector pointing out of the page with a dot (·).

a. $\vec{A}$ points right, $\vec{B}$ points up.

b. $\vec{A}$ points right, $\vec{B}$ points down.

c. $\vec{A}$ points up, $\vec{B}$ points right.

d. $\vec{A}$ points up, $\vec{B}$ points left.

e. $\vec{A}$ points down, $\vec{B}$ points in.

f. $\vec{A}$ points down, $\vec{B}$ points out.

g. $\vec{B}$ points up, $\vec{C}$ points right.

h. $\vec{B}$ points up, $\vec{C}$ points out.

i. $\vec{B}$ points in, $\vec{C}$ points left.

j. $\vec{B}$ points in, $\vec{C}$ points up.

k. $\vec{A}$ points up, $\vec{C}$ points right.

l. $\vec{A}$ points up, $\vec{C}$ points out.

m. $\vec{A}$ points out, $\vec{C}$ points left.

n. $\vec{A}$ points left, $\vec{C}$ points in.

⋆ **Problem 12.5.** Three vectors are related by $\vec{A} \times \vec{B} = \vec{C}$. Determine the possible directions of $\vec{C}$ for each of the following situations. Support your response with a diagram.

a. $\vec{A}$ and $\vec{B}$ lie on the xy plane.

b. $\vec{A}$ and $\vec{B}$ lie on the xz plane.

c. $\vec{A}$ and $\vec{B}$ lie on the yz plane.

d. $\vec{A}$ points along the $+x$ axis and $\vec{B}$ points along the $-y$ axis.

e. $\vec{A}$ points along the $+z$ axis and $\vec{B}$ points along the $+y$ axis.

⋆⋆ **Problem 12.6.** Determine $\vec{A} \times \vec{B}$ for each of the following situations. Support each response with a diagram.

a. $\vec{A} = 3.0\hat{x} + 4.0\hat{y}$, $\vec{B} = 5.0\hat{x} + 1.0\hat{y}$.

b. $\vec{A} = -3.0\hat{x} + 4.0\hat{y}$, $\vec{B} = 5.0\hat{x} - 1.0\hat{y}$.

c. $\vec{A} = 4.0\hat{y} + 4.0\hat{z}$, $\vec{B} = -2.0\hat{y} - 6.0\hat{z}$.

12.3 The Magnetic Force

★ **Problem 12.7.** A proton is traveling to the right as you observe an experimental device. The magnetic field is directed straight up. In which direction does the magnetic force acting on the proton point?

★★ **Problem 12.8.** Determine the force, in component form, acting on an electron for each situation below. In each case include a sketch showing all three vectors $\vec{v}$, $\vec{B}$, and $\vec{F}$.

    a. The electron's velocity is $\vec{v} = -v\hat{z}$ and it is immersed in a magnetic field $\vec{B} = B\hat{x}$.

    b. The electron's velocity is $\vec{v} = v\hat{x}$ and it is immersed in a magnetic field $\vec{B} = B\hat{x}$.

    c. The electron's velocity is $\vec{v} = v\hat{y}$ and it is immersed in a magnetic field $\vec{B} = B\hat{z}$.

    d. The electron's velocity is $\vec{v} = v\cos 45°\,\hat{x} + v\sin 45°\,\hat{z}$ and $\vec{B} = B\hat{y}$.

★★ **Problem 12.9.** Repeat Problem 12.8 where the particle is a proton rather than an electron.

★★ **Problem 12.10.** An antiproton is a particle with the mass of a proton but the charge of an electron (neat!). Consider an antiproton with a velocity $\vec{v} = -1500$ m/s $\hat{x}$ in a region of space that has a magnetic field 3.0 T $\hat{y}$ and an electric field 450 N/C $\hat{z}$.

    a. Sketch the situation, using a coordinate system where $\hat{x}$ is to the right, $\hat{y}$ is in to the page, and $\hat{z}$ is up the page.

    b. What is the antiproton's acceleration at this instant?

★ **Problem 12.11.** Suppose a cyclotron is intended to contain electrons traveling in a circular orbit at a speed of $0.15c$. If the largest magnetic field that can be readily achieved is 4.0 T, what radius of motion will the electrons have? What would the radius be if protons were used instead?

★★ **Problem 12.12.** A microwave oven uses magnetism! Electrons are forced to orbit in a magnetic field with a frequency of 2.4 GHz, which causes them to emit microwave radiation. (a) What is the strength of the magnetic field? (b) What is the kinetic energy of an electron orbiting in the field with a radius of 1.25 cm?

★★★ **Problem 12.13.** An electron enters a region of space with a magnetic field $\vec{B} = 0.050$ T $\hat{z}$. Initially, the electron's velocity is $\vec{v} = \left(1.0 \times 10^6 \text{ m/s}\right)\cos\left(45°\right)\hat{y} + \left(1.0 \times 10^6 \text{ m/s}\right)\sin\left(45°\right)\hat{z}$.

    a. The particle will move in a *helix*: in a circle along the xy plane while simultaneously moving at a constant velocity in the z direction. Clearly explain why.[25]

    b. What is the radius of the circular motion on the xy plane?

    c. How far will the electron move in the z direction for every revolution on the xy plane?

★★ **Problem 12.14.** Suppose a super heroine (a member of the anti-copyright infringement league) can change the size of her body. In one dramatic battle she goes inside an opponent's high-tech armor to try and disable it. Suppose her mass while shrunk is $m = 4.0 \times 10^{-6}$ kg and while running around in the suit she picks up an excess charge of $q = 5.0 \times 10^{-6}$ C. Also suppose that at one point she jumps into a hole with an upward magnetic field. As she falls, she is pulled into an orbit with a tangential velocity $v = 15$ cm/s and radius $r = 8.0$ mm. Figure 12.38 shows a top-down perspective of her motion (she's not very pleased with the situation).

    a. What is the strength of the magnetic field?

    b. Is the heroine moving clockwise or counterclockwise around the circle? Explain.

---

    [25]This helical motion explains the **aurora**, where charged particles from the Sun are "caught" by the Earth's magnetic field lines and spiral in to the where the field lines enter or exit the Earth's atmosphere near the magnetic poles. (The Earth's magnetic field doesn't map to "straight along the $\hat{z}$ axis" as we've considered here, but this captures the idea.) Interactions between the particles and the Earth's atmosphere results in the beautiful lights visible at night near the poles.

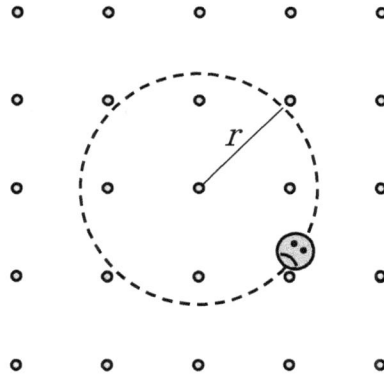

FIGURE 12.38
Problem 12.14

★★   **Problem 12.15.** A particle with a mass of 1.0 mg, charge of -0.50 C and speed $v = 2.0 \times 10^6$ m/s enters a region of space with magnetic fields as shown. Clearly sketch the subsequent path of the charge. If it eventually leaves the region, mathematically describe the exit location. If it stays inside the region forever, describe its path.

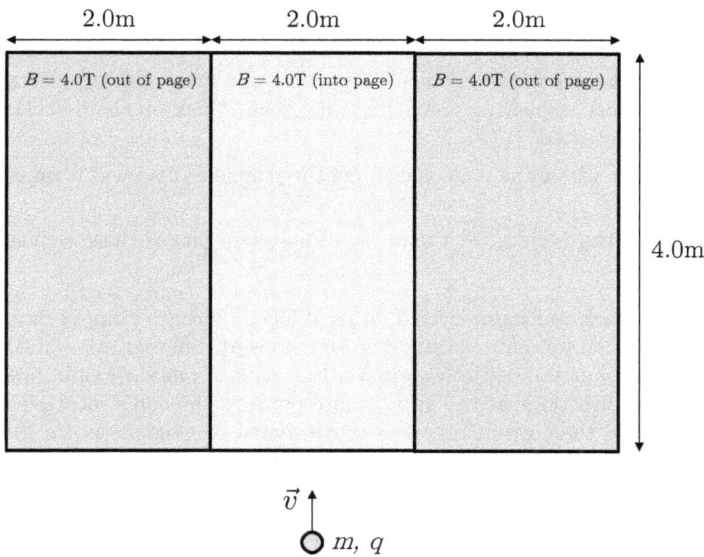

FIGURE 12.39
Problem 12.15

★★   **Problem 12.16.** A 0.20-m-long wire carries a current of 3.0 A directly into the page. A 0.50 T magnetic field is directed straight up the page. What force is acting on the wire (magnitude and direction)?

★★   **Problem 12.17.** A 10.0 V battery is connected to a 25 Ω resistor such that the current flows clockwise as you look down at the circuit. A 3.0 cm long piece of the circuit's wire on its right side is completely vertical and immersed in a 35 mT magnetic field that points into the page.

   a. Sketch the circuit and field.

   b. In what direction will the magnetic force experienced by the wire point?

   c. What magnitude force will the wire experience?

★★ **Problem 12.18.** In a mass spectrometer, a charged particle is accelerated through a potential difference of $\Delta V = 5.0$ kV such that it emerges with a velocity $\vec{v}$ in the $+x$ direction. It then enters a magnetic field $\vec{B} = 1.2\hat{z}$ T. Assume that the magnetic field fills a large rectangular region of space and that the particle enters perpendicular to one side of the rectangle.

  a. Sketch the situation and include an example of a trajectory for a positively charged particle and for a negatively charged particle.

  b. Set the origin of your coordinate system to be the particle's point of entry into the magnetic field. Where do you expect ionized oxygen, O^+, to exit the magnetic field? (Neutral oxygen consists of 8 protons, neutrons, and electrons, and ionized oxygen is missing one electron.)

  c. If a particle that carries a charge of magnitude e exits the field 5.4 cm below the entry point, what is its mass? Express your answer as the nearest integer multiple of the mass of a proton.

★ **Problem 12.19.** Carbon dioxide (CO_2) consists of a carbon atom bound to two oxygen atoms. Suppose that a supply of CO_2 molecules are broken apart (for instance, by being exposed to a beam of high energy electrons) and the remnants, which can include O^+ and CO^+, are directed into a mass spectrometer. What is the radius of motion of each of these particles if the mass spectrometer uses a 4.0 kV potential difference and a 1.5 T magnetic field?

★ **Problem 12.20.** Br_2 molecules can have slightly different masses because the number of neutrons in a Bromine atom can vary (i.e. Bromine has more than one stable *isotope*). The lightest Br_2 molecule has a mass $m_{min} = 2.62 \times 10^{-25}$ kg. Suppose such a molecule has been ionized so it has a net charge of $+e$. It is accelerated through a potential difference $\Delta V = 2.50 \times 10^3$ V and then enters a 2.00 T magnetic field directed in to the page.

  a. Sketch the subsequent path of the molecule assuming the molecule completes a half circle in the magnetic field before impacting the detection screen. Mark on the detection screen where the molecule will be detected.

  b. How far from the point where the molecule enters the magnetic field will it impact the detection screen?

  c. The heaviest mass of Br_2 is $m_{max} = 1.03 m_{min}$. Where will one of these molecules impact the detection screen?

★★ **Problem 12.21.** Refer back to Example 12.5 on page 594. There we pointed out that the torque experienced by the current loop points in opposing directions at different orientations of the loop in the magnetic field. If we want the coil to continuously rotate in one direction, one approach is to make the current change directions at the appropriate point in the coil's rotation. For what range of angular positions should the current be reversed compared to what is shown in Figure 12.13 to maintain a clockwise torque throughout the rotation? Your response should include clear diagrams!

12.4 The Magnetic Field: The Biot–Savart Law and Ampere's Law

★★ **Problem 12.22.** A proton is located at the origin and traveling with a velocity $\vec{v} = 5.0 \times 10^5 \hat{x}$ m/s. For each position listed below, determine the magnetic field.

  a. $\vec{r} = 1.0\hat{x}$ mm.

  b. $\vec{r} = (1.0\hat{x} + 1.0\hat{y})$ mm.

  c. $\vec{r} = (-4.0\hat{x} + 5.0\hat{y})$ mm.

  d. $\vec{r} = -2.0\hat{y}$ mm.

★★ **Problem 12.23.** Consider the three pieces of current-carrying wires shown in Figure 12.40. What is the net force (magnitude and direction) acting on

  a. the left wire?

  b. the center wire?

  c. the right wire?

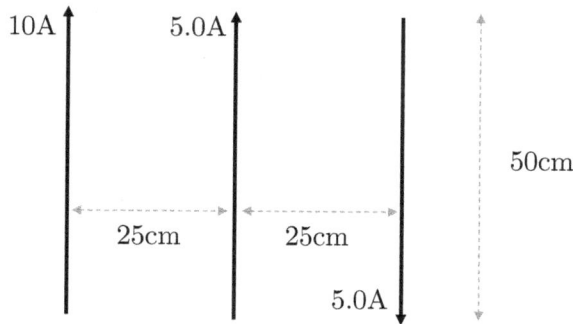

FIGURE 12.40
Problem 12.23

★★ **Problem 12.24.** A wire carries a 5.0 A current out of the plane of the page at $(x, y) = (3.0, 3.0)$ cm. A second wires carries a 5.0 A current in to the plane of the page at $(x, y) = (3.0, -3.0)$ cm. Sketch the situation and determine total magnetic field (magnitude and direction!) at each of the following positions:

a. $(x, y) = (0, 0)$ cm

b. $(x, y) = (3.0, 0)$ cm

c. $(x, y) = (6.0, 0)$ cm

★ **Problem 12.25.** The magnetic field at the center of a circular loop with 5.0 mm radius is 3.5 mT.

a. What is the current in the loop?

b. You measure the current in a long straight wire to be the same as what you found in (a). How far from the wire is the field 3.5 mT? Compare to the radius of the circular loop. Can you give an intuitive explanation for which is larger?

c. How many turns of wire would you need in a 5.0 cm long solenoid carrying the same current you found in (a) to achieve a 3.5 mT field at its center?

★★ **Problem 12.26.** Consider a current-carrying wire that has been twisted to form a half circle of radius r.

a. Use the Biot–Savart Law to set up an integral for the magnetic field at the center of the circle. Include a clear diagram.

b. Solve the integral you set up in (a). How does your result compare to Equation (12.15)?

★★ **Problem 12.27.** Two wires are parallel with the x axis; one is 5.0 cm directly above the other. The top wire carries a current $I = 5.0$ A in the $+\hat{x}$ direction and the bottom carries the same current I in the $-\hat{x}$ direction. Determine the total magnetic field at each of the following locations:

a. the midpoint between the wires.

b. 2.0 cm above the top wire.

c. 2.0 cm below the bottom wire.

★★ **Problem 12.28.** Consider Problem 12.27.

a. What is the magnetic force per unit length acting on the top wire due to the bottom wire?

b. What is the magnetic force per unit length acting on the bottom wire due to the top wire?

★★ **Problem 12.29.** Repeat Problems 12.27 and 12.28 in the situation where the top wire carries a current I in the $-\hat{x}$ direction and the bottom wire carries a current in the $+\hat{x}$ direction.

★ **Problem 12.30.** A circular ring carries a current that is (a) clockwise or (b) counter-clockwise as you look at it. In each case, determine the direction does the magnetic field at the center of the loop. Explain.

★ **Problem 12.31.** A long wire has been twisted to form a circular loop of radius R, but is otherwise perfectly straight (e.g. the straight section might lie along the x axis of your coordinate system, with the circular loop tangent to it). The wire carries a steady current of magnitude I.

  a. Sketch the situation (choose your own adventure and decide for yourself if the straight section is vertical, horizontal, etc. and which direction the current flows).

  b. What is the magnitude of the magnetic field at the center of the circular loop?

  c. In which direction does the magnetic field at the center of the wire point? (Your answer will depend on how you sketched the situation in (a)).

★ **Problem 12.32.** A long straight wire carries a 250 mA current into the page. A masochistic student draws a squiggly surface around the wire and declares, "I will determine $\oint \vec{B} \cdot d\vec{s}$!" What answer will he find (assuming he does it correctly)?

★ **Problem 12.33.** If $\oint \vec{B} \cdot d\vec{s}$ for some closed loop is 290×10^{-7} Tm, what current is passing directly through the interior of the loop?

★★ **Problem 12.34.** A solenoid (Figure 12.18 on page 599) consists of circular loops, or *turns*, of wire stacked into a column; the density of loops is N/L where N is the number of loops packed into a linear distance L. A current I runs through the wires. Figure 12.41 shows a cross-sectional view of a solenoid (the number of turns is arbitrary), with the current wrapping out of the page at the top and in to the page at the bottom.

  a. Use the right hand rule to determine the direction of the magnetic field at the center of the solenoid. Include at least 6 magnetic field vectors (from at least three pieces of wire each along the top and bottom).

  b. Use the right hand rule to determine the direction of the magnetic field outside the solenoid at a point directly above the midpoint. Again, use at least 6 vectors.

  c. Repeat (b) for a point outside the solenoid, directly below the midpoint.

  d. In an **ideal solenoid**, the magnetic field is considered to be a constant inside the solenoid (at points far from the edges, at least) and 0 outside. Does your work in (a)–(c) support this simplification? Explain.

  e. Assume we have an ideal solenoid. Use Ampere's law to determine an expression for the magnitude of the magnetic field inside the solenoid. (Hint: set your path to be a rectangle with one edge inside the solenoid, parallel with the magnetic field, and the opposite edge outside the solenoid.)

FIGURE 12.41
Problem 12.34

★ **Problem 12.35.** In the text I argued that Earth's geographic North Pole is actually its magnetic South pole. Explain how this must be the case, given that a compass needle's N pole points toward the (geographic) North Pole. Along with your written explanation, include a clear diagram that depicts both the compass needle and the Earth's magnetic field.

★ **Problem 12.36.** If a dipole is antiparallel with a magnetic field ($\vec{m}$ points in the $+\hat{z}$ direction and $\vec{B}$ in the $-\hat{z}$ direction, say), then what is the change in the dipole's energy if it rotates to align with the magnetic field (so they both point in the $-\hat{z}$ direction)?

★ **Problem 12.37.** Make a sketch where three dipoles are evenly spaced in a horizontal line. The left dipole is fixed with its orientation directed *up*, while the right is fixed with its orientation directed *down*. The middle dipole is free to rotate. What position(s) result in a minimum in potential energy? What is/are the corresponding torque(s)? (You don't need to do any laborious calculations here.)

★★ **Problem 12.38.** Consider two side-by-side dipoles separated by a distance d. The left dipole is fixed oriented up, and the right dipole is free to rotate.

a. First, determine the magnetic field at the location of the right dipole due to the left dipole. To do this you will need to make use of the equation for the magnetic field created by a dipole. In spherical polar coordinates (see the Appendix), with the dipole pointing along the $+z$ axis, the field is given by $\vec{B}_{\text{dip}} = \frac{\mu_0 m}{4\pi r^3} \left(2\cos\theta \hat{r} + \sin\theta \hat{\theta}\right)$.

b. Calculate the torque acting on the right dipole as a function of its counterclockwise rotation from a "straight to the right" orientation.

c. Calculate the potential energy of the interaction, again as a function of its counterclockwise rotation from a "straight to the right" orientation.

d. Based on your results, what orientation(s) for the right dipole correspond to stable equilibria? Do you find your results surprising?

★★ **Problem 12.39.** Some experiments in physics involve manipulating an "electron beam", which consists of many electrons traveling in the same direction with approximately the same kinetic energy. Suppose that at some instant two electrons, located at $(x, y) = (0, 0)$ and $(0, b)$ each have a velocity $\vec{v} = v\hat{x}$.

a. Determine the electric force (magnitude and direction) acting on each electron.

b. Determine the magnetic force (magnitude and direction) acting on each electron.

c. How do $\vec{v}$ and b determine which of the forces (electric or magnetic) is larger?

d. If you want to avoid a situation where the electrons are repelled from another (which would effectively destroy the beam), do you want to have a high energy beam or a low energy beam?

12.5 Magnetic Flux

★ **Problem 12.40.** A metallic circular hoop is in the plane of the page, and a magnetic field is directed into the page. What direction will current flow in the hoop if the field is (a) increasing (b) decreasing or (c) constant in magnitude?

★★ **Problem 12.41.** A wire is wrapped in place to form 200 loops of a 3.0 cm diameter circle. The coil is at rest with its axis vertical (i.e. the plane of the coil is horizontal so a line running perpendicular through the middle is vertical). A uniform magnetic field 65° from the vertical increases from 0.25 T to 1.75 T in 0.80 s. What is the induced emf in the coil?

★ **Problem 12.42.** A square loop of metal has an edge of 8.0 cm. It rests in the plane of the page. In 0.50 s, a magnetic field increases from 0 T to 0.75 T into the plane of the page. (a) What direction will current flow in the loop? (b) What average emf will be induced in the loop?

★ **Problem 12.43.** A circular hoop of radius r is immersed in a magnetic field of magnitude B; the orientation of the field relative to a line running through the middle of the ring is defined by the angle θ, as shown on the edge view in Figure 12.42. Determine an algebraic expression for the magnetic flux passing through the loop.

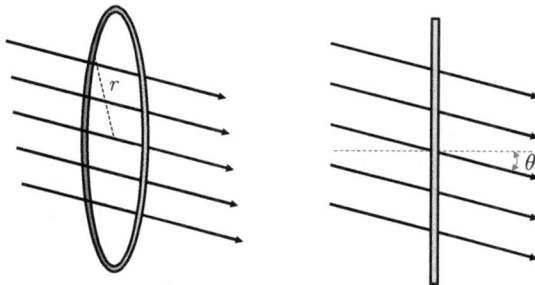

FIGURE 12.42
Problem 12.43

★   **Problem 12.44.** A square loop has an edge length of 5.0 cm. It lies flat on the xy plane in a region where there is a magnetic field $\vec{B} = (0.5\hat{x} - 0.5\hat{y})$ T. What is the total flux through the loop?

★★   **Problem 12.45.** Figure 12.43 shows a metal rectangular loop immersed in two different magnetic fields. (a) What is the magnetic flux through the hoop? (b) If the magnitude of the field on the left decreased, what direction would current flow in the hoop?

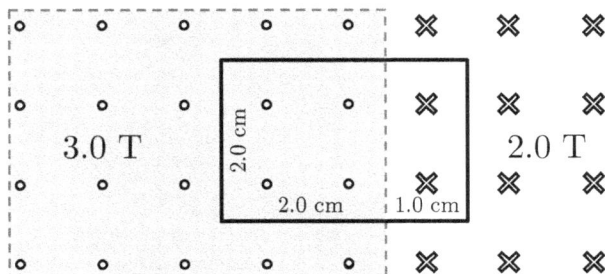

FIGURE 12.43
Problem 12.45

★★   **Problem 12.46.** A square loop of metal has a 12.5 cm edge length and a resistance of 0.65 Ω. A magnetic field is directed in to the plane of the page, and a 150 mA current is running counterclockwise around the hoop. (a) Is the magnitude of the field increasing or decreasing? (b) What is the rate of change of the field (in T/s)?

★   **Problem 12.47.** A battery with emf ε_1 is connected to a resistor R_1; the circuit is completed by wrapping the wire N_1 times around a cylindrical tube off to the side, as shown. A magnet is dropped through the tube with the north pole pointed down. How does the current through R_1 change as the magnet falls through the tube?

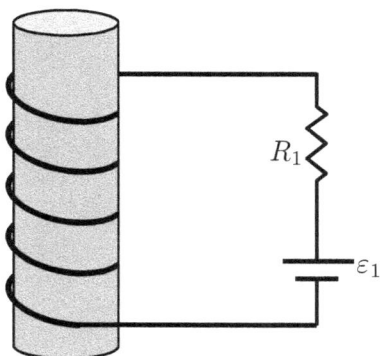

FIGURE 12.44
Problem 12.47

★★   **Problem 12.48.** Suppose a medical patient forgets to remove his gold wedding ring during an MRI, where the body is exposed to powerful and rapidly changing magnetic fields. The ring has a radius $R = 1.6$ cm and its cross-section is a semicircle of radius $r = 1.5$ mm. Suppose that during one portion of the procedure, the magnetic field is oriented directly into the page (Fig. 12.45) and increases from 0.75 T to 1.25 T in 2.0 s.

   a. In what direction (clockwise or counterclockwise in the diagram) will an electrical current flow through the ring?

   b. What is the magnitude of the current?

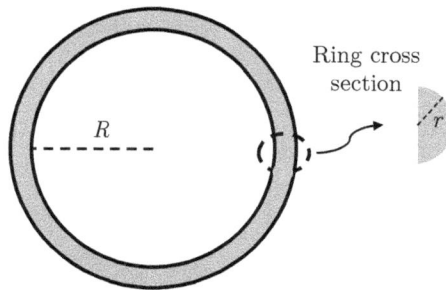

FIGURE 12.45
Problem 12.48

★★ **Problem 12.49.** We introduced magnetic flux by thinking of a magnet falling through a stationary loop of wire (Figure 12.32): as the magnet falls, the flux through the loop changes and so, according to Faraday's Law, a current is induced. Consider the same situation from a reference frame locked onto the magnet, where it appears that the magnet is stationary and the coil rises straight up such that the magnet is momentarily at the exact center of the loop.

   a. Sketch the situation. Consider an electron in the loop and indicate (by drawing vectors) its velocity, the magnetic field it is experiencing, and the magnetic force acting on it. Repeat for another electron in a different portion of the ring.

   b. Based on your work in (a), show that you expect a current to flow in the ring. Which direction will it flow when looking down on the ring from above?

   c. Analyze the situation using Lenz's Law and show that it predicts a current will flow in the same direction as you found in (b). (Evidently different mechanisms explain the same phenomenon based on different reference frames!)

★ **Problem 12.50.** If you place a metal ring on top of a powerful solenoid and then turn it on (in the diagram: touch the loose wire to the battery lead to allow a current to suddenly run through the solenoid), the ring will jump into the air. Why? You don't need to do any precise calculations, but invoke the relevant physical principles and clearly explain how they predict this behavior.

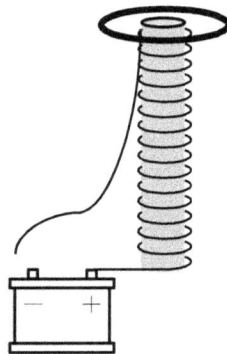

FIGURE 12.46
Problem 12.50

★ **Problem 12.51.** A strong bar magnet is placed on a table with the north pole facing up. You hold a solenoid in your hand and slam it down on the table such that the magnet will sit in the middle of the solenoid.

   a. Will a current flow in the solenoid as you slam it down? Explain. If yes, will it flow clockwise or counterclockwise as you look down on it from above?

   b. Will you feel any sort of magnetic force as you slam the solenoid? Explain. If yes, which direction will it point?

★★ *dx*　**Problem 12.52.** A circular loop of wire has radius r and total resistance R. It is immersed in a constant magnetic field with magnitude B and spins with angular speed ω about an axis perpendicular to the field. Figure 12.47 shows a perspective view (on the left) and looking down the axis of rotation (on the right).

　　a. Determine an expression for the flux as a function of the orientation (at some instant in time) θ.

　　b. If $\theta = 0°$ at $t = 0$, what is the induced emf in the loop as a function of time?

　　c. What is the current in the loop as a function of time?

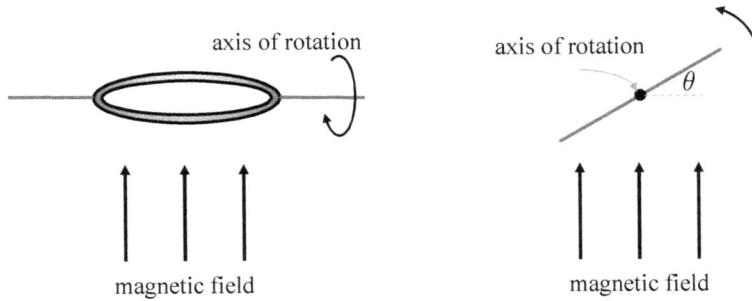

FIGURE 12.47
Problem 12.52

★★★　**Problem 12.53.** Figure 12.48 shows a long straight wire near a rectangular loop. The wire carries a linearly increasing current $I = \alpha t$ where α is a positive constant and t is the time.

　　a. Consider an infinitely narrow strip of the area inside the loop (width dx and height h, parallel to the wire and just inside the loop). What is the flux through this strip?

　　b. Use your result from (a) to set up an integral for the flux through the entire loop.

　　c. Solve the integral from (b).

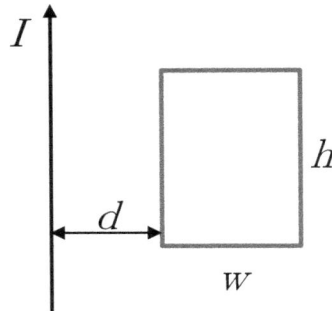

FIGURE 12.48
Problem 12.53

★　**Problem 12.54.** Consider a long copper tube that is held vertically. A bar magnet is dropped, N pole down, through the tube. Consider the time it takes for the magnet to fall all the way through the tube: how does it compare to the time it would take without the tube? Justify your response by clearly explaining what is happening inside the copper tube. (You don't need much math here, but be clear in your explanation!)

★★　**Problem 12.55.** Figure 12.49 shows a straight wire segment, with mass m and resistance R, that can slide along a second wire (with negligible resistance) that has been bent into the shape of a "U". The apparatus is arranged so the straight wire segment experiences a downward gravitational force. Neglect air resistance and assume the contacts between the wires are frictionless.

a. By applying a magnetic field to the entire region where the apparatus is located, it is possible to bring the horizontal wire to translational equilibrium (i.e. you can prevent it from accelerating forever). In what direction should the magnetic field be oriented for this to occur?

b. Determine an algebraic expression for the straight wire segment's terminal velocity.

c. Determine a numerical value for the terminal velocity when $B = 0.25$ T, $L = 10$ cm, $m = 15$ g, and $R = 20 \ \Omega$.

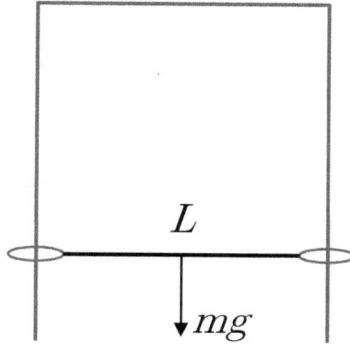

FIGURE 12.49
Problem 12.55

Additional Problems for Chapter 12

★ **Problem 12.56.** Suppose you have two identical circular loops of wire: they have the same radius and carry equal currents I. One loop sits on a table and the second loop is held a small distance directly over the first. In each case described below, determine if the loops are attracted to one another, repelled, or if they experience no force. Explain.

a. Looking down on the loops, both currents are clockwise.

b. Looking down on the loops, the top current is clockwise but the bottom current is counter-clockwise.

★★ **Problem 12.57.** A stiff horizontal wire is supported from below by 3 identical springs, each with a spring constant $k = 2.0 \times 10^1$ N/m. Initially, no current runs through the wire and the springs are at their equilibrium lengths. Then, a 0.75 T magnetic field pointing *in* is turned on in a 45 cm wide region centered on the springs. Simultaneously, a 5.0 A current turns on that moves from left to right through the wire.

a. Sketch the situation.

b. Will the springs compress or expand once the field and current are turned on?

c. By how much will each spring change its length?

★ **Problem 12.58.** A long straight wire carries a current I into the page. At the instant shown in the diagram, an electron is a distance r directly above the wire is traveling with velocity $\vec{v}$.

a. What is the direction of the magnetic field at the electron's position? Draw a vector on the diagram and explain.

b. What is the direction of the magnetic force on the electron? Draw a vector on the diagram and explain.

c. Determine an algebraic expression for the magnitude of the magnetic force on the electron.

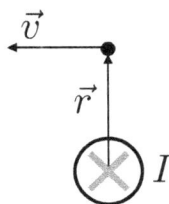

FIGURE 12.50
Problem 12.58

★   **Problem 12.59.** In a "hand crank electromagnetic generator" you turn a crank and to rotate a solenoid in place between two magnets. Figure 12.51 depict the rotation looking along the axis of rotation (from left to right, the solenoid rotates 90° in each frame; the solenoid is darker on one side to guide your eye for the rotation). Suppose that the solenoid has N turns of wire, depth d, and that each loop forms a square with edge length a. The solenoid rotations with a uniform angular velocity ω. Make your own sketch of this situate and draw one vector on each snapshot indicating the direction of the induced magnetic field in the coil at the instant shown. Be sure to clearly explain your logic.

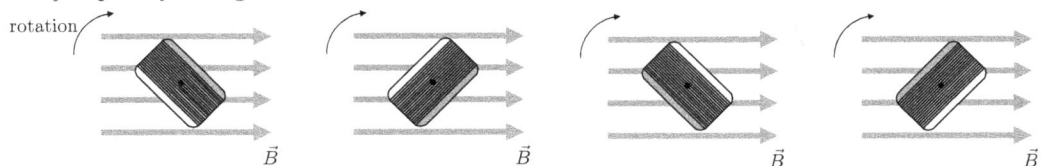

FIGURE 12.51
Problem 12.59

★★★   **Problem 12.60.** While a solenoid can create a uniform magnetic field along its axis, it is challenging to *see* inside the solenoid. An alternative is to use two circular coils of wire that share a common axis, as shown in Figure 12.52: the coils have equal radii R and consist of N loops carrying a current I. The distance between the coils is L.

     a. If the goal is to generate a magnetic field that points to the right at the middle of the device (marked with a dot), then which way should the current be oriented through the left coil? What about the right? Respond by indicating the direction of the current (in our out of the page) at the *top* of each coil. Briefly explain.

     b. Use the Biot–Savart Law to clearly set up an integral that could be evaluated to determine the magnitude of the magnetic field at the center of the device. As usual, simplify it to the point that it could be solved by someone unaware of the physical context of the problem.

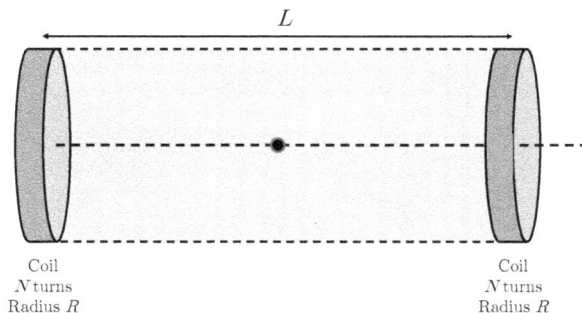

FIGURE 12.52
Problem 12.60

⋆∫ **Problem 12.61.** Solve the integral you set up in Problem 12.60.

⋆⋆ **Problem 12.62.** A simple model of the hydrogen atom considers an electron moving with speed v in a circular orbit around a stationary proton. The proton has charge $+e$ and also acts like a tiny bar magnet: it creates a magnetic field of magnitude B_0/r^3 that is directed in to the plane of the page along the orbit of the electron.[26]

    a. What is the direction of the magnetic force acting on the electron at the instant shown on the diagram? Explain.

    b. Set up an equation that *could* be solved for the radius of the electron's orbit considering both the electric force and the magnetic force between the electron and proton (the algebra is gross and you do **not** need to get it into "$r = ...$" form).

    c. Does including the magnetic field in your analysis suggest the electron's radius should be larger or smaller than if you consider only the electric force (assuming the velocity is unchanged)? How do you know?

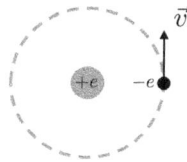

FIGURE 12.53
Problem 12.62

⋆⋆ **Problem 12.63.** A 12.6 V car battery is used to run a current through a resistor fixed above the battery, as shown in Figure 12.54.

    a. What force will the left vertical piece of wire experience after the circuit has been connected due to the magnetic field created by the right piece of wire? Specify both magnitude and direction. Ignore the resistance of the wire.

    b. Now suppose the 10 Ω resistor is removed and the only resistance in the circuit is due to the wires themselves. To calculate the resistance, suppose that the diameter if the wire is 4.6 mm and it is constructed from copper.

(In both cases use the "long straight wire" magnetic field and force, as usual, though clearly this is an approximation here because the wire isn't that long and the complicated edge effects would play a role.)

⋆⋆ **Problem 12.64.** A gas is being tested for the presence of ionized hydrogen (H_2^+), which has charge $+e$ and mass $2m_p$. Some of the gas is released into a chamber with walls held at a potential difference of 5.00×10^2 V, as shown. A narrow opening allows gas particles to move out of the chamber and into an area where there is a constant 0.350 T magnetic field directed into the plane of the page. A detector can be placed directly above or below the point of entry into the magnetic field to collect hydrogen gas particles.

    a. Should the detector be placed above or below the point of entry? Explain.

    b. How far above/below the point of entry should the detector be placed?

---

[26]Considering the magnetic interaction like this, in addition to the electric interaction, is an example of *hyperfine coupling*. A properly quantum-mechanical treatment of this effect is used in *paramagnetic nuclear magnetic resonance spectroscopy* (a type of "NMR"), which is used by chemists and material scientists to characterize certain materials.

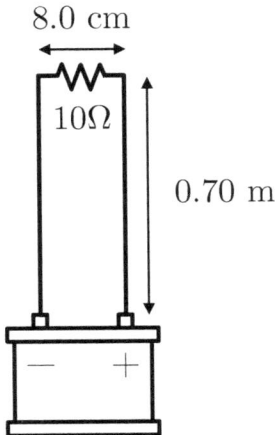

FIGURE 12.54
Problem 12.63

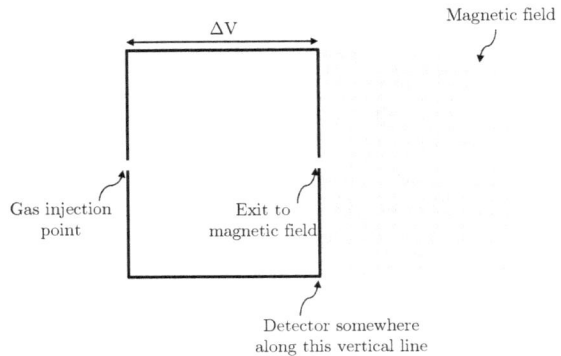

FIGURE 12.55
Problem 12.64

★★ **Problem 12.65.** An electron is accelerated from rest through a potential difference of 2.00×10^2 V such that its velocity is to the left. It then enters a 2.0 T magnetic field that is directed into the page.

  a. Determine the magnetic force experienced by the electron when it enters the magnetic field.

  b. Sketch the situation and indicate the subsequent motion of the electron. Explain. (Assume, as usual, that the magnetic field fills a large rectangular region and that the electron enters traveling perpendicular to one edge.)

★ **Problem 12.66.** A solenoid is connected to an ammeter and resistor and held horizontally in a region where there is a uniform horizontal magnetic field, as shown on the left side of Figure 12.56. It is then rotated 90° clockwise as shown on the right side of the figure. The solenoid has a cross-section area of 15 cm^2, a length of 0.30 m, and consists of 16 loops. The resistor is rated at 15 Ω and the magnetic field has a magnitude of 0.50 T.

  a. What average current will the ammeter register if the rotation is completed in 0.25 s? (Consider just the initial and final orientation of the solenoid.)

  b. Will the current travel up or down through the resistor as it is drawn in the second image? Clearly explain.

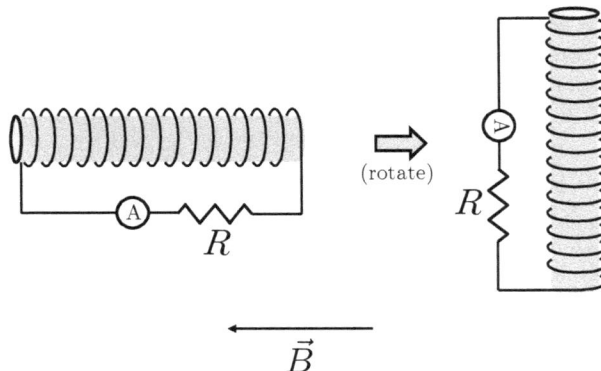

FIGURE 12.56
Problem 12.66

** **Problem 12.67.** A yo-yo can use electromagnetic induction to make LED lights turn on when the yo-yo is spinning. For simplicity, the diagram below shows just one magnet (connected to a stationary axle as the rest of the yo-yo spins) and one bulb connected to a simple loop of wire (which rotates about the yo-yo's axis with angular frequency ω). The loop passes just in front of the magnet as it spins. The loop and magnet both have radius r and the distance from the center of the yo-yo to the center of the loop is r_y. The magnet creates a magnetic field $\vec{B}$ out of the page in the shaded region, and the loop's resistance is R.

Determine an algebraic expression for the average power P dissipated in the bulb from the instant shown in the diagram to the instant when the loop has completely left the region of the magnetic field. Your expression should involve some or all of r, r_y, R, B, θ and ω.

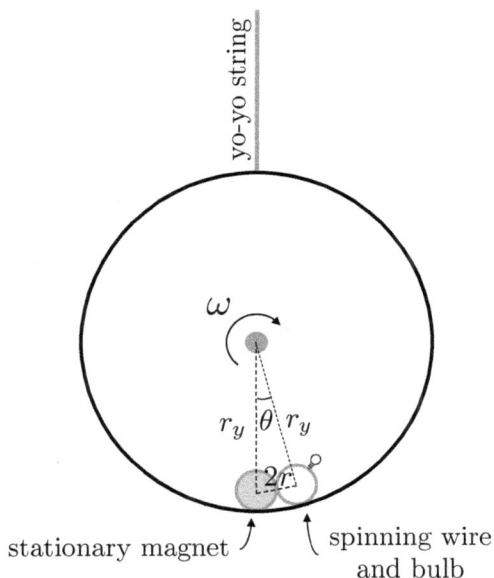

FIGURE 12.57
Problem 12.67

** **Problem 12.68.** A metal bar with resistance R is connected to a long U-shaped piece of wire, as shown in Figure 12.58. A magnetic field of magnitude B is directed into the page.

a. What force is necessary to pull the bar to the right at a constant speed v?

b. Does a current flow in the loop? Why or why not? If it does, in what direction will it travel? Explain *both* through Lenz's Law *and* by analyzing the forces acting on electrons in the bar.

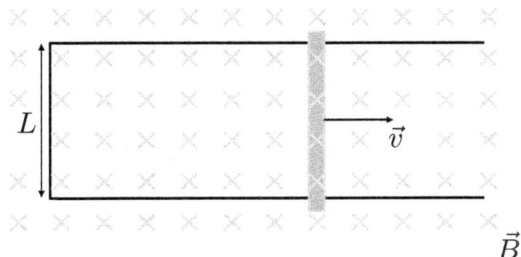

FIGURE 12.58
Problem 12.68

★★ **Problem 12.69.** Consider the "U" shaped rail of Problem 12.55, but this time suppose the rail has a length of 8.5 cm and the apparatus is held horizontally (so gravity plays no role), the magnetic field B is perpendicular to the rail, and the bar is being pulled with a steady force F. Suppose the total resistance of the bar is 0.45 Ω and when the bar is being pulled at 3.50 m/s the rail dissipates 4.0 W. What is (a) the magnitude of F and (b) the magnitude of B?

★★★ **Problem 12.70.** Consider the "U" shaped rail of Problem 12.55 again. This time, suppose the apparatus is held horizontally (so gravity plays no role), and instead of pulling on the bar you use a battery and a switch to drive a current through the rail and bar. Your goal is to drive the bar along the rail with the resulting magnetic force acting on the bar, which you can treat as a "long straight wire". (This is an electromagnetic projectile launcher!) The battery has an emf ε, the magnetic field throughout the entire region has a magnitude B, the bar has a length L, and the circuit has a total resistance R.

   a. Sketch the situation, paying special attention to the orientation of the battery (will it drive current clockwise our counterclockwise?) and the orientation of the magnetic field (into the page or out of the page?). Explain how your choices will result in a magnetic force that pushes the bar down the rail.

   b. What will the terminal velocity of the bar be? You should provide an algebraic expression in terms of necessary constants.

   c. Determine a numerical value for the terminal velocity of the var for $\varepsilon = 9.0$ V, $B = 0.50$ T, $L = 6.0$ cm, and $R = 0.20$ Ω.

★ **Problem 12.71.** A friend excitedly tells you that they've come up with an idea for a projectile launcher that relies on electromagnetism. The idea is sketched below: set up a solenoid at the base of the barrel, and allow the projectile, which is a metal ring, to sit next to it (but not touching). When the trigger is pulled, a current will flow through the solenoid and your friend predicts that the ring will be launched out of the barrel.

   a. Is your friend correct or incorrect? Clearly explain by referencing relevant physical principles.

   b. If they are correct, what physical quantities could be adjusted to increase the speed of the projectile? If they are incorrect, can the approach be modified so it will work?

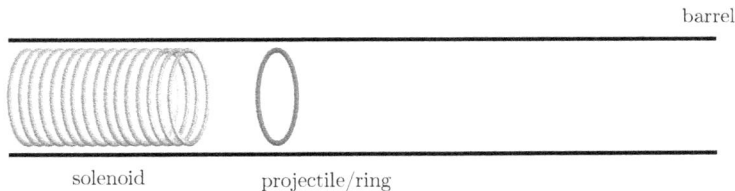

FIGURE 12.59
Problem 12.71

★ **Problem 12.72.** Two round magnets are attached to the terminals of a battery and placed on a sheet of conducting aluminum foil, as shown in the diagram. Sketch and label any magnetic field lines and electric currents. Will the battery and magnets move? If so, in what direction? Clearly explain your answer.

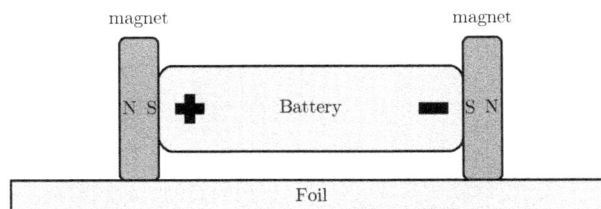

FIGURE 12.60
Problem 12.72

★★ **Problem 12.73.** Some microphones operate on the principle of electromagnetic induction: the speaker's voice moves a membrane; a magnet is connected to the membrane and moves with it. A coil consisting of N turns of wire of radius R is located near the magnet and does not move. The magnetic field created by the bar magnet in the coil has a magnitude given by $B = \mu_0 M / (2\pi r^3)$ where M is a constant that corresponds to the strength of the magnet and r is the distance of the magnet from the coil, as shown in Figure 12.61. Suppose a person's voice moves the magnet from an initial distance r_1 to a smaller distance r_2 in time Δt.

a. Determine an algebraic expression for the average emf generated in the coil in this interval.

b. Which way will the current flow in the coil as you look down at it from above?

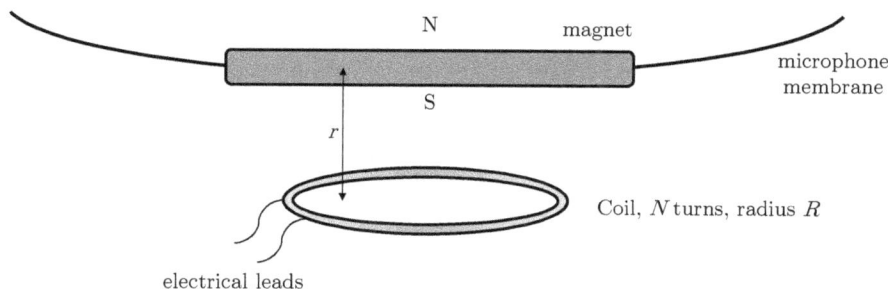

FIGURE 12.61
Problem 12.73

★★ **Problem 12.74.** In some sensitive experiments it is desirable to detect the motion of fast-moving charged particles down an observation tunnel. One way is set up a series of wire loops, each connected to an ammeter, near the path of the particle. Suppose that at some instant a proton is moving down the tunnel as shown in Figure 12.62.

a. What is the magnetic field at point P? Provide a numerical answer.

b. Will the coil surrounding point P register a current at the instant shown? Why or why not? If there is a current, clearly explain if you expect it to be clockwise (CW) or counterclockwise (CCW).

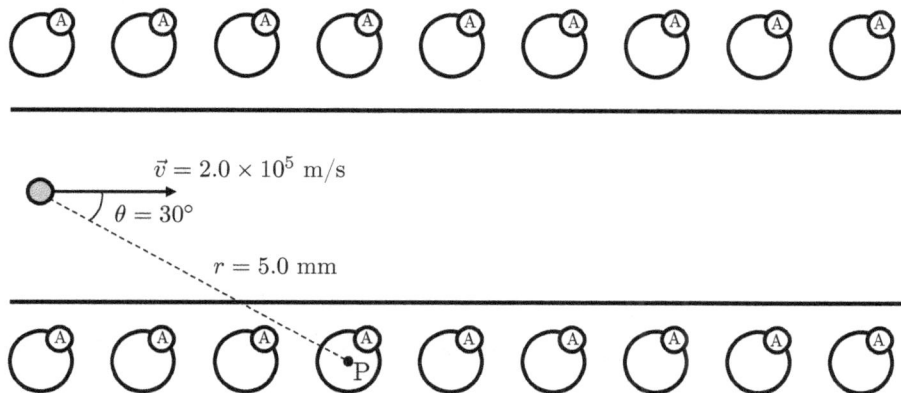

FIGURE 12.62
Problem 12.74

★ **Problem 12.75.** A lightning bolt involves electrons moving from the bottom of a cloud to the surface of the Earth. Refer to Problem 11.2 to determine a value for the average current of a lightning bolt. If we treat the bolt as a current-carrying wire traveling vertically down from the cloud, then what is the strength of the magnetic field a distance of 1, 10, or 100 m from the bolt?

Problem 12.76. Two parallel wires run through the plane of the page; one is directly above the other. The top wire carries a current $I = 3.5$ mA out of the page, while the bottom wire carries a current $I = 3.5$ mA in to the page. The wires are separated by 6.0 cm. What is the magnetic field (magnitude and direction) at the midpoint between the wires?

★★ **Problem 12.77.** A solenoid is formed by wrapping 100 loops of wire along a 3.3 cm-long cylinder with a 2.5 cm diameter. Separately, copper wire has been bent into a square with a 1.3 cm edge length. The square is placed inside the solenoid as shown in the left side of Figure 12.63 (in the right side the front of the solenoid blocks your view of the square). The solenoid is then attached to a power supply and the current increases from 0 A to 130 mA in 0.12 ms. What is the average emf induced in the copper square during this time?

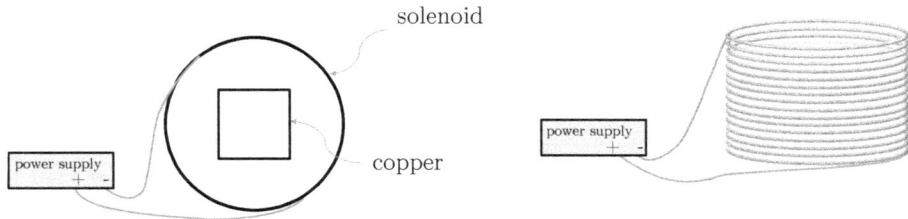

FIGURE 12.63
Problem 12.77

★★ **Problem 12.78.** When you charge an electric device, the adapter you plug into the wall typically contains a **transformer**. Figure 12.64 shows a simple model of a transformer, which consists of two insulated coils wrapped around opposite sides of an iron loop: the left side has N_L loops and the right side has N_R. The coil on the left is connected to the wall outlet, which delivers a time-varying electric potential $V(t) = V_0 \sin(\omega t)$ where V_0 is the maximum potential in the left coil and ω is the oscillation frequency. The right coil is connected to an electrical device (like your phone) with resistance R. At the instant shown, the current in the left coil is in the direction shown and is *increasing* in magnitude.

   a. What is the direction of the current in the right coil at the instant shown? Indicate your answer by drawing an arrow on the top *and* bottom wire on the right coil. Explain your logic.

   b. What is the maximum electric potential of the right coil? Express your answer in terms of V_0 and any other needed parameters. Explain your logic.

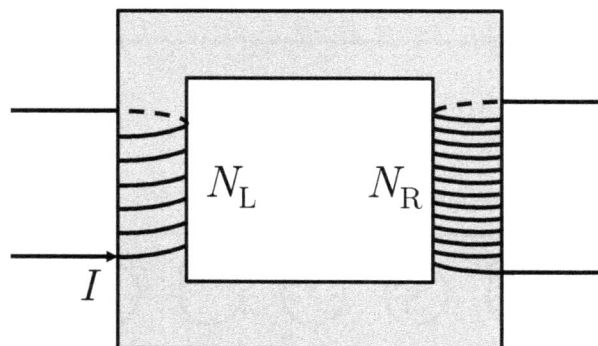

FIGURE 12.64
Problem 12.78

★★ **Problem 12.79.** A *particle selector* can separate the remnants of nuclear fission so particle physicists can study them. Suppose an atom (called the *parent*) has split into two identical nuclei (the *daughters*) each with mass m and charge $+Ze$. Initially, each daughter has a speed v_0 directed up the page and they have some small horizontal distance r between them. Later, they've separated to a much greater distance R (such that any interactions between the particles is negligible) and are about to enter a magnetic field of magnitude B, as shown in the figure.

a. At the instant shown at the bottom of the diagram, each daughter nucleus experiences a magnetic force due to the other nucleus. Determine an expression for the magnitude and the direction of the magnetic force acting on each nucleus.

For the rest of this problem, ignore the magnetic interaction between the particles (i.e. assume the electric interaction is much stronger).

b. Consider the electrical repulsion between the daughter nuclei to determine an expression for the magnitude of each daughter nucleus's velocity just before entering the magnetic field.

c. Sketch both velocity vectors from (a) on the diagram. Also sketch the magnetic force vectors that the nuclei will experience once they enter the magnetic field. Explain (no detailed calculations necessary).

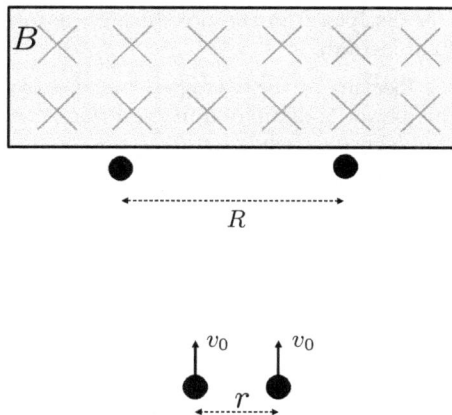

FIGURE 12.65
Problem 12.79

★★ **Problem 12.80.** In the plane that bisects the Earth's magnetic poles, the magnetic field points toward the magnetic South pole (see Problem 12.35), as shown in Figure 12.66. Suppose a negatively charged particle has been emitted by the sun and is on this plane with a velocity as shown on the right side of the figure.

a. Describe the direction of the force felt by the particle at the instant shown in the figure. (Sketch it and/or describe it in a phrase or sentence.)

b. The particle has mass m and charge $-q$. It is a distance r from the center of the Earth and it experiences a magnetic field of magnitude B. The particle's velocity has magnitude v and forms an angle θ with the horizontal, as shown. Determine an equation for the magnitude of the force at the instant shown in terms of some or all of these parameters.

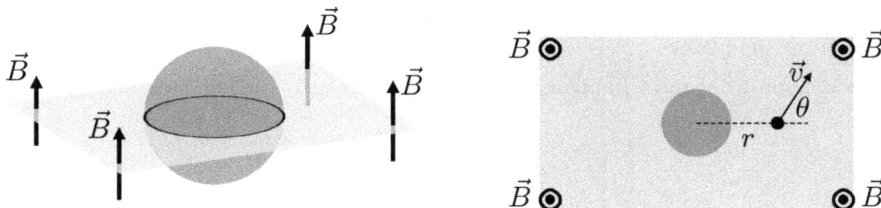

FIGURE 12.66
Problem 12.80

$\star\star\star$ **Problem 12.81.** If you need to boil a pot of water, you'd probably use a gas-powered or electric-powered stove top in your home. An alternative design is called an **induction cooktop**. The basic idea here is to run an alternating current through a coil of copper loops (an alternating current changes direction sinusoidally, i.e. $I = I_{max}\sin(\omega t)$). The pot of water sits just above the coil (though not touching), as sketched in Figure 12.67. To be specific and simplify the analysis, we'll make the following assumptions:

- The pot is a circular metal ring of resistance $R = 0.50$ Ω and radius $r_{pot} = 0.10$ m.
- Suppose that $I_{max} = 8.0$ A and $\omega = 1.5 \times 10^5$ rad/s.
- Treat the coil as having 2000 loops and a radius of $r_{coil} = 15$ cm.

a. At some instant the current through the coil is 0 A. A short time later, the current is a maximum and runs *clockwise* through the coil as viewed down from above. In what direction is the *induced current* flowing in the pot during this time? Clearly explain your answer.

b. What is the average power dissipated in the pot during the interval described in (a)? (Just consider the initial and final instants.)

c. The answer to (b) is also a measure of the average power dissipated in the pot over the entire period. Explain why. (This power is dissipated in the form of heat, which is the whole idea – you can boil water rather quickly this way!)

d. Using the two endpoints to calculate an average of a sine curve, as you did in (c), isn't quite correct because it assumes a linear, rather than sinusoidal, change between the two values. A more exact answer is $\pi^2/8$ multiplied by your result from (c). What is this updated value? (For added calculus-based joy, *prove* that this is the appropriate factor. The fact that the integral of $\sin^2(\omega t)$ or $\cos^2(\omega t)$ over a quarter period is $T/8$ will be useful.)

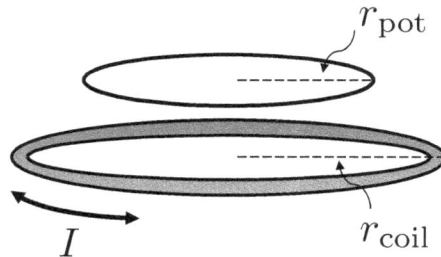

FIGURE 12.67
Problem 12.81

$\star\star$ **Problem 12.82.** Figure 12.68 shows part of a **solenoid boxer engine**.[27] Each solenoid contains a metal piston attached to a central crankshaft. The pistons are driven in and out of the solenoids by a current that runs through the solenoids; the net effect is that the crankshaft spins (think of an axle of a car). The cartoon sketches one solenoid and its piston mostly extended.

a. With a few sentences and a diagram, clearly explain how running a current through the solenoid can push out the piston.

b. Clearly explain what must happen for the piston to then be pulled back in.

---

[27]Images of these devices are readily available online.

FIGURE 12.68
Problem 12.82

★★ **Problem 12.83.** A popular "test your strength" carnival game involves slamming a hammer into a target on the ground, which then sends a little ball up a post: the harder the hit, the higher it goes. Suppose you are playing a version of this game where hitting the target instead makes a lightbulb glow. Suppose the target consists of 50 loop circular coil of wire with a 15 cm radius; its total resistance is 0.80 Ω. A powerful bar magnet is embedded in the tip of the hammer; we'll approximate the field just outside the magnet to be a steady 0.50 T.[28]

a. You take a swing and complete your stroke in 0.60 s. What power is dissipated in the bulb?

b. What is the mathematical relationship between the power and the amount of time required to complete the stroke? (For example, "power is proportional to time, so if you double time you double power.")

★★ **Problem 12.84.** A **diode** is an electrical device that only allows current to flow in one direction. Suppose a diode is connected to a rectangular loop of wire, as shown in Figure 12.69. You plan to drop a bar magnet through the loop to induce a current as it falls toward the plane of the loop.

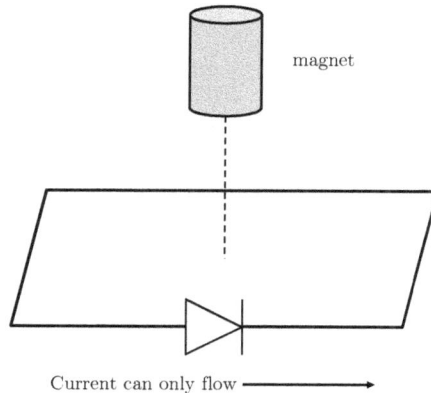

FIGURE 12.69
Problem 12.84

a. How should you orient the magnet: north pole up or north pole down? Explain.

b. Will current continue to flow after the magnet passes through the loop and continues to fall down and away from the loop?

★★ **Problem 12.85.** A circular hoop of metal is held horizontally with respect to the ground. A bar magnet is held above the hoop and dropped so it falls through the center of the hoop. Describe the flux through the hoop as the magnet falls to the ground if (a) the north pole is down, (b) the south pole is down, or (c) the magnet is held horizontally.

---

[28]Note that this isn't terribly practical because repeatedly striking a magnet can demagnetize it by reorienting the magnetic dipoles that form aligned domains in a magnet... unless of course you're an underhanded operator of the game and you don't want contestants to win!

★★ **Problem 12.86.** You are troubleshooting an MRI machine, which we'll model as a simple solenoid. You are about to put your hand into the middle of the machine, but your coworker stops you, concerned that the 2.0 cm diameter ring you're wearing will become hot and burn you. Suppose the magnetic field has a strength of 0.50 T, your ring has a resistance of $R = 1.0 \times 10^{-3}\Omega$, and you move it into the field in 0.25 s. Determine an upper limit for how much energy will be transferred to the ring. Do you think the ring will heat up to the point of being noticeable? Why or why not?[29]

★★ **Problem 12.87.** Consider the setup shown in Figure 12.70: a cylindrical piece of iron is free to move along the tube formed by three small solenoids that are stacked side by side. Each solenoid is connected to a switch that can be used to turn on/off a current I through the corresponding solenoid. The iron core is pulled left-to-right along the tube if the three switches are sequentially turned on, left-to-right.

    a. Explain this behavior. A complete answer should name and explain relevant principles of physics and describe, in words and with a diagram, how the principles apply to this situation.

    b. List three ways the force applied to the iron could be made stronger. Justify your responses by citing relevant mathematical relationships.

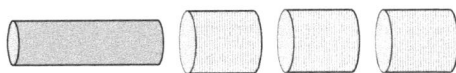

FIGURE 12.70
Problem 12.87

★★ **Problem 12.88.** Some emergency flashlights can be charged by rotating a handle that, in turn, spins a coil of wire immersed in a magnetic field. This drives a current that charges a battery (or capacitor) that can then be used to power the lights. Suppose a 750 turn circular coil of wire with a radius of 1.5 cm is immersed in a uniform 0.20 T magnetic field. What angular speed ω is needed to generate a 1.5 V emf (about equivalent to a AA battery)? Provide your answer in (a) rad/s and (b) rotations per minute (rpm). (You'll find that the average emf during a complete rotation is, strictly speaking, 0 V. Assume the 1.5 V refers to the average absolute value of the emf. Or, if you'd like to flex your calculus muscles, consider the instantaneous emf and suppose the 1.5 V refers to the **root mean square** potential, which is to say you should calculate the average of the emf squared over one rotation, then take the square root to find an expression for ε_{RMS}; if this quantity is to be 1.5 V, what is ω?)

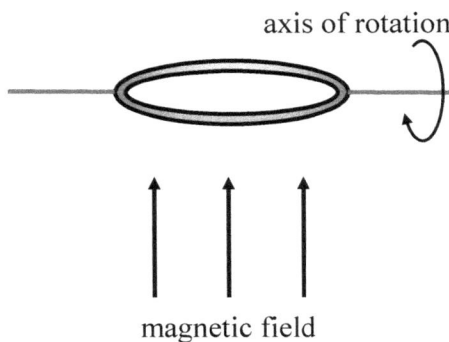

FIGURE 12.71
Problem 12.88

★★ **Problem 12.89.** A power supply is connected to two resistors and two long, flexible wires, as shown in Figure 12.72.

    a. What current will run through each wire?

---

[29] In practice, of course, you'd make sure the machine was safely shut down before working on it!

b. How much power is dissipated by the circuit?

c. What is the magnetic force per unit length acting on each wire? Provide the magnitudes and directions.

Problem 12.90. A square loop of wire is electrically connected to a small light bulb placed in a magnetic field and rotated about a central axis with an insulating handle (Fig. 12.73). Suppose the loop has sides of length $L = 4.0$ cm and the magnetic field has a strength of $B = 0.35$ T. If the period of revolution is 0.80 s and the bulb has resistance $R = 1.5$ Ω, what average power will the bulb dissipate? (This is a crude model of a generator.) *Hint:* The power equations are true for *instantaneous* potential and current; here they're changing as the bulb rotates. Write down the instantaneous flux, then calculate or look up the derivative needed to calculate the instantaneous emf. Then set up an integral and use $\int_0^T \sin^2(\omega t)\, dt = \int_0^T \cos^2(\omega t)\, dt = \frac{1}{2}T$ to find the power.

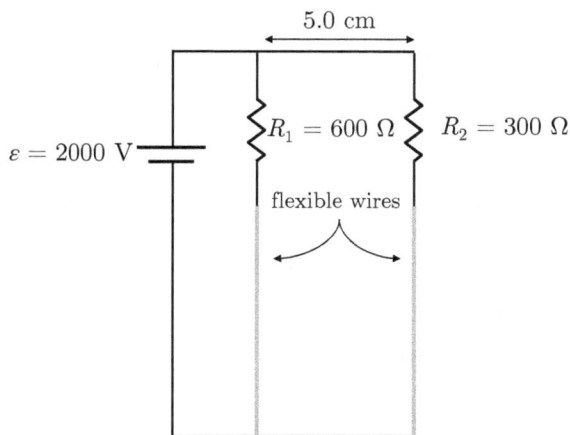

FIGURE 12.72
Problem 12.89

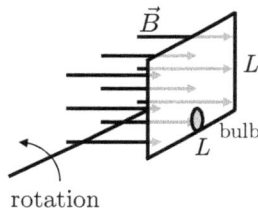

FIGURE 12.73
Problem 12.90

Problem 12.91. Here is an experiment that you can build at home and take to parties. Take a pair of powerful magnets and attach them to the terminals of a AAA battery so the same poles are touching the battery (so e.g. "NS-battery-SN"). Push the contraption (the "train") into a long coil of copper and be amazed as it is sucked in and moves along the "track". (Figure 12.74 shows a cross sectional view.) You can connect the ends of the copper coil so you have a closed loop and the device will keep moving until the battery runs dry. By thinking through how current will flow from one end of the battery to the other, explain why the battery and magnets will be pushed along the track. You don't need much math here, but a clear diagram and reference to the relevant physical principles is important.[30]

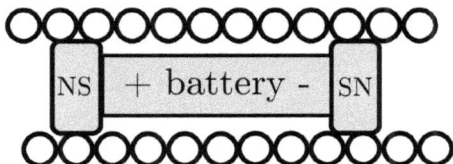

FIGURE 12.74
Problem 12.91

---

[30]It isn't too hard to find videos of this online. Try searching for "electric train with batteries and magnets" or something similar.

★★ **Problem 12.92.** Figure 12.75 shows a large magnet; suppose the N pole is facing up. A battery is connected to the side of the magnet. A flexible wire is connected to the edge of the battery. You hold the wire in place above the center of the magnet but leave enough of the wire dangling that it can make contact with the top of the magnet. At the instant shown, the wire is in contact with the edge of the magnet.

a. What is the direction of the magnetic force acting on the portion of the wire touching the magnet? Explain (words and equations required for full credit!).

b. If the portion of the wire below your hand is free to move, then what will happen to the wire? Explain.

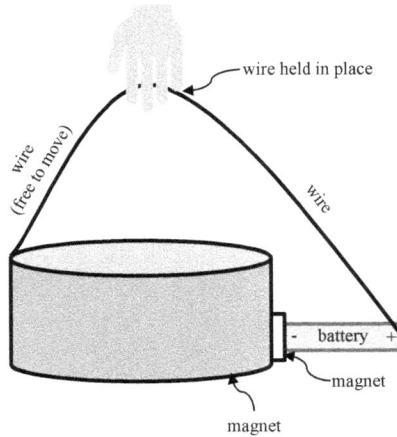

FIGURE 12.75
Problem 12.92

Part V

Modern Physics

Physics ultimately concerns itself with the laws of nature at their most fundamental level. (In this way physicists are somewhat like children who respond to every question with another "Why?" or "How?") In this last part of the book we will trace some developments in physics that push the frontiers of our understanding at the scale of the *small*.

First, we will see that, in contrast to our discussion of light as electromagnetic *waves*, in some cases light behaves like a *particle*. In parallel to this, we will see that fundamental *particles* (electrons, protons, and neutrons) can in some cases behave like a *wave*! Both of these ideas are important for our understanding of the atom, which has immense practical applications (including for instance the development of high-powered miscroscopes, cutting-edge electronics and the study of exotic materials such as graphene) and builds into our understanding of molecules and therefore chemistry and biology.

13

Modern Physics

Our goal in this chapter is to explore the atom, the modern understanding of which began to take shape in the late 1800s and exploded through the early 1900s and continues today. While we'll certainly grapple with some mathematics in this chapter, the full details get to be rather complicated and I will rely on descriptive language more than usual.

Indeed, my goal here is more of a *preview* than a thorough *introduction* to the so-called field of modern physics: I have attempted to write a book that fits neatly into a two-semester sequence of courses, and if you've been closely following the text so far, there simply isn't time to do more than scratch the surface and hopefully pique your curiosity for more. (If you continue beyond the introductory sequence, you'll likely take an entire course on *Modern Physics*.)

That said, in our consideration of the topic we will frequently refer to important experiments in physics, each of which prompted surprising (and often abstract) advances in our understanding of the fundamental nature of reality.

Learning Objectives

After reading this chapter, you should be able to:

- Describe the photoelectric effect and mathematically analyze the photoelectric effect experiment.

- Describe and mathematically analyze Bragg diffraction.

- Describe wave-particle duality for photons, electrons, protons, and neutrons, particularly in terms of the wave function and the probability distribution for an electron trapped in a well or traveling freely through space.

- Describe Rutherford scattering; in particular, explain how it showed that the atom has a positively charged nucleus.

- Describe the Bohr model and mathematically analyze its predictions for the spectrum of hydrogen.

- Mathematically analyze the spectrum of hydrogen in terms of the Balmer–Rydberg formula.

13.1 Light can Behave Like a Particle: The Photoelectric Effect

We saw in Chapter 12 that light is an *electromagnetic wave*: it consists of an electric field and a magnetic field, each oscillating at the same frequency f and wavelength λ, and propagating

DOI: 10.1201/9781003571568-18

with a velocity $\vec{v}$. If it is traveling through a vacuum then the speed is the speed of light c. Recall that these quantities are related according to

$$c = \lambda f \tag{13.1}$$

The wavelength and frequency characterize an electromagnetic wave: visible light, for instance, has a wavelength ranging from about 400 nm to 700 nm. We also characterize light by its *intensity*, which is related to the perceived *brightness* and the amplitude of the wave.[1] Intuitively, a brighter (or more intense) light carries more energy.

With this in mind, let me describe the **photoelectric effect**, which describes the interesting phenomenon where *light* can knock electrons off of a metal and create a *current*. Let me state right up front the "master equation" that describes the maximum kinetic energy a freed electron can have:

$$K_{\max} = hf - \phi \tag{13.2}$$

Before I explain what the terms on the right side of this equation mean, let me point out what *isn't* included: intensity. This is rather striking, because it means that the brightness of the light has nothing to do with whether or not an electron will actually become knocked loose from the metal. Instead, it depends on the frequency f, a metal-specific constant ϕ called the **work function**, and a constant of nature h called **Planck's constant:**[2]

$$h = 6.63 \times 10^{-34} \text{ J s} = 4.14 \times 10^{-15} \text{ eV s} \tag{13.3}$$

Let's unpack what Equation (13.2) says. Electrons in a metal normally stay there; they need to absorb a certain amount of energy (the metal-dependent constant ϕ) in order to break free. Think here of throwing a ball out of a pit, or well: you need to throw it with enough energy to reach the top and get it out (so in this analogy ϕ is proportional to the depth of the well; see Figure 13.1). Some work functions for typical metals are shown in Table 13.1.[3]

TABLE 13.1

Work functions

Material	Work function ϕ (eV)
Aluminum	4.20
Copper	5.10
Gold	5.47
Nickel	5.22
Silver	4.64
Zinc	3.63

Meanwhile, the *light* provides the electron with an amount of energy *exactly* equal to hf. If this quantity exceeds the metal's work function ϕ, then an electron is freed and can carry an amount of energy up to the difference between hf and ϕ. Returning to the ball in a bit analogy, if the ball needs, say, $mgh = 3$ J of energy to get to the top and you throw the ball up with initial kinetic energy 4 J, then when it reaches the top if will be left with $4 \text{ J} - 3 \text{ J} = 1$ J of energy remaining.[4]

Now it turns out (from the photoelectric effect and other experiments that we won't discuss) that the "energy packet" equal to hf is *indivisible*: in a very literal sense a beam of light is nothing more than an enormous quantity of these energy packets, which we call

---

[1]We won't worry about the mathematical details of this relationship.

[2]After Max Planck (1858–1947), who received the 1918 Nobel Prize in Physics. Planck's constant comes up in many experiments dealing with light and exceedingly small length scales.

[3]Remember "eV" stands for electronvolt; $1\text{eV} = 1.60 \times 10^{-19}\text{J}$. I should also note that work function values are somewhat sensitive, for instance to the cleanliness of the metal when a measurement is made. You will accordingly find variations if you look up work functions elsewhere.

[4]I said the electron receives energy "up to" $hf - \phi$ because ϕ is the *minimum* energy needed to release an electron: the details of how electrons are bonded to the metal are complex and some electrons can require more energy than others. In the "ball in a pit" example, we can say that some balls (electrons) are deeper than others and ϕ characterizes the minimum depth.

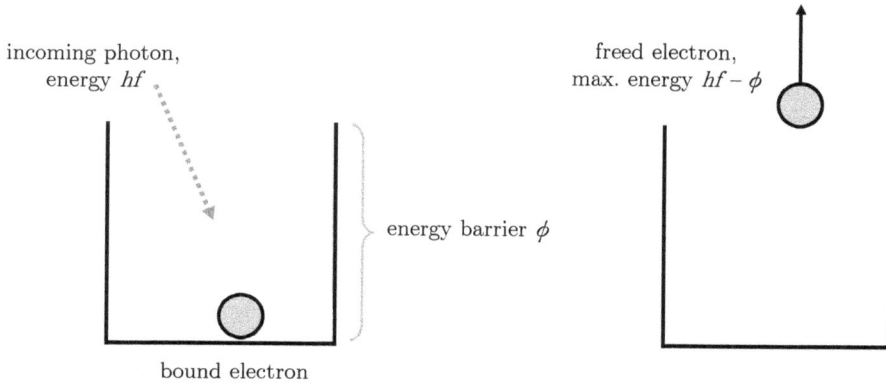

incoming photon,
energy hf

freed electron,
max. energy $hf - \phi$

energy barrier ϕ

bound electron

FIGURE 13.1

A schematic view of the photoelectric effect. A photon with energy hf interacts with an electron embedded in a metal in a "potential well" of depth ϕ: this simply means that the electron remains bound to the metal unless it absorbs the photon and $hf > \phi$. If the electron does escape, it carries up to the remaining energy as kinetic energy.

photons, all traveling together. Thus the energy of a photon is given by

$$E_{\text{photon}} = hf \tag{13.4}$$

where f is the photon's frequency. Interestingly, a photon also carries *momentum*, even though it has *no mass*![5] The energy and momentum of the photon are related by

$$E_{\text{photon}} = pc \tag{13.5}$$

"Whoa there!" you might say, "I thought light was a *wave*, meaning an oscillating electric field and an oscillating magnetic field!" Well, yes, but this wave carries *energy* (the Earth's dependence on the Sun for our energy should mean this is no surprise), and when the wave exchanges energy with its surroundings, it does so in these discrete (albeit tiny) chunks.

Photons are often referred to as *particles* because an isolated photon doesn't involve an oscillation in electric fields and magnetic fields in the same way as a beam of light (thus the "photon vs. wave" perspective of light is sometimes called **wave-particle duality**.) That being said, calling photons particles is a bit of a misnomer: a photon is *not* a little ball. In fact, even a single photon retains important wavelike characteristics: if you send photons through a double slit experiment, one by one, eventually they build up a diffraction pattern![6] Put another way: photons diffract, and so it is perhaps unsurprisingly that a laser beam (many photons) *also* diffracts.

That being said, the idea that many energy packets collectively make an oscillating electromagnetic wave is very far outside our intuition because there simply isn't any phenomenon on the scale of everyday life that behaves this way. It is *completely natural* to find this all a bit odd! Indeed, it took a real Einstein to work out the details.[7]

---

[5]This result falls rather elegantly out of a study of special relativity.

[6]This only makes sense if *each* photon passes through *both* slits: it is a wave rather than a particle with an infinitely precise location.

[7]Albert Einstein (1879–1955) to be precise. The 1921 Nobel Prize in Physics was awarded to Einstein for his work on the photoelectric effect.

The Photoelectric Effect Experiment

You may be wondering exactly how it is that we know all this. As it turns out, it isn't terribly challenging to explore the photoelectric effect experimentally (though the required equipment is very sensitive and therefore rather expensive). Figure 13.2 sketches the basic idea of the experiment, with the following features:

- Two metal plates are separated by a small distance; their opposite ends are connected by a wire.

- Along the wire there exists a power supply (to provide a user-determined potential difference ΔV) and an ammeter (to read the current in the wire).

- A nearby apparatus shines light on one of the metal plates. The experimenter is able to adjust both the frequency (or equivalently the wavelength) and the intensity (brightness) of the light.

If the light has a sufficiently high frequency (i.e. if $hf > \phi$), then some electrons will be knocked loose. If they make their way through the air to the *other* plate, they can travel through the wire and back to the original plate.[8] The ammeter will therefore detect a current.

Now the experimental approach is to turn on the power supply to create a potential difference that *opposes* this current. At the so-called *stopping potential* ΔV_s, the potential energy $e\Delta V_s$ between the plates perfectly matches the maximum kinetic energy of the electrons, *just barely* preventing the electrons from jumping the gap and causing a current to be registered by the ammeter. Mathematically, this occurs when

$$e\Delta V_s = K_{\max}$$
$$e\Delta V_s = hf - \phi \qquad (13.6)$$

Equation (13.6) is a very useful result and can be used in a variety of

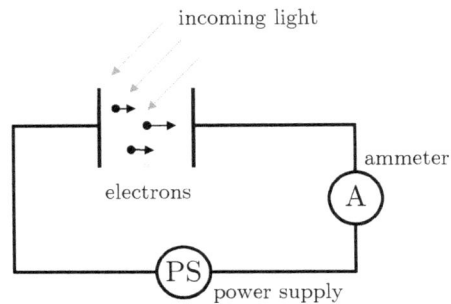

FIGURE 13.2

A circuit diagram for the photoelectric effect. Light is incident on one metal plate; if electrons are freed, they can move across the gap and create a current as they move through the wire and back to the original plate. A power supply can be adjusted to combat this current by setting up a potential difference ΔV across the plates and thereby slowing electrons traveling across the gap – or in the extreme case preventing them from completing the trip at all.

contexts. For instance, it can be used to determine an experimental value for h, because we control the frequency of light f, know ϕ (as long as we know the metal we're using), and experimentally determine ΔV_s. Problem 13.5 provides some sample data so you can do precisely this.

On the other hand, if we use the accepted value of h, then Equation (13.6) could be used, for instance, to identify an unknown work function ϕ. The following example demonstrates this approach.

---

[8]When the electrons leave the original plate, the plate is left with a positive charge, so a potential difference and electric field is established in the wire.

Example 13.1 The Photoelectric Effect ⋆

When working with a photelectric effect apparatus, a student uses ultraviolet light with $\lambda = 2.00 \times 10^2$ nm on an unknown metal. The current in the circuit reaches 0 when the power supply is toggled to 1.10 V. What is the work function of the metal? Does it correspond to any of the metals listed in Table 13.1?

We'll begin by converting from wavelength to frequency with Equation (13.1):

$$f = \frac{c}{\lambda} = \frac{3.00 \times 10^8 \text{ m/s}}{2.00 \times 10^{-7} \text{ m}} = 1.50 \times 10^{15} \text{ Hz}$$

Then we'll turn to Equation (13.6) to solve for the work function ϕ:

$$e\Delta V_s = hf - \phi$$
$$\phi = hf - e\Delta V_s$$
$$\phi = \left(6.63 \times 10^{-34} \text{Js}\right)\left(1.50 \times 10^{15} \text{ Hz}\right)$$
$$- \left(1.60 \times 10^{-19} \text{ C}\right)(1.10 \text{ V})$$
$$\phi = 8.19 \times 10^{-19} \text{ J} = 5.12 \text{ eV}$$

Comparing this to the table, we find that we are rather close to the value for copper.

13.2 Particles are Waves: Bragg Diffraction and the Wave Function

Before beginning this chapter you would probably claim that *light is a wave* and *electrons, neutrons, and protons are particles*. In the last section we saw that the first statement isn't true because a strict "light is a wave" perspective doesn't admit the possibility of the wave exchanging energy in discrete pieces (i.e. photons).

Well, it turns out the second statement isn't true either, because in certain cases electrons, protons, and neutrons behave like *waves!*[9] The classic experiment to see if something is a wave is to send it through a double slit apparatus, where the slits are narrow and narrowly spaced compared to the wavelength of the wave. It turns out that the wavelength of an electron is very small (we shall see that it depends on the electron's speed, but values as low as $\lambda = 0.01$ nm are readily achievable with modern equipment), and constructing a double slit apparatus to test the properties of electrons is technically challenging.

One way to sidestep this difficulty is to use naturally occurring crystals, where atom layers are very narrowly spaced. (In fact, the approach we will discuss here can be used to determine *how narrowly* they are spaced.) Some of the incoming wave (from a beam of electrons, say) will reflect off of the top layer of the crystal, but some will penetrate and then reflect off of the second (or deeper) layer. When the reflected waves rejoin at a common point (on an observation screen, say), they will have traveled slightly different distances – which,

---

[9]Thus the term *wave-particle duality* applies to electrons, protons, and neutrons (and even larger particles such as molecules, though we won't worry about those situations in this book) as well as light.

as you recall from our study of wave optics, is the entire idea![10] This phenomenon is referred to as **Bragg diffraction**.[11] Figure 13.3 shows the geometry.

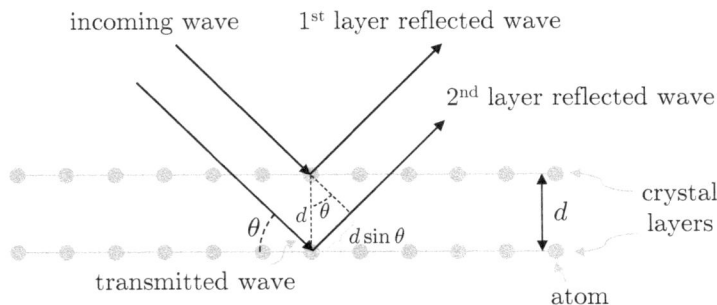

FIGURE 13.3
In Bragg diffraction, a wave impinges on the surface of a crystal. The portion of the wave that penetrates the top layer of the crystal travels an additional distance $d \sin \theta$ before interfacing with the second layer of the crystal. If the wave reflects, it travels another distance $d \sin \theta$ before it lines up with the portion of the wave that reflected off of the top layer. The total difference in path length is therefore $2d \sin \theta$. The figure shows the geometry on the right side of the reflection point, but the left side is symmetrically identical.

Indeed, if we analyze the geometry we see that the path length difference between beams reflecting from the top and second layers is given by $2d \sin \theta$. For constructive interference to occur on the screen, this (as usual) needs to be an integer multiple of the wavelength λ. Thus we have the so-called **Bragg condition**:

$$n\lambda = 2d \sin \theta \tag{13.7}$$

where n is a positive integer, d is the separation between crystal layers and θ is the *glancing angle* of the incoming wave (i.e. measured up from the plane of the crystal).

A particle's wavelength is related to its momentum according to the **de Broglie equation**:[12]

$$\lambda = \frac{h}{p} \tag{13.8}$$

where h is Planck's constant and $p = mv$ is the magnitude of the particle's momentum. As the following example demonstrates, this result is very useful when analyzing Bragg diffraction. Problem 13.11 provides some insight into *where* this comes from.

Example 13.2 Bragg Diffraction   ★★
Electrons are accelerated through a 3.00 kV potential difference and then impinge on a crystal with plane separation $d = 0.020$ nm. Find the lowest value of the glancing angle θ that will result in constructive interference on an observation screen.

---

[10]If the details are fuzzy, you may want to refer back to Chapter 8, but in brief the idea is that the waves will have traveled different distances and so be at different *phases*, admitting the possibility of constructive interference, destructive interference, or something in between.
[11]After William Henry Bragg (1862–1942) and William Lawrence Bragg (1890–1971), a father and son duo who shared the 1915 Nobel Prize in Physics for this discovery.
[12]After Louis de Broglie (1892–1987), who received the 1929 Nobel Prize in Physics for this work.

We'd like to know the wavelength of the electron so we can use Equation (13.7). This should draw our eye to Equation (13.8). However, we don't know the electron's momentum – just its accelerating voltage. Well, momentum and kinetic energy are related:

$$K = \frac{1}{2}mv^2 = \frac{p^2}{2m}$$

And we can relate the kinetic energy to the electric potential in the usual way:

$$e\Delta V = K$$

$$e\Delta V = \frac{p^2}{2m}$$

$$p = (2me\Delta V)^{1/2}$$

Plugging in, we find

$$p = \left(2\left(9.11 \times 10^{-31}\ \text{kg}\right)\left(1.60 \times 10^{-19}\ \text{C}\right)\left(3.00 \times 10^3\ \text{V}\right)\right)^{1/2}$$
$$p = 2.96 \times 10^{-23}\ \text{Ns}$$

We can now work out the wavelength:

$$\lambda = \frac{h}{p} = \frac{6.63 \times 10^{-34}\ \text{Js}}{2.96 \times 10^{-23}\ \text{Ns}} = 2.24 \times 10^{-11}\ \text{m} = 0.0224\ \text{nm}$$

With the wavelength we can finally turn to the Bragg condition to work out the values of θ:

$$n\lambda = 2d \sin\theta$$

$$\theta = \sin^{-1}\left(\frac{n\lambda}{2d}\right)$$

Plugging in λ, the given value of d, and $n = 1$, we find

$$\theta_1 = 0.59\ \text{rad} = 34°$$

Phew!

Bragg diffraction shows that electrons, protons, and neutrons have wavelike properties. But *what* exactly is it that is "waving"? In the case of light we now know that it is an electric field and a magnetic field, and in other situations the medium itself oscillates (air for sound, or in other cases perhaps a taut rope or sheet of rubber). When discussing the wavelike properties of protons, electrons, and neutrons, we refer to the quantum-mechanical **wave function**, which is expressed mathematically (depending on context) as Ψ or ψ.[13] In a very deep sense the wave function *is* the particle, and it is the wave function that oscillates: generally speaking, both in time (Ψ is different at a fixed location, instant by instant) and space (Ψ is different point by point, even at the same instant in time). Figure 13.4 sketches one way of visualizing the wave function for a traveling particle.

---

[13]The uppercase or lowercase Greek letter psi, pronounced "sigh". This is perhaps easy to remember because it is only natural to want to sigh now and again when reading about quantum mechanics.

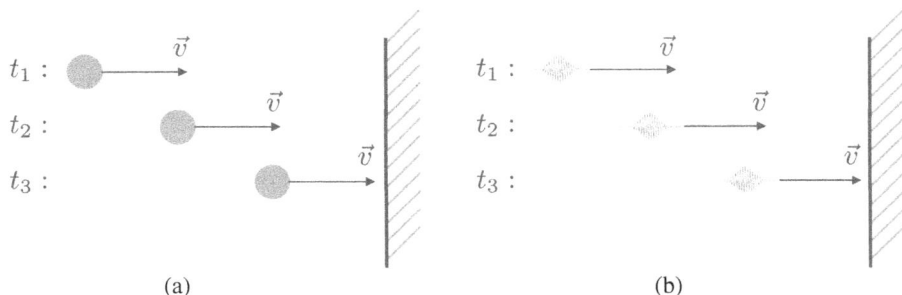

(a)   (b)

FIGURE 13.4
Two views of a traveling electron moving toward a detector. (a) The particle view: The electron is a little ball with a fixed position that moves, instant by instant. (b) The wave view: The electron is described by a wave, Ψ, that decays to 0 away from where the particle is expected to be classically. The electron can be said to exist with a non-exact location in the region where Ψ is not 0.

This oscillation is physically meaningful because if we know the complete mathematical form of a wave function for a particle then we can, at least in principle, determine anything that *can* be known about the particle (e.g. its momentum or energy) by going through the appropriate mathematics. The details are far beyond the scope of this book, but let me make a few descriptive comments.

First, the wave function can only tell us the *probability* of finding the particle at a certain location if we make a measurement. Another way of saying this is that if we want to pin down the location of an electron with increasing precision, at very small length scales we simply *can't*. Indeed, if you set up a series of *identical* experiments where an electron is trapped in a narrow region, then measure their positions, you'll get a range of possible outcomes that line up with the so-called *probability distribution* predicted from the wave function (Fig. 13.5). Now, you might think that our inability to say *exactly* where a given electron will be is due to some technical limitation with our equipment, but physicists have convincingly shown that this is instead because of the fundamental properties of the electron itself.[14]

Second, the details of a particle's wave equation depend, naturally enough, on its surroundings. The equation that fills in the details is called the **Schrödinger Equation**:[15] given the local potential energy (for example, due to a surrounding atomic lattice in a crystal or, more simply, a nearby nucleus) one can solve the Schrödinger Equation to determine the particle's wave function.[16] If you continue your study of quantum mechanics, you'll learn all about the mathematical techniques that are used to solve the Schrödinger Equation and analyze the wave function of particles in various important contexts.

For our present purposes, we shall keep in mind the wavelike nature of fundamental particles as we turn to how they interact in an atom.

---

[14]Or proton or neutron; I'm just referring to an electron here to be specific.

[15]After Erwin Schrödinger (1887–1961). Schrödinger received the 1933 Nobel Prize in Physics (with Paul Dirac, who you will learn more about if you continue to study quantum mechanics).

[16]This is somewhat analogous to solving Newton's Second Law in classical mechanics: once you know the net force on an object you can determine its acceleration and therefore characterize its subsequent motion, so long as you know its initial position and velocity.

Probability of measuring particle
to be at given location

edge of well   edge of well

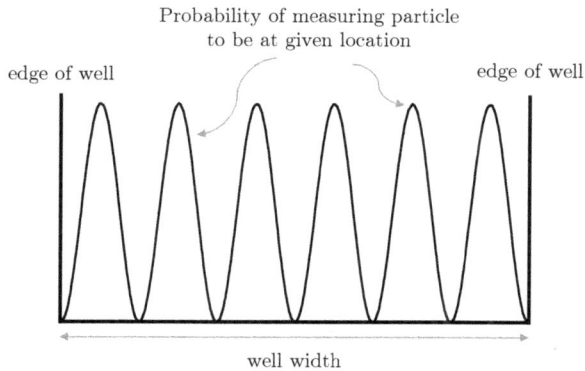

well width

FIGURE 13.5

A particle trapped in a narrow region (like an electron trapped in a crystal lattice) does not have a well-determined position. Instead, it is possible to determine the *probability* of measuring a certain position (when a measurement is made) from its wave function Ψ. As drawn, there are six positions that share the maximum probability, though this is just one possible configuration. (In general, take a sine wave scaled such that you have a node on either edge of the well, then *square* the sine wave so it is always positive. This is identical to saying you need an integer number of half-wavelengths in the original sine wave; as drawn there are six. Perhaps surprisingly, the original sine wave I am describing also describes a standing wave on a taut string! You will see how this solution arises from the mathematics of quantum mechanics if you continue to study the subject beyond this text.)

13.3 The Atom has a Nucleus: Rutherford Scattering

The Experiment

The fact that atoms have small, positively charged nuclei was discovered by an experiment that is rather simple to describe (even if it is challenging to carry out): send a narrow beam of positively charged particles toward a thin foil of gold, then detect where the particles impact a detector screen on the far side (Fig. 13.6). This experiment was first carried out in 1908 by Hans Geiger and Ernest Marsden, who were students of Ernest Rutherford.[17] The charged particles they used were **alpha particles**, which consist of two protons and two neutrons stuck together.[18]

One prevailing theory of the day was that atoms consist of negatively charged electrons that were embedded in a positively charged "paste".[19] Importantly, if atoms really *were*

---

[17]Geiger (1882–1945) went on to develop the **Geiger counter**, a device used to detect radiation. Marsden (1889–1970) was also involved in the study or radiation and nuclear physics for much of his career. Rutherford (1871–1937), who received the 1908 Nobel Prize in Chemistry, was a pioneer of nuclear physics and is widely regarded as one of the most influential physicists in history.

[18]These particles are emitted from radioactive substances during so-called **radioactive decay**. We won't study radioactivity in any depth in this book, but the basic idea is that nuclei are held together by the **strong force**, which overcomes the Coulombic repulsion between the positively charged protons in the nucleus. This balance isn't always perfect, and in some heavy elements the instability manifests in a number of ways – including alpha particles "chipping" off the parent nucleus.

[19]It isn't uncommon to hear analogies to fruit cake or jello salad here, since they both have bits of fruit (instead of electrons) embedded randomly inside. In fact, the British physicist J.J. Thomson (1856–1940), who received the 1906 Nobel Prize in Physics and proposed this model, called it the *plum pudding model*;

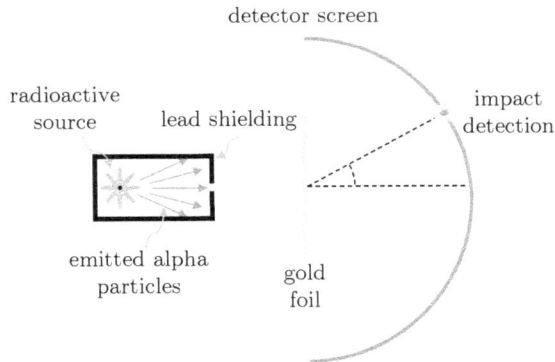

FIGURE 13.6

In the Rutherford experiment, a radioactive source is kept in a lead-lined container (radiation can be harmful to people and lead is a good absorber of radiation). A narrow hole allows radioactive particles to leave only if they have the appropriately aligned velocity (here, horizontal). The particles then travel through a thin foil of gold. Their path can be deflected as a result and their final path is detected based on their interactions with a detecting screen.

structured this way, then the electric field created by the (overall neutral) atom would, on average, be zero. This would suggest, in turn, that charged particles that passed even very close to atoms wouldn't be deflected (or at least not by much) due to a Coulombic interaction with the atom.

Instead, it was discovered that some of the incoming charged particles were deflected away from their original trajectory – some even up to 180°. Rutherford was able to show that the experimental results matched very nicely with what you would *expect* to find if the atom has a small positively charged nucleus: incoming particles that happened to travel close to the nucleus would experience significant deflection, whereas those that were farther away would experience very little.

Rutherford's Analysis

Suppose first that you have a positively charged particle Q (the nucleus) fixed in place, then fire a smaller positively charged particle q (and mass m) directly at it with some initial velocity $\vec{v}$. From our discussion of the Coulomb force and electric potential energy, we know that the traveling charge will slow down, stop, and then move away.

Indeed, we can determine the point where the traveling charge q briefly comes to rest, a distance $r_{\min}$ from charge Q. This occurs when the initial kinetic energy has been entirely converted into electric potential energy:

$$\frac{1}{2}mv^2 = q\Delta V$$

$$\frac{1}{2}mv^2 = q\left(\frac{KQ}{r_{\min}}\right)$$

$$r_{\min} = \frac{2KqQ}{mv^2} \tag{13.9}$$

---

this is still its most common name. (If you don't know what plum pudding is, I invite you to look into it.) To my American ear, the "brownies with nuts" model seems more fitting, though I doubt this name will catch on.

This all strikes you as reasonable, I hope (if not, perhaps refer back to Chapter 9!). Let's now make the problem somewhat more complicated: what if the incoming particle is instead shifted some perpendicular distance from its initial "collision course" alignment with the nucleus (Fig. 13.7)? We refer to this distance as the *impact parameter, b*. (This accounts for the fact that our beam of charged particles isn't perfectly narrow or perfectly aligned with a nucleus.)

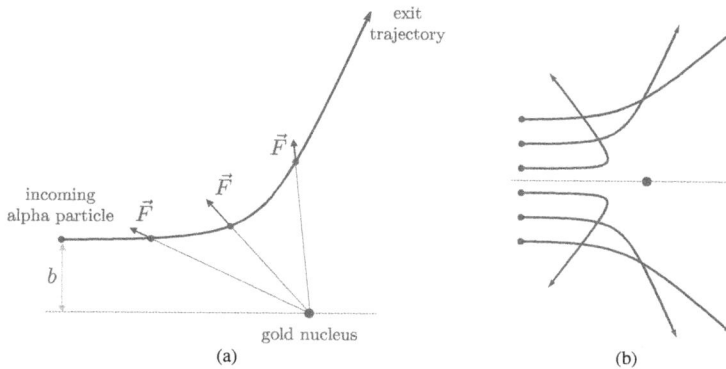

FIGURE 13.7

In Rutherford scattering, a stationary gold nucleus deflects the path of an incoming alpha particle. (a) A single alpha particle with an impact parameter b. The Coulomb force vector is indicated for a few positions along the alpha particle's path. (b) Multiple paths are shown for different impact parameters. Note that the deflection is more dramatic for smaller impact parameters.

Describing the precise path that charge q will take in this situation is quite challenging (especially when you recognize that the particles won't interact with just *one* nucleus as it passes through the foil, even if it is very thin), but the basic physics is still the same: the charges repel according to Coulomb's Law. Thus, as we've drawn the situation, charge q initially feels a force up and to the left, which means its horizontal velocity slows while it simultaneously picks up a vertical velocity. If we advance the position of the particle slightly and repeat the process, bit by bit we'll trace out a trajectory such as those shown in Figure 13.7.

Let's push this one step further. Suppose instead of a single charge q, you have whole *series* of charges (the experiment is run long enough for many charges to interact with the foil), each interacting with *many* nuclei (the foil is more than one atom thick, after all!). Through some more complicated mathematics, it is possible to work out how many particles you would expect to find deflected onto any given piece of a detecting screen on the far side of the foil. Carrying out this theoretical analysis – and then comparing it to the experimental data – is a lovely example of the synergy between theoretical and experimental physics.

The mathematical details of Rutherford's analysis are beyond what we shall consider in this book. That being said, the fact that atoms contain a positively charged nucleus is of fundamental importance. To go further, we will consider how *electrons* interact with the nucleus.

13.4 Atomic Electrons have Restricted Energies: Spectra

Perhaps the simplest way to model a negatively charged electron interacting with a positively charged nucleus is to suppose that the much lighter electrons orbit around the nucleus, much like planets orbits around a star (Fig. 13.8, top left). Indeed, we'll think through this in some mathematical detail shortly. However, we've already seen that electrons aren't tiny little planets but rather *waves* with indeterminate position, which suggests a more complicated view, perhaps along the lines of what is sketched in Figure 13.8, top right. This supposes that an electron exists as some sort of wave, but still at a fixed radius.

However, a full quantum-mechanical treatment with the Schrödinger equation shows that we can pin down *neither* the angular coordinate *nor* the radius: the electron exists as a sort of *probability cloud* around the nucleus. Moreover, there are multiple possible configurations, much like how any integer number of half-wavelengths can exist in the wave function that yields the probability distribution for an electron trapped in a well (Figure 13.5). The lowest two states for the electron in a hydrogen atom are shown in the bottom two panels of Figure 13.8.

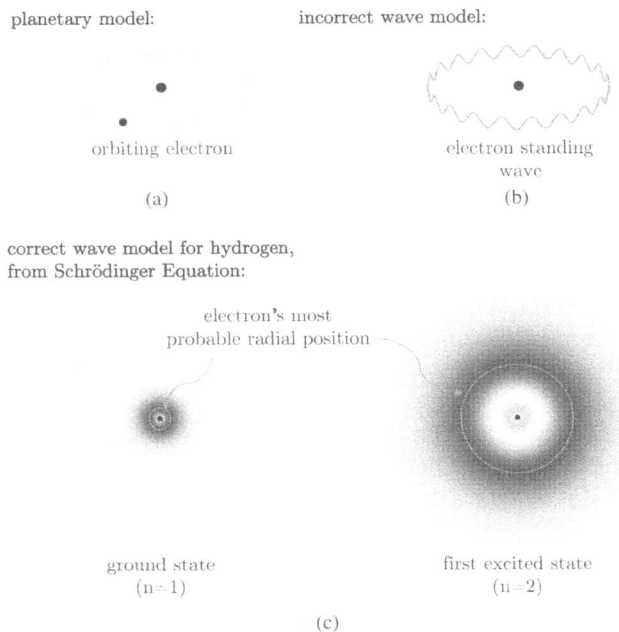

FIGURE 13.8

Models of the atom that correctly recognize the existence of the nucleus. (a) The (incorrect) *planetary model* where the electron is viewed as a particle orbiting around the nucleus like a planet around a star. (b) A (incorrect) *wave* model where the electron has an indeterminate position along the classical (planetary) orbit, similar to the situation discussed for an electron in a "well" as in Figure 13.5. (c) The Schrödinger equation gives the complete and correct solution. Here we show the simplest two states for the electron in hydrogen: the electron's probability distribution varies with radius as well as the angular coordinate (darker colors corresponds to higher probabilities of finding the electron at the given location if a measurement is made; the white dashed lines show the most probable radius). Both of these panels are actually cross-sections of a spherically symmetric distribution.

While we're not going to worry about solving the Schrödinger equation, we can actually capture some key results by returning to the planetary model. Our analysis will involve a few intuitive leaps initially proposed by Niels Bohr,[20] so we will be considering what is commonly referred to as the **Bohr model** of the atom.

The Bohr Model

Suppose an electron (mass m_e and charge $-e$) is orbiting around a proton (mass m_p and charge e) like a planet around a star – except here it is the Coulombic attraction between the particles that holds the electron in orbit, rather than the gravitational attraction between planetary bodies. This analysis can be a bit confusing at first, because we will need to analyze the situation from the perspective of angular momentum, potential energy, and kinetic energy – and then combine the results in a few different ways.

Before getting into the details, then, let me simply state the results. According to the Bohr model, the electron orbits the proton at only certain restricted radii given by

$$r_n = n^2 a_B, n = 1, 2, 3... \tag{13.10}$$

where a_B is the **Bohr radius** with numerical value

$$a_B = 5.29 \times 10^{-11} \text{ m} \tag{13.11}$$

Along with this, the Bohr model predicts that each allowed orbit has a corresponding energy given by

$$E_n = \frac{-E_R}{n^2}, n = 1, 2, 3... \tag{13.12}$$

where E_R is the **Rydberg energy**[21] with numerical value

$$E_R = 2.18 \times 10^{-18} \text{ J} = 13.6 \text{ eV} \tag{13.13}$$

Figure 13.9 sketches the allowed orbits in the Bohr model and Figure 13.10 sketches the energies. The success of the Bohr model is in describing what happens when the electron in hydrogen *transitions* between these allowed radii and energy levels. Before discussing this, however, let me describe how these equations are *derived*.

Analyzing the Bohr Model: Angular Momentum

Planck's constant has units of angular momentum (see Problem 13.16) and, as we have seen, Planck's constant comes up in the context of *quantization*. We will therefore suppose that **the angular momentum of an atomic electron is quantized.**[22] Specifically, we will suppose it can only have values given by

$$L = n \left(\frac{h}{2\pi} \right), n = 1, 2, 3... \tag{13.14}$$

---

[20](1885–1962) Bohr received the 1922 Nobel Prize in Physics for his work on atomic physics. He also contributed to the famous Manhattan Project during World War II.

[21]After Johannes Rydberg (1854–1919).

[22]You might be wondering "Why would we do this?", to which I respond "because it turns out to be give accurate predictions". We're glossing over the historical nuance here, but the Bohr model, which was formulated in 1913, was a first step in explaining experimental results for which no satisfying theory existed. The Schrödinger equation came later, in 1926.

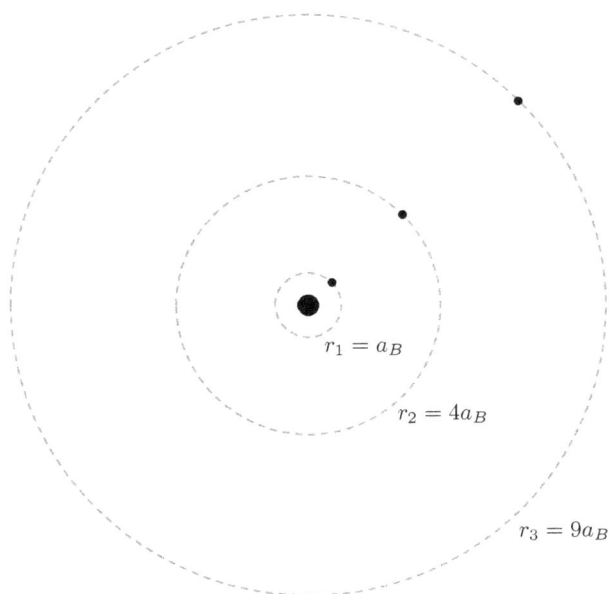

FIGURE 13.9
The first three electron orbits according to the Bohr model.

Let us combine this assumption with the classical equation for the angular momentum of a particle with linear momentum $\vec{p}$ and position given by the vector $\vec{r}$:

$$L = |\vec{r} \times \vec{p}| \tag{13.15}$$

If the object is in a circular orbit, then $\vec{r}$ and $\vec{p}$ are perpendicular, so

$$L = m_e vr \tag{13.16}$$

Combining Equations (13.14) and (13.16), we have

$$m_e vr = n\left(\frac{h}{2\pi}\right) \implies v = n\left(\frac{h}{2\pi m_e r}\right) \tag{13.17}$$

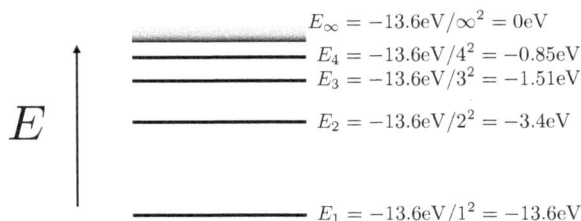

FIGURE 13.10
A schematic of the energy levels for hydrogen. Each level is represented with a horizontal line; the spacing between the lines corresponds to the difference in the energy levels. As n increases, the energy levels become increasingly close together, as indicated by the shading at the top of the figure. The fact that the energies are negative indicates that the electron is bound to the electron; the atom becomes *ionized* (i.e. it loses the electron) if the electron receives enough energy to reach the $E = 0$ state.

We will use this important result along with an analysis of the *energy* associated with the orbiting electron: first potential and then kinetic.

Analyzing the Bohr Model: Electric Potential Energy

The electric potential energy of the two charges (e and $-e$) a distance r from each other is given by

$$U = e\Delta V = \frac{1}{4\pi\varepsilon_0}\left(\frac{-e^2}{r}\right) \tag{13.18}$$

We have used the potential difference of a point charge and written the electrostatic constant in terms of the permittivity of free space ($K = 1/4\pi\varepsilon_0$) so we can reserve the symbol K for kinetic energy. OK, fair enough. Let's turn to kinetic energy.

Analyzing the Bohr Model: Kinetic Energy

To analyze the electron's kinetic energy, we'll start with *force* analysis, keeping in mind that it is the Coulomb force that keeps the electron in a circular orbit (where $a_c = m_e v^2/r$):

$$\vec{F}_{\text{NET}} = m_e \vec{a_c}$$

$$\frac{1}{4\pi\varepsilon_0}\frac{e^2}{r^2} = m_e \frac{v^2}{r}$$

$$\frac{1}{4\pi\varepsilon_0}\frac{e^2}{r} = m_e v^2 \tag{13.19}$$

In the last line I have cancelled a factor of r from the denominator on both sides. Now, if we recall $K = mv^2/2$, then our last line above can be rewritten

$$\frac{1}{4\pi\varepsilon_0}\frac{e^2}{r} = 2K \tag{13.20}$$

We're now in a position to combine our results: Equations (13.17), (13.18), (13.19), and (13.20).

Analyzing the Bohr Model: Allowed Radii

If we insert Equation (13.17) for v into Equation (13.19), we find

$$\frac{1}{4\pi\varepsilon_0}\frac{e^2}{r} = m_e\left(n\frac{h}{2\pi m_e r}\right)^2$$

$$r = n^2\left(\frac{h^2\varepsilon_0}{\pi m_e e^2}\right)$$

$$r = n^2 a_B \tag{13.21}$$

In this last line we have replaced the term in parentheses (which entirely includes known constants) with the Bohr radius a_B.

This is the result we stated above (Equation (13.10)); it says that an electron in a hydrogen atom can only "live" at certain allowed radii, which is determined by the allowed values of n ($= 1, 2, 3...$): $r_1 = a_B$, $r_2 = 4a_B$, $r_3 = 9a_B$, and so on.

Analyzing the Bohr Model: Allowed Energies

Note that the left side of Equation (13.20) is the *same* as the right side of Equation (13.18) except for the negative sign. In other words,

$$2K = -U$$

Now, the electron's *total* energy E is the sum of its electric and potential energies, so

$$E = K + U = \left(-\frac{U}{2}\right) + U = \frac{U}{2}$$

$$E = \frac{1}{8\pi\varepsilon_0}\left(\frac{-e^2}{r}\right)$$

But we have now derived the equation for the allowed values of r (Equation (13.10) or equivalently Equation (13.21)). Inserting, we find

$$E = \frac{1}{8\pi\varepsilon_0}\left(\frac{-e^2}{n^2 a_B}\right)$$

$$E = \frac{-E_R}{n^2} \tag{13.22}$$

In the last line I've grouped together the constants into the Rydberg energy E_R (check that the values work out if you like – Problem 13.17); we've arrived at Equation (13.12) proposed above.

Thus, just as the Bohr model claims that the electron can exist in only *certain orbits* (Equation (13.10)), it also claims that *each orbit has a certain energy* (Equation (13.12)).

The Bohr Model Predicts the Spectrum of Hydrogen

According to the Bohr model, the electron in a hydrogen atom is only *allowed* to orbit the proton with certain radii; each allowed radius has a corresponding energy. Now, as I stated above, the Bohr model is *wrong*; central to our analysis was the idea that the electron orbits the proton. Instead, as we saw in Figure 13.8, the electron exists as a kind of probability cloud around the proton. However, Equation (13.12) *is* correct: we get the same result with a full quantum-mechanical treatment.[23]

Now, the electron can *transition* between these allowed states (in the Bohr model, the electron would be changing the radius of its orbit; in the full Schrödinger model, the probability cloud is changing configurations). How does the electron's energy change when this happens? Well, according to Equation (13.12) we have

$$\Delta E = E_f - E_i$$

$$\Delta E = \left(\frac{-E_R}{n_f^2}\right) - \left(\frac{-E_R}{n_i^2}\right)$$

$$\Delta E = -E_R\left(\frac{1}{n_f^2} - \frac{1}{n_i^2}\right) \tag{13.23}$$

For instance, if $n_i = 1$ and $n_f = 2$, $\Delta E = 10.2\text{eV}$; the electron's energy *increases*. However, *energy must be conserved*. Where does this energy come from?

One answer is *from a photon*! Just as an electron can be freed from metal by absorbing a photon with sufficient energy, so too can an atomic electron absorb a photon and move to a higher energy level as a result.

Importantly, though, the photon *must* have an energy that corresponds exactly to the so-called *gap* between the electron's current energy level and a higher energy level

---

[23]Interestingly, Equation (13.10) has a bit of truth to it as well: the most probable radius in the ground state (the dashed circle in the lower left panel of Figure 13.8) is the Bohr radius a_B.

(Fig. 13.10). If the photon doesn't have an appropriate energy, it won't be absorbed.[24] Now, the absorption process we're describing can also run in *reverse*: an electron that has been excited to an energy level $n > 1$ can "fall" back down to a lower energy level (in any order, so for instance $3 \to 1$ is possible and so too is the two-step process $3 \to 2 \to 1$). According to conservation of energy, at every step a photon with the appropriate energy level is *emitted* rather that *absorbed*.

This can seem rather esoteric, so let me describe an important experiment involving this phenomenon. If you have a container of hydrogen gas and shine white light (containing a mixture of all wavelengths) through it, *light with the allowed wavelengths will be absorbed*, and all other light will pass through. This is referred to as the **absorption spectrum** of the gas. Similarly, if you *excite* the gas,[25] *light with the allowed wavelengths will be emitted*. This is referred to as the **emission spectrum** of the gas.

In 1888, the spectrum of hydrogen was shown to obey the mathematical relationship known as the **Balmer–Rydberg formula:**[26]

$$\frac{1}{\lambda} = R \left(\frac{1}{n_1^2} - \frac{1}{n_2^2} \right) \tag{13.24}$$

where $n_2 > n_1$ and both are positive integers (so $n_1 = 1, 2, 3...$ and $n_2 = 2, 3, ...$) and R is the **Rydberg constant** with numerical value

$$R = 1.10 \times 10^7 \text{ m}^{-1} \tag{13.25}$$

Does Equation (13.24) look somewhat familiar? It should! It doesn't take much work to show that the result from the Bohr Model (Equation (13.23)) matches this result (see Problem 13.18). Bohr's model, then, provides a theoretical grounding for the Balmer–Rydberg formula.

Example 13.3 The Spectrum of Hydrogen ⋆

The *Lyman series* of hydrogen refers to transitions to the $n_f = 1$ state from an excited state $n_i > 1$. Similarly, the *Balmer series* considers $n_f = 2$ and the *Paschen series* considers $n_f = 3$. What is the lowest and highest wavelength in each of these series? Which of the series overlaps with the visible spectrum (400 to 700 nm)?

We employ Equation (13.24), where n_1 is set equal to n_f (otherwise we would find a negative wavelength). The limiting cases are $n_2 = n_1 + 1$ (the closest allowed value) and $n_2 = \infty$ (the extreme case).

For instance, in the Lyman series, we have

$$\frac{1}{\lambda_{1,2}} = R \left(1 - \frac{1}{2^2} \right) \implies \lambda_{1,2} = 1.21 \times 10^{-7} \text{ m} = 121 \text{ nm}$$

and

$$\frac{1}{\lambda_{1,\infty}} = R \left(1 - 0 \right) \implies \lambda_{1,\infty} = 9.09 \times 10^{-8} \text{ m} = 90.9 \text{ nm}$$

This is entirely outside the visible spectrum. Going through the same process for the Balmer series, we find

$$\lambda_{2,3} = 655 \text{ nm}$$
$$\lambda_{2,\infty} = 364 \text{ nm}$$

---

[24]It isn't possible for the electron to absorb just some of the photon's energy because photons are indivisible.

[25]For instance, a neon light uses an electric potential difference across the gas tube to essentially jostle neon atoms into one another; it is the atomic collisions that excite the electrons to higher energy levels.

[26]After Johann Balmer (1825–1898) and Johannes Rydberg (1854–1919).

which overlaps with much of the visible spectrum (indeed, you can show that four of the lines in this series are in the visible spectrum, while the rest fall below it – see Problem 13.22). Finally, for the Paschen series

$$\lambda_{3,4} = 1870 \text{ nm}$$
$$\lambda_{3,\infty} = 818 \text{ nm}$$

which is again outside of the visible spectrum.

13.5   Summary

In this chapter we've seen that at the smallest scales, light and fundamental particles (electrons, neutrons, and protons) exhibit *wave-particle duality*. We've studied the structure of the atom and seen that the Bohr model is successful insofar as it arrives at the correct formula (Equation (13.12)) for the allowed energies of the electron in hydrogen – and this, in turn, correctly predicts the spectrum of hydrogen. However, the Bohr model *fails* to explain more complicated situations, for example atoms that include *multiple electrons*.

To go further, then, we need the full machinery of quantum mechanics (prominently including the Schrödinger equation). I would like to stress the practical importance of quantum mechanics because it can often strike people as weird, quasi-magical, and largely useless. However, at least the last of these is false. As just a few examples, quantum mechanics is immensely important for the following:

- *Chemistry*: The interatomic interactions that allow molecules to exist involves interactions between the electron "probability clouds" that we have been discussing, and indeed the exact structure of these clouds in a given atom has a lot to say about how it can interact with other atoms to form molecules. The field of *physical chemistry*, in particular, leverages quantum mechanics to understand complicated chemical phenomena – but you will likely see electron cloud diagrams even in introductory chemistry.

- *Computing*: Traditional computer processors are becoming so densely packed with electronic components (for instance, to increase the capabilities of a small phone) that quantum-mechanical interactions can be an important consideration. Perhaps more strikingly, the field of **quantum computing** is currently a major area of research. While traditional computers fundamentally use a "low Voltage = OFF" and "high voltage = ON" framework to store information and perform computations, quantum computers aim to use quantum mechanical properties that have more than two states and therefore stand to perform some very complicated tasks much (much much) more rapidly than traditional computers.

- *Imaging*: Advanced microscopes rely on quantum-mechanical phenomena including the ability for electrons to *tunnel* from a very small probe, through the air, and into some material that is being imaged. The resolution that can be achieved in this way is far beyond anything that can be performed with a traditional microscope, which uses light in the visible spectrum. In a very different context, quantum mechanics is essential to medical imaging technologies including Magnetic Resonance Imaging (MRI), which can safely scan a patient's organs.

Regardless of whether or not you go on to study quantum mechanics at a more advanced level, I hope you appreciate its importance! And even if this is the end of your formal training in physics, I encourage you to do some casual reading on any or all of these topics–and topics from earlier in this book, for that matter! Physics–which I might poetically define as the scientific study of the innate beauty of the Universe–speaks to something fundamental in all of us.

13.6 Problems for Chapter 13

13.1 Light can Behave Like a Particle: The Photoelectric Effect

⋆ **Problem 13.1.** Determine the energy (in eV and in J) of a photon with the following frequencies and wavelengths:

a. $\lambda = 5.00 \times 10^2$ nm (green light)

b. $f = 1.00 \times 10^2$ MHz (a radio wave on the FM band)

c. $f = 2.45$ GHz (the frequency of a microwave oven)

d. $\lambda = 2.00 \times 10^2$ nm (UV light)

e. $\lambda = 2.0$ nm (an X-ray)

f. $\lambda = 1.0 \times 10^{-12}$ m (a gamma ray)

⋆ **Problem 13.2.** For any metal, there are certain frequencies of light that *cannot* free an electron. What are these frequencies for (a) gold and (b) nickel?

⋆ **Problem 13.3.** A student shines ultraviolet light with $\lambda = 1.50 \times 10^2$ nm on an unknown metal. The current in the ammeter reaches 0 when the power supply is toggled to 2.0 V. What is the work function of the metal? Express your answer in (a) J and (b) eV.

⋆⋆ **Problem 13.4.** What stopping potential ΔV_s would just barely prevent a current from flowing in the photoelectric effect experiment when light with frequency 6.0×10^{15} Hz is shining on a zinc plate?

⋆⋆ **Problem 13.5.** You perform the photoelectric effect experiment with a block of silver. By varying the wavelength of light shining on the metal, you determine the potential that stops the flow of current through the ammeter. Your data is shown in Table 13.2.

a. Make a graph of ΔV_s vs. f.

b. Estimate an experimental value of h from the data. Compare to the accepted value of h.

c. Use a computer (or sufficiently sophisticated calculator) to calculate a line of best fit. From the fit, determine the experimental value for h and compare to the accepted value.

TABLE 13.2
Problem 13.5

Wavelength (nm)	ΔV_s (V)
200.0	2.60
150.0	4.63
100.0	8.83
50.0	21.23

⋆⋆   **Problem 13.6.** When performing the photoelectric effect experiment, one slowly increases the power supply voltage ΔV_p until no current flows, through the ammeter at which point the power supply is at the stopping potential: $\Delta V_p = \Delta V_s$. Is it possible to determine (or at least get a good estimate for) the stopping potential by instead recording the power supply potential and the ammeter current for a few low potentials? Explain how (or why not) using mathematics as well as a written explanation.

13.2 Particles are Waves: Bragg Diffraction and the Wave Function

⋆   **Problem 13.7.** An X-ray with a wavelength of 1.5 nm impinges on a crystal with crystal spacing of 3.0 nm. At what glancing angles will you observe the (a) $n = 1$ and (b) $n = 2$ points of constructive interference?

⋆   **Problem 13.8.** An X-ray with a wavelength of 0.20 nm impinges on a crystal. The $n = 1$ point of constructive interference is observed at an angle of 5.0°. What is the crystal plane spacing?

⋆⋆   **Problem 13.9.** The Bragg condition (Equation (13.7)) supposes the interference occurs between a wave reflecting off of the surface (i.e. the first layer) of the crystal and the second layer (Fig. 13.3). This is the most prominent interference because most of the wave's reflected energy is off of one of these layers. However, it is possible for some of the wave to reflect off of the *third* layer.

   a. What is the "Bragg condition" for interference between waves reflecting from the first and third layers of a crystal? (Below I refer to this as "$(1, 3)$" interference.)

   b. Repeat Problem 13.7 but use your result from (a) to assess the $(1, 3)$ interference (rather than the usual $(1, 2)$ interference assessed with Equation (13.7)).

   c. Is it possible for a $(1, 2)$ interference to occur at the same angle θ as a $(1, 3)$ interference for the same energy X-ray on the same crystal? If *yes*, how are the orders $n_{1,2}$ and $n_{1,3}$ related? If *no*, why not?

⋆   **Problem 13.10.** The de Broglie equation $\lambda = h/p$ (Equation (13.8)) is startling because there isn't anything in our everyday experience to suggest that particles have a wavelength. Indeed, the de Broglie equation suggests that the wavelength of particles on the scale of everyday life *shouldn't* be evident:

   a. What is the wavelength of a grain of sand ($m = 5.0 \times 10^{-5}$ g) moving at 1.0 m/s?

   b. What is the wavelength of a car ($m = 1.2 \times 10^3$ kg) moving at 22 m/s?

   c. What is the wavelength of an electron moving at $0.05c$?

⋆   **Problem 13.11.** Use Equations (13.4) and (13.5) to show that the de Broglie relation (Equation (13.8)) is true for photons. (Equations (13.4) and (13.5) follow for massless photons from special relativity, and the so-called **de Broglie hypothesis** supposed – correctly – that the result also holds for particles with mass.)

13.3 The Atom has a Nucleus: Rutherford Scattering

⋆   **Problem 13.12.** In a Rutherford scattering experiment, an alpha particle has an initial energy of 4.0 MeV. If the particle is moving directly toward the nucleus of a gold atom (with 79 protons):

   a. How close will it get to the nucleus before stopping?

   b. What is the magnitude of the force on the alpha particle when it is as close as it gets to the nucleus?

⋆   **Problem 13.13.** What initial energy for an alpha particle would result in the particle coming within 7.0×10^{-15} m of a gold nucleus (containing 79 protons), assuming the alpha particle is moving directly toward the nucleus?[27]

---

[27]This is the approximate radius of the nucleus. Around this approximate radius, Rutherford's analysis breaks down–which means bombarding nuclei with energetic particles like this can yield a sense of the radius of the nucleus. I should note though that there isn't a *hard* cutoff where Rutherford's formula breaks down because the nucleus is inherently quantum mechanical rather than a firm surface like, say, a bowling ball.

★ **Problem 13.14.** The point of using a heavy element like gold in the Rutherford scattering experiment is to increase the frequency with which alpha particles experience a large deflection. Why would gold lead to larger deflections than a metal such as, say, aluminum?

★★ **Problem 13.15.** Consider an alpha particle with initial kinetic energy K moving in the $+\hat{x}$ direction as it approaches a gold nucleus with an impact parameter $b > 0$. Suppose b is sufficiently large so the alpha particle passes by essentially undeflected (so θ is a *small angle* in Figure 13.6).

a. Set up an integral for the total impulse in the $\hat{x}$ direction as the alpha particle passes by the nucleus. Support your work with a clear diagram.

b. Set up an integral for the total impulse in the $\hat{y}$ direction as the particle passes by the nucleus. Support your work with a clear diagram.

c. Solve the integrals you set up in (a) and (b). Your answers will include fundamental constants, the atomic number, Z, of gold (i.e. the number of protons in a gold nucleus), and the velocity v of the alpha particle.

d. Use your results from (c) to determine a numerical value for b that yields a deflection of $1°$ if $K = 5.0$ MeV. Express your result as a multiple of the radius of a gold nucleus, $r_{Au} = 7.0 \times 10^{-15}$ m.[28]

13.4 Atomic Electrons have Restricted Energies: Spectra

★★ **Problem 13.16.** We claimed that Planck's constant has units of angular momentum. To show why:

a. Consider Equation (13.4), solve it for h, and assess the units on the other side of the equation.

b. Compare these units to the units for angular momentum according to Equation (4.33).

★ **Problem 13.17.** The Rydberg energy E_R is given by Equation (13.13):

$$E_R = 2.18 \times 10^{-18} \text{J} = 13.6 \text{ eV}$$

Surrounding Equation (13.22) we claimed $E_R = \frac{1}{8\pi\varepsilon_0} \left(\frac{e^2}{a_B} \right)$. Evaluate the right hand side of this equation to show that the result agrees with Equation (13.13).

★★ **Problem 13.18.** Show that Equations (13.23) and (13.24) are equivalent.

★ **Problem 13.19.** An electron in a hydrogen atom transitions from the $n = 1$ state to the $n = 3$ state.

a. Was energy absorbed by the electron or emitted by the electron in this process?

b. How much energy was absorbed/emitted?

★ **Problem 13.20.** An electron in a hydrogen atom is initially in the $n = 4$ state. A long time later it is back in the $n = 1$ state. List *all possible* energies that could be detected as this transition occurs.

★ **Problem 13.21.** The most prominent visible wavelength given off by the Sun corresponds to a transition of the electrons in hydrogen between the $n = 2$ and $n = 3$ levels.

a. What wavelength is this?

b. What color light is this (red, blue...)?

c. Which series (e.g. Lyman) is this?

★ **Problem 13.22.** Determine the wavelengths in the Balmer series for hydrogen that fall in the visible spectrum (i.e. between 400 nm and 700 nm).

---

[28]Your answer here will be an approximation only. Nonetheless, this approach gets rather close to a more exact answer while keeping the mathematics comparatively simple.

Additional Problems for Chapter 13

⋆⋆⋆ **Problem 13.23.** We mentioned special relativity in our discussion of the photon, and while we're not taking the time to delve too deeply into the subject, here we will consider one particularly famous result. Special relativity is important for understanding motion at speeds near c (these are appropriately called *relativistic speeds*). It turns out that the energy of a relativistic particle with mass m is given by

$$E = \gamma m c^2$$

where γ is the **Lorentz factor** that depends on the particle's speed v:

$$\gamma = \left(1 - \left(\frac{v}{c} \right)^2 \right)^{-1/2}$$

a. A particle that has mass cannot reach the speed of light. Why not?

b. For a particle moving much less than the speed of light ($v \ll c$), show that the energy is approximately given by $E \approx mc^2 + \frac{1}{2}mv^2$. (Thus a particle at rest still has **rest energy** $E = mc^2$, Einstein's famous equation that was of critical importance for the development of nuclear energy.)

c. We learned that the (kinetic) energy of a particle is $\frac{1}{2}mv^2$. Why doesn't the term $E = mc^2$ matter for "every day" physics?

Part VI

Appendices

A

Preliminaries

To effectively describe the world around us, we need to be able to use the precise language of mathematics. Before we get to the *physics*, therefore, this preliminary chapter presents an overview of some of the key background material that we will use more-or-less right out of the gate. Depending on your background, you may want to spend some time working carefully through these topics–or a quick skim might suffice as a refresher. In any event, as you work through the rest of the book you can refer back to this chapter as needed.

Learning Objectives

After reading this chapter, you should be able to:

- Identify the basic units from the International System of Units (SI).

- Convert numbers between standard notation and scientific notation.

- Identify the number of significant figures in a number and use significant figures when reporting the result of a mathematical calculation.

- Convert quantities between different metric prefixes (e.g. centimeters to meters) and different units (e.g. centimeters to inches).

- Solve an algebraic equation or a system of algebraic equations.

- Use basic trigonometric functions, the Pythagorean Theorem, and principles of geometry to analyze geometric configurations, especially right triangles.

- Describe common methods of representing scientific data (e.g. in a table or with a diagram).

- Describe effective problem solving strategies in physics.

- Define what is meant by a *physical model*.

A.1   Representing Physical Quantities

Units

In characterizing the world around us, we make many *types* of measurements: what you can measure with a ruler (a *length* or distance) is quite different from what you can measure with a thermometer (a *temperature*)! Moreover, even for the same type of measurement there are many different *units*: for example, a ruler might report a distance in centimeters, inches, and/or feet.

In this book we will almost exclusively use the International System of Units, which is abbreviated "SI".[1] The base units of the SI system are shown in Table A.1. These base units can be combined to form other *derived* units that represent other physical quantities of interest. We will explore many of them as we proceed through this book.[2]

TABLE A.1

SI quantities and units

Quantity	Unit	Abbreviation
Length	Meter	m
Mass	Kilogram	kg
Time	Second	s
Electric Current	Ampere	A
Amount of Substance	Mole	mol
Temperature	Kelvin	K
Luminous Intensity	Candela	cd

Example A.1 SI Units   ⋆

The SI unit for *force* is the Newton (N). In terms of the basic quantities in the SI system, the Newton has units of "mass × length / time^2". What are the basic SI *units* for each of these quantities? How can a Newton be expressed in terms of basic SI abbreviations?

Referring to Table A.1, we see that "mass × length / time^2" has units of kg m/s^2. Thus

$$1 \text{ N} = 1\frac{\text{kg m}}{\text{s}^2}$$

Scientific Notation

In exploring the physical world, we will deal with very small and very large numbers, and it can be tedious to write (and hard to read) them in **standard notation**. For instance, the mass of the Earth is about

$$m_{\text{Earth}} = 5,972,000,000,000,000,000,000,000 \text{ kg}$$

A much more concise way to write this is

$$m_{\text{Earth}} = 5.972 \times 10^{24} \text{ kg}$$

Here we are saying "multiply 5.972 by ten raised to the twenty-fourth power", which amounts to moving the decimal to the right 24 times to get back to the standard form of the number.[3]

Numbers written in this way are in **scientific notation**. Formally, a number written in scientific notation:

1. Starts with one digit (which can't be 0) before the decimal point. There can be any number of digits (including none at all) to the right of the decimal point.

2. Ends with "×10^n" where n is an integer that tells you the number of places to move the decimal to get back to the standard form of the number.

---

[1]The use of SI rather than IS comes from the French, *Systeme international d'unites*. The SI system is coordinated by the International Bureau of Weights and Measures.

[2]For instance, the SI unit of pressure is the Pascal (which has units of force per area or mass / $\left(\text{length} \times \text{time}^2\right)$), and the SI unit of charge is the Coulomb (which has units of current × time).

[3]For example, 3.6×10^1 is just 36.

If $n > 0$, the decimal place is moved to the right (you're multiplying by powers of ten). If $n < 0$ the decimal place is moved to the left (you're *dividing* by powers of ten[4]).

Example A.2 Scientific and Standard Notation ★

Determine if each of the following numbers is in scientific notation, standard notation, or neither. Express each number in both forms. (a) 412 (b) 30.20×10^{-3} (c) 0.00290×10^4 (d) 1.65×10^2.

(a) is in standard notation; note that it breaks both of the rules for scientific notation. Rewriting in scientific notation involves moving the decimal to the left two places:

$$412 = 4.12 \times 10^2$$

(b) is in neither standard nor scientific notation: there are *two* digits to the left of the decimal (so it isn't proper scientific notation) but there is still a multiplicative power of ten (so it isn't standard notation). We need to slide the decimal one more spot to the left to get it into scientific notation, which increases the power of ten by one:

$$30.20 \times 10^{-3} = 3.020 \times 10^{-2} = 0.03020$$

(c) is not in scientific notation because the digit to the left of the decimal is a 0; like (b) it is not in standard notation because of the multiplicative power of ten. To get into scientific notation we need to move the decimal to the right three spots, which lowers the power of 10 from 4 to 1:

$$0.00290 \times 10^4 = 2.90 \times 10^1 = 29.0$$

(d) is in scientific notation; moving the decimal two spots to the right puts it in standard notation:

$$1.65 \times 10^2 = 165$$

Metric Prefixes

Writing numbers in scientific notation is concise, but reading them out loud can be a mouthful. It is therefore convenient to use metric prefixes, which absorb the "powers of ten". For instance, one kilogram is equal to 10^3 grams:

$$1 \times 10^3 \text{ g} = 1 \text{ kg}$$

Table A.2 Lists common metric prefixes.

Example A.3 Converting Prefixes ★

How many picoseconds are in a second? How many seconds are in a picosecond?

If you're not comfortably with prefixes, it is easy to get these mixed up. By definition, we have

$$1 \text{ ps} = 10^{-12} \text{ s}$$

This answers the second question: 1 picosecond is 10^{-12} seconds, or in other words there are 10^{-12} seconds in a picosecond. To answer the *first* question we can multiply

---

[4]Because $10^{-n} = 1/10^n$. For example, $4 \times 10^{-2} = 4/10^2 = 4/100 = 0.04$.

both sides of this equation by 10^{12} to yield

$$10^{12} \text{ ps} = 1 \text{ s}$$

In other words, there are 10^{12} picoseconds in 1 second.

Unit Conversions

While we'll mostly work in the SI system of units, other systems of units certainly exist and it is important to be able to translate; to do so we need to know the relevant conversion factor. For instance, distances can be measured in inches (in), and the conversion factor to the SI unit of length is

$$1 \text{ in} = 2.54 \text{ cm} = 2.54 \times 10^{-2} \text{ m}$$

Now, for any equation $x = y$ it follows $x/y = 1$, which means that we can carry out a unit conversion by multiplying by the conversion factor ratio.[5] For instance, to determine the height of a 42-inch-tall child in centimeters:

$$42 \text{ in} \times \frac{2.54 \text{ cm}}{1 \text{ in}} = 110 \text{ cm}$$

Notice that the conversion factor is expressed with inches in the denominator, so the inches unit algebraically cancels: in $\times$ cm/in = cm.[6] Importantly, we have found that 42 inches *carries the same meaning* as 110 cm: they represent the *same* length, just with different units. Common unit conversions are provided in Appendix B; for now we'll consider one more example before moving on.

TABLE A.2

Prefix	Symbol	Factor
peta-	P	10^{15}
tera-	T	10^{12}
giga-	G	10^{9}
mega-	M	10^{6}
kilo-	k	10^{3}
hecto-	h	10^{2}
deka-	da	10^{1}
deci-	d	10^{-1}
centi-	c	10^{-2}
milli-	m	10^{-3}
micro-	μ	10^{-6}
nano-	n	10^{-9}
pico-	p	10^{-12}
femto-	f	10^{-15}

Example A.4 Cubic Centimeters in a 2 Liter Bottle ★

A liter (L) is defined to be the volume occupied by 1 kg of water under standard temperature and pressure (often abbreviated "STP"; we'll learn more about this in Chapter 5), which is equivalent to 1×10^{-3} m^3. How many cubic centimeters are present in a 2.0 liter bottle of your favorite carbonated beverage?

We'll carry this out in two steps: first from liters to cubic meters, then from cubic meters to cubic centimeters. The first conversion gives

$$2.0 \text{ L} \times \frac{1 \times 10^{-3} \text{ m}^3}{1 \text{ L}} = 2.0 \times 10^{-3} \text{ m}^3$$

The second conversion gives

$$\left(2.0 \times 10^{-3} \text{ m}^3\right) \times \left(\frac{100 \text{ cm}}{1 \text{ m}}\right)^3 = 2.0 \times 10^3 \text{ cm}^3$$

---

[5]For instance, for some equation $A + B = C$, it is also the case that $A\frac{x}{y} + B = C$ if $\frac{x}{y} = 1$.

[6]If you punch 42×2.54 into your calculator you will find 106.68: I rounded the answer to two significant figures because 42 has two significant figures; see the discussion of significant figures starting on page 673. The conversion factor ratio is *exact* so it doesn't affect the precision of the result.

Notice that the conversion from meters to centimeters is performed three times (the conversion factor ratio is cubed): because we're starting with m^3, if we carried out just one conversion we'd have units of $m^2 \times cm$ ("square meters multiplied by centimeters"), which is clearly not what we're after.

Significant Figures, Precision, and Error

Suppose we are measuring the width of the rectangular block of wood shown in Figure A.1. The ruler is marked in units of cm and there are marks indicating each tenth of a centimeter: clearly the block is wider than 1.5 cm but smaller than 1.6 cm. It is common practice to *estimate* the next digit, so we might report our result as 1.55 cm. However, we *shouldn't* report, say, 1.550 cm. Why not? Well, when we report 1.55 cm we are saying that we are confident that the actual width of the block is 1.55 cm *when rounded to the nearest hundredth of a centimeter*. In other words, we're claiming that the *actual* width is between 1.545 cm and 1.554 cm. If, on the other hand, we claimed the length is 1.550 cm, then we're indicating we know the width when rounded to the nearest *thousandth* of a centimeter. As scientists, we must be careful to report our measurements with the appropriate level of precision.

Take it to a more extreme example: you certainly *shouldn't* respond "1.55000000 centimeters" because then we'd be saying "the block's width definitely *isn't* 1.55000001 centimeters". The difference between these two numbers (which is 1×10^{-8} cm) is about the width of a single hydrogen atom; there is no way your ruler can give such an exact number!

We formalize the precision of a number by determining how many **significant figures** it has according to the following rules:

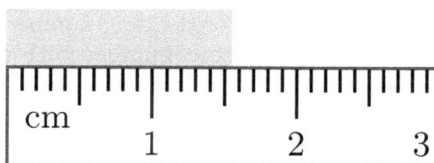

FIGURE A.1
A ruler, marked in centimeters, measuring one dimension of a rectangle.

- Non-zero digits are significant

- Digits between significant digits are significant

- If the number ends with one or more zeros to the right of the decimal (so-called *trailing zeroes*), they are significant

- If the number ends in a decimal, all digits to the left of the decimal are significant

- All other digits are not significant

Conveniently, when a number is expressed in scientific notation, *every* digit is significant. Thus we can say that the **precision** of an instrument corresponds to the number of significant digits with which we can report results obtained with the instrument.

Example A.5 Significant Figures   ⋆
How many significant figures are there in each of the following numbers? (a) 120 (b) 305.8 (c) 0.0420 (d) 300.02.
   (a) has two significant figures because the final 0 is not significant. (But if we specified "120". then all three digits would be significant.)
   (b) has four (the 0 is between non-zero digits, and so is included).

(c) has three (the final zero is a *trailing* zero and is therefore significant).

(d) has five; all of the zeros are between non-zero digits.

Keeping track of significant figures is important when performing calculations. (For example, we might wish to calculate the area of a rectangle by multiplying its length by its width.) The number of significant figures we should use when reporting the result depends on the number of significant figures in the values we're manipulating *and* the type of calculation we're performing. The general principle is that we should never report a result with more confidence than is afforded by the quality of the underlying measurements.[7]

Of course, we don't *only* work with numbers that we have measured. Some of our formulas will have numerical factors that are *exact*: for instance, the kinetic energy of an object is $K = \frac{1}{2}mv^2$ and there is no uncertainty associated with the factor of $\frac{1}{2}$. In other cases we will use *defined* quantities. For instance, we will find that objects near the surface of the Earth accelerate down toward the ground at 9.8 m/s^2 and we will generally take this as a defined value that does not affect the number of significant figures in our results. In other words, defined values and numerical factors can be treated as if they have arbitrarily high precision.

That being said, here are the rules for mathematically dealing with significant figures for measured values (the first two are the most important as you get started; the others are included for the sake of completeness and to serve as a useful reference):

- When multiplying or dividing numbers, the number of significant figures in the result is the same as the input number with the fewest significant figures. So if your numbers have, say, 3 significant figures and 4 significant figures, then the product or quotient should have 3 significant figures.

- When adding or subtracting numbers, the number of digits to the right of the decimal should be the same as the number being added or subtracted with the fewest digits to the right of the decimal. So if you're adding one number specified to the tenths to a number specified to the hundredths, the final answer should be rounded to the tenths.

- When calculating trigonometric functions like $\sin\theta$ (see Section A.3), the result should be reported with the same number of significant figures as the value of θ.

- When calculating logarithms of the form $\log x = y$, the number of significant figures in x equals the number of significant figures to the right of the decimal point in y.

- When calculating antilogarithms of the form $x = 10^y$, the number of significant figures to the right of the decimal in y equals the number of significant figures in x.

We will frequently perform multi-step calculations (e.g. $ab + c$ involves a product and *then* a sum). In these cases you shouldn't round intermediate calculations to avoid so-called *rounding error*. Instead, keep additional digits and apply the rules described above when reporting your final answer.

Example A.6 Calculations with Significant Figures   ⋆

Carry out the following calculations, taking care to express the result with the correct number of significant figures.

a. $702/5.5$

---

[7]In my first physics class, our teacher introduced this idea by having students use rulers to calculate the volume of a variety of blocks of wood. Each team wrote their answer on the chalkboard, and our teacher dramatically marveled at the wide variety of significant figures in the answers.

b. $14.55 + 3.5$

c. $\cos{(53°)}$

d. $\ln{(5.20)}$

e. $10^{2.21}$

f. $(3.0)(2.56) + 5.73$

In (a) we have two numbers, with three significant figures and two significant figures, so we report the result with two significant figures: $702/5.5 = 130$

In (b) the result in our calculator is 18.05, but we round to one digit after the decimal: $14.55 + 3.5 = 18.1$.

In (c) we report the answer with two significant figures: $\cos{(53°)} = 0.60$

In (d), we are taking the log of a number with three significant figures, so we specify the result with three significant digits to the right of the decimal: $\ln{(5.20)} = 0.716$.

In (e), We are raising 10 to a number with two significant digits to the right of the decimal, so we report our answer with two significant figures: $10^{2.21} = 160$.

Finally, in (f) the result in our calculator yields 13.41. However, the product $(3.0)(2.56)$ has two significant figures, so we would report that value as 7.7. Added to 5.73, we see we should keep just one digit past the decimal place, so we round our final answer to 13.4.

Finally, we should discuss how we *compare* numbers. If we make many repeated measurements with the same instrument and obtain identical or nearly identical results, then the instrument is precise,[8] or, equivalently, there isn't much uncertainty. To quantify this for any two measurements we define the **percent difference**, which compares two values x_1 and x_2 (usually representing different measurements of the same quantity) against one another:

$$\% \text{ difference} = \frac{|x_1 - x_2|}{(x_1 + x_2)/2} \times 100\% \tag{A.1}$$

In words, this says to divide their difference by their average, then multiple by 100 to express the result as a percentage. The absolute value guarantees that the percent difference is a positive value and it doesn't matter which value is treated as x_1 vs. x_2. Note that if we have two identical values then there is 0% difference, as you would expect.

In some cases we know what we *should* find: for instance, suppose we are attempting to measure the speed of light, and we would like to get as close to the known value as possible. The relevant term here isn't precision but rather **accuracy**, which refers to how close a measurement is to the accepted value: if the measurement is *accurate* then there isn't much error.[9] We quantify this with the **percent error**, which compares one experimental value x_{exp} to an expected theoretical value x_{thy}:

$$\% \text{ error} = \frac{|x_{\text{exp}} - x_{\text{thy}}|}{x_{\text{thy}}} \times 100\% \tag{A.2}$$

Note that if we obtained the expected value we would find a 0% error, again as you would expect.

---

[8]Note the subtly different use of the term compared to above, where it referred to how many significant figures can be used to describe measurements with the instrument.

[9]Suppose we are consistently off from the actual value by the same amount: the results are *precise* but not *accurate*.

Example A.7 Percent Difference and Percent Error   ⋆

The theoretical value for the acceleration of an object falling under the influence of gravity (and only gravity) is 9.8 m/s^2. You and a friend both perform some experiments; you find a value of 9.6 m/s^2 and your friend finds 9.9 m/s^2. What is the percent difference between you and your friend's experimental values? What are each of the percent errors compared to the expected value?

First, the percent difference is given by

$$\% \text{ difference} = \frac{|9.6 - 9.9|}{(9.6 + 9.9)/2} \times 100\% = 3.1\%$$

The percent errors are

$$\% \text{ error (you)} = \frac{|9.6 - 9.8|}{9.8} \times 100\% = 2.0\%$$

and

$$\% \text{ error (friend)} = \frac{|9.9 - 9.8|}{9.8} \times 100\% = 1.0\%$$

A.2 Algebra and Solving Systems of Equations

The laws of nature are most concise, effective, and beautiful when expressed mathematically. We'd like to express these laws in their most general form, which is to say *algebraically*: in a certain situation I might be very concerned about what happens to a specific box with a mass of 15.0 kg, but in general I'd like to characterize what happens to an arbitrary box of some mass m: characterizing a situation in a general case allows us to recognize relationships between quantities (like between a mass and a *force*). Moreover, if we work in algebraic notation, then our results can be used for as many particular numerical situations as we might like (whereas if we start in the numeric case we'd have to repeat our work for every other situation we care to analyze, even if all that has changed is a slight change to one numerical value).

In this book, then, we will frequently deal with algebraic equations, and we should be comfortable manipulating them to isolate any variable. You should already have experience doing so when we're dealing with a single equation. For instance, if we have

$$ax = (by + c)^2$$

and we would like to isolate y, we proceed as follows:[10]

$$(ax)^{1/2} = by + c$$
$$(ax)^{1/2} - c = by$$
$$\frac{(ax)^{1/2} - c}{b} = y$$

---

[10]Take the square root of both sides, subtract c, and then divide by b.

Given numerical values for a, b, c, and x, it would now be straightforward to calculate a numerical value for y. More, we can see from this result how these quantities affect y: for example, if b increases while the other quantities are held fixed, then y decreases (because we'd be dividing the fixed numerator by a larger denominator).

Similarly, for equations that are *quadratic* in the quantity of interest (here, x):

$$ax^2 + bx + c = 0 \tag{A.3}$$

we can use the **quadratic formula** to determine the solutions:

$$x = \frac{1}{2a}\left(-b \pm \left(b^2 - 4ac\right)^{1/2}\right) \tag{A.4}$$

More challenging is the situation where you have two equation and two unknowns (a so-called *system of equations*). Consider the following system:

$$5x = 3y + 1$$
$$2x = 2y - 2$$

Beyond the brute force "guess and check" method of inserting values and adjusting them (which is only effective in simple cases), there are two general strategies you should be aware of when dealing with situations such as these. First is the **substitution** method:

1. Solve one equation for one of the variables, so you have one variable *in terms of the other*. (For instance, manipulate one equation to the form $x = f(y)$ where $f(y)$ is some function of y.)

2. Substitute the first equation into the second equation. The second equation thereby has only one variable, so you can solve it in the usual way. (For instance, replace x with the function $f(y)$ in the second equation, then solve it for y.)

3. Once you have a final value for one variable from step 2, insert it into one of the original equations to solve for the other variable.

We'll demonstrate with the example given above. We solve the first equation for x (step 1) and find

$$x = (3y + 1)/5$$

Inserting into the second equation and solving for y (step 2) yields

$$2(3y + 1)/5 = 2y - 2$$
$$2(3y + 1) = 10y - 10$$
$$6y + 2 = 10y - 10$$
$$12 = 4y$$
$$3 = y$$

Then we insert this result into either of the original equations to find x (step 3). We'll use the first equation:

$$5x = 3y + 1$$
$$5x = 3(3) + 1$$
$$x = 2$$

Thus the system of equations has the solution $x = 2$, $y = 3$.

The second method to solving systems of equations involves **equation arithmetic**. First, note that if you have two equations $a = b$ and $c = d$, then it is also true that $a + c = b + d$: you can add the left hand side of the equations and set it equal to the sum of the right hand sides of the equations. Similarly, you can subtract the equations ($a - c = b - d$) and/or multiply one or both of the equations by a constant first (e.g. $4a - 3c = 4b - 3d$).

This is useful because if we're clever, we can eliminate one of our variables by an appropriate manipulation of our equations. Here, for instance, we can multiply the first equation by 2 and the second equation by 5 (so the coefficient on x is 10 for both equations), then subtract:

$$2\,(5x) - 5\,(2x) = 2\,(3y + 1) - 5\,(2y - 2)$$
$$0 = 6y + 2 - 10y + 10$$
$$4y = 12$$
$$y = 3$$

We find the same result for y, as we should, and we could then solve for x as with the substitution method.

The choice of which method to apply is largely one of taste, though with a bit of practice you will find that equation arithmetic can sometimes lead you to a solution more quickly than substitution. As Example A.10 demonstrates, these techniques can also be extended to systems of more than two equations.[11]

Example A.8 Systems of Equations (Equation Arithmetic)   ★★
Solve the following system of equations for x, y, and z:

$$2x + y - z = 2$$
$$4x - 2y + 3z = 12$$
$$y - z = -1$$

We can use equation arithmetic to eliminate x: double the first equation and subtract it from the second:

$$4x - 2y + 3z - 2\,(2x + y - z) = 12 - 2\,(2)$$
$$-4y + 5z = 8$$

If we solve the third equation for z (yielding $z = y + 1$) and insert into this result, we find

$$-4y + 5\,(y + 1) = 8 \implies y = 3$$

From here, the third equation yields $z = y + 1 = 4$ and (as you can check for yourself) the first or second equation can be solved to yield $x = 1.5$.

Thus we conclude $x = 1.5$, $y = 3$, and $z = 4$.

---

[11]I should perhaps mention that if you study **linear algebra** you will learn other techniques for solving systems of equations. These techniques will become very important if you continue your study physics beyond the level of this book.

Example A.9 Systems of Equations (Substitution Method) ★★

Solve the following system of equations for F_1, F_2, and F_3:

$$F_2 - F_3 = -1$$
$$2F_1 + F_2 - F_3 = 2$$
$$4F_1 - 2F_2 + 3F_3 = 12$$

Here I'll exclusively use the substitution method. We'll start by solving the first equation for F_2:

$$F_2 = F_3 - 1$$

Inserting this into the second equation yields

$$2F_1 + (F_3 - 1) - F_3 = 2$$
$$2F_1 = 3$$
$$F_1 = 1.5$$

If we insert this result and the previous result for F_2 into the third equation, we find

$$4F_1 - 2(F_3 - 1) + 3F_3 = 12$$
$$4(1.5) - 2(F_3 - 1) + 3F_3 = 12$$
$$F_3 = 4$$

If $F_3 = 4$ and $F_2 = F_3 - 1$, then $F_2 = 3$. Thus we have $(F_1, F_2, F_3) = (1.5, 3, 4)$. Look familiar? This is actually the same system of equations as the previous example! All I did was switch the names of the variables and reorder the equations. I did this on purpose to emphasize the fact that the *same mathematics* can come up in different scenarios and look, at first glance, like they might be rather different (a good example of this in physics are the linear kinematic equations and the angular kinematic equations: they come up in different contexts and use entirely different symbols but are mathematically identical). Learning to recognize underlying patterns is challenging, but it is well worth the effort!

Example A.10 Systems of Equations (Algebraic) ★★

Use the following system of equations to determine an algebraic expression for F_p in terms of m, a, g, and μ (the Greek letter mu):

$$F_p - F_f = ma$$
$$F_N - mg = 0$$
$$F_f = \mu F_N$$

The question amounts to asking us to eliminate F_f in the first equation, but we can't simply insert the third equation because then we'd be introducing F_N, which is not one of the allowed variables. Instead, let's substitute the third equation into the second to eliminate F_N:

$$\frac{F_f}{\mu} - mg = 0 \implies F_f = mg\mu$$

Then we can substitute this expression into the first equation:

$$F_p - mg\mu = ma$$
$$F_p = ma + mg\mu$$
$$F_p = m\left(a + g\mu\right)$$

This last expression is what we were after: F_p in terms of the requested variables. The choice of symbols here may strike as you rather odd, but I pulled this from a physical example: these expression describe the forces acting on an object being pushed over a horizontal surface in the presence of friction, and our final result expresses the magnitude of the push in terms of the mass of the object, m, its acceleration, a, the acceleration due to gravity, g, and the coefficient of friction, μ, which quantifies the roughness of the object-surface interface. I sometimes refer to *setting up* these equations (which we'll cover in Chapter 3) as "doing the physics" and then performing the algebra to acquire whatever result we're after–here, solving for F_p in terms of specific variables–as "doing the math".

A.3   Geometry and Trigonometry

You should be comfortable calculating the areas of simple shapes such as a rectangle ($A = bh$ for base b and height h), circle ($A = \pi r^2$ for radius r), and right triangle ($A = \frac{1}{2}bh$ for base b and height h). Similarly, you are likely familiar with the fact that the interior angles of a triangle add to $180°$, and if the triangle is a *right* triangle, one of the interior angles is $90°$ (so the remaining two angles add to $90°$) and the longest side of the triangle is called the hypotenuse (Fig. A.3).

There are two standard units used for angles: **degrees** and **radians**. The SI unit is the radian; the conversion is often provided in terms of the angular measure of a complete circle:

$$360° = 2\pi \text{ rad} \tag{A.5}$$

Strictly speaking, the radian is a *unitless* quantity that serves as a proportionality factor. What do I mean by this? Well, consider the equation for the circumference of a circle, C, of radius r:

$$C = 2\pi r$$

If we're interested in a smaller portion of perimeter of the circle we can use the more general equation for the arc length, s (Fig. A.2):

$$s = \theta r$$

Plugging in $\theta = 2\pi$ corresponds to the entire circle and yields the circumference. But θ, which we're measuring in radians, has to ultimately carry *no units* because we have "length = radians × length"! Thus the angle serves as a proportionality factor that measures the ratio of the arc length to the radius of the circle. Conversely, a radian tells us how far along the circle we've gone such that 2π rad indicates "we've gone around once". We will see that this is a useful concept for analyzing, for instance, circular motion and oscillations.

You may be less familiar with the trigonometric functions sine, cosine, and tangent, which relate these interior angles (commonly labeled θ and ϕ, which are respectively the

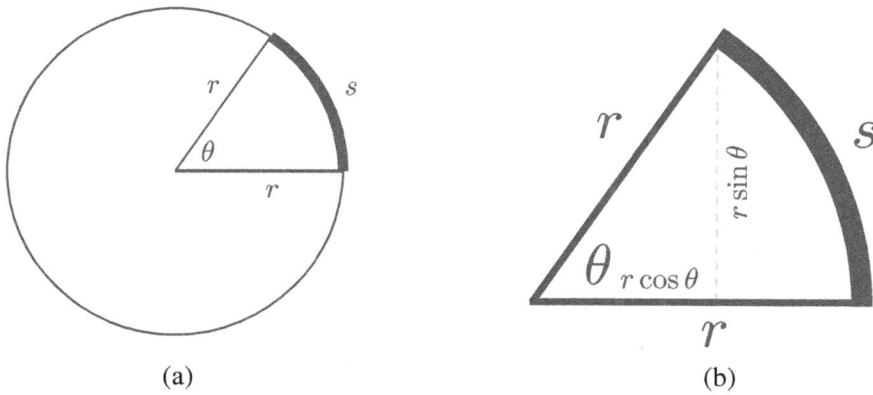

FIGURE A.2
(a) Some portion of the perimeter of a circle has an arc length s, which is related to the circle's radius r and the angle subtended by the arc length, θ. (b) Some additional detail: and if one forms a right triangle by including the vertical dashed line, the non-hypotenuse sides can be determined with the trigonometric functions sine and cosine (see Fig. A.3).

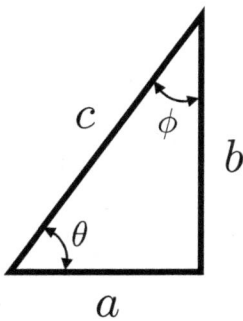

$$\sin(\theta) = \frac{\text{opposite}}{\text{hypotenuse}} = \frac{b}{c}$$
$$\sin(\phi) = \frac{\text{opposite}}{\text{hypotenuse}} = \frac{a}{c} \quad \text{SOH}$$

$$\cos(\theta) = \frac{\text{adjacent}}{\text{hypotenuse}} = \frac{a}{c}$$
$$\cos(\phi) = \frac{\text{adjacent}}{\text{hypotenuse}} = \frac{b}{c} \quad \text{CAH}$$

Pythagorean Theorem:
$$a^2 + b^2 = c^2$$

$$\tan(\theta) = \frac{\text{opposite}}{\text{adjacent}} = \frac{b}{a}$$
$$\tan(\phi) = \frac{\text{opposite}}{\text{adjacent}} = \frac{a}{b} \quad \text{TOA}$$

FIGURE A.3
An overview of right-triangle trigonometry and the Pythagorean Theorem. The mnemonic "SOHCAHTOA" can help you remember the definition of the trigonometric functions. For instance, sin ("S") is defined as the ratio of the length of the side opposite the angle ("O") over the length of the hypotenuse ("H").

lowercase Greek letters theta and phi) to the lengths of the triangle's sides:

$$\sin\theta = \frac{\text{opposite}}{\text{hypotenuse}} \tag{A.6}$$

$$\cos\theta = \frac{\text{adjacent}}{\text{hypotenuse}} \tag{A.7}$$

$$\tan\theta = \frac{\text{opposite}}{\text{adjacent}} \tag{A.8}$$

Similarly, the *inverse* trigonometric functions can be used to determine and angle θ given the sides of the triangle:

$$\theta = \sin^{-1}\left(\frac{\text{opposite}}{\text{hypotenuse}}\right) \tag{A.9}$$

$$\theta = \cos^{-1}\left(\frac{\text{adjacent}}{\text{hypotenuse}}\right) \tag{A.10}$$

$$\theta = \tan^{-1}\left(\frac{\text{opposite}}{\text{adjacent}}\right) \tag{A.11}$$

When evaluating trigonometric functions in a calculator, be careful to use the correct setting so the calculator understands if you are providing an angle in degrees or radians. Once evaluated, the trigonometric functions return a number with no units. (For example, $\sin(0 \text{ rad}) = 0$.) The inverse trigonometric functions reverse this process by taking a unitless argument and returning an angle. (For example, $\cos^{-1}(0.5) = \pi/3$ rad.) We will see situations where the argument of a trigonometric function is more complex than a simple angle, and if we are working in SI units then the argument will reduce to radians, as Example A.11 demonstrates.

Example A.11 Unit Analysis ⋆

Consider the equation $y = A\sin(kx)$. If y and x are measures of length, what are the fundamental units of A and k?

Let's first consider k. The product kx must yield the SI unit of an angle, which is the radian. Thus k has units of radians / length: when multiplying by x the units of length cancels, leaving us with radians. Now $\sin(kx)$ will return a number with no units, which means the units of y and the units of A must agree. Thus A has units of length.

While the hypotenuse is fixed for a given triangle, the meaning of "opposite" and "adjacent" depends on *which angle* we're talking about. Importantly, it follows from these definitions that if the interior angles (other than the right angle) are θ and ϕ then $\sin\theta = \cos\phi$ and $\sin\phi = \cos\theta$ (Fig. A.3).

Meanwhile, the **Pythagorean Theorem** relates the length of the right triangle's sides without explicit reference to the angles:

$$a^2 + b^2 = c^2 \tag{A.12}$$

where c is the length of the hypotenuse and the other two sides have lengths a and b.

We shall frequently come across situations where we need to apply these tools to analyze right triangles; one simple example is a *ramp*. Example A.12 demonstrates the idea. Just as importantly, right triangle trigonometry is crucial to understanding (and converting between) Cartesian and polar coordinates, as we shall see in Example A.13.

Example A.12 Right Triangle Trigonometry ⋆

The triangle shown in Figure A.4 has a hypotenuse of length 5.0 and $\theta = 30.0°$. What are the lengths of the other two sides of the triangle? What is the value of ϕ?

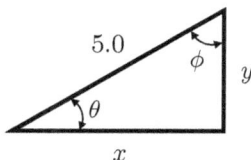

FIGURE A.4

We can apply the sine function to θ to determine the length of the side opposite θ, which is labeled y:

$$\sin\theta = \frac{y}{5.0} \implies y = 5.0\sin 30.0° = 2.5$$

Similarly, the cosine function can be used to determine the length of the adjacent side, x:

$$\cos\theta = \frac{x}{5.0} \implies x = 5.0\cos 30.0° = 4.3$$

This is not the only way to solve this problem. For example, we could have noted that $\theta + \phi = 90° \implies \phi = 60.0°$, and then evaluated the trigonometric functions with ϕ rather than θ. Or, once we have determined one of the non-hypotenuse sides, we could have determined the third by using the Pythagorean Theorem.

Example A.13 Describing a Hiker's Position ⋆

Figure A.5 shows a hiker's camp at the center of a compass coordinate system that shows North, South, East, and West. The hiker's position through the day is shown with a sequence of dots that are connected by thin lines to guide the eye. At the end of the hike the hiker is 5.0 mi East and 12.0 mi North of camp. The arrow points from the camp to the hiker's last position. What is the length of the arrow, and what angle does it form with the horizontal compass line marking the East-West direction?

The arrow can be drawn as the hypotenuse of a right triangle where the other two sides have lengths of 5.00 mi and 12.0 mi (Fig. A.6). From the Pythagorean Theorem it follows that the hypotenuse has length h given by

$$h = \left(5.00^2 + 12.0^2\right)^{1/2} \text{ mi} = 13.0 \text{ mi}$$

The angle θ in Figure A.6 has a value given by (Equation (A.11)):

$$\theta = \tan^{-1}\left(\frac{12.0}{5.00}\right) = 67.4°$$

We shall see that the arrow described here is an example of a *vector*, which defined both by a quantity (represented by the length of the arrow) and a direction (specified by the angle it forms with the reference axis).

Note also that we have two complete and equivalent ways of describing where the hiker ended up compared to our reference point: "5.0 mi East and 13 mi North", *or* "13.0 mi away directed 67.4° North of East". The first approach, where we specify a component along each axis, is in **Cartesian coordinates** and the second "magnitude angle" form is called **polar coordinates**. Both shall prove useful as we proceed.

Finally, while it isn't immediately relevant for this example, it is worth noting that the lines that connect our points shouldn't be interpreted as guarantees that the hiker actually took the straight line path from each dot to the next. All we *know* are the locations of the hiker at certain times.

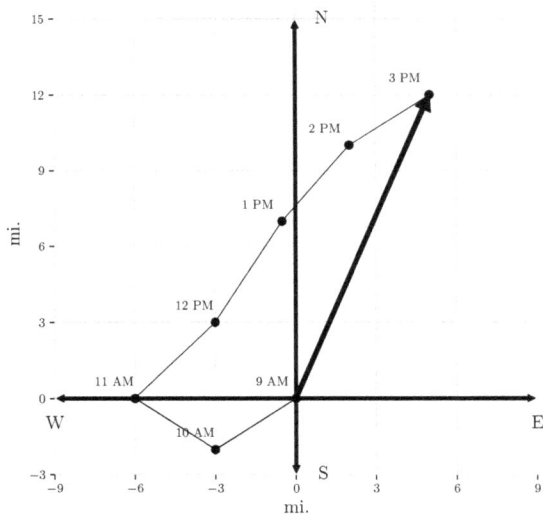

FIGURE A.5

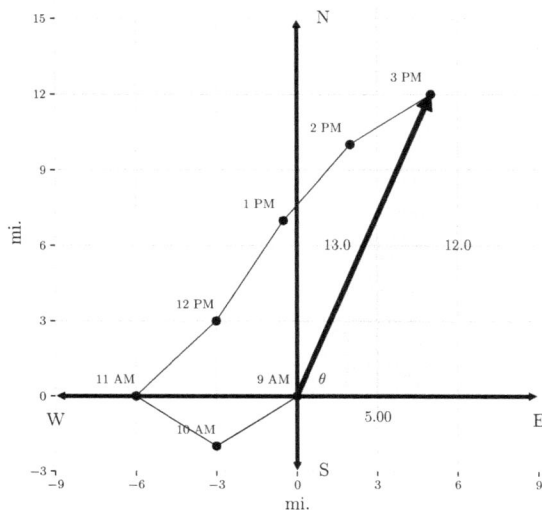

FIGURE A.6

What we've described above summarizes the principles of *right triangle trigonometry*. We'll sometimes use these principles in isolation, but sometimes we'll apply them to larger

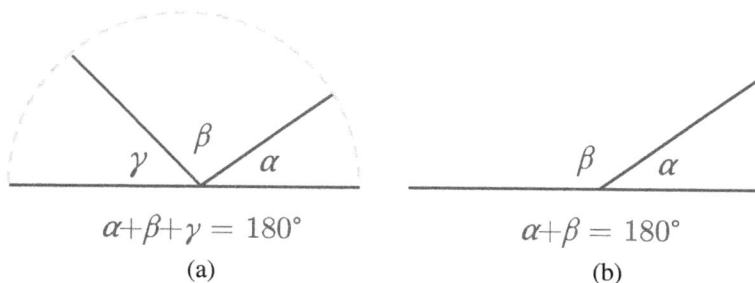

$$\alpha+\beta+\gamma = 180°$$
(a)

$$\alpha+\beta = 180°$$
(b)

FIGURE A.7
Angles on one side of a line (here, the top half of a horizontal line) add to 180°. (a) A semicircle is included to emphasize how the angles form a half circle. (b) The principle holds even without a circle included. Here the region is split into two regions so the angles α and β add to 180° and are said to be *supplementary*.

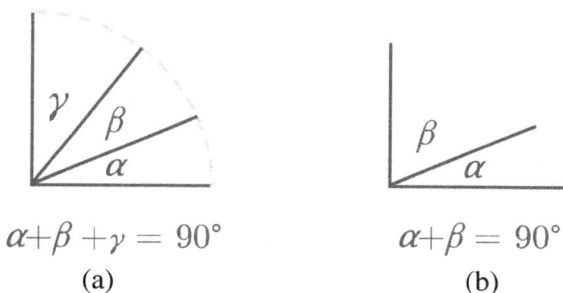

$$\alpha+\beta+\gamma = 90°$$
(a)

$$\alpha+\beta = 90°$$
(b)

FIGURE A.8
Angles between perpendicular lines add to 90°. (a) An arc is included to emphasize how the angles form a quarter circle. (Right) The principle holds even without an arc included. Here the region is split into two regions so the angles α and β add to 90° and are said to be *complementary*.

problems that also require us to relate lengths and angles in more complicated situations. Here are four of the most useful principles:

- Because a complete circle consists of 360°, the total angle on either side of a line is 180°. So if two or more lines[12] meet at a point, the angles on one side of a line must add to 180°. If there are two such angles, they are said to be **supplementary angles**. Figure A.7 demonstrates the idea.[13]

- Similarly, if a right angle (which measures 90°) is split into two or more regions, they must add to 90°. If there are two such angles, they are said to be **complementary angles**. Figure A.8 demonstrates the idea.

- If two lines cross at a point, the space around the point is split into four regions. Opposite angles have the same value and adjacent values add to 180° (Fig. A.9 (a)).

---

[12]Formally we mean here lines *or line segments*. A line has infinite length, which never occurs in a physical system, though we sometimes approximate a "very long" line as infinitely long.

[13]The symbols α, β, and γ are respectively the lower case Greek letters alpha, beta, and gamma.

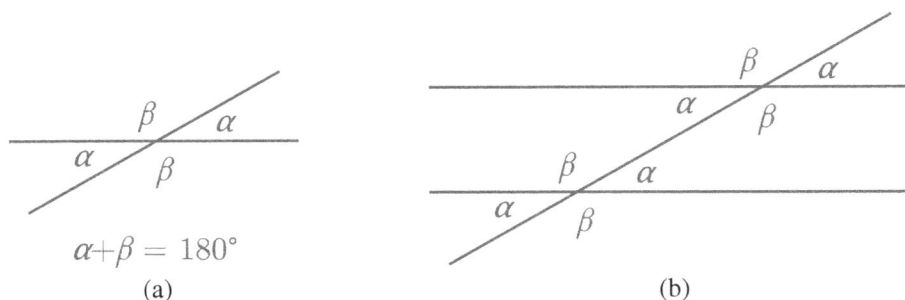

$$\alpha + \beta = 180°$$

(a)

(b)

FIGURE A.9
(a) If two lines intersect at a point, opposite angles are equivalent and adjacent angles add to 180°. (b) If two parallel lines intersect a third, then you're left with two copies of the left side of the figure, so knowing an angle from one copy allows you to determine angles from the other copy.

- If two parallel lines are crossed by a third line, we can invoke symmetry and the preceding point to make a number of statements about the values of the angles, as summarized in the right side of Figure A.9. For instance, the two angles marked α closer to the middle of the figure are called "alternate interior angles".

  Examples A.14–A.16 demonstrates the use of several of these ideas.

Example A.14 Describing a Boat's Position   ⋆
A ship is 40.0 mi East and 30.0 mi South of its dock, which is at the center of the compass shown in Figure A.10. How far is the ship from the dock? Determine a numerical value for the angle α. Then find a numerical value for β two different ways. Are α and β complementary, supplementary, or neither?

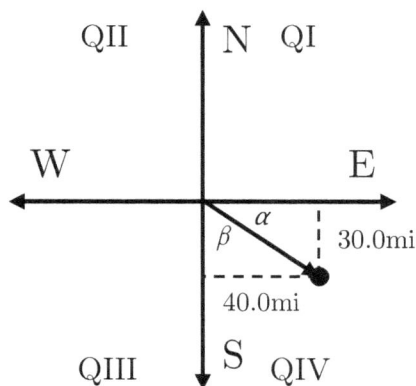

FIGURE A.10

We'll begin with two general comments. First, notice that the North-South and East-West axes split the figure into four regions. For any diagram with four regions like this, we refer to each region as a **quadrant**, labeled QI to QIV starting counterclockwise from the top right. Second, the arrow that indicates the ship's position is an example of a **vector**, which refers to any quantity defined by both a magnitude (how long is the arrow?) and a direction (which way does it point?). We will have quite a bit to say about vectors beginning in Chapter 1, but for now we can think of the vector simply as the hypotenuse of a triangle with non-hypotenuse sides of length 30.0 mi and 40.0 mi.

OK, on to the details of the problem. The length of the vector, which we'll call r, is given by the Pythagorean Theorem:

$$r = \left(30.0^2 + 40.0^2\right)^{1/2} \text{ mi} = 50.0 \text{ mi}$$

We can determine α with any of the inverse trigonometric functions. I'll use inverse tangent since it relies directly on the values given in the problem statement (it doesn't matter so long as we use sufficient precision in any calculated values we use, but using the given values eliminates the possibility of accidentally introducing rounding error):

$$\alpha = \tan^{-1}\left(\frac{30.0 \text{ mi}}{40.0 \text{ mi}}\right) = 36.9°$$

Finally, to determine β we can use inverse tangent again, being careful to note from the geometry that the opposite and adjacent values have switched:

$$\beta = \tan^{-1}\left(\frac{40.0 \text{ mi}}{30.0 \text{ mi}}\right) = 53.1°$$

Alternatively, if we note from the geometry that $\alpha + \beta = 90°$ (the angles are *complementary*), it follows that

$$\beta = 90° - \alpha = 90° - 36.9° = 53.1°$$

as we found above.

Example A.15 Parallel Lines Angle Analysis ⋆
Identify the remaining angles in Figure A.11. The two angled lines are parallel.

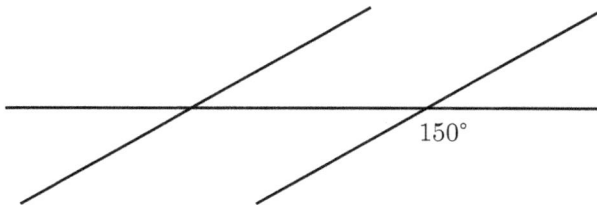

FIGURE A.11

If we start at the marked angle and work counterclockwise, we find that the first missing angle is 30° (it and the first angle must add to 180°; they are supplementary angles). The next two angles are 150° and 30° because they are opposite known angles (Fig. A.9 (a)).

For the angles on the left side of the figure, we can invoke symmetry now that the right side is fully determined. The result is shown in Figure A.12.

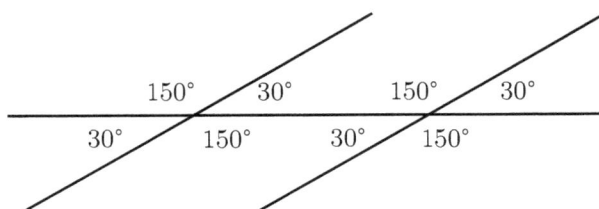

FIGURE A.12

Example A.16 Additional Angle Analysis   ★★

Find the angles (in terms of α) marked "?" in Figure A.13. The horizontal lines are parallel and three right angles are marked with a □ symbol.

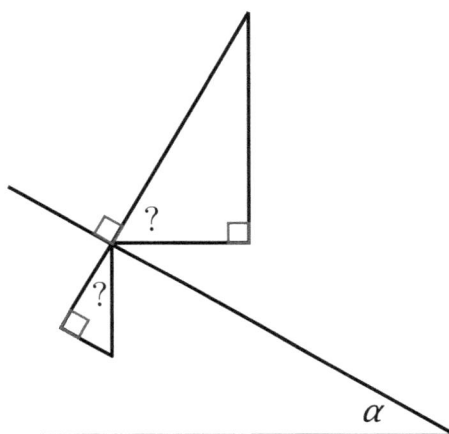

FIGURE A.13

We proceed in the steps labeled in Figure A.14. The angle labeled α in step 1 follows because it is an alternate interior angle with the original angle α (see Figure A.9). The rest of the problem involves recognizing different right angles. The two angles marked $90° - \alpha$ in steps 2 and 3 each form a $90°$ angle with the angle α from step 2.

Finally, the two angles marked in steps 3 and 4 form *another* $90°$ angle, so if the angle in step 3 is $90° - \alpha$, the final angle in step 4 must equal α. We can check our work for consistency by adding all four angles together and verifying that they sum to $180°$, as they must (they build up the half plane to the bottom right of the long diagonal line that goes to the top of the figure).

This example might seem very contrived, but we'll see that this geometry is very relevant when we analyze forces acting on objects on an incline plane (such as a car parked on a hill).

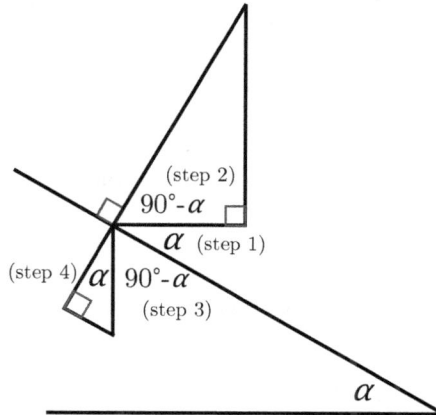

FIGURE A.14

A.4 Representing Data

Some measurements can be analyzed in isolation: "The temperature is 60° F". However, it is often the case that data are associated with other data: "The temperature at 9:00 AM was 60° F, at 10:00 AM it was 62° F.". For all but the simplest situations, we have to choose how to represent our data–and we need to be comfortable interpreting others' data however *they* choose to represent it. Here are a few options:

- With a movie or sequence of images. We could set a clock by our thermometer, take several pictures throughout the day (or just record it all day), and then provide the pictures (or recording) and say "here is our data". (This is a bit extreme in this case, but if we're analyzing, say, the *motion* of some object then a recording might be an excellent idea.)

- With a diagram. We might draw a sequence of thermometers, side by side, each labeled with the time and corresponding temperature.

- With words. We could just write a sequence of sentences of the form "At *some time* the temperature was *some value*", much like how I wrote the example at the start of this discussion.

- In a table, where each set of data (e.g. a time and the corresponding temperature) is arranged in a row.

- With a graph that visualizes trends in numerical data.

You've likely seen all of these at some point already in your education; I list them out here to emphasize that data can be conveyed in myriad ways and we should be comfortable interpreting data however it is presented and translating between different formats. I'll focus here on the final two formats described above: tables and graphs. Some example data for the outside temperature over a 24 hour period is shown in Table A.3.

We can also show this data on a **Cartesian coordinate system** where perpendicular axes are used to represent two sets of data: typically, the horizontal axis is arbitrarily labeled the x axis and the vertical axis is labeled the y axis, though in practice the axes can represent any quantities of interest and the labels can be chosen to be more informative.[14] Regardless of what you call the axes, one set of measurements (here, a time and the corresponding temperature) is called an **ordered pair** and can be marked on the Cartesian coordinate system at the intersection of the appropriate positions on the axes (Fig. A.15).

When we have data represented in this way, we can infer mathematical relationships. For instance, it should be evident that the daily temperature doesn't change *linearly* with the temperature because if we take a ruler and try to draw a straight line through the data, we'll do a terrible job (much of the data will be far from the line regardless of how we draw it). For these data, it turns out we can generate a good **fit** with a sine curve (Fig. A.15).

TABLE A.3

Example data for the outside temperature over a 24 hour period.

Time	Temperature (°F)
12:00 AM	54
2:00 AM	47
4:00 AM	45
6:00 AM	42
8:00 AM	37
10:00 AM	41
12:00 PM	45
2:00 PM	48
4:00 PM	46
6:00 PM	39
8:00 PM	34
10:00 PM	29

This realization can lead us to other questions: is this a general result for temperature data collected anywhere in the world? How do the properties of the fit change over the course of the year and based on where on the planet the data are taken? What do the specific details of the fit mean, physically?[15] Can we explain this behavior by thinking about the brightness of the sun, the angle of the sun over horizon based on the location and time of the year, etc.?

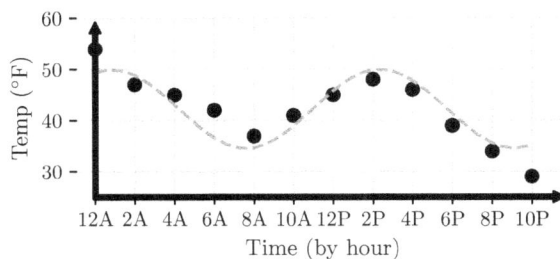

FIGURE A.15

Example data for the outdoor temperature over the course of a day (Table A.3). The black circles are data points and the dashed curve shows a sine curve fitted to the data: the mathematical form of the fit is $T = A \sin(Bt + C) + D$ where A, B, C, and D are constants chosen to make the curve match the data as closely as possible.

These are just example questions to make the point that meaningful information–which can stimulate interesting questions–can be extracted from visual representations of data.

---

[14] For example, if you're measuring the height of a plant every day as it grows, you might label the horizontal axis t for time and the vertical axis h for height.

[15] For instance, in this context you might expect the time required for a complete cycle, called the **period**, to be a complete day. You're asked to think this through in the context of the data shown in Figure A.15 in Problem A.47.

Example A.17 Graphing Positional Data ★

Consider the data included in the table below, which shows the horizontal and vertical positions (x and y, respectively) of an object as it moves. Make a graph of the position data on a Cartesian coordinate system. Can you describe in words what this object might be doing?

time (s)	x (m)	y (m)
0.0	0.0	0.0
0.2	0.1	0.0
0.4	0.2	0.0
0.6	0.3	0.0
0.8	0.4	0.0
1.0	0.5	0.0
1.2	0.6	−0.2
1.4	0.7	−0.8
1.6	0.8	−1.8
1.8	0.9	−3.1

The graph in shown in Figure A.16. The *time* isn't explicit, but if we refer back to the table we see that x is always increasing, so the motion is from the left to the right side (we could label the points on the graph with "$t = 0.0$ s", "$t = 0.1$ s", etc. if we wanted to embed the timing explicitly). So the ball is moving to the right and, after the first second, it begins to move down as well. One possible situation where this might occur is an object moving across a table and then tumbling over the edge.

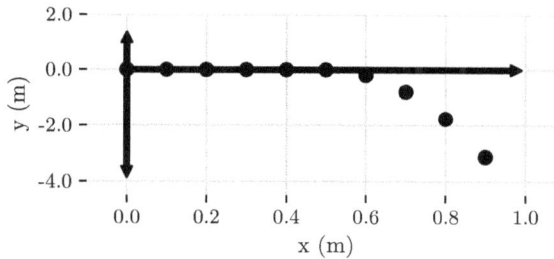

FIGURE A.16

Of course, if we'd like to fit different kinds of curves to our data (or if we'd like to visualize the laws of nature in their algebraic form), we should be familiar with what typical curves look like, and how their mathematical form maps to their visual form. For now we'll consider the two types of equations we'll encounter most frequently as we get under way: *linear* and *quadratic*.

Linear Equations

The general form of a linear equation is

$$y = ax + b \tag{A.13}$$

where a is the *slope* and b is the *y-intercept* (Fig. A.17).[16] A good example of a linear equation in physics considers the motion of an object moving at a constant velocity: $x = vt + x_0$.

---

[16]You may have seen this equation written as $y = mx + b$; I am using the parameters a and b here to be a bit more generic. We'll see the same symbols used to represent different things as we go–this is unavoidable

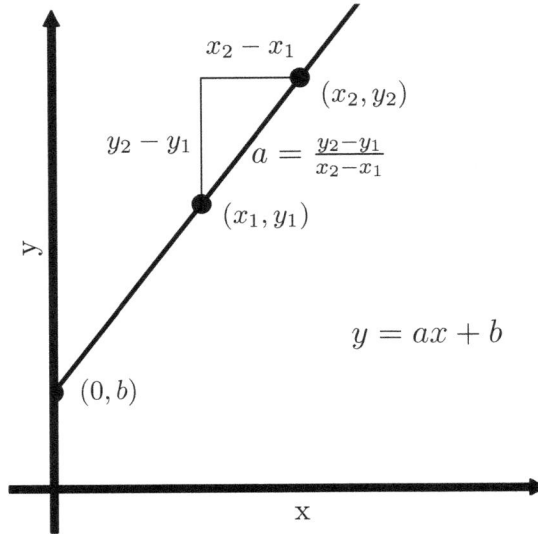

FIGURE A.17
A graph of a linear equation $y = ax + b$. If $a < 0$, the line slopes down and to the right (instead of up and to the right) and if $b < 0$ the y-intercept is below the horizontal axis.

Here the object's position, x, is determined by its initial position x_0, the magnitude of its velocity, v, and the amount of time it has been moving, t.

If you consider two ordered pairs on the line, (x_1, y_1) and (x_2, y_2) then you have the system of equations

$$y_1 = ax_1 + b$$
$$y_2 = ax_2 + b$$

If you subtract the first equation from the second you find

$$y_2 - y_1 = a(x_2 - x_1)$$

or, solving for the slope,

$$a = \frac{y_2 - y_1}{x_2 - x_1} \tag{A.14}$$

This is the mathematical statement of "the slope is the rise over the run", i.e. the change in the vertical coordinate divided by the change in the horizontal coordinate.

Quadratic Equations

The general form of a quadratic equation is

$$y = ax^2 + bx + c \tag{A.15}$$

---

when we talk about a lot of diverse phenomena, so you'll want to pay attention to the context of what we're discussing. This can be frustrating, I know, but it is also unavoidable so we might as well get used to it early!

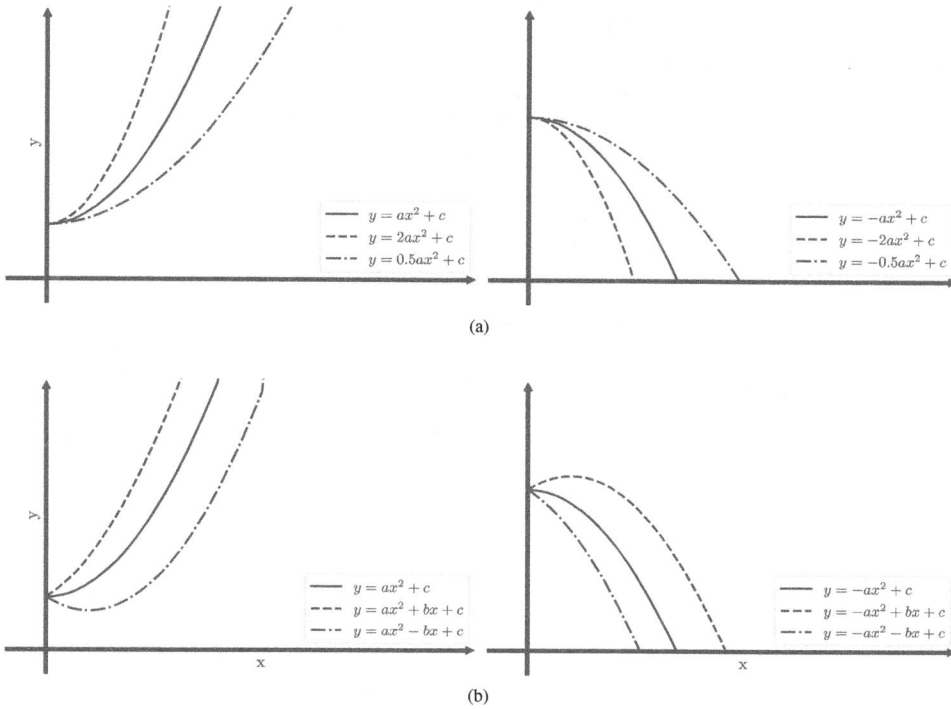

(a)

(b)

FIGURE A.18

(a) Quadratic equations of the form $y = ax^2 + c$, i.e. parabolas that cross the y axis at $y = c$. The left and right panels respectively show the effect of positive and negative coefficients on x^2 and the different curves in each panel show the effect of different magnitudes of the coefficients on x^2. (b) The parabolic cases (solid black lines) are compared to quadratic equations of the form $y = ax^2 + bx + c$.

where a, b, and c are constants, just like with the linear case.[17] An example of a quadratic equation in physics expands our previous example in the case where an object is accelerating at some rate a rather than moving at a constant velocity: $x = \frac{1}{2}at^2 + v_0 t + x_0$. The other symbols have the same meaning as above except v_0, which refers to the initial velocity (before we generically used v because it was understood to be a constant; here we specify v_0 since the velocity v is understood to be changing).

If we first consider the case where $b = 0$ then we have a *parabola* with curvature defined by a and a vertical shift defined by c (Fig. A.18, top). If $b \neq 0$, the maximum (if $a < 0$) or minimum (if $a > 0$) is no longer on the y axis; we can characterize the shape of the curve based on whether a and b have the same sign (i.e. positive or both negative) or not.

If a and b have the same sign then the curve grows more dramatically than if $b = 0$, but if the signs are *different*, then the sign of b dominates for $x \approx 0$ and then the sign of a dominates for $x \gg 0$. For example, if $a > 0$ and $b < 0$, then $y = ax^2 + bx + c$ starts at c for $x = 0$, decreases from c for small positive values of x, then eventually increases (Fig. A.18, bottom left).[18] If $a < 0$, the same arguments apply but the slope is negative for $x \gg 0$ (Fig. A.18, bottom right).

---

[17]Though note that a, as the coefficient of the highest power of x, is the coefficient of the quadratic term here, but the coefficient of the linear term in $y = ax + b$. Context!

[18]This is because $x \geq x^2$ for $x \leq 1$ but $x < x^2$ for $x > 1$.

Example A.18 Graphing Temporal Data   ⋆

The table shows the position of a car driving down a straight, flat road and the corresponding times (measured by a stopwatch) when the car was at each position. Make a graph of the position (on the vertical axis) vs. the time (on the horizontal axis) and calculate the slope of the line running through the data points. Interpret the *units* of the slope.

Time (s)	Position (m)
0.00	0.00
1.00	22.00
2.00	44.00

These data are shown in Figure A.19. Because the points lie on the same line, it doesn't matter which two we use to calculate the slope. Here we'll use the first and last (note that I am using the subscripts 1, 2, and 3 for the data in the corresponding rows of the table; this is arbitrary–we could just as well use a, b, and c if we like–but a common convention):

$$\text{Slope} = \frac{x_3 - x_1}{t_3 - t_1} = \frac{(44.00 - 0.00)\ \text{m}}{(2.00 - 0.00)\ \text{s}} = 22.0\ \text{m/s}$$

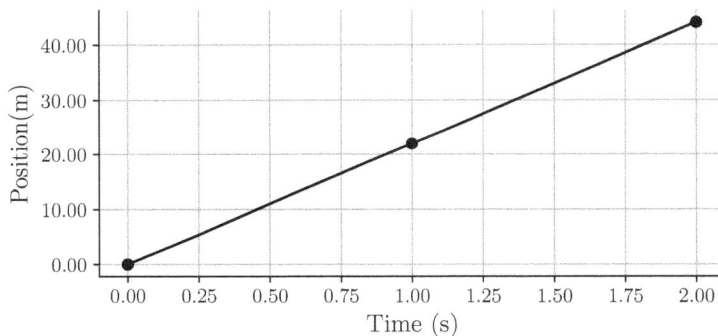

FIGURE A.19

I invite you to check that you obtain the same result if you use a different pair of data points (that is, 1 and 2 or, alternatively, 2 and 3). The *units* of our slope correspond to the units of the vertical axis, distance (in meters) divided by the units of the horizontal axis, time (in seconds). Thus we can say that the *rate of change* of the position over time is measured in m/s. As we discuss in Chapter 1, these are the SI units of *velocity*.

A.5   Calculus Concepts

If you jump off of a diving board into a swimming pool, we can think of your position above the water *at every instant* from when we start considering the motion until when we stop.[19] This is the language of *continuous change*, which is the language of *calculus*.

---

[19]If we use a stopwatch that measures time down to the millisecond then we might not have data measured *between* every millisecond we record, but you are moving nonetheless!

You may have already studied calculus, you might be studying calculus for the first time as you're studying physics, or you may have never studied calculus. Regardless of which category you're in, this book is for you. Here we will consider the basic *language* of calculus, so you'll be comfortable with the notation as we proceed through the book. If you've studied (or are studying) calculus, I additionally invite you to try the examples and problems marked with a dx or $\int$ symbol.

The key terms we should be comfortable with are the **derivative** and **integral**, which we will consider in turn.

Derivatives (slopes)

First, consider Equation (A.14), which describes the slope of a line as

$$a = \frac{y_2 - y_1}{x_2 - x_1} = \frac{\Delta y}{\Delta x}$$

Here I've added the last step and introduced the uppercase Greek letter Delta (Δ), which is a convenient shorthand for "the change in" whatever follows:

$$\Delta x \equiv x_2 - x_1 \text{ and } \Delta y \equiv y_2 - y_1$$

For a line, Δy is the same regardless of what points we use to calculate Δx; in other words, the slope is *constant*. For a more complicated function, however, our choice of Δx can yield different values for Δy, which is to say that the slope is *not* a constant. (So for any function other than a line, Equation (A.14) yields only an *average* slope for whatever interval Δx we use.)

Instead of considering the slope of a line, then, in general we're concerned with the slope of a function $f(x)$ *at a certain value of x*.[20] Graphically, this is the slope of the line **tangent** to the curve at x: a tangent line touches $f(x)$ at x and at no other (nearby) locations (Fig. A.20).

Mathematically, we compute the slope of the tangent line at some point $x = x_0$ by setting the values of x_1 and x_2 arbitrarily close to x_0. This wipes away any nearby variations in the function and changes the *average slope* from x_1 to x_2 into the *instantaneous rate of change*, or **derivative** at $x = x_0$. When x_2 and x_1 are arbitrarily close together, we shift the notation from Δ (which implies an average) and use a lowercase d instead:

$$\frac{\Delta y}{\Delta x} \rightarrow \left.\frac{dy}{dx}\right|_{x=x_0} = \left.\frac{df(x)}{dx}\right|_{x=x_0} \tag{A.16}$$

The mathematical expression $df(x)/dx$ is read "the derivative of $f(x)$ with respect to x", and if we specify $x = x_0$ then we say we "evaluate the derivative at $x = x_0$". Note also that I shifted notation here from y to $f(x)$: these are both common ways of describing a curve and you don't want to be thrown by the notation. (Indeed, when we get to the physics we'll label the curve in accordance with the physical quantities we're describing, so we'll see many more choices than y or $f(x)$.)

The key point here is that the derivative of a function $f(x)$ tells you the rate at which $f(x)$ is changing; visually it is given by the slope of a line tangent to $f(x)$.

---

[20] An example from physics involves position as a function of time: the derivative tells us the rate at which the position is changing at a certain instant, e.g. "At $t = 2.0$ s, position is changing at 1.5 m/s".

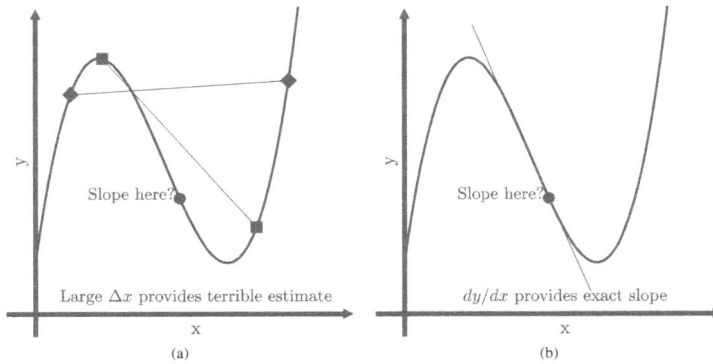

FIGURE A.20

(a) Estimating the slope of a function with $\Delta y/\Delta x$. The two diamonds are equidistant (in x) from the point marked with a circle, and the slope of the line connecting them clearly does a poor job estimating the slope at the circle. The squares are somewhat closer and do a somewhat better job. (b) Squeezing the points used for Δx arbitrarily close together (we can also say that the separation becomes infinitesimally small) yields the derivative of y, dy/dx, which provides the exact slope.

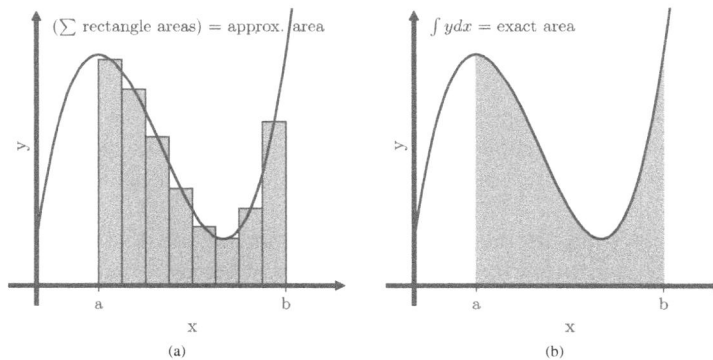

FIGURE A.21

(a) Estimating the area bounded by a curve and the horizontal axis between two x coordinates a and b using a series of rectangles. Note that the rectangles are sometimes above and sometimes below the curve; making the rectangles narrower (and necessarily increasing how many there are) will reduce the error. (b) In the limit of infinitely narrow rectangles we obtain the integral $\int y dx$, which provides the exact area.

Integrals (Areas)

An integral is a continuous version of a sum; in physics it can be helpful to think of it as the *cumulative effect* of whatever quantity we're considering".[21] Just as a derivative at some point x_0 is visually the slope of the tangent line at $x = x_0$, the integral is the *area bounded by $f(x)$ and the x axis* (Fig. A.21).

---

[21]An example from physics involves velocity as a function of time: the integral tells us the object's change in position, e.g. "From $t = 2.0$ s to $t = 5.0$ s the runner moved 20.0 m closer to the finish line".

If our curve is a straight line, we can evaluate the area geometrically by splitting the area into rectangles and triangles, but if we have a more complicated function we need to rely on the techniques of so-called **integral calculus.** The basic idea is still the same: break the area under the curve into a sequence of rectangles with some small width Δx, and compute the area by multiplying each rectangle by its height, then add them all up.

What heights do we use, though? For any finite rectangular width Δx the function could be changing dramatically from the left edge to the right edge, so whatever value for the height we use would be approximate. We get around this by setting the width of the rectangle to be infinitely narrow, i.e.

$$\Delta x \to dx$$

so for a rectangle at $x = x_0$ the height of the rectangle is just $f(x_0)$. The notation is

$$\sum (\text{rectangle height}) (\text{rectangle width}) \to \int f(x)\,dx$$

The integral symbol $\int$ is therefore the continuous version of the summation symbol $\sum$, which is used for a discrete number of terms to be added together.

If we wish to specify the starting and ending values of x, we include the so-called *bounds of integration*:

$$\int_a^b f(x)\,dx$$

where a is the starting value of x and b is the final value of x.

The key point is that an integral measures the *cumulative effect* of some function; visually it is given by the area between $f(x)$ and the x axis.

A.6   Calculating with Calculus

In the last section we introduced the main ideas from calculus–derivatives and integrals–that we'll be using in this book. We did not, however, discuss how one actually *evaluates* a derivative or an integral. If you don't intend to engage with any calculus-based examples or problems in this book, then you can safely skip this section. If you do, however, then here we will present a quick review of some basic methods for differential and integral calculus.[22]

Derivatives and Integrals of Polynomials

A polynomial is of the form

$$f(x) = a + bx + cx^2 + dx^3 + \ldots$$

where in general the series can continue to an arbitrarily high power. In practical situations the series is finite, and the highest order power is said to be the *power* of the series (e.g. the quadratic polynomials considered in Fig. A.18 are of order 2).

---

[22]For the sake of brevity I make no pretenses of *deriving* the results in this section.

When taking the derivative of any function involving a sum, we can distribute the derivative and thereby split the derivative of a sum into a sum of derivatives:[23]

$$\frac{d}{dx}f(x) = \frac{d}{dx}\left(a + bx + cx^2 + dx^3 + \ldots\right)$$

$$= \left(\frac{d}{dx}(a) + \frac{d}{dx}(bx) + \frac{d}{dx}(cx^2) + \frac{d}{dx}(dx^3) + \ldots\right)$$

Thus any polynomial derivative can be broken down into a sequence of terms that look like

$$\frac{d}{dx}(ax^n)$$

for an arbitrary coefficient a and power n. The rule for how to evaluate such a derivative is "multiply the coefficient by the power, then subtract 1 from the power":

$$\frac{d}{dx}(ax^n) = anx^{n-1} \tag{A.17}$$

As a particular example, this means that the derivative of a *constant* is 0 (the initial power of x is already 0 because $a = ax^0$, so the derivative yields $0ax^{-1} = 0$). Returning to the general polynomial we started with, we find

$$\frac{d}{dx}\left(a + bx + cx^2 + dx^3 + \ldots\right) = b + 2cx + 3dx^2 + \ldots \tag{A.18}$$

And the *second* derivative is simply the derivative of the derivative:

$$\frac{d^2}{dx^2}\left(a + bx + cx^2 + dx^3 + \ldots\right) = 2c + 6dx + \ldots \tag{A.19}$$

The first derivative evaluated at a point tells use the *rate of change* at that point; graphically, this is the slope of the tangent line at that point, as we described above. Similarly, the second derivative evaluated at a point tells us the **concavity** of the curve: a positive second derivative is "concave up" and a negative second derivative is "concave down" (see Fig. A.18). Combining these two statements allows one to mathematically identify the local minima (where the first derivative is 0 and the second derivative is positive) and local maxima (where the first derivative is 0 and the second derivative is negative).

Example A.19 Minima and Maxima $\star dx$
Find the local minima and maxima for the equation $x = 4t^2 - 4t + 5$.

The first derivative is

$$\frac{dx}{dt} = 8t - 4$$

---

[23]While we won't make use of other notations for derivatives in this text, it is worth mentioning that the first derivative is sometimes written $\dot{x}$ or x' (read "x dot" or "x prime" respectively) and the second derivative is sometimes written $\ddot{x}$ or x'' ("x double dot" or "x double prime". The "dot" notation is sometimes called Newton's notation and the "prime" notation is sometimes called Lagrange's notation. The d/dx notation that we'll use is called "derivative operator" notation.

Setting this equal to 0 and solving for t, we find $t = 0.5$. Meanwhile, the second derivative is

$$\frac{dx^2}{dt^2} = 8$$

Because this is positive (for all values of t), we have a local minimum at $t = 0.5$. The curve is shown in Figure A.22.

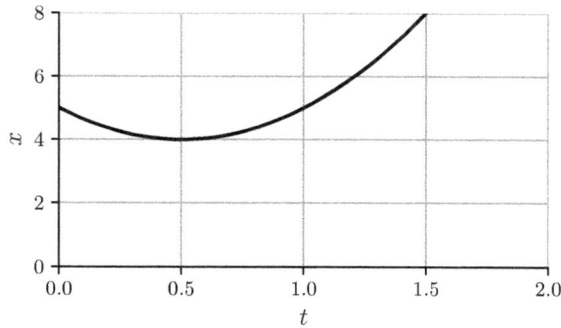

FIGURE A.22

Let's turn to *integrating* polynomials. We can distribute the integral to each term just as we did with derivatives, so in general we're concerned with a single term; we find:

$$\int_{x_i}^{x_f} = ax^n dx = \frac{a}{n+1}x^{n+1}\Big|_{x_i}^{x_f} = \frac{a}{n+1}\left(x_f^{n+1} - x_i^{n+1}\right)$$

Here we "add 1 to the power, then divide by the new power". The $\big|_{x_i}^{x_f}$ in the middle term says "evaluate from x_i to x_f", i.e. set $x = x_f$ and then *subtract* the case where $x = x_i$, as is shown in the last step.

Because we have specified or *definite* bounds of integration, this is called a *definite integral*. In some cases we wish to evaluate an integral without specifying the bounds. In this case we do not need to set x equal to anything but we do add an arbitrary *constant of integration*, C:

$$\int ax^n dx = \frac{a}{n+1}x^{n+1} + C$$

Because there are no bounds, this is called an *indefinite integral*. A numerical value of C can be determined, if desired, by specifying a *boundary value*, as Example A.21 demonstrates. First, in Example A.20 we consider an example drawn from physics and review the process of evaluating minima and maxima of functions.

Example A.20 Calculus with a Ball Toss $\star dx$

An object tossed vertically into the air that is subject only to gravity obeys

$$y(t) = y_i + v_i t - \frac{1}{2}gt^2$$

Where $y(t)$ is the object's height, y_i is its initial height, v_i is its initial velocity, t is the time elapsed since the ball was thrown, and g is the acceleration due to gravity (all these quantities are positive). What is the ball's maximum height, and how long does it take for the ball to return to its original height?

Note first from the functional form of this equation that its graph looks like the dashed curve in the bottom right of Figure A.18. We can calculate the maxima of this function by checking where the derivative is equal to 0:

$$\frac{dy}{dt} = \frac{d}{dt}\left(y_i + v_i t - \frac{1}{2}gt^2\right)$$

$$\frac{dy}{dt} = v_i - gt$$

Notice that the *second* derivative of $y(t)$ is always negative (it is simply $-g$) so we will only find maxima rather than minima (as we expect from Figure A.18).

Setting the *first* derivative equal to 0 and solving for t allows us to locate the maxima:

$$0 = v_i - gt$$

$$t = \frac{v_i}{g}$$

We can insert this value for t into the original function for $y(t)$ to find the maximum height:

$$y_{\text{max}} = y_i + v_i\left(\frac{v_i}{g}\right) - \frac{1}{2}g\left(\frac{v_i}{g}\right)^2$$

$$= y_i + \frac{v_i^2}{g} - \frac{v_i^2}{2g}$$

$$= y_i + \frac{v_i^2}{2g}$$

Thus the object travels a distance $v_i^2/2g$ above its initial height y_i.

Example A.21 Derivatives and Integrals   ★

Consider the function

$$f(x) = 5x^3 + 2x + 3$$

(a) Compute the derivative of $f(x)$ and evaluate it at $x = 3$. (b) Integrate this function and evaluate it from $x = 0$ to $x = 5$. (c) Determine the indefinite integral of this function, $g(x) = \int f(x)\,dx$. What is the constant of integration if $g(0) = 10$?

We will first compute the derivative for (a):

$$\frac{df(x)}{dx} = (3 \times 5)x^2 + 2x^0 + 0 = 15x^2 + 2$$

Evaluating this for $x = 3$ yields $15 \times 3^2 + 2 = 137$.

Next, the integral for (b):

$$\int_0^5 \left(5x^3 + 2x + 3\right) dx = \left(\frac{5}{4}x^4 + \frac{2}{2}x^2 + 3x\right)\Big|_0^5$$

$$= \left(\frac{5}{4}(5)^4 + 5^2 + 3(5)\right)$$

$$\approx 821 \tag{A.20}$$

Finally if we leave the integral in indefinite form for (c) we find

$$g(x) = \int \left(5x^3 + 2x + 3\right) dx = \frac{5}{4}x^4 + x^2 + 3x + C$$

If $g(0) = 10$, then $C = 10$ since every other term in $g(x)$ vanishes for $x = 0$.

Example A.22 Tangent Lines ★

Figure A.23 shows a graph of the function $y(t) = 4t^2$ and a line that is tangent to the function at $t = 2$. What is the equation of the tangent line?

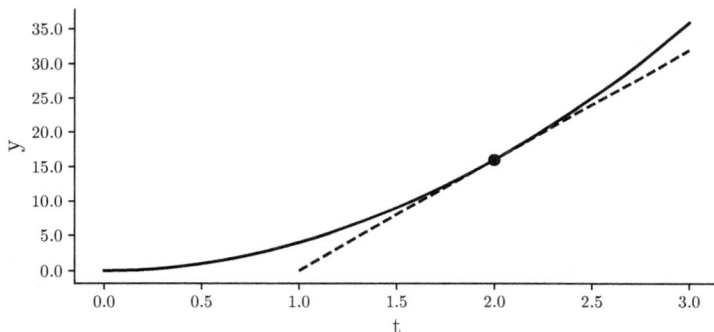

FIGURE A.23

The slope is given by the derivative of the function:

$$\frac{dy}{dt} = \frac{d}{dt}\left(4t^2\right) = 8t$$

Thus at $t = 2$ the slope has a value 16. To get this in to the standard form $y = at + b$ we need the value of b, which is to say the y-intercept. Now, we already have one point on the line, because if $t = 2$ we know $y = 4(2)^2 = 16$. To determine the value of y when $t = 0$, we can apply the definition of slope:

$$\text{slope} = a = \frac{y_2 - y_1}{t_2 - t_1}$$

$$a(t_2 - t_1) = y_2 - y_1$$

$$y_1 = y_2 - a(t_2 - t_1)$$

$$y_1 = 16 - 16(2 - 0)$$

$$y_1 = -16$$

So, finally, we have the equation of the tangent line is $y = 16t - 16$. Because we're working with relatively small, whole numbers, it isn't too hard to arrive at this answer without the last bit of math: once we know the point on the line is at $(t, y) = (2, 16)$ and that the slope has a value of 16, we can "walk" toward the y axis: if we decrease t by 1, we decrease y by 16 (since that is the value of the slope). Thus $(t, y) = (1, 0)$ is also on the tangent line. (In fact, this is the leftmost point drawn on Figure A.23.) Applying the same procedure one more time moves along the line to the point $(t, y) = (0, -16)$, as we found above.

Trigonometric Functions

Table A.4 provides the derivative and integral of the common trigonometric functions (for brevity the constant of integration is omitted). In isolation these are fairly straightforward, but often the argument of the trigonometric function is more complicated, for instance

$$y = \sin(3\theta + \pi)$$

For such a function we take the derivative by applying the chain rule: take the derivative of the trigonometric function while leaving the argument alone, then multiply by the derivative of the argument:

$$\frac{dy}{d\theta} = 3\cos(3\theta + \pi)$$

TABLE A.4

Function	Derivative	Integral
$\sin\theta$	$\cos\theta$	$-\cos\theta$
$\cos\theta$	$-\sin\theta$	$\sin\theta$
$\tan\theta$	$\sec^2\theta$	$-\ln\lvert\cos\theta\rvert$

If we're instead taking the *integral* of such a function, we often make use of u substitution: introduce a function $u(x)$ and change the variable of integration from dx (or $d\theta$ or whatever other variable we're considering) to du.[24] Example A.23 demonstrates the procedure.

Example A.23 Integrals with Trigonometry   ⋆

Compute the indefinite integral of $15\cos(3\theta + \pi)$.

We can't make easy use of Table A.4 here since the argument of the cosine function is more complicated than a simple x (or θ). But we can move to that form if we set the argument of the cosine function equal to a new parameter u, then evaluate du/dx:

$$u = 3\theta + \pi \implies \frac{du}{d\theta} = 3 \implies \frac{1}{3}du = d\theta$$

We can then make a substitution in our integral to eliminate both θ and $d\theta$ in favor of u and du, respectively (we've thereby made a so-called *change of variables*):

$$\int 15\cos(3\theta + \pi)\,dx = \int \frac{15}{3}\cos(u)\,du$$

$$= 5\sin(u) + C$$

$$= 5\sin(3\theta + \pi) + C$$

---

[24]The choice of u rather than some other symbol is arbitrary, but this is the most common convention.

> In the last line I swapped back to θ from u so we express our result in terms of the original variable. (If we were taking a definite integral, we could then evaluate the expression at the bounds expressed in terms of θ.)

A.7 Problem Solving Strategies and Physical Models

Students in physics courses often find that *getting started* is the hardest part of a problem: you've read about some physical principles, worked through some examples, and perhaps even done some introductory problems on the topic, but then you see something new and can't help but think, "What does this have to do with anything I am supposed to be learning?"[25]

To combat this, it is helpful to have a formal procedure to guide you through the problem solving process:

1. **Make a visual representation of the situation.** Often this involves making a labeled diagram (e.g. "The ball starts *here* but we know it rolls off the table, so it should land on the floor *here*.."), but if, say, you are presented with a table of data, it might help to make a graph.

2. **List and label what you know.** In general, whatever information is presented in the problem statement should be represented on your visualization, or neatly listed nearby. Often, some relevant facts are left unstated in a problem, but any that seem relevant should be included as well.[26]

3. **Identify relevant assumptions and principles.** This requires you to have a handle on the physical models that we will be discussing, and to have a sense of when they are appropriate. For instance, momentum conservation is relevant when isolated objects are colliding or exploding, and mathematically applying the principle involves the masses and velocities of the objects. So, if objects are banging together or flying away from one another (and especially if you know and/or are asked to determine the masses or velocities of the objects), it is worth your time to consider using the principle of momentum conservation as you analyze the situation. If you are feeling your way through a problem and aren't sure how to proceed, listing out *potentially relevant* principles and taking some preliminary steps is an entirely valid way to get your bearings: it is OK to engage in some reflective "play" without having a clear sense of how to get to the desired result.

4. **Proceed carefully through your analysis.** Don't gloss over conceptual steps: imagine explaining to a peer *why* you're proceeding as you are, and double check any mathematical steps.

---

[25] Physics is all about getting at the fundamental principles that explain how the Universe operates, so we shouldn't be surprised when we end up applying these principles to situations that seem, at first glance, to be quite different! Getting better at anything takes effort, but knowing how to break a problem down into its fundamental components and then meaningfully analyzing it is a general skill that will serve you well in your career.

[26] For instance, we will find that if a ball is thrown across the room, it has an acceleration with a fixed value of 9.8 m/s^2 directed down toward the ground. Because this is *always* the case (at least near the surface of the Earth and neglecting air resistance) it often isn't stated explicitly, but it is usually relevant when analyzing thrown objects.

5. **Check the realism of your answer.** This can require some practice, especially for areas of physics for which you don't have much natural intuition. But if you find, say, that the Moon orbits the Earth at a distance of 1 meter, you should catch yourself and say "Well, perhaps that isn't quite right".

6. **Review your work.** Even if you're confident you've gotten it right, it is worth your time to re-check your work with fresh eyes (i.e. a day or so after you initially finish it). Ask yourself, "How did I determine the appropriate strategy based on the information provided?" and double check your steps. If your initial solution is messy, rewrite it in a neat, logical way: include your visualizations, your mathematics, and explain key steps with words. This will solidify your understanding and help you catch mistakes.

7. **Start early and don't spin your wheels.** Even if you're being thoughtful and deliberate, it is entirely normal to get stuck now and again. Assuming you're working problems for a course with due dates, give yourself the courtesy of being able to walk away and then return to a problem with a fresh set of eyes well before it is due; this gives you the opportunity to go to your instructor (or peers) prepared to say, "Here is what I have so far, and here is where I am stuck". Doing this effectively will improve your learning, decrease stress, and help you to work productively when you *are* working on your own.[27]

Let me close by emphasizing a point I made in step 3 above: "Identify relevant assumptions". As we proceed through this book, we will establish many principles of physics, but to make the analysis tractable as we get our bearings, we will often neglect some details. For example, we will discuss how balls move through the air, but (at least at first) we will ignore the fact that the ball is moving *through the air*. And, when we analyze electrical circuits, we will often ignore the resistive effect of the wires because their effect tends to be minimal compared to other components of the circuit.

These simplifying assumptions characterize our **model** of a physical situation.[28] It is important to keep the assumptions that underly our models in mind: in an experimental setting (be it for a lab you're completing for a class or in your career) the details that we omit for the sake of simplicity can explain experimental deviations from what we expect, and in some cases these deviations can help us to build (and properly contextualize) better models. Of course, the further you go in your education in physics, the more comfortable you will be in handling more complicated and realistic models of the Universe.

A.8   Problems for Appendix A

A.1 Representing Physical Quantities

⋆   **Problem A.1.** The SI unit for *energy* is the Joule (J). In terms of the basic quantities in the SI system, the Joule is "mass × length^2 / time^2". What are the basic SI *units* for each of these quantities? How can a Joule be expressed in terms of basic SI abbreviations?

---

[27]Yes, I know there is a good chance that after reading this you thought something along the lines of "Whatever, I'll just do the homework the night before it is due". I get it... but keep this advice in mind in case you find that strategy doesn't work out as well as you'd like.

[28]A standard joke is that if you ask a physicist how to get a cow to the top of a ramp, they will begin by responding, "First, assume the cow is a sphere..".

★ **Problem A.2.** The SI unit for *electric charge* is the Coulomb (C). In terms of the basic quantities in the SI system, the Coulomb is "current × time". What are the basic SI *units* for each of these quantities? How can a Coulomb be expressed in terms of basic SI abbreviations?

★★ **Problem A.3.** The SI unit for *power* is the watt (W). 1 watt is the same as 1 J/s (see Problem A.1). How can a watt be expressed in terms of basic SI abbreviations?

★ **Problem A.4.** What are the SI base units for an equation of the form v^2/a (velocity squared divided by acceleration)?

★★ **Problem A.5.** Categorize each of the following numbers as being in *standard notation, scientific notation,* or *neither*. Rewrite each in both standard and scientific notation.

a. 10200

b. 2.040×10^3

c. 80.0050×10^{-2}

d. 6.56×10^1

e. 0.032

★★ **Problem A.6.** Determine the number of significant figures in each number provided in Problem A.5.

★★ **Problem A.7.** Categorize each of the following numbers as being in *standard notation, scientific notation,* or *neither*. Rewrite each in both standard and scientific notation. How many significant figures are in each number?

a. 7.59×10^3

b. 45,000,000

c. 80 million

★★ **Problem A.8.** Perform each of the requested conversions. Report your answers in scientific notation and be sure to report your answer with the appropriate number of significant figures. Unit conversions are provided in Appendix B.

a. Convert 5.00 km to meters.

b. Convert 5.00 km to miles.

c. Convert 4.2 ly to kilometers. (1 ly is one "light year", i.e. the distance that light travels through a vacuum in one year. The closest star to Earth, other than our sun, is about 4.2 ly away.)

★ **Problem A.9.** The volume of a sphere is calculated as 1 in³. What is the volume in cm³?

★ **Problem A.10.** If the area of the football field is 5.76×10^4 square feet, what is the SI equivalent area?

★ **Problem A.11.** Right now, the International Space Station is traveling around the Earth at about 8.0 kilometers every second. What is this speed in miles per hour (mph)?

★ **Problem A.12.** Right now, the International Space Station is traveling around the Earth at about 8.0 kilometers every second. What is this speed in miles per hour (mph)?

★★ **Problem A.13.** A large swimming pool might hold 6.0×10^5 gallons of water.

a. What is this volume in m³? (1 gallon is approximately 3.79 liters, and 1 liter is 10^{-3} m³.)

b. A typical density of a neutron star is 4.0×10^{17} kg/m³. If the swimming pool was filled in with matter from a neutron star, what would be the total mass, in kg? (Density is an object's mass divided by its volume.)

c. Express your answer to (b) as a multiple of the mass of the Earth. In other words, if you call your answer to (b) m and the mass of the Earth is M_E, then what is α in $m = \alpha M_E$? Does this impressive to you? Why or why not?

d. The average density of the Earth is about 5.5×10^3 kg/m^3. How many swimming pools would you need to contain the mass you found in (b) if it is to have this density? Does this seem impressive to you? Why or why not?

★ **Problem A.14.** The exhaust fan on a typical kitchen stove pulls 600 CFM (cubic feet per minute) through the filter. How many cubic meters per second is this?

★★ **Problem A.15.** Write each of the below quantities in scientific notation using base SI units.

a. 365 days

b. 0.5 in/month (a standard rate at which hair grows)

c. 1.97×10^8 sq. miles (the surface area of Earth)

★ **Problem A.16.** I sometimes talk with my children about distances in terms of their height. For example, "That tree is about 4 Joeys tall!" This isn't very scientific but it can be a good (or at least amusing) way to get a sense of scale. For the sake of this problem, suppose 1 Joey = 48 inches. Determine the height (or length) of the following objects in units of Joeys.

a. A 100-yard-long football field.

b. The Eiffel tower, which is about 1,080 feet tall.

c. The circumference of the Earth, which is about 24,900 miles.

d. Repeat (a)–(c) using a unit that corresponds to your name and height.

★★ **Problem A.17.** In SI, a torque is expressed as a "Newton meter", (Nm) whereas in the Imperial system of units a torque is expressed in as a "foot pound" (ft-lb). How many Newton meters is 3.0×10^2 ft-lb? (Answer by converting pounds to Newtons, then feet to meters. The conversion factors are in Appendix B.)

★ **Problem A.18.** Calculate the following values. Express your answer to the correct number of significant figures. Briefly explain *why* your answer has the number of significant figures that it does.

a. $1.29 + 40.3$

b. $837 - 37.1$

c. $5.0 \times 10^2 + 3.2 \times 10^{-3}$

★ **Problem A.19.** Calculate the following values. Express your answer to the correct number of significant figures. Briefly explain *why* your answer has the number of significant figures that it does.

a. $42.0/7$

b. 3×3.5

c. $17.9/5.24$

d. $\left(4.6 \times 10^1\right) \times \left(3.58 \times 10^{-2}\right)$

★ **Problem A.20.** Calculate the following values. Express your answer to the correct number of significant figures. Briefly explain *why* your answer has the number of significant figures that it does.

a. $\cos\left(32°\right)$

b. $\sin\left(5.00°\right)$

c. $\tan\left(85.00°\right)$

★ **Problem A.21.** Calculate the following values. Express your answer to the correct number of significant figures. Briefly explain *why* your answer has the number of significant figures that it does.

a. $\log_{10}\left(3.5\right)$

b. $\ln\left(4.724\right)$

c. $10^{4.56}$

d. $e^{0.533}$

A.2 Algebra and Solving Systems of Equations

★ **Problem A.22.** Consider the equation $a = 3\left(b + c\right)/d$. Algebraically solve it for c (i.e. get it to the form $c = ...$). Show at least one intermediate step of the algebra.

★ **Problem A.23.** Determine a numerical value for x in each of the following equations.

a. $5 = 3\left(x + 2\right)$

b. $10 = 9x + 1$

c. $0 = \left(x + 2\right)\left(x - 2\right)$

d. $5x^2 + 3x - 8 = 0$

★★ **Problem A.24.** Determine an algebraic expression for A in terms of B for each of the following equations. Then, go through again and determine an algebraic expression for B in terms of A.

a. $3A = \frac{7B^2}{6}$

b. $5A + 3B = A^2 - 12$

c. $10^A = 15B$

d. $4\cos\left(8A\right) = 12B$

★ **Problem A.25.** Solve this system of equations for F_1 and F_2. Do it twice, once with substitution and once with equation arithmetic.

$$F_1 + F_2 = 8$$
$$3F_1 - F_2 = 8$$

★★ **Problem A.26.** Solve this system of equations for x and y:

$$5x - 12y = 122$$
$$yx = -60$$

★ **Problem A.27.** Solve this system of equations for x and y:

$$3x - 5y = -11$$
$$6x + 2y = 30$$

★★ **Problem A.28.** Solve this system of equations for r_1 and r_2 subject to the constraint that r_1 and r_2 are both integers:

$$r_1^2 + r_2^2 = 10$$
$$r_1\left(4 + 5r_2\right) = 27$$

★★ **Problem A.29.** Solve this system of equations for F_1, F_2, and F_3:

$$3F_1 + 4F_2 - 7F_3 = -102$$
$$-3F_1 - 2F_2 + 7F_3 = 118$$
$$5F_1 - 2F_2 + 3.5F_3 = 64$$

★ **Problem A.30.** Solve this system of equations for T_1 in terms of w and θ:

$$T_1 \sin\theta = T_2 \sin\theta$$
$$T_1 \cos\theta + T_2 \cos\theta = w$$

★★ **Problem A.31.** Solve this system of equations for v in terms of m_1, m_2, v_{1xi}, v_{1yi}, v_{2xi}, and v_{2yi}.

$$m_1 v_{1xi} + m_2 v_{2xi} = \left(m_1 + m_2\right) v_{fx}$$
$$m_1 v_{1yi} + m_2 v_{2yi} = \left(m_1 + m_2\right) v_{fy}$$
$$v^2 = v_{fx}^2 + v_{fy}^2$$

★ **Problem A.32.** Solve this system of equations for θ in terms of F_f and F_N:

$$F_N = mg\cos\theta$$
$$F_f = mg\sin\theta$$

A.3 Geometry and Trigonometry

★ **Problem A.33.** You walk 1/3 of the way around a circular track. If you've walked 5.00×10^2 m, what is the radius of the track?

★ **Problem A.34.** You wrap a 5.0 cm-long string around a cylindrical can of soup with a radius of 4.0 cm. What fraction of the circumference is covered by the string?

★ **Problem A.35.** A dog is wearing a leash tied to a stake in the ground. Consider two cases:

  a. The leash is 10.0 m long

  b. The leash is 5.0 m long

In each case suppose that the dog walks 1.2 circular laps as far from the stake as possible. What distances (in meters) will the dog have walked? What angular distances will the dog have walked (in radians)? Include a sketch of each situation and compare your answers in the two cases.

★ **Problem A.36.** A right triangle has a hypotenuse of length 10.0 and one interior angle measures $35°$. Determine the length of all three sides and all three interior angles. Include a sketch of the triangle where all sides and angles are labeled.

★ **Problem A.37.** A cart is sliding up a hill that is inclined relative to the horizontal at an angle of $35°$. If the cart is moving with a speed of 2.5 m/s, what is the horizontal component of the speed? What is the vertical component?

★ **Problem A.38.** A sailboat is accelerating to the East at 1.5 m/s^2 and simultaneously accelerating to the South at 2.5 m/s^2. What is the total acceleration, and which way does it point? (*Hint:* think of the two provided accelerations as the lengths of the non-hypotenuse sides of a triangle. The total acceleration is then length of the hypotenuse.)

★★ **Problem A.39.** Two of a right triangle's sides have lengths 5.0 and 6.0. What are the possible lengths of the third side? For each possible length, sketch the triangle and label the length of each side and the value of each interior angle.

★ **Problem A.40.** Find the lengths of the unmarked sides and the value of the interior angles for the right triangle shown in Figure A.24.

★ **Problem A.41.** Find the lengths of the unmarked sides and the value of the interior angles for the right triangle shown in Figure A.25.

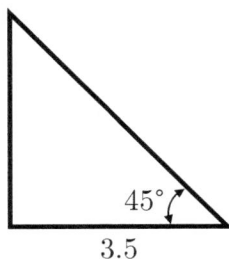

FIGURE A.24
Problem A.40

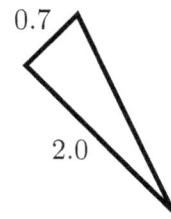

FIGURE A.25
Problem A.41

★★ **Problem A.42.** If a right triangle has a side of length 4.5 and a side of length 5.0, then what is the length of the third side? Sketch and label the triangle, and determine the interior angles. If there is more than one possible length for the third side (*Hint:* there is), go through the analysis for every option.

★★ **Problem A.43.** A diamond has edge length L and opposing interior angles θ as shown in Figure A.26. What is the distance from the top point to the bottom point? Give your answer algebraically in terms of L and θ and numerically in the case where $L = 1.5$ and $\theta = 30°$.

★ **Problem A.44.** Determine the angle marked "?" in terms of θ in Figure A.27.

★ **Problem A.45.** Determine the angles marked "?" in terms of θ in Figure A.28. The horizontal lines are parallel.

	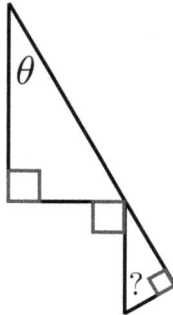	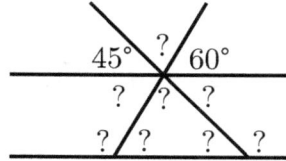
FIGURE A.26	**FIGURE A.27**	**FIGURE A.28**
Problem A.43	Problem A.44	Problem A.45

A.4 Representing Data

★ **Problem A.46.** Come up with an experiment where you'd end up with a series of ordered pairs of data. Describe the experiment and describe three different methods you might use to represent the data. Discuss the pros and cons of each of the three methods.

★ **Problem A.47.** In Figure A.15 a sine curve is fitted to example data showing the outside temperature during a day. How much time passes before the sine curve has gone through one complete cycle and begins to repeat the same behavior? Is this what you would expect? Why or why not? If it *isn't* what you expected, can you come up with a possible explanation?

★★ **Problem A.48.** You measure the height of a sunflower every 7 days and find the heights (in m) are 0, 0.18, 0.36, 0.70, and 1.00. Make a table and a graph of these data. Estimate the height of the sunflower on day 10. Estimate its height after 60 days assuming the rate of growth stays consistent with the first 28 days. Explain your reasoning.[29]

★ **Problem A.49.** Graph each of the following equations:

a. $y = 5x + 3$

b. $x = 1.5t - 2$

c. $F = -2t + 5$

d. $v_f = -2t - 4$

★ **Problem A.50.** Make a graph of the following equations. Describe (in words) what would change about your graphs if the sign of x^2 changed.

a. $y = x^2 + 2$

b. $y = 2x^2 - 1$

c. $y = -3x^2 + 3$

---

[29]The last part of this question demonstrates the danger of *extrapolating*, i.e. using data to predict the behavior outside of where you *have* data. As a general rule, one shouldn't assume that trends will continue indefinitely! In contrast, estimating the height of the plant on day 10 is an example of *interpolating*, which is generally safer... though even then you're implicitly assuming that whatever you're measuring doesn't vary wildly between instances where you've obtained some data (here, between days 7 and 14).

★★ **Problem A.51.** Table A.5 shows sample data for the mass of a male baby over their first year after being born. Make a graph of this data (mass vs. age). Draw in a line connecting the first and last data points. What is the slope of the line? Describe the data: does it seem like the data are linear?

★ **Problem A.52.** Determine the equation of each of the lines shown in Figure A.29 (that is, determine numerical values for a and b in $y = ax + b$).

TABLE A.5
Problem A.51

Age (mo.)	Mass (kg)
0.0	3.3
2.0	5.5
4.0	6.8
6.0	7.9
8.0	8.5
10.0	9.1
12.0	9.5

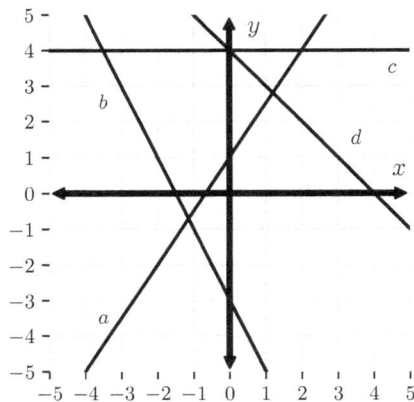

FIGURE A.29
Problem A.52

★ **Problem A.53.** Consider the graph in Figure A.30.

   a. Make a table of these data (be careful to estimate appropriately and report your values with the appropriate number of significant digits).

   b. The axes don't include units and no context has been provided for these data. Come up with a scenario where these data could plausibly be relevant. What does each axis represent (and with what units) in your example?

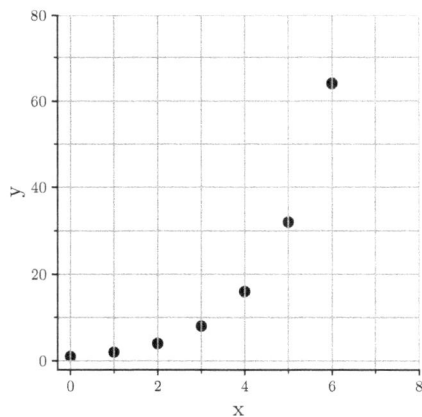

FIGURE A.30
Problem A.54

★ **Problem A.54.** Determine the equation for each of the quadratic curves shown in Figure A.31. There are no *linear* terms in any of these, so your equations will be of the form $y = ax^2 + c$.

★ **Problem A.55.** Figure A.32 shows two curves of the form $y = ax^2 + bx + c$ For each, determine the *sign* (positive or negative) of each coefficient a, b, and c. (None of them are equal to 0.) Explain your reasoning.

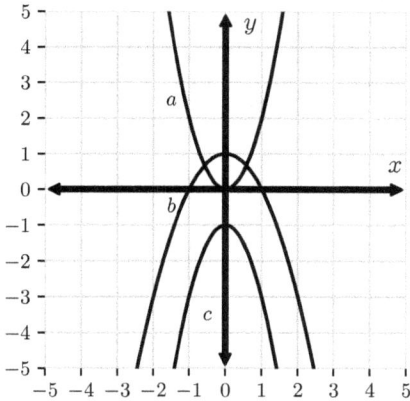

FIGURE A.31
Problem A.54

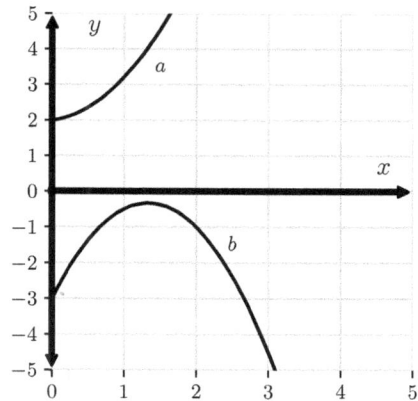

FIGURE A.32
Problem A.55

A.6 Calculating with Calculus

★ **Problem A.56.** Take the derivative of each of the following polynomials:

a. $y = 4$

b. $y = 20x^2$

c. $y = 3x^3 - 4x^2$

d. $y = 10x^5 - 5x + 2$

★★ **Problem A.57.** Consider each of the equations in Problem A.56. For each:

a. Find the slope of the tangent line at $x = 5$.

b. Find the concavity at $x = 3$.

c. Find all local minimal and local maxima for $x \geq 0$.

★ **Problem A.58.** Evaluate each of the following integrals:

a. $\int 3dx$

b. $\int 4.5x^{1/2}dx$

c. $\int_0^2 \left(2x^3 - 5\right) dx$

d. $\int_{-5}^5 -xdx$

★ **Problem A.59.** Take the derivative of each of the following equations:

a. $y = \sin(3\theta)$

b. $y = \cos(2\theta - \pi)$

c. $y = 5\sin(\theta + 2\pi)$

d. $y = 8\cos\left(\frac{1}{2}\theta + 6\right)$

★★ **Problem A.60.** Evaluate each of the following integrals:

a. $\int_0^\pi \sin(2\theta) d\theta$

b. $\int \cos(\theta/2) d\theta$

c. $\int 10\sin(10\theta + 1.5\pi) d\theta$

Additional Problems for Appendix A

★   **Problem A.61.** For an equation to be physically valid, the fundamental units on the left side of the equation must be the same as on the right side of the equation: we can say "5280 feet $=1$ mile" because both feet and miles are units of *length*, but it is nonsense to say any number of feet (a measure of length) is equal to any number of kilograms (a measure of mass).

Consider the fundamental units of the following quantities, then determine if each of the suggested equations in (a)–(c) can be physically valid.

- Velocity is length / time
- Force is mass $\times$ length / time^2
- Energy is mass $\times$ length^2 / time^2
- Momentum is mass $\times$ length / time
- Impulse is mass $\times$ length / time

a. Force $\times$ Length $=$ Energy

b. Mass $\times$ Velocity $=$ Impulse

c. Impulse / Time $=$ Force

★★   **Problem A.62.** Consider each of the equations below. Determine the units for each instance of the variables α, β, and γ (the lowercase Greek letters alpha, beta, and gamma). The definitions provided in Problem A.61 will be useful.

a. $y = \alpha + \beta \Delta t + \gamma (\Delta t)^2$, where y is a measure of length and Δt is a measure of time.

b. $v^2 = \alpha + \beta (\Delta x)$, where v is a measure of velocity and Δx is a measure of length.

c. $K = \alpha v^2$ where K is a measure of energy and v is a measure of velocity.

d. $W = \alpha x$ where W is a measure of energy and x is a measure of length.

e. $y = A \sin(\alpha x + \beta t + \gamma)$, where y, A, and x are measures of length and t is a measure of time.

★★   **Problem A.63.** We can attempt to construct physically valid equations by raising terms on either side to a power, then determining the numerical values of the powers that balance the units on either side of the equation. For instance, we might wonder if we can combine a mass and velocity to yield an energy (the fundamental units of these quantities are provided in Problem A.61):

$$[\text{mass}]^\alpha \left(\frac{[\text{length}]}{[\text{time}]} \right)^\beta = \left(\frac{[\text{mass}]\,[\text{length}]^2}{[\text{time}]^2} \right)^\gamma$$

(Of course, a unitless constant might exist on either side of the equation, so this method only helps to establish the relationship between quantities *with* units.) Looking first at mass, we see that we have $[\text{mass}]^\alpha = [\text{mass}]^\gamma$, which indicates $\alpha = \gamma$. Repeat this argument for the other units (length and time). If a mass and velocity *can* be multiplied together to yield units of energy, then what are the values of α, β, and γ? If they *can't*, why not?

★★   **Problem A.64.** We can extend the approach described in Problem A.63 to infer a wide variety of physically meaningful relationships. The period of a pendulum is the time it takes the pendulum to swing back and forth once. If the only dimensional quantities that the period depends on are the acceleration of gravity, g, and the length of the pendulum, L, what combination of g and L must the period be proportional to?

★★★ **Problem A.65.** We can extend the approach described in Problem A.63 to infer a wide variety of physically meaningful relationships. Here, consider the fact that a skydiver will speed up as they plummet toward the ground until they reach a *terminal velocity* (units of length/time). Suppose that we reason that this velocity should depend only on the density of air (units of mass / length^3), the weight of the skydiver (units of mass $\times$ length / time^2), the surface area of the front of the skydiver (units of length^2), and a unitless coefficient.[30]

---

[30] It turns out that the coefficient depends on the *shape* of the object, e.g. a sphere vs. a bullet.

a. Determine an equation for the terminal velocity that involves the product of each of these terms. To what power must each term be raised to ensure the units balance?

b. Use your answer to (a) to sketch the relationship between the terminal velocity (on the y axis) and the weight of the skydiver (on the x axis) assuming the other quantities are held constant.

c. If the weight of a light skydiver is w_L and their terminal velocity is v_L, then what is the terminal velocity of a heavier skydiver with a weight $w_H = 3w_L$, assuming no other quantities differ? (Your answer will be a multiple of v_L.)

d. Assume that the unitless coefficient has a value of about 1.4 (a reasonable choice), so your answer is in the form $v = 1.4 \left(w^\alpha \rho^\beta A^\gamma \right)$ where w is the person's weight, ρ is the density of air, and A is the surface area of the front of the person (and α, β, and γ are the powers you worked out in (a)). The density of air is $\rho = 1.2$ kg/m^3, and typical values for w and A are 750 N and 0.70 m^2, respectively.[31] Use these values to estimate a person's terminal velocity (your answer will be in units of m/s).

★★ **Problem A.66.** We can extend the approach described in Problem A.63 to infer a wide variety of physically meaningful relationships. In this problem we'll consider a historical example: several years after the first explosion of an atomic bomb in 1945, some high-speed photographs of the expanding mushroom cloud were published; they were each annotated with how long since the detonation they were taken and a length scale. From this the physicist G.I. Taylor was able to estimate the energy contained in the bomb. We'll take a dimensional analysis approach: suppose that the radius of the explosion depends only on the energy of the bomb, the density of the surrounding air, and the time since the explosion. (Density has units of mass / length^3 and the units of energy are provided in Problem A.61.)

a. Can the radius, density, and time each be raised to some power and then multiplied together to yield units of energy? If *yes*, determine the value of each power. If no, check your reasoning and try again, because the answer is *yes*.

b. The density of air is approximately 1.2 kg/m^3, and at $t = 0.006$ s after the explosion the radius was approximately 75 m. Use these values and your result from (a) to determine an estimate for the energy in the bomb (to avoid going too far into the proverbial weeds, assume the dimensionless coefficient has a value of 1). Your answer will be in the SI unit of energy called Joules (J).

c. Convert your answer to (b) from Joules to "tons of TNT" by using the conversion

$$1 \text{ ton of TNT} = 4.18 \times 10^9 \text{ J}$$

d. If a second bomb with twice the energy was detonated, what radius would the cloud have at the same time used above, $t = 0.006$ s?

---

[31]1 N is "one Newton", the SI unit of weight. While not relevant for this problem, you may be curious to know that 1N is 0.225 pounds.

B

Reference Information

B.1 Constants

$$g = 9.8 \text{ m/s}^2$$ free fall acceleration on Earth

$$c = 3.00 \times 10^8 \text{ m/s}$$ speed of light (vacuum)

$$v_{\text{sound}} = 343 \text{ m/s}$$ speed of sound (air at 20°C)

$$G = 6.67 \times 10^{-11} \text{ Nm}^2/\text{kg}^2$$ gravitational constant

$$M_E = 5.98 \times 10^{24} \text{ kg}$$ mass of the Earth

$$R_E = 6.37 \times 10^6 \text{ m}$$ radius of the Earth

$$R = 8.31 \text{ J}/(\text{mol K})$$ gas constant

$$k_B = 1.38 \times 10^{-23} \text{ J/K}$$ Boltzmann's constant

$$N_A = 6.02 \times 10^{23} \text{ particles/mole}$$ Avogadro's constant

$$P_{\text{atm}} = 1.013 \times 10^5 \text{ Pa}$$ atmospheric pressure

$$\varepsilon_0 = 8.85 \times 10^{-12} \text{ C}^2/\text{Nm}^2$$ permittivity of free space

$$K = 8.99 \times 10^9 \text{ N m}^2/\text{ C}^2$$ electrostatic constant

$$m_p = 1.673 \times 10^{-27} \text{ kg}$$ mass of the proton

$$m_n = 1.675 \times 10^{-27} \text{ kg}$$ mass of the neutron

$$m_e = 9.11 \times 10^{-31} \text{ kg}$$ mass of the electron

$$e = 1.60 \times 10^{-19} \text{ C}$$ fundamental unit of charge

$$\mu_0 = 4\pi \times 10^{-7} \text{ N/A}^2$$ permeability of free space

$$h = 6.63 \times 10^{-34} \text{ Js}$$ Planck's constant

$$R = 1.10 \times 10^{-2} \text{ nm}^{-1}$$ Rydberg constant

$$E_R = 13.6 \text{ eV}$$ Rydberg energy

B.2   Conversion Factors

Length:

$$1 \text{ ft} = 30.48 \text{ cm}$$
$$1 \text{ yd} = 3 \text{ ft}$$
$$1 \text{ in} = 2.54 \text{ cm}$$
$$1 \text{ mi} = 1609 \text{ m}$$
$$1 \text{ ly} = 9.46 \times 10^{15} \text{ m}$$

Volume:

$$1 \text{ L} = 10^{-3} \text{ m}^3$$

Weight and Mass:

$$1 \text{ lb (on Earth)} \rightarrow 0.454 \text{ kg}$$
$$1 \text{ lb (on Earth)} = 4.45 \text{ N}$$
$$1 \text{ amu} = 1.66 \times 10^{-27} \text{ kg}$$

Speed:

$$1 \text{ mph} = 0.447 \text{ m/s}$$
$$1 \text{ kph} = 0.278 \text{ m/s}$$

Time:

$$1 \text{ day} = 8.64 \times 10^4 \text{ s}$$
$$1 \text{ yr} = 3.16 \times 10^7 \text{ s}$$

Temperature:

$$T_F = \frac{9}{5} T_C + 32$$
$$T_K = T_C + 273.15$$

Energy:

$$1 \text{ cal} = 4.184 \text{ J}$$
$$1 \text{ eV} = 1.60 \times 10^{-19} \text{ J}$$

B.3   Common Metric Prefixes

< 10^0			> 10^0		
Prefix	Symbol	Factor	Prefix	Symbol	Factor
deci-	d	10^{-1}	deka-	da	10^1
centi-	c	10^{-2}	hecto-	h	10^2
milli-	m	10^{-3}	kilo-	k	10^3
micro-	μ	10^{-6}	mega-	M	10^6
nano-	n	10^{-9}	giga-	G	10^9
pico-	p	10^{-12}	tera-	T	10^{12}
femto-	f	10^{-15}	peta-	P	10^{15}

B.4   Greek Letters Used in This Book

Name	Symbol	Name	Symbol
alpha	α	rho	ρ
beta	β	Sigma	Σ
gamma	γ	sigma	σ
Delta	Δ	tau	τ
delta	δ	Phi	Φ
epsilon	ε	phi	ϕ
eta	η	Psi	Ψ
theta	θ	psi	ψ
lambda	λ	Omega	Ω
mu	μ	omega	ω
pi	π		

B.5   Series Expansions

Taylor Series:

$$f\left(x + \delta\right) = f\left(x\right) + \left(\frac{d}{dx}f\left(x\right)\right)\delta + \left(\frac{1}{2!}\frac{d^2}{dx^2}f\left(x\right)\right)\delta^2 + \left(\frac{1}{3!}\frac{d^3}{dx^3}f\left(x\right)\right)\delta^3 + \cdots$$

Common Expansions:

$$\left(1 + \delta\right)^n = 1 + n\delta + \frac{1}{2}n\left(n - 1\right)\delta^2 + \cdots$$

$$\sin\left(\delta\right) = \delta - \frac{\delta^3}{6} + \frac{\delta^5}{120} - \cdots$$

$$\cos\left(\delta\right) = 1 - \frac{\delta^2}{2} + \frac{\delta^4}{24} - \cdots$$

$$\tan\left(\delta\right) = \delta + \frac{\delta^3}{3} + \frac{2\delta^5}{15} + \cdots$$

$$e^\delta = 1 + \delta + \frac{\delta^2}{2} + \frac{\delta^3}{6} + \cdots$$

$$\ln\left(1 + \delta\right) = \delta - \frac{\delta^2}{2} + \frac{\delta^3}{3} - \cdots$$

B.6 Common Trigonometric Identities

$$\sin(-\theta) = -\sin\theta$$

$$\cos(-\theta) = \cos\theta$$

$$\cos\theta = \sin\left(\frac{\pi}{2} - \theta\right)$$

$$\sin\theta = \cos\left(\frac{\pi}{2} - \theta\right)$$

$$\sin^2\theta + \cos^2\theta = 1$$

$$\sin 2\theta = 2\sin\theta\cos\theta$$

$$\cos 2\theta = \cos^2\theta - \sin^2\theta$$

$$\sin\frac{\theta}{2} = \pm\left(\frac{1}{2}(1 - \cos\theta)\right)^{1/2}$$

$$\cos\frac{\theta}{2} = \pm\left(\frac{1}{2}(1 + \cos\theta)\right)^{1/2}$$

$$\sin(\theta \pm \phi) = \sin\theta\cos\phi \pm \cos\theta\sin\phi$$

$$\cos(\theta \pm \phi) = \cos\theta\cos\phi \mp \sin\theta\sin\phi$$

$$\sin\theta \pm \sin\phi = 2\sin\left(\frac{\theta \pm \phi}{2}\right)\cos\left(\frac{\theta \mp \phi}{2}\right)$$

$$\cos\theta + \cos\phi = 2\cos\left(\frac{\theta + \phi}{2}\right)\cos\left(\frac{\theta - \phi}{2}\right)$$

$$\cos\theta - \cos\phi = -2\sin\left(\frac{\theta + \phi}{2}\right)\sin\left(\frac{\theta - \phi}{2}\right)$$

$$\sin\theta\cos\phi = \frac{1}{2}(\sin(\theta + \phi) + \sin(\theta - \phi))$$

$$\cos\theta\cos\phi = \frac{1}{2}(\cos(\theta + \phi) + \cos(\theta - \phi))$$

$$\sin\theta\sin\phi = \frac{1}{2}(\cos(\theta - \phi) - \cos(\theta + \phi))$$

B.7 Coordinate Systems

Cartesian

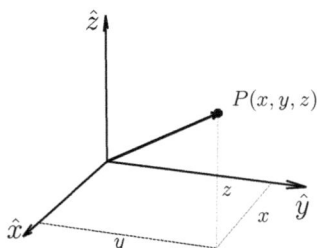

Cylindrical

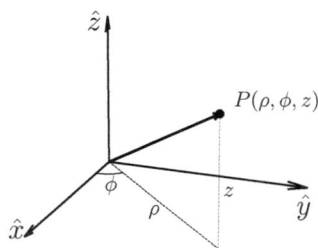

Spherical Polar

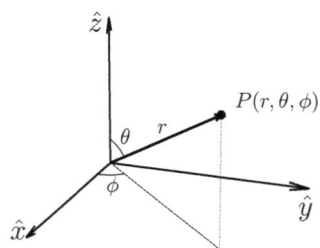

B.8   Solar System Data

Name	Mass (10^{24} kg)	Average radius		Period (day)	
		Body (km)	Orbit (10^6 km)	Rotation	Orbit
Sun	1,990,000	695,700	-	-	-
Mercury	0.3301	2,440	57.91	58.65	87.97
Venus	4.868	6,052	108.2	243.0	224.7
Earth	5.972	6,371	149.6	0.9958	365.2
Moon	0.07348	1,737	0.3844	27.32	27.32
Mars	0.6418	3,390	228.0	1.026	687.0
Jupiter	1,898	69,910	778.5	0.4135	4331
Saturn	568.3	58,230	1,432	0.4484	10,750
Uranus	86.81	25,360	2,867	0.7167	30,590
Neptune	102.4	24,620	4,515	0.6708	59,800

Index

For Product Safety Concerns and Information please contact our EU
representative GPSR@taylorandfrancis.com
Taylor & Francis Verlag GmbH, Kaufingerstraße 24, 80331 München, Germany